Handbook of Statistics

Volume 40

Integrated Population Biology and Modeling, Part B

Handbook of Statistics

Series Editor

C.R. Rao
C.R. Rao AIMSCS, University of Hyderabad Campus,
Hyderabad, India

Handbook of Statistics

Volume 40

Integrated Population Biology and Modeling, Part B

Edited by

Arni S.R. Srinivasa Rao

Augusta University, Augusta, GA, United States

C.R. Rao

AIMSCS, University of Hyderabad Campus, Hyderabad, India

ELSEVIER

N·H North-Holland

An imprint of Elsevier

North-Holland is an imprint of Elsevier
Radarweg 29, PO Box 211, 1000 AE Amsterdam, Netherlands
The Boulevard, Langford Lane, Kidlington, Oxford OX5 1GB, United Kingdom

Notices
Knowledge and best practice in this field are constantly changing. As new research and experience broaden our understanding, changes in research methods, professional practices, or medical treatment may become necessary.

Practitioners and researchers must always rely on their own experience and knowledge in evaluating and using any information, methods, compounds, or experiments described herein. In using such information or methods they should be mindful of their own safety and the safety of others, including parties for whom they have a professional responsibility.

To the fullest extent of the law, neither the Publisher nor the authors, contributors, or editors, assume any liability for any injury and/or damage to persons or property as a matter of products liability, negligence or otherwise, or from any use or operation of any methods, products, instructions, or ideas contained in the material herein.

ISBN: 978-0-444-64152-6
ISSN: 0169-7161

For information on all North-Holland publications
visit our website at https://www.elsevier.com/books-and-journals

Publisher: Zoe Kruze
Acquisition Editor: Sam Mahfoudh
Editorial Project Manager: Peter Llewellyn
Production Project Manager: Vignesh Tamil
Cover Designer: Alan Studholme

Typeset by SPi Global, India

Contents

Section VII
Stochastic Complexity and Structural Dynamics

Section VIII
GWAS, Species Divergence and Bayesian Item Response Theory

Contributors

Numbers in Parentheses indicate the pages on which the author's contributions begin.

Viswanathan Arunachalam (157), Department of Statistics, Universidad Nacional de Colombia, Bogotá, Colombia

Narayan Behera (405), Department of Physics, Dayananda Sagar University, Bengaluru, India

Erin N. Bodine (3), Department of Mathematics & Computer Science, Rhodes College, Memphis, TN, United States

Nicolas Brouard (435), National Institute for Demographic Studies (INED), Paris, France

James R. Carey (487), Department of Entomology, University of California, Davis; Center for the Economics and Demography of Aging, University of California, Berkeley, CA, United States

Carlos Castillo-Chavez (157), Simon A. Levin Mathematical Computational Modeling Science Center, School of Human Evolution and Social Change, Arizona State University, Tempe, AZ, United States

Glenn R. Flierl (595), Department of Earth, Atmospheric, and Planetary Science, MIT, Cambridge, MA, United States

Paul Georgescu (85), Department of Mathematics, Technical University of Iaşi, Iaşi, Romania

Vishwesha Guttal (551), Centre for Ecological Sciences, Indian Institute of Science, Bengaluru, India

Jitesh Jhawar (551), Centre for Ecological Sciences, Indian Institute of Science, Bengaluru, India

Ruth King (33), School of Mathematics, University of Edinburgh, Edinburgh, United Kingdom

Christopher Kribs (157), Department of Mathematics, The University of Texas at Arlington, Arlington, TX, United States

Daniel Maxin (85), Department of Mathematics and Statistics, Valparaiso University, Valparaiso, IN, United States

Rachel McCrea (33), School of Mathematics, Statistics and Actuarial Science, University of Kent, Canterbury, United Kingdom

Moirangthem Hemanta Meitei (501), Department of Economics, Manipur University, Imphal, Manipur, India

Casey E. Middleton (3), Department of Mathematics & Computer Science, Rhodes College, Memphis, TN, United States

Richard G. Morris (551), National Centre for Biological Sciences, TIFR, Bengaluru, India

Anuj Mubayi (157), Simon A. Levin Mathematical Computational Modeling Science Center, School of Human Evolution and Social Change, Arizona State University, Tempe, AZ, United States

Piyush Kant Rai (195), Department of Statistics, Banaras Hindu University, Varanasi, India

Katabathula Ramachandra Murthy (355), Case Comprehensive Cancer Center, Case Western Reserve University, Cleveland, OH, United States

Shelby M. Scott (3), Department of Ecology & Evolutionary Biology, University of Tennessee, Knoxville, TN, United States

Laurentiu Sega (85), Department of Mathematics, Augusta University, Augusta, GA, United States

Salendra Singh (355), Case Comprehensive Cancer Center, Case Western Reserve University, Cleveland, OH, United States

David A. Spade (287), Department of Mathematical Sciences, University of Wisconsin–Milwaukee, Milwaukee, WI, United States

Arni S.R. Srinivasa Rao (487), Augusta University, Augusta, GA, United States

David Steinsaltz (133), Department of Statistics, University of Oxford, Oxford, United Kingdom

David Tuck (355), Curis, Inc., Lexington, MA, United States

Shripad Tuljapurkar (133), Department of Biology, Stanford University, Stanford, CA, United States

Vinay Varadan (355), Case Comprehensive Cancer Center, Case Western Reserve University, Cleveland, OH, United States

Ram Chandra Yadava (195), Department of Statistics, Banaras Hindu University, Varanasi, India

Hong Zhang (85), Department of Financial Mathematics, Jiangsu University, Zhenjiang; School of Economics and Management, Changzhou Institute of Technology, Changzhou, Jiangsu, PR China

Preface

Handbook of Statistics: Integrated Population Biology and Modeling, Part B, Volume 40 brings a comprehensive collection of recently developed concepts and materials on the subject in an accessible and research adaptable way. We took great care in finalizing the chapters which invite readers to understand the content and motivate for further investigations. The subject "population biology" has its theoretical foundations since last few centuries, yet there have been useful extensions of the subject with the advancement of understanding of various ecological principles those govern our lives. We felt that there is a need to understand biological process in an integrated approach and it is important to bring a collection for our readers to give a taste of recent advancements in this integrated approach. Increasingly newer models have been developed that integrates one or more of the human–environment, human–animal, biogenomics, and socio-demographic systems into unified models. Stability principles are relooked for revised mathematical models and stationarity principles of certain human populations are studied again with the newly available animal movement experiments. Still there are unresolved questions on timings of stationarity in populations and exact quantified outcomes of impact of human–environment interactions and stochastics. Hence, we thought there is a justification to bring details of all these developments in several structured chapters in the two part volumes (volume 39 and volume 40) such that advanced readers and beginners enjoy equally.

The subject *Integrated Population Biology* is not very old in comparison to the subject *population biology* as explained in the previous paragraph and this has evolved from the rich data that was made available by the experimental, field, and laboratory biologists during recent decades. Increasing number of theoretical researchers across the world started utilizing newer experimental results and could able to come-up with modified hypotheses and newer theoretical models. Newer models could able to resolve deeper questions due to multilevel data and evolution of computational techniques.

These two volumes offer very delicately handled complex and precise realities of quantifying modern and traditional methods of understanding populations and population dynamics. The topics range from collective behaviors to multi-species mutualism, theories of stationary populations and association with life lived and life left, prey–predator animal models to population biology through data science, evolutionary biology computations to stochastic processes and population dynamics models, the mechanisms for extinction, persistence and resonance to human longevity analysis, biogeochemical modeling analysis, etc.,

We feel that these two volumes have brought compelling material for statisticians, applied mathematicians, theoretical biologists, stochastic modelers, and computational biologists which they can use for their research and teaching. We have divided 14 chapters of the volume 40 into five sections:

Section *VI*: Agent Based Models, Capture–Recapture Methods and Multispecies Mutualism
Section *VII*: Stochastic Complexity and Structural Dynamics
Section *VIII*: GWAS, Species Divergence, and Bayesian Item Response Theory
Section *IX*: Aging and Age-structured Population Dynamics.
Section *X*: Collective Behaviors in Ecology.

Section *VI* contain three chapters: the first chapter by S. M. Scott, C. E. Middleton, and E. N. Bodine brings a very practical application of agent based models to understand the dynamics of Santa Cruz Island Fox in the United States; the second chapter by R. King and R. McCrea provide detailed account of capture–recapture theory applied to population size estimations and recent breakthroughs in this direction were described; and the third chapter by P. Georgescu, D. Maxin, L. Sega and H. Zhang write a detailed description, modeling, and analysis of multispecies mutualism from a deterministic approaches stand point.

Section *VII* contains three chapters: the first chapter by S. Tuljapurkar and D. Steinsaltz introduce modeling and theory behind stochastic structured population dynamics; the second chapter by A. Mubayi, C. Kribs, V. Arunachalam, and C. Castillo-Chavez introduce stochastic models introduction and development of complex ecosystems; and the third chapter by R. C. Yadava and P. K. Rai provide a detailed account of development of stochastic birth interval models and their analysis to understand human fertility.

Section *VIII* contains three chapters: the first chapter by David A. Spade writes a detailed chapter on detection of quantitative trait loci of Genome-Wise Association Studies from the both Bayesian and non-Bayesian approaches; the second chapter by K. R. Murthy, S. Singh, D. Tuck, and V. Varadan describes Bayesian item response theory and useful applications for cancer biomarker discovery; and the third chapter by N. Behera writes on modeling the effects of phenotypic plasticity and unpredictability on species divergence.

Section *IX* contains three chapters: the first chapter by N. Brouard brings a detailed approach in modeling aging societies through backward probabilities; the second chapter by A. S. R. Srinivasa Rao and J. R. Carey introduce two dimensional stationary population identity to study the role of truncation in population age-structure estimation through captive cohort studies; and the third chapter by M. H. Meitei writes a substantial description of demographic situation in the state of Manipur in India.

Section *X* contains two chapters: the first chapter by J. Jhawar, R. G. Morris, and V. Guttal bring a detailed narration on models of collective behavior for finite populations and overview of two methods for deriving mesoscopic descriptions of microscopic models of collective behaviors and the second chapter by G. R. Flierl write a technical account of collective behavior and ecology from the very basics to the advanced readers, for example, individual based and continuum models, aggregations dynamics and pre-predator dynamics, etc.

Our special thanks to Mr. Sam Mahfoudh, acquisition editor (Elsevier and North-Holland) for his overall administrative support throughout the preparation of this volume. His valuable involvement in time to time handling of authors' queries is also appreciated. We also thank Mr. Peter J. Llewellyn, Editorial Project Manager (Elsevier) for excellent assisting of editors and authors in various technical editorial aspects throughout preparation and to provide valuable assistance towards the proofs and production. Our thanks are also to Mr. Vignesh Tamilselvvan (Elsevier) for leading the production and printing activities and assistance to the authors. Our sincere thanks and gratitude to all the authors for writing highly resourceful and wonderful chapters in their respective research topics.

We very much convinced that this collection will be equally resourceful for the beginners and advanced scientists in the theory and modeling of population biology.

Arni S.R. Srinivasa Rao
C.R. Rao

Section VI

Agent Based Models, Capture-Recapture Methods and Multi-Species Mutualism

Chapter 1

An Agent-Based Model of the Spatial Distribution and Density of the Santa Cruz Island Fox

Shelby M. Scott*, Casey E. Middleton† and Erin N. Bodine[1,†]
**Department of Ecology & Evolutionary Biology, University of Tennessee, Knoxville, TN, United States*
†*Department of Mathematics & Computer Science, Rhodes College, Memphis, TN, United States*
[1]*Corresponding author: e-mail: bodinee@rhodes.edu*

ABSTRACT

During the 1990s, the endemic Santa Cruz island (SCI) fox (*Urocyon littoralis santacruzae*) population experienced heavy predation from invasive golden eagles (*Aquila chrysaetos*) and nearly went extinct. The Island Fox Recovery Program instituted in 2002 has since reestablished the population to normal levels. We have constructed an agent-based model (ABM), which simulates the population dynamics of the SCI foxes in both the absence and presence of golden eagle predation. Geographic information system (GIS) data are integrated into the model to replicate the distribution of multiple vegetation types on the island, and the location of golden eagles' nests and hunting territories. Using the model, we determine how both island vegetation and predation by golden eagles impacted the spatial density and distribution of the foxes across SCI. Model analysis shows the 10 golden eagles known to reside on the island during the 1990s and 2000s were sufficient to have caused this rapid population decline even without the presence of other factors such as disease or competition. Additionally, model analysis shows that while removal of the golden eagles was necessary for the survival of the SCI fox population, the remaining population of 75–100 foxes had a 90% survival probability without aid. The model we present provides a powerful tool for understanding the spatial and temporal dynamics of the SCI fox population, can be used and adapted to facilitate the further management of the population, and provides a working example in integrating GIS data with ABMs.

Keywords: Predator–prey dynamics, Santa Cruz island fox, Golden eagle, Spatial dynamics, Agent-based model, Numerical simulation, Geographic information system data

Handbook of Statistics, Vol. 40. https://doi.org/10.1016/bs.host.2018.10.001

1 INTRODUCTION

Many species coexist within the varied vegetation of the California Channel Islands, yet few are endemic (Collins and Latta, 2006). One such species is a small canid known as the island fox (*Urocyon littoralis*). This derivative of the mainland gray fox likely traveled to Santa Cruz island (SCI) by land bridges and adapted a dwarfed stature over thousands of years (Collins, 1991). During mating season (late January to March), mated pairs form between individuals within close geographic proximity (Moore and Collins, 1995). Island foxes are monogamous with mated pairs maintained until one member of the pair dies (Coonan, 2003). Once a mated pair is established, they attempt to find a suitable territory. This territory must be both unoccupied (by another fox mated pair) and habitable (within grassland or mixed vegetation, see Fig. 6). Island fox territory size depends on the vegetation in which the mated pair settles. If the potential territory is mainly grassland, then the territory size will be $\sim$0.87 km^2, but if the potential territory contains more mixed vegetation then the territory size will be $\sim$0.36 km^2 (Coonan, 2003; Roemer et al., 2001). Mated island foxes will attempt to reproduce during breeding season (Moore and Collins, 1995). Gestation lasts 50–53 days, and typical litter sizes are between one and three pups (Moore and Collins, 1995). Pups remain in their natal territory for the first year of their lives, after which they search for their own mates and territories (Coonan, 2003).

Invasive to the island were a population of golden eagles (*Aquila chrysaetos*) (Collins and Latta, 2009). Initially bald eagles (*Haliaeetus leucocephalus*) were the apex predator of SCI, mainly eating fish rather than preying on terrestrial species (Clark, 2009). However, by the 1950s there were no more breeding bald eagles on the island due to DDT in the water causing eagles egg shells to thin; all remaining bald eagles migrated off the island in the 1960s (Clark, 2009). The departure of bald eagles from SCI opened a niche for golden eagles to migrate from the mainland and establish territory on the island (Collins and Latta, 2006). Golden eagles are large raptors with a wingspan of up to two meters and weighing between 3.6 and 5.7 kg (Todd, 2000). Due to their large size, golden eagles require 200–300 g of food daily (Todd, 2000). They are opportunistic specialists in their feeding habits, often exhibiting a consistent preference for a specific prey until that prey diminishes, at which point they will feed on more diverse prey (Collins and Latta, 2006; Todd, 2000). Mated pairs of golden eagles hunt within territories they establish around their nests (known as eyries) with territories areas ranging from 25 to 145 km^2 where the territories of mated pairs can marginally overlap (see Fig. 1) (Todd, 2000). However, during eagle breeding season (March to August), eagles hunt within a significantly reduced area (1.6–40 km^2) centered around their eyries, and these reduced hunting regions do not overlap with other mated eagle pairs (Todd, 2000). We will refer to these smaller regions as *breeding territories* and the larger regions as *hunting territories* (Todd, 2000). Golden eagle mated pairs maintain multiple eyries (Collins and Latta, 2006; Todd, 2000). Each breeding season, eagles choose one of their eyries for mating, at which point it becomes the center of the mated pair's breeding territory (see Fig. 7) (Collins and Latta, 2006; Todd, 2000). Over several years, a mated pair will cycle through each of its eyries (Todd, 2000). Based on observation and nest excavation, five mated pairs once occupied SCI, with first sightings occurring around 1995 (Collins and Latta, 2006). Fig. 1 shows the nest locations and hunting territories of each of the five mated pairs on SCI using data from Collins and Latta (2006) and Latta (2004).

In the early 1990s, conservationists observed a rapid population decline of island fox populations. In 1994, the island fox population on SCI was estimated to be 1465. By 2004, that number had declined to between 75 and 100 foxes (see Fig. 2) (Coonan, 2003). The main reason for this decline was

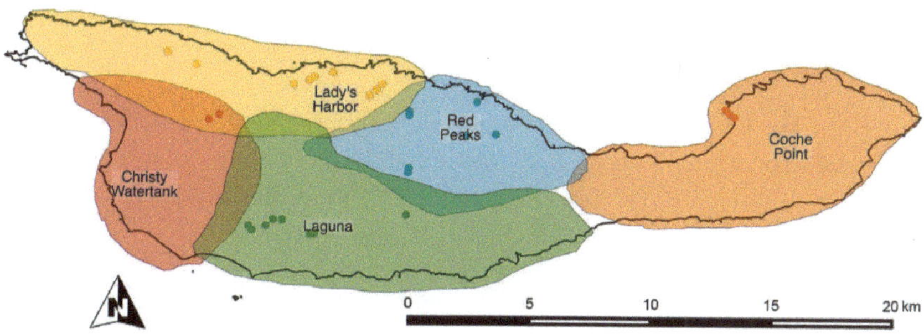

FIG. 1 Map of eagle nest locations and hunting territories on Santa Cruz island using data from Collins and Latta (2006) and Latta (2004). Each *shaded* region represents eagle mated pair hunting territory. Each corresponding *colored dot* represents a mated pair nest location. During the annual breeding season, paired eagles will choose a nest within their hunting territory at random and create a breeding territory encompassing this nest.

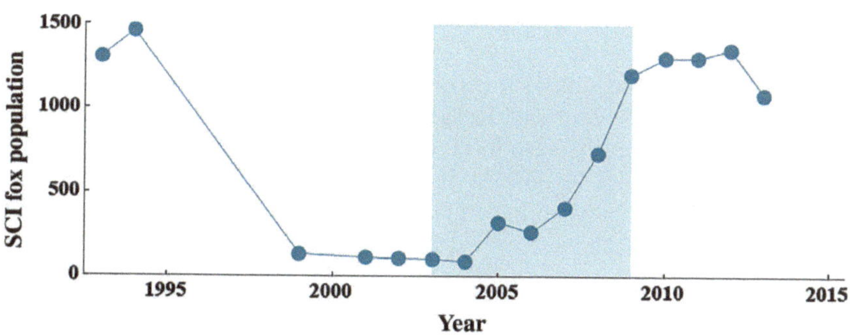

FIG. 2 The SCI island fox population size in the 1990s–2000s. The data points are taken from Island Fox Recovery Annual Reports for 1999–2013 and other observational studies (Bakker et al., 2003; Coonan, 2003, 2008, 2009, 2010, 2011, 2012, 2013, 2014; Coonan and Dennis, 2006, 2007; Coonan and Rutz, 2001; Coonan et al., 2010; Roemer et al., 2001, 2002; Schwemm, 2008). The *blue shaded* area represents times when recovery efforts took place.

predation by golden eagles, though disease and competition may have also had an effect (Coonan, 2003; Scott et al., 2014). In a previous study, we developed an agent-based model (ABM) that demonstrated the existence of an emergent Allee effect within the SCI fox population (Scott et al., 2014). Furthermore, we determined that the presence of a detrimental recessive allele that reduces the probability of successful reproduction could exacerbate negative per capita growth rates at low population sizes (Scott et al., 2014).

The model presented here focuses on the population dynamics of the island fox under golden eagle predation. Classical mathematical models of predator–prey interactions use the Lotka–Volterra method or other systems of equations (i.e., Kermack–McKendrick or Jacob–Monod models) (Hoppensteadt, 2006). These models are mainly useful for relatively large populations of both predators and prey, which is not the condition for SCI (Hoppensteadt, 2006). Because we assume the population sizes of both the predator and the prey are small, the results of stochastic decisions and individual interactions heavily impact population dynamics and viability. Therefore, agent-based models (ABMs) are more appropriate.

ABMs are a class of mathematical and computational models in which individuals (or agents) are unique and autonomous entities that can locally interact with other individuals, as well as with their environment (Railsback and Grimm, 2005, 2011; Scott et al., 2014). This class of models is best used for scientific questions in which interactions between agents are complex, when spatial considerations are crucial, when heterogeneous and complex topological interactions between agents are present, and when the agents included exhibit complex behavior (Bonabeau, 2002; Heppenstall et al., 2012). While equation-based models can be used in these cases, their assumptions often mask the inherent fluctuations and heterogeneity of the systems in question (Bonabeau, 2002). For example, aggregate flow equations may be used when looking at movement of collectives, but the equations usually assume homogeneous mixing (Bonabeau, 2002). In reality, the interaction network of the individuals within the collective can lead deviations from expected aggregate behavior (Bonabeau, 2002). The inclusion of stochastic processes and heterogeneity allows for emergent behaviors to be observed (Railsback and Grimm, 2011). An important feature of ABMs is the ability to inform environmental characteristics. Not only can these characteristics be defined directly by the user, but many ABM softwares also have the ability to integrate GIS data, enhancing the heterogeneous spatial realism of the model (Heppenstall et al., 2012). ABMs are applicable at a wide range of scales and in a wide range of fields, from human behavior to cellular dynamics and ecological systems (Bonabeau, 2002; Cannata et al., 2013; Conner et al., 2008; Gaff, 2011). The ABM we have developed incorporates both fox and eagle behavior and interactions, along with spatial data specific to SCI, to create a realistic, spatially explicit, multiagent-based model.

The remainder of this chapter presents a detailed description of our fox–eagle ABM and the results of several different model analyses. Specifically,

Section 2 describes the types of GIS data used and how the GIS data were incorporated into the model. Within the section we provide some discussion about considerations that must be made specifically when importing and exporting GIS data in the ABM software NetLogo (Wilensky, 1999). Section 3 describes the model in detail using the Overview, Design Concepts, and Details (ODD) Protocol proposed by Grimm et al. (2010), Railsback and Grimm (2005), and Railsback and Grimm (2011). In Section 4, the results from simulations of the model under various conditions are presented. Lastly, Section 5 discusses what can be concluded from the simulations and discusses the implications of the model.

2 INCORPORATING GIS DATA INTO AN ABM

Geographic Information Systems (GIS) are used to manipulate, analyze, and present geographic and spatial data. Numerous platforms with varying functionalities and price-ranges exist to perform these geographic analyses. ArcMap and QGIS are two commonly used software platforms that can be used to edit and analyze geospatial data (ESRI, 2011; QGIS Development Team, 2009). QGIS software was used for all geospatial analyses in this project. The purpose of these analyses using GIS techniques was to create and manipulate data that can inform the agent-based model presented in Section 3 and to analyze geospatial data that was generated from model simulations (see Section 4).

2.1 GIS Data Types

Two types of data are often used in GIS analyses: raster and vector data. Raster data is defined by matrices of pixels, such as satellite imagery or hand-drawn maps, while vector data is defined by points, lines, and polygons. In order to be imported into the ABM software NetLogo, all data must to be in vector form (Wilensky, 1999). Additionally, each vector item (point, line, or polygon) can have attributes associated with it in addition to its geographic coordinates. For example, the fox location data not only includes the latitude and longitudes of the foxes, but it also includes the year and week at which the point is recorded, allowing us to look at the changes in spatial distribution over time (see density maps generated from simulations shown in Section 4).

For the fox–eagle ABM we import data on vegetation type, which is composed of polygons (see Fig. 3 for a zoomed in region of the vegetation map showing the polygons more clearly) and data on the locations of golden eagle eyries and their hunting territories (see Fig. 1). Note the locations of the eyries are stored as points while the hunting territories are stored as polygons.

The vegetation data were taken from a spatial database created by Cohen et al. (2009). Aerial photos were broken down by vegetation type, fennel category, and a cover rating (Cohen et al., 2009). To determine vegetation type,

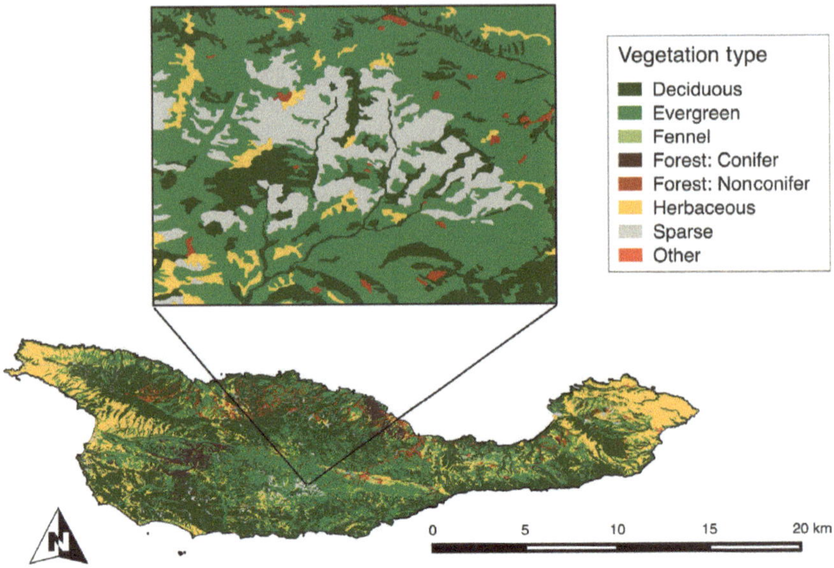

FIG. 3 The vegetation composition of Santa Cruz island classified by vegetation type. The zoomed in portion more clearly shows the polygons defining the area of each category. Data used to create this map was from Cohen et al. (2009).

the photos were broken down into eight vegetation groups: deciduous forest, evergreen forest, conifer woodlands, nonconifer woodlands, fennel, herbaceous, sparse, and an "other" category that accounts for nonvegetative areas (see Fig. 3) (Cohen et al., 2009). For our purposes, we were only concerned with the eight vegetation groups because fox territoriality depends on vegetation type (Cohen et al., 2009). We therefore divided these eight vegetation groups into grassland, mixed, or uninhabitable vegetation types. The forest and woodland groups were considered to be "mixed" vegetation, while the fennel and herbaceous groups were considered to be "grasslands." The sparse and other groups were considered to be "uninhabitable" for foxes (see Table 1 and Fig. 6) (Cohen et al., 2009).

The eagle hunting territories were taken from Latta's report on golden eagle translocation (Latta, 2004). The map of territory boundaries was overlaid with the map of SCI and a polygon layer was created using the Freehand Editing plugin (QGIS Development Team, 2009). Eagle breeding territories were created by creating circles with area 25 km^2 (radius 2.8217) around each nest site and creating a new polygon layer of each territory (see Figs. 1 and 7) (Collins and Latta, 2006; Todd, 2000).

After the model simulations are run, we create and load the location data into QGIS and run the Heatmap plugin. The produced raster images are then rectified onto the map of the island and used to show the spatial density of foxes both with and without golden eagle predation (see density maps generated from simulations shown in Section 4) (QGIS Development Team, 2009).

2.2 Projections and Coordinate Systems for GIS Data

Projections and Coordinate Systems vastly affect the way data are displayed and measured in GIS. Geographic coordinate systems are systems of latitude and longitude, which use the equator as zero degrees and form parallel lines of latitude North and South and longitude East and West. Projections are used to "project" the three-dimensional surface of the Earth onto a two-dimensional plane by converting geographic coordinates into an (X, Y) coordinate system, where X is longitude and Y is latitude, or by projecting the surface of a sphere onto a cone or cylinder, and then cutting and unrolling the cone or cylinder into a flat surface. Distortions will always exist when projecting the surface of a sphere onto a plane, so an appropriate projection must be chosen for every map to minimize distortion in the area of interest.

In a *conic projection*, the surface of the Earth (approximated as a sphere) is projected on to the surface of a cone. The projection is created by intersecting a cone with the globe. In a *tangent conic projection* the cone creates only one curve of intersection (see Fig. 4A). The dimensions of the cone are chosen such that the curve of intersection forms a circle that represents a single latitude (or an ellipse that crosses multiple latitudes if using an oblique conic projection). Fig. 4A shows the former case where the curve of intersection occurs at a single latitude, 30 degrees above the equator (i.e., 30 degrees N) in this case (shown in green). When the surface of the Earth is then projected onto the cone, there is no distance distortion at the latitude of intersection. The further away points are from the latitude of intersection, the greater the distortion of distance between them. In a *secant conic projection*, the cone intersects the globe at two latitudes called *standard parallels* (see Fig. 4B), thus creating two latitudes where no distance distortion occurs. Once the surface of the globe is projected onto the surface of the cone, the cone is cut and unfurled to create a flat surface containing the projected map (see Fig. 4C).

There are a variety of transformations which describe how the points on the globe are mapped to the surface of the cone. Two commonly used projections are *Albers Conic Equal-Area Projection* and *Lambert Conformal Conic*. Both methods use secant conic projections. The Albers projection was formulated to make all areas on the projected map have proportionally the same area as on Earth (the globe) (ESRI, 2011); see Weisstein (n.a.-a) for transformation equations. The Lambert projection, on the other hand, was formulated to maintain angles between latitude and longitude lines which minimizes shape distortion over small areas (ESRI, 2011); see Weisstein (n.a.-b) for transformation equations. The projected map in Fig. 4C uses the Albers conic equal-area projection with standard parallels of 20 degrees N (red) and 50 degrees N (blue).

For our purposes, a projection was needed that could minimize the distortion of Santa Cruz island off the coast of California. However, we also had to consider the compatibility of the agent-based modeling software, NetLogo, to handle projections. NetLogo does not support *all* projected coordinate systems. A full list of the supported projected coordinate reference systems

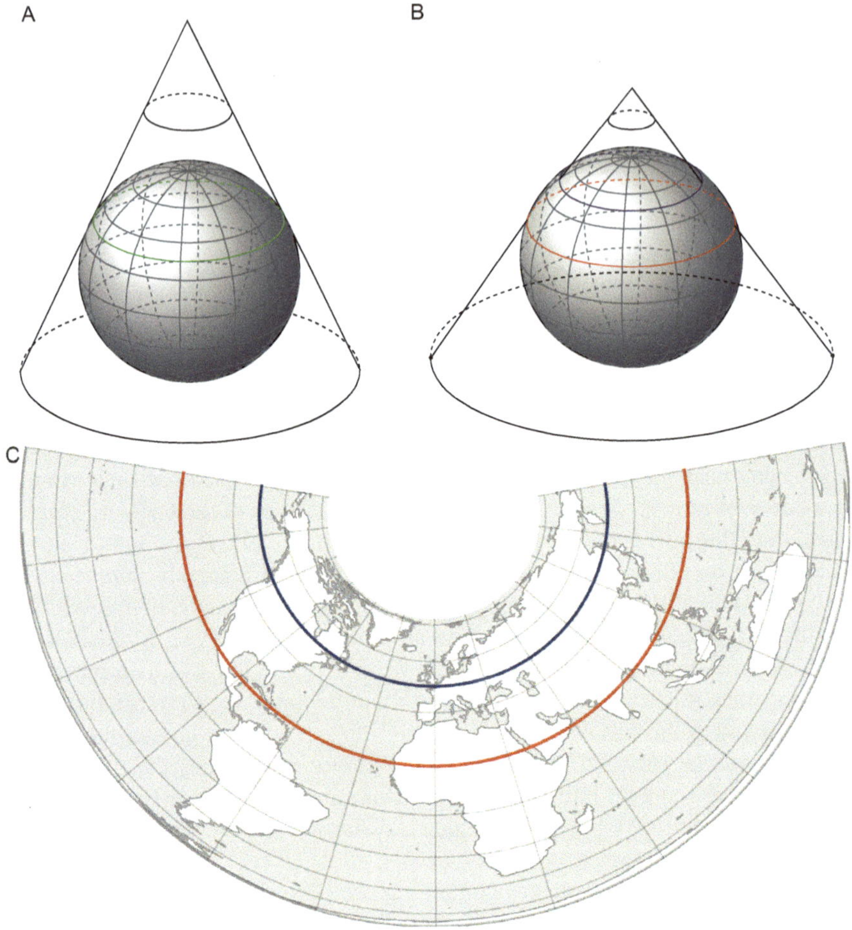

FIG. 4 A conic projection of the earth projects the surface of the globe (approximated as a sphere) onto the surface of a cone. Along the curve(s) where the cone and sphere intersect, there is no distortion in measures of distance in the projection. In the examples shown here, the tangent curve of intersection occurs at latitude 30 degrees (*green*) above the equator in (A), and the secant curves of intersection occur at the standard parallels of 20 degrees (*red*) and 50 degrees (*blue*) above the equator in (B) and (C). The projected surface in (C) uses an Albers (conic) projection.

(CRS) can be found the NetLogo documentation (Wilensky, 1999). The conic projections supported by NetLogo include

- Albers conic equal area
- Lambert conformal conic 2SP
- Polyconic
- Equidistant conic

We used the Albers conic equal-area projection with CRS NAD83 for North America (EPSG:4269) for all maps other than the vegetation map. For this map, we used the Albers conic equal-area projection with CRS NAD83/ California Albers (EPSG:3310). These projections were chosen due to their abilities to minimize data distortion while remaining compatible with NetLogo software.

3 MODEL DESCRIPTION

The description of the agent-based model presented here uses the Overview, Design Concepts, and Details (ODD) Protocol as described in Grimm et al. (2010), Railsback and Grimm (2005), and Railsback and Grimm (2011).

3.1 Overview

3.1.1 Purpose

The purpose of this model is to simulate the population dynamics of the SCI fox and to determine the resulting effects on the island fox spatial distribution and population viability under golden eagle predation. This allows us to observe the impact of golden eagle predation on the population of island foxes. The model incorporates golden eagle and island fox behaviors to realistically replicate individual decisions and species interactions. Additionally, the model makes use of GIS data of SCI vegetation and golden eagle nest sites and territory ranges. We therefore realistically constrain the establishment of island fox territories and the hunting behaviors of the golden eagles based on this data. We use the model to (i) simulate the population decline seen on the island in the early 1990s, (ii) determine population size-dependent population viability of the island foxes in the presence and absence of golden eagle predation, and (iii) observe spatio-temporal distribution of island foxes under golden eagle predation.

3.1.2 Entities, State Variables, and Scales

The agents in this model are male foxes, female foxes, eagles, and alternative food sources (AFS) for the eagles. Both male and female foxes have state variables for their age, number of offspring, whether they have established a territory, the area (in km^2) of their territory (set to 0 if they have not established a territory), the number of attempts at forming a territory, and their mass (in kg). Each female fox additionally has a state variable for her gestation week (set to 0 if not currently gestating offspring). Each eagle has state variables for the mass of prey (in kg/wk) consumed, the number of hunting attempts during the current time step, an id number for their hunting territory and breeding territory, probability of consuming foxes over AFS, the number of foxes consumed per week, and a Boolean variable as to whether they have or have not consumed at least one fox within a given week. Finally, each AFS has a variable determining its mass (in kg).

Each patch within the model represents a physical location with area $0.0049\ km^2$ ($0.07\ km \times 0.07\ km$), which is determined through integration of GIS data into the model (Cohen et al., 2009). Each patch has state variables for the patch vegetation type and vegetation group, which defines the majority of vegetation type present in that patch, based on GIS data (Collins and Latta, 2006). Patches also have an owner variable that indicates the fox that has made a territory there, and a variable indicating whether the patch is land or water. Lastly, each patch has Boolean variables determining whether the patch is part of a specific eagle hunting or breeding territory. GIS data for island vegetation type and eagle hunting and breeding territories are imported in the initialization procedure (see Section 3.3.1 for details; Cohen et al., 2009).

The model's global parameters include GIS datasets for the island, the eagle breeding territories, the eagle hunting territories, and the island vegetation. There are also variables for the number of patches on the island and several temporary variables that allow GIS data to be integrated into the model based on the associated attribute tables. Global parameters for foxes include successful reproduction rate, reproductive age, territory search radius, size, a counter for the number of foxes, a current mate variable, and a count of the overall fox offspring produced. There are also global variables pertaining to AFS, which include proportions of the total island biomass each type of AFS composes, the percent of each AFS type that composes an eagle's weekly diet, the number of each type of AFS created, the size of the AFS agents, and the total biomass of AFS present on the island. The model includes observational global variables for the number of foxes present at week 25 each year (for yearly population counts), the per capita growth rate of the fox population from the initial population size, and the yearly per capita growth rate of the fox population. There are also week and year counters.

Lastly, each time step (tick) represents 1 week and the simulation runs until there are no remaining foxes or 200 years (10,400 weeks) have passed.

3.1.3 Process Overview

The model is initialized with a user-defined number of male and female foxes on the island. The foxes are distributed across the island randomly. Those of reproductive age attempt to form mated pairs and successful mated pairs are placed into territories. Ten eagles (consistent with the number of verified nesting pairs on the island) are initialized and pairs are placed in each hunting territory (see Fig. 1) (Collins and Latta, 2006; Latta, 2004).

At each tick, foxes attempt to mate with the opposite sex (see DISPERSE and MAKE-FOX-TERRITORY submodels). If a successful mated pair is formed, the female fox will attempt to produce offspring based on the successful reproduction rate (see BREED-FOXES submodel). If the female is successful, she will produce offspring (see BIRTH submodel). Impregnated female foxes increase their gestation by one and the age of all foxes (counted in weeks) is incremented. AFS are distributed around the island as alternative food sources for the eagles (see DISTRIBUTE-FOOD submodel). The eagles then set their energy to zero and begin to feed until they reach their minimum weekly energy requirements (see FORAGE submodel). The week counter is incremented and, once 52 weeks have passed, the year counter is incremented. The foxes die naturally based on a stochastic decision (see DEATH submodel). Lastly, the total number of foxes is calculated at the end of each time step; if there are no remaining foxes or the model has run for 200 years, then the simulation will stop. Fig. 5 shows how each procedure leads to the next within this model.

3.2 Design Concepts

3.2.1 Basic Principles

The basic principle of this simulation is to replicate Santa Cruz island fox dynamics in the presence of golden eagle predation. With 10 golden eagles and an initial population size of 1464 foxes, we observe similar dynamics to those seen in the early 1990s (Coonan, 2003). The number of eagles initially present and preying on island foxes is based on verified nests on the island (Collins and Latta, 2006). We model eagle feeding habits and diet composition using data collected from nest excavations (Collins and Latta, 2006). The fox populations reproduce and behave realistically based on data collected through behavioral and ecological field studies (Coonan, 2003; Moore and Collins, 1995; Roemer et al., 2001; Scott et al., 2014; Vissman, 2004).

3.2.2 Emergence

When golden eagle predation is not included in this model and there is an initial population of 40 male foxes and 40 female foxes, we see the emergence of logistic growth with a carrying capacity around 1440 (see Section 4.1.1). This is similar to the population size observed in 1994 (see Fig. 2) (Coonan, 2003).

3.2.3 Adaptation

Male and female foxes determine the size of their territory based on surrounding vegetation (Vissman, 2004). The smaller territory radius is 0.36 km, while the larger territory radius is 0.87 km (Vissman, 2004). If mated foxes are within an area that is more than 30% grassland, then they will create the larger territory size, if not, they will create the smaller territory size (Vissman, 2004). In the model, foxes have a limited attempts to find an unoccupied, habitable territory within the boundaries of the island. For details, see MAKE-FOX-TERRITORY submodel in Section 3.3.2. In this way, as the island becomes more crowded, the territory sizes become smaller and the chances of finding a suitable territory within the allotted trials decrease. If a fox mated pair is unable to find a suitable territory within the five allotted trials, then they will break their mated pair and

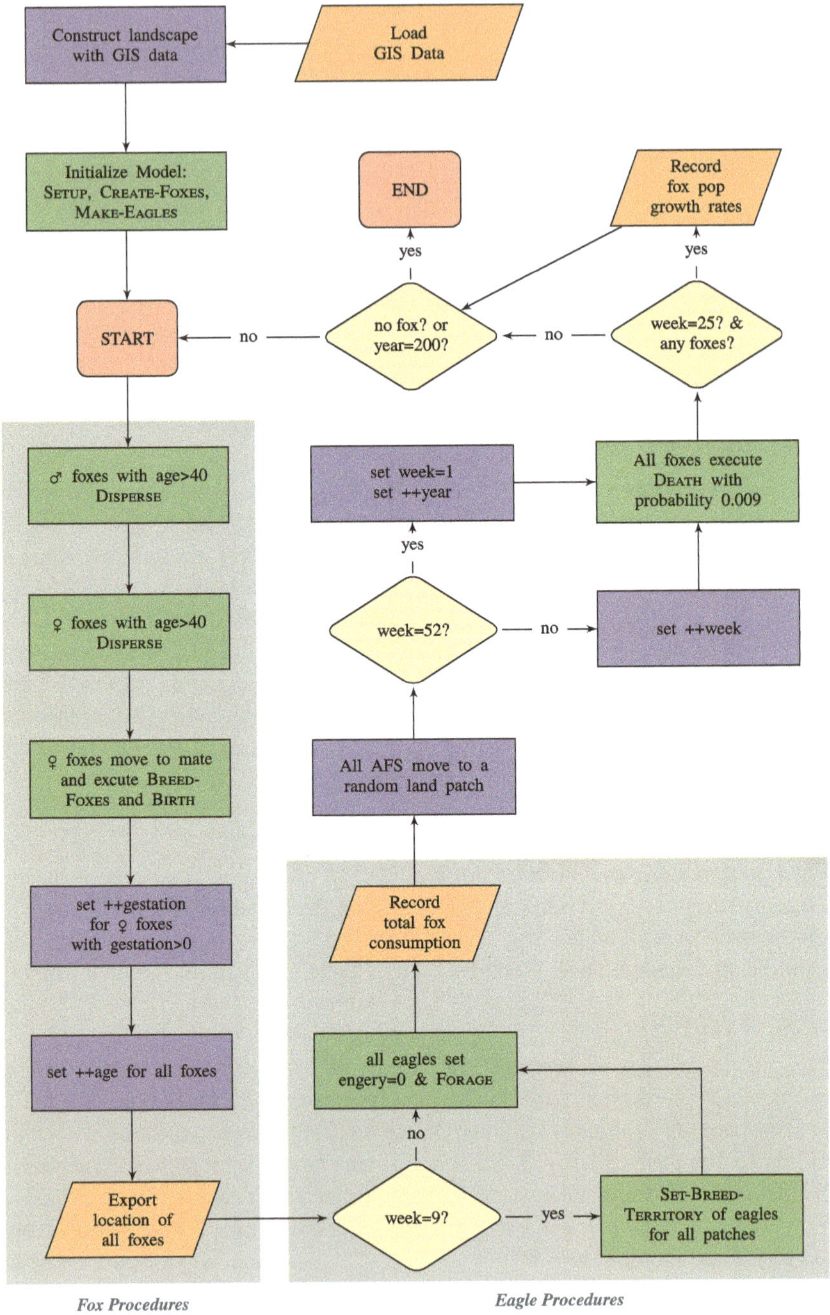

FIG. 5 Flow diagram of the model. Procedures in *green* represent specific submodels (see Section 3.3.2). The loop represents the order of procedures during the execution of one time step.

attempt to mate with other individuals. After the birth of fox pups, foxes will not move from their natal territory until they are of reproductive age (40 weeks). If there are no available territories due to high fox population densities, fox pups may remain in their natal territories their entire lives, even after reaching reproductive age (Coonan, 2003).

3.2.4 Objectives

The objective of a fox is to mate and successfully reproduce. Their territory decisions are based on the vegetation around them. Mating decisions depend on availability of mates and whether they are of reproductive age and are unmated. An eagle's objective is to reach their weekly energy requirements within the number of allowed trials. The AFS have no objectives in this model.

3.2.5 Learning

The foxes, eagles, and AFS have no learning capacity in this model.

3.2.6 Prediction

The foxes, eagles, and AFS have no predictive capacity in this model.

3.2.7 Sensing

Foxes can sense the age, gender, and mated-status of foxes within their search radius, which directs movement and mating. They can also sense whether patches are occupied by other foxes and the vegetation type there. Foxes can sense breeding season (weeks 7–12) and will attempt to reproduce if they are mated (Moore and Collins, 1995). Eagles sense their own breeding season (weeks 9–32) and whether patches are within their hunting or breeding territories (Todd, 2000).

3.2.8 Interaction

Male and female foxes interact to establish mated pairs and to form territories on the island. Foxes and eagles have an antagonistic interaction when the eagles hunt and feed. Eagles and AFS have a similar antagonistic interaction. Foxes and AFS have no interaction.

3.2.9 Stochasticity

Within this system, there are many processes dependent upon stochastic decisions. Whenever a set of entities (agents or patches) executes a command or procedure, the order in which each entity within the set are selected to execute the task is randomized. Additionally, there are many decisions made by the agents or patches which are stochastic. Within the framework of the NetLogo software, stochastic decisions by entities are determined by random number generation or random selection of an agent or patch from a set of entities with equal probability.

3.2.9.1 Decisions Using Random Number Generation

Let X be a decision that occurs in the model with probability P, and does not occur with probability $1 - P$. Each time decision X is to be made, we sample the uniform distribution $\mathcal{U}[0,1]$ to generate the value $y \in \mathbb{R}$; the event X occurs if $y \leq p$, and does not occur otherwise. Here, we provide the specific probabilities used for stochastic model decisions.

- Each fox pair that is not currently gestating offspring will successfully reproduce with probability $p = 0.61$ in a given time step during the fox breeding season (weeks 7–12) (Scott et al., 2014).
- At each time step each fox will die naturally with probability $p = 0.009$ (Scott et al., 2014).
- The number of offspring born to any pregnant female fox is one with $p_1 = 0.558$, two with $p_2 = 0.356$, and three with $p_3 = 0.086$ (note $p_1 + p_2 + p_3 = 1$) (Coonan, 2003). In this case given the value y sampled from the uniform distribution $\mathcal{U}[0,1]$, the pregnant female fox will produce a litter of

$$\begin{cases} 1 \text{ pup} & \text{if} \quad y \leq p_1 \\ 2 \text{ pups} & \text{if} \quad p_1 > y \leq p_1 + p_2 \\ 3 \text{ pups} & \text{if} \quad y > p_1 + p_2 \end{cases}$$

Other stochastic events require sampling from uniform distributions over wider or smaller ranges.

- Upon the creation or birth of each fox it is given a randomly generated mass. The mass is sampled from the uniform distribution $\mathcal{U}[1.200, 2.700]$ kg for male foxes and $\mathcal{U}[1.070, 2.722]$ kg for female foxes (Moore and Collins, 1995).
- Each eagle has an individual preference for choosing a fox as its prey over an AFS (when both are present within its territory). For each eagle this preference value is sampled (during initialization) from the uniform distribution $\mathcal{U}[0.457, 0.577]$ (Collins and Latta, 2009).

3.2.9.2 Decisions Using Random Selection of Entities

When entities (agents and patches) interact, typically one entity is stochastically selecting the entity or subset of entities with whom it will interact. The selecting entity makes its selection randomly from a group where each entity within the group has an equal probability of being selected. Here, we provide the details on who the selecting entities are and from which group of entities they are selecting.

- When mating, each fox chooses an individual of the opposite sex and of reproductive age randomly from within their search radius (see MAKE-FOX-TERRITORY submodel in Section 3.3.2).
- At each breeding season, the eagles randomly choose which nest to occupy for that year.

- In the case where breeding territories of two eagles overlap (see Fig. 7), the eagle pair that gets use of the shared space is chosen randomly.
- When an eagle is foraging, if foxes are present, the eagle selects a fox at random from within its territory with probability equal to the eagle's individual preference for choosing a fox as its prey; otherwise the eagle selects an AFS at random from within its territory (see FORAGE submodel in Section 3.3.2). Note, the eagle's territory is restricted to its breeding territory during breeding season (weeks 9–32).
- Each time step the AFS move to a randomly selected land patch (see DISTRIBUTE-FOOD submodel in Section 3.3.2).
- If a fox is unable to find a mate (see the DISPERSE submodel in Section 3.3.2), the fox will move to a randomly selected patch within its search radius that is a land patch and does not contain a fox of the opposite sex.

3.2.10 Collectives

The collectives of this model include the mated pairs of female and male foxes. Once fox pups are birthed, they will remain in their natal territory and add to the collective until they are of reproductive age and can establish their own territory. Eagles are also in mated pairs and forage only within their hunting or breeding territories, which are shared with a mate. AFS do not form collectives.

3.2.11 Observation

At each time step, we record the fox population size, the number of foxes consumed, the current week, and the current year. At week 25 of each year, we record the per capita growth rate over the previous year as well as the per capita growth rate from the initial population size (see Scott et al. (2014) for associated formulas). We also record the total fox population and the number of eagles feeding on foxes each week.

3.3 Details

3.3.1 Initialization

A data set giving the spatial distribution of vegetation types on SCI is used to define the patch variables of `vegetation-type` and `vegetation-group`, and to determine if a patch is land or water (Cohen et al., 2009). Table 1 and Fig. 6 show the vegetation composition and distribution of SCI. Location data were rectified onto an SCI map and used to define the five golden eagle hunting territories and eagle nest locations (Collins and Latta, 2009). Eagle breeding territories were defined with 25 km^2 buffer regions around each nest location (Todd, 2000). This GIS data set is used to define eagle hunting and breeding territories.

During the setup procedure, global variables are initialized, all GIS data sets are loaded, and corresponding patch variables defined. An initial fox

TABLE 1 SCI Composition by Vegetation Type and Group (Cohen et al., 2009)

Vegetation Type	% of Patches	Vegetation Group	% of Patches
Fennel Herbaceous	1.4 39.8	Grassland	41.2
Deciduous Evergreen Forest-conifer Forest-nonconifer	18.0 27.5 3.3 9.4	Mixed	58.2
Sparse Other	0.1 0.5	Uninhabitable	0.6

FIG. 6 Vegetation map of SCI. In *yellow* is grassland where fox territory size is larger. In *dark green* is mixed vegetation where fox territory is smaller. In *gray* is uninhabitable vegetation where foxes cannot form territories (Cohen et al., 2009).

population is generated with an even sex ratio and the initial age of each fox is drawn from a uniform distribution over the range [0,312] weeks (note 312 weeks is $\sim$6 years) (Coonan, 2003). The initial fox population is randomly distributed over the island, a pair of eagles is placed on within each golden eagle hunting territory (a total of 10 eagles), and the AFS are distributed across the island (see the Distribute-Food submodel in Section 3.3.2). If there exists a suitable mate for a fox, they will form a mated pair and make a territory. The territory size of each mated pair is dependent upon the vegetation they inhabit (see Make-Fox-Territory submodel in Section 3.3.2 for details) (Coonan, 2003).

3.3.2 Submodels

The various submodels of the ABM are described here in detail. Note, submodels are also referred to as *procedures* through the ODD description. The italicized text in each procedure's description identifies which set of entities executes the procedure. When a set of entities executes a procedure, the order in which each entity of the set is selected to execute the procedure is random.

Forage

This procedure is executed by eagles. First, eagles set their hunting trials to zero. Until they have reached their required weekly energy level and while they have remaining trials (of the allotted 7), eagles will look for prey in their hunting territory. During eagle breeding season (weeks 9–32), eagle hunting territory is limited to one of the breeding territories, which surround a randomly chosen nest (Collins and Latta, 2006). If there are any foxes within the hunting territory, then the eagles will consume one of the foxes within its territory. This occurs with a probability defined by the eagle's variable of their preference for eating foxes ($p \in [0.457, 0.577]$). If the eagle does not feed on a fox, it will randomly select an AFS within their territory. Once the eagle has selected its prey, it increments its energy by the mass of the fox or AFS selected. If a fox was consumed, the global number of foxes consumed is incremented by one and the eagle sets the Boolean variable determining that they have consumed a fox in that time step to "true." If there are no foxes in the eagle's territory, then the eagle will randomly select an AFS within their hunting territory and increment its energy by the energy of the AFS. The amount of energy obtained from each prey type is determined by the weight of each prey (see Table 2). The eagle will set its `trials` variable equal to zero after successful foraging. If there are no AFS or foxes in the eagle's territory, it will increment its trials by one. If the `energy` variable of the eagle is still below 1.68 kg of food and the `trials` variable of the eagle is still below 7, the eagle will make another attempt to forage.

Disperse

This procedure is executed by male and female foxes of reproductive age. If there are any unmated foxes of reproductive age and of the opposite sex within the executing fox's search radius, then a mated pair will be created and the male will attempt to form a territory (see the Make-Fox-Territory procedure). If there are no unmated foxes of reproductive age and the opposite sex in the search radius, then the fox will randomly move to one of the island patches within its search radius that has no foxes of the opposite sex.

Make-Fox-Territory

This procedure is executed by newly mated males. The large and small fox territory sizes are set to 0.87 km^2 and 0.36 km^2, respectively (Vissman, 2004).

TABLE 2 Composition of a Golden Eagle's Diet

Common Name	Avg. Individual Weight (kg)	% of Eagle Diet
Deer mouse	0.020	0.15
Western spotted skunk	0.560	1.04
European mouflon sheep	2.300	12.77
Double-crested/Brandt's cormorant	1.962	29.05
Herring gull	1.135	4.20
California gull	0.607	2.25
Western/glaucous-winged gull	0.875	21.05
California quail	0.173	0.64
Common raven	1.199	28.85

The original data comes from a study which excavated four golden eagle nests on SCI to determine the species of discarded remains (Collins and Latta, 2006). For our model, only mammalian and bird species were included as only trace amounts of fish, invertebrate, and crustacean remains were found in the excavated nests.

Then, the fox's territory size is initially set to the smaller territory centered at the fox's current location. Next, the `percent-grassland` variable is set as the number of land patches within the fox's territory radius that have `vegetation-group` *grassland* divided by the total number of patches within this radius, that is

$$\texttt{percent-grassland} = \frac{\#\text{ of grassland patches in territory}}{\text{territory size}}.$$

If `percent-grassland` $<30\%$, then the fox's territory remains as the small territory size, however if `percent-grassland` $\geq 30\%$, then the fox's territory size will change to the larger territory size. Next, the fox initializes its number of trials to establish a territory with its mate to 0. The fox then searches for a viable territory until he either establishes a suitable territory or reaches five trials.

While there are any unsatisfactory patches (patches that are occupied or of uninhabitable vegetation) within the male fox territory, the male fox will try to find a suitable territory. He will change his territory size by a small factor with each successive attempt. The specifics of the territory trial process are as follows: if there are any land patches within his territory size that are unoccupied, he will randomly move to one of those patches, increase his number of trials by 0.4 and multiply his territory size by a factor of 0.98. If there are no unoccupied, land patches within the male territory, he will randomly move to one of the patches in his territory size, increment his territory trials by 1, and multiply his territory size by a factor of 0.95. Note, depending on how densely populated the island is with foxes, the male fox may have between 5 and 12 trials to establish a territory; the more densely populated the island the fewer trials the fox is likely to have. In this way, when there are still viable patches, foxes have more attempts to decrease their territory size slightly in order to find a suitable territory.

Once the male fox has a set his territory, he checks to be sure the territory has at least 45 land patches (0.2205 km^2, a lower bound on territory size). If so, then he is successful in forming a territory and will set his Boolean `territorial?` variable to "true" and will set the owner of the patches within his territory to his ID number. Additionally, he will set his mate's Boolean `territorial?` variable to "true" and define her territory size as the same as his. In the case where a territory cannot be made, the male fox will set his `territorial?` variable to "false", break his mated pair, and the male fox will move to a random location on the island.

Set-Eagle-Breeding-Territories

This procedure is not executed by any entities; instead it is called by the main procedure at week 9 each year (i.e., once every 52 time steps). Note, there are five different eagle territories: Christy's Watertank (CW), Coche Point (CP), Lady's Harbor (LH), Laguna (La), and Red Peaks (RP) (Collins and Latta, 2006). The locations of the eagle territories and eyrie locations are accessed through imported GIS data (see Fig. 1) into the model during initialization (Collins and Latta, 2006). Each patch in the landscape has Boolean variables `CW-BT?`, `CP-BT?`, `LH-BT?`, `La-BT?`, and `RP-BT?`, which will be set to "true" when that patch is a part of the associated breeding territory for that year's breeding season. At the beginning of this procedure, all five of these Boolean variables are set to "false". Next, one nest for each mated eagle pair is chosen; this will be the center of each pair's breeding territory for the current breeding season. Using the imported GIS data, the appropriate Boolean variables are set to "true" to indicate which patches are a part of each pair's breeding territory for this breeding season. Depending on which eyrie locations were randomly chosen, there may be overlap between (1) Christy's Watertank and Lady's Harbor, (2) Lady's Harbor and Red Peaks, (3) Christy's Watertank and Laguna, and (4) Laguna and Red Peaks; see Fig. 7 for an example of one selection of breeding territories showing the potential overlaps (1) and (4). If any two breeding territories overlap, one of the territories is randomly chosen to occupy the overlapping area for the current breeding season.

Breed-Foxes

This procedure is executed by female foxes. If a female fox is of reproductive age (>40 weeks old), not already pregnant, is part of a territorial mated pair, and it is between weeks 7 and 12 of the current model year, then the female

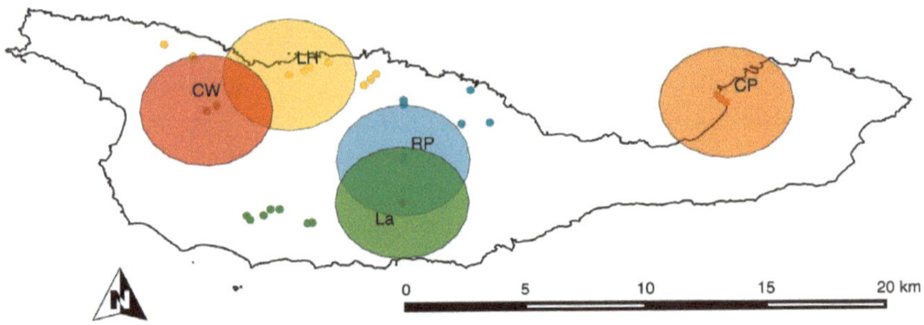

FIG. 7 An example of breeding territories for the five eagle pairs: Christy's Watertank (CW, *red*), Coche Point (CP, *orange*), Lady's Harbor (LH, *yellow*), Laguna (La, *green*), and Red Peaks (RP, *blue*) (Collins and Latta, 2006). The Eyrie locations are shown as points in corresponding *colors.*

fox will conceive a litter with probability $p = 0.61$ (i.e., there is a 61% chance of successful reproduction) (Sanchez and Hudgens, 2011).

Birth

This procedure is executed by female foxes. If the gestation period of a female fox has reached 7 weeks, she gives birth. The number of offspring born is randomly chosen with a 55.8% chance of producing 1 offspring, a 35.6% chance of producing 2 offspring, and a 8.6% chance of producing 3 offspring (Coonan, 2003). Next, male and female foxes of the mated pair increment their numbers of progeny by the number of offspring birthed. Male and female fox pups are produced with an equal chance due to an even sex ratio (Coonan, 2003). The offspring's age is set to zero, gestation (if female) is set to zero, Boolean variable `territorial?` is set to "false" indicating the pups have not created a mated pair and established territory yet, and mass (or energy) is set to random value in the range [1.200, 2.700] kg if male and [1.070, 2.722] kg if female. Lastly, the mother female fox resets her gestation to zero (Moore and Collins, 1995).

Death

This procedure is executed by foxes at random with probability 0.009 (the natural death rate) (Scott et al., 2014). Any patches within the territory of the dead fox will be reset to have no owner. If the dead fox has a mate, the link establishing the mated pair will be broken and the mate's Boolean `territorial?` variable will be set to "false" (allowing the widowed fox to find a new mate).

Distribute-Food

This procedure is called by the setup procedure during initialization. Based on prey remains collected from eleven golden eagle nests on the Channel Islands and using the *matrix* extension in NetLogo, we develop a matrix of a typical SCI golden eagle's diet composition (see Table 2) (Collins and Latta, 2009;

Wilensky, 1999). There are nine main species within Channel Island eagle diets, excluding foxes. We determine eagle diet composition based on island biomass; SCI biomass is 152.172 kg with SCI foxes and 144.028 kg without SCI foxes (Collins and Latta, 2006). The model creates the number of each AFS based on the percentage of the overall biomass (without foxes) each species composes. Each AFS is also given an average amount of energy (equal to its mass in kg) an eagle will gain from consumption based on the fresh weight of the species (Collins and Latta, 2006).

4 RESULTS

In the following sections, we show the results of the fox population dynamics with and without the inclusion of golden eagle predation.

4.1 Results of Model Analysis Without Golden Eagle Predation

4.1.1 SCI Fox Carrying Capacity

Starting with 40 female and 40 male foxes of varying ages distributed randomly across the island, we simulated the SCI fox population dynamics over 200 years without eagle predation. This simulation experiment was repeated 100 times to capture the range in population dynamics due to model stochasticity. For each simulation the size of the population was recorded at week 25 each year after the fox breeding season. The mean and standard deviation in population size were calculated for each year over all 100 simulations and are shown as the gray curve and shaded region, respectively, in Fig. 8. For each of the 100 simulations, the population increases to a carrying capacity near the 1465 carrying capacity estimated

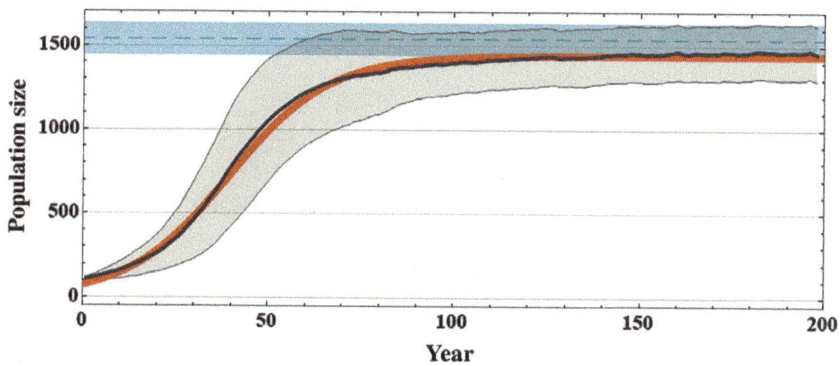

FIG. 8 This graph plots the average population size over 200 years. The blue horizontal region is the carrying capacity for the island as estimated by Coonan (2003) with the dashed line at 1465. The thick *gray line* represents the mean population size of 100 simulated populations. The surrounding *gray* region represents ±1 standard deviation of the mean population size over 100 simulations. The thick *orange line* is the logistic growth curve fitting Eq. (1).

in 1993 (see the dashed blue line) (Coonan, 2003). The simulated SCI fox population size over time can be approximated by a logistic growth curve with the equation:

$$N(t) = \frac{KN_0 e^{rt}}{K + N_0(e^{rt} - 1)}, \tag{1}$$

where $N(t)$ is the SCI fox population size at year t with $N_0 = N(0)$ (the initial population size), r is the intrinsic growth rate, and K is the carrying capacity (Edelstein-Keshet, 2005). For the simulated 100 SCI fox populations, $N_0 = 80$ since each simulation started with 80 foxes. Using the `FindFit` function in Wolfram Mathematica (Wolfram Research, 2010) the logistic growth curve parameters for the SCI fox population are estimated to be $K = 1442.33$ ($p < 0.01$) and $r = 0.07400$ ($p < 0.01$).

4.1.2 SCI Fox Population Viability

The model simulations to determine carrying capacity all started with the same population size: 80 foxes. However, because previous modeling had shown the emergence of an Allee effect which led to population decline at low population size, we next tested the population viability of the SCI foxes by beginning with varying initial fox population sizes, always with an even sex ratio (Scott et al., 2014). Starting with an initial population of x foxes, we ran 100 simulations until either the fox population was extinct or the population had grown to $\frac{1}{2}K$ (where $K = 1442.33$ as estimated from the previous analysis, see Fig. 8). For each set of 100 simulations we calculated the proportion of simulations where the population survived to $\frac{1}{2}K$. This calculation was repeated for each initial population size x. We found that the viability of island foxes (in the absence of golden eagle predation) increased as initial population size increased according to a Holling type III functional response curve (Kot, 2001). The probability of reaching $\frac{1}{2}K$ is given by:

$$P\left(\text{reach}\,\frac{K}{2} \mid N_0 = x\right) = \frac{\alpha x^n}{1 + \alpha x^n}. \tag{2}$$

Using the `FindFit` function in Wolfram Mathematica and the simulated data, the functional response curve parameters were estimated as $n = 3.545$ and $\alpha = 7.53 \times 10^{-6}$, fitting the model with standard error of 0.0344 (Wolfram Research, 2010). The simulated data and parameterized functional response curve are shown in Fig. 9. Given the parameterized model, $P(\text{reach } K/2 \mid N_0 = x) = 0.99$ when $x = 102$. Thus, once the initial population size is at least 102 foxes, we would expect, with a 99% probability, that the population would grow to at least half-carrying capacity. Presumably, once the population has reached half-carrying capacity it will be able to continue to grow toward carrying capacity, though the speed at which the population is growing would slow (assuming logistic growth dynamics).

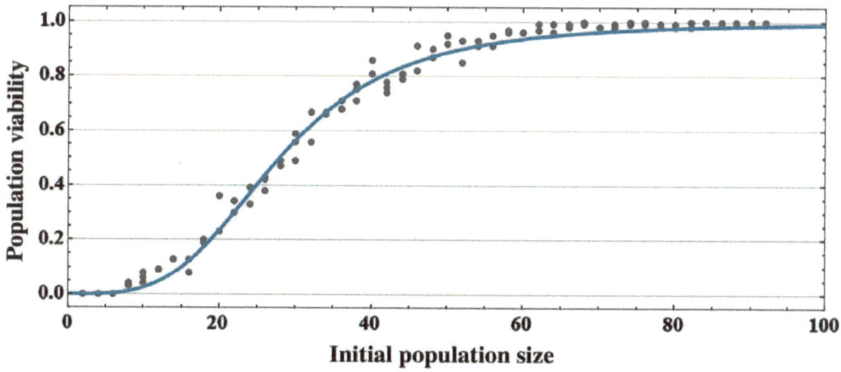

FIG. 9 Population viability based on varying initial population sizes. Each data point represents the proportion of 100 simulations, starting with x foxes, that survived to reach $K/2$ (half of carrying capacity, ~721 foxes, as seen in Fig. 8). The *blue curve* represents Eq. (2) with $n = 3.545$, and $\alpha = 7.53 \times 10^{-6}$.

4.1.3 SCI Fox Spatial Distribution

After running simulations, we also mapped spatial data of fox locations onto GIS maps to show the impact of including vegetation data (see Fig. 6 for vegetation distribution). SCI fox territory size depends on the local composition of vegetation, with smaller territories occurring in the mixed vegetation areas, while larger territories occur in areas with grassland vegetation. Larger territories therefore occupy the outer edges, with smaller territories toward the island's center.

Fig. 10 maps this density over time (1–130 years) without golden eagle predation, as predicted by our model, when starting with an initial population size of 80 foxes (40 male, 40 female). The location (using CRS NAD83 for North America; EPSG:4269 coordinates) of each fox in the simulation is recorded at week 25 each year. We then map each fox onto SCI and use the `heatmap` function of QGIS to observe the change in density over time. This QGIS plugin uses kernel density estimation to estimate the fox density values at every point on the island. Kernel density estimation is a nonparametric density estimator that smooths the contribution of each observed data point over a local neighborhood of that individual data point. The contribution of data point x_i to the estimate at x^* depends on the distance between x_i and x^* (Bozdogan, 2018). The QGIS kernel density estimation plugin uses a quartic kernel in order to estimate density. To create the heatmaps in Figs. 10 and 12, we use a radius of 2000 m, giving an area of ~12.57 km^2. Therefore, if the density value at a certain point is 12.57 foxes, there is one fox per km^2 in that location. The legend in Figs. 10 and 12 has been rescaled in terms of foxes per km^2 for ease of interpretation.

Fig. 10 shows that without predation pressure, the island fox population forms more dense populations in the center of the island, consistent with the

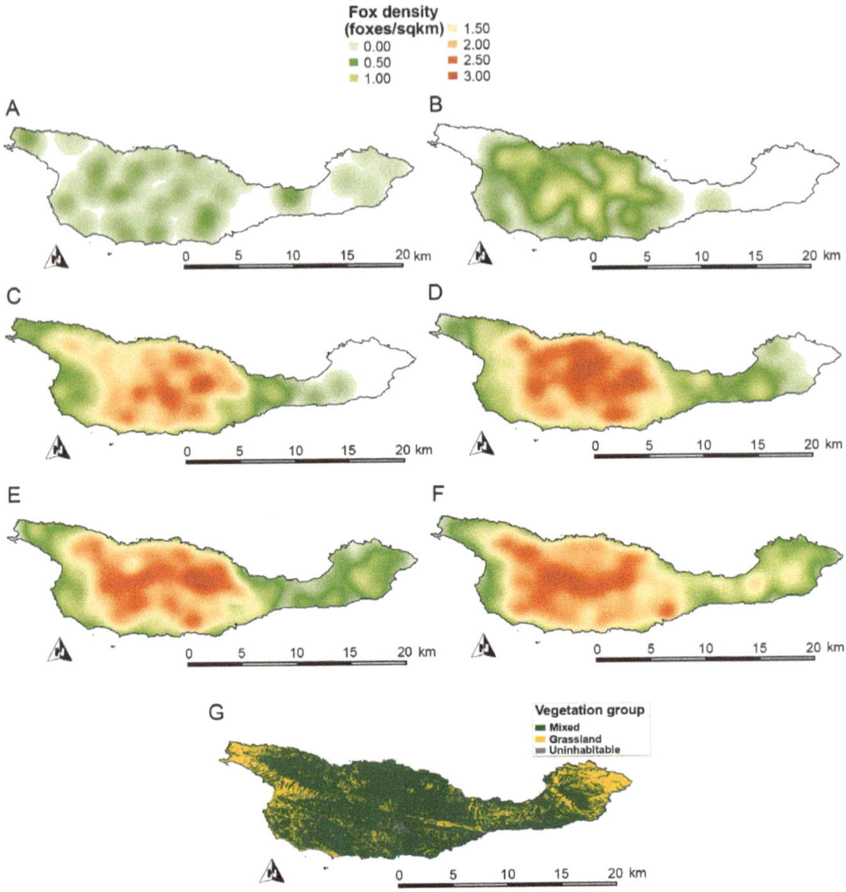

FIG. 10 These figures represent the density of foxes without predation on SCI at years 1 (A), 25 (B), 50 (C), 75 (D), 100 (E), and 130 (F). The *red* areas represent regions of high fox density, the *yellow* areas represent regions of medium fox density, and the *green* areas represent regions of low fox density; *white* regions have no foxes. The vegetation group map from Fig. 6 is shown in (G) for comparison of vegetation distribution and fox distribution.

vegetation. As time continues and population size grows, the island fox population spreads to cover the island, but the highest density regions remain in the island's center where mixed vegetation allows for smaller territory sizes and thus greater density.

4.2 Model Results with Golden Eagle Predation

We also simulated the SCI fox population with golden eagle predation. At each time step, we took a measure of the number of island foxes present on

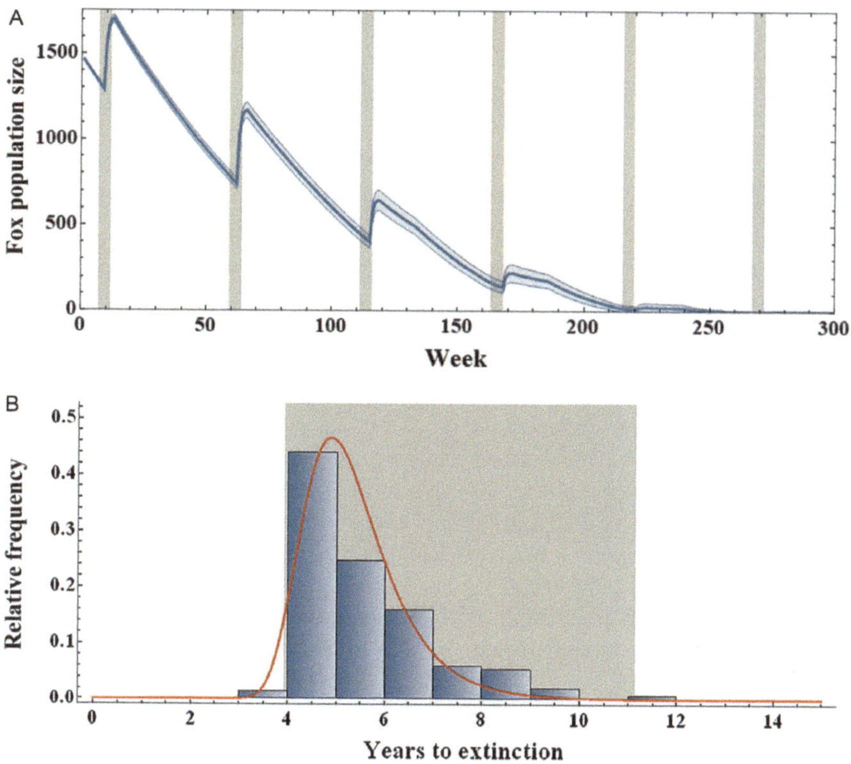

FIG. 11 (A) The average population size of SCI foxes at each time step over 150 simulations with the influence of golden eagle predation. The *gray* regions represent fox breeding season (weeks 9–12); the *shaded blue* region around the curve shows ±1 standard deviation from the mean. (B) A histogram of the time to fox population extinction over 150 simulations; the full range of the years to extinction is shown by the *gray shading*. The data are fit with an extreme value distribution curve given by Eq. (3) where $\alpha = 4.89$ and $\beta = 0.791$.

the island. The population declined steadily, but during fox breeding season, there were spikes in population size. Fig. 11A shows the fox population size over time with golden eagle predation averaged over 150 simulations. Note that Fig. 11A shows simulations from weeks 1–300 even though 45 of the 150 simulations had nonzero population sizes after week 300. However, the maximum fox population size over all 150 simulations at week 301 was 7 foxes.

We additionally determined the average number of years a fox population could subsist on SCI under the influence of golden eagle predation. Using the same 150 simulations, we recorded the first time step at which the fox population size dropped below 1 fox. The weeks to extinction were then scaled to years to extinction and plotted in a histogram shown in Fig. 11B. The mean number of years to fox extinction is 5.40, the median is 5.02 years, and the standard deviation is 1.25 years. The distribution for the years to extinction can be approximated by an extreme value distribution (EVD) (Hosking et al., 1985) given by equation

$$P(x) = \exp\left(\frac{\alpha - x}{\beta} - \exp\left(\frac{\alpha - x}{\beta}\right)\right), \tag{3}$$

where $\alpha = 4.89$ and $\beta = 0.791$. Given the null hypothesis that the years to extinction data is fit by the extreme value distribution with $\alpha = 4.89$ and $\beta = 0.791$ and an acceptance level of $p = 0.001$, the null hypothesis is rejected by the Pearson χ^2 test ($p \ll 0.001$), but accepted by the Anderson–Darling test ($p = 0.0064$) and the Cramer–von Mises test ($p = 0.0074$).

Using the same procedure from Fig. 10, we map the spatial distribution of the fox population under eagle predation for three separate simulations, shown in Fig. 12. We then create a time-lapse video of the SCI fox distribution over the length of the three simulations. Each frame in the video contains three SCI fox distribution heat maps (one for each simulation) generated for one time step (1 week) in the model. The Video clip 1 https://doi.org/10.1016/bs.host.2018.10.001 is available in the Supplementary Material. For the first years of the simulations, fox breeding season allows the population to partially rebound during the fox breeding season following eagle

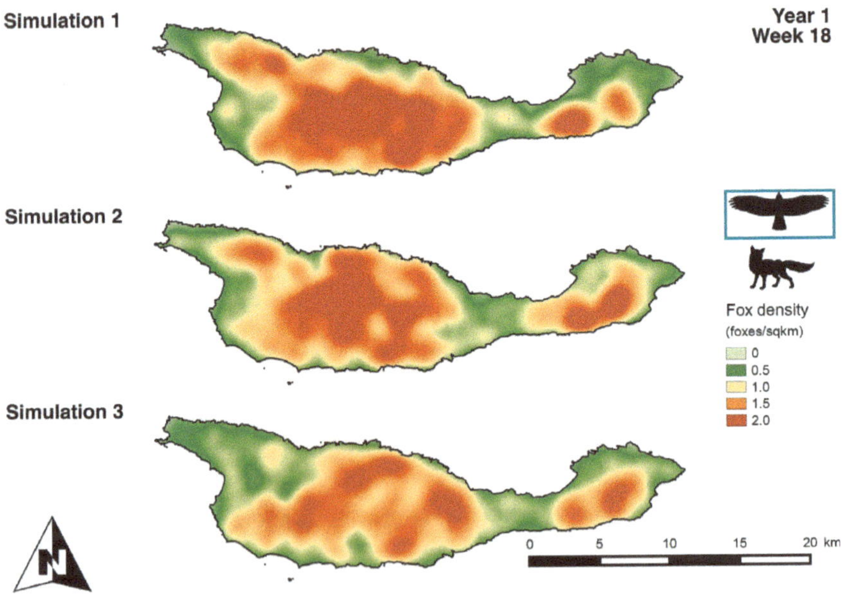

FIG. 12 Three separate simulations showing the spatial distribution of foxes under golden eagle predation. During the breeding season of each species, the icon on the right is highlighted to indicate changes in behavior or dynamics.

predation. As the fox population size becomes smaller, the number of pups added after breeding season does not replenish those lost to predation. Additionally, the spatial distribution of the foxes changes over time. Early in the simulation there is higher fox density in the center of the island. As the population is subject to predation, the foxes are distributed to the outer edges of the island. In all three of the simulations shown in the time-lapse, the last remaining foxes have territories on the edges of the island.

5 CONCLUSIONS AND DISCUSSION

From the above results, we draw the following conclusions. First, in analyzing our model without the presence of golden eagle predation, we see that our model correctly predicts variable SCI fox territory size based on vegetation type. On the outer edges of the island where larger territories form due to grassland habitats, we see lower fox densities. Conversely, in the island's center where there are areas of mixed vegetation, we see higher fox densities. Additionally, while our model does not explicitly include a carrying capacity for the SCI fox population, due to our inclusion of fox behavior decisions regarding the establishment of territories and the inclusion of vegetation data for SCI, our model closely predicts the SCI fox carrying capacity estimated by previous research and field data (Coonan, 2003; Scott et al., 2014).

Furthermore, we also show this ABM model can produce logarithmic growth of the fox population when the population size is recorded yearly thus ignoring the temporal dynamics of the birth pulse. A population starting with 80 foxes (40 male, 40 female) will reach half-carrying capacity between 40 and 50 years later (see Fig. 8). Reaching half-carrying capacity always ensured survival to carrying capacity according to our analysis of SCI fox population viability. In fact, our results show that an SCI fox population starting with slightly more than 100 foxes (with an even sex ratio) all but guarantees survival to half-carrying capacity, and then to carrying capacity after that.

Population viability analysis allow us to determine the minimum population size that will produce carrying capacity, if the population is unadulterated by golden eagle predation or other population decline factors (see Fig. 9). Similarly, if our GIS maps and data are correct, we can pinpoint locations on the island where we would expect foxes to exist at high densities and low densities, as well as eagle nest proximities to fox populations (see Fig. 10).

Under the predation of 10 golden eagles comprising 5 mated pairs, our model predicted the SCI fox population to experience a rapid decline. It took the eagles around 5 years to eradicate the fox population ($\mu = 5.41$). This concurs with data collected on the SCI fox population in the 1990s–2000s. From 1994 to 1999 (a 5-year span), the SCI fox population fell from 1465 to 135 foxes, by 2004 (10 years later) the fox population reached a low of 89 foxes. Based on results from our model, while there is a 58% chance the SCI fox population would have survived more than 5 years before going extinct, there is less than a 1% chance that the SCI fox population would have survived more 9 years before going extinct. This indicates that the SCI fox population was either very "lucky" or other factors like intensified surveillance of the SCI foxes and the beginnings of conservation measures helped ensure their survival in otherwise unfavorable circumstances. The conservation efforts of the National Park Service and Nature Conservancy were therefore fortuitous.

From the number of years to extinction, we can determine the average number of years that the fox population can exist before going extinct in the absence of conservation efforts (see Fig. 11). There is a significant drop in fox consumption during eagle breeding season, meaning conservation efforts may be more successful during this time, when there is less of a threat of eagle predation.

The research presented here shows the fragility of island populations and the utility of including carefully curated GIS data in agent-based models used to study such fragile populations. Using models similar to this, we can improve the predictive capacity of conservation efforts on the Channel Islands and in other regions of isolated populations.

ACKNOWLEDGMENTS

The authors would like to acknowledge Rhodes College Department of Mathematics and Computer Science, University of Tennessee, Knoxville Department of Ecology and Evolutionary Biology, NIH Grant #R25GM086761, the National Defense Science and Engineering Graduate Fellowship and the Associated Colleges of the South for funding portions this research.

APPENDIX

Supplemental Resources

The authors have provided the model code (in NetLogo) and associated GIS data files which are imported into the model. Additionally, a video clip of the SCI fox population density under golden eagle predation for three simulated fox populations (referenced in Fig. 12) is provided. The model code, data files, and Video clip 1 https://doi.org/10.1016/bs.host.2018.10.001 are available in the Supplementary Material online.

REFERENCES

Bakker, V.J., Garcelon, D.K., Aschehoug, E.T., Crooks, K.R., Newman, C., Schmidt, G.A., Van Vuren, D.H., Woodroffe, R., 2003. Current status of the Santa Cruz island fox (Urocyon littoralis santacruzae). In: 6th California Islands Symposium.

Bonabeau, E., 2002. Agent-based modeling: methods and techniques for simulating human systems. Proc. Natl. Acad. Sci. 99 (suppl 3), 7280–7287.

Bozdogan, H., 2018. Modern statistical theory and inference: an informational modeling process with Matlab. (in preparation).

Cannata, N., Corradini, F., Merelli, E., Tesei, L., 2013. Agent-based models of cellular systems. Methods Mol. Biol. 930, 399–426.

Clark, H.O., 2009. Species at risk: golden eagle predation on arid-land foxes. Endangered Species Update 26 (1 & 2), 10–14.

Cohen, B., Corry, C., Menke, J., Hepburn, A., 2009. A spatial database of Santa Cruz island vegetation. In: Proceedings of the 7th California Islands Symposium.

Collins, P.W., 1991. Interaction between island foxes *Urocyon littoralis* and Indians on islands off the coast of Southern California. J. Ethnobiol. 11 (1), 51–81.

Collins, P.W., Latta, B.C., 2006. Nesting Season Diet of Golden Eagles on Santa Cruz and Santa Rosa Islands, Santa Barbara County, California. Santa Barbara Museum of National History. Technical Reports No. 3.

Collins, P.W., Latta, B.C., 2009. Food habits of nesting golden eagles *Aquila chrysaetos* on Santa Cruz and Santa Rosa Islands, California. In: Proceedings of the 7th California Islands Symposium.

Conner, M.M., Ebinger, M.R., Knowlton, F.F., 2008. Evaluating coyote management strategies using a spatially explicit, individual-based, socially structured population model. Ecol. Model. 219 (1), 234–247.

Coonan, T.J., 2003. Recovery Strategy for Island Foxes on the Northern Channel Islands. National Park Service: Channel Islands National Park.

Coonan, T.J., 2008. Island Fox Captive Breeding Program: 2007 Annual Report. National Parks Service: Channel Islands National Park, Ventura, CA. 08-01.

Coonan, T.J., 2009. Eleventh Annual Meeting: Island Fox Working Group. National Parks Service, Sheraton Four Points Hotel, Ventura, CA.

Coonan, T.J., 2010. Twelfth Annual Meeting: Island Fox Working Group. National Parks Service, Sheraton Four Points Hotel, Ventura, CA.

Coonan, T.J., 2011. Thirteenth Annual Meeting: Island Fox Working Group. National Parks Service, Sheraton Four Points Hotel, Ventura, CA.

Coonan, T.J., 2012. Fourteenth Annual Meeting: Island Fox Working Group. National Parks Service, Sheraton Four Points Hotel, Ventura, CA.

Coonan, T.J., 2013. Fifteenth Annual Meeting: Island Fox Working Group. National Parks Service, Sheraton Four Points Hotel, Ventura, CA.

Coonan, T.J., 2014. Sixteenth Annual Meeting: Island Fox Working Group. National Parks Service, Sheraton Four Points Hotel, Ventura, CA.

Coonan, T.J., Dennis, M., 2006. Island Fox Captive Breeding Program: 2005 Annual Report. National Parks Service: Channel Islands National Park, 1901 Spinnaker Drive, Ventura, CA. 06-02.

Coonan, T.J., Dennis, M., 2007. Island Fox Captive Breeding Program: 2006 Annual Report. National Parks Service: Channel Islands National Park, 1901 Spinnaker Drive, Ventura, CA, 07-04.

Coonan, T., Rutz, K., 2001. Island Fox Captive Breeding Program: 2001 Annual Report. National Parks Service: Channel Islands National Park, 1901 Spinnaker Drive, Ventura, CA. 02-01.

Coonan, T.J., Schwemm, C.A., Garcelon, D.K., 2010. Decline and Recovery of the Island Fox: A Case Study for Population Recovery. Cambridge University Press.

Edelstein-Keshet, L., 2005. Mathematical Models in Biology. Society for Industrial and Applied Mathematics, Philadelphia, PA, USA, ISBN: 0898715547.

ESRI, 2011. ArcGIS Desktop (Version 10.3). http://www.esri.com/software/arcgis.

Gaff, H., 2011. Preliminary analysis of an agent-based model for a tick-borne disease. Math. Biosci. Eng. 8, 463–473.

Grimm, V., Berger, U., DeAngelis, D.L., Polhill, J.G., Giske, J., Railsback, S.F., 2010. The ODD protocol: a review and first update. Ecol. Model. 221, 2760–2768.

Heppenstall, A., Crooks, A., See, L., Batty, M., 2012. Agent-Based Models of Geographical Systems, first ed. Springer, Netherlands.

Hoppensteadt, F., 2006. Predator-prey model. Scholarpedia 1 (10), 1563.

Hosking, J.R.M., Wallis, J.R., Wood, E.F., 1985. Estimation of the generalized extreme-value distribution by the method of probability-weighted moments. Technometrics 27 (3), 251–261. https://doi.org/10.1080/00401706.1985.10488049.

Kot, M., 2001. Elements of Mathematical Ecology. Cambridge University Press. https://doi.org/10.1017/CBO9780511608520.

Latta, B.C., 2004. Channel Islands Golden Eagle Translocation Program Summary Report. The Nature Conservancy.

Moore, C.M., Collins, P.W., 1995. Urocyon littoralis. Mamm. Species 489, 1–7.

QGIS Development Team, 2009. QGIS Geographic Information System (Version 3.0). https://qgis.org.

Railsback, S.F., Grimm, V., 2005. Individual-Based Modeling and Ecology. Princeton Series in Theoretical and Computational Biology, Princeton University Press, Princeton, NJ.

Railsback, S.F., Grimm, V., 2011. Agent-Based and Individual-Based Modeling: A Practical Introduction. Princeton University Press.

Roemer, G.W., Smith, D.A., Garcelon, D.K., Wayne, R.K., 2001. The behavioural ecology of the island fox (Urocyon littoralis). J. Zool. 255 (1), 1–14. ISSN 1469-7998. https://doi.org/10.1017/S0952836901001066.

Roemer, G.W., Donlan, C.J., Courchamp, F., 2002. Golden eagles, feral pigs, and insular carnivores: how exotic species turn native predators into prey. Proc. Natl. Acad. Sci. 99 (2), 791–796.

Sanchez, J., Hudgens, B., 2011. Spatial Ecology of the Island Fox. Department of Defense Legacy Resource Management Program.

Schwemm, C.A., 2008. Tenth Annual Meeting: Island Fox Working Group. National Parks Service, Sheraton Four Points Hotel, Ventura, CA.

Scott, S.M., Bodine, E.N., Yust, A., 2014. An agent-based model of Santa Cruz island foxes (*Urocyon littoralis santacruzae*) which exhibits an Allee effect. Lett. Biomath. 1 (1), 97–109.

Todd, C.S., 2000. Golden Eagle Assessment. Maine Department of Inland Fisheries and Wildlife.

Vissman, S., 2004. Island Fox: Management Guidelines for Species at Risk on Department of Defense Installations. NatureServe.

Weisstein, E. W., n.a.-a. Albers Equal-Area Conic Projection. From MathWorld-A Web Resource, http://mathworld.wolfram.com/AlbersEqual-AreaConicProjection.html.

Weisstein, E. W., n.a.-b. Lambert Conformal Conic Projection. From MathWorld-A Web Resource, http://mathworld.wolfram.com/LambertConformalConicProjection.html.

Wilensky, U., 1999. Netlogo. Center for Connected Learning and Computer-Based Modeling, Northwestern University, Evanston, IL.

Research, Wolfram, 2010. Mathematica 8.0.

Chapter 2

Capture–Recapture Methods and Models: Estimating Population Size

Ruth King*,[1] and Rachel McCrea†

*School of Mathematics, University of Edinburgh, Edinburgh, United Kingdom

†School of Mathematics, Statistics and Actuarial Science, University of Kent, Canterbury, United Kingdom

[1]Corresponding author: e-mail: ruth.king@ed.ac.uk

ABSTRACT

This book chapter describes ecological capture–recapture studies and associated models often fitted to capture–recapture data to obtain estimates of total population size. Such estimates can be important for numerous reasons, including, for example, conservation and management purposes. We focus on different forms of heterogeneity that may affect the propensity of individuals to be observed within the study period. Failing to account for such heterogeneity can lead to significant bias in the population estimates. We focus on different types of heterogeneity corresponding to recorded (discrete-valued) covariates/characteristics of individuals that are observed within the study period; in addition to unobserved heterogeneity in the form of mixture distributions. The different models are motivated and discussed, including the specification of the likelihood functions, before being applied to a real dataset. Finally we conclude with a discussion including the modern challenges which are arising due to technological advances.

Keywords: Abundance, Capture histories, Discrete covariates, Heterogeneity, Likelihood, Mixtures, Models

1 INTRODUCTION

Population sizes are of interest across a wide spectrum of disciplines and geographical scales. For example, at the macro-scale, such as the number of individuals on the planet; to the micro-scale, such as the number of specific animals that inhabit a localized area; and everything in between. Many different data collection techniques are used to look at population sizes and their trends over time. These include, for example, national censuses sent to all

Handbook of Statistics, Vol. 40. https://doi.org/10.1016/bs.host.2018.09.006

households (carried out periodically in many countries, such as the United Kingdom and United States every 10 years) or local subsampling of individual animals within a small survey area. In practice, observing all members of a given population is often infeasible for a combination of resource and pragmatic issues. For example, marine animals primarily spend their time underwater and traverse a very wide location so that survey boats may not observe the animals despite being in the same area; alternatively problem drug users live in communities and their drug use may be well hidden. Thus, reliable estimation of total population sizes, for hidden or difficult to observe populations, is an important aspect in many different areas. For example, within wildlife conservation the IUCN Red List of Threatened Species (http://www.iucnredlist.org) classifies species and plants in relation to their conservation status and uses, among other criteria, population size (and trends over time of these) in categorizing the different species. Alternatively, assessing the impact of different governmental policies on populations such as the homeless or problem drug users will require robust estimation of population sizes in order to detect associated changes over time; similarly financial socio-economic planning may be dependent on these subpopulations, with areas with a higher level of problem drug use requiring additional healthcare resources. See, for example, Böhning et al. (2018) for further discussion and associated examples of this type.

In this book chapter we focus on ecological populations, and the estimation of wildlife population sizes. There are many different ways that data can be collected on such populations. The most accurate, in terms of a complete census count of all individual animals within the population, is typically not possible due to the nature of the population combined with time and financial resource limitations. In such circumstances partial observation and enumeration of the population under study may be possible. These may take multiple different forms including, for example, distance sampling (where for line transect sampling observers traverse some transect and record the number of animals observed and their distance to the transect; Buckland et al., 2004); aerial survey data over a given region (i.e., a snapshot of individuals within the given location and time; Fleming and Tracey, 2008); and nets/traps to capture animals that pass through a given area (Ralph and Dunn, 2004). Typically, multiple types of surveys may be conducted and a combination of both spatial and temporal replications used.

We focus on repeated surveys over time, such that there are a series of "capture" occasions where researchers go into the field at distinct times, and subsequently identify and record all individuals observed from the population of interest. Such studies are generally referred to as capture–recapture (or mark–recapture/resight) studies (see, for example, McCrea and Morgan, 2014 and King, 2014). We note that for the analogous case within epidemiological studies, the capture occasions correspond to different lists of individual names, such as hospital records and police records; the data are consequently often referred to as multilist data, and the approach of using such data to estimate the total population size "multiple systems estimation" (see, for example, Bird and King (2018) for a review).

For standard capture–recapture studies it is assumed that individuals are uniquely identifiable at each capture occasion. This assumption permits the construction of individual capture histories for each individual observed within the study, detailing which capture occasions they are (or are not) observed. Traditionally, to uniquely identify individuals, marks are applied to an individual the first time it is observed, such as a tag attached (e.g., sheep, seals) or a ring applied to its leg (e.g., common for many bird species). However, new techniques for uniquely identifying individuals include the use of DNA (for hair/scat samples; Lukacs and Burnham, 2005a) or natural unique markings and photographic identification (e.g., great crested newts, seals, tigers; Royle et al., 2009; Smout et al., 2011; Worthington et al., 2018b)—the latter becoming increasingly common with the use of motion sensor traps. There are several attractions for the use of these new technologies as they do not disturb the animals by the addition of an artificial mark and the associated marks cannot generally be lost; however, there is a trade-off as uniquely identifying individuals can be more difficult and can lead to additional uncertainty in the capture histories (see Section 5 for further discussion of modern techniques).

The primary parameter of interest within such capture–recapture studies is the total population size. However, in estimating this we also need to consider the capture probabilities of the animals, i.e., the probability that a given individual is observed at a given capture occasion—though these parameters are essentially regarded as nuisance parameters. The capture probabilities may be dependent on a number of different factors such as the effort expended at the given capture occasion (the longer amount of time, or increased number of observers, the more likely an individual may be seen); the time of day that animals are attempted to be observed (animals may be more active at different times of day, such as dawn or dusk, dependent on the species); geographical location (seeing an individual in areas with long grass is typically much harder than seeing an individual in open short grassland); and many other additional factors including individual characteristics. Despite the capture probabilities being nuisance parameters, and typically not of interest themselves, the dependence structure of the capture probabilities may have an appreciable impact on the overall estimate of total population size—see, for example, Cubaynes et al. (2010). We discuss different forms of heterogeneity that may affect the capture probabilities, and hence estimate of the total population size. We focus on models associated with discrete forms of heterogeneity (both at the global and individual levels), which may be either observed or unobserved. In this book chapter we focus on the models themselves and for illustration apply a classical approach to the data. For a discussion of the use of Bayesian methods, see, for example, King et al. (2009) and King (2012).

In Section 2 we give a brief history of the use of capture–recapture data in terms of the number of capture occasions used, from the simplest, minimum, case of just two occasions to multiple occasions and associated modeling issues, focusing on closed populations where it is assumed that there no births/deaths or migrations within the study period. In Section 3 we discuss in more detail the different forms of heterogeneity (observed and unobserved; time-invariant and time-varying) that may be present in the system, particularly in reference to the capture probabilities, and the associated impact on the statistical analysis. We extend the ideas to open populations in Section 4. In Section 5 we provide a general discussion and briefly describe further extensions to the models described within this book chapter. Throughout these sections we primarily focus on a simple capture–recapture example relating to deer mice for comparison of the different models.

2 CLOSED POPULATION CAPTURE–RECAPTURE STUDIES

Within this section we begin by describing the simplest form of capture–recapture studies, where there are just two sampling occasions, before discussing issues relating to this two-sample approach. Some of these shortcomings, and necessary associated assumptions, can be addressed by increasing the number of capture occasions within the study. We define the multiple occasion approach and discuss in further detail the associated heterogeneities that may arise, that, if unaccounted for, can lead to significant bias within the estimates of the total population size. Models that account for different types of heterogeneity given the available information within the data are discussed. For the different models described, we fit the models to data from a capture–recapture study of mice and discuss the corresponding results.

2.1 Simple Beginnings: Lincoln–Petersen Estimator

We begin by describing the simplest capture–recapture study where there are just two capture occasions. We assume that the population is closed so that the population size is fixed over the study period with no entries to, or exits from, the population. The capture history for a given individual observed within the study corresponds to the pair (x_1, x_2), where $x_t = 0, 1$ corresponding to whether the individual was not observed ($x_t = 0$) or observed ($x_t = 1$) at time $t = 1, 2$. There are three possible observable capture history combinations:

$$(1,1) \quad (1,0) \quad (0,1).$$

The first pair corresponds to an individual being observed at both capture occasions; the second to only being observed at the first occasion; and the final pair to only being observed at the second occasion. Clearly there is a further *unobservable* capture history: (0, 0).

The data can be summarized as the number of individuals with each given observed capture history (or pair). Let n_{ij} denotes the number of individuals with capture history (i, j) for $i, j = 0, 1$. The observed data correspond to n_{11}, n_{10} and n_{01}—the number of individuals with each distinct combination of observable capture histories. The number of individuals not observed at either capture occasion, denoted by n_{00}, is unknown. The total number of observed individuals is given by $n = n_{11} + n_{10} + n_{01}$. The true total population size is given by $N = \sum_{i=0}^{1} \sum_{j=0}^{1} n_{ij} = n_{00} + n$ and is the primary parameter of interest to be estimated.

The observed data are often usefully represented in the form of an incomplete (contingency) table, where the cell entries correspond to the number of individuals with the given capture history—see Table 1.

The pattern of the observed cell entries permits the estimation of the total population size, N (or equivalently the number of individuals in the population not observed at either occasion, n_{00}). For example, the following argument can be applied, given the assumption that the probability an individual is observed at occasion 2 is independent of whether or not they are observed at occasion 1. In this case we would expect the ratio of observed individuals to unobserved individuals at occasion 2 to be approximately equal for the two subsets of animals corresponding to those (i) observed at occasion 1 and (ii) not observed at occasion 1. Mathematically, this means that we would expect:

$$\frac{n_{11}}{n_{10}} \approx \frac{n_{01}}{n_{00}}.$$

Rearranging the equation we obtain the corresponding estimates:

$$\hat{n}_{00} = \frac{n_{01} n_{10}}{n_{11}}; \text{ and } \hat{N} = \frac{(n_{11} + n_{10})(n_{11} + n_{01})}{n_{11}}.$$

This estimate can be obtained using analogous arguments by equating different ratios or proportions of the cell entries which should be (approximately)

TABLE 1 An Example of the Observed Data of a Capture–Recapture Study With Two Capture Occasions

		Capture Occasion 2	
		0	1
Capture	0	$n_{00} = ?$	n_{01}
Occasion 1	1	n_{10}	n_{11}

The cell n_{00} corresponding to the number of individuals not observed at either occasion is unobserved and hence unknown.

equal if the probability an individual is observed on the second occasion is independent of whether or not they were observed on the first capture occasion.

This estimate for the total population size, $\hat{N}$, is typically referred to as the Lincoln–Petersen estimate (Lincoln, 1930; Petersen, 1896). For a discussion of the history of the estimate see, for example, Goudie and Goudie (2007). The estimate predates both Lincoln and Petersen by several hundred years, dating back to at least the 1600s, when Graunt used a similar idea to estimate the size of the population of England and the effect of the plague. However, the most famous early application of the estimate is typically attributed to Laplace (1786)—and the application of these ideas to estimate the total population of France using "occasions" corresponding to birth records across France and local complete municipality census records (Bird and King, 2018).

The Lincoln–Petersen estimate is also the maximum likelihood estimate (MLE) for the total population size so that it is a consistent estimate, however, it is not unbiased. This led to the proposal of the Chapman estimator (Chapman, 1951) that is less biased and given by

$$\hat{N} = \frac{(n_{11} + n_{10} + 1)(n_{11} + n_{01} + 1)}{n_{11} + 1} - 1.$$

2.1.1 Example: Mice Data

We consider an example of capture–recapture data relating to deer mice (*Peromyscus maniculatus*) collected using traps over a period of 6 consecutive days at East Stuart Gulch, Colorado by V. Reid. See Huggins (1991) for further discussion of the data. For the purposes of demonstrating the application of the two-sample estimators, we simplify the data by defining capture occasion 1 to be days 1–3 and capture occasion 2 to be days 4–6. Thus an animal caught in a trap on any of the days 1, 2, or 3 was recorded as being observed at capture occasion 1; and similarly an animal caught in a trap on any of the days 4, 5, or 6 was recorded as being observed at capture occasion 2. Note that we will consider the data more fully in Sections 2.2 and 3, and in particular we will remove the combining of days implemented here. The corresponding, simplified, two-sample capture–recapture data are provided in Table 2.

Applying the Lincoln–Petersen estimator we obtain an estimate of the total population size of 38.41. The corresponding 95% confidence interval for the total population using a nonparametric bootstrap approach is given by (38.19, 38.81). We note that applying the Chapman estimator provides a very similar estimate of 38.64 and associated 95% nonparametric bootstrap confidence interval of (38.00, 40.09). Both estimates suggest that the majority of individuals have been observed within the study with little uncertainty.

TABLE 2 Observed Mice Data Using a Two-Sample Capture–Recapture Study

		Capture Occasion 2	
		0	1
Capture	0	?	9
Occasion 1	1	2	27

Throughout this chapter we apply a nonparametric bootstrap approach for calculating the confidence intervals of the parameters. To calculate the associated confidence interval of the parameters, within each bootstrap replication, we simulate "new" data by sampling with replacement over the set of observed capture histories and calculate the associated MLEs of the model parameters. Thus, we take the capture histories as the sampling level. We use a percentile bootstrap confidence interval by taking the lower and upper 2.5% quantiles of the associated MLEs obtained over the bootstrap replicates. The nonparametric bootstrap approach is used for several reasons: (i) the bootstrap approach does not rely on the asymptotic normality assumption which is perhaps questionable in this case given the amount of available data; (ii) the estimates of population size are close to the boundary leading to unreliable confidence intervals (using the delta method to obtain standard errors of transformed estimates even when this may appear reasonable may lead to naive confidence intervals with bounds outside the permissible range); (iii) for ease of use for calculating associated confidence intervals of functions of the estimated parameters (see, for example, Section 3.2 where we individually estimate the population sizes of subgroups corresponding to males and females, and wish to also obtain an estimate of the total population size).

For the above observed data, with only two capture occasions, the estimates rely on a number of assumptions. For example, this includes that the population is homogeneous with regard to capture probabilities, so that all individuals have the same capture probability at a given occasion. In particular this implies that the capture occasions are independent of each other so that an individual observed at the first capture occasion does not affect the associated probability that the individual is observed on the second capture occasion—recall the rationale underlying the Lincoln–Petersen estimator. In other words the probability that an individual is observed at the second capture occasion is independent of whether or not they are observed at the first capture occasion. Deviations in the system from such assumptions will lead to unreliable and biased estimates of the total population size. However, unfortunately, for two-sample capture–recapture studies it is not possible to test such assumptions—we have three observed data values permitting the estimation of three parameters: the total population size, the probability an individual is observed at capture occasion 1, and the probability an individual is observed at capture occasion 2. There are no additional degrees of freedom to model or test for independence or additional heterogeneity. We have however simplified the data for this example for comparability with later analyzes. A simple visual inspection of the more detailed data suggests that these assumptions are likely to be violated. For example, some mice are captured on all six occasions while others are captured only once, leading us to question the assumption of equal detectability of all individuals. An obvious way to investigate such issues and address them within the modeling process is to extend the study design and consider multiple capture occasions (as is the case with the more detailed mice data).

2.2 Multiple Capture Occasions

Expanding the number of capture occasions is a natural extension to the two-sample case and associated Lincoln–Petersen estimator. Suppose that there are a total of T capture occasions. The data collection process extends immediately so that on each capture occasion $t = 1, \ldots, T$ individuals are uniquely identified and recorded before being released. A number of assumptions are typically made with regard to the population, including

- The population is closed (though we discuss how this may be relaxed in Section 4);
- Animals cannot lose or change their unique identifying marks so that there is no misidentification within the study (i.e., there are no false positives or false negatives with regard to matching individuals across sightings);
- All individuals behave independently of each other.

For further discussion of these (and additional) assumptions, see, for example, McCrea and Morgan (2014).

2.2.1 *Schnabel Census*

Early studies to look at multiple (>2) capture occasions simply recorded the number of times that each individual was observed within the study period, and such data are typically referred to as a Schnabel census (Schnabel, 1938). This can be thought of as similar to a binomial coin-tossing example—where we simply record the number of "successes" (i.e., heads or tails) but not where the successes occur in the T coin tosses. The data were recorded as the number of individuals observed a total of t times within the study period, denoted by f_t, for $t = 1, \ldots, T$. For notational simplicity we let $\boldsymbol{f} = \{f_1, \ldots, f_T\}$. The unobserved value f_0 corresponds to the number of individuals not observed within the study. The summary statistics f_t for $t = 1, \ldots, T$ are the minimal sufficient statistics for individual heterogeneity models which we will discuss in Section 3.1. We note that, as for a binomial experiment, it is assumed that for each individual the associated capture (or noncapture) at each capture occasion are assumed to be independent Bernoulli trials with the same probability of "success" (i.e., capture probability). Recording only the number of times an individual is observed means that we cannot investigate any time dependence on the capture probabilities (as we do not know at what times individuals are observed). To more fully explore dependence structures we need the full encounter (or capture) histories of individuals over all the capture occasions.

2.2.2 Capture Histories

For T capture occasions there are a total of $2^T - 1$ distinct observable capture histories (though in general we may not observe all of these possible histories within a given dataset) corresponding to an individual being observed or not at each possible capture occasion. Thus, here we consider the exact capture occasions that an individual is observed, rather than simply the summary of how many times an individual is observed as in the Schnabel census. Notationally, we let:

$$x_{it} = \begin{cases} 0 & \text{if individual } i \text{ is not observed at time } t; \\ 1 & \text{if individual } i \text{ is observed at time } t. \end{cases}$$

The capture history for individual i is denoted by $\boldsymbol{x}_i = \{x_{i1}, \ldots, x_{iT}\}$. For notational convenience, we let $\boldsymbol{x}_0$ denote the unobserved capture history, i.e., $\boldsymbol{x}_0 = \{0, \ldots, 0\}$. The set of all observed capture histories is denoted by $\boldsymbol{x} = \{\boldsymbol{x}_1, \ldots, \boldsymbol{x}_D\}$, where D denotes the number of individuals observed within the study. The raw data (x_{it} values) for the individuals observed in the study can be displayed in the form:

	Capture Occasion (t)				
Unique ID	*1*	*2*	*3*	...	*T*
1	1	1	0	...	1
2	0	1	0	...	1
3	0	1	0	...	0
⋮					
D	0	0	0	...	1

The "unique ID" corresponds to the unique identifier (in this case number) assigned to each individual; and a total of D individuals are observed within the study period. Thus D provides a lower bound on the total population size, but of course does not provide any indication regarding the number of individuals not observed within the study. Alternatively these data can be summarized in the form of an incomplete 2^T contingency table, analogous to Table 1 extended to multiple capture occasions, where the cell corresponding to not being observed at any capture occasion is again unknown.

2.2.3 Model Formulations

There are two common formulations for the general model that arise for capture–recapture data corresponding to the number of individuals observed by each combination of capture occasions (i.e., contingency table cell): a Poisson model; and a multinomial model. Both of these formulations are examples of a generalized linear model for the data.

Poisson Model

Mathematically, we index the set of possible contingency table cells by $j = 0, 1, \ldots, 2^T - 1$ and for simplicity assume that cell $j = 0$ corresponds to the unobserved capture history. The associated cell entries are denoted by $n_0, \ldots, n_{2^T-1}$. For notational simplicity we let $\boldsymbol{n} = \{n_0, \ldots, n_{2^T-1}\}$, so that the number of observed individuals is given by $D = \sum_{j=1}^{2^T-1} n_j$. We assume that for $j = 0, \ldots, 2^T - 1$,

$$n_j | \mu_j \sim Poisson(\mu_j),$$

independently, where μ_j denotes the corresponding mean cell count. The mean cell counts μ_j for $j = 1, \ldots, 2^T - 1$ are assumed to be a function of a number of model parameters, such as an underlying prevalence level and propensity of being observed at each capture occasion. In general, we let the mean cell count model parameters be denoted by $\boldsymbol{\theta}$.

We note that only the cell entries for $j = 1, \ldots, 2^T - 1$ are observed. We let the set of observed cell entries be denoted by $\boldsymbol{n}^{obs} = \{n_1, \ldots, n_{2^T-1}\}$. The number of individuals in the population but not observed in the study, denoted n_0, is unknown. However, since the cell entries are *independent* given the model parameters, we can apply a two-step estimation process. First we fit the model to the observed cell entries $j = 1, \ldots, 2^T - 1$ and estimate the model parameters. The corresponding likelihood for the *observed* data is given by

$$L(\boldsymbol{\theta}; \boldsymbol{n}^{obs}) = \prod_{j=1}^{2^T-1} \frac{\exp(-\mu_j)(\mu_j)^{n_j}}{n_j!}.$$

For example, we may obtain the maximum likelihood estimates (MLEs) $\hat{\boldsymbol{\theta}}$ and associated MLE for the mean number of individuals not observed, denoted, $\hat{\mu}_0$. Second, we estimate the unobserved cell entry using $\hat{n}_0 = \hat{\mu}_0$, via the property of the MLE for a Poisson random variable (recall that the cell entries are assumed to be independent Poisson random variables). The overall estimate of the total population size is given by

$$\hat{N} = \hat{n}_0 + \sum_{j=1}^{2^T-1} n_j = \hat{n}_0 + D.$$

Using the same form of two-step process a Bayesian approach can be implemented and a posterior distribution for $\boldsymbol{\theta}$ obtained up to proportionality. A sample from the posterior distribution can be obtained using standard techniques, such as Markov chain Monte Carlo; and hence a sample from the posterior distribution of n_0 can be obtained by subsequently simulating a Poisson random variable with given mean (calculated from the sampled $\boldsymbol{\theta}$ values).

Multinomial Model

Alternatively we can consider a multinomial model formulation. Suppose that we randomly select an individual from the population of interest—that individual must have one of the possible 2^T capture histories. The probability associated with capture history $j = 0, \ldots, 2^T - 1$ is denoted by q_j; and we set $\boldsymbol{q} = \{q_0, \ldots, q_{2^T-1}\}$. Summing over the number of individuals in the population leads to,

$$\boldsymbol{n}|N, \boldsymbol{q} \sim Multinomial(N, \boldsymbol{q}).$$

The corresponding likelihood of the capture histories, conditional on the total population size, is given by

$$L(N, q; \boldsymbol{n}^{obs}) = \frac{N!}{\prod_{j=0}^{2^T-1} n_j} \prod_{j=0}^{2^T-1} q_j^{n_j},$$

where $N = n_0 + D$ (or conversely $n_0 = N - D$) and N corresponds to the total population size. This likelihood can be directly maximized in order to obtain unconditional MLEs of the parameters, and in particular the total population size.

We note that we can factorize the multinomial likelihood as follows:

$$\begin{aligned} L(N, \boldsymbol{q}; \boldsymbol{n}^{obs}) &= \frac{N!}{D!(N-D)!} q_0^{N-D} (1 - q_0)^D \times \frac{D!}{\prod_{j=0}^{2^T-1} n_j} \prod_{j=0}^{2^T-1} \left(\frac{q_j}{1 - q_0} \right)^{n_j} \\ &= \frac{N!}{D!(N-D)!} q_0^{N-D} (1 - q_0)^D L(q; \boldsymbol{n}^{obs}), \end{aligned} \tag{1}$$

where q_0 denotes the probability that an individual is not observed within the study; so that $1 - q_0$ corresponds to the probability an individual is observed at least once within the study. The first term is a binomial component corresponding to the number of individuals not observed (or conversely observed at least once), given the total population size N; the second term, $L(q;\boldsymbol{n}^{obs})$ is a conditional multinomial component corresponding to the number of individuals observed by each combination of sources (i.e., cells $j = 1, \ldots, 2^T$), conditional on an individual being observed at least once within the study.

Relationship Between Poisson and Multinomial Models

It is straightforward to show that for observations from independent Poisson distributions, if we condition on the sum of the observations, then the distribution of the observations has a multinomial distribution. The associated probabilities of the multinomial distribution are equal to the Poisson mean for that cell divided by the sum of all Poisson mean values. Thus, suppose that for $j = 0, \ldots, 2^T - 1$ we let:

$$\mu_j = Np_j.$$

In other words the Poisson mean cell counts are equal to the total population size multiplied by the corresponding cell probabilities. Conditional on the total population size the cell counts have a multinomial distribution with probabilities given by

$$\frac{\mu_j}{\sum_{j=0}^{2^T-1}\mu_j} = \frac{Np_j}{N\sum_{j=0}^{2^T-1}p_j} = p_j.$$

The Poisson model can then be factorized (after algebra) as follows:

$$\begin{aligned} L(N, \boldsymbol{q}; \boldsymbol{n}) &= \frac{\exp(-N(1-q_0))(N(1-q_0))^D}{D!} \times \left\{ \frac{D!}{\prod_{j=1}^{2^T-1} n_j!} \prod_{j=1}^{2^T-1} \left(\frac{q_j}{1-q_0}\right)^{n_j} \right\} \\ &= \frac{\exp(-N(1-q_0))(N(1-q_0))^D}{D!} \times L(q; \boldsymbol{n}^{obs}). \end{aligned}$$

The first term corresponds to a Poisson component for the total number of observed individuals, D (recall that the sum of independent Poisson random variables is again Poisson with mean equal to the sum of the means); and the second component corresponds to the distribution of the cell counts given the number of observed individuals (which is thus independent of the total population size N). Comparing to Eq. (1) we can see that this Poisson likelihood shares the same conditional multinomial component, $L(q; \boldsymbol{n}^{obs})$ (corresponding to the likelihood of the observed cell counts conditional on only the observed individuals)—the difference lies with the first term—either a Poisson term for the total number of individuals observed; or a binomial component for the number of observed individuals (or equivalently unobserved individuals) given the total population size. Sandland and Cormack (1984) showed that the MLEs of these multinomial and Poisson formulations are the same. For further discussion see, for example, McCrea and Morgan (2014).

Note that within this chapter we will focus on the multinomial formulation as we will consider heterogeneity at the individual level, and the most intuitive way of specifying the multinomial cell probabilities corresponds to individuals (or groups of individuals) within the population. In particular, we can rewrite the multinomial formulation as follows. Let p_{it} denote the probability that individual i is observed at time t; and let the set of capture probabilities be given by $\boldsymbol{p} = \{p_{it} : i = 1, \ldots, D; t = 1, \ldots, T\}$. In addition we let p^* denote the probability that an individual is observed at least once within the study period; so that $1 - p^*$ is the probability that an individual is not observed within the study period. Thus $p^* = 1 - q_0$ in the previous notation. The general multinomial form of the unconditional likelihood, specified at the individual level, is given by

$$L(N, \boldsymbol{p}; \boldsymbol{x}) \propto \frac{N!}{(N-D)!} \prod_{i=1}^{D} \Pr(\boldsymbol{x}_i | \boldsymbol{p}) \times \Pr(\boldsymbol{x}_0 | \boldsymbol{p})^{N-D} \tag{2}$$

where $\Pr(\boldsymbol{x}_i|\boldsymbol{p})$ denotes the probability of capture history $\boldsymbol{x}_i$ given the set of capture probabilities $\boldsymbol{p}$; and $\Pr(\boldsymbol{x}_0|\boldsymbol{p})$ the probability of an encounter history with no captures. For notational simplicity, here and throughout the remainder of this chapter, we drop the dependence on the capture probabilities and simply write $\Pr(\boldsymbol{x}_i)$ and $\Pr(\boldsymbol{x}_0)$ instead of $\Pr(\boldsymbol{x}_i|\boldsymbol{p})$ and $\Pr(\boldsymbol{x}_0|\boldsymbol{p})$, respectively. We note that we can write

$$\Pr(\boldsymbol{x}_i) = \prod_{t=1}^{T} p_{it}^{x_{it}} (1 - p_{it})^{(1-x_{it})};$$

and

$$\Pr(\boldsymbol{x}_0) = 1 - p^*.$$

In general we specify restrictions on the capture probabilities, to represent different possible dependencies. For example, we may assume that $p_{it} = p$ for all individuals i and times t—so that the capture probability is constant over time and individuals; in this case $(1 - p^*) = (1-p)^T$. Alternatively the capture probabilities may vary across capture occasions, but be consistent across individuals, so that $p_{it} = p_t$ for all individuals i; and $(1 - p^*) = \prod_{t=1}^{T}(1 - p_t)$.

We discuss some of these models in the context of the mice data considered previously, before discussing these dependencies in more detail in Section 3.

2.2.4 Example: Mice Data

We return to the mice data discussed in Section 2.1.1. For this study there were six separate trapping occasions on consecutive days—one for each day of the study. Previously we reduced the data to two capture occasions by combining the trapping occasions on days 1–3 and days 4–6. Now we consider each day of trapping as a separate capture occasion. The data are provided by Huggins (1991)—we consider a subset of the full data (removing further covariate information of the individual animals) and present these in Table 3.

A total of 38 individuals are observed over the study period such that of these 17 are female and 21 are male. We ignore this additional individual characteristic data for the moment—but will return to this in Section 3.2. Of the 38 individuals observed only 1 individual is observed at every capture occasion; with 9 individuals observed only once. However, the number of individuals in the study population but not observed is unknown.

The previous estimate of the total population size using the Lincoln–Petersen estimator on the reduced two-sample data provided an estimate of 38.41 with 95% confidence interval (38.19, 38.81). We now extend the statistical analysis and use the individual days to be six separate capture occasions. In general, there is no closed form expression for the MLE of the total population size or capture probability, and a numerical optimization algorithm is applied to maximize the (log-)likelihood to obtain the associated MLEs and associated confidence intervals. We consider two different models here (i) the model with a constant capture probability over time (denoted model M_0); and (ii) the model with a different capture probability for each capture occasion (denoted model M_t)—we discuss temporal heterogeneity in further detail in Section 3.2.

Model M_0

The simplest model for such multiple capture occasion data assumes a constant capture probability, p, for all individuals. Thus we have that $p_{it} = p$ for all individuals i and time t. The likelihood can be simplified to be of the form:

$$L(N, p; R, D) \propto \frac{N!}{(N-D)!} p^R (1-p)^{NT-R}, \tag{3}$$

where,

$$R = \sum_{i=1}^{D} \sum_{t=1}^{T} x_{it} = \sum_{t=1}^{t} tf_t;$$

TABLE 3 The Capture Histories of the 38 Deer Mice Observed (Arbitrarily Numbered) Over the 6 Capture Occasions

Individual Number	Capture Occasion (t)						Individual Number	Capture Occasion (t)					
	1	**2**	**3**	**4**	**5**	**6**		**1**	**2**	**3**	**4**	**5**	**6**
1	1	0	0	1	1	1	18	1	1	1	1	1	1
2	1	1	1	0	1	1	19	1	1	0	0	1	1
3	1	0	0	1	0	0	20	1	1	0	1	1	1
4	0	1	0	0	1	0	21	1	1	1	1	1	1
5	0	1	1	0	0	1	22	1	1	0	1	1	1
6	0	1	0	1	0	1	23	1	1	1	1	1	0
7	0	1	0	0	0	1	24	1	1	1	0	0	1
8	0	0	1	0	0	0	25	1	1	1	1	1	1
9	0	0	1	1	1	1	26	1	1	0	1	1	1
10	0	0	1	0	1	1	27	1	1	0	1	1	1
11	0	0	1	1	1	1	28	1	1	1	1	1	1
12	0	0	1	0	1	0	29	1	0	1	1	1	0
13	0	0	1	0	0	0	30	0	1	0	0	0	1
14	0	0	0	1	0	0	31	0	1	0	1	0	1
15	0	0	0	1	1	0	32	0	1	1	0	1	0
16	0	0	0	0	1	0	33	0	1	0	0	1	1
17	0	0	0	0	0	1	34	0	0	0	1	1	1

Continued

TABLE 3 The Capture Histories of the 38 Deer Mice Observed (Arbitrarily Numbered) Over the 6 Capture Occasions—Cont'd

Individual Number	Capture Occasion (t)						Individual Number	Capture Occasion (t)					
	1	2	3	4	5	6		1	2	3	4	5	6
							35	0	0	0	0	1	0
							36	0	0	0	0	1	0
							37	0	0	0	0	0	1
							38	0	0	0	0	0	1

A "0" indicates the individual is unobserved at the given capture occasion; and a "1" that the individual is observed at the given capture occasion. Individuals 1–17 are female (left-hand side); individuals 18–38 are male (right-hand side).

such that $D = 38$ and $T = 6$. Recall, from Section 2.2.1, f_t denotes the number of animals observed at t different capture occasions within the study. We note that $R = 120$ simply corresponds to the number of captures observed within the study period.

For model M_0 the summary statistics R and D are minimally sufficient. The likelihood can be easily interpreted as a form of binomial expression where, for each individual in the population, we have a series of simple independent and identical Bernoulli trials corresponding to whether or not the individual is observed at each of the capture occasions. Under this binomial model (i.e., model M_0) we obtain an MLE for N of 38 and associated 95% nonparametric bootstrap confidence interval of (38.00, 38.22). We note that this estimate is slightly smaller than the previous Lincoln–Petersen estimator with a reduced 95% confidence interval width. It is perhaps not surprising that the confidence interval is narrower, since we are estimating the total population size using all six capture occasions (and not aggregated to simply two occasions), so that we are using a greater amount of information and have only two parameters to estimate (i.e., N and p). We also note that the estimate of the total population is on the boundary of the parameter space. Further the MLE of the (common) capture probability is 0.526 with 95% nonparametric bootstrap confidence interval (0.434, 0.614).

We note that the Lincoln–Petersen estimator does not assume the same capture probability at each of the two capture occasions, simply that each individual has the same probability of capture within a single capture occasion. The capture probabilities themselves are assumed to be different. In particular, the MLEs (95% nonparametric bootstrap confidence intervals) of the capture probabilities are 0.729 (0.595, 0.889) for capture occasion 1 and 0.989 (0.856, 0.994) for capture occasion 2. Thus this suggests that assuming a constant capture probability may not be appropriate, for example, as sampling conditions can vary over time. Of course the Lincoln–Petersen estimator has simplified the data by aggregating the capture occasions, but, for consistency, the same number of capture occasions are aggregated within each of the two sampling periods.

We now consider the six capture occasions in more detail. Under the assumption of a constant capture probability we would expect approximately the same number of individuals to be observed at each capture occasion. We let R_t denote the number of individuals observed at capture occasion $t = 1, \ldots, T$, and set $\boldsymbol{R} = \{R_1, \ldots, R_T\}$. For the mice data we have the following number of observed captures at each occasion:

	Capture Occasion (t)					
	1	*2*	*3*	*4*	*5*	*6*
Number observed (R_t)	15	20	16	19	25	25

These data would perhaps suggest that the capture probability is not constant over time—so that we consider an alternative model allowing for different capture probabilities at each capture occasion.

Model M_t

For this model we assume that each of the capture occasions have a different capture probability (but that the capture probabilities are the same for each individual at a given capture occasion). We let p_t denote the capture probability at time $t = 1, \ldots, T$, and for notational simplicity we let the set of all capture probabilities be denoted by $\boldsymbol{p} = \{p_1, \ldots, p_T\}$. The multinomial likelihood formulation for this model can be reduced to:

$$L(N, \boldsymbol{p}; D, \boldsymbol{R}) \propto \frac{N!}{(N-D)!} \prod_{t=1}^{T} p_t^{R_t} (1-p_t)^{N-R_t}, \quad (4)$$

where for the mice data $D = 38$; $T = 6$; and the R_t values are given above. Using a numerical optimization algorithm we obtain the MLE (95% nonparametric bootstrap confidence interval) of N to be 38.00 (38.00, 38.05). This estimate is similar to the previous estimate for model M_0, but with a slightly shorter confidence interval (i.e., less uncertainty), most likely due to the improved fit of the model to the data due to the greater flexibility of the additional parameters in the model. The associated MLEs and 95% nonparametric bootstrap confidence intervals of the capture probabilities are

	MLE	95%
Capture Occasion	$\hat{p}_t$	*Confidence Interval*
1	0.395	(0.237, 0.553)
2	0.526	(0.364, 0.684)
3	0.421	(0.263, 0.579)
4	0.500	(0.342, 0.658)
5	0.658	(0.500, 0.816)
6	0.658	(0.500, 0.816)

Thus we note that there is some variability with regard to the estimated capture probabilities, although the confidence intervals are overlapping for all capture occasions.

To formally compare the models, M_0 and M_t (and additional models throughout this chapter), we will consider Akaike's Information Criterion (AIC; Akaike, 1974). The AIC statistic is defined to be equal to the deviance ($-2 \times$ log-likelihood evaluated at the MLE of the parameters) plus $2 \times$ the number of parameters. For further discussion of model selection techniques, and in particular with reference to the AIC statistic, see, for example, Burnham and Anderson (2002). The associated AIC statistics of model M_0 and M_t are 113.51 and 113.67, respectively. The AIC statistic is a relative goodness-of-fit measure, and so in the calculation of the deviance within this expression, we only use the multinomial combinatorial terms containing the unknown population size (the other combinatorial terms are constants). There is very little difference between the models, M_0 and M_t with regard to the AIC statistic, and the additional time dependence of the capture probabilities have not significantly increased the fit of the model to the data, allowing for the additional model complexity (i.e., number of parameters)—this is perhaps not surprising given the overlapping confidence intervals of the capture probabilities. Further model refinements could be considered such as considering "low" catchable times (occasions 1–4) and "high" catchable times (occasions 5–6)—however, this is fairly arbitrary and without some ecological justification regarding knowledge of the data collection process is difficult to sensibly interpret. Clearly such results can lead to useful discussions with the data collectors and ecologists to understand (and possibly improve) future study protocols or additional data being recorded (such as effort expended at each capture occasion; or weather conditions). Thus for simplicity we will generally assume no temporal heterogeneity for these data.

As previously mentioned, the AIC statistic is a *relative* goodness-of-fit measure—given the set of models considered which model fits the data best. In order to assess *absolute* goodness-of-fit (does the given model fit the observed data well?) it is possible to compare the observed and expected values for the number of individuals with each possible capture history (i.e., cell entries) and apply a Pearson chi-squared test, for example. Considering such tests will typically lead to issues of sparsity with small observed and expected values. However, for some models there are sensible ways of pooling observations to overcome this issue. For example, consider the Schnabel census for the mice data—here the data can be summarized via the use of the sufficient statistics $\boldsymbol{f} = \{9, 6, 7, 6, 6, 4\}$. The expected values of $\boldsymbol{f}$ under model M_0 are $\{2.9, 8.0, 11.9, 9.9, 4.4, 0.8\}$, thus by simple inspection there would appear to be a large discrepancy between the observed and expected values, under the given model. Formally, applying Pearson's chi-squared goodness-of-fit test we obtain an observed test statistic of 30.27; and an associated p-value of 1.2×10^{-6}, providing strong evidence of a lack of fit under model M_0. Thus for these data there appear to be other factors that have not been accounted for—we discuss next different forms of heterogeneity that may impact the capture probabilities (and hence total population size).

3 HETEROGENEITY

Heterogeneity can arise in many different forms and can have a noticeable impact on the data observed and hence a significant effect on the estimate for the total population size. Let us temporarily return to the two-sample Lincoln–Petersen experiment. Suppose capture heterogeneity exists within the population, so that a proportion of the population is much more likely to be captured on each occasion than the rest of the population. This means that on the second sampling occasion the proportion of already marked individuals is likely to be higher (since marked individuals have a higher probability overall of being caught than unmarked individuals), and the resulting population size estimate will be negatively biased (i.e., an underestimate of the population size). Conversely suppose that the marking of an individual is an unpleasant experience, which changes the behavior of the individual so that they avoid future captures. This means that on the second capture occasion a marked individual has a lower probability of being observed than an unmarked individual—often referred to as a "trap shy" response. This subsequently leads to a population size estimate that is positively biased (i.e., an overestimate of the population size).

The manifestation of heterogeneity in capture is not always simply due to study design, and therefore it is not always possible to predict prior to model fitting that you need to consider models incorporating capture heterogeneity. A relatively simple diagnostic test, proposed in Jeyam et al. (2018), can be applied to capture–recapture data with multiple occasions to detect whether the capture histories exhibit capture heterogeneity. The test outcome can guide which models should be considered as potential candidate models for the given data set. The test compares the previous number of captures to the number of future captures and calculates a nonparametric test to test for positive association between these values. No model structure is assumed for the computation of this test. The cause of capture heterogeneity is not always clear. For example, some individuals may be more likely to be captured because of their role within a population. For other populations it is not obvious what the differences are and they may be as subtle as some individuals showing no fear to being captured and therefore would be attracted to baited traps; or some animals being more curious than others when humans approach.

Otis et al. (1978) identified and described three primary sources of heterogeneity corresponding to temporal (t); behavioral (b); and individual heterogeneity (h). We briefly discuss each of these different forms of heterogeneity:

Temporal heterogeneity: Temporal heterogeneity naturally reflects changes in capture probabilities across the different capture occasions—as previously discussed in Section 2.2.7. Such changes may reflect, for example, different weather conditions that affects the behavior of individuals or search strategy used by the observers; or different efforts that may be expended within the capture occasions—so that the capture probabilities at the different capture occasions are not constant but vary.

Behavioral heterogeneity: Behavioral effects remove the independence assumption between captures and allows for *trap happy* or *trap shy* responses. In other words the behavior of an individual may change following their initial capture—they may become more likely to be observed at future capture occasions if the experience is profitable such as providing available food (a *trap happy* response); or less likely to be observed at future capture occasions if the initial capture experience is unpleasant such as being physically handled by an observer or unable to escape a trap (a *trap shy* response). In the simplest case there are two distinct capture probabilities corresponding to the initial capture probability of an individual and the subsequent recapture probability of the individual.

Individual heterogeneity: Individual heterogeneity assumes that the population is not homogeneous and so the capture probabilities of individual animals differ. This may be a result of many different factors including, for example, natural variability of individual behavior (some individuals may simply be more adventurous than others); relationship between the study region and an individual's home range; or characteristics that may affect behavior, for example, males and females often display different behaviors.

Models may incorporate any combination of these heterogeneities. To represent the different possible models the notation M with the set of subscripts t, b and h (used in this order presented) is typically used to denote the set of heterogeneities present. For example, M_b denotes the model with only behavioral heterogeneity present; the model M_{th} the model with both temporal and individual heterogeneity present. We note that the model notation does not correspond to a single unique model—but simply denotes the heterogeneities present in the model—there are many possible model formulations for the different heterogeneities. For example, for M_t we may specify arbitrary capture probabilities for each capture occasion (to be estimated) or a catch-effort model where the capture probabilities are written as a function of the recorded effort used at each capture occasion. Alternatively for model M_b, the trap response may be permanent, or the behavioral effect may decrease over time since last capture, as an individual "forgets" its previous capture.

In this chapter we focus on individual heterogeneity models where different individuals in the population may have different capture probabilities (though note that in Section 3.3 that a behavioral response can be regarded as a special case of individual time-varying heterogeneity). In particular, within this section we explore three particular types of heterogeneity: unobserved heterogeneity; observed (time-invariant) heterogeneity; and time-varying observed heterogeneity; and in each case focus on where the heterogeneity can be discretely categorized.

3.1 Unobserved Heterogeneity

We begin by considering unobserved heterogeneity where we have no additional information on the observed capture histories but recognize that some individual animals may be more "catchable" than others. For example, the catchability of an individual may be related to the closeness of the individual's home range to the study area; or the position of the individual in the hierarchical social structure of the population which may influence their behavior and/or movement patterns. We focus on the case where we assume that the overall population is heterogeneous, but is composed of an (unknown) number of homogeneous subpopulations. We consider the model $M_{h(k)}$ denoting the heterogeneity model with k homogeneous (unobserved) subpopulations.

3.1.1 Model $M_{h(k)}$

The model can be mathematically described as a finite mixture model on the individuals, where it is unknown which mixture, or subpopulation, each individual belongs to—such models were initially proposed by Pledger (2000). We assume that there are a total of k mixture components, such that an individual belongs to component (or subpopulation) j with probability $\pi(j)$, where $0 \leq \pi(j) \leq 1$ and $\sum_{j=1}^{k} \pi(j) = 1$. We set $\boldsymbol{\pi} = \{\pi(1), \ldots, \pi(k)\}$. In other words the different subpopulations correspond to a partition such that each individual in the population belongs to one and only one subpopulation. Each individual in subpopulation $j = 1, \ldots, k$ has a capture probability $p(j)$, assumed to be constant over time. The set of capture probabilities is given by $\boldsymbol{p} = \{p(1), \ldots, p(k)\}$. For this model, the sufficient statistics correspond to the Schnabel census, $\boldsymbol{f} = \{f_1, \ldots, f_T\}$, where f_t corresponds to the number of animals observed a total of t times within the study period.

In this case the likelihood of the multinomial formulation given in Eq. (2) simplifies to be of binomial form:

$$L(N, \boldsymbol{p}, \boldsymbol{\pi}; \boldsymbol{f}) \propto \frac{N!}{(N-D)!} \prod_{t=1}^{T} \Pr(\text{individual observed } t \text{ times})^{f_t} \times \Pr(\boldsymbol{x}_0)^{N-D}.$$

To calculate the probability that an individual is seen t times (for $t = 0, \ldots, T$) we use the basic laws of probability. In particular we sum over all possible subpopulations that an individual may belong to of the joint probability of belonging to the given subpopulation and having the given capture probability. Mathematically, we have that, for $t = 0, \ldots, T$,

$$\begin{aligned}
\Pr(\text{individual observed } t \text{ times}) &= \sum_{j=1}^{k} \Pr(\text{individual observed } t \text{ times and belongs to group } j) \\
&= \sum_{j=1}^{k} \Pr(\text{in group } j) \times \Pr(\text{observed } t \text{ times} \mid \text{in group } j) \\
&= \sum_{j=1}^{k} \pi(j) \times p(j)^t (1 - p(j))^{T-t}.
\end{aligned}$$

We note that this likelihood is simply a mixture of binomial components.

Although finite mixture models are structurally straightforward to construct there are some computational challenges with their use in practice. Commonly referred to as a label-switching problem, the finite mixture model can have a local identifiability issue if the parameter space of the proportion parameter is not restricted; see Kim and Lindsay (2015) for a full discussion of this issue. Further, there exist some challenges in the context of model selection. The issues (which are discussed in full within Pledger (2000)) arise because if the models being compared differ in the number of groups of animals, the parameters at the boundary under the null model mean that conditions of the standard likelihood ratio test are not met. However, a nonstandard likelihood ratio test can be used—see Self and Liang (1987) for details. In practice Cubaynes et al. (2012) showed that the AIC statistic (or other information criteria approaches), work reasonably well.

Alternative approaches for the modeling of unobserved heterogeneity exist. In particular, Dorazio and Royle (2003) proposed the use of an infinite mixture model, by considering a beta-binomial distribution, and Morgan and Ridout (2008) suggested combining the use of a finite and infinite mixture. Alternative infinite mixture models have been proposed, specifying the individual capture probabilities in the form of a logistic regression with an individual (normal) random effect component and using a numerical integration approach to evaluate the likelihood (Coull and Agresti, 1999) or a Bayesian data augmentation (sometimes referred to as a complete-data likelihood) approach (King and Brooks, 2008; King et al., 2016). For further discussion of the strengths and weaknesses of some of these approaches, see, for example, Pledger (2005) and the associated response.

3.1.2 *Example: Mice Data*

For the mice data the corresponding Schnabel census sufficient statistics for model $M_{h(k)}$ are given by

t	1	2	3	4	5	6
f_t	9	6	7	6	6	4

We fit models of the form $M_{h(k)}$ (i.e., mixtures of k binomial components). In particular, we consider the cases $k = 2$ and $k = 3$. The estimates of the parameters, associated 95% nonparametric bootstrap confidence intervals and AIC statistics are provided in Table 4. The analyzes indicate that the model with $k = 2$ is preferred over the model where $k = 3$ using the AIC statistic. We note further that these are nested models (although the null model has parameters on the boundary) and considering only the fit of the model to the data in terms of the deviance (or equivalently the log-likelihood evaluated at the MLE of the parameters), there is very little difference between these two models. The difference in AIC statistics is thus dominated by the number of parameters associated with each model (4 parameters

TABLE 4 Model Estimates From Fitting Binomial Mixture Models, $M_{h(2)}$ (Two Components) and $M_{h(3)}$ (Three Components) to the Mice Data

	Model $M_{h(2)}$		Model $M_{h(3)}$	
AIC	*100.767*		*104.479*	
		95%		*95%*
Parameter	*MLE*	*Confidence Interval*	*MLE*	*Confidence Interval*
N	40.52	(38.00, 83.59)	41.91	(38.11, 149.62[a])
$\pi(1)$	0.577	(0.351, 0.885)	0.467	(0.219, 0.840)
$\pi(2)$	—	—	0.387	(0.016, 0.675)
$p(1)$	0.290	(0.038, 0.475)	0.223	(0.015, 0.452)
$p(2)$	0.771	(0.609, 0.976)	0.629	(0.414, 0.875)
$p(3)$	—	—	0.890	(0.563, 1.000)

[a]*Note that the upper 2.5% quantile may not have completely converged due to the very long right-hand tail of this distribution.*

when $k = 2$ and 6 parameters when $k = 3$). Thus there is clear support for $k = 2$ and we focus our discussions on this model.

For the case $k = 2$, there clearly appears to be a large difference in the detectability of the two different subpopulations assumed to be present in the population. The associated AIC statistic (100.767) is significantly smaller than for model M_0 fitted in Section 2.2.7, indicating some level of unobserved heterogeneity present within the population. The estimate for the total population size is 40.52, suggesting that not quite all members of the population were observed within the study (recall that a total of 38 individuals were observed). We note that there appears to be a very long right-hand tail for the estimate of the total population size. This is often observed in such models, particularly where the capture probability of one of the components (with reasonable mixture weight) is small—capture probabilities and population size (for the given subpopulation) are inversely proportional to each other, so that variations in the small capture probabilities lead to generally large variability of the estimate of the total population size. (We note that this is particularly marked for the case $k = 3$ where we appear to be over-estimating the number of subpopulations.) The expected values of f under the $k = 2$ model are $\{7.5, 8.0, 6.0, 6.1, 6.7, 3.7\}$. Applying Pearson's chi-squared goodness-of-fit test we obtain a test statistic of 1.07; and an associated p-value of 0.3, suggesting that the observed and expected values are not significantly different from one another. We note that sparsity of the available data means that it is often not straightforward to assess absolute goodness-of-fit due to the requirement for all expected values to be greater than 5, and even for this fairly straightforward heterogeneity model we have an expected value of 3.7. Therefore for the more complex models we consider in the remainder of this chapter we do not consider absolute goodness-of-fit further. This issue of sparsity of data for goodness-of-fit assessment is well-documented in many areas of ecological modeling—see, for example, Stanley and Burnham (1999) and Amstrup et al. (2005).

For the heterogeneity studied here we assume that we do not observe the subpopulations. These subpopulations may relate to physical characteristics of the individuals such as gender, breeding status, age, etc. In many studies, such characteristics may also be recorded when individuals are observed. These may be subsequently used to create observable subpopulations to incorporate such heterogeneity.

3.2 Observed (Time-Invariant) Heterogeneity

We now consider the case where additional information may be recorded on individuals observed within the study. For example, this may correspond to gender, age group or breeding status. We let $\mathcal{G}=\{1,\ldots,G\}$ denote the set of G mutually exclusive and exhaustive individual physical characteristics (or groups) within a population. Here we assume that the group membership of an individual is constant within the study period—we relax this assumption in Section 3.3. When individuals are captured they are identified as belonging to group $g \in G$ without error, and we assume that a total of $D(g)$ individuals in group g are observed within the study period. For notational convenience we let $\boldsymbol{D} = \{D(1), \ldots, D(G)\}$ denote the set of observed animal numbers corresponding to each group. As before, a total of $D=\sum_{g\in\mathcal{G}} D(g)$ unique individuals are observed within the study period. We wish to estimate the total subpopulation sizes for each group denoted by $\boldsymbol{N} = \{N(1), \ldots, N(G)\}$ and associated overall total population size $N=\sum_{g\in\mathcal{G}} N(g)$. For the observed heterogeneity models, we use the notation $h(\mathcal{G})$ to denote individual heterogeneity attributed to the set of characteristics $\mathcal{G}$.

3.2.1 *Model* $M_{0(\mathcal{G})}$

For model $M_{0(\mathcal{G})}$ we include the additional covariate information, and condition on this when forming the likelihood, but assume that the population is homogeneous across the different subpopulations or groups, i.e., the capture probabilities are independent of group membership. The associated likelihood expression is given by

$$L(\boldsymbol{N}, p; R, \boldsymbol{D}) \propto \left[\prod_{g\in\mathcal{G}} \frac{N(g)!}{(N(g)-D(g))!}\right] \times p^R (1-p)^{NT-R},$$

where we recall that R corresponds to the number of observed sightings within the whole study period (and so in this case across all groups); and $N = \sum_{g \in \mathcal{G}} N(g)$.

The expression is very similar to that given in Eq. (3) corresponding to model M_0 when there is no additional observed covariate information. However, the initial combinatorial terms are different. For model M_0 there is no information relating to the groups and so all individuals can be simply reordered as the order is unimportant in the likelihood calculation, leading to a single multinomial coefficient in the likelihood expression. However, for model $M_{0(\mathcal{G})}$ we also condition on the observed group that each individual belongs to so that individual animals are distinguishable between groups, though the order that they are numbered within the groups are exchangeable, thus leading to a multinomial coefficient term for each individual group within the likelihood expression. The two models, M_0 and $M_{0(\mathcal{G})}$, are using different levels of information—with M_0 taking an "unconditional" approach with regard to gender and model $M_{0(\mathcal{G})}$ conditioning on the additional covariate information within the likelihood. Consequently we cannot directly compare such models with regard to AIC statistics or likelihood ratio tests.

3.2.2 Model $M_{h(\mathcal{G})}$

We consider the model $M_{h(\mathcal{G})}$, where the only heterogeneity acting on the system corresponds to the observed group membership. For any given capture occasion, we let $p(g)$ denote the probability that an individual from group $g \in \mathcal{G}$ is observed, i.e., $p(g)$ is the capture probability of an individual in group $g \in \mathcal{G}$. For notational simplicity we let $\boldsymbol{p} = \{p(1), \ldots, p(G)\}$ denote the set of capture probabilities across all the different groups. Conditioning on the observed group membership of each individual we can extend Eq. (3) and express the likelihood of the data as a product of the likelihood contributions of the capture histories of the individuals associated with each group $g \in G$ and write

$$L(\boldsymbol{N}, \boldsymbol{p}; \boldsymbol{R}, \boldsymbol{D}) \propto \prod_{g \in \mathcal{G}} \frac{N(g)!}{(N(g) - D(g))!} p(g)^{R(g)} (1 - p(g))^{N(g)T - R(g)}, \quad (5)$$

where $\boldsymbol{R} = \{R(1), \ldots, R(G)\}$, such that $R(g)$ corresponds to the number of captures of individuals in subpopulation $g \in \mathcal{G}$ and can be expressed as

$$R(g) = \sum_{t=1}^{T} t f_t(g);$$

such that $f_t(g)$ denotes the number of individuals from group g captured t times within the study; and the parameters of primary interest, $N(g)$, correspond to the (unknown) population size of group $g \in \mathcal{G}$. For each individual group $f_0(g)$ denotes the number of individuals from group g that were not observed within the study, so that $f_0(g) = N(g) - D(g)$ and $\sum_{g \in \mathcal{G}} f_0(g) = N - D$.

We note that in this approach we essentially stratify the individuals into the separate groups and analyze each group independently of each other. The estimates of $N(g)$ for each group can then be combined to obtain an overall estimate for N. Constructing the confidence intervals using a bootstrap approach means that it is trivial to obtain an associated bootstrap confidence interval for N, using the relationship that $N = \sum_{g \in \mathcal{G}} N(g)$ applied to the MLEs of $N(g)$ for each bootstrap replicate.

3.2.3 *Model* $M_{th(\mathcal{G})}$

Here we consider further both temporal and observed individual heterogeneity, denoted by the model notation $M_{th(\mathcal{G})}$. We extend the above group stratification, so that the capture probabilities for each group are also time-dependent. The data can be most easily summarized in terms of the set of the total number of individuals observed in each group, $\boldsymbol{D} = \{D(1), \ldots, D(G)\}$, and the number of individuals from group g that are observed at time t, denoted $\boldsymbol{R} = \{R_t(g) : t = 1, \ldots, T; g \in \mathcal{G}\}$. The corresponding likelihood can be expressed in the form:

$$L(\boldsymbol{N}, \boldsymbol{p}; \boldsymbol{R}, \boldsymbol{D}) \propto \prod_{g \in \mathcal{G}} \frac{N(g)!}{(N(g) - D(g))!} \prod_{t=1}^{T} p_t(g)^{R_t(g)} (1 - p_t(g))^{N(g) - R_t(g)}, \quad (6)$$

where $p_t(g)$ denotes the capture probability at time $t = 1, \ldots, T$ for group $g \in \mathcal{G}$ and $\boldsymbol{p} = \{p_t(g) : t = 1, \ldots, T; g \in \mathcal{G}\}$. We once again note the similarity with Eq. (4) allowing for the additional group membership.

Arbitrarily estimating $p_t(g)$ for each time and group once again essentially stratifies the data into the different groups and leads to independent analyzes for each of the groups. However, this may lead to small sample sizes for each of the different groups with a relatively large numbers of parameters to be estimated—in this case there are $(T + 1)G$ parameters to estimate (TG recapture probabilities and G population sizes) and consequently leading to potentially poor precision of the estimates. However, alternative capture probabilities can be specified that reduces the number of parameters to be estimated and permits the borrowing of information across the different groups. For example, we may specify an additive logistic model for the capture probabilities, such that,

$$\text{logit}\, p_t(g) = \alpha(g) + \beta_t,$$

where $\alpha(g)$ denotes the effect associated with group $g \in \mathcal{G}$ and β_t the temporal effect for $t = 1, \ldots, T$. For identifiability we typically set one of the parameters to be equal to 0, for example, without loss of generality, we may set $\alpha(1) = 0$. The remaining $\alpha(g)$ terms, for $g = 2, \ldots, G$ are then interpreted as the effect of group g *relative* to group 1. This model assumes that the capture probabilities associated with each group display a similar temporal pattern over the capture occasions—but allowing for different levels of catchability for the different groups. The model also assumes that these differences in catchability across groups are additive on the logistic scale—taking the logistic link function is often primarily for mathematical convenience to ensure that the capture probabilities are constrained to the interval [0,1]. Alternative relationships can also be specified, such as $p_t(g) = \beta_t^{\alpha(g)}$. However, additional care needs to be taken in this case to ensure that $p_t(g) \in [0, 1]$ for all $g \in \mathcal{G}$ and $t = 1, \ldots, T$. Thus, for simplicity we retain the additive logistic models described above. For this model there are a total of $2G + T - 1$ parameters (G population sizes; T temporal effects and $G - 1$ group effects to estimate, since we specify the constraint $\alpha(1) = 0$ for identifiability); this is compared to $(T + 1)G$ for the case above with arbitrary group and time effects.

This logistic model for the capture probabilities no longer stratifies the data into the separate subpopulations and considers each stratified dataset independently. Instead the model uses information from all of the groups to estimate the temporal model parameters, borrowing information across each of the different groups, but with the given assumed structure across times and groups.

3.2.4 Model $M_{t\times h(\mathcal{G})}$

The final model that we consider is of a similar form to model $M_{th(\mathcal{G})}$ with the associated likelihood expression given in Eq. (6). However, here we consider an interaction term in the logistic model specified for the capture probabilities with regard to the group and time effects, expressed as model $M_{t\times h(\mathcal{G})}$. Thus for $g \in \mathcal{G}$ and $t = 1, \ldots, T$, we specify

$$\operatorname{logit} p_t(g) = \alpha(g) + \beta_t + \delta_t(g),$$

with the standard constraints for identifiability (for example, $\alpha(1) = 0$ and $\delta_1(1) = 0$). We note that this model is equivalent to the previous model discussed above with arbitrary capture times $p_t(g)$ to be estimated—and hence again removes this borrowing of information across the different groups, and associated issues discussed above. The sufficient statistics are again the number of individuals from group $g \in \mathcal{G}$ that are observed at time $t = 1, \ldots, T$, and denoted by, $\boldsymbol{R} = \{R_t(g) : t = 1, \ldots, T; g \in \mathcal{G}\}$.

3.2.5 Example: Mice

We return to the mice example, and now consider the additional information of gender within our analyzes. There are naturally two groups corresponding to females and males. Thus we set $\mathcal{G} = \{f, m\}$ with obvious notation (so that in the previous group labeling $1 \equiv f$; and $2 \equiv m$). We consider four different models:

- Model $M_{0(\mathcal{G})}$—capture probabilities are constant over all times and groups.
- Model $M_{h(\mathcal{G})}$—capture probabilities are different for each group but constant over time.

- Model $M_{th(\mathcal{G})}$—capture probabilities logistically regressed with additive group and time effects.
- Model $M_{t\times h(\mathcal{G})}$—capture probabilities logistically regressed with additive group and time effects with interactions—equivalent to arbitrary capture probabilities for each group and time.

For model $M_{h(\mathcal{G})}$ the sufficient statistics are given by $R(f) = 41$; $R(g) = 79$; $R = 120$; $D(f) = 17$; and $D(m) = 21$. For models $M_{th(\mathcal{G})}$ and $M_{t\times h(\mathcal{G})}$ the sufficient statistics are given by

t	1	2	3	4	5	6
$R_t(f)$	3	5	8	7	9	9
$R_t(m)$	12	15	8	12	16	16

We note that there does appear to be some variability over capture occasions with regard to the number of females and males observed—with a general mild positive correlation (correlation coefficient = 0.19) between the number of observed captures for females and males over capture occasions. However, the correlation is difficult to assess with the relatively few data points being considered.

The associated estimates of the population sizes and AIC statistics are provided in Table 5. In the calculation of the nonparametric bootstrap confidence intervals we apply a stratified bootstrap to preserve the proportion of male and female histories within each bootstrap replication (i.e., for these data we sample with replacement 17 female capture histories from the set of all observed female histories; and 21 male capture histories from the set of all observed male histories). The model identified as optimal with regard to the AIC statistic is model $M_{th(\mathcal{G})}$ with additive group and time effects—although we note that $\Delta AIC = 0.31$ for the nested time-independent model $M_{h(\mathcal{G})}$. The estimates of the capture probabilities are given in Table 6 for $M_{th(\mathcal{G})}$. We note that the gender difference (male–female) on the logit scale is estimated to be 0.984, with 95% nonparametric bootstrap confidence interval (0.299, 1.814), signifying a difference between the capture probabilities for males and females, such that males are more likely to be observed than females. Similarly, for the second ranked model, $M_{h(\mathcal{G})}$, with no temporal heterogeneity, the corresponding MLE (and 95% confidence interval) for the capture probability of females is 0.394 (0.261, 0.500) and for males is 0.627 (0.500, 0.746). This again suggests a substantial gender difference, assuming that there is no temporal heterogeneity.

In this example we observe group membership of all individuals. However, if group information is missing for some individuals an EM algorithm (Little and Rubin, 2002) or Bayesian data augmentation approach (Tanner and Wong, 1987) can be used to impute the group membership within the computational algorithms. Alternatively, a mixture modeling approach can

TABLE 5 The MLEs and Associated 95% Nonparametric Bootstrap Confidence Intervals (CIs) for Population Sizes and Associated AIC Statistic for the Four Fitted Models

	$N(f)$		$N(m)$		N[a]		AIC
Model	*MLE*	*95% CI*	*MLE*	*95% CI*	*MLE*	*95% CI*	
$M_{0(\mathcal{G})}$	17.00	—[b]	21.00	—[b]	38.00	—[b]	163.67
$M_{h(\mathcal{G})}$	17.34	(17.00, 19.79)	21.00	—[b]	38.34	(38.00, 40.79)	154.03
$M_{th(\mathcal{G})}$	17.23	(17.00, 19.04)	21.00	(21.00, 21.01)	38.34	(38.00, 40.04)	153.72
$M_{t\times h(\mathcal{G})}$	17.00	(17.00, 17.21)	21.00	—[b]	38.00	(38.00, 38.21)	156.74

[a]*Note that N is a derived parameter from N(f) and N(m) calculated by the simple relationship* $N = N(f) + N(m)$.
[b]*Denotes that the confidence interval was essentially ill-defined with the parameter lying on the boundary of the parameter space with the width of the 95% confidence interval less than 0.01.*

TABLE 6 The Derived MLEs and 95% Nonparametric Bootstrap Confidence Intervals (CIs) for the Capture Probabilities at Each Capture Occasion for Females and Males for Model $M_{th(\mathcal{G})}$

Parameter	MLE	95% CI	Parameter	MLE	95% CI
$p_1(f)$	0.268	(0.151, 0.407)	$p_1(m)$	0.494	(0.305, 0.683)
$p_2(f)$	0.391	(0.240, 0.549)	$p_2(m)$	0.632	(0.447, 0.798)
$p_3(f)$	0.291	(0.147, 0.457)	$p_3(m)$	0.523	(0.341, 0.711)
$p_4(f)$	0.363	(0.210, 0.527)	$p_4(m)$	0.605	(0.405, 0.787)
$p_5(f)$	0.533	(0.330, 0.729)	$p_5(m)$	0.753	(0.605, 0.881)
$p_6(f)$	0.533	(0.330, 0.729)	$p_6(m)$	0.753	(0.605, 0.881)

be implemented, combining the theory from Sections 3.1 and 3.2—see McCrea et al. (2013) for an application of this procedure in the related field of ring-recovery modeling for individuals with missing age-class information.

3.3 Time-Varying Observed Individual Heterogeneity

So far we have only focused on "fixed" forms of individual heterogeneities—in other words observed or unobserved characteristics of the individuals that are static over the period of the study. For example, characteristics such as gender or species which are invariant over all time; or other factors that do not vary over the study period itself, although are not invariant over longer periods of time, such as age or breeding status. However, other individual level factors leading to heterogeneity in the capture probabilities may vary over the time of the study period itself, such as location or group size. For simplicity, we refer to these as covariates. If an individual is observed, the associated time-varying covariate is generally observed (though this may not always be the case); however, if an individual is not observed the corresponding time-varying covariate cannot be recorded and is thus "missing." An exception to this is where the covariate can be inferred without error from the observed data—this includes, for example, where the covariate corresponds to age or for describing behavioral responses (i.e., trap happy and trap shy) as an observed individual time-varying covariate (we consider this as a special case in Section 3.3.2). We initially consider this latter, simpler, case before the more general case where we account for the missing covariate values explicitly within the likelihood expression.

3.3.1 Deterministic Covariate Values

Here we consider the case where the covariates vary deterministically over time. For closed populations (typically over a short study period) this situation is not very common, except for the special case described in Section 3.3.2 corresponding to behavioral effects. However, we describe the situation in this general case for completeness and for motivation for Section 3.3.9 where we extend the idea to stochastic time-varying covariate values. For deterministic covariate values, if the covariate is observed at the initial capture of an individual (and possibly conditional on the given observed capture history) the corresponding covariate values are known for all capture occasions for that individual. We let the set of possible covariate values be denoted by $\mathcal{Z}=\{1,\ldots,Z\}$. Further, we let, z_{it} denote the known covariate value for individual $i = 1, \ldots, D$ at capture occasion $t = 1, \ldots, T$; and set $z = \{z_{it} : i = 1, \ldots, D; t = 1, \ldots, T\}$.

We again specify the capture probabilities to be a function of the covariate values. In particular, assuming no additional heterogeneities, the capture probability for observed individual $i = 1, \ldots, D$ at time $t = 1, \ldots, T$ is given by

$$\text{logit}\, p_{it} = \alpha(z_{it}).$$

The covariate values for the unobserved individuals are, however, unknown. Thus, in this general case we specify a distribution for the initial covariate value for all individuals at time $t = 1$. We note that since the covariate values are deterministic, given the observed capture history, the choice of time is arbitrary—we could have chosen the final state distribution, for example, at time $t = T$, but this, in general, is less intuitive. For observed individuals, $i = 1, \ldots, D$, we specify

$$z_{i1} \sim Multinomial(1, \boldsymbol{\pi}),$$

where $\boldsymbol{\pi} = \{\pi(1), \ldots, \pi(Z)\}$ such that $\pi(z)$ denotes the probability that an individual is initially in state $z \in \mathcal{Z}$ at time $t = 1$.

The likelihood expression given in Eq. (2) can be extended to incorporate these deterministic covariate values to give

$$L(N, \boldsymbol{p}, \boldsymbol{\pi}; \boldsymbol{x}, z) \propto \frac{N!}{(N-D)!}\prod_{i=1}^{D}\pi(z_{i1})\prod_{t=1}^{T}p_{it}^{x_{it}}(1-p_{it})^{1-x_{it}} \times (Pr(\boldsymbol{x}_0))^{N-D}.$$

The likelihood function is a product of terms corresponding to the multinomial coefficient; initial state distribution and capture history probabilities associated with observed individuals; and finally the contribution associated with unobserved individuals. In particular, assuming that the unobserved individuals have the same initial state distribution as the observed individuals, we have that

$$\begin{aligned}\Pr(\boldsymbol{x}_0) &= \sum_{z\in\mathcal{Z}}\Pr(\text{initial state}=z)\Pr(\text{unobserved in study} \mid \text{initial state}=z)\\ &= \sum_{z\in\mathcal{Z}}\pi(z)\prod_{t=1}^{T}(1-p_t(z)),\end{aligned}$$

where $p_t(z)$ corresponds to the capture probability at time t, given that the initial covariate value of an individual at time $t = 1$ is equal to z. Suppose that we let $h(z, t)$ denote the covariate value at time t, given that the initial covariate value at time $t = 1$ is equal to z. Then, we have that

$$\text{logit}\, p_t(z) = \alpha(h(z,t)).$$

Recall that the covariate values are deterministic, conditional on the capture histories, so for any given initial covariate value at time $t = 1$, all other covariate values as times $t = 2, \ldots, T$ are a deterministic function. We note that the above likelihood can be extended to the case where we do not necessarily observe the covariate values for all observed individuals. In this case, for those individuals, we can apply a similar argument to that of the unobserved individuals and sum over all possible initial covariate values within the associated likelihood expression.

As discussed previously, recording deterministic time-varying covariate values that may influence the capture probabilities is not particularly common for capture–recapture studies over a short period of time where it is reasonable to assume that the population is closed (i.e., there are no births/deaths/migrations within the study period). The most common examples of such covariates are often age- or stage-related, that simply increases deterministically over time. One particular example where the covariate is time-varying and known for all individuals and capture occasions (given the observed capture histories) is with regard to behavioral effects.

3.3.2 *Behavioral Heterogeneity*

Behavioral heterogeneity permits an individual's capture probability to change following its initial capture—either increasing (a trap happy response) or decreasing (a trap shy response) the capture probability at subsequent capture occasions. The response is typically assumed to be a permanent response effect, but can be easily extended to allow for the memory of the capture to decrease over time. In the simplest model, M_b, there are just two capture probabilities: an initial capture probability; and a recapture probability (for individuals who have been observed at least once in the study). In this model only the time to initial capture has any effect on the associated capture probability of the individual—subsequent multiple recaptures do not lead to any further modification.

For this model, we set the individual time-varying covariate values to be equal to $z_{it} = 0$ up to and including the initial capture of the individual; and $z_{it} = 1$ for all times after initial capture for individual $i = 1, \ldots, N$. Thus these values are known for *all* individuals both those observed within the study and those unobserved—for those unobserved the covariates values are equal to 0 for all capture times. The capture probability of individual i at time t is specified to be of the form

$$\text{logit}\, p_{it} = \begin{cases} \text{logit}\, p = \beta & \text{for } z_{it} = 0; \\ \text{logit}\, c = \beta + \gamma & \text{for } z_{it} = 1. \end{cases}$$

The term p correspond to the initial capture probability; and c to the recapture probability. The model is parameterized in the above form for interpretability of the difference between initial and subsequent capture probabilities (on the logit scale); and for consistency with the extension to include additional forms of heterogeneity, such as temporal heterogeneity which we discuss below. The parameter γ directly corresponds to the behavioral response effect: if $\gamma > 0$ there is a trap happy response; if $\gamma < 0$ there is a trap shy response. The corresponding likelihood simplifies to

$$L(N,p,c|R,U,D) \propto \frac{N!}{(N-D)!} p^D (1-p)^{U+(N-D)T} c^{R-D} (1-c)^{DT-R-U},$$

where

$$R = \sum_{i=1}^{D} \sum_{t=1}^{T} x_{it}; \quad \text{and} \quad U = \sum_{i=1}^{D} \sum_{t=1}^{T} (1 - z_{it}) - D.$$

We note that R corresponds to the total number of captures within the study period; thus $R - D$ corresponds to the total number of recaptures (excluding the initial capture of each observed individual). Similarly U corresponds to the total number of noncaptures of individuals that occur prior to the individuals being initially observed (unobserved individuals are not included in this calculation since they are not initially observed within the study); $(N - D)T$ the number of noncaptures relating to unobserved individuals; and $DT - R - U$ the total number of noncaptures following the initial capture of the individuals (equal to the total number of possible captures for individuals observed within the study minus the number of captures minus the number of noncaptures prior to initial capture). We note that for this model, the likelihood can be partitioned into two distinct components relating to (i) up to and including initial capture (if any); and (ii) after initial capture. It is only the initial captures (and associated initial capture probability, p) that affects the estimate of the total population size ($N - D$ and c are independent; similarly p and c are independent of each other). For this simplest model, M_b, the statistics D, R, and U are the minimal sufficient statistics.

As discussed previously, there can be multiple heterogeneities present within the same system. For example, consider the model M_{tb}, with both temporal and behavioral effects present. We note that there is no unique model specification for M_{tb}, with several different parameterizations possible. We consider an additive logistic model here, such that the (logit of the) capture probability of an individual at time t is specified to be of the form

$$\operatorname{logit} p_{it} = \begin{cases} \operatorname{logit} p_t = \beta_t & \text{for } z_{it} = 0; \\ \operatorname{logit} c_t = \beta_t + \gamma & \text{for } z_{it} = 1. \end{cases}$$

The term p_t corresponds to the initial capture probability at time t; and c_t the recapture probability of an individual at time t, given that they have already been observed within the study period. The corresponding sufficient statistics for the model are given by

$$R_t = \sum_{i=1}^{D} x_{it}; \; A_t = \sum_{i=1}^{D} x_{it}(1 - z_{it}); \quad \text{and} \quad U_t = \sum_{i=1}^{D}(1 - z_{it}) - A_t.$$

for $t = 1, \ldots, T$. The statistics can be easily interpreted as follows: R_t corresponds to the total number of captures at time t; A_t the number of initial captures at time t (by definition $A_1 = R_1$ and $\sum_{t=1}^{T} A_t = D$); and U_t the number of individuals unobserved at time t that have not been observed at any previous occasion. For notational convenience we set $\boldsymbol{R} = \{R_1, \ldots, R_T\}, \boldsymbol{A} = \{A_1, \ldots, A_T\}$ and $\boldsymbol{U} = \{U_1, \ldots, U_T\}$. The corresponding likelihood follows similarly to the above for M_b, but allowing for the additional temporal dependence. In particular we obtain

$$L(N, \boldsymbol{p}|\boldsymbol{R}, \boldsymbol{U}, \boldsymbol{A}, D) \propto \frac{N!}{(N-D)!} \prod_{t=1}^{T} p_t^{A_t}(1 - p_t)^{U_t + (N-D)} c_t^{R_t - A_t}(1 - c_t)^{D - R_t - U_t}. \tag{7}$$

Similarly additional individual heterogeneity can be incorporated into such models—such as unobserved or observed individual heterogeneity. We see an example of this for the mice data, where we consider the observed covariate of gender.

3.3.3 Example: Mice

We return to the mice data and consider models with additional behavioral effects. Given the previous importance of including the observed (static) covariate of gender, we again condition on the gender within all of the models. We consider a number of competing models corresponding to the additional inclusion of behavioral and/or temporal effects present in the model.

Model $M_{bh(0)}$

In this model we assume that there is a behavioral effect, but no gender effects on the capture probabilities, denoted $M_{bh(0)}$. The sufficient statistics for this model are given by $D(f) = 17$ (number of females observed); $D(m) = 21$ (number of males observed); $U = 56$ and $R = 120$. The corresponding likelihood expression is given by

$$L(\boldsymbol{N}, p, c|R, U, \boldsymbol{D}) \propto \prod_{g \in \mathcal{G}} \frac{N(g)!}{(N(g) - D(g))!} p^D (1-p)^{U + (N-D)T} c^{R-D}(1-c)^{DT-R-U},$$

where $D = D(f) + D(m)$ (and corresponds to the total number of individuals observed). We note that the only difference between this expression and that given in Section 3.3.2 (with no gender effects) corresponds to the multinomial coefficient terms for the different groups.

Model $M_{bh(\mathcal{G})}$

Here there is both a behavioral effect, specified to be additive on the logistic scale, and a gender effect for the capture probabilities. Thus we assume that the behavioral effect is independent of gender. In particular we specify the capture probability for an individual of gender $g \in \mathcal{G}$ at time t to be of the form

$$\operatorname{logit} p_{it}(g) = \begin{cases} \operatorname{logit} p(g) = \alpha(g) & \text{for } z_{it} = 0; \\ \operatorname{logit} c(g) = \alpha(g) + \gamma & \text{for } z_{it} = 1. \end{cases}$$

As usual for notational convenience we set $\boldsymbol{p} = \{p_t(g) : g \in \mathcal{G}\}$ and $\boldsymbol{c} = \{c_t(g) : g \in \mathcal{G}\}$. The corresponding sufficient statistics are given by $\boldsymbol{D} = \{D(g) : g \in \mathcal{G}\}$; $\boldsymbol{U} = \{U(g) : g \in \mathcal{G}\}$; and $\boldsymbol{R} = \{R(g) : g \in \mathcal{G}\}$, with obvious notation (conditioning on the gender of the individual animals). For the mice data, $U(f) = 31$; $U(m) = 25$; $D(f) = 17$; $D(m) = 21$; $R(f) = 41$; and $R(m) = 79$. The corresponding likelihood is given by

$$L(\boldsymbol{N}, \boldsymbol{p}, \boldsymbol{c} | \boldsymbol{R}, \boldsymbol{U}, \boldsymbol{D}) \propto \prod_{g \in \mathcal{G}} \frac{N(g)!}{(N(g) - D(g))!} p(g)^{D(g)} (1 - p(g))^{U(g) + (N(g) - D(g))T}$$
$$\times c(g)^{R(g) - D(g)} (1 - c(g))^{D(g)T - R(g) - U(g)}.$$

This model assumes that the behavioral response, denoted by the additive γ term is the same for all observed covariate values (in this case gender). This may not be the case, for example, if the different observed subpopulations respond differently to the capture process. In this case we can consider an interaction between the behavioral effect and observed covariate.

Model $M_{b \times h(\mathcal{G})}$

Within this model we assume that the behavioral response is gender specific. The capture probabilities are specified similarly to model $M_{bh(\mathcal{G})}$, but where the behavioral response is now a function of the observed covariate:

$$\operatorname{logit} p_{it}(g) = \begin{cases} \operatorname{logit} p(g) = \alpha(g) & \text{for } z_{it} = 0; \\ \operatorname{logit} c(g) = \alpha(g) + \gamma(g) & \text{for } z_{it} = 1. \end{cases}$$

The sufficient statistics and associated likelihood expression are of the same form as for $M_{bh(\mathcal{G})}$—it is simply the specification of the capture probabilities that differ (and the estimation of an additional parameter). We note that in this case (as for the model $M_{t \times h(\mathcal{G})}$) the likelihood once again decomposes into separate, and independent, components for the distinct groups $g \in \mathcal{G}$.

Model $M_{tbh(\mathcal{G})}$

Here we consider temporal, behavioral and observed heterogeneity via an additive logistic model. The capture probabilities are specified in the form

$$\text{logit}\, p_{it}(g) = \begin{cases} \text{logit}\, p(g) = \alpha(g) + \beta_t & \text{for } z_{it} = 0; \\ \text{logit}\, c(g) = \alpha(g) + \beta_t + \gamma(g) & \text{for } z_{it} = 1, \end{cases}$$

and we set $\alpha(f) = 0$ for identifiability. The corresponding likelihood function is an extension of Eq. (7) following the previous idea of extending to include group membership by producting over groups and specifying the parameters (capture probabilities and total population sizes) and sufficient statistics to be conditional on the observed groups. The sufficient statistics are given by $\boldsymbol{R} = \{R_t(g) : t = 1,\ldots,T; g \in \mathcal{G}\}$; $\boldsymbol{A} = \{A_t(g) : t = 1,\ldots,T; g \in \mathcal{G}\}$; $\boldsymbol{U} = \{U_t(g) : t = 1,\ldots,T; g \in \mathcal{G}\}$; and $\boldsymbol{D} = \{D(g) : g \in \mathcal{G}\}$.

Results

The four models described above are fitted to the data, and the associated results (in terms of MLEs, 95% nonparametric confidence intervals and AIC statistics) are presented in Table 7. The model deemed optimal has both behavioral and individual heterogeneity effects in terms of observed group, i.e., model $M_{bh(\mathcal{G})}$, with an associated AIC statistic of 145.38. Recall that the model previously identified as optimal via the AIC statistic without behavioral effects was model $M_{th(\mathcal{G})}$ with an associated AIC statistic of 153.72 (see Table 5). Thus, this suggests that the behavioral effects are more important than arbitrary temporal effects. Further, we note that including the behavioral effects consistently leads to a significantly improved fit of the model to the data, irrespective of the other effects that are present in the model. The behavioral response for this model (and for all models fitted with a behavioral effect present) is positive, corresponding to a "trap happy" response. In other words an individual who is initially observed at a given capture time is more likely to be observed at future capture occasions. Finally, we note that the estimates of the total population size are also generally dependent on the model being fitted—with greater variability with regard to the female subpopulation.

3.3.4 *Stochastic Covariate Values*

In practice many individual covariates may change over time in a stochastic manner, such as their breeding status, hunger levels, health, and location. The temporal scale acting on the changes is dependent on the actual covariate—for example, breeding status (breeding/not breeding) may change at an annual level; whereas hunger levels may vary over hours or days, dependent on the species. The capture probability of an individual may be highly dependent on the covariate—animals that are hungry are more likely to forage for food and hence may be more likely to be seen during a given capture

TABLE 7 The MLE and Associated 95% Nonparametric Bootstrap Confidence Intervals (CIs) for Population Sizes and Associated AIC Statistic for the Four Fitted Models

	$N(f)$		$N(m)$		N^a		AIC
Model	*MLE*	*95% CI*	*MLE*	*95% CI*	*MLE*	*95% CI*	
$M_{b(\mathcal{G})}$	17.61	(17.00, 22.03)	21.88	(21.00, 27.33)	39.49	(38.00, 49.36)	155.53
$M_{bh(\mathcal{G})}$	21.25	(17.51, 34.62)	21.00	—[b]	42.25	(38.51, 55.62)	145.38
$M_{b\times h(\mathcal{G})}$	19.59	(17.00, 48.60)	21.00	(21.00, 21.01)	38.00	(38.00, 69.60)	147.17
$M_{tbh(\mathcal{G})}$	20.95	(17.04, 33.45)	21.00	(21.00, 21.13)	41.95	(38.04, 54.54)	151.00

[a]*Note that N is a derived parameter from N(f) and N(m) calculated by the simple relationship* $N = N(f) + N(m)$.
[b]*Denotes that the confidence interval was essentially ill-defined with the parameter lying on the boundary of the parameter space with the width of the 95% confidence interval less than 0.01.*

occasion, compared to an individual who is satiated and hence stays within some home location. Alternatively, the location of an animal may also influence the probability that they are seen—with individuals more likely to be observed in open grassland, compared to dense forest. Here we consider such covariates where changes in their values may occur in individuals at a finer scale than the study period.

Notationally, we let $\mathcal{Z} = \{1,\ldots,Z\}$ denote the set of possible discrete covariate values (or states). Recall that we let z_{it} denote the covariate value for individual $i = 1, \ldots, D$ at time $t = 1, \ldots, T$. For simplicity, we initially assume that if an individual is observed, the corresponding covariate value is observed without error. However, if an individual is not observed, the corresponding covariate value is also unknown. We note that the model can be extended to allow for misclassification of states, and such a model is referred to as the multievent model (Pradel, 2005), and a related model exists for partially observed data (King and McCrea, 2014).

We define the set of observed covariate values, denoted, $z^{obs} = \{z_{it} : z_{it}$ is known$\}$. Further we set

$$\mathcal{Z}_{it}^* = \begin{cases} z_{it} & \text{for } z_{it} \text{ known;} \\ \mathcal{Z} & \text{for } z_{it} \text{ unknown.} \end{cases}$$

Thus $\mathcal{Z}_{it}^*$ corresponds to the set of values that the value z_{it} may be, given the observed data. If the covariate value is observed then as we assume that these values are correctly recorded so that the state of the individual is known without error; else if the covariate is unknown then it may take any possible covariate value, i.e., any value in $\mathcal{Z}$. The set of possible covariate values over the capture occasions for individual i, given the observed data are given by $\mathcal{Z}_i^* = \{\mathcal{Z}_{it}^* : t = 1,\ldots,T\}$.

We specify the capture probabilities to be a function of the covariate values, and other possible effects. For example, we may specify the capture probabilities to be of the following forms:

- Model M_h: capture probabilities of the form logit $p_t(z_{it}) = \alpha(z_{it})$—dependence only on state;
- Model M_{th}: capture probabilities of the form logit $p_t(z_{it}) = \alpha(z_{it}) + \beta_t$—dependence on state and time assumed to be additive on logistic scale (for identifiability we set $\alpha(1) = 0$);
- Model $M_{t \times h}$: capture probabilities of the form logit $p_t(z_{it}) = \alpha(z_{it}) + \beta_t + \delta_t(z_{it})$—dependence on state and time with an interaction (i.e., arbitrary time and state dependence) (for identifiability we set $\alpha(1) = 0$ and $\delta_1(1) = 0$).

The above models, and associated capture probabilities, can be extended, for example, to allow for group effects (observed deterministic effects) or behavioral effects in a similar way.

We also need to specify some form of model for the change in state over time. We assume a first-order Markov model, so that the state at time t depends only on the state at time $t - 1$ (and no previous states). We define the $Z \times Z$ state transition probabilities at time t to be,

$$\begin{aligned}\psi_t(z,z^\dagger) &= \Pr(\text{individual in state } z^\dagger \text{ at time } t+1 | \text{ in state } z \text{ at time } t)\\ &= \Pr(z_{it+1} = z^\dagger | z_{it} = z).\end{aligned}$$

Notationally, we let the set of all transition probabilities be denoted by $\boldsymbol{\psi} = \{\psi_t(z,z^\dagger) : t = 1,\ldots,T-1;\ z,z^\dagger \in \mathcal{Z}\}$. Finally, we define the parameters corresponding to the initial state distribution, and let $\pi(z)$ denote the probability that an individual is in state z at time $t = 1$. We note that $\sum_{z\in\mathcal{Z}}\pi(z) = 1$ (an individual must belong to one of the possible states at time $t = 1$). The set of initial state probabilities is given by $\boldsymbol{\pi} = \{\pi(1), \ldots, \pi(Z)\}$.

The corresponding likelihood of the models can be constructed, summing over all possible unknown states, and taking into account the transition probabilities between the possible states:

$$\begin{aligned}L(N, \boldsymbol{p}, \boldsymbol{\pi}, \boldsymbol{\psi}; \boldsymbol{x}, z^{obs}) \propto \frac{N!}{(N-D)!}\prod_{i=1}^{D}\sum_{z_i\in\mathcal{Z}_i^*}&\left(\pi(z_{i1})p_1(z_{i1})^{x_{i1}}(1-p_1(z_{i1}))^{1-x_{i1}}\right.\\ &\left.\times\prod_{t=1}^{T-1}\psi_t(z_{it},z_{it+1})p_{t+1}(z_{it+1})^{x_{it+1}}(1-p_{t+1}(z_{it+1}))^{1-x_{it+1}}\right)\\ &\times \Pr(\boldsymbol{x}_0)^{N-D},\end{aligned}$$

where,

$$\Pr(\boldsymbol{x}_0) = \sum_{z_1\in\mathcal{Z}}\cdots\sum_{z_T\in\mathcal{Z}}\pi(z_1)(1-p_1(z_1))\prod_{t=1}^{T-1}\psi_t(z_t,z_{t+1})(1-p_{t+1}(z_{t+1})).$$

Summing over all the possible missing covariate values will often be computationally expensive, or even infeasible, due to the number of possible combinations of unobserved values. For example, for the probability of not being observed, $\Pr(\boldsymbol{x}_0)$, the summation is over Z^T possible values—this is the most "expensive" capture history in terms of the calculation, but the number of unique capture histories will also typically be "large" for which the associated probability will need to be calculated.

The (first-order) Markovian assumption for the state transitions, however, allows us to significantly simplify the likelihood expression, so that it can be written in the form of an efficient hidden Markov model (see Zucchini et al. (2016) for an introduction to general hidden Markov models). Due to the first-order Markov structure we can consider "pairs" of state values at times t and $t + 1$. Intuitively, this is due to the memoryless property of the Markov process—what happens after time t only depends on the state of an individual at time t (and not anything that has happened before). Thus we do not need to consider how an individual arrives at its state at time t (i.e., its previous covariate values or capture history) but simply what its state is at time t.

The subsequent likelihood is most easily expressed in matrix notation. We describe the three different components corresponding to (i) the initial covariate (or state) distribution; (ii) observation process; and (iii) transition process between the different possible discrete covariate values (or states). We consider each in turn.

(i) Initial state distribution: We consider two different cases, corresponding to whether or not the initial state of the individual is known or not, i.e., for individual i whether the covariate value is known, so that $\mathcal{Z}^*_{i1} = z_{i1} \in \mathcal{Z}$; or unknown so that $\mathcal{Z}^*_{i1} = \mathcal{Z}$. We define the initial state distribution row vector:

$$\boldsymbol{\pi}(\mathcal{Z}^*_{i1}) = \begin{cases} (\pi(1)I(z_{i1}=1),\ldots,\pi(Z)I(z_{i1}=Z)) & \text{for } \mathcal{Z}^*_{i1} = z_{i1}\,(\text{i.e., } z_{it} \text{ is known}); \\ (\pi(1),\ldots,\pi(Z)) & \text{for } \mathcal{Z}^*_{i1} = \mathcal{Z}(\text{i.e., } z_{it} \text{ is unknown}), \end{cases}$$

where $I(\cdot)$ denotes the indicator function.

(ii) Observation process: We define the observation matrix for individual $i = 1, \ldots, D$, at time $t = 1, \ldots, T$ to be of the form

$$\boldsymbol{P}_t(\mathcal{Z}^*_{it}) = \begin{cases} \text{diag}(p_t(1)I(z_{it}=1),\ldots,p_t(Z)I(z_{it}=Z)) & \text{for } \mathcal{Z}^*_{it} = z_{it} \in \mathcal{Z}; \\ \text{diag}(1-p_t(1),\ldots,1-p_t(Z)) & \text{for } \mathcal{Z}^*_{it} = \mathcal{Z}. \end{cases}$$

As for the initial state distribution, the observation matrix depends on whether the individual is observed or not at time t. We assume that if an individual is observed then the associated covariate value is known without error. This means that if an individual is observed the contribution to the likelihood is simply $p_t(z_{it})$; if an individual is not observed, the individual may be in any of the possible states with associated contribution to the likelihood corresponding to the probability of not being observed (conditional on the covariate value).

(iii) Transition process: Finally the state transition matrix is specified as follows, dependent on whether or not the covariate values are observed or not at time t and $t + 1$ for individual $i = 1, \ldots, D$. We initially consider the case where the covariate value is unknown at times t and $t + 1$, so that $\mathcal{Z}^*_{it} = \mathcal{Z}^*_{it+1} = \mathcal{Z}$, then:

$$\boldsymbol{\Psi}_t(\mathcal{Z}^*_{it}, \mathcal{Z}^*_{it+1}) = \begin{pmatrix} \psi_t(1,1) & \ldots & \psi_t(1,Z) \\ \vdots & & \vdots \\ \psi_t(Z,1) & \ldots & \psi_t(Z,Z) \end{pmatrix}.$$

Now if the covariate value is known at time t, so that z_{it} is known ($\mathcal{Z}^*_{it} = z_{it}$), but the covariate value is unknown at time $t + 1$ ($\mathcal{Z}^*_{it+1} = \mathcal{Z}$), the corresponding transition matrix is the same as $\boldsymbol{\Psi}_t(\mathcal{Z}^*_{it}, \mathcal{Z}^*_{it+1})$ but with 0 replacing all the values in the rows except for the z_{it}th row. Similarly if the covariate value at time $t + 1$ is known ($\mathcal{Z}^*_{it+1} = z_{it+1}$) but the covariate value at time t is unknown ($\mathcal{Z}^*_{it} = \mathcal{Z}$), the transition matrix is the same $\boldsymbol{\Psi}_t(\mathcal{Z}^*_{it}, \mathcal{Z}^*_{it+1})$ but with 0 replacing all values except for the z_{it+1}th column. Finally if both z_{it} and z_{it+1} are both known, the matrix reduces such that all values are set equal to 0 except $\psi_t(z_{it}, z_{it+1})$.

The corresponding likelihood can be expressed in the form:

$$L(N, \boldsymbol{p}, \boldsymbol{\pi}, \boldsymbol{\psi}; \boldsymbol{x}, z^{obs}) \propto \frac{N!}{(N-D)!} \prod_{i=1}^{D} \left(\boldsymbol{\pi}(\mathcal{Z}_{i1}^*) \mathbf{P}_1(\mathcal{Z}_{i1}^*) \prod_{t=1}^{T-1} \left[\boldsymbol{\Psi}_t(\mathcal{Z}_{it}^*, \mathcal{Z}_{it+1}^*) \boldsymbol{P}_{t+1}(\mathcal{Z}_{it+1}^*) \right] \mathbf{1}_{\mathcal{Z}}^T \right)$$

$$\times \Pr(\boldsymbol{x}_0)^{N-D},$$

where $1_{\mathcal{Z}}^T$ corresponds to the column vector with all elements equal to unity; and

$$\Pr(\boldsymbol{x}_0) = \pi(\mathcal{Z}) \mathbf{P}_1(\mathcal{Z}) \prod_{t=1}^{T-1} \boldsymbol{\Psi}_t(\mathcal{Z}, \mathcal{Z}) \boldsymbol{P}_{t+1}(\mathcal{Z}) \mathbf{1}_{\mathcal{Z}}^T.$$

The likelihood expression is of the same form as a hidden Markov model, but allows for known observed covariate values (for a hidden Markov model typically no true states are observed). A similar approach is taken by Langrock and King (2013) for capture–recapture data for an open population with continuous time-varying individual covariates using a discrete approximation (and allowing for covariate values being unknown when an individual is observed); and King and Langrock (2016), permitting semi-Markov transition processes for multistate capture–recapture data (and where covariate values may be observed with error), removing the first-order Markov assumption. Worthington et al. (2018a) describe the associated summary statistics for these data, and an efficient likelihood specification, using an alternative recursive formulation to the matrix approach.

4 OPEN POPULATIONS

We have, so far, considered closed populations where it is assumed that the population does not change over the duration of the study period, i.e., there are no births/deaths or emigration/immigration acting on the system. In other words all individuals in the study are available for capture at every capture occasion within the study. In order to satisfy this assumption, the study typically covers only a short duration of time to minimize the chances of this assumption being violated—for example, for the mice data studied throughout, the capture–recapture study was undertaken over 6 consecutive days. It is possible to perform a test for closure: Otis et al. (1978) present a test for closure which is unaffected by the presence of heterogeneous capture probabilities, while Stanley and Burnham (1999) propose a test for closure for cases where capture probabilities vary over time. We have previously seen how failing to account for heterogeneity in capture may lead to biased estimates of population size. Similarly, if closed population models are fitted to data from populations which exhibit birth/immigration and death/emigration, capture probabilities will generally be underestimated and hence population size overestimated. The literature of open population capture–recapture models is extensive (for a review see, for example, King (2014) and McCrea and Morgan (2014)). However, for such open population models, the focus has generally been on the estimation of demographic parameters such as survival, and the factors that may affect the survival probabilities, for example, to gain an understanding of the system and the primary factors driving the population which may be important for conservation and management purposes.

The original open capture–recapture model which allowed inference to be made on population size was the Jolly–Seber model (Jolly, 1965; Seber, 1965). However for this original model the births in the populations were not directly incorporated into the model but were inferred from the relationship between survival and the numbers of unmarked individuals captured on consecutive capture occasions. The population size estimate was obtained through the use of a Horvitz–Thompson-like estimate (Borchers et al., 2002; Horvitz and Thompson, 1952) as a ratio of the number of captured individuals and estimated probability that an individual is observed within the study. The disadvantage of such an approach is that it is not possible to constrain births to be positive, and hence negative birth rates can be obtained. In addition it is not possible to formally model the birth rates, for example, as a function of population size and/or additional covariate information.

Schwarz and Arnason (1996) proposed an alternative approach for open population models introducing the idea of a "super-population" from which individuals may enter and leave the study area. The extended Jolly–Seber-type model builds on the likelihood given in Eq. (2), permitting arrivals to the study area (after which individuals can be subsequently observed); and departures from the study site (from which time individuals are no longer able to be observed). It is assumed that once each individual arrives at the study site they remain there until they depart and cannot return thereafter (i.e., there is no temporary migration).

To account for the additional processes (arrivals and departures), we define the additional parameters:

β_t : the probability that an individual in the super-population and available for capture at some point within the study period is first available for capture at occasion $t+1$;

ϕ_t : the probability an individual present in the study area at occasion t remains in the study area until occasion $t+1$.

We note that by definition, $\sum_{j=0}^{T-1}\beta_j = 1$, since we condition on the individual being available for capture at some point within the study period. For notational convenience we set $\boldsymbol{\beta} = \{\beta_0, \ldots, \beta_{T-1}\}$ and $\boldsymbol{\phi} = \{\phi_1, \ldots, \phi_{T-1}\}$.

The likelihood expression can be written in the form:

$$L(N, \boldsymbol{p}, \boldsymbol{\beta}, \boldsymbol{\phi}; \boldsymbol{x}) \propto \frac{N!}{(N-D)!} \prod_{i=1}^{D} \Pr(\boldsymbol{x}_i) \times \Pr(\boldsymbol{x}_0)^{N-D},$$

where $\Pr(\boldsymbol{x}_i)$ denotes the probability of the capture history of individual i; and $\Pr(\boldsymbol{x}_0)$ the associated probability of not being observed within the study (given the model parameters). Both of these probability terms need to account for the arrival and departure of the individuals—an individual cannot be recaptured before they arrive at the study site; or after they leave the study site (or alternatively the associated capture probabilities at these times are equal to 0). For notational convenience, we let g_i, and l_i, denote the occasions where individual i is observed for the first time, and last time, respectively. Recall that we assume that there is no temporary migration, so that under this assumption, an individual must be present (and available for capture) for times $g_i, \ldots, l_i$. However, we do not necessarily know when the individual arrives—it could be that the individual is first available for capture at time g_i, when it was observed; or the individual may have arrived earlier but have been unobserved until time g_i. Similarly, we know that the individual is in the population and available for capture between times g_i and l_i, inclusive; but after l_i it is again unknown whether the individual leaves the study area before capture occasion $l_i + 1$ (and so cannot be observed), or remains in the study area available for capture but is simply unobserved until it leaves at a future time (or remains in the study area until the end of the study). Thus we need to account for the unknown arrival and departure times (if any) from the study—this can be done by summing over all possible arrival and departure times. For example, for model M_t, with simply a temporal dependence on the capture probabilities, the probability of the capture history of individual $i = 1, \ldots, D$, denoted $\boldsymbol{x}_i$, is given by

$$\Pr(\boldsymbol{x}_i) = \sum_{b=1}^{g_i} \beta_{b-1} \sum_{d=l_i}^{T} (1 - \phi_d) \prod_{t=b}^{d-1} \phi_t \prod_{t=b}^{d} p_t^{x_{it}} (1 - p_t)^{1-x_{it}},$$

where we define $\phi_T = 0$; and apply the mathematical convention that the product over the empty set is equal to unity, i.e., $\prod_{t=b}^{b-1} \equiv 1$. The first term relates to an individual joining the study area; the second term to the individual leaving the study area; the third term for remaining in the study from arrival to departure; and the final term to the capture of the individual given they are in the study area. Similarly, we let p^* denote the probability that an individual is observed at least once within the study. Then, $\Pr(\boldsymbol{x}_0) = 1 - p^*$, such that

$$1 - p^* = \sum_{b=1}^{T} \beta_{b-1} \sum_{d=b}^{T} (1 - \phi_d) \prod_{t=b}^{d-1} \phi_t \prod_{t=b}^{d} (1 - p_t).$$

This model has been successfully extended to account for the probability that an animal leaves the study area as a function of its arrival time (or equivalently the duration of time at the study location; Pledger et al., 2009). The model is particularly useful for animals stopping over at breeding sites—see Matechou et al. (2013, 2014) for some applications—and is thus often referred to as a *stopover* model. Additional behavioral and/or individual heterogeneity effects can be incorporated into the model, by extending the definition of the capture probabilities. For further discussion of these models, including the extension to discrete time-varying individual heterogeneity, see Worthington et al. (2018b); and for a Bayesian approach of fitting a model with discrete time-varying heterogeneity see, for example, Dupuis and Schwarz (2007).

The theory of robust design (Kendall et al., 1995; Pollock, 1982) combined both open and closed population models by defining primary periods of the study which are open and secondary sampling periods which occur in a primary period within which the population is assumed to be closed. This study design facilitates the estimation of abundance, while allowing for the estimation of important demographic parameters; however, the assumption of closure is again required within the secondary sampling periods, which may be unrealistic. The open robust design model Kendall and Bjorkland (2001) removed this closure assumption for the secondary periods, by conditioning on the number of individuals capture in each primary occasion, which in turn removed the ability to estimate abundance directly. The concept of robust design has been extended to the related area of removal modeling (Zhou et al., 2018) which is a special case of the behavioral model discussed in this chapter. Worthington et al. (2018b) extend the robust design models in two ways: by removing the conditional requirement of robust design models by directly modeling all arrival and departure times to the study site within both the primary and secondary sampling periods; and introducing additional time-varying discrete individual covariates, using the analogous approach to Section 3.3.9.

5 DISCUSSION

The accurate estimation of population size is an important aspect of conservation and wildlife management. There are several common methods used to obtain estimates of population size, including capture–recapture studies. The associated capture–recapture models rely on a number of assumptions—we briefly discussed the implications of the assumption of closure in Section 4 and described extended models to allow for such open populations. The failure of the study to satisfy the model assumptions can have a significant impact on the estimates of total population size. This has led to the development of numerous capture–recapture models, such as those described in Section 3 to allow for different forms of heterogeneity that may arise. These models have been constructed based on an understanding of the different factors that may influence the systems: the possible effects on the system are traditionally described in terms of temporal, behavioral, and individual heterogeneity. These models are generally fairly rich in nature and can be adapted to incorporate specific knowledge of the system under study. For example, catch-effort models can be applied when the known resources expended at each capture occasion is recorded—in such cases it may be appropriate to reparameterize the model in terms of the recapture probability per unit effort (see, for example, King and Brooks (2004) for an example of such a parameterization).

Traditionally, the capture–recapture models that have been developed rely on the data collection techniques involving a number of observers going into the field and observing (and subsequently marking) individual animals. However, recent technological advances have had a significant impact on the way that data may be collected—for example, using photographic identification rather than physically adding a mark to the animal to minimize disruption/impact on the animals; using alternative nonvisual cues such as scat/hair samples; or even motion-sensor cameras instead of physical traps. These advances in data collection techniques provide new forms of data and associated challenges that the traditional models in their current form do not address. We discuss some of these challenges and the associated statistical developments to combat these.

In the construction of the capture histories of the individuals observed within the study, it is assumed that each individual carries a unique identification, and that these are recorded without error when an individual is observed. However, within mark-resight studies it may be the case that an individual can be observed to have a mark, but that the unique identifier cannot be read. Thus, the individual capture histories cannot be constructed completely and a given number of observations of marked individuals can be recorded but not assigned to individuals. To address this issue McClintock et al. (2014) consider a Bayesian data augmentation technique (and discuss previous approaches). Alternatively, modern techniques for identifying individuals, such as DNA matching from scat/hair samples or photo-identification can lead to nonignorable mismatches within the study—including false negatives when a single individual is observed at two separate capture occasions but is not matched and hence recorded as two different individuals; or false positives where two different individuals seen at different capture occasions are incorrectly matched to be the same individual. The nature of the scheme for identifying individuals, and associated protocol, provides information on whether false positives and/or false negatives may occur. For example, when using DNA to match individuals, due to genotyping errors that may occur in the process of the given sample, false negatives may occur, but not false positives. For further discussion of issues using DNA for matching individuals, see, for example, Lukacs and Burnham (2005a, b) and Wright et al. (2009).

The problem of not being able to form individual capture histories arises when individuals are not able to be uniquely identified. A traditional marking method where this arises relates to *batch* marking. This is simply where an individual is marked such that its initial capture time is recorded, for example, a color-coded system may be used such that the first time an individual is observed a (permanent) marker/paint may be used—with a different color applied at each capture occasion. The color is only applied the first time an individual is observed; if an individual already has a color no additional color is added at a future capture occasion. At each capture occasion, the number of individuals are recorded that were first observed at each previous capture occasion—but we do not have the individual capture histories of each individual observed within the study (just the total number of individuals observed that were first observed at each previous capture occasion). Estimates of the total population size can be obtained using, for example, estimating equations and a pseudo-likelihood (Huggins et al., 2010) or hidden Markov model approach (Cowen et al., 2017).

Location data may also be recorded in addition to an individual being observed/not observed. For example, capture–recapture studies may involve an array of traps over some spatial region—the capture histories of individuals will then typically detail not only whether an individual was observed at a capture occasion—but also in which trap the individual was observed. The traditional, standard, capture–recapture models collapse these data to simply observed/not observed, discarding the associated spatial information. However, incorporating the spatial information permits the estimation of spatial densities of animals, in addition to the estimation of total population size (Borchers and Efford, 2008; Efford, 2004; Royle and Young, 2008). The development of models to explicitly incorporate this spatial information has exploded in the last decade—and is now referred to as spatially explicit capture–recapture (SECR); or spatial capture–recapture. Typically, for these models the capture probability of an individual at a given time and trap location is expressed as a function of the distance of the given trap to the (unobserved) home range center of the individual. The unobserved home range center can be regarded as a two-dimensional individual (continuous) random effect—the likelihood is calculated by integrating out these unobserved locations. A common aspect of modern SECR data relates to the captures being observed in continuous time—with the spatial array of traps corresponding to motion sensors, so that the data recorded relates to both the individual and the exact time that the individual is observed by the sensor. Two issues immediately arise with regard to correctly identifying individuals from the photographs/videos and the continuous time recordings as opposed to the traditional discrete observation times. For discussion of many of these issues, and associated models, see, for example, Royle et al. (2013) and Borchers et al. (2019).

The technological advances associated with the collection of capture–recapture-type data will continue. However, there is an inevitable delay between the advanced technological tools for collecting the data and the necessary statistical tools to fully analyze the data to maximize the information that can be obtained. In addition, not only may the types of data change over time, but also the amount of available data, for example, multiple types of data may be collected on the same species/ecosystem; or the number of individuals on which data are collected may increase from 10s or 100s of individuals to 1000s of individuals. This leads to additional statistical challenges such as developing integrated models that incorporate different types of data within a single robust analysis which may be particularly challenging when the datasets are not independent of each other; or where the use of traditional methods are computationally infeasible due to the number of individuals within the study. Although the types of data, and associated size and/or and complexity are changing, the need for accurate estimation of population size across many different areas of application will continue to drive forward the statistical innovation in developing appropriate models and corresponding efficient computational model-fitting techniques.

REFERENCES

Akaike, H., 1974. A new look at the statistical model identification. IEEE Trans. Autom. Control 19, 716–723.

Amstrup, S.C., McDonald, T.L., Manly, B.F.J., 2005. Handbook of Capture-Recapture Analysis. Princeton University Press.

Bird, S.M., King, R., 2018. Multiple systems estimation (or capture-recapture estimation) to inform public policy. Annu. Rev. Stat. Appl. 5, 95–118.

Böhning, D., van Der Heijden, P.G.M., Bunge, J., 2018. Capture-Recapture Methods for the Social Sciences. Chapman and Hall, pp. 429.

Borchers, D.L., Efford, M.G., 2008. Spatially explicit maximum likelihood methods for capture-recapture studies. Biometrics 64, 377–385.

Borchers, D.L., Buckland, S.T., Zucchini, W., 2002. Estimating Animal Abundance. Springer, Closed Populations, pp. 314.

Borchers, D.L., Stevenson, B.C., Howe, E.J., Distiller, G., Glennie, R., 2019. Spatial Capture-Recapture by Maximum Likelihood. Springer.

Buckland, S.T., Anderson, D.R., Burnham, K.P., Laake, J.L., Borchers, D.L., Thomas, L., 2004. Introduction to Distance Sampling: Estimating Abundance of Biological Populations. Oxford University Press, pp. 448.

Burnham, K.P., Anderson, D.R., 2002. Model Selection and Multimodel Inference. A Practical Information-Theoretic Approach, second ed. Springer, pp. 488.

Chapman, D.G., 1951. Some Properties of the Hypergeometric Distribution With Applications to Zoological Sample Censuses. University of California Press.

Coull, B., Agresti, A., 1999. The use of mixed logit models to reflect heterogeneity in capture-recapture studies. Biometrics 55, 294–301.

Cowen, L.L.E., Besbeas, P., Morgan, B.J.T., Schwarz, C.J., 2017. Hidden Markov models for extended batch data. Biometrics 73, 1321–1331.

Cubaynes, S., Pradel, R., Choquet, R., Duchamp, C., Gaillard, J.-M., Lebreton, J.-D., Marboutin, E., Miquel, C., Reboulet, A., Poillot, C., Taberlet, P., Gimenez, O., 2010. Importance of accounting for detection heterogeneity when estimating abundance: the case of the French wolves. Conserv. Biol. 24, 621–626.

Cubaynes, S., Lavergne, C., Marboutin, E., Gimenez, O., 2012. Assessing individual heterogeneity using model selection criteria: how many mixture components in capture-recapture models? Methods Ecol. Evol. 3, 564–573.

Dorazio, R.M., Royle, J.A., 2003. Mixture models for estimating the size of a closed population when capture rates vary among individuals. Biometrics 55, 294–301.

Dupuis, J.A., Schwarz, C.J., 2007. A Bayesian approach to the multistate Jolly-Seber capture-recapture model. Biometrics 63, 1015–1022.

Efford, M.G., 2004. Density estimation in live-trapping studies. Oikos 106, 598–610.

Fleming, P., Tracey, J., 2008. Some human, aircraft and animal factors affecting aerial surveys—how to enumerate animals from the air. Wildl. Res. 35, 258–267.

Goudie, I.B.J., Goudie, M., 2007. Who captures the marks for the Petersen estimator? J. R. Stat. Soc. Ser. A 170, 825–839.

Horvitz, D.G., Thompson, D.J., 1952. A generalization of sampling without replacement from a finite universe. J. Am. Stat. Assoc. 47, 663–685.

Huggins, R., 1991. Some practical aspects of a conditional likelihood approach to capture experiments. Biometrics 47, 725–732.

Huggins, R., Wang, Y., Kearns, J., 2010. Analysis of an extended batch marking experiment using estimating equations. J. Agric. Biol. Environ. Stat. 15, 279–289.

Jeyam, A., McCrea, R.S., Pradel, R., 2018. A test of positive association for detecting heterogeneity in capture for capture-recapture data. J. Agric. Biol. Environ. Stat. 23, 1–19.

Jolly, G.M., 1965. Explicit estimates from capture-recapture data with both death and immigration stochastic model. Biometrika 52, 225–247.

Kendall, W.L., Bjorkland, R., 2001. Using open robust design models to estimate temporary emigration from capture-recapture data. Biometrics 57, 1113–1122.

Kendall, W.L., Pollock, K.H., Brownie, C., 1995. A likelihood-based approach to capture-recapture estimation of demographic parameters under the robust design. Biometrics 51, 293–308.

Kim, D., Lindsay, B.G., 2015. Empirical identifiability in finite mixture models. Ann. Inst. Stat. Math. 67, 745–772.

King, R., 2012. A review of Bayesian state-space modelling of capture-recapture-recovery data. Interface Focus 2, 190–204.

King, R., 2014. Statistical ecology. Ann. Rev. Stat. Appl. 1, 410–426.

King, R., Brooks, S.P., 2004. A classical study of catch-effort models for Hector's dolphins. J. Am. Stat. Assoc. 90, 325–333.

King, R., Brooks, S.P., 2008. On the Bayesian estimation of a closed population size in the presence of heterogeneity and model uncertainty. Biometrics 64, 816–824.

King, R., Langrock, R., 2016. Semi-Markov Arnason-Schwarz models. Biometrics 72, 619–628.

King, R., McCrea, R.S., 2014. A generalized likelihood framework for partially observed capture-recapture-recovery data. Stat. Methodol. 17, 30–45.

King, R., Morgan, B.J.T., Gimenez, O., Brooks, S.P., 2009. Bayesian Analysis for Population Ecology. CRC Press, pp. 456.

King, R., McClintock, B., Kidney, D., Borchers, D.L., 2016. Abundance estimation using a semi-complete data likelihood approach. Ann. Appl. Stat. 10, 264–285.

Langrock, R., King, R., 2013. Hidden Markov-type maximum likelihood estimation of mark-recapture-recovery models in the presence of continuous covariates. Ann. Appl. Stat. 7, 1709–1732.

Laplace, P.S., 1786. Sur les naissances, les mariages et les morts. In: Histoire de LAcademie Royale des Sciences, pp. 1783. Paris.

Lincoln, F.C., 1930. Calculating Waterfowl Abundance on the Basis of Banding Returns. US Department of Agriculture Circular 118, pp. 1–4.

Little, R.J.A., Rubin, D.B., 2002. Statistical Analysis With Missing Data. John Wiley & Sons, pp. 389.

Lukacs, P.M., Burnham, K.P., 2005a. Review of capture-recapture methods applicable to noninvasive genetic sampling. Mol. Ecol. 14, 3909–3919.

Lukacs, P.M., Burnham, K.P., 2005b. Estimating population size from DNA-based closed capture-recapture data incorporating genotyping error. J. Wildl. Manag. 69, 396–403.

Matechou, E., Morgan, B.J.T., Pledger, S., Callazo, J.A., Lyons, J.E., 2013. Integrated analysis of capture-recapture-resighting data and counts of unmarked birds at stop-over sites. J. Agric. Biol. Environ. Stat. 18, 120–135.

Matechou, E., Dennis, E., Freeman, S.N., Brereton, T., 2014. Monitoring abundance and phenology in (multivoltine) butterfly species: a novel mixture model. J. Appl. Ecol. 51, 766–775.

McClintock, B.T., Hill, J.M., Fritz, L., Chumbley, K., Luxa, K., Diefenbach, D.R., 2014. Mark-resight abundance estimation under incomplete identification of marked individuals. Methods Ecol. Evol. 5, 1294–1304.

McCrea, R.S., Morgan, B.J.T., 2014. Analysis of Capture-Recapture Data. Chapman and Hall, pp. 314.

McCrea, R.S., Morgan, B.J.T., Cole, D.J., 2013. Age-dependent models for recovery data on animals marked at unknown age. J. R. Stat. Soc. Ser. C Appl. Stat. 62, 101–113.

Morgan, B.J.T., Ridout, M.S., 2008. A new mixture model for capture heterogeneity. J. R. Stat. Soc. Ser. C Appl. Stat. 57, 433–446.

Otis, D., Burnham, K., White, G., Anderson, D., 1978. Statistical inference from capture data on closed animal populations. Wildl. Monogr. 62, 3–135.

Petersen, C.G.J., 1896. The yearly immigration of young plaice into the Limfjord from the German sea. Dan. Biol. Station 6, 5–84.

Pledger, S., 2000. Unified maximum likelihood estimates for closed capture-recapture models using mixtures. Biometrics 56, 434–442.

Pledger, S., 2005. The performance of mixture models in heterogeneous closed population capture-recapture. Biometrics 61, 868–876.

Pledger, S., Efford, M., Pollocj, J., Callazo, J., Lyons, J., 2009. Stopover duration analysis with departure probability on unknown time since arrival. Environ. Ecol. Stat. 3, 1071–1082.

Pollock, K.H., 1982. A capture-recapture design robust to unequal probability of capture. J. Wildl. Manag. 46, 752–757.

Pradel, R., 2005. Multievent: an extension of multistate capture-recapture models to uncertain states. Biometrics 61, 442–447.

Ralph, C.J., Dunn, E.J. (Eds.), 2004. In: Monitoring bird populations using mist nets, Studies in Avian Biology No. 29. Cooper Ornithological Society.

Royle, J.A., Young, K.V., 2008. A hierarchical model for spatial capture-recapture data. Ecology 89, 2281–2289.

Royle, J.A., Karanth, K.U., Gopalaswamy, A.M., Kumar, N.S., 2009. Bayesian inference in camera trapping studies for a class of spatial capture-recapture models. Ecology 90, 3233–3244.

Royle, J.A., Chandler, R.B., Sollmann, R., Gardner, B., 2013. Spatial Capture-Recapture. Academic Press, pp. 612.

Sandland, R.L., Cormack, R.M., 1984. Statistical inference for Poisson and multinomial models for capture-recapture experiments. Biometrika 71, 27–33.

Schnabel, Z.E., 1938. The estimation of total fish population of a lake. Am. Math. Mon. 45, 348–352.

Schwarz, C.J., Arnason, A.N., 1996. A general methodology for the analysis of capture-recapture experiments in open populations. Biometrics 52, 860–873.

Seber, G.A.F., 1965. A note on the multiple-recapture census. Biometrika 52, 249–259.

Self, S.G., Liang, K.-Y., 1987. Asymptotic properties of maximum likelihood estimators and likelihood ratio tests under nonstandard conditions. J. Am. Stat. Assoc. 82, 605–610.

Smout, S.C., King, R., Pomeroy, P.P., 2011. Estimating demographic parameters for capture-recapture data in the presence of multiple mark types. Environ. Ecol. Stat. 18, 331–347.

Stanley, T.R., Burnham, K.P., 1999. A closure test for time-specific capture-recapture data. Environ. Ecol. Stat. 6, 197–209.

Tanner, M., Wong, W., 1987. The calculation of posterior distributions by data augmentation (with discussion). J. Am. Stat. Assoc. 81, 528–550.

Worthington, H., McCrea, R.S., King, R., Griffiths, R., 2018a. Estimating abundance from multiple sampling capture-recapture data via a multi-state multi-period stopover model (Technical Report). University of St Andrews, arXiv. http://arxiv.org/abs/1810.10292.

Worthington, H., McCrea, R.S., King, R., Griffiths, R., 2018b. Estimating population size for multi-state closed capture-recapture data (Technical report). University of St Andrews, arXiv. https://arxiv.org/abs/1708.00348.

Wright, J.A., Barker, R.J., Schofield, M.R., Frantz, A.C., Byrom, A.E., Gleeson, D.M., 2009. Incorporating genotyping uncertainty into mark-recapture-type models for estimating abundance using DNA samples. Biometrics 65, 833–840.

Zhou, M., McCrea, R.S., Matehcou, E., Cole, D.J., Griffiths, R.A., 2018. Removal models accounting for temporary emigration. Biometrics. https://doi.org/10.1111/biom.12961. (in press).

Zucchini, W., MacDonald, I.L., Langrock, R., 2016. Hidden Markov models for time series—an introduction using R, second ed. Chapman and Hall/CRC, pp. 370.

Chapter 3

Mathematical Modeling of Multispecies Mutualism: From Particular Models Toward a Generalization of the Concept

Paul Georgescu*, Daniel Maxin†,1, Laurentiu Sega‡ and Hong Zhang§,¶

*Department of Mathematics, Technical University of Iaşi, Iaşi, Romania
†Department of Mathematics and Statistics, Valparaiso University, Valparaiso, IN, United States
‡Department of Mathematics, Augusta University, Augusta, GA, United States
§Department of Financial Mathematics, Jiangsu University, Zhenjiang, Jiangsu, PR China
¶School of Economics and Management, Changzhou Institute of Technology, Changzhou, Jiangsu, PR China
[1]Corresponding author: e-mail: daniel.maxin@valpo.edu

ABSTRACT

Mutualism refers to any species interaction in which the presence of one species benefits the others. While many specific models of two-species mutualism have been developed and analyzed, comparatively, more attention has been devoted to antagonistic interactions among species (i.e., predation and competition). However, recent work revisited the problem of modeling mutualistic interactions with a focus on two basic objectives: an attempt at unifying existing particular models into a smaller set of general models and at providing a qualitative analysis of their solutions in order to reveal the fundamental role of mutualism in the dynamics of the population under consideration. In this chapter we provide an overview on the most important results concerning general models of mutualism. In particular, we place the development of these models in the general context of establishing realistic "building blocks" in ecological models. To this end, particular attention is paid on the mathematical conditions necessary to separate bounded solutions from unbounded ones and, whenever possible, on the conditions leading to a unique, globally stable, coexistence equilibrium. We show how these results can be extended to an unspecified number of mutualistic species and how one can identify a subset of these species most likely to cause unbounded solutions. Last but not least, we argue for the importance of developing general models with carefully selected assumptions that are not too restrictive but, at the same time, still useful in potential experimental research.

Handbook of Statistics, Vol. 40. https://doi.org/10.1016/bs.host.2018.09.001

Keywords: Mutualism, Generalization, Bounding thresholds, Global stability, Coexistence equilibrium

1 INTRODUCTION

All species, including humans, evolve in ecosystems in which interactions between different species play significant roles, shaping the future evolution of these environments. To distinguish whether these interactions are beneficial, detrimental, or neutral, a general classification can be made to account for the perceived effect of the interaction on each species. Holland and Bronstein (2008) label with (+), (0), and (–) a positive, neutral, or negative effect, respectively, and distinguish between the following possible cases: *mutualism* (+ +), *predation* (+ –), *commensalism* (+ 0), *neutralism* (0 0), *ammensalism* (0 –), and *competition* (– –) whenever two-species interact. Following this classification, a mutualistic interaction between two species is an association which is beneficial for both.

Most mathematical modeling literature is focused more on the antagonistic interactions (i.e., predation and/or competition) than on the mutualistic ones. This is, partially, because the effect of one species on the other is more obvious in these cases. Any realistic model of antagonistic species must include predation as the main source of growth. Furthermore, it is generally accepted that any species is subject to finite resources and predation and/or competition are possible manifestations of gathering these resources. The paradigm of finite resources is the primary motivation behind the logistic assumptions in the models which, generally speaking, state that the population growth slows down as the size of the population increases. Without these assumptions the models will produce unbounded increasing solutions which are unrealistic in the long run.

In this chapter, we argue that building robust models of mutualistic interactions is also important for the same overall goal: having realistic and useful mathematical models. While mutualism is not always evident, it certainly plays an important a role. The classical example of flowers and pollinators is an obvious kind of mutualism. However, after a closer look, an argument for mutualism can be made in pretty much all biological systems. For example, the carnivorous species that prey on herbivorous ones may provide, with their presence, a mutualistic benefit for certain types of plants. Since no species evolves in a vacuum, it is clear that mutualism should be included in a framework of a realistic multispecies models alongside competition/predation. From a mathematical perspective, this immediately raises the question on how to account for mutualism without stepping back into the problem of unrealistic population growth which is guaranteed to happen if the mutualistic effects are not constrained. In other words, we have a similar problem to the one solved by the logistic assumptions: if the mutualistic benefits are not constrained then it is possible to have, again, unbounded population growth. Taken together, we arrive at two principles concerning the role of competition/predation and mutualism:

- *competition/predation* is the default justification for logistic assumptions in the models, i.e., making the removal rate increase with the total population size;
- *mutualism* by its own nature has a positive effect on the growth rate and any model will require biological self-limiting assumptions on its effect in order to avoid unrealistic population growth.

The focus of this chapter is how to model mutualistic interactions between two or more species, with a special attention to the problem of building models that are both general and useful enough to serve as a blueprint for more detailed and specific, concrete species centered models and how to analyze them from a stability viewpoint. Particular attention will be devoted to the principle stated above referring to the necessary self-limiting nature of any mutualistic interaction.

This chapter is structured as follows: in the next section we discuss various types of mutualism that appear in nature and how they affect the modeling framework, along with the motivation behind the generalization paradigm employed in this chapter, which strikes a balance between generality and specificity. Section 3 provides a brief history of particular two-species mutualistic models, together with some of the questions that arose from them that motivated later work on generalizations of these models. Section 4 describes the results of generalized two-species models and the mathematical techniques used on proving these results. In Section 5 we show how the results from two-species models can be naturally extended to a framework for a general model with an unspecified number of species. We conclude with some thoughts on the limitations of these models and the related research questions that are still open, which may provide avenues for further research.

2 MUTUALISM WITHIN THE FRAMEWORK OF A GENERAL MODEL

Before elaborating on the research concerning the general models, we provide, in this section, more details on the biological aspects of mutualistic interactions that are of importance when building these models. We also briefly touch upon the issue on how generic the components of these models should be. As we shall see, the answer to this last question is not always obvious, in the sense that, in an attempt to be as general as possible, a model may be less useful than expected.

2.1 Mechanisms Behind Mutualisms

The biological mechanisms on which a mutualistic interaction is based can take many forms and manifest in various ways. They may involve the *exchange of resources* between species when each of them has easy access to one hard to obtain by another. For example, most terrestrial plants rely on fungi for the uptake of phosphorus providing carbon in return. Plants also provide nectar and fruits to insects, while insects are crucial in pollination and/or seed dispersal, their mutualistic support involving thereby *transportation*. Note that, in these examples, the mutualism appears as a positive term in the modeling of each species while, in other cases, it plays a mortality-reducing role. This happens when mutualism involve *protection* from natural enemies or from unfavorable environmental conditions. For example, acacia ants (*Pseudomyrmex ferruginea*) defend the bull-horn acacia (*Acacia cornigera*) from herbivores, providing in exchange nutrients and sheltering (Palmer et al., 2010). Other associations of a mutualistic type are those between cleaner species and their hosts (Floeter et al., 2007).

These are just several simple examples, illustrating specific types of mutualisms. However, it is important to point out that all living organisms belong in a mutualistic interaction to some extent, some in as many as hundreds associations of this type (Bronstein et al., 2003). Further, mutualistic associations such as those between the cells of ancient eukaryotic organisms and the independent microorganisms such as mitochondria or plasmids were of crucial importance, driving the very evolution of life forms on Earth (Margulis, 1971). Other mutualisms contribute to the rise of certain species to ecological dominance, being known that over a thousand plant species have an exclusive insect pollinator, who lays eggs in the flowers it pollinates (Thompson et al., 2013). The insect then has the survival of its larval stages guaranteed, while the plant can rely on a specialized pollinator, at the cost of a few seeds. This suggest that herbivore–pollinator interaction was responsible for the diversification of both plants and insect pollinators. It can then be argued that mutualisms are widespread, occurring frequently and at multiple spatial and temporal scales, in and between living organisms, food chains, and even social systems.

Humans are, of course, not an exception. Indeed, crucial processes like the digestion are completely dependent on a symbiosis with beneficial bacteria. Serious illnesses like IBD (inflammatory bowel disease) can be understood as a disturbance in the gut microbiota and restoring it is a way to treat this illness. Finally, the bisexual reproduction system itself (specific to many species, including humans) can be thought, in some situations, as a mutualistic interaction between females and males.

2.2 To Be or Not to Be Independent

Mutualisms can be further classified into two types: *obligatory*, when neither species can survive without the other and *facultative*, when each species can survive on their own. A classical example of the latter is a plant that is subject to both self-pollination and cross-pollination while also being pollinated by multiple insects. It is believed that the majority of mutualisms that occur naturally are facultative, being argued by Bronstein (1994) that the outcomes of facultative mutualisms should be more diverse and variable than the outcome of more obligate mutualisms. Furthermore, when building a mathematical model, a compromise must be reached when it comes to the number of equations. It may be the case that a mutualistic system involves dozen of species, however, in order to keep the model amenable for analysis, one must focus on only a subset of these species. In this case, the facultative feature of that mutualistic interaction may represent the species absent from the model that, nevertheless, contribute to the growth of those present in the model. For example, if we want to study a two-species mutualism model representing bees and a specific flower plant, it is likely we would model the two in a facultative mutualistic interaction since bees do not depend on a specific flower to gather food nor a specific flower is pollinated by bees only. For these reasons, we argue that robust models of *facultative mutualism* are more important than obligatory ones.

A realistic modeling approach should also acknowledge that mutualisms involve not only benefits but also costs. For instance, in protection mutualisms there is a net benefit for both species involved only in the presence of natural enemies or in detrimental environmental conditions. Otherwise, the mutualism may yield negative effects for one or both species, due to the inherent costs of the mutualistic interaction and the nature of the entire interaction may shift in a continuum between mutualism and parasitism. To deal with this issue, mutualisms can be thought as being consumer–resource interactions in which benefits outweigh costs (Holland and DeAngelis, 2010). In the unifying framework of Holland and DeAngelis (2010), most mutualisms are recognized to belong to one of the following categories: two-way consumer–resource interactions, one-way consumer–resource interactions and indirect mutualisms via a third species consumer or resource, and to fit one of the six food web topologies given in Holland and DeAngelis (2010).

2.3 The Generalization Paradigm

Because mutualisms play a crucial role in the evolution of many species and are, at the same time, ubiquitous in nature, an important goal is to build models that are realistic but also comprehensive enough to be used in a variety of situations. Mathematical modeling of population dynamics, in its most general sense, is a trade-off between *simplicity* and *realism*. In other words, as more details are added to the model, its predictions become more credible at a cost of increased difficulty from an analysis standpoint.

This trade-off between simplicity and realism has deep historical roots. In 1798, Malthus (2018) pointed out in his famed essay that a simple population model with constant per capita birth and death rate, of the form

$$\text{(Malthusian model)} \quad \frac{dP}{dt} = rP, \quad r > 0,$$

P being the population size, will be subject to an exponential growth, far surpassing the physical space available, provided that living resources are unlimited.

This simple but fundamental observation led to the famous logistic equation established by Verhulst (1838) and further studied by Pearl and Reed (1920), which assumes that the growth rate is negatively dependent on the population size so that, as the population increases, the overall population size approaches a finite limit:

$$\text{(Verhulst logistic equation)} \quad \frac{dP}{dt} = r\left(1 - \frac{P}{K}\right)P, \quad K > 0, \quad r > 0.$$

This is a classical example of a trade-off in which the simplest possible framework (the Malthusian model) is not realistic enough (except for very short time spans at low values of P) since the exponential quickly increases to completely unrealistic/impossible values. The logistic equation is far more realistic (it predicts that the population growth will eventually level off), with an acceptable cost increase in difficulty (the logistic equation can still be easily solved explicitly).

Of course, the logistic equation is still far from being satisfactory for a number of reasons. First of all, it assumes a linear decline of the per capita growth rate $r\left(1 - \frac{P}{K}\right)$ which may or may not be appropriate for the species that is modeled. This is because such assumption would imply that the logistic effects (i.e., competition for finite resources) manifests itself even at very low population levels. An argument can be made to the contrary: that, initially, when the population is low, more individuals promote a bigger growth rate since, at that level, resources are plentiful and reproduction will be more effective if the population increases. This leads to the so-called Allee effect which will be discussed later. Second (and most relevant to our considerations), the logistic equation does not separate various biological factors that are at play in the growth rate: natural birth, natural death, competition for resources with other species, increased reproduction with higher population levels, and so on.

Thus, generally, in building a model and specifically in constructing one that is suited to our purpose, there are two main directions: either build a very specific model that is good only for the biological system to be investigated, or build a model as general and as useful as possible that serves as a blueprint for a larger class of specific biological processes. An approach leaning toward the latter direction for the logistic equation yields:

$$\frac{dP}{dt} = f(P)P,$$

where $f(P)$ is a decreasing function in P that inherits the properties of the linear function it replaces, i.e., $f(0) = r > 0, f(K) = 0, f(P) > 0$ for $0 < P < K$ and $f(P) < 0$ for $P > K$, K being the carrying capacity of the environment.

While this approach has a significant generality, it may leave too much work to be done if one wants to use this model and adapt it to a more specific situation. In other words, the model says nothing in particular about any fundamental features such as birth or death rates, since these are all lumped together in the function $f(P)$. If this framework is used to build a mutualism model, it becomes obvious that using just a single generic function for each species per capita growth is, in fact, too limiting. To illustrate this point, consider the following approach presented by Brauer and Castillo-Chavez (2012). The authors provide a generic two-species mutualism model by imposing several assumptions on a Kolmogorov-type model where the per capita growth rates are functions of the population sizes:

$$\begin{aligned} x' &= xf(x,y) \\ y' &= yg(x,y). \end{aligned} \tag{1}$$

This model assumes that, without interaction, each species is modeled by a single per capita growth function, i.e., $x' = xf(x, 0)$ and $y' = yg(0, y)$. The assumptions on f are

$$\frac{\partial f}{\partial x} < 0 \text{ and } \frac{\partial f}{\partial y} > 0.$$

The first condition represents growth saturation (i.e., logistic effect) and the second is the mutualistic effect of species y for species x. The analogous conditions on g are

$$\frac{\partial g}{\partial y} < 0 \text{ and } \frac{\partial g}{\partial x} > 0.$$

With these assumptions, the authors show that any possible equilibrium points are either saddle points or attracting nodes. From a biological perspective, this means that, in the long run, the population will approach various limit points depending on its initial size. However, as mentioned earlier, this model is very general and, without further assumptions on f and g, one cannot say anything about the actual number of limit points. In particular, the existence of a unique attracting point is of special interest because, when that happens, it indicates a situation where the two species evolve toward stable coexistence. Furthermore, it is in fact possible that there will be no attracting point, with important consequences when it comes to the biological validity of the model (i.e., the possibility of unbounded growth). Another problem with this framework is that we cannot explicitly model whether the mutualism manifest itself as a positive contribution to the growth rate or as a decreasing mortality effect on the benefiting species. Finally, by not separating the mutualistic effect from the rest of the model component, the assumptions listed above can be in fact too strong. It is possible, as already mentioned, that, for certain low population values, the birth rate to actually increase with the total population due to an increase on mating opportunities.

We argue that a better approach is to strike a balance between *generality* (i.e., using generic functions for most or all species interactions) and *specificity* (i.e., using an explicit function/term). In the case of the logistic equation, this compromise could take the form

$$\frac{dP}{dt} = [a(P) - f(P)]P,$$

where $a(P)$ is the birth rate or, more generally, any positive influence in the growth rate while $f(P)$ is the death/removal rate. Specific assumptions will still be made on $a(P)$ and $f(P)$ later. We will present research done using this base-line approach for modeling each species within the mutualistic interaction. The mutualistic assumptions on either the source term $a(P)$ and/or the removal term $f(P)$ will be introduced in Section 4. While no model is perfect, we argue that this approach is more useful in the continuum between a very specific model and a very generic model as the one from Brauer and Castillo-Chavez (2012).

3 MODELING CONSIDERATIONS: FROM SPECIFIC MODELS TO A MORE GENERAL FRAMEWORK

Any mathematical model is a crude approximation of the real population aimed to be studied. It is, perhaps, no surprise that various models of mutualistic interactions have been proposed. Each of them aims to strike the usual balance between mathematical simplicity and realism. In this section we provide an overview of these models and various attempts on improving them. We will also emphasize the common questions addressed when it comes to general properties of their solutions which, later, motivated the generalization of these models.

3.1 The Lotka–Volterra Legacy

An array of formally distinct models have been proposed to describe the complexity and diversity of mutualistic interactions, starting with a two-species mutualistic version of the classic Lotka–Volterra model with logistic growth, in which both interspecies interaction terms have positive signs and for which the per capita effects of a (facultative) mutualism are density independent, of the form

$$\begin{aligned} x_1' &= r_1 x_1\left(1 - \frac{x_1}{K_1}\right) + \frac{r_1 b_{12}}{K_1} x_1 x_2 = r_1 x_1\left(1 + \frac{b_{12} x_2}{K_1} - \frac{x_1}{K_1}\right) \\ x_2' &= r_2 x_2\left(1 - \frac{x_2}{K_2}\right) + \frac{r_2 b_{21}}{K_2} x_1 x_2 = r_2 x_2\left(1 + \frac{b_{21} x_1}{K_2} - \frac{x_2}{K_2}\right). \end{aligned} \tag{2}$$

Note that, as it can be seen below for this model, each species can experience unlimited per capita growth rates due to the mutualistic support it receives from the other species when the population size of the mutualistic partner grows unbounded. Also, if one species is missing, the other is not condemned to extinction but grows in a logistic fashion and consequently the model describes a facultative mutualism.

This model, proposed by Gause and Witt (1935), has been employed, among others, by Vandermeer and Boucher (1978) to describe the dynamics of a mutualistic model which is subject to constant harvesting. Probably as a tribute paid to this pioneering model, many of its successors focusing on facultative mutualisms consist of two structurally identical equations, which reduce more often than not to the logistic equation when the mutualistic counterpart of the species being modeled is not present (Holland and DeAngelis, 2009, 2010; Vázquez et al., 2015). Although this may be a legitimate assumption in many concrete situations, it is certainly a restricting one if the mutualistic partners are significantly different, for instance if one is a plant and the other one is an animal, as it happens for pollination/seed dispersal mutualisms, or if there is a drastic size imbalance, as it happens for several host/cleaner mutualisms. In such a situation, the beneficial effects of the mutualism may impact different fitness components for each mutualistic partner, leading to interaction terms which are structurally different. Consequently, any sufficiently general modelling framework which is suitable for mutualistic models should also allow for equations that are structurally different and cover cases in which different fitness components are enhanced in different species.

Let us briefly discuss the dynamics of (2), which will highlight an important issue when modelling mutualisms: the looming unboundedness. Setting the right-hand sides of (2) equal to 0 and solving for x_2, one obtains the following nullclines:

$$x_2 = \frac{1}{b_{12}} x_1 - \frac{K_1}{b_{12}} \qquad \text{(the } x_1\text{-nullcline)}$$

$$x_2 = K_2 + b_{21} x_1 \qquad \text{(the } x_2\text{-nullcline).}$$

Both nullclines are half-lines, the x_1-nullcline starting below, at $\left(\frac{K_1}{b_{12}}, 0\right)$, and being of slope $\frac{1}{b_{12}}$ and the x_2-nullcline starting above, at $(0, K_2)$, and being of slope b_{21}. If $b_{12}b_{21} > 1$ (that is, $b_{21} > \frac{1}{b_{12}}$), then the nullclines do not meet each other since the slope of the x_2-nullcline is larger than the slope of the x_1-nullcline and there is a region in the phase plane where the solutions become unbounded. That is, if the mutualistic interaction is of enough strength, then the reinforcing effects of the mutualistic benefits lead to unbounded growth, suggestively named "an *orgy of mutual benefaction*" in May (Malthus, 2018).

If $b_{12}b_{21} < 1$ (that is, if the mutualistic interaction is weak or asymmetric), then there is a unique positive equilibrium, which is asymptotically stable. In this case, the mutualistic interaction actually helps toward stabilizing the dynamics of the system. A similar linear example is provided in Brauer and Castillo-Chavez (2012) that includes both facultative and obligatory mutualism cases.

Apart from mutualisms not adhering to the Darwinian theory of natural selection, the unboundedness issue probably helped conveying the ideas that mutualisms are of less importance than antagonistic interactions such as predation and competition, which in turn led to mutualisms receiving much less attention than the antagonistic counterparts. Consequently, mechanisms to limit the positive feedback received by each species from the mutualistic partner, such as saturating functional responses, need to be built into models, perhaps along with a more accurate description of single-species dynamics.

3.2 Saturating Functional Responses and More General Frameworks

Relying upon the same logistic-facultative paradigm (Wolin and Lawlor, 1984), introduced the following model to accommodate the situation in which a (facultative) mutualism has a higher per capita impact at higher densities:

$$
\begin{aligned}
x_1' &= r_1x_1 - \frac{r_1x_1^2}{K_1 + b_{12}x_2} = r_1x_1\left(1 - \frac{x_1}{K_1 + b_{12}x_2}\right) \\
x_2' &= r_2x_2 - \frac{r_2x_2^2}{K_2 + b_{21}x_1} = r_2x_2\left(1 - \frac{x_2}{K_2 + b_{21}x_1}\right).
\end{aligned}
\tag{3}
$$

In both models (2) and (3) (which we shall often reference in what follows), r_i is the intrinsic growth rate and K_i is the carrying capacity of species $i = 1, 2$ when not accompanied by its mutualistic partner. In addition, b_{12} and b_{21} are positive constants that scale the beneficial effect of one species on the other. However, the mutualistic association is manifested differently in models (2) and (3), which translates into different signs of the interaction terms. In the latter, the mutualism is manifested itself as a reduction of mortality rate.

In particular, while model (2) is a special case of a general model

$$
\begin{aligned}
x_1' &= r_1x_1[f_1(x_1, x_2) - a_1(x_1)] \\
x_2' &= r_2x_2[f_2(x_2, x_1) - a_2(x_2)]
\end{aligned}
\tag{4}
$$

for some positive functions a_1, a_2, f_1, f_2, model (3) can rather be generalized as

$$
\begin{aligned}
x_1' &= r_1x_1[a_1(x_1) - f_1(x_1, x_2)] \\
x_2' &= r_2x_2[a_2(x_2) - f_2(x_2, x_1)],
\end{aligned}
\tag{5}
$$

the functions a_1, a_2, b_1, b_2 being subject to further conditions to be stated later.

Other established models in widespread use can also be rearranged to fit either (4) or (5) for different choices of a_i and f_i. The following model

$$
\begin{aligned}
x_1' &= r_1 x_1\left(1-\frac{x_1}{K_1}\right)+c_1 x_1\left(1-e^{-\alpha_2 x_2}\right)\\
x_2' &= r_2 x_2\left(1-\frac{x_2}{K_2}\right)+c_2 x_2\left(1-e^{-\alpha_1 x_1}\right),
\end{aligned}
\tag{6}
$$

proposed by Graves et al. (2006), can be rearranged in the form

$$
\begin{aligned}
x_1' &= x_1\left[(r_1+c_1)-\left(\frac{r_1 x_1}{K_1}+c_1\left(1-e^{-\alpha_2 x_2}\right)\right)\right]\\
x_2' &= x_2\left[(r_2+c_2)-\left(\frac{r_2 x_2}{K_2}+c_2\left(1-e^{-\alpha_1 x_1}\right)\right)\right],
\end{aligned}
\tag{7}
$$

to fit (5). As for its derivation, the model (6), termed in Graves et al. (2006) a "limited per capita growth rate (LGR) mutualism model" to acknowledge the contrast with the much more generous per capita mutualistic benefits yielded by the model of Gause and Witt, assumes that each species enhances the growth rate of the other without affecting its carrying capacity, unlike the Wolin–Lawlor model. The main assumption made in Graves et al. (2006) in order to derive (6) is that the marginal rate of change for the per capita growth rate of each species due to the increase of the other species is proportional to the difference between the maximum growth rate and the current growth rate. An example of a mutualistic interaction which is adequately described by this model is the symbiosis between plants and nitrogen-fixing microorganisms, a detailed argument in this direction being given in Graves et al. (2006).

The following model of mutualism which incorporates the handling of the mutualistic partner

$$
x' = x\left[r(1-cx)+\frac{\beta y}{\alpha+y}\right]
\tag{8}
$$

has been proposed in Wright (1989). In this form, the mutualistic term appears as a positive contribution to the growth rate of the benefiting species. However, similarly to what happens for the model (6), this contribution is limited even for large populations of the mutualist, as $\frac{\beta y}{\alpha+y} \leq \beta$ for $y \geq 0$. Let us consider now its two-species version proposed in Georgescu et al. (2017), given by

$$
\begin{aligned}
x_1' &= x_1\left[r_1(1-c_1 x_1)+\frac{\beta_1 x_2}{\alpha_1+x_2}\right]\\
x_2' &= x_2\left[r_2(1-c_2 x_2)+\frac{\beta_2 x_1}{\alpha_2+x_1}\right]
\end{aligned}
\tag{9}
$$

which can be rearranged in the form

$$\begin{aligned} x_1' &= x_1\left[r_1+\beta_1-\left(r_1c_1x_1+\frac{\beta_1\alpha_1}{\alpha_1+x_2}\right)\right] \\ x_2' &= x_2\left[r_2+\beta_2-\left(r_2c_2x_2+\frac{\beta_2\alpha_2}{\alpha_2+x_1}\right)\right] \end{aligned} \tag{10}$$

to fit again (5).

A general model of mutualism, in the form

$$\begin{aligned} x_1' &= a_1(x_1)+f_1(x_1)g_1(x_2) \\ x_2' &= a_2(x_2)+f_2(x_2)g_2(x_1), \end{aligned} \tag{11}$$

has been proposed and analyzed in Georgescu and Zhang (2014), under several sets of monotonicity and sign conditions for the functions a_1, a_2, f_1, f_2, g_1, g_2 (not necessarily all positive) and expressions involving them, the main focus being finding sufficient conditions for the global stability of the unique coexistence equilibrium. Although (11) is not associated to any specific biological interaction in particular, it represents a framework which is comprehensive enough to encompass the models (2) and (3) as its special cases.

3.3 An Unlikely Marriage: Two-Sex Reproduction in the Mutualistic Framework

Population models that separate the individuals by sex have a long history in the mathematical biology literature. All models, whether purely demographic or extended to larger ones as in the case of epidemic models, come with increased mathematical challenges. First, the number of equations doubles which makes the model far more difficult to analyze. Second, as shown below, the portion of the model that describes the two-sex reproduction/mating is usually not differentiable at the origin which makes standard stability analysis techniques unusable when trying to establish population extinction/persistence threshold conditions.

These models evolved into their own branch of mathematical biology, i.e., they are not usually presented as cases of mutualism for reasons that will be discussed shortly. However, in some cases, they still fit the mutualistic framework that is the subject of this chapter. Consider the following two-sex model in which F and M denote the female and male populations at time t (for a comprehensive survey of two-sex modeling see Iannelli et al. (2005) and references within):

$$\begin{aligned} F' &= \beta\gamma_f\mathcal{M}(F,M)-\mu_fF \\ M' &= \beta\gamma_m\mathcal{M}(F,M)-\mu_mM. \end{aligned} \tag{12}$$

The meaning of the parameters are: μ_f and μ_m are the mortality rates for females and males respectively, γ_f and γ_m are the probabilities of a newborn being a female or male respectively (hence $\gamma_f + \gamma_m = 1$), β is the per-mating birth rate and, finally, the central component of the model, $\mathcal{M}(F, M)$ is the number of matings per unit of time. It is this last term than can be interpreted as a mutualistic interaction between F and M. While demographers do not completely agree on what specific form $\mathcal{M}$ should take, there is, in general, agreement on a set of properties that this function should satisfy:

- positivity: $\mathcal{M}(F, M) \geq 0$ whenever $F \geq 0$ and $M \geq 0$.
- heterosexuality: $\mathcal{M}(F, 0) = \mathcal{M}(0, M) = 0$. This indicates that the number of matings is zero if individuals of one gender are absent. In the language of mutualistic interactions this would correspond to an obligatory mutualism.
- degree one homogeneity: $\mathcal{M}(\alpha D, \alpha M) = \alpha \mathcal{M}(F, M)$ for every $\alpha > 0$. This ensures the preservation of sex ratio if the number of singles in each gender changes by a factor α.
- monotonicity: $\mathcal{M}(F, M)$ is increasing in F and M. This means that the number of matings increases with the number of individuals of each gender which, obviously, corresponds to the mutualistic assumption in the two-species models.
- consistency: $\mathcal{M}(F, M) \leq kF$ and $\mathcal{M}(F, M) \leq kM$ for some constant $k > 0$ and for every F and M. This ensures that the number of matings is limited by the number of available individuals of a certain gender. Put it differently, this is a scarcity argument: if one gender is over-represented, the number of matings is controlled by the opposite (and scarce) gender. In the context of mutualistic models this correspond to a saturation effect in which the benefit from the other species approaches a certain limit if that species size increases to infinity.

Since all demographic models designed for long-term population modeling must contain some form of logistic assumptions on the natural mortality rate, we see that these models will not be suitable to the two-species mutualistic framework. Nevertheless, two-sex models can also be used for short-term modeling in which case the logistic effects on the mortality may be neglected. In such a case, the models do fit the mutualistic framework. Typical examples of the mating function analyzed in the literature are: the harmonic mean $\frac{FM}{F+M}$, geometric mean $\sqrt{FM}$ or the generalized weighted mean $\left(aF^b + (1-a)M^b\right)^{1/b}$.

One important additional reason why two-sex models are not typically treated as particular cases of mutualism is that, in many cases, the role of individuals of each gender does not appear *only* as a positive contribution to the other. Indeed, in the case of using the linear logistic mortality, the model (12) becomes

$$\begin{aligned} F' &= \beta\gamma_f \mathcal{M}(F,M) - [\mu_f + b(F+M)]F \\ M' &= \beta\gamma_m \mathcal{M}(F,M) - [\mu_m + b(F+M)]M. \end{aligned} \tag{13}$$

In this case, F and M, in addition to their positive contribution in the reproduction terms, they also increase the mortality due to logistic effect. As such, model (13) does not fit the mutualistic interaction framework discussed in this chapter.

3.4 Getting Away From the Logistic Growth: Allee Effects

Apart from discussing the properties of the interspecies interaction terms, that is, the functional responses of the interspecific interaction, it is also worth searching for a possible replacement for the usual linear form of the logistic per capita growth rate. A modification of the Verhulst logistic equation, in the form

$$\text{(Richards growth equation)}\quad \frac{dP}{dt} = r\left[1 - \left(\frac{P}{K}\right)^p\right]P,\quad K>0,\ \ r>0,\ \ p>0,$$

has been proposed by Richards (1959) in order to better deal with populations with different types of birth and death rates. In the above, p is an additional parameter which controls the location of the point where the growth rate is maximal. Apart from its primary modelling purpose, Richards equation has been employed in models used to predict the severity of a disease outbreak (Hsieh, 2008), for which the inflection point represents the turning point of the outbreak. With a high degree of generality, the Richards growth term encompasses as particular or limit cases other commonly used growth rates (von Bertalanffy and Gompertz), apart from the logistic growth rate. In this regard, versions of (2) and (3) with logistic growth terms replaced by Richards growth terms have been introduced and investigated from a stability viewpoint in Georgescu et al. (2016).

Another way to look for possible modeling enhancements of the logistic growth rate is to investigate what happens at small species densities. As mentioned earlier, there are clear biological reasons to associate a positive dependence of the growth rate with the population size when the population density is low. The most common ones are increasing benefits arising from group defense or cooperative behaviors and more reproductive success due to easier mate finding. This is known in the biological literature as the Allee effect. To put it another way, in a population which is subject to an Allee effect, the per capita population growth rate declines with decreasing population density when the population is rare.

By their severity, the Allee effects can be classified as *weak* or *strong*. If there is a critical population size below of which extinction is guaranteed, then the Allee effect is called *strong*. In this case, per capita growth rate becomes negative once the population density falls below a critical value, which is called the *Allee threshold*. If there is no such critical threshold and, in spite of the population decline, the per capita growth rate never becomes negative in low-density populations, then the Allee effect is called *weak*.

For instance, relabeling the classical logistic model as

$$x' = rx\left(1 - \frac{x}{K}\right),$$

it is seen that the per capita growth rate is $f(x) = r\left(1 - \frac{x}{K}\right)$, decreasing for all population densities. Consequently, the logistic model exhibits no Allee effects. On the contrary, the model

$$x' = rx\left(\frac{x}{A} - 1\right)\left(1 - \frac{x}{K}\right),$$

for which the per capita growth rate is $f(x) = r\left(\frac{x}{A} - 1\right)\left(1 - \frac{x}{K}\right)$, negative for $x \in (0, A)$, but with derivative $f'(x) = -\frac{2x}{AK} + \left(\frac{1}{A} + \frac{1}{K}\right)$, positive for small x, exhibits a strong Allee effect with threshold equal to A. Finally, the model

$$x' = rx^p\left(1 - \frac{x}{K}\right), \quad p > 1,$$

for which the per capita growth rate is $f(x) = rx^{p-1}\left(1 - \frac{x}{K}\right)$, positive for small x, with derivative $f'(x) = rx^{p-2}\left(p - 1 - \frac{px}{K}\right)$, also positive for small x, exhibits a weak Allee effect.

Using nonrestrictive assumptions on the growth and self-limiting functions employed in (4) and (5), established in Maxin et al. (2017) necessary and sufficient conditions on the boundedness of model solutions and proved the global stability of a unique coexistence equilibrium whenever it exists, the cases of strong and weak Allee effects being separately discussed. In this regard, a better, more inclusive theoretical framework allows for a much greater flexibility and complexity in model formulation. As seen in (4) and (5), the species interaction can now be more complex (i.e., do not need to factorize as a product of factors, each describing the dynamics of a single species, which opens the door for the modeling and treatment of ratio-dependent interactions). Further, each species can be modeled under different assumptions (e.g., one can benefit from a reduction in mortality and the other from an increase in fertility), which allows for the investigation of mutualistic systems in which the interacting species are deeply dissimilar. Also, since Allee effects (and especially weak Allee effects) are known to be relatively common (see, for instance, Courchamp et al., 1999) many more systems than before are now known to be stable and that the presented models may become more adequate descriptions of mutualisms in models of more complex food webs.

4 TWO-DIMENSIONAL MODELS: A STABILITY ANALYSIS

Virtually all theoretical models analyzed in the literature are subject to a mathematical analysis concerned with two main questions:

(1) Is the model valid and/or realistic from a biological perspective?

(2) Are there any locally and/or globally stable equilibrium points? Under what conditions are they stable? Under what conditions such a stable equilibrium point is unique?

It is obvious that question (1) must be addressed first since a negative answer to it renders the model practically useless. This is not always obvious when reading typical mathematical biology papers since, for many models, the validity is trivial to prove and often skipped from the exposition. Take for example the positivity invariance requirement (i.e., all solutions starting with positive initial values should remain bounded away from zero). Failure of this condition predicts negative population sizes which is, obviously, not possible. In a typical multispecies model in x_1, x_2,…,x_n consider the i-th equation

$$x_i' = f_i(x_1, x_2, \ldots, x_n).$$

To show positivity invariance one looks at the limit of the right-hand side as $x_i \to 0$ and show that this limit is strictly positive which indicates that x_i will never cross into negative region. Alternatively, if x_i is a factor of f_i, $f_i = x_i g_i$, this shows that $x_i(t)$ can be written as

$$x_i(t) = x_i(0)\exp\left[g_i(x_1, x_2, \ldots, x_n)\right]$$

and, thus, positive for all time t.

While failure of positivity invariance makes the model invalid, a failure of self-limiting growth (such as in the case of an exponential vs a logistic growth) makes the model clearly not realistic. When it comes to mutualistic interactions, in addition to a logistic assumption for each species, the mechanism of mutualism itself should be self-limiting in order to prevent the so-called *orgy of mutual benefactions* in which the population becomes unbounded. In the attempt to provide a useful generalization of mutualism models, the work on generalized models focused primarily on satisfying the boundedness condition (in addition to the usual requirements of a population model) as a prerequisite for asking further questions on the existence and the stability of equilibrium points.

The generalized models developed were heavily dependent on the mathematical tools employed in their analysis. There were two directions in this regard:

(1) *The LaSalle Invariance Principle.* For a two-species model in x_1 and x_2 this method requires finding a suitable Lyapunov functional $V(x_1, x_2)$ which satisfies several conditions such as: defined on a compact set which is positively invariated by the model trajectories, $\dot{V}(x_1, x_2)$ is negative along the model trajectories and $\dot{V}(x_1, x_2) = 0$ has only a positive equilibrium point as an invariant set. It then follows that all trajectories converge to this equilibrium point. The required compactness is often ensured checking that V is unbounded for very large or very small values of x_1 or x_2.

(2) *The Dulac Criterion.* This method, which is only suitable for two-species models, states that if the trajectories are bounded in a compact domain, periodic solutions are ruled out and there exist a unique equilibrium point in that domain, then all trajectories converge to that point. In particular, the possibility of periodic solutions is ruled out by finding a function $\varphi(x_1, x_2)$ with the property that

$$\frac{\partial}{\partial x_1}(\varphi(x_1,x_2)\dot{x}_1)+\frac{\partial}{\partial x_2}(\varphi(x_1,x_2)\dot{x}_2)\neq 0 \quad \text{on a compact domain.}$$

We point out that, in using this criterion, one needs to establish a priori conditions for the boundedness of solutions. These conditions are also more valuable, to an extent, since they can be established even in the case of an unspecified number of species (i.e., an n-dimensional model).

In the case of LaSalle Invariance Principle, the difficulty resides on finding a suitable Lyapunov function and proving the negativity of $\dot{V}$ condition. As we shall present below, this difficulty often limits the generality of the model and forces additional technical hypotheses. In the second case, applying the Dulac Criterion is, usually, easier when it comes to finding the suitable function φ. On the other hand, the boundedness of solutions must be established beforehand. By dissociating the bounding conditions from the stability conditions it is possible to obtain better generalizations with fewer technical assumptions.

4.1 Stability Results Using the LaSalle Invariance Principle

Georgescu and Zhang (2014) consider the following two-species mutualism model

$$\begin{aligned} x_1' &= a_1(x_1)+f_1(x_1)g_1(x_2)\\ x_2' &= a_2(x_2)+f_2(x_2)g_2(x_1). \end{aligned} \tag{14}$$

a_1, a_2, f_1, f_2, g_1, and g_2 are real continuous function defined on $(0,\infty)$ satisfying several assumptions as follows.

- f_1 and f_2 are positive on $(0,\infty)$ and g_1, g_2 are nonzero on $(0,\infty)$
- g_1 and g_2 are strictly increasing on $(0,\infty)$

Notice that these assumptions do not assume a particular sign for the interaction terms f_1g_1 and f_2g_2 which means this generalization can be applied in both cases of a mutualism affecting the source or a removal term in each equation. Another set of assumptions is added to the two above for each stability result. In each case a particular Lyapunov functional is used following those assumptions. In their main stability result, Georgescu and Zhang (2014) showed that the equilibrium point (x_1^*, x_2^*) is *globally asymptotically stable* whenever it exists and is unique and when one of the following sets of additional assumptions is satisfied.

Each set of hypotheses follows from a particular choice of the Lyapunov function. In the first one, several monotonicity assumptions are needed for the model component functions:

(1) The functions $\frac{a_1}{f_1} + g_2$ and $\frac{a_2}{f_2} + g_1$ are decreasing with at least one strictly decreasing.

(2) $\int_{x_i^*}^{\xi} \frac{g_j(\theta)-g_j(x_i^*)}{f_i(\theta)} d\theta = \infty$ for $\xi \in \{0, \infty\}$ and $(i, j) \in \{(1, 2), (2, 1)\}$.

(3) The Lyapunov functional is

$$V(x_1, x_2) = \int_{x_1^*}^{x_1} \frac{g_2(\theta) - g_2(x_1^*)}{f_1(\theta)} d\theta + \int_{x_2^*}^{x_2} \frac{g_1(\theta) - g_1(x_2^*)}{f_2(\theta)} d\theta.$$

Notice that the first condition essentially indicates a self-limiting property when it comes to both terms specific to the growth of each species (f_i and a_i) but also to the effect of mutualism on the other species (g_i). The fact that this monotonicity condition appears in this specific form has to do as well with the specific steps in the stability proof. The second condition expresses the fact that the functional is unbounded when populations become too large or too small and, since the functional decreases along solutions, those solutions can not stray away to the boundary of the domain. In other words these are technical assumptions that can be partially explained by biological considerations.

Another set of hypotheses leading to the same stability result is as follows:

(1) The functions $a_2 g_1$ and $a_1 g_2$ are strictly negative on $(0, \infty)$.

(2) The functions $\frac{f_1}{a_1} + \frac{1}{g_2}$ and $\frac{f_2}{a_2} + \frac{1}{g_1}$ are monotone, both increasing if g_1 and g_2 have the same sign or both decreasing if they keep a different sign.

(3) $\int_{x_i^*}^{\xi} \left(1 - \frac{g_j(\theta)}{g_j(x_i^*)}\right) \frac{1}{a_i(\theta)} d\theta = \infty$ for $\xi \in \{0, \infty\}$ and $(i, j) \in \{(1, 2), (2, 1)\}$.

(4) The Lyapunov functional is

$$V(x_1, x_2) = \int_{x_1^*}^{x_1} \left(1 - \frac{g_2(\theta)}{g_2(x_1^*)}\right) \frac{1}{a_1(\theta)} d\theta + \int_{x_2^*}^{x_2} \left(1 - \frac{g_1(\theta)}{g_1(x_2^*)}\right) \frac{1}{a_2(\theta)} d\theta.$$

In Georgescu and Zhang (2014) the global stability result is also proved without the restrictive monotonicity assumptions:

(1) The functions g_1 and g_2 are strictly positive on $(0, \infty)$.

(2) $\frac{f_1}{a_1}(0) > 0, \frac{a_1}{f_1}(0) > 0, \frac{f_2}{a_2}(0) > 0, \frac{a_2}{f_2}(0) > 0$, and $\frac{a_1}{f_1}(x_1) < \frac{a_1}{f_1}(0)$ for all $x_1 \geq 0$, $\frac{a_2}{f_2}(x_2) < \frac{a_2}{f_2}(0)$ for all $x_2 \geq 0$.

(3) $\int_{x_i^*}^{\xi} \left(1 - \frac{g_j(x_i^*)}{g_j(\theta)}\right) \frac{1}{f_i(\theta)} d\theta = \infty$ for $\xi \in \{0, \infty\}$ and $(i, j) \in \{(1, 2), (2, 1)\}$.

(4) The Lyapunov functional is

$$V(x_1,x_2) = \int_{x_1^*}^{x_1} \left(1 - \frac{g_2(x_1^*)}{g_2(\theta)}\right) \frac{1}{f_1(\theta)} d\theta + \left[\int_{x_2^*}^{x_2} \left(1 - \frac{g_1(x_2^*)}{g_1(\theta)}\right) \frac{1}{f_2(\theta)} d\theta\right] \frac{g_1(x_2^*)}{g_2(x_1^*)}.$$

While these conditions are certainly less strict on the basic properties of f_i, g_i, and a_i, the global stability result proved in Georgescu and Zhang (2014) does require several additional technical assumptions that we do not reproduce here.

Despite the presence of rigid technical assumptions, the generalization provided in Georgescu and Zhang (2014) includes a large variety of particular models already analyzed in the literature prior to that paper. One only needs to carefully rearrange the model equations and check to see if a set of hypotheses in the stability results above apply to that particular model. We provide below an example and refer the reader to Georgescu and Zhang (2014) for more.

$$
\begin{aligned}
x_1' &= r_1 x_1\left(A_1 - \frac{x_1}{K_1}\right) + r_1\frac{b_{12}}{K_1}x_1x_2 \\
x_2' &= r_2 x_2\left(A_2 - \frac{x_2}{K_2}\right) + r_2\frac{b_{21}}{K_2}x_1x_2.
\end{aligned}
\tag{15}
$$

We choose

$$
a_1(x_1) = r_1x_1\left(A_1 - \frac{x_1}{K_1}\right),\quad a_2(x_2) = r_2x_2\left(A_2 - \frac{x_2}{K_2}\right),\quad f_1(x_1) = r_1\frac{b_{12}}{K_1}x_1,
$$

$$
g_1(x_2) = x_2,\quad f_2(x_2) = \frac{r_2x_2}{K_2},\quad g_2(x_1) = b_{21}x_1.
$$

With these choices, the model fits the first stability result which requires the monotonicity conditions. Notice that

$$
\left(\frac{a_1}{f_1} + g_2\right)(x_1) = \frac{K_1A_1}{b_{12}} - \frac{x_1}{b_{12}}(1 - b_{12}b_{21}),\quad \left(\frac{a_2}{f_2} + g_1\right)(x_2) = K_2A_2.
$$

The existence and uniqueness condition for the coexistence equilibrium is $b_{12}b_{21} < 1$ and, in this case, the above functions are nonincreasing (i.e., the stability result requires only one of them to be strictly decreasing).

While clearly useful, the generalization provided in Georgescu and Zhang (2014) is far from comprehensive. Georgescu et al. (2016) showed in the follow-up paper that the same class of Lyapunov functionals employed in Georgescu and Zhang (2014) can be used to analyze mutualistic models that do not, otherwise, satisfy the hypotheses in Georgescu and Zhang (2014). In Georgescu et al. (2016), two particular models are analyzed. One with restricted growth rates (6), proposed by Graves et al. (2006), and one with Richard growth function instead of the classical logistic one shown below:

$$
\begin{aligned}
x_1' &= r_1 x_1\left[A_1 - \left(\frac{x_1}{K_1}\right)^p\right] + r_1\frac{b_{12}}{K_1}x_1x_2 \\
x_2' &= r_2 x_2\left[A_2 - \left(\frac{x_2}{K_2}\right)^p\right] + r_2\frac{b_{21}}{K_2}x_1x_2.
\end{aligned}
\tag{16}
$$

It is easy to see that neither of these two models satisfy the stability conditions in Georgescu and Zhang (2014). Yet, the stability results themselves still hold which can be proven, as seen in Georgescu et al. (2016), using different Lyapunov functionals. We should also point out that another limitation is given by the mutualistic interaction term itself, which has to be factored into two functions depending each on a single variable. As we will see later, there are many mutualistic models where this property does not hold.

4.2 Stability Results Using the Dulac Criterion: Threshold-Like Parameters

Maxin et al. (2017) present a different mathematical approach in the analysis of mutualistic models. Before we describe the main results, we point out that these different approaches are motivated primarily by the possibility to have much greater flexibility in how one designs the model equations. The global stability result presented above requires structurally identical equations (i.e., the intrinsic growth rate and the mutualistic interaction are given by similar functions).

In many cases the mutualists have completely different ecological properties, such as when representing a plant and an animal (Vázquez et al., 2015). The interaction terms differ structurally for each mutualist. In addition, beneficial effects of mutualism may impact different fitness components in each species which implies using different functions for the mutualistic interaction in the same model. As an example consider the oxpeckers and the zebras or rhinos (Boucher et al., 1982). By consuming parasitic bugs on the zebras' or rhinos' skin, the oxpeckers acquire food and the zebras or rhinos get rid of pests. Hence the mutualistic benefit is a reduction in mortality for zebras and rhinos and a source of food for the oxpeckers. The main results in Maxin et al. (2017) are presented in more details for symmetrical models first and then later are expanded to a more general interaction terms since the mathematical proofs differ only slightly.

Each model in Maxin et al. (2017) starts with several assumptions on the underlying model for each species in the absence of mutualism. Specifically, each species follows a Kolmogorov-type equation

$$x' = rx[a(x) - f(x)], \tag{17}$$

with the following assumption:

(G): There exists $K > 0$ such that $a(x) - f(x) > 0$ for $x < K$, $a(x) - f(x) < 0$ for $x > K$ and $a(K) - f(K) = 0$. This is a self-limiting (i.e., logistic type) assumption which includes the possibility of a weak Allee effect (i.e., the concave up shape of the growth rate function at low population densities). K denotes here the carrying capacity, i.e., the population level after which the population growth rate declines. Consider now a mutualistic model

$$x_1' = r_1 x_1 [a_1(x_1) - f_1(x_1, x_2)],$$
$$x_2' = r_2 x_2 [a_2(x_2) - f_2(x_2, x_1)]. \tag{18}$$

It is further assumed that both $a_1(x_1) - f_1(x_1, 0)$ and $a_2(x_2) - f_2(x_2, 0)$ satisfy assumption (G).

The following, so-called, *consistency assumptions* are

(C1) $f_1(x_1, x_2)$ is increasing in x_1 and $f_2(x_2, x_1)$ is increasing in x_2.
(C2) $\frac{a_1(x_1)}{f_1(x_1,x_2)}$ is strictly decreasing in x_1 and $\frac{a_2(x_2)}{f_2(x_2,x_1)}$ is strictly decreasing in x_2.
(C3) $\frac{a_1(x_1)}{f_1(x_1,\alpha x_1)}$ and $\frac{a_2(x_2)}{f_2(x_2,\alpha x_2)}$ are eventually decreasing for all $\alpha > 0$. That is,

$$\frac{d}{dx_1}\left(\frac{a_1(x_1)}{f_1(x_1,\alpha x_1)}\right) < 0 \text{ for } x_1 > M_1,$$

$$\frac{d}{dx_2}\left(\frac{a_2(x_2)}{f_2(x_2,\alpha x_2)}\right) < 0 \text{ for } x_2 > M_2,$$

where M_1 and M_2 are positive real numbers that may depend on α.

The meaning of (C1) and (C2) is that if one species is kept constant, then the other still follows a logistic-like in which the ratio between the growth and self-limiting functions decreases as the population increases. The meaning of (C3) is similar but in the case that the ratio between the population densities is kept constant. We point out that assumption (C3) holds only for population sizes greater than a given size (i.e., M_1 and M_2). This is an intentionally weaker monotonicity property which allows the boundedness results shown below to hold also for mutualism models including strong Allee effects.

Finally, the following mutualism-specific assumption is imposed:

(M) $f_1(x_1, x_2)$ is decreasing in x_2 and $f_2(x_2, x_1)$ is decreasing in x_1.

With these assumptions, it is easy to see that, for all $\alpha > 0$ there exist finite nonnegative numbers $\mathcal{R}_1(\alpha)$, $\mathcal{R}_2(\alpha)$ given by

$$\mathcal{R}_1(\alpha) = \lim_{x_1\to\infty}\frac{a_1(x_1)}{f_1(x_1,\alpha x_1)}, \quad \mathcal{R}_2(\alpha) = \lim_{x_2\to\infty}\frac{a_2(x_2)}{f_2(x_2,\alpha x_2)}.$$

4.2.1 A Comparison With Epidemiologic Reproductive Numbers

The main stability results established in Maxin et al. (2017) are given in terms of conditions on the threshold-like quantities $\mathcal{R}_1$ and $\mathcal{R}_2$. To some extent, these quantities emulate the concept of a *demographic reproductive number*. In its simplest form, when the equation for a population contains only the two vital terms for birth and death, the reproductive number is essentially the product between the birth rate and the reciprocal of the death rate. The latter approximates the life time of a typical individual. Hence the reproductive number represents the number of offspring that an individual will have during his life time. This is the reason why, when compared to 1, the reproductive number indicates either population growth or decline. Of course the meaning of "birth" and "death" is far more general in more complicated models as these terms can include other factors besides natural birth/death rate. Indeed, the mutualism effect itself can be part of the "birth" (better called a "source" term) or of the "death" (better called a "removal" term). In the case of (18), the mutualism effect is a reduction in mortality for each species.

The term "reproductive number" also appears in the epidemiological models. It estimates the number of secondary cases of infections caused by an infectious individual in a healthy population (the so-called patient zero). Thus, it indicates whether an epidemic occurs or not. Its analogy with "reproduction" is due to the fact that it views new infections as "offspring" of a given infectious individual when the disease is rare (i.e., near a state when the number of infectious individuals is close to zero). Unlike the epidemic reproduction numbers, the threshold-like parameters defined above are not computed at near-extinction situation, but at high population densities, under given species proportionality. The reason is that, as far as the validity of the model is concerned, what is important is not the extinction of species, but their blow-up (caused by unconstrained mutualistic effects). Also, a model of mutualism has an entirely different structure, not exhibiting the asymmetry which is characteristic to disease propagation models, to which the behavior of the system around the disease-free equilibrium, as expressed by the basic reproduction number, is of paramount importance. A single threshold parameter, based on the dynamics of a single species or compartment, may consequently not be enough to describe the behavior of solutions for a model of mutualism. Thus, in proving the results in Maxin et al. (2017), the authors employ one reproductive ratio per species to introduce the boundedness conditions.

4.2.2 The Main Results From Maxin et al. (2017)

The threshold conditions for bounded and unbounded solutions of (18) are as follows:

(1) If there is $\alpha > 0$ such that $\mathcal{R}_1(\alpha) < 1$ and $\mathcal{R}_2\left(\frac{1}{\alpha}\right) < 1$, then solutions of the model (18) are uniformly bounded in a compact region of the positive quadrant and there is a coexistence equilibrium.

(2) Under the same assumptions listed earlier, if there is $\alpha > 0$ such that $\mathcal{R}_1(\alpha) > 1$ and $\mathcal{R}_2\left(\frac{1}{\alpha}\right) > 1$, then solutions of the model (18) are unbounded if the initial population density is large enough.

Notice that α represents the ratio between the size of the second species x_2 and the first one x_1. The first result indicates that, in order to ensure the boundedness of solutions, there should be a certain ratio between the two species which, if maintained, it will cause the population net growth rate to decline as it gets larger and larger. The second result mirrors the first one and it provides a condition when the solutions become unbounded. As mentioned earlier, this will be an important condition in order to avoid the unrealistic mutualistic effect, also know as "the orgy of mutual benefactions." As we will see in the examples below, the two conditions above can often be combined into a single threshold-like condition in which case the value of α satisfying the inequalities above is not unique and it becomes irrelevant. In those cases, the bounding condition more closely reflects the typical results attached to reproductive numbers in which a single quantity determines the behavior of the model solutions depending on whether it is less or greater than one.

If the first (bounding) condition is satisfied and the coexistence equilibrium is unique then it is shown that it is *globally asymptotically stable* using the Dulac Criterion with the following Dulac function: $\varphi(x_1,x_2)=\frac{1}{x_1x_2a_1(x_1)a_2(x_2)}$. The more mathematically relevant part of proving this result lies on showing the bounding argument. This proof depends on controlling which species has a greater population size at a given time t. Since increasing/decreasing one species affects the growth of the other one, proving all bounding results relies on considering max/min functions of $\{\alpha x_1(t),\ x_2(t)\}$. This way, it is proved that these functions are bounded/unbounded which then implies the same, individually, for $x_1(t)$ and $x_2(t)$.

We make the important observation that the conditions for boundedness and unboundedness are mutually exclusive. Due to the monotonicity assumptions made earlier, $\mathcal{R}_1$ and $\mathcal{R}_2$ are increasing and

$$[\mathcal{R}_1(\alpha)-\mathcal{R}_1(\beta)]\left[\mathcal{R}_2\left(\frac{1}{\alpha}\right)-\mathcal{R}_2\left(\frac{1}{\beta}\right)\right]\leq 0 \quad \text{for all } \alpha,\beta\geq 0.$$

Therefore, if there is any α such that $\mathcal{R}_1(\alpha)<1$ and $\mathcal{R}_2(\frac{1}{\alpha})<1$, there cannot be any β such that $\mathcal{R}_1(\beta)>1$ and $\mathcal{R}_2(\frac{1}{\beta})>1$, and vice versa. Nevertheless, the bounding theorems cannot always be combined into a single necessary and sufficient condition that separates a bounded from an unbounded system. For all we know, it may be possible to have a situation in which neither the two conditions hold.

4.2.3 When Do We Have a Single Threshold?

For virtually all forms of known mutualistic models, the two results presented earlier do combine into a threshold-like condition. To see this, let us first consider a subclass of models for which the limits $\mathcal{R}_1(\alpha)$ and $\mathcal{R}_2(\alpha)$ are powers of α, having the form

$$\mathcal{R}_1(\alpha)=C_1\alpha^{p_1},\quad \mathcal{R}_2(\alpha)=C_2\alpha^{p_2},\quad p_1,p_2>0,\quad C_1,C_2\geq 0. \tag{19}$$

In particular, as it will be seen below, the model (2) of Gause and Witt, the model (3) of Wolin and Lawlor and the model (6) of Graves et al. all belong to this class. The boundedness condition becomes equivalent to finding $\alpha>0$ such that

$$C_1\alpha^{p_1} < 1, \quad C_2\left(\frac{1}{\alpha}\right)^{p_2} < 1.$$

If one of the constants C_1, C_2 is 0, then the boundedness condition is satisfied regardless of the value of the other constant. For example, if $C_1 = 0$ then $\mathcal{R}_1(\alpha) = 0$ for any α and $\mathcal{R}_2\left(\frac{1}{\alpha}\right)$ can be made less than 1 by choosing α large enough. Similarly, if $C_2 = 0$ then $\mathcal{R}_2\left(\frac{1}{\alpha}\right)$ is always zero and $\mathcal{R}_1(\alpha)$ can be made less than 1 by choosing α small enough. If both $\mathcal{R}_1$ and $\mathcal{R}_2$ are identically zero, then the boundedness follows without any condition on the existence of a suitable α.

Finally, if $C_1, C_2 > 0$, then

$$C_1\alpha^{p_1} < 1, \quad C_2\left(\frac{1}{\alpha}\right)^{p_2} < 1 \Longrightarrow C_2^{\frac{1}{p_2}} < \alpha < \left(\frac{1}{C_1}\right)^{\frac{1}{p_1}},$$

condition which is satisfied if $C_1^{\frac{1}{p_1}} C_2^{\frac{1}{p_2}} < 1$. It follows that the boundedness condition is satisfied for all values of C_1, C_2, null or not, such that $C_1^{\frac{1}{p_1}} C_2^{\frac{1}{p_2}} < 1$.

Similarly, the unboundedness condition is equivalent to finding $\alpha > 0$ such that

$$C_1\alpha^{p_1} > 1, \quad C_2\left(\frac{1}{\alpha}\right)^{p_2} > 1.$$

Obviously, the unboundedness condition cannot be satisfied if any of C_1, C_2 is null. If $C_1, C_2 > 0$, then

$$C_1\alpha^{p_1} > 1, \quad C_2\left(\frac{1}{\alpha}\right)^{p_2} > 1 \Longrightarrow \left(\frac{1}{C_1}\right)^{\frac{1}{p_1}} < \alpha < C_2^{\frac{1}{p_2}},$$

condition which is satisfied if $C_1^{\frac{1}{p_1}} C_2^{\frac{1}{p_2}} > 1$. Notice, as mentioned earlier, that the particular value of α is not of importance. There are, obviously infinitely many α's satisfying the bounding results in this case.

As we can see, in the particular case defined by (19), the bounding conditions are threshold-like. Specifically, if we define

$$\mathcal{R} = C_1^{\frac{1}{p_1}} C_2^{\frac{1}{p_2}}, \tag{20}$$

then the boundedness condition can be expressed as $\mathcal{R} < 1$, while the unboundedness condition can be expressed as $\mathcal{R} > 1$.

4.2.4 Additional Flexibility in Modeling Mutualistic Interactions

The main results from Maxin et al. (2017) can be adapted for models that are less restrictive. A first example is provided by asymmetric mutualism models and/or models with a different type of mutualistic interaction. The results are stated in the same way with only small modifications on the assumptions made and on the way the limits $\mathcal{R}_1(\alpha)$ and $\mathcal{R}_2(\alpha)$ are defined. For example, consider the following model where the mutualism affects the source term.

$$\begin{aligned} x_1' &= r_1 x_1 [f_1(x_1, x_2) - a_1(x_1)], \\ x_2' &= r_2 x_2 [f_2(x_2, x_1) - a_2(x_2)]. \end{aligned} \tag{21}$$

The *consistency assumptions* are now as follows:

(C1) $a_1(x_1)$ is increasing in x_1 and $a_2(x_2)$ is increasing in x_2.

(C2) $\frac{f_1(x_1,x_2)}{a_1(x_1)}$ is decreasing in x_1 and $\frac{f_2(x_2,x_1)}{a_2(x_2)}$ is decreasing in x_2.

(C3) $\frac{f_1(x_1,\alpha x_1)}{a_1(x_1)}$ and $\frac{f_2(x_2,\alpha x_2)}{a_2(x_2)}$ are eventually decreasing for all $\alpha > 0$,

and the *mutualism-specific assumption* is

(M) $f_1(x_1, x_2)$ is increasing in x_2 and $f_2(x_2, x_1)$ is increasing in x_1.

Hence, the limits $\mathcal{R}_1(\alpha)$, $\mathcal{R}_2(\alpha)$ are now given by

$$\mathcal{R}_1(\alpha) = \lim_{x_1 \to \infty} \frac{f_1(x_1, \alpha x_1)}{a_1(x_1)}, \quad \mathcal{R}_2(\alpha) = \lim_{x_2 \to \infty} \frac{f_2(\alpha x_2, x_2)}{a_2(x_2)}.$$

Obviously, if we have a mixture of the two in which the mutualism affects the removal rate for x_1 and the source term for x_2, then the assumptions above will also be modified accordingly.

A second example concerns species undergoing a strong Allee effect. Using (18) as a framework, the assumptions made for each species are

(S1) $a_i(x) - f_i(x) < 0$ for $0 < x < A_i$,
(S2) $a_i(x) - f_i(x) > 0$ for $A_i < x < K_i$,
(S3) $a_i(x) - f_i(x) < 0$ for $x > K_i$,
(S4) $a(A_i) - f(A_i) = a(K_i) - f(K_i) = 0$,

where $i = \{1, 2\}$. Notice that K_i maintains its meaning as the logistic threshold while A_i is the Allee threshold which indicates the level under which the population declines. When it comes to a strong Allee effect, the bounding theorems hold with no change. However, the existence, uniqueness, and the global stability of the coexistence equilibrium need a separate discussion. This is due to the fact that, if both species exhibit a strong Alee effect, it can be shown mathematically that the extinction equilibrium is always locally stable. In other words, the mutualism cannot prevent extinction of both species if their population size drops below the Allee threshold. Since the origin is always locally stable, obviously, any coexistence equilibrium can only be locally stable. However, if only one species undergoes a strong Allee effect it is possible to have, in that case, a unique globally stable coexistence equilibrium. This provides a very useful ecological test to see under what conditions the mutualistic effect of one species can prevent the extinction of another species that is subject to the Allee effect.

Suppose that x_1 undergoes a strong Allee effect with the Allee threshold A_1 and the carrying capacity K_1, while x_2 obeys the growth assumption (G) with the carrying capacity K_2. In the presence of mutualism and assuming all other consistency conditions, if there exist a unique coexistence equilibrium of the model (18), then this equilibrium will be globally stable provided that all solutions are uniformly bounded in the positive quadrant. The boundedness arguments described earlier hold also here, with the caveat that one need to establish a condition which ensures that x_1 does not go extinct due to the strong Allee effect. Using standard linearization around the boundary equilibrium $(0, K_2)$, this condition becomes

$$a_1(0) - f_1(0, K_2) > 0. \tag{22}$$

This represents the necessary and sufficient condition that species x_2 prevents the extinction of species x_1 and annihilates its strong Allee effect.

4.3 Examples

We illustrate the advantages of the generalization presented above with several examples. All models presented below do not satisfy entirely the assumptions necessary for the analysis using the LaSalle Invariance Principle. On the other hand, they are suited to the bounding theorems proven with the max/min functions discussed earlier as well as the Dulac Criterion whenever we have a unique coexistence equilibrium. A more detailed discussion on these examples is provided in Maxin et al. (2017).

4.3.1 *Species With a Weak Allee Effect*

Consider the following q-logistic Wolin–Lawlor model with weak Allee effect:

$$\begin{aligned} x_1' &= r_1 x_1^{p_1}\left(1 - \frac{x_1^{q_1}}{K_1^{q_1} + b_{12} x_2^{n_1}}\right), \\ x_2' &= r_2 x_2^{p_2}\left(1 - \frac{x_2^{q_2}}{K_2^{q_2} + b_{21} x_1^{n_2}}\right), \end{aligned} \tag{23}$$

assuming that $K_i > 0$, $p_i \geq 1$, $q_i > 0$, and $n_i > 0$ for $i = 1, 2$.

The constants p_i model the weak Allee effect (i.e., the per capita growth rates are increasing at low population densities), while q_i allow the possibility to manipulate the point at which the logistic effect takes place (i.e., the inflection point, which indicates a slower population growth at high densities).

For this model, we have

$$a_1(x_1) = x_1^{p_1 - 1}, \quad a_2(x_2) = x_2^{p_2 - 1},$$

$$f_1(x_1, x_2) = \frac{x_1^{p_1 + q_1 - 1}}{K_1^{q_1} + b_{12} x_2^{n_1}}, \quad f_2(x_2, x_1) = \frac{x_2^{p_2 + q_2 - 1}}{K_2^{q_2} + b_{21} x_1^{n_2}},$$

and consequently

$$\frac{a_1(x_1)}{f_1(x_1,\alpha x_1)} = \frac{K_1^{q_1} + b_{12}\alpha^{n_1}x_1^{n_1}}{x_1^{q_1}}, \quad \frac{a_2(x_2)}{f_2(x_2,\alpha x_2)} = \frac{K_2^{q_2} + b_{21}\alpha^{n_2}x_2^{n_2}}{x_2^{q_2}}.$$

The consistency condition (C3) requires that $n_1 \leq q_1$ and $n_2 \leq q_2$. If $n_1 < q_1$ then $\mathcal{R}_1(\alpha) = 0$ for all $\alpha > 0$, while if $n_1 = q_1$ we get $\mathcal{R}_1(\alpha) = b_{12}\alpha^{q_1}$. Similarly, if $n_2 < q_2$ then $\mathcal{R}_2(\alpha) = 0$ for all $\alpha > 0$, while if $n_2 = q_2$ then $\mathcal{R}_2(\alpha) = b_{21}\alpha^{q_2}$. Therefore, if $n_1 < q_1$ or $n_2 < q_2$ the boundedness condition is satisfied for all other parameter values, while if $n_1 = q_1$ and $n_2 = q_2$ the boundedness condition is satisfied provided that

$$\mathcal{R} = (b_{12})^{1/q_1}(b_{21})^{1/q_2} < 1.$$

We now reproduce a numerical example from Maxin et al. (2017). Parameter values are chosen for the benefit of a clearer picture and are not necessarily related to actual biological data. Let $p_1 = p_2 = 1 + 10^{-4}$, $q_1 = n_1 = 3$, and $q_2 = n_2 = 1$. For $K_1 = 15$, $K_2 = 35$, $b_{12} = 0.75$, and $b_{21} = 0.7$ we have $\mathcal{R} = 0.64$. There is also a unique equilibrium, which is globally stable. On the contrary, for $K_1 = 40$, $K_2 = 55$, $b_{12} = 2$, and $b_{21} = 1$ we have $\mathcal{R} = 1.38$ and the solutions are unbounded (i.e., the case of the so-called orgy of mutual benefactions). In Fig. 1 we show the phase portraits corresponding to these two cases, with $r_1 = 3 \times 10^{-5}$ and $r_2 = 1.05 \times 10^{-4}$ in panel (A) and $r_1 = 8 \times 10^{-5}$ and $r_2 = 1.65 \times 10^{-4}$ in panel (B).

In Fig. 2 we show the region of a selected two-parameter space in which the solutions of the model (23) are bounded and we have a unique globally stable equilibrium. The shaded region corresponds to $\mathcal{R} < 1$. In panel (A), we compare the relative effect of the terms describing the mutualism effect, b_{12} and b_{21}. In panel (B), we compare the effect of a limiting resource term of one species (q_1) with the mutualistic effect of the other species (b_{12}). All other parameters are as considered for Fig. 1.

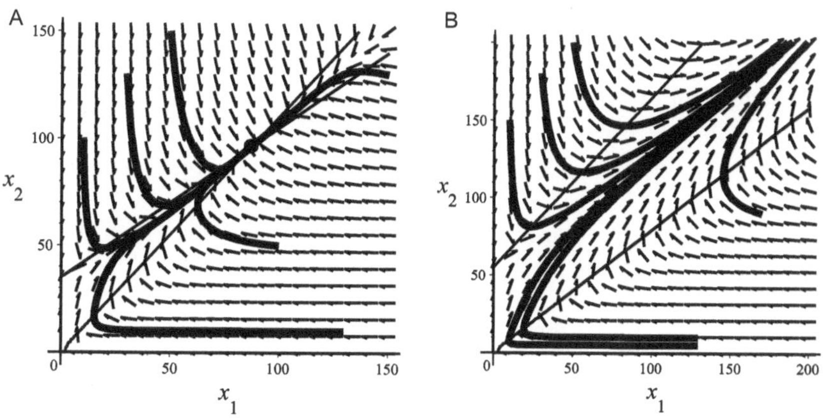

FIG. 1 Phase portrait of model (23). (A) $\mathcal{R} = 0.64$. (B) $\mathcal{R} = 1.38$.

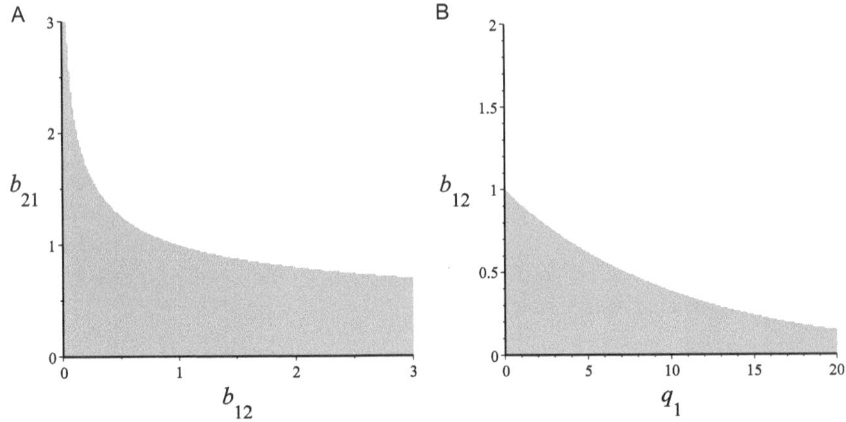

FIG. 2 Global stability region for model (23). (A) $\mathcal{R}(b_{12}, b_{21}) < 1$. (B) $\mathcal{R}(q_1, b_{12}) < 1$.

4.3.2 *Nonpolynomial Thresholds*

For most mutualism models published prior to the generalization results from Maxin et al. (2017), the limits $\mathcal{R}_1$ and $\mathcal{R}_2$ are powers of α. These need not be the case and the results from Maxin et al. (2017) allow for models in which these limits have a different form, as seen in the following example:

$$\begin{aligned} x_1' &= r_1 x_1 \left[1 - \frac{K_1 x_1}{1+x_1}\left(1+\frac{1}{1+x_2}\right)^{x_1}\right], \\ x_2' &= r_2 x_2 \left[1 - \frac{K_2 x_2}{1+x_2}\left(1+\frac{1}{1+x_1}\right)^{x_2}\right], \end{aligned} \tag{24}$$

with $K_1 < 1$ and $K_2 < 1$. Then

$$\mathcal{R}_1(\alpha) = \frac{1}{K_1} e^{-\frac{1}{\alpha}}, \quad \mathcal{R}_2(\alpha) = \frac{1}{K_2} e^{-\frac{1}{\alpha}}.$$

The existence of a real number α such that $\mathcal{R}_1(\alpha) < 1$ and $\mathcal{R}_2\left(\frac{1}{\alpha}\right) < 1$ is now equivalent to

$$\mathcal{R} := \ln K_1 \ln K_2 < 1$$

and the solutions of model (24) are unbounded if the opposite inequality holds.

Consider, again, a numerical example. If $K_1 = 0.3$ and $K_2 = 0.55$ then $\mathcal{R} = 0.72$, the solutions are bounded and we have a unique globally stable equilibrium (Fig. 3A). If $K_1 = 0.23$ and $K_2 = 0.43$ and all the other parameters are the same then $\mathcal{R} = 1.24$ and all solutions are unbounded (Fig. 3B). We have $r_1 = 0.002$ and $r_2 = 0.0033$ in panel (A) and $r_1 = 0.0026$ and $r_2 = 0.0042$ in panel (B).

In Fig. 4 we plot the region of global stability when K_1 and K_2 vary.

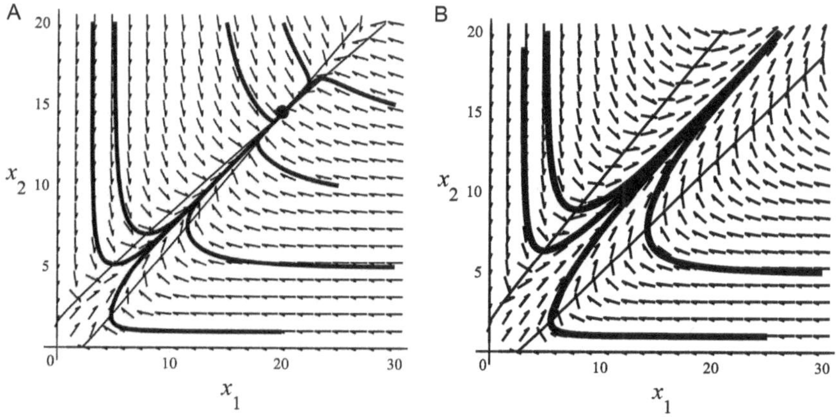

FIG. 3 Phase portrait of model (24). (A) $\mathcal{R} = 0.72$. (B) $\mathcal{R} = 1.24$.

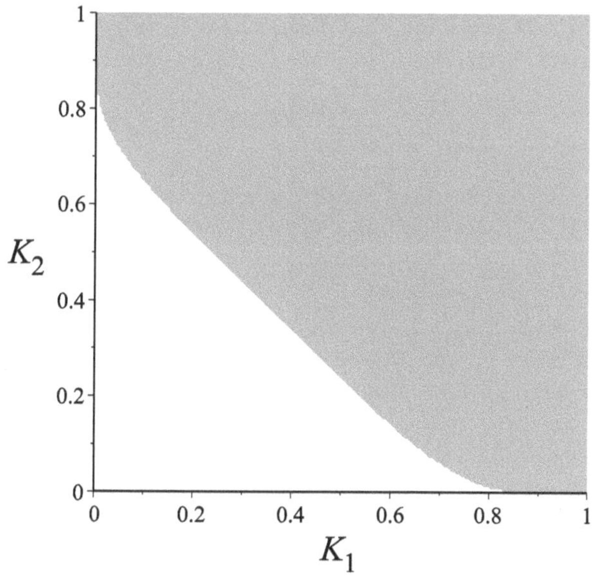

FIG. 4 Global stability region for model (24), in which $\mathcal{R}(K_1, K_2) < 1$.

4.3.3 Species With Strong Allee Effect

In the last example of this section we illustrate the advantages of the generalization from Maxin et al. (2017) when modeling mutualism models subject to a strong Allee effect. In other words, we assume that the species undergoing a strong Allee effect will be modeled by the following equation in the absence of the mutualist:

$$x' = rx\left(\frac{x}{A} - 1\right)\left(1 - \frac{x}{K}\right),$$

As mentioned earlier, a globally stable coexistence equilibrium is not possible if both species are subject to a strong Allee effect. But even when just one of the species is subject to a strong Allee effect, a beneficial effect of the other species need not always prevent that Allee effect as seen in the following example:

$$\begin{aligned} x_1' &= r_1 x_1 \left[\left(\frac{1}{A_1} + \frac{1}{K_1} \right) x_1 - \frac{A_1 K_1 + x_1^2}{A_1 K_1 + b_{12} x_2} \right], \\ x_2' &= r_2 x_2 \left(1 - \frac{x_2}{K_2 + b_{21} x_1} \right). \end{aligned} \tag{25}$$

In this example, species x_1 is the one with the strong Allee effect.

In this setting,

$$a_1(x_1) = \left(\frac{1}{A_1} + \frac{1}{K_1} \right) x_1, \; f_1(x_1, x_2) = \frac{A_1 K_1 + x_1^2}{A_1 K_1 + b_{12} x_2},$$

and it is clear that condition (22) cannot hold. However, the boundedness results still apply. Since

$$R_1(\alpha) = \frac{(K_1 + A_1) b_{12} \alpha}{A_1 K_1}, \quad R_2(\alpha) = b_{21} \alpha,$$

it follows that model (25) has bounded solutions whenever

$$\mathcal{R} := \frac{b_{12} b_{21} (A_1 + K_1)}{A_1 K_1} < 1.$$

On the other hand, if $\mathcal{R} > 1$ then all solutions are unbounded except those starting in the basin of attraction of the boundary equilibrium (0, K_2), which describes the presence of a strong Allee effect in spite of the mutualism. In other words, species x_1 still declines to zero if its population size is too low.

On the other hand, in the following example, the mutualism can overcome an existing strong Allee effect, condition (22) holds and there is a unique coexistence equilibrium that is globally stable:

$$\begin{aligned} x_1' &= r_1 x_1 \left(A_1 e^{x_1} - \frac{B_1 + x_1 e^{x_1}}{1 + b_{12} x_2} \right), \\ x_2' &= r_2 x_2 \left(1 - \frac{x_2}{K_2 + b_{21} x_1} \right). \end{aligned} \tag{26}$$

In this case,

$$a_1(x_1) = A_1 e^{x_1}, \; f_1(x_1, x_2) = \frac{B_1 + x_1 e^{x_1}}{1 + b_{12} x_2}.$$

The per capita growth rate of the first species in the absence of mutualism, $a_1(x_1) - f_1(x_1, 0)$, has a critical point at $x_1 = A_1 - 1$. Hence, in order to satisfy the strong Allee effect assumptions, we need to set the parameters A_1 and B_1 so that

$$e^{A_1-1} > B_1 > A_1 > 1. \tag{27}$$

This ensures that, in the absence of mutualism, the first species is subject to a strong Allee effects and has a hump-shaped per capita growth rate that equals zero at two different positive values of x_1. Since

$$R_1(\alpha) = A_1 b_{12}\alpha, \quad R_2(\alpha) = b_{21}\alpha,$$

model (26) has bounded solutions if and only if

$$\mathcal{R} := b_{12}b_{21}A_1 < 1.$$

In addition, the strong Allee effect is eliminated in the first species if and only if condition (22) holds, which in this particular case requires that

$$A_1 > \frac{B_1}{1 + b_{12}K_2}. \tag{28}$$

It should be noted that these examples (the last one in particular) are built for the purpose of illustrating the advantages of the general models and are not correlated with biological data. It will be primarily the job of a theoretical ecologist to choose particular functional forms that fit the particular aspects of mutualistic interactions found in nature.

From a chronological perspective, we see that the main results concerning building general models of two-species mutualism were established in two stages. In the first stage, using LaSalle Invariance Principle, the general models covered most of the existing specific two-species models while still keeping somewhat restrictive assumptions on the components of these models. In the second stage, threshold conditions for the existence of bounded or unbounded solutions have been established, independent on the existence and stability of equilibrium points. In the case of the latter, the Dulac Criterion can be used for two-species models to prove the asymptotic global stability of a possible unique equilibrium point. This second stage also covers previous specific models but it considerably adds to the flexibility in building other models not previously analyzed. Moreover, as seen in the following section, the boundedness and unboundedness thresholds can be established for models with an unspecified number of species.

5 *n*-DIMENSIONAL MODELS: BOUNDEDNESS AND UNBOUNDEDNESS

While many existing studies on mutualistic interactions restrict themselves to the two-species case (see, for instance, Addicott, 1981; Graves et al., 2006; Malthus, 2018; Vargas-De-León, 2012; Wolin and Lawlor, 1984; Wright, 1989), comparatively fewer are dedicated to the case of multispecies communities, although ecosystems rarely consist in sets of isolated two-species interactions. This happens in spite of biological findings which demonstrate that mutualistic interactions between two species may be greatly affected by species external to the mutualism (Bacher and Friedli, 2002), which may make the consideration of pairwise interactions without their environmental context somewhat inconclusive. In this regard, it has been argued in Ringel et al. (1996) that mutualisms have a stabilizing effect when they are embedded in realistic multispecies real-world communities, such as the four-species model of a pollination mutualism proposed in Ringel et al. (1996). It has also been observed by Addicott (1981) that the consideration of most kinds of mutualism and most kinds of stability criteria shows that a mutualism is more stable than the corresponding system without mutualism. In this sense, Goh (1979) has proved that in a two-species Lotka–Volterra model of mutualism local stability implies global stability.

Further, a network perspective has been introduced to the study of mutualistic communities (Bascompte et al., 2003; Jordano et al., 2003), being observed that mutualistic networks are often of a heterogeneous nature. That is, while most of the participating species are only involved in a couple of interactions, a few "privileged" species are much more connected than the others (Jordano et al., 2003), one of the causes being that longer-lived species often interact with different short-lived partners at distinct stages of their lives (Husband et al., 2002; Palmer et al., 2010). Also, it has been found that high mutualistic to antagonistic ratios generate significantly more diversity than found in the randomized networks (Melián et al., 2009).

Multispecies mutualisms appear sometimes in quite unexpected circumstances. One such example is provided by reef-building corals, which are associated with a diverse array of microbes. Long time considered a single species, a class of symbionts called *zooxanthellar* (found in all reef-building corrals, see also Knowlton and Forest (2003) and references therein), is now recognized as consisting in several genetically diverse groups that live in a single colony thus forming a multispecies mutualism interaction. As mentioned in Knowlton and Forest (2003), coral bleaching (the loss of zooxanthellae) and coral diseases, both increasing over the last decades, may be examples of mutualistic instability.

Georgescu et al. (2017) are shown that the results concerning the thresholds that separate boundedness from unboundedness of model solutions can be extended to an unspecified number of species. We present these results and some details on the proofs in what follows.

5.1 A Generic Framework

Consider the n-species mutualism system

$$x_i' = x_i[a_i(x_i) - f_i(x_1, x_2, \ldots, x_n)], \quad 1 \leq i \leq n, \tag{29}$$

in which the functions a_i and f_i, $1 \leq i \leq n$, are assumed to be positive and continuous. The basic assumptions on each species are

(L) Logistic assumption. For all $1 \leq i \leq n$, there exist $K_i > 0$ such that $a_i(K_i) - f_i(0, .., 0, K_i, 0, \ldots 0) = 0$ and

$$(a_i(x_i) - f_i(0,..,0, x_i, 0, \ldots 0))(x_i - K_i) < 0 \text{ for } x_i \neq K_i.$$

(M) Mutualism-specific assumption. For all $1 \leq i \leq n$, $f_i(x_1, .., x_n)$ is decreasing in x_j for $j \neq i$.

(C1) Consistency assumption. For all $1 \leq i \leq n$, $f_i(x_1, \ldots, x_n)$ is increasing in x_i.

(C2) Consistency assumption. For all $1 \leq i \leq n$, $\frac{a_i(s_i x)}{f_i(s_1 x, \ldots, s_n x)}$ is eventually decreasing for all $s_i > 0$, i.e.,

$$\frac{d}{dx}\left(\frac{a_i(s_i x)}{f_i(s_1 x, \ldots, s_n x)}\right) < 0 \text{ for } x > M_i$$

where M_i are positive real constants that may depend on $s_1, s_2, \ldots, s_n$.

Notice that these are analogous to the ones for the two-species models presented earlier with some small differences in notation for better readability. For example the role of α which indicated the ratio between x_1 and x_2 is replaced by the coefficients $s_1, s_2, \ldots, s_n$ with a similar role (see below). Since the f_i's represent removal rates, assumption (M) represents the fact that the mutualistic support the species give to each other leads to a decrease in the death rates. Assumptions (C1) and (C2) essentially describe the effects of saturation. Also, assumption (C2) ensures that the following limits are defined:

$$\mathcal{R}_i(s_1, s_2, \ldots, s_n) := \lim_{x \to \infty} \frac{a_i(s_i x)}{f_i(s_1 x, s_2 x, \ldots, s_n x)}.$$

It is easy to see that the $\mathcal{R}_i$'s are invariated by a scalar multiplication of all s_i's, in the sense that

$$\mathcal{R}_i(cs_1, cs_2, \ldots, cs_n) = \mathcal{R}_i(s_1, s_2, \ldots, s_n), \quad \text{for } c > 0.$$

Consequently, s_i, $1 \leq i \leq n$, are to be understood as characterizing relative population sizes and $\mathcal{R}_i(s_1, s_2, \ldots, s_n)$ characterizes the ability of species i to reproduce at high population sizes under the given species proportionality. Under the stated form of (C2), x is to be understood as a common measuring unit of all populations and does not bear an immediate biological meaning. It is true that (C2) can be stated (and later used) only for parameters s_i, $1 \leq i \leq n$ with sum 1, in which case x represents the total (of all species) population size and s_i represent the percentage held by the i-th population. However, this approach has the disadvantage of lumping all species (which may or may not have similar characteristics) into a total population size, although it should also be mentioned that if the x_i's represent adimensionalized quantities, then this approach may become reasonable enough.

It is also possible to state (C2) (and later the boundedness and unboundedness results) in such a way to involve $n-1$ parameters instead of n (much closer to the definition given in the previous section for the two-species case). However, this can be done only at the expense of using significantly more cumbersome and less transparent statements.

Just as in the case of two-species models, the $\mathcal{R}_i$'s represent birth-to-death ratios, or, perhaps more aptly, growth-to-loss ratios, since a_i and f_i are only defined modulo functions of x_i and consequently a_i may or may not contain terms pertaining to the removal of species i. The $\mathcal{R}_i$'s also measure the amount of mutualistic support each species receive from the others.

5.2 Boundedness vs Unboundedness: A Tale of *n* Parameters

Assuming (M) and (L), one may prove that the solutions of the system (29) are positivity-preserving, in the sense that the solutions which start with positive initial data are bounded from below by positive constants that may depend on the initial data.

The validity and utility of the $\mathcal{R}_i$'s as threshold-like parameters to characterize the behavior of the system (29) remains the same as in the previous section, considering their roles in proving the boundedness and unboundedness of the solutions, respectively, as well as in discussing the existence of a positive equilibrium.

We now summarize the results proved in Georgescu et al. (2017). Assuming then (M) and (C2), if there are $\alpha_1, \alpha_2, \ldots \alpha_n > 0$ such that

$$\mathcal{R}_i(\alpha_1, \alpha_2, \ldots, \alpha_n) < 1, \quad \text{for all } 1 \leq i \leq n,$$

then the solutions of (29) are bounded from above by positive constants that may depend on the initial data. The proof relies on establishing the boundedness of the function

$$h(t) = \max\left\{\frac{x_1(t)}{\alpha_1}, \frac{x_2(t)}{\alpha_2}, \ldots, \frac{x_n(t)}{\alpha_n}\right\},$$

and is, otherwise, similar to the one employed for two-species models in Maxin et al. (2017). Having obtained the boundedness of the solutions from both above and below, it is possible to establish the existence of a positive equilibrium via a fixed point argument. Consequently, one obtains the following result. Assuming (L), (M), (C1), and (C2), if there are $\alpha_1, \alpha_2, \ldots, \alpha_n > 0$ such that

$$\mathcal{R}_i(\alpha_1, \alpha_2, \ldots, \alpha_n) < 1, \quad \text{for all } 1 \leq i \leq n,$$

then there is a positive equilibrium $\mathbf{E}^* = (x_1^*, x_2^*, \ldots, x_n^*)$ of (29).

A converse (in some sense) result holds as well, under slightly stronger monotonicity hypotheses on the functions which appear in (C2). That is, suppose that (C2) is replaced by

(C2)′ For all $1 \le i \le n$, $\frac{a_i(s_i x)}{f_i(s_1 x, \ldots, s_n x)}$ is decreasing as a function of x for all s_1, s_2, …, $s_n > 0$, i.e.,

$$\frac{d}{dx}\left(\frac{a_i(s_i x)}{f_i(s_1 x, \ldots, s_n x)}\right) < 0 \ \text{ for } \ x > 0.$$

If there is a positive equilibrium $\mathbf{E}^*$, then there exist α_1, α_2, …, $\alpha_n > 0$ such that

$$\mathcal{R}_i(\alpha_1, \alpha_2, \ldots, \alpha_n) < 1, \quad \text{for all } 1 \le i \le n.$$

If the opposite situation happens, that is, if the growths exceed the losses, then it is natural to expect that the population sizes grow unbounded. Consequently, the following result also holds. Assuming (M) and (C2), if there are α_1, α_2, …, $\alpha_n > 0$ such that

$$\mathcal{R}_i(\alpha_1, \alpha_2, \ldots, \alpha_n) > 1, \quad 1 \le i \le n,$$

then the solutions of (29) are unbounded if the initial population sizes are high enough. Again, analogous to the two-species case, the proof relies on establishing the unboundedness of the function

$$h(t) := \min\left\{\frac{x_1(t)}{\alpha_1}, \frac{x_2(t)}{\alpha_2}, \ldots, \frac{x_n(t)}{\alpha_n}\right\}.$$

It is worth noting that for the n-species model (29) one could not use the $\mathcal{R}_i$'s to obtain more than boundedness or unboundedness, the stability of the positive equilibrium being now out of reach. This is because the behavior of a n-dimensional ODE model is much richer than that of a two-dimensional one and ruling out periodic solutions would not translate anymore into stability properties. Primarily, the $\mathcal{R}_i$'s are tools to obtain boundedness (or lack thereof), being qualitatively different from $\mathcal{R}_0$, the basic reproduction number of Mathematical Epidemiology, which could be classified as a stability tool.

5.3 Particular Growth Conditions: A Single Parameter to Rule Them All

The boundedness and unboundedness results from Georgescu et al. (2017) are tailored to the higher generality of the framework established in the system (29). It is possible to specialize them to higher dimensional versions of established two-dimensional models. Georgescu et al. (2017) provide a framework to this end. First we shall review some notions of matrix theory. For given matrices A, B, we shall write $A \le (\ge) B$ if $a_{ij} \le (\ge) b_{ij}$ for all i, j and $A \ll (\gg) B$ if $a_{ij} < (>) b_{ij}$ for all i, j. A square matrix A will be called a Metzler matrix if it has nonnegative off-diagonal entries, that is, $a_{ij} \ge 0$ for all $i \ne j$. For a square matrix A, we shall denote the spectrum of A by $\sigma(A)$ (i.e., the set of its eigenvalues) and the spectral abscissa of A by $\mu(A)$ (i.e., the maximum among the real part of the elements of $\sigma(A)$), denoted as

$$\mu(A) = \max\{\Re(\lambda); \lambda \in \sigma(A)\}.$$

We subsequently say that a square matrix A is Hurwitz stable if $\mu(A) < 0$, while A will be called Hurwitz unstable if $\mu(A) > 0$.

Now suppose that the $\mathcal{R}_i$'s are given in the following particular form.

$$\mathcal{R}_i(s_1, s_2, \ldots, s_n) = \sum_{j=1, j\neq i}^{n} C_{ij}\left(\frac{s_j}{s_i}\right)^p, \quad 1 \leq i \leq n, \tag{30}$$

with

$$C_{ij} \geq 0 \quad \text{for all } 1 \leq i \neq j \leq n.$$

Then the boundedness conditions

$$\mathcal{R}_i(\alpha_1, \alpha_2, \ldots, \alpha_n) < 1, \quad 1 \leq i \leq n,$$

reduce to

$$\sum_{j=1, j\neq i}^{n} C_{ij} v_j - v_i < 0, \quad 1 \leq i \leq n,$$

with $v_i = \alpha_i^p$, $1 \leq i \leq n$. This means that the vector $v = (v_1, v_2, \ldots, v_n)^T$ verifies

$$Cv \ll 0,$$

where C is the Metzler matrix given by

$$C = \begin{pmatrix} -1 & C_{12} & \ldots & C_{1n} \\ C_{21} & -1 & \ldots & C_{2n} \\ \vdots & \vdots & & \vdots \\ C_{n1} & C_{n2} & \ldots & -1 \end{pmatrix}. \tag{31}$$

With the same notations, the unboundedness conditions reduce to

$$Cv \gg 0.$$

One obtains the following result of Georgescu et al. (2017), to which the reader is referred for further details. Suppose that the growth conditions (30) are satisfied. The following properties hold true, the matrix C being defined in (31).

(1) All solutions of (29) are bounded if and only if C is Hurwitz stable.

(2) If C is Hurwitz unstable and irreducible, then (29) has unbounded solutions.

Note that the matrix C is, in some sense, a limiting quantity, being defined in terms of $\mathcal{R}_i$'s, and is not tied to any particular equilibrium. Also, the $n\mathcal{R}_i$'s reduce to a single threshold quantity, the spectral abscissa of a given matrix, which successfully separates boundedness from unboundedness.

Suppose the $\mathcal{R}_i$'s are given in the exponential form.

$$\mathcal{R}_i(s_1,s_2,\dots,s_n)=M_i\exp\left(-\frac{s_i^p}{\sum\limits_{j=1,j\neq i}^{n} d_{ij}s_j^p}\right),\quad 1\leq i\leq n, \tag{32}$$

with

$$M_i>0,\, d_{ij}\geq 0 \quad \text{for all } 1\leq i\neq j\leq n.$$

Via similar considerations, one then arrives at the following Metzler matrix of interest

$$D=\begin{pmatrix} -1 & d_{12}\ln M_1 & \dots & d_{1n}\ln M_1 \\ d_{21}\ln M_2 & -1 & \dots & d_{2n}\ln M_2 \\ \vdots & \vdots & & \vdots \\ d_{n1}\ln M_n & d_{n2}\ln M_2 & \dots & -1 \end{pmatrix}. \tag{33}$$

which may replace C in the statement of the above result.

Below we provide several particular cases of n-species mutualism where these bounding results can be applied.

5.4 Mutualism as Reduction of Mortality for the Benefiting Species

Let us consider the model

$$x_i'=r_ix_i\left[A_i-\frac{x_i^p}{K_i+\sum\limits_{j\neq i} b_{ij}x_j^p}\right],\quad 1\leq i\leq n,\quad p>0, \tag{34}$$

introduced here as an immediate extension of the model considered in Wolin and Lawlor (1984) and Malthus (2018), which can be obtained from (34) for $n=2$ and $p=1$. Here, "reduction of losses" means that the mutualism acts toward a reduction of the negative term, although the derivation of the model in Wolin and Lawlor (1984) assumes that it is the birth rate that is affected by the mutualistic interaction.

For this model, it is seen that

$$a_i(x_i)=r_iA_i,\quad f_i(x_1,x_2,\dots,x_n)=\frac{r_ix_i^p}{K_i+\sum\limits_{j\neq i} b_{ij}x_j^p}$$

and consequently

$$\frac{a_i(s_i x)}{f_i(s_1 x, s_2 x, \ldots, s_n x)} = A_i \left[\frac{K_i}{(s_i x)^p} + \sum_{j \neq i} b_{ij} \left(\frac{s_j}{s_i} \right)^p \right].$$

It then follows that

$$\mathcal{R}_i(s_1, s_2, \ldots, s_n) = A_i \left[\sum_{j \neq i} b_{ij} \left(\frac{s_j}{s_i} \right)^p \right],$$

the corresponding matrix C being given by

$$C = \begin{pmatrix} -1 & A_1 b_{12} & \ldots & A_1 b_{1n} \\ A_2 b_{21} & -1 & \ldots & A_2 b_{2n} \\ \vdots & & & \\ A_n b_{n1} & A_n b_{n2} & \ldots & -1 \end{pmatrix}$$

For $n = 2$, the model (34) reduces to the following model:

$$\begin{aligned} x_1' &= r_1 x_1 \left(A_1 - \frac{x_1^p}{K_1 + b_{12} x_2^p} \right), \\ x_2' &= r_2 x_2 \left(A_2 - \frac{x_2^p}{K_2 + b_{21} x_1^p} \right). \end{aligned} \qquad (35)$$

Since in this case $C = \begin{pmatrix} -1 & A_1 b_{12} \\ A_2 b_{21} & -1 \end{pmatrix}$, its leading principal minors being

$$\Delta_1 = -1, \quad \Delta_2 = 1 - A_1 A_2 b_{12} b_{21},$$

it follows that the solutions of (35) are bounded provided that $A_1 A_2 b_{12} b_{21} < 1$. This condition ensures not only the boundedness of the solutions (and, as seen above, the existence of the positive equilibrium), but also the global stability of the positive equilibrium, via the use of the Dulac Criterion.

5.4.1 A Privileged Species

Let us now suppose that (34) has a "privileged" species, the n-th one, which interacts with all other, while species 1, 2, …, $n - 1$ interact with species n, but do not interact with each other (see Vargas-De-León, 2015 for a related model). An usual example is provided by plant–pollinators systems where the species interaction may form a bipartite graph (i.e., insects interact with plants but not among themselves). The example below represents a subset of this network focused on one insect (the privileged species) that pollinates multiple plant species. In some cases plants do depend on only one single pollinator as is the case with certain orchid species that are exclusively pollinated by the male orchid bee (see also Ramirez, 2009 and references therein). In this case, the model (34) reduces to

$$\begin{aligned}
x_1' &= r_1 x_1 \left(A_1 - \frac{x_1^p}{K_1 + b_{1n} x_n^p} \right), \\
x_2' &= r_2 x_2 \left(A_2 - \frac{x_2^p}{K_2 + b_{2n} x_n^p} \right), \\
&\vdots \\
x_n' &= r_n x_n \left(A_n - \frac{x_n^p}{K_n + b_{n1} x_1^p + \cdots + b_{nn-1} x_{n-1}^p} \right)
\end{aligned} \tag{36}$$

and consequently

$$C = \begin{pmatrix} -1 & 0 & \dots & 0 & A_1 b_{1n} \\ 0 & -1 & \dots & 0 & A_2 b_{2n} \\ \vdots & & & & \\ 0 & 0 & \dots & -1 & A_{n-1} b_{n-1n} \\ A_n b_{n1} & A_n b_{n2} & \dots & A_n b_{nn-1} & -1 \end{pmatrix}.$$

Since

$$\det C = (-1)^{n-1} [-1 + A_n (A_1 b_{1n} b_{n1} + A_2 b_{2n} b_{n2} + \cdots + A_{n-1} b_{n-1n} b_{nn-1})],$$

the other leading minors of order p, $1 \le p \le n-1$, being equal to $(-1)^p$, one finds that the bounding condition is in this case

$$A_n (A_1 b_{1n} b_{n1} + A_2 b_{2n} b_{n2} + \cdots + A_{n-1} b_{n-1n} b_{nn-1}) < 1.$$

5.5 Mutualism as a Positive Contribution to the Fertility Rate of the Benefiting Species

In what follows, we shall consider the multispecies version of the model (6) proposed by Graves et al. (2006), given by

$$x_i' = r_i x_i \left(1 - \frac{x_i}{K_i} \right) + c_i x_i \left(1 - e^{-\sum_{j \ne i} \alpha_{ij} x_j} \right), \quad 1 \le i \le n. \tag{37}$$

Let us first rearrange (37) in the form

$$x_i' = x_i \left[(r_i + c_i) - \left(\frac{r_i x_i}{K_i} + c_i e^{-\sum_{j \ne i} \alpha_{ij} x_j} \right) \right], \quad 1 \le i \le n. \tag{38}$$

For this model, it is seen that

$$a_i(x_i) = r_i + c_i, \quad f_i(x_1, x_2, \ldots, x_n) = \frac{r_i x_i}{K_i} + c_i e^{-\sum_{j \ne i} \alpha_{ij} x_j}$$

and consequently

$$\frac{a_i(s_i x)}{f_i(s_1 x, s_2 x, \ldots, s_n x)} = \frac{r_i + c_i}{\frac{r_i s_i}{K_i} x + c_i e^{-\left(\sum_{j \neq i} \alpha_{ij} s_j\right) x}},$$

which is eventually decreasing. It then follows that

$$\mathcal{R}_i(s_1, s_2, \ldots, s_n) = 0, \quad 1 \leq i \leq n,$$

the corresponding matrix C being given by

$$C = -I_n,$$

which is Hurwitz stable. This ensures the boundedness of the solutions of (38).

Let us consider, after some renotations, the multispecies version of the model (8) proposed by Wright (1989), given by

$$x_i' = r_i x_i \left(1 - \frac{x_i}{K_i}\right) + x_i \sum_{j \neq i} \frac{\beta_{ij} x_j}{\alpha_{ij} + x_j}, \quad 1 \leq i \leq n. \tag{39}$$

One may remark that this multispecies version is related to the models considered in Jelle Lever et al. (2014) and Rohr et al. (2014), the differences being that no competition terms are considered and there is no functional distinction between two categories of species (plants and pollinators in Jelle Lever et al. (2014), plants and animals in Rohr et al. (2014)). Lastly, our model (39), unlike the model in Jelle Lever et al. (2014), assumes that there is no immigration.

Let us first rearrange (39) in the form

$$x_i' = x_i \left[(r_i + \sum_{j \neq i} \beta_{ij}) - \left(\frac{r_i x_i}{K_i} + \sum_{j \neq i} \frac{\beta_{ij} \alpha_{ij}}{\alpha_{ij} + x_j} \right) \right], \quad 1 \leq i \leq n. \tag{40}$$

For this model, it is seen that

$$a_i(x_i) = r_i + \sum_{j \neq i} \beta_{ij}, \quad f_i(x_1, x_2, \ldots, x_n) = \frac{r_i x_i}{K_i} + \sum_{j \neq i} \frac{\beta_{ij} \alpha_{ij}}{\alpha_{ij} + x_j}$$

and consequently

$$\frac{a_i(s_i x)}{f_i(s_1 x, s_2 x, \ldots, s_n x)} = \frac{r_i + \sum_{j \neq i} \beta_{ij}}{\frac{r_i s_i x_i}{K_i} + \sum_{j \neq i} \frac{\beta_{ij} \alpha_{ij}}{\alpha_{ij} + s_j x_j}},$$

which is eventually decreasing. It then follows that

$$\mathcal{R}_i(s_1, s_2, \ldots, s_n) = 0, \quad 1 \leq i \leq n,$$

the corresponding matrix C being given by

$$C = -I_n,$$

which is Hurwitz stable. This, again, ensures the boundedness of the solutions of (40).

We observe that (39) includes the versions without competition of the models of mutualism in Jelle Lever et al. (2014) and Rohr et al. (2014) (for Rohr et al., 2014, also without immigration), the functional distinction being a matter of setting the appropriate β_{ij}'s equal to 0. Via a comparison argument, one may see that the solutions of the models in Jelle Lever et al. (2014) and Rohr et al. (2014) are bounded if one discounts immigration, which is not surprising, since both are models with limited per capita benefits from the mutualism.

As a consequence, it is seen that these boundedness results can be applied not only for mutualisms which act by reducing removal rates, such as (34), but also for mutualisms which enhance growth rates, although, at a glance, the particular form of our system (29) seems to tell otherwise.

5.6 Which Species Is Responsible for Unboundedness?

In the examples provided above, the n conditions necessary for the bounding results are actually reduced to a single threshold quantity, the spectral abscissa of a given matrix, which separates boundedness from unboundedness. A further question to ask is whether the limits $\mathcal{R}_i$ have separate roles in the bounding conditions or, from a biological perspective, whether the outcome (bounded vs unbounded) is caused by the mutualistic interaction of all species together. It can be shown that, with some additional hypotheses, the above results can be caused by a subset of the species in the system. This is important since, if the goal is to develop a realistic mutualistic model, the attention can be restricted only to the species likely to produce an unbounded solution.

To this end, suppose that the n species are separated into two subgroups: group 1 made of x_1, …, x_m on the one hand and group 2 made of x_{m+1}, …, x_n on the other, with $m < n$.

Furthermore, suppose that group 1 obtains its mutualistic benefits within their own group while group 2 exhibits an upper bound on the mutualistic benefits from group 1. Mathematically, this translates into the following additional hypotheses:

$$\begin{aligned} f_i(x_1,\ldots,x_m,x_{m+1},\ldots,x_n) &\le g_i(x_1,\ldots,x_m), \quad 1 \le i \le m, \\ f_i(x_1,\ldots,x_m,x_{m+1},\ldots,x_n) &\ge h_i(x_{m+1},\ldots,x_n), \quad m+1 \le i \le n, \end{aligned}$$

the functions g_i and h_i being assumed to satisfy the same hypotheses assumed for f_i in the previous results. Denoting

$$\mathcal{R}_i(s_1,s_2,\ldots,s_m) := \lim_{x\to\infty} \frac{a_i(s_i x)}{g_i(s_1 x, s_2 x, \ldots, s_m x)}, \quad 1 \le i \le m,$$

and

$$\mathcal{R}_i(s_{m+1},s_{m+2},\dots,s_n) := \lim_{x\to\infty} \frac{a_i(s_i x)}{h_i(s_{m+1}x,s_{m+2}x,\dots,s_n x)}, \quad m+1 \le i \le n.$$

The following result from Georgescu et al. (2017) shows that only a subset of the total number of species experience unbounded growth, while the others remain bounded. Assuming (C1) and (C2) for the functions g_i and h_i, if there are $\alpha_1, \alpha_2, \dots \alpha_m > 0$ such that

$$\mathcal{R}_i(\alpha_1,\alpha_2,\dots,\alpha_m) > 1, \quad 1 \le i \le m,$$

and there are $\alpha_{m+1}, \alpha_{m+2}, \dots, \alpha_n > 0$ such that

$$\mathcal{R}_i(\alpha_{m+1},\alpha_{m+2},\dots,\alpha_n) < 1, \quad m+1 \le i \le n,$$

then $x_1, \dots, x_m$ are unbounded if the initial population sizes are large enough while $x_{m+1}, \dots, x_n$ remain bounded.

The techniques used in proving this result are very similar to the ones presented so far. The main difference consists in using a minimum function for the first m species only and a maximum function for the remaining species as follows:

$$h(t) = \min\left\{\frac{x_1(t)}{\alpha_1}, \frac{x_2(t)}{\alpha_2}, \dots, \frac{x_m(t)}{\alpha_m}\right\},$$

in order to show that $x_1, x_2, \dots, x_m$ are unbounded, and

$$h(t) = \max\left\{\frac{x_{m+1}(t)}{\alpha_{m+1}}, \frac{x_{m+2}(t)}{\alpha_{m+2}}, \dots, \frac{x_n(t)}{\alpha_n}\right\}$$

to show that $x_{m+1}, x_{m+2}, \dots, x_n$ are bounded.

To illustrate this result, consider the following model

$$\begin{aligned} x_1' &= r_1 x_1\left(A_1 - \frac{x_1^p}{K_1 + b_{12}x_2^p + b_{13}x_3^p}\right), \\ x_2' &= r_2 x_2\left(A_2 - \frac{x_2^p}{K_2 + b_{21}x_1^p + b_{23}x_3^p}\right), \\ x_3' &= r_3 x_3\left[A_3 - \left(1 + \frac{x_3^p}{K_3 + b_{31}x_1^p + b_{32}x_2^p}\right)x_3\right], \end{aligned} \tag{41}$$

with

$$p > 0, \quad b_{ij} \ge 0, \quad A_1 A_2 b_{12} b_{21} > 1.$$

We see that

$$x_1' \ge r_1 x_1 A_1 - \frac{r_1 x_1^{p+1}}{K_1 + b_{12}x_2^p}, \quad x_2' \ge r_2 x_2 A_2 - \frac{r_2 x_2^{p+1}}{K_2 + b_{21}x_1^p}$$

with

$$g_1(x_1,x_2) = \frac{r_1 x_1^{p+1}}{K_1 + b_{12}x_2^p}, \quad g_2(x_1,x_2) = \frac{r_2 x_2^{p+1}}{K_2 + b_{21}x_1^p}.$$

Also,

$$x_3' \leq r_3 x_3 A_3 - r_3 x_3^2,$$

with

$$h_3(x_3) = r_3 x_3^2.$$

Then

$$\mathcal{R}_1(s_1, s_2) = A_1 b_{12} \left(\frac{s_2}{s_1}\right)^p, \quad \mathcal{R}_2(s_1, s_2) = A_2 b_{21} \left(\frac{s_1}{s_2}\right)^p, \quad \mathcal{R}_3(s_3) = 0.$$

If $A_1 A_2 b_{12} b_{21} > 1$, there are α_1, $\alpha_2 > 0$ such that $\mathcal{R}_1(\alpha_1, \alpha_2) > 1$, $\mathcal{R}_2(\alpha_1, \alpha_2) > 1$ and obviously $\mathcal{R}_3(s_3) < 1$. Therefore x_3 remains bounded, while x_1 and x_2 grow unbounded (in this case, for all initial population sizes).

6 FINAL CONSIDERATIONS

In this chapter, we have provided an overview on the development of models of mutualism with an emphasis on the most important attempts to shape these models into a general framework, amenable to a unified treatment. Motivated by the widespread nature of mutualistic interactions, these general models were analyzed in such a way that they could serve as a blueprint for more elaborate population models. To this end, the derivation of boundedness conditions plays the most important role in this work since bounded solutions are a fundamental requirement to any realistic population model used in long-term predictions. The general models developed by Georgescu and Zhang (2014) and Maxin et al. (2017) used different mathematical techniques that affect the way their main results are obtained but, taking together, they unify most of the existing particular two-species models of mutualism and greatly increase the flexibility of future models, allowing the inclusion of other important features such as the weak and strong Allee effects and different mechanisms of mutualism for each species.

It is important to point out several limitations of these results since they shed a light also on the possible avenues of future research.

First, there is no general criterion concerning the existence of equilibria. This is not unexpected, since we are dealing with the solvability of a nonlinear system in a rather general form. In particular, there are no necessary and sufficient conditions in the general models that ensure the existence and uniqueness of the coexistence equilibrium. Nearly all global stability results assume the existence of this type of equilibrium a priori. On the one hand, this is to be expected as the price to pay for having a high level of generalization. In other words, without assuming anything beyond what is needed for the boundedness, the existence of equilibria is an open problem. There are plenty of examples in which there are, in fact, multiple equilibrium points alternating from stable to unstable. In some sense, this is a mild shortcoming since, whenever a concrete model replaces the general one, the conditions on the existence of equilibria can be derived without problems, at least in the case of two-species models.

Furthermore, the terms α and $s_1, s_2, \ldots, s_n$ used in defining $\mathcal{R}_i$'s are somewhat artificial and they are a direct consequence of the particular mathematical technique used in proving the bounding thresholds. The first indicates the ratio between the two population sizes in the case of two-species models, while the latter are scaled quantities describing the relative sizes of each population in the case of the $n-$species model. As seen in all particular examples, they are not needed whenever the bounding condition reduces to a single threshold. As such, it is an open question whether one can derive similar results without these ratios in which the bounding conditions reduce to a threshold even for the most general model.

Finally, it is important to point out that there will always be limitations with any attempt toward a generalization since there is no universal format for a mutualistic model. As an example, the settings from Georgescu et al. (2017) do not include any model of mutualism derived within the modeling framework of Levin's colonization-extinction metapopulation model, such as those of evolution of coviruses (Nee, 2000) and plant–pollinator (Nee et al., 1997), due to the sign pattern of the interaction terms. Actually, in both Nee (2000) and Nee et al. (1997) (and generally in any colonization-extinction metapopulation model) the variables represent proportions of patches, not population sizes or densities as in Georgescu et al. (2017), leading to constant total size models. Consequently, the growth of all but one variables (for instance) is guaranteed to have a negative effect on the remaining variable, and consequently the resulting model cannot adhere to the spirit of the mutualistic assumption presented (Georgescu et al., 2017). However, for the models presented in Nee (2000) and Nee et al. (1997), boundedness of solutions is a feature of the model and does not need to be studied separately.

As a final thought toward future research, we believe that a worthwhile extension of the work presented in this chapter is to consider further generalizations of multigroup models containing both mutualism and the interplay between mutualisms and other types of interactions such as predation and competition. In the ecological literature many authors point out that the distinction between competition/predation/mutualism is not always strict (often a species plays all of these roles during various stages of its growth). As a consequence, a useful abstract framework containing all types of interactions is desirable.

ACKNOWLEDGMENT

The authors thank the reviewers who read this chapter and provided valuable feedback.

REFERENCES

Addicott, J.F., 1981. Stability properties of 2-species models of mutualism: simulation studies. Oecologia 49, 42–49.

Bacher, S., Friedli, J., 2002. Dynamics of a mutualism in a multi-species context. Proc. Biol. Sci. 269, 1517–1522.

Bascompte, J., Jordano, P., Melián, C.J., Olesen, J.M., 2003. The nested assembly of plant-animal mutualistic networks. Proc. Natl Acad. Sci. USA 100, 9383–9387.

Boucher, D.H., James, S., Keeler, K.H., 1982. The ecology of mutualism. Annu. Rev. Ecol. Syst. 315–347.

Brauer, F., Castillo-Chavez, C., 2012. Mathematical Models in Population Biology and Epidemiology, vol. 1. Springer, New York.

Bronstein, J.L., 1994. Conditional outcomes in mutualistic interactions. Trends Ecol. Evol. 214–217.

Bronstein, J.L., Wilson, W.G., Morris, W.F., 2003. Ecological dynamics of mutualist/antagonist communities. Am. Nat. 162 (Suppl.), S24–S39.

Courchamp, F., Clutton-Brock, T., Grenfell, B., 1999. Inverse density dependence and the Allee effect. Trends Ecol. Evol. 14, 405–410.

Floeter, S.R., Vázquez, D.P., Grutter, A.S., 2007. The macroecology of marine cleaning mutualisms. J. Anim. Ecol. 76, 105–111.

Gause, G.F., Witt, A.A., 1935. Behavior of mixed populations and the problem of natural selection. Am. Nat. 69, 596–609.

Georgescu, P., Zhang, H., 2014. Lyapunov functionals for two-species mutualisms. Appl. Math. Comput. 226, 754–764.

Georgescu, P., Zhang, H., Maxin, D., 2016. The global stability of coexisting equilibria for three models of mutualism. Math. Biosci. Eng. 13, 101–118.

Georgescu, P., Maxin, D., Zhang, H., 2017. Threshold boundedness conditions for n-species mutualisms. Nonlinearity 30, 4410–4427.

Goh, B.S., 1979. Stability in models of mutualism. Am. Nat. 113, 261–275.

Graves, W.G., Peckham, B., Pastor, J., 2006. A bifurcation analysis of a differential equations model for mutualism. Bull. Math. Biol. 68, 1851–1872.

Holland, J.N., Bronstein, J.L., 2008. Mutualism. In: Jorgensen, S.E., Fath, B.D. (Eds.), Population Dynamics, Vol 3 of Encyclopedia of Ecology. Elsevier, pp. 2485–2491.

Holland, J.N., DeAngelis, D.L., 2009. Consumer-resource theory predicts dynamic transitions between outcomes of interspecific interactions. Ecol. Lett. 12, 1357–1366.

Holland, J.N., DeAngelis, D.L., 2010. A consumer-resource approach to the density-dependent population dynamics of mutualism. Ecology 91, 1286–1295.

Hsieh, Y.H., 2008. Richards model: a simple procedure for real-time prediction of outbreak severity. In: Ma, Z., Wu, J., Zhou, Y. (Eds.), Modeling and Dynamics of Infectious Diseases, in Contemporary Applied Mathematics (CAM), vol. 11. Higher Education Press, pp. 218–239.

Husband, R., Herre, E.A., Turner, S.L., Gallery, R., Young, J. P. W., 2002. Molecular diversity of arbuscular mycorrhizal fungi and patterns of host association over time and space in a tropical forest. Mol. Ecol. 11, 2669–2678.

Iannelli, M., Martcheva, M., Milner, F.A., 2005. Gender-Structured Population Modeling: Mathematical Methods, Numerics, and Simulations. vol. 31. SIAM.

Jelle Lever, J., van Nes, E. H., Scheffer, M., Bascompte, J., 2014. The sudden collapse of pollinator communities. Ecol. Lett. 17, 350–359.

Jordano, P., Bascompte, J., Olesen, J.M., 2003. Invariant properties in coevolutionary networks of plant-animal interactions. Ecol. Lett. 6, 69–81.

Knowlton, N., Forest, R., 2003. Multispecies microbial mutualisms on coral reefs: the host as a habitat. Am. Nat. 162 (Suppl.), S51–S62.

Malthus, T.R., 2018. An Essay on the Principle of Population as It Affects the Future Improvement of Society. The Economics of Population. Routledge 41–50.

Margulis, L., 1971. Symbiosis and evolution. Sci. Am. 225, 49–55.

Maxin, D., Georgescu, P., Sega, L., Berec, L., 2017. Global stability of the coexistence equilibrium for a general class of models of facultative mutualism. J. Biol. Dyn. 11, 339–364.

Melián, C.J., Bascompte, J., Jordano, P., Krivan, V., 2009. Diversity in a complex ecological network with two interaction types. Oikos 118, 122–130.

Nee, S., 2000. Mutualism, parasitism and cooperation in the evolution of coviruses. Phil. Trans. R. Soc. Lond. B 355, 1607–1613.

Nee, S., May, R.M., Hassell, M.P., 1997. Two-species metapopulation models. In: Hanski, I., Gilpin, M.E. (Eds.), Metapopulation Biology. Academic Press, pp. 123–147.

Palmer, T.M., Doak, D.F., Stanton, M.L., Bronstein, J.L., Kiers, E.T., Young, T.P., Goheen, J.R., Pringle, R.M., 2010. Synergy of multiple partners, including freeloaders, increases host fitness in a multispecies mutualism. Proc. Natl. Acad. Sci. USA 107, 17234–17239.

Pearl, R., Reed, L.J., 1920. On the rate of growth of the population of the United States since 1790 and its mathematical representation. Proc. Nat. Acad. Sci. 6, 275–288.

Ramirez, S.R., 2009. Orchid bees. Curr. Biol. 19, 1061–1063.

Richards, F.J., 1959. A flexible growth function for empirical use. J. Exp. Bot. 10, 290–300.

Ringel, M., Hu, H., Anderson, G., 1996. The stability and persistence of mutualisms embedded in community interactions. Theor. Popul. Biol. 50, 281–297.

Rohr, R.P., Saavedra, S., Bascompte, J., 2014. On the structural stability of mutualistic systems. Science 345 (6195), 1253497.

Thompson, J.N., Schwind, C., Guimarães, P.R., Friberg, M., 2013. Diversification through multitrait evolution in a coevolving interaction. Proc. Natl. Acad. Sci. USA 110, 11487–11492.

Vandermeer, J., Boucher, D., 1978. Varieties of mutualistic interaction in population models. J. Theor. Biol. 74, 549–558.

Vargas-De-León, C., 2012. Lyapunov functions for two-species cooperative systems. Appl. Math. Comput. 219, 2493–2497.

Vargas-De-León, C., 2015. Global stability for multi-species Lotka-Volterra cooperative systems: one hyper-connected mutualistic-species. Int. J. Biomath. 8, 9. 1550039.

Vázquez, D.P., Ramos-Jiliberto, R., Urbani, P., Valdovinos, F.S., 2015. A conceptual framework for studying the strength of plant-animal mutualistic interactions. Ecol. Lett. 18, 385–400.

Verhulst, P.F., 1838. Notice sur la loi que la population suit dans son accroissement. Corr. Math. et Phys. 10, 113–121.

Wolin, C., Lawlor, L., 1984. Models of facultative mutualism: density effects. Am. Nat. 124, 843–862.

Wright, D.H., 1989. A simple, stable model of mutualism incorporating handling time. Am. Nat. 134, 664–667.

Section VII

Stochastic Complexity and Structural Dynamics

Chapter 4

Stochastic Models for Structured Populations

Shripad Tuljapurkar*[,1] and David Steinsaltz[†]
[*]*Department of Biology, Stanford University, Stanford, CA, United States*
[†]*Department of Statistics, University of Oxford, Oxford, United Kingdom*
[1]*Corresponding author: e-mail: tulja@stanford.edu*

ABSTRACT

This chapter deals with stochastic models for structured populations whose dynamics depend crucially on individual characteristics such as age, size, or location. We deal with linear stochastic models, and their analysis is also essential to the analysis of non-linear stochastic models, particularly their boundary behavior which determines invasion, extinction, and persistence. The models discussed here operate in discrete time, and describe individuals who can be in one of the many finite stages. The analyses described here also apply to other discrete-time models, known as integral population models, under suitable conditions. These models are naturally framed in terms of products of random matrices, and so are connected to models used in a number of other areas of science. We explain the assumptions and structure of the models, how they are constructed, and how they may be analyzed. We focus on two kinds of random environments, those that are independently identically distributed (IID), and finite-state Markov chains, but many of the results also apply to other environmental processes. We give examples to illustrate the theory. We also discuss the computation of derivatives of the stochastic growth rate for our models. Code (in R) to compute these derivatives will soon appear as a package; until then, the code is available from the authors.

Keywords: Structured populations, Stochastic population dynamics, Stochastic vital rates, Lyapunov exponent, Random matrix product, Long-run growth rate, Derivatives of stochastic growth rate, Small-noise approximation

1 INTRODUCTION

This chapter deals with stochastic models for structured populations that describe dynamics over discrete-time intervals, such as months or years. Structured populations consist of individuals organized into discrete classes, here called stages, that are defined by features such as age, size, or location.

Handbook of Statistics, Vol. 40. https://doi.org/10.1016/bs.host.2018.11.005

Thus the models here are discrete-time models for movements of individuals between discrete stages, and so are naturally matrix models. The reader may be familiar with models for age-structured populations that have long been described using Leslie matrices (Leslie, 1966). But the theory we describe applies also to populations where we track size instead of age (Lefkovitch, 1965), or track both age and size (Coulson et al., 2010), or track some more general combination such as age, size, and developmental state (Steiner et al., 2014). A core feature of all such models is a specification of *vital rates*—rates of transition between discrete stages in the life of an individual. As such rates must be nonnegative quantities, the theory can exploit known properties of nonnegative matrices. The matrix entries in our models—the transition rates—vary randomly over time, so we need to draw upon the mathematical theory that describes products of random nonnegative matrices. The stochasticity sets these models outside the bounds of familiar linear methods.

While we confine our attention here to linear stochastic models, their properties underlie the analysis of nonlinear stochastic models, particularly their boundary behavior, which determines invasion, extinction, and persistence (Benaïm and Schreiber, 2009), as we explain below. Many aspects of our theory also apply to models in which the stages are continuous, so that transitions are described by a nonnegative integral operator instead of a nonnegative matrix. The latter are known in biology as integral population models and Ellner and Rees (2006) show how they can be described by a generalization of the present theory.

We do not provide extensive historical background or a review of past work, but describe here briefly the origins of the subject and, toward the end of the paper, mention a few connections with other areas of science. The biological literature on the subject began with the important paper by Lewontin and Cohen (1974) that examined a population in discrete time whose numbers changed according to random (scalar) multiplicative growth rates. They showed that the population's long-run size depends on the expected value of the logarithm of the growth rate, and the population size has an asymptotically lognormal distribution.

The mathematics of stochastic growth indeed goes back much further, with the matrix case already having originated, it would seem, in a thought-provoking paper by Bellman et al. (1954). This inspired the seminal paper of Furstenberg and Kesten (1960), whose very general solution defined the framework for all later work on the subject. These authors were stimulated by general questions related to the effects of iterated transformations that could be represented by random matrix products acting on some initial vector. They showed that, just as in the case of random scalar multiplication, the long-run size of such a vector (measured by, for example, the length) may be described by a deterministic logarithmic limit, now called the *leading Lyapunov exponent* of the random matrix product (Arnold et al., 1986); the size of the vector also has an asymptotically lognormal distribution. It was natural to expect that the mathematical theory might apply to a population described by a vector (whose entries list, for example, population counts by age class) and whose change depended on a matrix of vital rates. And indeed the demographic generalization of the mathematical work was established by Cohen in a series of papers starting in 1976 (Cohen, 1976). A useful general discussion of the population mathematics is given by Lange and Holmes (1981). Cohen's work has since been extended and applied in many contexts. Biological applications and developments are discussed in the books by Tuljapurkar (1990a) and Caswell (2001), and in recent papers by Benaīm and Schreiber (2009), Ellner and Rees (2007), and Steinsaltz et al. (2011).

Here we summarize the main features of the theory of stochastic population models for structured populations and illustrate the theory with examples. We explain the structure of the models, how they are constructed, and how they are analyzed. The discussion focuses on two kinds of environments, independent identically distributed (IID) and finite-state Markov chains, but many of the results also apply to other environmental processes. As a concrete example we describe the application to one time-series model.

2 INTRODUCTION: POPULATION DYNAMICS IN A CONSTANT ENVIRONMENT

As an elementary introduction to the nature of the models we are concerned with, we begin by looking at the growth of a structured population in a constant environment. The populations we consider have K discrete stages, determined by one or more of: age, size, developmental stage, and spatial locations. At each time t such a population is described by a $K \times 1$ vector $\mathbf{N}_t$, with total population size $|\mathbf{N}_t|$, where we define the notation

$$|\mathbf{u}| = \text{Sum of components of vector } \mathbf{u}. \tag{1}$$

The population structure is just the vector of proportions $\mathbf{U}_t$, defined as

$$\mathbf{U}_t = \frac{\mathbf{N}_t}{|\mathbf{N}_t|}. \tag{2}$$

In a constant environment, transitions among the K stages are described by vital rates: for any pair of stages i, j, a nonnegative number $b_{ij} \geq 0$ measures the number of individuals in stage i at time $t + 1$ resulting from a single individual in stage j at time t. These rates can be organized into a nonnegative matrix $\mathbf{B}$, and so determine the dynamics of the population by the equations

$$\mathbf{N}_{t+1} = \mathbf{B}\,\mathbf{N}_t, \tag{3}$$

$$\mathbf{U}_{t+1} = \frac{\mathbf{B}\,\mathbf{U}_t}{\lambda_{t+1}}, \tag{4}$$

$$\lambda_{t+1} = |\mathbf{B}\,\mathbf{U}_t|. \tag{5}$$

The number λ_{t+1} is the one-period growth rate of total population, from $|\mathbf{N}_t|$ to $|\mathbf{N}_{t+1}|$,

$$\lambda_{t+1} = \frac{|\mathbf{N}_{t+1}|}{|\mathbf{N}_t|} = \frac{|\mathbf{B}\,\mathbf{N}_t|}{|\mathbf{N}_t|} = |\mathbf{B}\,\mathbf{U}_t|. \tag{6}$$

The K discrete stages describe a life cycle. For example, let the stages be age-classes 1, 2, 3, with newborns in age-class $i = 1$. The transition rates between ages i and $i + 1$ are simply survival rates, $0 < p_1, p_2 < 1$, and transitions between age-classes 2, 3 and age-class 1 are rates of reproduction, $0 < f_2, f_3$. The matrix of vital rates is thus a Leslie matrix,

$$\mathbf{B} = \begin{pmatrix} 0 & f_2 & f_3 \\ p_1 & 0 & 0 \\ 0 & p_2 & 0 \end{pmatrix}. \tag{7}$$

For a more involved example we might let stages 1, 2, 3 represent small, medium, and large newborns, and represent the corresponding sizes in age-classes 2 and 3 by stage numbers 4,5,...,9. A possible vital rate matrix here is

$$\mathbf{B} = \begin{pmatrix} 0 & 0 & 0 & f_4 & f_5 & f_6 & f_7 & f_8 & f_9 \\ p_1g_1 & p_1h_1 & p_1k_1 & 0 & 0 & 0 & \ldots & 0 & \\ p_1g_2 & p_1h_2 & p_1k_2 & 0 & 0 & & 0\ldots & 0 & \\ p_1g_3 & p_1h_3 & p_1k_3 & 0 & 0 & & 0\ldots & 0 & \\ 0 & 0 & 0 & p_2G_1 & p_2H_1 & p_2K_1 & 0 & 0 & 0 \\ 0 & 0 & 0 & p_2G_2 & p_2H_2 & p_2K_2 & 0 & 0 & 0 \\ 0 & 0 & 0 & p_2G_3 & p_2H_3 & p_2K_3 & 0 & 0 & 0 \\ \vdots & \vdots & \vdots & \vdots & \vdots & \vdots & \vdots & \vdots & \vdots \end{pmatrix}. \tag{8}$$

As in the simpler example (7), here the f_i are fertilities, and the p_i are survival rates. But

$$(g_1 + g_2 + g_3) = (h_1 + h_2 + h_3) = \ldots(k_1 + k_2 + k_3) = 1,$$

and are transition rates from small offspring to small (g_1), medium (g_2), or large (g_3) adults in age-class 2 and similarly for the other transitions.

For any of the structured models we analyze, the matrix $\mathbf{B}$ is nonnegative, so the Perron–Frobenius theorem (Seneta, 2006) tells us that (for primitive irreducible $\mathbf{B}$) the population structure and growth rate converge to limits,

$$\mathbf{U}_t \to \mathbf{u}_0, \lambda_t \to \lambda_0, \tag{9}$$

as $t \uparrow \infty$. Here $\mathbf{u}_0$ is the right eigenvector of $\mathbf{B}$ corresponding to its dominant (largest) eigenvalue λ_0. Thus, as time passes, the population size $|\mathbf{N}_t|$ grows asymptotically at rate λ_0. When this limiting behavior results from any initial population vector $\mathbf{N}_0$ we say that the vital rates $\mathbf{B}$ satisfy *strong demographic ergodicity*.

We note that time reversal plays a hidden role here. Let us define a vector $\mathbf{Y}_t$ with K nonnegative elements, and construct a backward-in-time version of (3) by writing

$$\mathbf{Y}_{t-1}^T = \mathbf{Y}_t^T \mathbf{B}. \tag{10}$$

where superscript T indicates a transpose. The corresponding vector of proportions $\mathbf{V}_t = \mathbf{Y}_t/|\mathbf{Y}_t|$ changes as

$$\mathbf{V}_{t-1}^T = \frac{\mathbf{V}_t^T \mathbf{B}}{|\mathbf{V}_t^T \mathbf{B}|}. \tag{11}$$

For a fixed t we can use (11) to generate a sequence of vectors $\mathbf{V}_{t-m}$ for increasing $m > 0$. The conditions that ensure strong demographic ergodicity in (9) also ensure that as $m \uparrow \infty$,

$$\mathbf{V}_{t-m} \to \mathbf{v}_0, \tag{12}$$

where $\mathbf{v}_0$ is the left eigenvector of $\mathbf{B}$ corresponding to the dominant eigenvalue λ_0.

The vectors $\mathbf{Y}$ and $\mathbf{v}_0$ play two useful roles. First, suppose that $\mathbf{Y}_t$ is a vector of the contributions to the future (i.e., to all times $> t$) of the different possible stages. Then (10) is clearly a backward iteration yielding $\mathbf{Y}_{t-1}$ as a vector of the contributions to all times $> (t-1)$ of the different possible stages. In demography, the vector $\mathbf{Y}_t$ is called the vector of *stage-specific reproductive values*. Clearly these reproductive values grow as time recedes further into the past, so it makes sense to use a relative reproductive value, which is just $\mathbf{V}_t = \mathbf{Y}_t/|\mathbf{Y}_t|$ as in (11). Demographic strong ergodicity means that the vector of relative reproductive values converges to the constant vector $\mathbf{v}_0$.

The other useful role of reproductive values is contained in the well-known perturbation result (Caswell, 2001) that

$$\frac{\partial \lambda_0}{\partial b_{ij}} = \frac{v_0(i)\, u_0(j)}{\mathbf{v}_0^T \mathbf{u}_0}. \tag{13}$$

In the numerator above the quantity $v_0(i)$ is the ith element of $\mathbf{v}$, and the denominator is just the scalar product of the vectors. Reproductive value has a similar definition, and similar uses, in the time-dependent case of this chapter.

3 STRUCTURED POPULATIONS IN STOCHASTIC ENVIRONMENTS

3.1 The Environment

Equal discrete-time intervals are marked by integer values $t = 0, 1, \ldots, t, t+1, \ldots$. In each time interval (e.g., between 0 and 1) all the environmental conditions that matter to the population are referred to as an environmental "state." The simplest random environments are IID at each time. Two other kind of environmental sequences, where successive environments are not independent, are also often used.

3.1.1 Environments Follow a Finite-State Markov Chain

One class of models assumes there is a finite number M of environmental states, and transitions between the states follow a Markov chain. For example, the temperature may be hot (H) or cold (C), and the rainfall may be wet (W) or dry (D), yielding the environmental states, HW, HD, CW, and CD, so $M = 4$. We use the letters c, d to indicate environmental states. Suppose that the environmental state at time t (i.e., in the interval $t - 1$ to t) is c. In a Markov chain, the next environmental state at $t + 1$, call it E_{t+1}, is randomly determined.

A Markovian environment is specified by conditional probabilities

$$\Pr[E_{t+1} = d | E_t = c] = p(c,d), \text{ with } \sum_d p(c,d) = 1 \text{ for every } c. \tag{14}$$

This is the mathematical convention: to make clear the direction of transitions, it helps to write (or imagine) an arrow,

$$p(c,d) = p(c \rightarrow d).$$

The $p(c, d)$ are elements of a Markov transition matrix $\mathbf{P}$, and $0 \leq p(c, d) \leq 1$. In a simulation we pick a value for c, start from state $E_1 = c$, and use (14) repeatedly to generate a random environmental sequence $E_1, E_2, \ldots, E_t$ of any length. We label a random sequence by an index ω.

We assume that the Markov chain is irreducible and aperiodic (Seneta, 2006) and so has an ergodic distribution $\boldsymbol{\pi} = \{\pi(1), \pi(2), \ldots\}^T$ where $\pi(a)$ is the long-run probability of observing state a and, of course, $\sum_a \pi(a) = 1$. Our assumptions on $\mathbf{P}$ guarantee that the limiting distribution $\boldsymbol{\pi}$ is approached, starting from any initial distribution.

We also can specify a time-reversed chain that will be useful when we analyze reproductive value. Given $E_{t+1} = d$, choose $E_t = c$ with probability $\hat{p}(d,c)$, say. We define the time-reversed chain in the stationary state, with $\boldsymbol{\pi}$ being the equilibrium probability for both forward and backward chains, so

$$\hat{p}(d,c) = \frac{\pi(c)\, p(c,d)}{\pi(d)}, \text{ are elements of } \hat{\mathbf{P}}. \tag{15}$$

Using these transition probabilities, we can start with an initial state at some future time $t + T$, say $E_{t+T} = d$, and generate a backward random sequence $E_{t+T}, E_{t+T-1}, \ldots, E_t$. For large T we can generate such a sequence backward using $\hat{\mathbf{P}}$, or forward using $\mathbf{P}$.

3.1.2 Environments are Continuously Distributed and Markovian

More general environmental processes include, e.g., time-series models, and models with continuous variation in variable such as temperature or pressure. For example, random variation in a vital rate F_t can be modeled using

$$F_t = \overline{F}\,(1 + s\,\mathcal{F}_t), \tag{16}$$

with a continuous bounded random variable

$$\mathcal{E}\mathcal{F}_t = 0, \mathrm{Var}\mathcal{F}_t = 1, 0 < f_- \leq \mathcal{F}_t \leq f_+ < \infty. \tag{17}$$

Under these conditions, the environmental state tends to a stationary distribution. In general, any stochastic process that is mixing and converges to an (ergodic) limit can be used (Lange and Holmes, 1981). This framework allows us the freedom to model any desired pattern of correlations between rates. Thus suppose that we have a random factor $\mathcal{F}_t$ affecting fertility rates, and another, $\mathcal{G}_t$, affecting survival rates. We may have a correlation between the two factors at each time, so that $\mathrm{Cov}\{\mathcal{F}_t\,\mathcal{G}_t\} \neq 0$. We may also have serial correlations so that one or more of the covariances

$$\mathrm{Cov}\{\mathcal{F}_t\,\mathcal{F}_{t+m}\}, \mathrm{Cov}\{\mathcal{G}_t\,\mathcal{G}_{t+m}\}, \mathrm{Cov}\{\mathcal{F}_t\,\mathcal{G}_{t+m}\}, m \neq 0,$$

are nonzero. It is common to have random rates with correlations at one time but no serial correlations.

3.2 Population Dynamics

As before, at time t a population is described by a $K \times 1$ vector $\mathbf{N}_t$ with a total population size $|\mathbf{N}_t|$, and the population structure $\mathbf{U}_t = \frac{\mathbf{N}_t}{|\mathbf{N}_t|}$.

For every environmental state c, changes in the population stages are described by the elements of a corresponding nonnegative population projection matrix $\mathbf{B}_c$. At time t say the environmental state is $E_t = c$ and the population structure is $\mathbf{U}_t$. Given the next environment E_{t+1}, we have

$$\mathbf{N}_{t+1} = \mathbf{B}_{E_{t+1}}\,\mathbf{N}_t, \tag{18}$$

$$\mathbf{U}_{t+1} = \frac{\mathbf{B}_{E_{t+1}}\,\mathbf{U}_t}{\lambda_{t+1}}, \tag{19}$$

$$\lambda_{t+1} = |B_{E_{t+1}}\,\mathbf{U}_t|. \tag{20}$$

The number λ_{t+1} is the one-period growth rate of total population, from $|\mathbf{N}_t|$ to $|\mathbf{N}_{t+1}|$,

$$\lambda_{t+1} = \frac{|\mathbf{N}_{t+1}|}{|\mathbf{N}_t|} = \frac{|\mathbf{B}_{E_{t+1}}\,\mathbf{N}_t|}{|\mathbf{N}_t|}.$$

Since values of $(E_t, \mathbf{U}_t)$ and the probability rules generate $(E_{t+1}, \mathbf{U}_{t+1})$, this pair follows a Markov process. We call $(E_t, \mathbf{U}_t), (E_{t+1}, \mathbf{U}_{t+1}), \ldots$ the **forward sequence**. In the case where environments follow a Markov chain, we have say $E_t = c$, and use the transition matrix $\mathbf{P}$ to generate E_{t+1}.

3.2.1 Demographic Weak Ergodicity

We assume demographic weak ergodicity of the matrices $\mathbf{B}_c$ for all states c; in the case of a Markov chain we have a countable set of matrices with $c = 1, \ldots, M$ (Cohen, 1976, 1977). Demographic weak ergodicity means that products of these matrices, in any order, will eventually converge to a matrix with all elements positive. This assumption is easily satisfied if the individual $\mathbf{B}_c$ are primitive, which is often true. If not, it is sometimes straightforward to test the condition using products of the signature matrices of the $\mathbf{B}_c$. Otherwise there is no general test.

3.2.2 Example: Age-Structured Population

We consider a three age-group population whose dynamics in the absence of stochasticity are described by the Leslie matrix

$$\mathbf{P} = \begin{bmatrix} 0 & 1.2 & 0.8 \\ 0.3 & 0 & 0 \\ 0 & 0.9 & 0 \end{bmatrix}. \tag{21}$$

In this matrix the first row contains the age-specific fertility rates, and the subdiagonal elements are survival rates. Fig. 1 shows that when there is no environmental variation we have strong demographic ergodicity, meaning that the population structure rapidly converges to a stable limit that is independent of the starting population structure; the three panels show the three components of the age structure.

Now let us suppose that the fertility and survival rates vary randomly over time, with random variation modeled as in Eq. (16). In the examples used here, the same random factor $\mathcal{F}_t$ is applied to each of the three fertility rates, and a different random factor $\mathcal{G}_t$ is applied to the survival rates. We assume that there is no serial correlation in either random factor, but in each time interval the random factors can have a correlation ρ. Such a correlation can easily arise when the vital rates respond to the same (or to correlated) environmental factors. The responses can be in the same or opposite directions: for example, a negative correlation arises when increasing temperature increases fertility but decreases survival.

In Figs. 2 and 3, the same distinct starting population structures are used as in Fig. 1, but with fertility and survival rates chosen randomly over time according to Eq. (17). Fig. 2 demonstrates demographic weak ergodicity when $\rho = 0$: Starting from different initial values, the age structure rapidly converges to the same time-varying sequence of values. In contrast, Fig. 3 shows

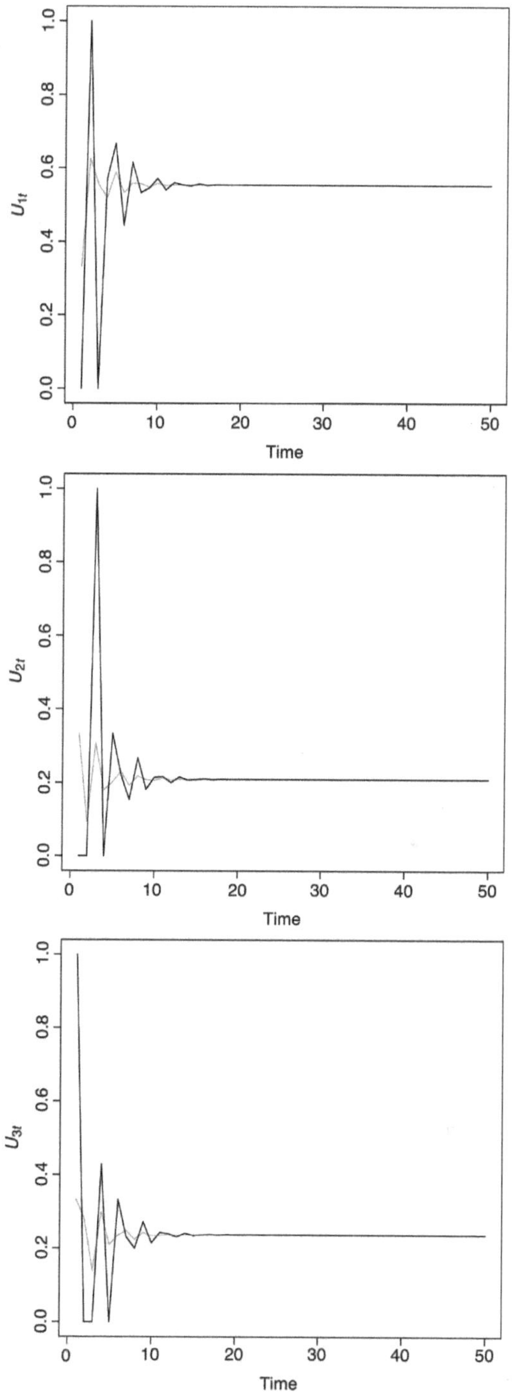

FIG. 1 Strong demographic ergodicity. With no randomness, population structure converges rapidly when started from very different initial values.

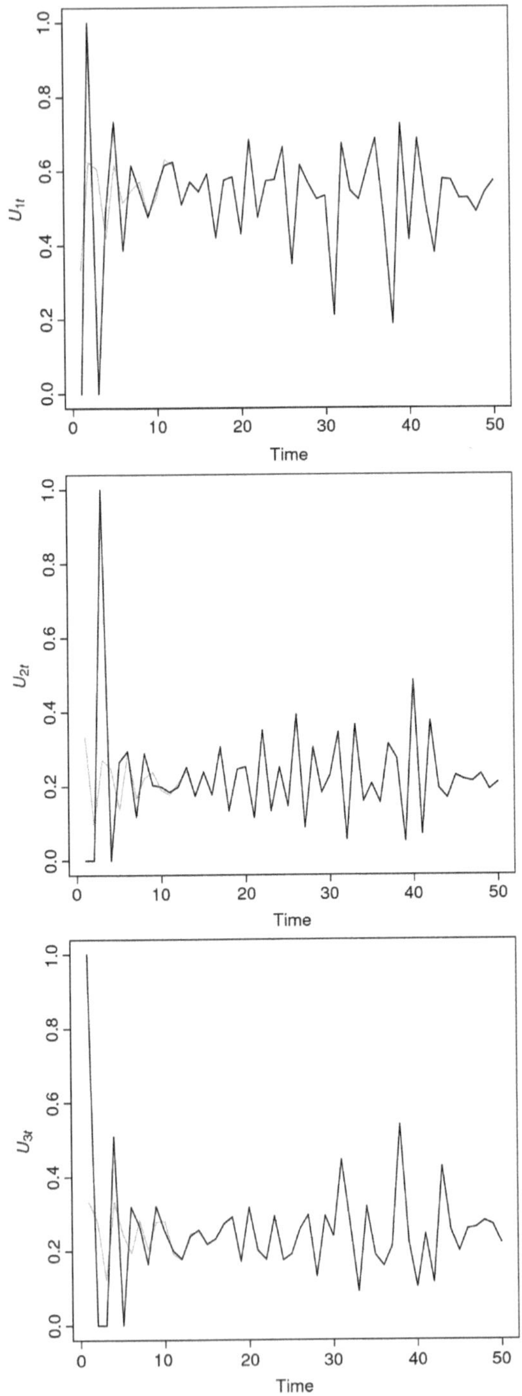

FIG. 2 Demographic convergence with random variation, different initial structures, fertility and survival uncorrelated.

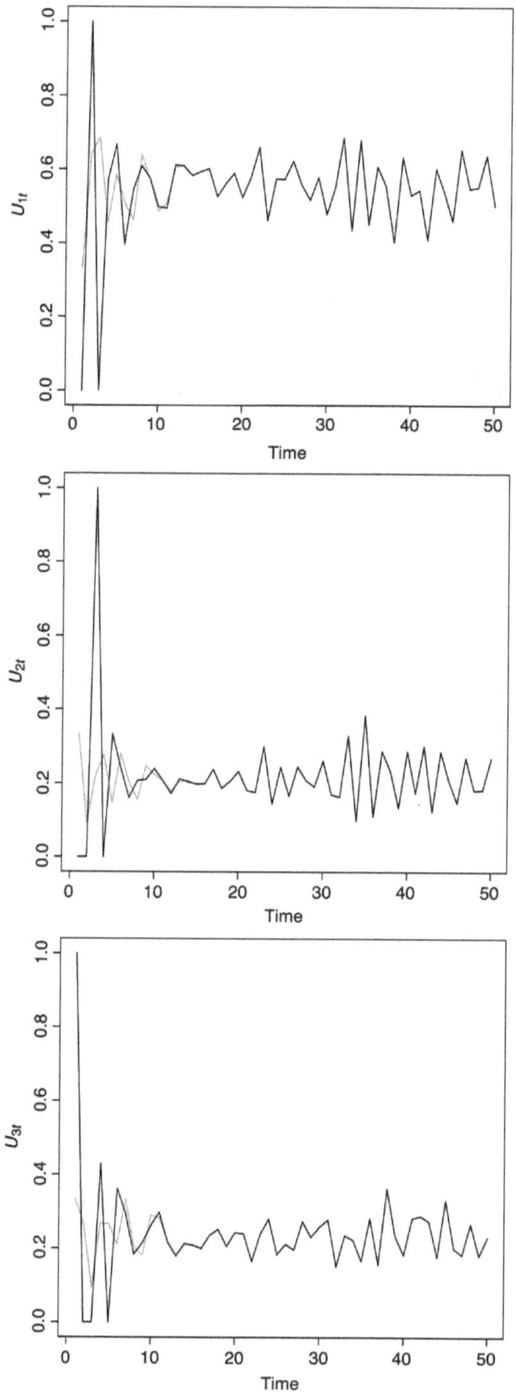

FIG. 3 Demographic convergence with random variation, different initial structures, fertility and survival negatively correlated.

demographic convergence when there is extremely negative correlation between survival and fertility, $\rho = -0.9$; again, there is rapid convergence but the limiting trajectory for the youngest age is noticeably less volatile than in Fig. 2. The difference between Figs. 2 and 3 makes biological sense, because the negative correlation acts to attenuate the variability in net reproduction.

3.3 Limiting Distributions

Together with the ergodicity assumptions on the stochastic process, weak demographic ergodicity means that the joint distribution of (E_t, $\mathbf{U}_t$) converges to a limit, call it $\mathbf{\Gamma}$. To describe this limiting distribution, start by noting that the population structure vector sums to $|\mathbf{U}_t| = 1$, so the structure vector has only $K - 1$ independent components. Thus, any population structure can be described by a vector of the form $\mathbf{x} = (x_1,\ldots,x_{K-1},x_K)$ with components in [0, 1] with $(x_1 + x_2 + \cdots + x_{K-1}) = 1 - x_K$. Hence we can just consider vectors of the form $\mathbf{x} = (x_1,\ldots,x_{K-1})$, with positive coordinates summing to no more than 1. Such vectors are points in a space called the K-simplex (with ($K - 1$) dimensions). Let $\mathcal{A}$ be a region in this simplex; e.g., pick a vector and define a rectangular region around it. Then there is a limiting probability $\mathbf{\Gamma}(E_t = c, \mathbf{U}_t \in \mathcal{A})$. We can estimate this by simulation and in rare cases analytically (Tuljapurkar, 1990a).

For a simulation, start with fixed values of (E_0, $\mathbf{U}_0$), choose some moderate integer T, and generate a large number of independent forward estimates of (E_T, $\mathbf{U}_T$). Fix c, and consider the points ($E_T = c$, $\mathbf{U}_T$)—we can translate each of these vectors into a point $\mathbf{y}_T$, say, on the simplex. The cloud of these points on the simplex describes the joint distribution $\mathbf{\Gamma}(c, \mathbf{y})$; an example is given below.

As we have discussed earlier, time reversal plays an important role, because it allows us to define and analyze reproductive values. Backward sequences of environments can be generated for many kinds of stationary stochastic processes. As we have noted above, it is straightforward to generate backward sequences in the case when environments follow a finite-state Markov chain. For this case, we use the backward transition probabilities $\widehat{\mathbf{P}}$, start with $E_t = c$, say, and generate a backward sequence $E_t = c, E_{t-1}, E_{t-2}, \ldots$. Define a vector $\mathbf{Y}_t$ with K nonnegative elements, a corresponding vector of proportions $\mathbf{V}_t = \mathbf{Y}_t/|\mathbf{Y}_t|$, and compute (superscript T indicates a transpose),

$$\mathbf{Y}_{t-1}^T = \mathbf{Y}_t^T \mathbf{B}_{E_t} = \mathbf{Y}_t^T \mathbf{B}_c, \tag{22}$$

$$\mathbf{V}_{t-1}^T = \frac{\mathbf{V}_t^T \mathbf{B}_{E_t}}{|\mathbf{V}_t^T \mathbf{B}_{E_t}|}. \tag{23}$$

Then $(E_t, \mathbf{V}_t), (E_{t-1}, \mathbf{V}_{t-1}^T), \ldots$ comprise the *backward sequence*.

In demographic terms, the $\mathbf{U}_t$ are the stochastic population structures, with time t increasing forwards. On the other hand, the $\mathbf{V}_t$ are stochastic reproductive values with time t decreasing backwards. For further discussion, see Tuljapurkar and Wiener (2000).

3.4 A Key Limit: Stochastic Growth Rate

An important result here is that the long-run population growth rate converges to a (deterministic) limit. For any random sequence ω, the cumulative growth rate from $t = 0$ to $t = T$ is the product of single-period growth rates,

$$\Lambda_T(\omega) = \lambda_T \, \lambda_{T-1} \ldots, \lambda_1. \tag{24}$$

For every random sequence (except sequences of zero probability), there is a limit (Furstenberg and Kesten, 1960), here called the stochastic growth rate

$$a = \lim_{T \to \infty} \frac{1}{T} \log \Lambda_T. \tag{25}$$

Clearly this is also an average over sequences,

$$a = \lim_{T \to \infty} \frac{1}{T} \mathcal{E}[\log \Lambda_T(\omega)]. \tag{26}$$

Even more remarkable, the convergence of cumulative growth rates to the limit (25) follows a normal distribution: there is a deterministic value $\sigma > 0$ such that

$$\frac{\log \Lambda_T - aT}{\sqrt{T}} \to \mathcal{N}(0, \sigma) \text{ in distribution.} \tag{27}$$

For more general stochastic environments, including stationary time-series models, the limits in (26) and (27) also hold.

The biological significance of a can be made clear by considering first the situation when the environment is constant. The fixed environment determines just one stage transition matrix $\mathbf{B}$ dominant eigenvalue λ is the one-period growth, and long-term growth rate is $r = \log \lambda$. With a Markovian environment, there is a stochastic growth computable as the limit a above, and an equivalent stochastic "period growth" given by $\lambda_S = \exp(a)$. Notational point: Cohen (1976) probably introduced λ_S, used and popularized by Caswell (2001); Tuljapurkar and Orzack (1980) used a, which seemed then and now to be a more obvious notation (besides being easier to write). Other symbols are also used in mathematics and physics.

The logarithm of total population size develops over time as a random walk (or Brownian motion) with constant drift a. The biological interpretation of σ derives from its role as the diffusion constant in this random walk. Consequently the extinction probability that population size will reach low levels will depend on the value of σ, as first discussed by Tuljapurkar and Orzack (1980). The use of extinction probabilities based on this analysis is discussed in several papers and books (Lande and Orzack, 1988; Lande et al., 2003; Morris and Doak, 2002).

3.4.1 Example: Stochastic Growth Rate

We continue with the example based on Eq. (21) and random variation with a correlation between survival and fertility. Intuitively, we expect that the stochastic growth rate should be higher when this correlation is negative, as illustrated by the difference in variability between Figs. 2 and 3. Simulation estimates of the stochastic growth rate are shown in the top panel of Fig. 4 and may be seen to decline with increasing correlation between survival and fertility at each time. The error bars shown are standard statistical estimates of sampling error, and demonstrate that while simulation estimates of the stochastic growth rate are imprecise, the qualitative pattern of behavior can be robustly detected. The bottom panel of Fig. 4 shows how the limiting variance σ^2, defined in Eq. (27), changes with the correlation between rates. Estimates of this variance are clearly noisy. As the lower panel of Fig. 4 suggests that there is little change as the correlation increases from -0.9 to 0, but the variance rises steadily as the correlation becomes increasingly positive.

3.4.2 Example: Markovian Environments

For a different example, suppose there are three environmental states with Markov transition matrix

$$\mathbf{P} = \begin{bmatrix} 0.4 & 0.3 & 0.3 \\ 0 & 0.5 & 0.5 \\ 0.1 & 0.2 & 0.7 \end{bmatrix}, \quad (28)$$

with corresponding population matrices

$$B_1 = \begin{bmatrix} 0.00 & 3.0 & 1.00 \\ 0.07 & 0.5 & 0.00 \\ 0.00 & 0.7 & 0.01 \end{bmatrix}, B_2 = \begin{bmatrix} 0.00 & 3.0 & 1.50 \\ 0.11 & 0.5 & 0.00 \\ 0.00 & 0.7 & 0.01 \end{bmatrix}, B_3 = \begin{bmatrix} 0.00 & 3.83 & 1.58 \\ 0.183 & 0.50 & 0.00 \\ 0.00 & 0.70 & 0.01 \end{bmatrix}. \quad (29)$$

Note that these are not Leslie matrices, but they are general matrices representing stage-structured life cycles. The matrices include stasis, in which an individual can stay in the same stage (thus the nonzero (2,2) elements in (29)), and the continued survival of older adults (thus the nonzero (3,3) elements in (29)).

We conduct 500 simulations for 50 periods of the forward Markov chain yielding a sequence of states E_t. In each we compute the corresponding forward sequence $\mathbf{U}_t$, using initial states $E_1 = 2$ and $\mathbf{U}_1 = (1, 1, \ldots)^T / K$. We do

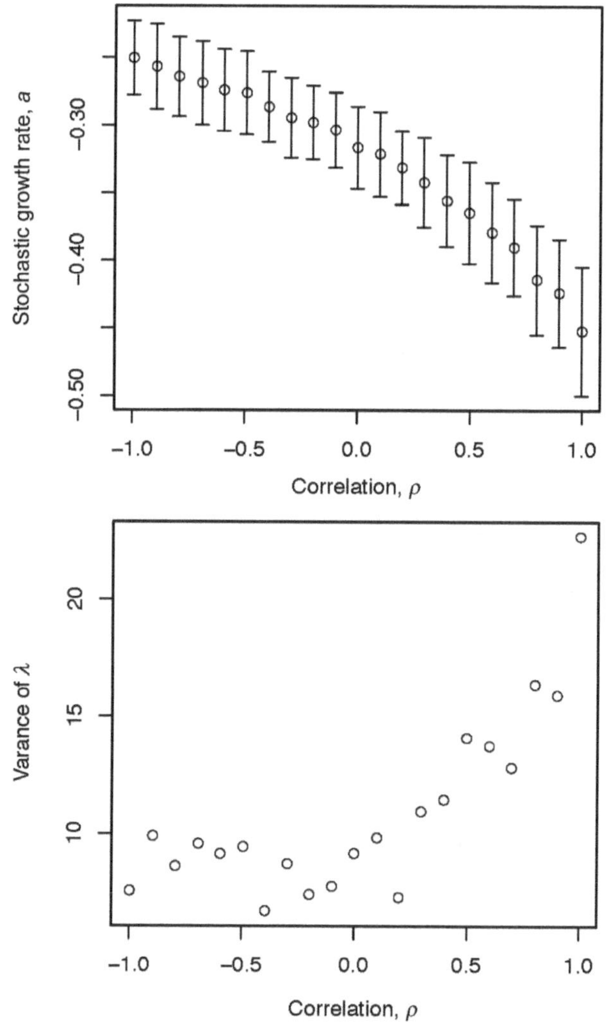

FIG. 4 Stochastic growth rate (*top*), variance (*bottom*), fertility and survival negatively correlated.

the same for the backward chain. Fig. 5 uses the components $U_t(1)$, $U_t(2)$ to plot the final values ($t = 50$) and the final environments: the points shown estimate the equilibrium distribution $\Gamma(E_t, \mathbf{U}_t)$. The estimated limit distribution shown appears fractal, a common feature in these models (Tuljapurkar, 1990a). For the same 500 forward simulations we show in Fig. 6 estimated values of stochastic growth rate a, computed using Eq. (25). The figure uses simulations that run for only 50 time periods each, yet clearly manifests the

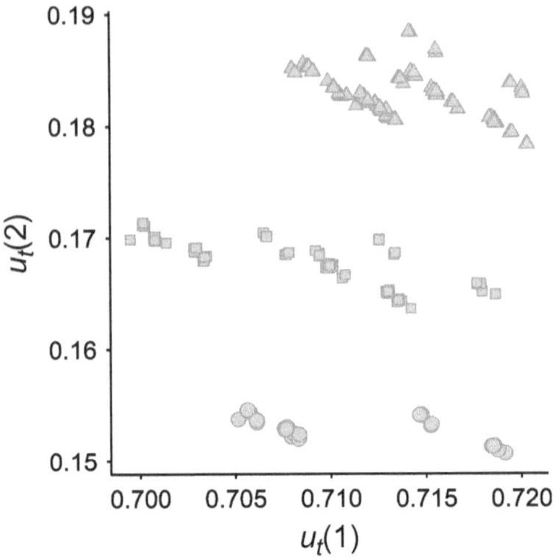

FIG. 5 The distribution $\Gamma(E_t = c, \mathbf{U}_t \in \mathcal{A})$ of environments and population structure, estimated by forward simulation. The distribution shown appears fractal, a common feature in these models.

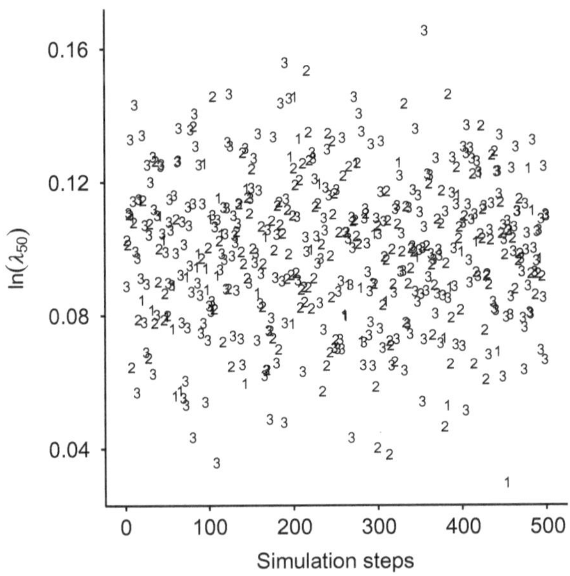

FIG. 6 Stochastic growth rates, $a = \log \lambda_s$, for each of 500 forward simulations, each of length 50.

convergence expected from Eq. (26). A histogram of empirical growth rates (Fig. 7) demonstrates the limiting normal distribution we expect from (27). Finally, Fig. 8 shows a simulated equilibrium distribution for the backward chain; this also appears fractal.

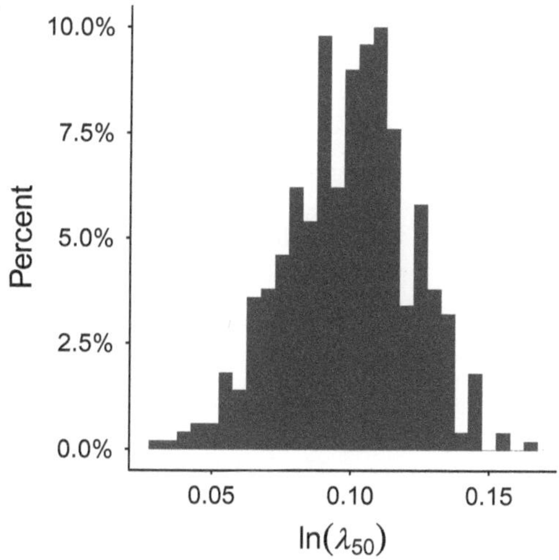

FIG. 7 Histogram of stochastic growth rates (from Fig. 6). Note the approximately normal distribution as expected from the limiting behavior.

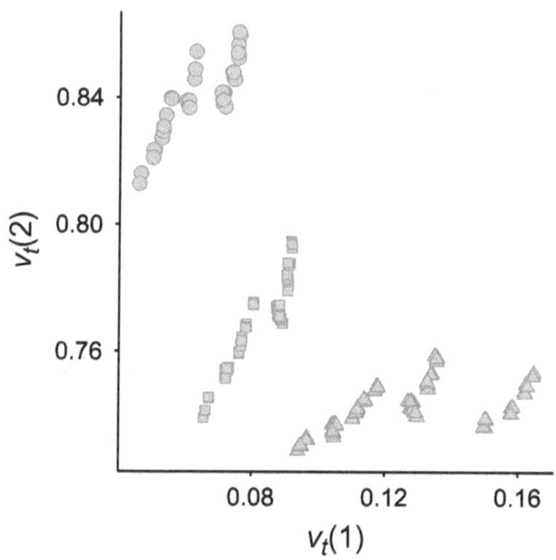

FIG. 8 The distribution $\Gamma(E_t = c, \mathbf{U}_t \in \mathcal{A})$ of environments and reproductive value, estimated by backward simulation. The distribution shown also appears fractal (compare Fig. 5).

4 STOCHASTIC GROWTH RATE: SMALL-NOISE APPROXIMATION

The stochastic growth rate is also well known as the leading Lyapunov exponent of a product of random matrices (Arnold et al., 1986; Cohen et al., 1986). Exact analytical expressions for the stochastic growth rate are difficult to obtain except in the simplest cases (Tuljapurkar, 1990b), and the bounds obtained so far, some of which are discussed in the same book, have limited applicability.

Many decades ago, Tuljapurkar (1982) obtained a general analytical approximation to the stochastic growth rate, which holds for small levels of noise (randomness). In practical experience this approximation has been found to remain reasonably accurate in many cases even for high levels of noise, and so offers considerable qualitative and quantitative insight. It has been widely applied in ecology (Caswell, 2001).

In Eq. (20), assume that the random matrices have the form

$$B_t = \mathbf{b} + \epsilon\, \mathbf{H}_t, \tag{30}$$

$$\mathcal{E}(\mathbf{H}_t) = 0, \tag{31}$$

$$\mathcal{E}(B_t) = \mathbf{A}. \tag{32}$$

(We use t directly and suppress E_t.) The stochastic process is ergodic, so the random matrix $\mathbf{H}_t$ which contains the random variation in (31) has a stationary variance-covariance matrix,

$$\mathbf{\Sigma}(0) = \mathcal{E}\{\mathbf{H}_1 \otimes \mathbf{H}_1\}, \tag{33}$$

and stationary lags

$$\mathbf{\Sigma}(t) = \mathcal{E}\{\mathbf{H}_{t+1} \otimes \mathbf{H}_1\}. \tag{34}$$

The average matrix $\mathbf{b}$ is assumed primitive and nonnegative, so we can write $\mathbf{v}_0$ and $\mathbf{u}_0$ as the left and right eigenvectors of $\mathbf{b}$ corresponding to its dominant eigenvalue λ_0. We normalize these eigenvectors so $\mathbf{v}_0^T \mathbf{u}_0 = 1$, where the superscript indicates a transpose. The small-noise expansion is

$$a \approx \log \lambda_0 - W_1 + W_2, \tag{35}$$

where W_1 is the effect due to within-period variability and W_2 is the effect due to autocorrelation. The contribution from variability within each time interval is

$$W_1 = \frac{1}{2\lambda_0^2}\ (\mathbf{v}_0 \otimes \mathbf{v}_0)'\ \Sigma_0\ (\mathbf{u}_0 \otimes \mathbf{u}_0), \tag{36}$$

The expression for W_1 can be rewritten in terms of the sensitivities (derivatives) of the dominant eigenvalue λ_0 of the average matrix. To do this, note the perturbation theory result (Caswell, 2001; Kato, 1976) that

$$\frac{\partial \lambda_0}{\partial b(i,j)} = v_0(i)\, u_0(j). \tag{37}$$

Using this in (36) gives the interpretable expression

$$W_1 = \frac{1}{2\lambda_0^2} \sum_{ij,kl} \frac{\partial \lambda_0}{\partial b(i,j)} \frac{\partial \lambda_0}{\partial b(k,l)} \operatorname{Cov}\{H_1(i,j)\, H_1(k,l)\}. \tag{38}$$

This is especially useful for Leslie matrices and age-structured populations, since the partial derivatives are well understood. There have been many applications of (38), e.g., Morris and Doak (2002) and Tuljapurkar et al. (2009).

The serial correlations in (34) are contained in the term W_2 in (35). The expression for W_2 is more involved and we refer to the original general derivation (Tuljapurkar, 1982), and the much more detailed discussion of the Markovian case (Tuljapurkar and Haridas, 2006). The latter reference presents detailed discussions of the qualitative and quantitative effects of serial correlation.

Finally we summarize here qualitative properties of a that follow from the small-noise approximation.

(a) The small-noise approximation separates the effect of within-period variability, $W_1 \geq 0$ from the effect of autocorrelation, W_2. The latter is a weighted average of environmentally driven time-lagged correlations between vital rates. In our stochastic models, all correlations between rates at times t and $t + k$ decrease to zero as $|k|$ increases.

(b) When environments are IID W_2 vanishes, and our approximation reduces to $a \approx \log \lambda_0 - W_1$, which yields the familiar inequality $a < \log \lambda_0$. This result is usually derived by applying Jensen's inequality. In autocorrelated environments, a positive W_2 increases stochastic growth rate a, and vice versa. Thus stochastic growth rate in a correlated environment can be larger or smaller than that in the IID case, e.g., Orzack (1985). Furthermore autocorrelation can result in a stochastic growth rate $a > \log \lambda_0$. Jensen's inequality does not prevent this because in autocorrelated environments the growth rate of the average population size is not λ_0.

(c) The eigenvalues of the transition matrix **P** measure the magnitude of autocorrelation. As discussed by Tuljapurkar and Haridas (2006) the environmental autocorrelation is modulated by the demographic damping. Thus the (effective) contribution to a from autocorrelation depends on the dynamics of stage-structure.

(d) Orzack (1985) used a two-state environment and a range of five-stage life histories to compare the effect of autocorrelated variability in fertility at different ages. He showed that W_2 can vary substantially, depending on which vital rate varies with the environment.

5 STOCHASTIC GROWTH RATE: DERIVATIVES

The stochastic growth rate is determined by the stochastic process that drives environments over time, and by the mapping from environmental states to vital rate matrices that determine structured population dynamics. Ruelle (1979) showed that the stochastic growth rate is an analytic function of that map. In the present context, this result allows us to define derivatives of a with respect to the elements of the matrices $\mathbf{B}_c$. These elements describe how vital rates depending on the environmental process.

Let us now assume that the environmental states follow a Markov chain, as discussed previously. Peres (1992) analyzed the properties of a in this case, and showed that the stochastic growth rate is also a real analytic function of the elements of the transition matrix $\mathbf{P}$. In the present context, we can define derivatives of a with respect to the transition probabilities between environmental states.

5.1 Derivatives With Respect to Vital Rates

In a fixed environment, sensitivity is unambiguously defined by the (partial) derivatives of the growth rate with respect to the vital rates. In a varying environment, a vital rate for transition (i, j) may have different values in each habitat state. Population growth rate is a function of the each of these values, i.e., a function of the phenotypic norm of response of the trait over environments. Therefore the "derivative" of growth rate with respect to the (i, j) trait is not a single number, but rather a collection of numbers (one for each habitat state) that make up a functional derivative. When we consider how growth rate changes in response to a perturbation of a mean, variance, or any other statistic, we are summarizing these derivatives.

We can define derivatives for stochastic growth rate, using exact formula derived in Tuljapurkar (1990b). First use Eqs. (19) and (22) to obtain stationary sequences of population structures $\mathbf{U}_t$ and reproductive value vectors $\mathbf{V}_t$. Now the growth rate $a = a(\mathbf{B}_t)$ is a function of the sequence of random projection matrices. To examine elasticity is to examine the effect of a small change in the vital rates, which amounts to a change in $\mathbf{B}_t$ for each t. Suppose that for the (i, j) life history transition, we change the vital rate from $B_t(i, j)$ to $B_t(i, j) + \delta\mathbf{C}_t(i, j)$, where $\delta > 0$. A proportional derivative of a corresponding to the perturbations $\mathbf{C}$ can now be defined as the limit

$$D(i,j) = \lim_{\delta \to 0} \left(\frac{1}{\delta}\right) \; [a(\mathbf{B} + \delta\,\mathbf{C}) - a(\mathbf{B})]. \tag{39}$$

Tuljapurkar (1990b) showed that

$$D(i,j) = \lim_{T \to \infty} \left(\frac{1}{T}\right) \sum_{t=1}^{T} \frac{V_t + 1(i)\; C_t(i,j)\; U_{t-1}(j)}{\lambda(t) \;\; (\mathbf{V}_{t+1}, \mathbf{U}_t)}. \tag{40}$$

The implementation and application of this result with a number of different biological examples is described by Tuljapurkar et al. (2003).

5.2 Derivatives With Respect to Transition Matrix Elements

As discussed above, the stochastic growth rate is an analytic function of the elements of **P**. Steinsaltz et al. (2011), henceforth referred to as STH, provided computable formulas and bounds for the stochastic growth rate and its derivatives with respect to life history transition rates and to environmental transition rates.

We note that STH shows for the first time how to bound the systematic bias and the sampling error in the simulation estimates that are used in most biological cases. Consequently, the algorithmic parts of STH are accompanied by computable bounds that make it possible to say, for example, that a lies in a given range with probability 90%. In addition, STH provides rigorous proofs for previously known but heuristically derived results for a, and derivatives of a with respect to life history transition rates. Most importantly, STH proves a general formula for derivatives of a with respect to environmental transition rates—a notable and important development.

We do not explain the mechanics of the formulas and calculations presented in STH, because they are more than a little complicated. But we note that the analytical results in STH are the basis for R code that makes the computations easily accessible. This code will soon be made available as a package; until then, please write to the authors. For the purposes of this article it will suffice to explain that the results in STH provide algorithms to efficiently estimate derivatives of the form $(\partial a/\partial \mathbf{P})$. It is important to remember that the elements of the transition matrix are constrained as in Eq. (14), which means that we are only able to compute relative magnitudes for these derivatives.

Biological applications of these results are just beginning. In an era in which of changing climate, the environmental transition matrices relevant to biological systems will also be changing. The tools provided by STH are thus increasingly relevant to problems that we will be facing.

6 DISCUSSION

The methods described here are useful in studying structured populations that are subject to environmental variability. Uncertainty in environmental progression is of course ubiquitous in nature, and many of the early applications of these methods were aimed at the ecological and evolutionary effects of such uncertainty.

As climate changes, empirical data and models agree that there will be significant changes in environmental variability affecting ecological systems at all spatial scales. Some aspects of such changes can be described using the methods of this paper, at least on the level of comparative statics. That is to say, we can compare the dynamics under current environmental regimes with future dynamics under projected environmental regimes. We can also use the sensitivity analyses discussed here to study the effects of small departures from current environmental patterns. Both types of analysis can yield valuable information about the nature of ecological responses to changing environments.

These tools and techniques have found wide application has in the study of conservation biology, particularly population viability analysis and analysis of extinction probabilities. Extinction can be thought of as the result of a "large" deviation, in which a population is reduced by an unfavorable string of environments. The methods described here are fundamental to the analysis of such deviations, and to computing the probability of such extreme atypical contractions. Also, as mentioned before, the persistence of nonlinear ecological dynamics is determined in large part by "boundary" behavior. Of course persistence of populations ultimately depends on nonlinear ecological dynamics, not on linear dynamics. But the crucial boundary behavior—whether the system tends to be repelled or absorbed (extinction) when population numbers get very small—will be analyzed, in the first instance, through linearized stochastic dynamics, using the methods described here.

These random matrix models appear naturally in many other fields. Some examples are the physics of disordered media, wave propagation in random media, portfolio theory, and percolation theory (Arnold et al., 1986; Bougerol et al., 2012; Ruelle, 1979). Whereas the general theory leading to the existence of a Lyapunov exponent and the stochastic growth rate was established early on, the qualitative and quantitative properties of the limit have taken many years to understand, and remain a subject of both theoretical and practical interest.

REFERENCES

Arnold, L., Kliemann, W., Oeljeklaus, E., 1986. Lyapunov exponents of linear stochastic systems. In: Arnold, L., Wihstutz, V. (Eds.), Lyapunov Exponents. Springer, pp. 85–125.

Bellman, R., et al., 1954. Limit theorems for non-commutative operations. I. Duke Math. J. 21 (3), 491–500.

Benaı-m, M., Schreiber, S.J., 2009. Persistence of structured populations in random environments. Theor. Popul. Biol. 76 (1), 19–34.

Bougerol, P., et al., 2012. Products of Random Matrices With Applications to Schrödinger Operators. vol. 8. Springer Science & Business Media.

Caswell, H., 2001. Matrix Population Models: Construction, Analysis and Interpretation, second ed. Sinauer Associates, Sunderland, MA.

Cohen, J.E., 1976. Ergodicity of age structure in populations with Markovian vital rates, I: countable states. J. Am. Stat. Assoc. 71 (354), 335–339.

Cohen, J.E., 1977. Ergodicity of age structure in populations with Markovian vital rates, III: finite-state moments and growth rate; an illustration. Adv. Appl. Probab. 462–475.

Cohen, J.E., Kesten, H., Newman, C.M., 1986. Random Matrices and Their Applications: Proceedings of the AMS-IMS-SIAM Joint Summer Research Conference Held June 17–23, 1984, With Support From the National Science Foundation. vol. 50. American Mathematical Society.

Coulson, T., Tuljapurkar, S., Childs, D.Z., 2010. Using evolutionary demography to link life history theory, quantitative genetics and population ecology. J. Anim. Ecol. 79 (6), 1226–1240.

Ellner, S.P., Rees, M., 2006. Integral projection models for species with complex demography. Am. Nat. 167, 410–428.

Ellner, S.P., Rees, M., 2007. Stochastic stable population growth in integral projection models: theory and application. J. Math. Biol. 54 (2), 227–256.

Furstenberg, H., Kesten, H., 1960. Products of random matrices. Ann. Math. Stat. 31 (2), 457–469.

Kato, T., 1976. Perturbation Theory for Linear Operators. vol. 132. Springer Science & Business Media.

Lande, R., Orzack, S.H., 1988. Extinction dynamics of age-structured populations in a fluctuating environment. Proc. Natl. Acad. Sci. 85 (19), 7418–7421.

Lande, R., Engen, S., Saether, B.E., 2003. Stochastic populated dynamics in ecology and conservation. Oxford University Press.

Lange, K., Holmes, W., 1981. Stochastic stable population growth. J. Appl. Probab. 18 (2), 325–334.

Lefkovitch, L.P., 1965. The study of population growth in organisms grouped by stages. Biometrics 21, 1–18.

Leslie, P.H., 1966. The intrinsic rate of increase and the overlap of successive generations in a population of guillemots (Uria aalge Pont). J. Anim. Ecol. 35 (2), 291–301.

Lewontin, R.C., Cohen, D., 1974. On population growth in a randomly varying environment. Proc. Natl. Acad. Sci. USA 62, 1056–1060.

Morris, W.F., Doak, D.F., 2002. Quantitative Conservation Biology Sinauer. Sinauer Associates, Sunderland, MA.

Orzack, S.H., 1985. Population dynamics in variable environments. V. The genetics of homeostasis revisited. Am. Nat. 550–572.

Peres, Y., 1992. Analytic dependence of Lyapunov exponents on transition probabilities. In: Arnold, L. (Ed.), Lecture Notes in Math, vol. 1486, pp. 64–80.

Ruelle, D., 1979. Analycity properties of the characteristic exponents of random matrix products. Adv. Math. 32, 68–80.

Seneta, E., 2006. Non-Negative Matrices and Markov Chains. Springer, New York.

Steiner, U.K., Tuljapurkar, S., Coulson, T., 2014. Generation time, net reproductive rate, and growth in stage-age-structured populations. Am. Nat. 183 (6), 771–783.

Steinsaltz, D., Tuljapurkar, S., Horvitz, C., 2011. Derivatives of the stochastic growth rate. Theor. Popul. Biol. 80 (1), 1–15.

Tuljapurkar, S.D., 1982. Population dynamics in variable environments. III. Evolutionary dynamics of r-selection. Theor. Popul. Biol. 21 (1), 141–165.

Tuljapurkar, S., 1990. Population Dynamics in Variable Environments. Springer-Verlag, New York.

Tuljapurkar, S.D., 1990. Population Dynamics in Variable Environments. Lecture Notes in Biomathematics, vol. 85. Springer-Verlag, New York.

Tuljapurkar, S., Haridas, C.V., 2006. Temporal autocorrelation and stochastic population growth. Ecol. Lett. 9 (3), 327–337.

Tuljapurkar, S.D., Orzack, S.H., 1980. Population dynamics in variable environments. I. Long-run growth rates and extinction. Theor. Popul. Biol. 18, 314–342.

Tuljapurkar, S., Wiener, P., 2000. Escape in time: stay young or age gracefully? Ecol. Model. 133 (1–2), 143–159.

Tuljapurkar, S., Horvitz, C., Pascarella, J., 2003. The many growth rates and elasticities of populations in random environments. Am. Nat. 162 (4), 489–502.

Tuljapurkar, S., Gaillard, J.M., Coulson, T., 2009. From stochastic environments to life histories and back. Philos. Trans. R. Soc. Lond. Ser. B, Biol. Sci. 364 (1523), 1499–1509.

Chapter 5

Studying Complexity and Risk Through Stochastic Population Dynamics: Persistence, Resonance, and Extinction in Ecosystems

Anuj Mubayi[1,*], Christopher Kribs[†], Viswanathan Arunachalam[‡] and Carlos Castillo-Chavez*

[*]*Simon A. Levin Mathematical Computational Modeling Science Center, School of Human Evolution and Social Change, Arizona State University, Tempe, AZ, United States*

[†]*Department of Mathematics, The University of Texas at Arlington, Arlington, TX, United States*

[‡]*Department of Statistics, Universidad Nacional de Colombia, Bogotá, Colombia*

[1]*Corresponding author: e-mail: amubayi@asu.edu*

ABSTRACT

Stochasticity provides a way to evaluate risk in complex population dynamics. This chapter describes how basic stochastic processes and algorithms are used to build models of ecological populations which can then derive probability distributions for population persistence and extinction, explain sustained oscillations generated by stochastic resonance, and measure emergent properties of a biological system, all as functions of demographic or environmental stochastic effects.

Keywords: Stochastic processes, Dynamical systems, Stochastic algorithms, Agent-based models

1 INTRODUCTION

The dynamics of real systems are often influenced by internal and external factors which are not completely understood and, therefore, cannot be described precisely. In order to understand such systems, deterministic models (which explain broad overall behavior and growth patterns) are modified to incorporate complex variations in the mechanisms underlying the system. These variations, defined in terms of probabilities which evolve over time

Handbook of Statistics, Vol. 40. https://doi.org/10.1016/bs.host.2018.11.001

alongside the populations, result in new structures referred to as *stochastic* models. Deterministic models capture the mean behavior of a system and serve as a framework for formulating the corresponding stochastic models. Stochastic models are designed to capture the ways that a system's behavior may deviate from the mean. Stochastic effects (variations) may arise either from differences among individuals, termed *demographic stochasticity*, or from fluctuations in the environment (due to forces such as weather or differences in landscape which are beyond the model's ability to describe), termed *environmental stochasticity*. Stochastic models may be preferable when studying a small community (where the law of large numbers, which pegs a population size to its mean behavior, does not apply), or when addressing questions related to *variations* in population (regardless of community size) caused by demographic or environmental stochasticity.

A stochastic model has one or more random elements or components, and, unlike deterministic models, a unique input leads to different outputs for each run, due to the random component of the modeled process—each sample path represents a different random draw (or set of draws) for the stochastic elements influencing the population growth. The behavior of stochastic dynamical systems can be described using several different types of stochastic processes, such as Poisson, renewal, discrete-time Markov chains, continuous-time Markov chains, Brownian motion, and diffusion. The difference between types of stochastic processes is loosely analogous to the difference among probability distributions for simpler random variables. In simulating a stochastic model, a random number is normally generated by some method at every time step. Such a simulation is called a Monte Carlo method or Monte Carlo simulation. Multiple runs are used to estimate probability distributions. The same set of parameter values and initial conditions will lead to an ensemble of different outputs, and a large enough set of outputs produces a probability distribution for any measurable model output.

In population biology, a primary question of interest has been to know the mechanisms for population persistence—in other words, the factors that may lead to its extinction. When stochasticity of any type is added to a population dynamic model, the moment-to-moment fluctuations in the population size increase in general, and hence the likelihood of extinction may increase too. The factors related to risk for extinction of a population include interactions with other populations, its fitness in the environment and numerous stochastic processes that may influence its declines in population size (Davidson et al., 2009; Fieberg and Ellner, 2000; Fritz et al., 2009). There have been various modeling approaches to quantify a population's extinction risk. Stochastic models are natural for studying questions related to extinction. However, computing extinction risk or time to extinction mathematically is difficult and, hence, most studies focus on providing approximations for corresponding distributions.

In a typical stochastic model, the extinction state is an *absorbing state*: that is, the population once become zero cannot recover. The long-time behavior of a stochastic model with an absorbing state, conditioned on nonextinction, can be approximated using tools from probability theory and is referred to as a *quasi-stationary* distribution (QSD) of the stochastic process. The extinction time (time to reach the absorbing state) can be large and the population size can fluctuate for a large amount of time before extinction actually occurs. The existence and uniqueness properties of QSDs for various processes have been studied in detail. The approximation of QSDs for subcritical Galton-Watson processes was first derived by Yaglom (Méléard et al., 2012; Yaglom, 1947). Darroch and Seneta provided the initial approximation of QSDs for discrete and continuous Markov processes on finite state spaces (Darroch and Seneta, 1965, 1967).

Another major focus of study involving stochastic models involves how fluctuations in a population's environment can sustain oscillations which would normally die out. The solutions of some nonlinear deterministic models for interacting populations show damped oscillations toward equilibrium values. However, simulations produced with corresponding stochastic models usually present sustained oscillations which preserve the natural frequency of the damped oscillations of the deterministic model but show nonvanishing amplitudes. Environmental variations can significantly impact population dynamics, that is, can alter the spread and persistence of population. Understanding how the timing and causes of seasonality (periodic forcing) impact population fluctuations, which may range from simple annual cycles to more complex multiyear, is critical to understanding its dynamics. Noise induced oscillations in the system can be of two types: stochastic resonance and coherence resonance. Stochastic resonance occurs in response to an external periodic forcing when stochasticity is present in the system whereas coherence resonance is observed due to the noise and intrinsic nonlinear dynamics of the system. The power spectral density can describe and explain the impacts of stochasticity. Predicted spectral peaks represent periodic phenomena induced or modified by stochasticity in the model.

Finally, some stochastic models incorporate demographic (individual-to-individual) stochasticity by maintaining a separate description for each individual in the population. Rather than simply counting the number of individuals which have certain characteristics (*compartmental* models), some models keep detailed registers of each individual's location and other characteristics, allowing simulation and tabulation of specific interactions and personal histories (*agent-based* models). These latter models require significantly more information to parametrize and complicate population counting beyond one's ability to describe their behavior analytically, but in exchange they allow one to build up detailed portraits of individual behavior which, when simulated in populations of many individuals, may reveal resulting behaviors at the system level which would not be observable at the individual level. Such properties are called *emergent* properties, and are a major reason why agent-based models are used in studies.

TABLE 1 Potential Differences Between Behaviors of Deterministic and Stochastic Models

Deterministic Model	Stochastic Model
Variables are functions of time only	Variables depend on time and probability
All mechanisms are described precisely	Process could have two sources of variability (demographic and environmental)
Captures only mean characteristics of process	Captures variations from mean behavior
Deterministic path provides expected value	Stochasticity leads to variances and covariances
Trajectory is fixed between simulations	Variability between simulations
For given parameter set, one simulation is sufficient	Needs many simulations
Behavior entirely governed by parameters	Allows "chance" to play a role
If we knew perfectly the present state, we could predict future states accurately	If we knew perfectly the present state, model assigns only a probability distribution to future states
Perfect reproducibility	Each realization is different
Deterministic dynamics, in general, have equilibrium behavior	Stochasticity can excite system to sustained oscillations (resonance) or can drive system to extinction
Mathematically easier	Often harder to analyze mathematically

Some of the properties are shown in Fig. 1.

This chapter proposes to give enough basic definitions in the area of stochastic processes and population dynamics to explain how stochastic models have been used to gain insights into the areas described above. Table 1 summarizes some of the main differences between deterministic and stochastic models, while Fig. 1 illustrates some of the elements and properties of stochastic models.

1.1 Definitions

Stochasticity is captured in a model using *noise* (a random process). There are different types of noise: observational/measurement and process noise. Process noise can be constant, scaled as a random variable, or described using an external parameter. Understanding how noise defines and impacts stochasticity for a system requires understanding certain definitions.

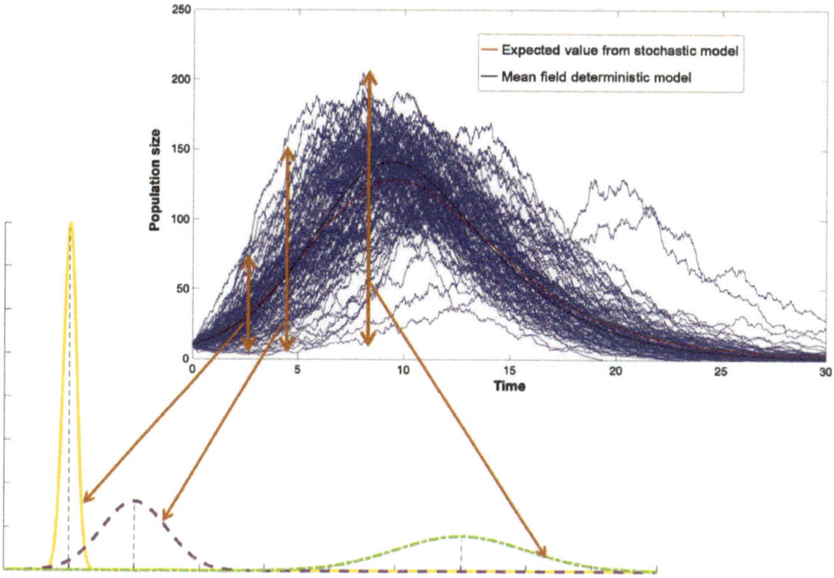

FIG. 1 Stochastic process: Collection of random variables, sample paths, mean pattern, and its variance.

- *Colored noise*: Refers to temporal autocorrelation structure of the noise associated with environmental stochasticity. White noise is uncorrelated in time, so that future environmental conditions do not depend on earlier environmental conditions. Different colors refer to situations where the environmental conditions vary slower (red or pink noises, positive correlation) or faster (blue noise, negative correlation) than for white noise.
- *Demographic heterogeneity*: Variation in the intrinsic birth and death rates among individuals.
- *Demographic stochasticity*: Random variation in the number of births and deaths in a population caused by the discrete nature of individuals and stochastic character of these processes. Present even if all individuals have identical birth and death rates.
- *Environmental stochasticity*: Irregular variation in environmental conditions that affect the birth and death rates of a population.
- *Quasi-stationary probability distribution*: The limit of the probability distribution at large time, conditional on the population being not yet extinct. In models that eventually lead to extinction, the stationary probability distribution describes the state where the population is extinct, whereas the quasi-stationary distribution describes the shape of the distribution of the population sizes long after an initial transient but before extinction.

1.2 Examples of Basic Stochastic Processes

Often analysis of complex stochastic processes involves systematic simplification of stochastic model to any basic stochastic process. We provide some basic stochastic processes that are referred in the analysis of complex stochastic models introduced in this chapter. A flow chart of stochastic process that includes these basic stochastic processes is shown in Fig. 2.

- A *random walk* is an example of a stochastic process in discrete time, where a particle starts at the origin at time 0 and moves one distance left with probability p or one distance unit right with probability $1 - p$ at each time unit. The random variable X_n then denotes the position of the particle at time step n.
- The *Wiener process (or Brownian Motion)* is an example of a stochastic process in continuous time. In 1828 the Scottish botanist Robert Brown observed that pollen grains suspended in water moved in an apparently random way, changing direction continuously, which was due to the pollen grains being bombarded by water molecules. The precise mathematical formulation of this phenomenon was later given by Norbert Wiener in 1923. A Wiener process is defined as a stochastic process $W(t)_{t \geq 0}$ which satisfies the following characteristics:

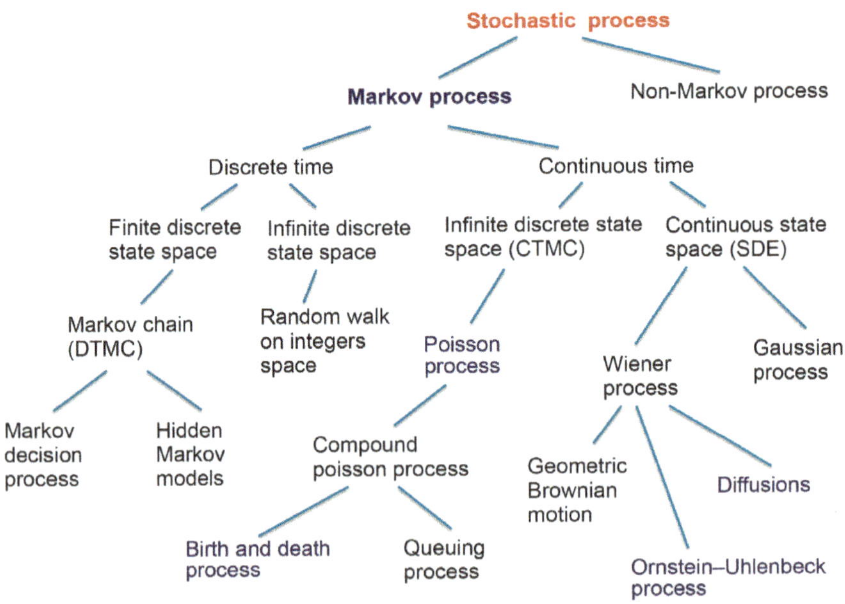

FIG. 2 A flowchart showing different types of stochastic processes. Processes lower in the chart are special cases of the processes hierarchical above them. DTMC, CTMC, and SDE stands for discrete-time Markov chain, continuous-time Markov chain, and stochastic differential equations, respectively. Example of processes that are provided in this chapter are shown in *blue color*.

(i) $W(0) = 0$

(ii) $W(t)_{t \geq 0}$ has independent increments (i.e., $W_{t_1}, W_{t_2} - W_{t_1}, \ldots, W_{t_2} - W_{t_1}$ are independent random variables for all $0 \leq t_1 < t_2 \ldots < t_k$)

(iii) $W_{s+t} - W_s \sim N(0, t)$ for all $t > 0$

Remark 1. This stochastic process is a building block of many of the models in the literature. It can be obtained as the limit of a random walk when the time steps and the jump sizes go to 0 in a suitable way.

Remark 2. The Wiener process is often called standard Brownian motion process or Brownian motion due to its connection with the physical process known as Brownian movement or Brownian motion. It is one of the best known Lévy processes (stochastic processes with stationary independent increments).

A Gaussian process is a simple stochastic process which satisfies the following property: For any finite set of indices $t_1, \ldots, t_k$ the vector of random variables $(X(t_1), \ldots X(t_k))$ follows a k-dimensional normal distribution.

Remark 3. Any continuous-time stochastic process with independent increments and finite second moments $(E(X^2(t)) < \infty$ for all t) is a Gaussian process provided that $X(t_0)$ is Gaussian for some t_0.

Remark 4. The Wiener process is a Gaussian process. The Wiener process is continuous with mean zero $(E(W(t)) = 0)$ and variance proportional to the elapsed time $(Var(W(t)) = t)$.

A *stationary stochastic process* is a specific type of process in which $X(t)$ has the same distribution as $X(t + h)$ for all $h > 0$.

Remark 5. The Wiener process cannot be stationary since the variance increases with t. The auto-covariance function is given by $Cov(W_t, W_s) = min(s, t)$.

Remark 6. The sample paths of a Wiener process are nowhere differentiable as their total variation on a closed interval of the real line is infinite.

- A *stochastic differential equation* can be defined as a stochastic process, $X_t = X(t)$, satisfying the following equation:

$$dX_t = f(X_t, t)dt + g(X_t, t)dW_t \tag{1}$$

which can be also written as

$$X_t = X_{t_0} + \int_{t_0}^{t} f(X_s, s)ds + \int_{t_0}^{t} g(X_s, s)dW_s, \tag{2}$$

where $W_t = W(t)$ is the Wiener process, X_{t_0} is a random variable (independent of the Wiener process), $f(\cdot)$, the drift part or the deterministic component, and $g(\cdot)$ is the diffusion part or the stochastic component.

Remark 7. If $f(\cdot)$ and $g(\cdot)$ do not depend on t, the process is called time-homogeneous.

Remark 8. Suppose $g(X_t, t) = c$ (i.e., constant function) then the second integral becomes $c(W(t) - W(t_0))$, which is a random variable with expectation 0 since the increments of a Wiener process have expectation 0.

- The limiting diffusion processes can be obtained as solutions of stochastic differential equations of the form

$$dX(t) = \mu(X(t))dt + \sigma(X(t))dW(t),$$

where W denotes a standard Wiener process. $X(t)$ then has the well-known properties

$$\lim_{h\to 0} \frac{E_x\{X(t+h) - X(t)|X(t) = x\}}{h} = \mu(x), \tag{3}$$

$$\lim_{h\to 0} \frac{E_x\{[X(t+h) - X(t)]^2|X(t) = x\}}{h} = \sigma(x), \tag{4}$$

where E_x denotes expectation given $X(t) = x$. These properties show that the mean increment of the stochastic process $X(t)$ during a small time interval (h), given that the process has reached x, is the same as the increment of the solution of the deterministic ordinary differential equation

$$\frac{dX(t)}{dt} = \mu(X), \tag{5}$$

and the variability of the process is completely described by a function σ^2.

- A *Wiener process with drift* is a stochastic process, X_t, which satisfies the following equation:

$$dX_t = \mu dt + \sigma dW_t,$$

and its solution is

$$X_t = \mu t + \sigma W_t.$$

Hence, this X_t process is normally distributed with mean μt and variance $\sigma^2 t$.

- A *Geometric Brownian motion* (stochastic version of Malthusian model or Black-scholes model) is a stochastic process which satisfies

$$dX_t = \mu X_t dt + \sigma X_t dW_t,$$

and its solution is

$$X_t = X_0 \exp\left(\left(\mu - \frac{1}{2}\sigma^2\right)t + \sigma W_t\right).$$

Hence, the process only takes positive values, and follows a log-normal distribution with parameters $\left(\mu - \frac{1}{2}\sigma^2\right)$ and $\sigma^2 t$.

Remark 9. The parameter μ represents the mean growing rate per capita and σ the intensity of the random effects on the growing rate.

- An *Ornstein–Uhlenbeck process* is a process that is attracted to some constant level but is also continuously perturbed by noise. It satisfies the following equation:

$$dX_t = -\left(\frac{X_t - \alpha}{\tau}\right)dt + \sigma dW_t$$

where τ has units of time, and the autocorrelation of the process is *corr* $(X_t, X_{t+s}) = e^{-s/\tau}$ (autocorrelation decreases with a factor of $1/e$ after τ units of time). Its solution is

$$X_t = X_0 e^{-t/\tau} + \alpha(1 - e^{-t/\tau}) + e^{-t/\tau}\int_0^t e^{s/\tau}\sigma dW_s.$$

Remark 10. The Ornstein–Uhlenbeck process is a stochastic process that describes the velocity of a Brownian particle under the influence of friction. The process is a stationary Gauss–Markov process, that is, it is a Gaussian process, a Markov process, and is homogeneous, which tends to drift toward its long-term mean (i.e., mean-reverting).

Remark 11. Note, the amplitude of the noise does not change over time for the Wiener and the Ornstein–Uhlenbeck process, whereas for Geometric Brownian motion the amplitude of the noise is proportional to the state variable. When the diffusion term does not depend on the state variable X_t as in the Wiener process with drift and the Ornstein–Uhlenbeck process, we say that it has *additive noise* and in the case of Geometric Brownian motion, we say that it has *multiplicative noise.*

Remark 12. Since a Wiener process is nondifferentiable, *Itô's formula* is often used to find the explicit solution of simple stochastic differential equations with a Wiener process. When no explicit solution is available, we can approximate different characteristics of the process by simulation, such as sample paths, moments, qualitative behavior, etc. Usually such simulation methods are based on discrete approximations of the continuous solution to a stochastic differential equation. Some of the principal discrete schemes are Euler–Maruyama and Milstein.

In this chapter, we provide examples of various types of analysis that can be carried out for understanding dynamics of some stochastic models (Fig. 3).

2 PERSISTENCE IN STOCHASTIC MODELS

There are various techniques through which general solution of a stochastic model can be approximated. Here, we provide some examples in which either explicit general (Section 2.1) or special approximate (quasi-stationary distribution) solution (Section 2.2) is computed.

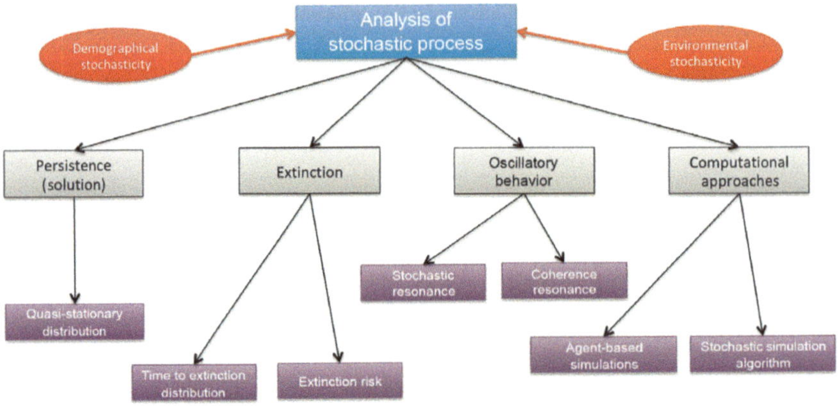

FIG. 3 Different types of analysis of a stochastic process collected in this chapter are shown here in schematic flow diagram.

2.1 Stochastic Logistic Growth Model for a Single Population Species

The *stochastic 1-D logistic equation* (also called the Verhulst equation) models the rate of growth of a single species of population whose rate of growth decreases as the population starts to compete (among themselves) for resources. The equation is given by

$$dX_t = X_t(\Lambda - X_t)dt + \sigma X_t dW_t \tag{6}$$

with $X(0) = X_0$. Here the first, deterministic term, referred to as *drift*, matches the deterministic logistic equation, while the second term, called *diffusion*, represents the stochastic variability in the population growth.

The explicit solution of this equation (Nåsell, 2001) is

$$X_t = \frac{x_0 \exp(\Lambda - 0.5\sigma^2)t + \sigma W_t}{1 + X_0 \int_0^t \exp((\Lambda - 0.5\sigma^2)s + \sigma W_s)ds}. \tag{7}$$

The *stochastic 1-D generalized logistic equation* model with a multiplicative noise term is given by the stochastic differential equation:

$$dx_t = bx_t\left(1 - \left(\frac{x_t}{F}\right)^m\right)dt + cx_t dw_t \tag{8}$$

with $x(0) = x_0$. In order to find its solution we use a so-called *reduction method* (Kloeden and Platen, 1991; Kloeden et al., 2012). Dividing both sides by F and substituting $y_t = x_t/F$ we get

$$dy_t = by_t(1 - (y_t)^m)dt + cy_t dw_t \tag{9}$$

Now let $z_t = y_t^{-m}$, and then applying Itô's formula to the transformation function z_t, we obtain

$$dz_t = \left(\left(\frac{c^2}{2} m(m+1) - bm \right) z_t + bm \right) dt - cmz_t dw_t \tag{10}$$

This is a linear autonomous stochastic differential equation and its solution is given by

$$z_t = \Phi_t \left\{ z_0 + bm \int_0^t \Phi_s^{-1} ds \right\} \tag{11}$$

where $\Phi_t = \exp\{(-bm + c^2 m/2)t - cmw_t)\}$. Since $y_t = z_t^{-1/m}$ and $x_t = Fy_t$, we can obtain the solution as

$$x_t = F\Phi_t^{-1/m} \left\{ (x_0/F)^{-m} + bm \int_0^t \Phi_s^{-1} ds \right\}^{-1/m} \tag{12}$$

2.2 Quasi-Stationarity (QSD) of a Birth–Death Process

In cases where model equations cannot be solved outright, we may nevertheless compute the QSD of a process to describe the corresponding population's persistence. To illustrate this procedure, let us consider a birth–death process $\{X(t),\ t \geq 0\}$ with finite state space $\{0, 1, \ldots, N\}$. Transitions in the process are only allowed to neighboring states. The rate of transition from state n to state $n + 1$ (the population birth rate) is denoted λ_n, and the rate of transition from the state n to the state $n - 1$ (the population death rate) is denoted by μ_n. It is convenient to also have notation for the sum of λ_n and μ_n: we put $\kappa_n = \lambda_n + \mu_n$. We assume that μ_0 and λ_N are equal to zero, to be consistent with the assumption that the state space is limited to $\{0, 1, \ldots, N\}$. Furthermore, we assume that λ_0 also equals zero, to be consistent with the assumption that the origin is an absorbing state. All other transition rates are assumed to be strictly positive.

The Kolmogorov forward equations for the state probabilities $p_n(t) = P\{X(t) = n\}$ can be written

$$p_n'(t) = \mu_{n+1} p_{n+1}(t) - \kappa_n p_n(t) + \lambda_{n-1} p_{n-1}(t) \quad n = 0, 1, \ldots, N. \tag{13}$$

Note, we put $\mu_{N+1} = \lambda_{-1} = p_{N+1}(t) = p_{-1}(t) = 0$. The state probabilities depend on the initial distribution $p_n(0)$. An alternative way of writing this system of equations is in the form

$$\mathbf{p}' = \mathbf{p}\mathbf{Q} \tag{14}$$

where $\mathbf{p}(t) = (p_0(t), p_1(t), \ldots, p_N(t))$ is the row vector of state probabilities and the matrix $\mathbf{Q}$ contains the transition rates as follows: The nondiagonal element

q_{mn} of this matrix equals the rate of transition from state m to state n, and the diagonal element q_{nn} equals the rate κ_n multiplied by -1. The matrix **Q** can be written as follows:

$$\begin{pmatrix} -\kappa_0 & \lambda_0 & 0 & \dots & 0 \\ \mu_1 & -\kappa_1 & \lambda_1 & \dots & 0 \\ 0 & \mu_2 & -\kappa_2 & \dots & 0 \\ \dots & \dots & \dots & \dots & \dots \\ 0 & 0 & 0 & \dots & -\kappa_N \end{pmatrix}$$

Note that **Q** is a traditional matrix with all row sums equal to 0. Thus the determinant of **Q** equals zero, so **Q** is noninvertible. Note also that $\lambda_0 = \kappa_0 = 0$, so the first row of **Q** is a row of zeros. A stationary distribution is found by setting the time derivative equal to zero, i.e., by solving $\mathbf{pQ} = 0$. It is readily shown that the process $\{X(t)\}$ has a degenerate stationary distribution $\mathbf{p} = (1, 0, \dots, 0)$. The distribution of $\{X(t)\}$ approaches the stationary distribution as time t approaches infinity. This says that ultimate absorption is certain. In many cases, the time to absorption is long. It is therefore of interest to study the distribution of $\{X(t)\}$ before absorption has taken place.

Let the quasi-stationary distribution of the process $\{X(t)\}$ be represented by **q**. Suppose the state space is divided into two subsets, one containing the absorbing state 0, and the other equal to the set of transient states $\{1, 2, \dots, N\}$. Corresponding to this partition, we write Eq. (14) in block form. The vector $\mathbf{p}(t)$ is expressed as $\mathbf{p}(t) = (p_0(t), \mathbf{p_T}(t))$, where $\mathbf{p_T}(t) = (p_1(t), \dots, p_N(t))$ is the row vector of state probabilities in the set of transient states. The corresponding block form of the matrix **Q** contains four blocks. The first row of **Q** gives rise to two blocks of row vectors of zeroes. The remaining two blocks contain a column vector **R** of length N and a square matrix **T** of order N. The first entry of **R** equals μ_1, while all other entries are equal to 0. The matrix **T** is formed be deleting the first row and the first column from **Q**. With this notation Eq. (14) becomes

$$(p_0'(t), \mathbf{p_T}'(t)) = (p_0(t), \mathbf{p_T}(t)) \begin{pmatrix} 0 & 0 \\ \mathbf{R} & \mathbf{T} \end{pmatrix}.$$

By carrying out the product on the right-hand side and equating the block components of the two sides of the equation, we are led to the following two differential equations:

$$p_0'(t) = \mathbf{p_T}(t)\mathbf{R} = \mu_1 p_1(t) \tag{15}$$

and

$$\mathbf{p}_\mathbf{T}'(t) = \mathbf{p_T}(t)\mathbf{T}. \tag{16}$$

Before absorption, the process takes values in the set of transient states. The state of the process at time t is restricted to this set if two conditions are fulfilled. One is that the initial distribution is supported on this set, i.e., that $P\{X(0) > 0\} = 1$. The second condition is that absorption at the origin has not occurred at time t, i.e., that $X(t) > 0$. The corresponding conditional state probabilities are denoted $\tilde{q}_n(t)$. It is also useful to put $\tilde{\mathbf{q}}(t) = (\tilde{q}_1(t), \ldots, \tilde{q}_N(t))$ to denote the row vector of conditional state probabilities. We note that $\tilde{\mathbf{q}}(t)$ depends on the initial distribution $\tilde{\mathbf{q}}(0)$. The conditioning on nonabsorption at time t leads to the relation

$$\tilde{q}_n(t) = P\{X(t) = n | X(t) > 0\} = \frac{p_n(t)}{1 - p_0(t)}. \tag{17}$$

Hence the vector of conditional state probabilities $\tilde{\mathbf{q}}(t)$ can be determined from the vector $\mathbf{p_T}(t)$ of state probabilities on the set of transient states via the relation

$$\tilde{\mathbf{q}}(t) = \frac{\mathbf{p_T}(t)}{1 - p_0(t)}. \tag{18}$$

By differentiating this relation and using Eqs. (15) and (17) we get

$$\tilde{\mathbf{q}}'(t) = \frac{\mathbf{p_T}'(t)}{1 - p_0(t)} + \mu_1 \tilde{q}_1(t) \frac{\mathbf{p_T}(t)}{1 - p_0(t)}$$

By using Eqs. (16) and (18) we get the following differential equation for the vector of conditional state probabilities $\tilde{\mathbf{q}}$:

$$\tilde{\mathbf{q}}'(t) = \tilde{\mathbf{q}}(t)\mathbf{T} + \mu_1 \tilde{q}_1(t) \tilde{\mathbf{q}}(t)$$

The quasi-stationary distribution $\mathbf{q}$ can now be defined. It is the stationary solution of this equation. Thus, it satisfies the equation

$$\mathbf{qT} = -\mu_1 q_1 \mathbf{q}. \tag{19}$$

This shows that the quasi-stationary distribution $\mathbf{q}$ is a left eigenvector of the matrix $\mathbf{T}$ corresponding to the eigenvalue $-\mu_1 q_1$. One can show that the eigenvalue $-\mu_1 q_1$ is the maximum eigenvalue of the matrix $\mathbf{T}$. This result can be evaluated using numerical computations (Nåsell, 2011).

2.3 Stochastic Model for Interacting Population Species: Native and Invading Species

The successful establishment of introduced species has been intensively studied for a variety of organisms in the field of invasion ecology, including plants (Corbin and D'Antonio, 2004; Vilà et al., 2011), insects (Kenis et al., 2009), fishes (Kolar and Lodge, 2002), and birds (Blackburn et al., 2009). There are number of factors (inter and intra species, and environmental factors) on which the establishment of an introduced species may depend

(Duncan et al., 2014; Romanuk et al., 2009; Sax et al., 2007). The population size of an introduced species is small in the initial stages and, hence, demographic stochasticity plays a critical role in its establishment (Drake et al., 2006). The main focus of studies in the literature has been to address why some introduced species become established while other species die out. In particular, how a species's demographic stochasticity, traits, and competitive interaction strengths affects its establishment rate is fundamental to understanding the mechanisms of invasion (Lande et al., 2003).

A stochastic version of the generalized Lotka–Volterra model for multiple species competition incorporating demographic stochasticity and with large resident species population has been developed via a continuous-time birth and death process (Black and McKane, 2012; Nåsell, 2001; Palamara et al., 2016). Singular perturbation theory is used to analyze the model, and the probability of invasion of a small introduced species is derived via a diffusion approximation (Ross et al., 2009). The probability of extinction for an introduced species is given by

$$p(\text{extinction by time } t) = 1 - p(\text{establishment by time } t),$$

where establishment means that the introduced species has not yet gone extinct in spite of the presence of demographic stochasticity.

Suppose there is a community of S interacting species. Then the Lotka–Volterra type deterministic model (Ebenman et al., 2004) can be described as:

$$\frac{dn_i}{dt} = r_i n_i \left(1 - \frac{n_i + \sum_{j \neq i}^{S} \alpha_{ij} n_j}{K_i} \right), \tag{20}$$

where $n_i(t)$ is the population density of species i, r_i is its per capita growth, K_i is its carrying capacity, and α_{ij} is the per capita effect of species j on species i ($\alpha_{ii} = 1$). For an introduced species I, similar but stochastic birth and death processes using the following birth and death rates can be constructed

$$B(n_I) = \lambda_I n_I \left[1 - \frac{n_I + \sum_{j \neq I}^{S} \alpha_{Ij} n_j}{N_I} \right], \tag{21}$$

$$D(n_I) = \mu_I n_I, \tag{22}$$

where λ_I and μ_I are the per capita intrinsic birth and death rates of the introduced species ($r_I = \lambda_I - \mu_I$) and N_I is the population size at which there is zero probability of births for the introduced species (Nåsell, 2001). Note, N_I

and K_I are related by the intrinsic birth and death rates, which define the maximum size of the demographic fluctuations at equilibrium ($K_I = r_I N_I/\lambda_I$).

Considering the population density of the introduced species n_I and omitting terms of order $O(1/S)$ or smaller, the equation for introduced species, I, becomes

$$\frac{dn_I}{dt} \approx \frac{r_I n_I}{K_I}\left(K_I - n_I + \alpha_{IR} n_R + \sum_{j \neq I}^{S} (S-1) cov(n_j, \alpha_{Ij})\right), \tag{23}$$

where r_I and K_I are the growth rate and carrying capacity of the introduced species, $\alpha_{IR} = \sum_{j \neq I} \alpha_{Ij}$ is the overall effect of the species in the resident community on the introduced species, n_R is the average species density of the resident community species, and $cov(n_j, \alpha_{Ij})$ is the covariance measured over the densities of the resident species and the introduced species' interaction strengths.

In order to decouple the system of this last equation, we assume that the species in the resident community remain at their average density, that is, at carrying capacity, K_R. This assumption holds by considering the earlier stages of an introduction, where it can be shown that the resident species remain close to the average density. Hence, the process can be written as a single species birth–death process for the introduced species with modified birth and death rates (Nåsell, 2001; Palamara et al., 2016):

$$B(n_I) \approx \overline{\lambda}_I n_I \left[1 - \frac{n_I}{N_I}\right], \tag{24}$$

$$D(n_I) = \mu_I n_I \tag{25}$$

where $\overline{\lambda}_I = \lambda_I \left(1 - \frac{\alpha_{IR} K_R}{N_I}\right)$.

If we can set this system in the form of the logistic equation or in the form of the birth–death process, an approximate general (Section 2.1 for logistic model) or quasi-stationary solution (Section 2.2 for birth–death process) of the system can be computed, respectively, using similar steps.

2.4 Stochastic Model for Heterogeneity in Population of a Single Type of Cells

As discussed in the earlier section on the population growth of a species, we turn our attention to the study of cell populations, with particular reference to cell proliferation which plays a central role in most microbiological processes (Smith and Martin, 1973). Many biologists have used Gompertz curves to fit the growth of cell populations. In many cell populations, not all cells are actively growing, but some are in a resting or quiescent phase. Quiescence is a state used by a cell population to control its growth. An important aspect of the proliferating cell population is the number of offspring or parity of cells in the population (Pollard, 1975). Our interest here is to model the number of cells that have divided a certain number of times in which cells could be in a proliferating or quiescent state. A computational model for the proliferation dynamics of a structured population is presented in Hasenauer et al. (2012). We shall deal with a cell population model in which the birth and death rates of proliferative cells are parity dependent. Let $X_i(t)$ be the number of proliferative cells of parity i at time t. Also let $X(t)$ and $Y(t)$, respectively, denote the number of proliferative and quiescent cells at time t. Then

$$X(t) = \sum_{k=0}^{\infty} X_k(t). \tag{26}$$

We assume that initially $X_k(0) = b_k$, $k = 0, 1, 2, \ldots$ and

$$X(0) = \sum_{k=0}^{\infty} b_k = b, \qquad Y(0) = a. \tag{27}$$

We also assume that the population evolves in accordance with the following principles:

- the subpopulations generated by any two coexisting cells are statistically independent of one another;
- a proliferative cell of parity k at time t produces another proliferative cell in $(t, t + \Delta)$ with probability $\rho_1\lambda_k\Delta + o(\Delta)$, while the new cell is quiescent with probability $\rho_2\lambda_k\Delta + o(\Delta)$, where $\rho_1 + \rho_2 = 1$;
- a proliferative cell of parity k has a probability $\mu_k\Delta + o(\Delta)$ of dying in the time interval $(t, t + \Delta)$, and a quiescent cell has a probability of $\mu'\Delta + o(\Delta)$ of dying in the same interval.

Define the generating function

$$G(\Theta; t) = E\left[\theta^{Y(t)} \prod_{k=0}^{\infty} \theta_k^{X_k(t)}\right] \tag{28}$$

By considering various possible transitions in $(t, t + \Delta)$, we construct the following forward Kolmogorov equation:

$$\begin{aligned} \frac{\partial G(\Theta;t)}{\partial t} = \mu'(1-\theta)\frac{\partial G(\Theta;t)}{\partial \theta} + \sum_{k=0}^{\infty} [\theta_0\theta_{k+1}\rho_1\lambda_k + \theta\theta_{k+1}\rho_2\lambda_k \\ -(\lambda_k + \mu_k)\theta_k + \mu_k]\frac{\partial G(\Theta;t)}{\partial \theta_k} \end{aligned} \tag{29}$$

with initial condition

$$G(\theta_i = \theta = 1, \ \forall \ i; t) = 1. \tag{30}$$

Eq. (29) appears to be intractable. However, it can be used to obtain the moments of the process. For example, if $m_k(t)=E[X_k(t)]$, and $y(t)=E[Y(t)]$, then from (29) we obtain

$$\frac{dm_0(t)}{dt}=-(\mu_0+(1-\rho_1)\lambda_0)m_0(t)+\rho_1\sum_{k=1}^{\infty}\lambda_k m_k(t), \tag{31}$$

$$\frac{dm_k(t)}{dt}=\lambda_{k-1}m_{k-1}(t)-(\lambda_k+\mu_k)m_k(t), \qquad k=1,2,\ldots, \tag{32}$$

$$\frac{dy(t)}{dt}=-\mu' y(t)+\sum_{k=0}^{\infty}\rho_2\lambda_k m_k(t). \tag{33}$$

Eqs. (31)–(33) can be solved using Laplace transforms. Thus if $m_k^*(s)$ and $y^*(s)$ are the Laplace transforms of $m_k(t)$ and $y(t)$, respectively, we have

$$(s+\mu_0+(1-\rho_1)\lambda_0)m_0^*(s)=b_0+\rho_1\sum_{k=1}^{\infty}\lambda_k m_k^*(s), \tag{34}$$

$$(s+\lambda_k+\mu_k)m_k^*(s)=\lambda_{k-1}m_{k-1}^*(s), \tag{35}$$

$$(s+\mu')y^*(s)=\sum_{k=0}^{\infty}\rho_2\lambda_k m_k^*(s). \tag{36}$$

The above equations can be solved recursively to yield

$$m_k^*(s)=\frac{1}{\lambda_k}\left[\sum_{j=0}^{k-1}b_{k-j}\prod_{r=0}^{j}\frac{\lambda_{k-r}}{s+\lambda_{k-r}+\mu_{k-r}}\right.$$
$$\left.+(s+\lambda_0+\mu_0)m_0^*(s)\prod_{r=0}^{k}\frac{\lambda_{k-r}}{s+\lambda_{k-r}+\mu_{k-r}}\right] \tag{37}$$
$$k=1,2,\ldots.$$

Substituting (37) in (34) and (36), respectively, we finally obtain

$$m_0^*(s)=\frac{b_0+\sum_{j=1}^{\infty}\sum_{k=0}^{j-1}b_{j-k}\prod_{r=0}^{k}\frac{\lambda_{j-r}}{s+\lambda_{j-r}+\mu_{j-r}}}{s+(1-\rho_1)\lambda_0+\mu_0-(s+\lambda_0+\mu_0)\sum_{j=1}^{\infty}\prod_{r=0}^{j}\frac{\lambda_{j-r}}{s+\lambda_{j-r}+\mu_{j-r}}}, \tag{38}$$

$$y^*(s)=\frac{a}{s+\mu'}+\frac{\rho_2}{s+\mu'}\sum_{k=0}^{\infty}\lambda_k m_k^*(s).$$

Also if $m(t) = E[X(t)]$ then from (26) and System (31)–(33) we obtain

$$\frac{dm(t)}{dt} = \sum_{k=0}^{\infty}(\rho_1\lambda_k - \mu_k)m_k(t) \tag{39}$$

or

$$sm^*(s) = b + \sum_{k=0}^{\infty}(\rho_1\lambda_k - \mu_k)m_k^*(s) \tag{40}$$

Since $m_k^*(s)$ are all known, $m^*(s)$ and $y^*(s)$ can be easily obtained. To illustrate the model we consider the following examples.

Example 1. Suppose the birth and death rates are constants: $\lambda_k = \lambda$; $\mu_k = \mu$, $\forall\ k$. For this choice

$$\lambda m_k^*(s) = \sum_{j=0}^{k-1} b_{k-j}\frac{\lambda^j}{(s+\lambda+\mu)^{j+1}} + \frac{\lambda^k}{(s+\lambda+\mu)^k}m_0^*(s), \tag{41}$$

$$y^*(s) = \frac{a}{s+\mu'} + \frac{\rho_2\lambda m^*(s)}{(s+\mu')}, \tag{42}$$

$$m_0^*(s) = \frac{b_0(s+\mu)+(b-b_0)}{(s+\mu)(s+\mu+(1-\rho_1)\lambda)-\lambda^2}, \tag{43}$$

and

$$m^*(s) = \frac{b}{s-(\rho_1\lambda-\mu)}. \tag{44}$$

Upon inversion the above equations yield

$$\begin{aligned} m_k(t) &= e^{-(\lambda+\mu)t}\sum_{j=0}^{k-1} b_{k-j}\frac{(\lambda t)^j}{j!} \\ &+ \frac{\lambda^k e^{-\mu t}}{(r_1-r_2)\Gamma(k)}\left[\frac{\gamma[k,(r_1+\lambda)t]}{r_1+\lambda}e^{r_1 t} - \frac{\gamma[k,(r_2+\lambda)t]}{r_2+\lambda}e^{r_2 t}\right], \end{aligned} \tag{45}$$

$$y(t) = ae^{-\mu' t} + \frac{\rho_2\lambda_2 b}{\rho_1\lambda-(\mu+\mu')}\left[e^{(\rho_1\lambda-\mu)t} - e^{-\mu' t}\right], \tag{46}$$

$$m_0(t) = \frac{e^{-\mu t}}{r_1-r_2}[[b_0(r_1-\lambda)+b\lambda]e^{r_1 t} - [b_0(r_2-\lambda)+b\lambda]e^{r_2 t}], \tag{47}$$

and

$$m(t) = be^{(\rho_1\lambda-\mu)t}, \tag{48}$$

where r_1 and r_2 are the roots of the polynomial $s^2 + (1 - \rho_1)\lambda s - \lambda^2$, and $\Gamma(n)$ and $\gamma[n, x]$, respectively, are the gamma and incomplete gamma functions.

Example 2. In this example we assume that $X_0(0) = 1$, $X_j(0) = 0$, $j \geq 1$, $\mu_k = \mu$, and the λ_k are given by

$$\lambda_k = \begin{cases} \lambda_0, & k = 0; \\ \lambda, & k \geq 1. \end{cases}$$

Following the approach outlined above, after inversion we have

$$m_k(t) = \frac{\lambda^k e^{-\mu t}}{(W_1 - W_2)\Gamma(k)} \left[\frac{\gamma[k, (W_1 + \lambda)t]}{W_1 + \lambda} e^{W_1 t} - \frac{\gamma[k, (W_2 + \lambda)t]}{W_2 + \lambda} e^{W_2 t} \right], \tag{49}$$

$$y(t) = ae^{-\mu' t} + \frac{\rho_2 \lambda_2}{(\mu' + W_1)(\mu' + W_2)} e^{-\mu' t} + \frac{\rho_2 \lambda_2}{W_1 - W_2} \left[\frac{e^{W_1 t}}{W_1 + \mu'} + \frac{e^{W_2 t}}{W_2 + \mu'} \right], \tag{50}$$

$$m_0(t) = \frac{e^{-\mu t}}{W_1 - W_2} \left[e^{W_1 t} - e^{W_2 t} \right], \tag{51}$$

where W_1 and W_2 are roots of $s^2 + s(1 - \rho_1)\lambda_0 - \lambda^2$.

3 TYPES OF STOCHASTICITY AND EXTINCTION IN STOCHASTIC MODELS

In this section, we provide examples of stochastic models to compute two different metrics of extinction of population (time to extinction and extinction risk).

3.1 Time to Extinction of a Birth–Death Process

The time to extinction T is a random variable and it depends on an initial distribution. Often it is easy to compute the distribution of time to extinction, T_Q, if the initial distribution $p(0)$ is the quasi-stationary distribution $\mathbf{q}$. If process $X(t)$ is absorbed at time t, then the time to extinction T is at most equal to t and the state of the process $X(t)$ is equal to 0 (that is, the events $\{T \leq t\}$ and $\{X(t) = 0\}$ are identical). Hence, we need to compute

$$P\{T \leq t\} = P\{X(t) = 0\} = p_0(t).$$

We compute here T_Q rather than for given arbitrary initial distribution. Note from previous Section 2.2, we can write

$$p_T'(t) = -\mu_1 q_1 p_T(t), \quad p_T(0) = \mathbf{q}$$

and its solution is

$$p_T(t) = \mathbf{q}\exp(-\mu_1 q_1 t). \tag{52}$$

This solution can be used to determine $p_0(t)$. Since Eq. (15) gives a differential equation for p_0 in terms of p_1, with the initial value $p_0(0) = 0$, hence

$$p_0'(t) = \mu_1 p_1(t), \quad p_0(0) = 0,$$

with solution

$$p_0(t) = 1 - \exp(-\mu_1 q_1 t). \tag{53}$$

Since there is no explicit expression for q_1, we do not have an explicit solution. Hence, the time to extinction from the quasi-stationary distribution, T_Q, has an exponential distribution with the expectation (Nåsell, 2001, 2011):

$$E[T_Q] = \frac{1}{\mu_1 q_1}.$$

That is, the distribution of T_Q is completely determined from the probability q_1. The distribution of time of extinction can be similarly computed for more complex population dynamics models such as for models analyzed in Mubayi and Greenwood (2013) and Bani et al. (2013).

3.2 The Role of Stochasticity and Heterogeneity on Extinction Risk

The risk for extinction depends on factors such as demographic stochasticity (birth and death rates), environmental stochasticity, sex of individuals, and demographic heterogeneity (Melbourne and Hastings, 2008). In this section we will show how each factor is incorporated in a model and derive a well-known stochastic model referred to as the stochastic Ricker Model. Finally, extinction risk is evaluated as a function of stochasticity and heterogeneity in birth and death related mechanisms and is compared numerically for the different models discussed below.

3.2.1 Demographic Stochasticity

Suppose there is a species whose population size is N_t at time t. The adults lay eggs at the starting of a year but also may cannibalize eggs and juveniles. There are different sources of *demographic stochasticity* in the model including births, density-dependent mortality, density-independent mortality, and sex determination (Melbourne and Hastings, 2008). Suppose $B_{i,t}$ represents number of eggs or young produced by an adult i at time t and assume that

$$B_{i,t} \sim Poisson(\beta).$$

This is referred as the *Poisson Ricker model*, where β is the average birth/egg-laying rate. The Ricker model assumes that an adult individual eats eggs or young ones randomly with constant probability of eating with no handling time. Hence, the probability that an individual offspring will not be eaten by adult i (represented by c_i) is given by

$$c_i = e^{-\alpha}$$

where α represents the adult search rate for eggs/young ones. Therefore, the probability that an individual offspring is not eaten by any adults can be described by a joint probability distribution, given by

$$c = \prod_{i}^{N_t} c_i = e^{-\alpha N_t}.$$

We can now describe the probability of survival of eggs from all forms of mortality can be written as:

$$s = (1 - m)c$$

where m is the probability of dying due to natural causes and c survival from not being eaten.

Hence, the number of offspring from adult i that survive to adult stage (represented by $S_{i,t+1}$) can be modeled using a binomial distribution as follows:

$$S_{i,t+1} \sim Binomial(B_{i,t}, s)$$

Note, since $B_{i,t}$ follows a Poisson distribution, $S_{i,t+1}$ will have a compounded binomial-Poisson distribution, which using the law of total probability reduces to a Poisson distribution with mean, $Re^{-\alpha N_t}$, as follows:

$$S_{i,t+1} \sim Poisson(Re^{-\alpha N_t})$$

where R is the finite rate of population increase of the deterministic Ricker model and is equivalent to $\beta(1 - m)$. The total number of offspring surviving to become adults is given by

$$N_{t+1} = \sum_{i}^{N_t} S_{i,t+1} \sim Poisson(N_t R e^{-\alpha N_t}).$$

That is, N_{t+1} follow a Poisson distribution with mean equal to the deterministic Ricker model.

3.2.2 Demographic Stochasticity + Environmental Stochasticity

Environmental stochasticity is captured in the model via external sources and it can change stochastically in time, space, or both. Suppose that the birth rate $\beta_{t,x}$ now varies randomly due to environmental factors. Since summation of Poisson distributed random variables is also Poisson distributed, we obtain following equation:

$$B_{t,x} = \sum_{i}^{N_t} B_{i,t,x} \sim Poisson(N_t \beta_{t,x})$$

Let variation in $\beta_{t,x}$ follows gamma distribution with mean β (mean rate in time or space) and additional parameter k_E, $\beta_{t,x} \sim Gamma(\beta, k_E)$. Since the distribution is conditional on $N_{t,x}$, we get distribution of $N_{t,x}\beta_{t,x}$ also gamma but with different mean:

$$N_{t,x}\beta_{t,x} \sim Gamma(N_{t,x}\beta, k_E).$$

Since distribution of $B_{t,x}$ is a gamma mixture of Poisson distributions, it can be easily shown that the form of this distribution will be of a negative binomial distribution:

$$B_{t,x} \sim NegBinom(N_{t,x}\beta, k_E).$$

Note, that from the Poisson Ricker model, survival of offspring is binomial: $S_{t+1,x} \sim Binomial(B_{t,x}, s)$. Since $B_{t,x}$ is negative binomial, the number of surviving offspring from time t at location x has a compounded (binomial-negative binomial) distribution, which can be reduced to a negative binomial distribution:

$$N_{t+1,x} \sim NegBinom(N_{t,x}Re^{-\alpha N_{t,x}}, k_E),$$

where $N_{t,x} = S_{t,x}$ and $R = \beta(1 - m)$. This model combined the effects of demographic and environmental stochasticity.

3.2.3 Demographic Stochasticity + Demographic Heterogeneity

Let adult i give birth randomly at a constant rate β_i specific to that adult. The number of offsprings produced by adult i at the beginning of the life cycle is given by

$$B_{i,t} \sim Poisson(\beta_i)$$

where β_i is the mean number of births for adult i, which is the tendency of that individual to produce more or less offspring than other adults (e.g., a large adult individual produce more offspring than a small one). Let $\beta_i \sim Gamma(\beta, k_D)$ (representing variation in birth rate between individuals) with mean β and variance determined by shape parameter k_D then

$$B_{i,t} \sim NegBinom(\beta, k_D)$$

Since survival is Binomial, i.e.,

$$S_{i,t+1} \sim Binomial(B_{i,t}, s) \sim NegBinom(Re^{-\alpha N_t}, k_D)$$

where $R = \beta(1 - m)$. Since the sum of independent negative binomials with the same k parameter is also a negative binomial, then

$$N_{t+1} = \sum_{i}^{N_t} S_{i,t+1} \sim NegBinom(N_t Re^{-\alpha N_t}, k_D N_t).$$

Note, variance in final abundance for this model is density dependent.

3.2.4 Demographic Stochasticity + Demographic Heterogeneity + Environmental Stochasticity

In this model, we combine the assumptions of the previous three models, that is, the number of offspring produced by an adult is a Poisson random variable (demographic stochasticity), variation in birth individuals within a population (demographic heterogeneity), β_i, follows a gamma distribution with mean $\beta_{t,x}$, variation in the population mean at different times and locations, $\beta_{t,x}$ follows a gamma distribution with mean β, and survival is binomial. With these assumptions, the distribution of $N_{t+1,x}$ is a compound negative binomial-gamma distribution (gamma mixture of negative binomials) with mean equal to the deterministic Ricker model and two variance parameters, k_D and k_E.

3.2.5 Summary

The models (shown in Table 2) include demographic stochasticity in all stages of the lifecycle (births as well as density-independent and density-dependent mortality), whereas environmental stochasticity and demographic heterogeneity are only included in births. Mortality is also subject to environmental stochasticity and demographic heterogeneity; however, their impact does not change the dynamical conclusions. Moreover, demographic sources of stochasticity are the prominent cause of "variability in most natural populations" and "extinction risk" rather than environmental variability. In the models considered above, demographic stochasticity arises from the number of eggs laid by individuals and survival of individual eggs from predation by adults. Demographic heterogeneity arises from the variation in number of offspring between individuals. For the same total variance, extinction risk is enhanced more by demographic heterogeneity than by environmental stochasticity.

TABLE 2 Models Incorporating Demographic Stochasticity (DS), Environmental Stochasticity (ES), and Demographic Heterogeneity (DH)

DS	ES	DH	Model (i.e., $N_{t+1} \sim$)
Y	—	—	Poisson-Demographic
Y	Y	—	Neg-Binomial-Environmental
Y	—	Y	Neg-Binomial-Heterogeneity
Y	Y	Y	Neg-Binomial-Gamma-All

4 OSCILLATIONS AND RESONANCE IN THE STOCHASTIC MODELS

Quasi-cycles may exists in stochastic model where demographic noise of the population excites the system, resulting in noisy cycles in the dynamics. The power spectrum using power spectral analysis of a sample path will be peaked with a finite width, indicating a range of excited frequencies. Although, all frequencies of the spectrum are excited in the time-series, the position of the peak in the spectrum indicates the dominant frequencies that are amplified the most. The quasi-cycles found in a stochastic model do not have a single period, but a distribution of periods centered about an average value. Moreover, there also exists a fluctuations in amplitude of the cycles in the time series of a model.

It is well known that demographic stochasticity could maintain oscillations by constantly perturbing the system away from its endemic steady state (Aparicio and Solari, 2001). Trajectories of stochastic model with a large population typically follow the damped trajectory of the corresponding deterministic model for a certain initial time, after which the stochastic path remains oscillatory (Kuske et al., 2007). This natural oscillatory behavior of a stochastic sample path could further increase or decrease due to introduction of artificial/forced oscillations in one or more of the model parameters. The phenomenon, in which stochastic fluctuations sustain nearly sustained oscillations in a system which has a stable constant equilibrium in the deterministic limit, has been called coherence resonance or autonomous stochastic resonance. In this section, we provide examples of natural and forced sustained oscillations in the stochastic model.

4.1 Sustained Oscillations via Coherence Resonance

We study here epidemic SIR model with demography and derived it as a system of stochastic differential equations using (Greenwood and Gordillo, 2009) and is given below:

$$dS = \left(\mu(N-S) - \beta\frac{SI}{N}\right)dt + \sqrt{\mu(N+S)}dW_1 - \sqrt{\beta\frac{SI}{N}}dW_2 \tag{54}$$

$$dI = \left(\beta\frac{SI}{N} - (\gamma+\mu)I\right)dt + \sqrt{\beta\frac{SI}{N}}dW_2 - \sqrt{(\gamma+\mu)I}dW_3 \tag{55}$$

We introduce dimensionless variables:

$$u = \frac{S-S^*}{S^*}, \quad v = \frac{I-I*}{I^*}, \quad \tau = \Omega t, \quad \Omega = \sqrt{\beta\frac{\mu}{R_0}(R_0-1)}$$

where $(S^*, I^*) = (N/R_0, N\mu(R_0-1)/\beta)$ is the stable endemic equilibrium and $R_0 = \beta/(\gamma + \mu)$ is the reproduction number of the system. This change of variables transform the system such that the new stable equilibrium is at (0, 0) and damped oscillations have frequency 1. Note, the choice of dimensionless time is based on trial and error, but this transformation yields a system with oscillations of unit frequency. The linearize system about $(u, v) = (0, 0)$ (setting $(u, v) = (0, 0)$ also in the noise coefficients) is given by

$$d\begin{bmatrix} u \\ v \end{bmatrix} = M\begin{bmatrix} u \\ v \end{bmatrix} d\tau + G\begin{bmatrix} dW_1 \\ dW_2 \\ dW_3 \end{bmatrix} \tag{56}$$

where

$$G = \begin{bmatrix} \sqrt{\frac{\mu}{\Omega S_{eq}^2}(N+S_{eq})} & -\sqrt{\frac{\beta I_{eq}}{\Omega N S_{eq}}} & 0 \\ 0 & \sqrt{\frac{\beta S_{eq}}{\Omega N I_{eq}}} & -\sqrt{\frac{\gamma+\mu}{\Omega I_{eq}}} \end{bmatrix} = \begin{bmatrix} g_1 & -b^2 g_2 & 0 \\ 0 & g_2 & -g_2 \end{bmatrix}.$$

The eigenvalues of

$$M = \begin{bmatrix} -\mu R_0/\Omega & -\mu(R_0-1)/\Omega \\ \beta/\Omega R_0 & 0 \end{bmatrix}$$

are $\lambda = -\epsilon^2 \pm \sqrt{\epsilon^4 - 1}$, where $\epsilon^2 = \mu^2 R_0/(2\Omega)$. For $\epsilon^2 \ll 1$, the leading order solution of the nondimensionalized corresponding deterministic model near the equilibrium is given by slowly damped oscillations with frequency near unity, that is, its solution is approximated by

$$\begin{bmatrix} u \\ v \end{bmatrix} \sim C_1 \exp(-\epsilon^2 t)\begin{bmatrix} b\ \cos(t) \\ \sin(t) \end{bmatrix} + C_2 \exp(-\epsilon^2 t)\begin{bmatrix} b\ \sin(t) \\ -\cos(t) \end{bmatrix}$$

where $b^2 = \frac{I_{eq}}{S_{eq}}$.

The resonance of the stochastic variations with the frequency of the deterministic system leads to an amplification of the otherwise damped oscillations with a slowly varying stochastic modulation. Hence, we provide an approximation which captures both the multiple scales behavior and the resonance with the deterministic frequency. We consider a proposed approximation for the solution of the linearized stochastic model (Kevorkian and Cole, 2012):

$$\begin{bmatrix} u \\ v \end{bmatrix} = A(T)\begin{bmatrix} b\ \cos(t) \\ \sin(t) \end{bmatrix} + B(T)\begin{bmatrix} b\ \sin(t) \\ -\cos(t) \end{bmatrix} \tag{57}$$

where $A(T)$ and $B(T)$ incorporate both the slow decay and the stochastic part of the solution, where T varies slowly relative to t, $T = \epsilon^2 t$, $\epsilon \ll 1$.

Substituting Eq. (57) into Eq. (56), we get following system:

$$du = (-2b\epsilon^2[A\cos t + B\sin t] + b[-A\sin t + B\cos t])dt + g_1 dW_1 - b^2 g_2 dW_2, \tag{58}$$

$$dv = [A\cos t + B\sin t]dt + g_2 dW_2 - g_2 dW_2 \tag{59}$$

Since function A and B are stochastic coefficients in Eq. (57), which approximately solve the diffusion system (56), we can assume that they are driven by Brownian motions. A reasonable form for the amplitude functions equation is stochastic equation with constant diffusion coefficients:

$$\begin{bmatrix} dA \\ dB \end{bmatrix} = \begin{bmatrix} f_1(A,B) \\ f_2(A,B) \end{bmatrix} dT + \begin{bmatrix} dN_1(T) \\ dN_2(T) \end{bmatrix} \tag{60}$$

where noise terms N_1 and N_2 satisfy

$$\begin{bmatrix} dN_1(T) \\ dN_2(T) \end{bmatrix} = \Sigma_1 \begin{bmatrix} d\xi_{11}(T) \\ d\xi_{12}(T) \end{bmatrix} + \Sigma_2 \begin{bmatrix} d\xi_{21}(T) \\ d\xi_{22}(T) \end{bmatrix} + \Sigma_3 \begin{bmatrix} d\xi_{31}(T) \\ d\xi_{32}(T) \end{bmatrix} \tag{61}$$

where the ξ_{ij}'s are independent standard Brownian motions on the slow time-scale T, and the Σ_j's are constant matrices. The drift coefficients, f_1 and f_2 and the diffusion coefficients Σ_1, Σ_2, and Σ_3 are unknowns and must be derived by relating (60) and (61).

For a solution $(u, v)'$ (Eq. 57), Itô's formula can be used to obtain a second expression for du and dv (the first expression is given by System (58)). Equating these two expressions of du and dv, we can compute the drift coefficients f_i's and diffusion coefficients Σ_k's as follows:

$$f_1 = -A, \quad f_2 = -B$$

and

$$\Sigma_1 = \frac{g_1}{2b\epsilon}\begin{bmatrix} 1 & 0 \\ 0 & 1 \end{bmatrix}, \qquad \Sigma_2 = \frac{g_2}{2\epsilon}\begin{bmatrix} -b & 1 \\ -1 & -b \end{bmatrix}, \qquad \Sigma_3 = \frac{g_2}{2b\epsilon}\begin{bmatrix} 0 & -b \\ b & 0 \end{bmatrix}, \tag{62}$$

Substituting coefficients from (62) in (60), we can obtain an updated amplitude system:

$$dX = ZXdT + \Gamma dW(T)$$

where $X = (A\ B)^T, Z = \begin{bmatrix} -1 & 0 \\ 0 & -1 \end{bmatrix}, \Gamma = \frac{1}{2b\epsilon}\begin{bmatrix} g_1 & -b^2 g_2 & b g_2 & -b g_2 & 0 & 0 \\ 0 & -b g_2 & -b^2 g_2 & 0 & g_1 & b g_2 \end{bmatrix}$,
and $dW(T) = (dw_{11}, dw_{21}, dw_{22}, dw_{32}, dw_{12}, dw_{31})$.

This resulting system is an Ornstein–Uhlenbeck-type process for which the stationary distribution (for A and B) can be easily computed. This stationary distribution follows bivariate normal distribution, $N(0, K)$ (where K is the covariance matrix), and provides information on the parameter regions where coherence resonance can occur. More details on computations can be found in Kuske et al. (2007).

4.2 Sustained Oscillations via Stochastic Resonance

It has been analyzed in the literature, how introduction of seasonality (forced oscillations in parameters) in deterministic models can cause population fluctuations with varying periods (Greenman et al., 2004). For example, models have been used to identify mechanisms related to environmental seasonality (periodic forcing; impacting variation in host and pathogen biology) on the diseases dynamics (Altizer et al., 2006; Grassly and Fraser, 2006; He and Earn, 2007). If demographic stochasticity is introduced in the deterministic model with seasonality (via periodic forcing), we see stochastic perturbations around the limit cycles, resulting in sustained oscillations (Black and McKane, 2012).

Consider the deterministic systems of three interacting populations (u, v, and w, for example, in the epidemic model u, v, and w can represents fractions of susceptible, infected and recovered) (Mata et al., 2018):

$$u' = -\beta(t)uv - \eta u + \alpha v + \alpha w \tag{63}$$

$$v' = \beta(t)uv + \eta u - \alpha v - \gamma v \tag{64}$$

$$w' = \gamma v - \alpha w \tag{65}$$

where $u + v + w = 1$ and $\beta(t) = \beta(t + T)$. Since population is constant we can analyze two dimensional system given below:

$$u' = -\beta(t)uv - \eta u + \alpha(1-u) \tag{66}$$

$$v' = \beta(t)uv + \eta u - (\alpha+\gamma)v \tag{67}$$

The corresponding stochastic differential model for limiting diffusion process is

$$ds = [-\beta(t)si - \eta s + \alpha(1-s)]dt + \frac{1}{\sqrt{N}}(-G_1(t)dW_1 + G_2(t)dW_2) \tag{68}$$

$$di = [\beta(t)si + \eta s - (\alpha+\gamma)i]dt + \frac{1}{\sqrt{N}}(-G_1(t)dW_1 - G_3(t)dW_3) \tag{69}$$

Suppose the limit cycle of the deterministic system (66) and (67) is represented by $(\bar{u}, \bar{v})$. For large t

$$s(t) = \bar{u} + \frac{1}{\sqrt{N}}\xi_1(t) \tag{70}$$

$$i(t) = \bar{v} + \frac{1}{\sqrt{N}}\xi_2(t) \tag{71}$$

where $\xi(t) = (\xi_1(t), \xi_2(t))^T \in \mathfrak{R}^2$ is the stochastic perturbation process. Linearizing around the limit cycle, we obtain

$$\dot{\xi}(t) = \mathbf{A}(t)\xi(t) + \mathbf{f}(t) \tag{72}$$

where

$$\mathbf{A}(t) = \begin{bmatrix} -\beta(t)\overline{v} - \eta - \alpha & -\beta(t)\overline{u} \\ \beta(t)\overline{v} + \eta & \beta(t)\overline{u} - \gamma - \alpha \end{bmatrix},$$

$$\langle \mathbf{f}(t)\mathbf{f}(t') \rangle = G(t)\delta(t - t'),$$

$$G(t) = \begin{bmatrix} \beta(t)\overline{uv} + \eta\overline{u} + \alpha(1 - \overline{u}) & -\beta(t)\overline{uv} - \eta\overline{u} - \alpha\overline{v} \\ -\beta(t)\overline{uv} - \eta\overline{u} - \alpha\overline{v} & \beta(t)\overline{uv} + \eta\overline{u} + (\alpha + \gamma)\overline{v} \end{bmatrix}.$$

The linear stochastic differential equation is therefore

$$d\xi(t) = \mathbf{A}(t)\xi(t)dt + C(t)d\mathbf{W}(t) \tag{73}$$

where $C(t) = G(t)^{1/2}$.

Now we will obtain an approximate solution of $\xi(t)$ using Floquet Theory [Grimshaw,1991] for the corresponding deterministic model. Let $\xi(t) = Q(t)y(t)$, then the system becomes

$$dy(t) = Fy(t)dt + Q^{-1}(t)C(t)d\mathbf{W}(t) \tag{74}$$

where

$$F = \begin{bmatrix} -\lambda & \omega \\ -\omega & -\lambda \end{bmatrix}.$$

Let $y(t) = R_{-\omega t}z(t)$; then

$$dz(t) = -\lambda z(t)dt + R_{\omega t}Q^{-1}(t)C(t)d\mathbf{W}(t). \tag{75}$$

Assume $\widetilde{C}_t = Q^{-1}(t)C(t)$ as well as define $\sigma^2(t) = \frac{1}{2}Tr(\widetilde{C}_t\widetilde{C}_t^*)$ and $\overline{\sigma}^2 = \frac{1}{T}\int_0^T \sigma^2(t)dt$.

Let $\mathbf{U}(t) = \frac{\sqrt{\lambda}z(t/\lambda)}{\overline{\sigma}}$; then

$$d\mathbf{U}(t) = -\mathbf{U}(t)dt + R_{\omega t/\lambda}D(u/\lambda)d\widetilde{\mathbf{W}}(t) \tag{76}$$

where $D(t/\lambda) = \frac{1}{\overline{\sigma}}Q^{-1}(t/\lambda)C(t/\lambda)$ and $\widetilde{\mathbf{W}}(t) = \sqrt{\lambda}\mathbf{W}(t/\lambda)$.

Using a theorem in Baxendale and Greenwood (2011), for fixed bounded time interval [0, T], the distribution of $\mathbf{U}(t)$ satisfying Eq. (76) converges to the distribution of the bivariate standard Ornstein–Uhlenbeck (mean-reverting) process ($dS(t) = -S(t)dt + dW(t)$) if $0 < \lambda \ll \omega$.

That is, when $0 < \lambda \ll \omega$, the stochastic perturbations can be written as:

$$\xi(t) = Q(t)y(t) = Q(t)[R_{-\omega t}z(t)] = \frac{\overline{\sigma}}{\sqrt{\lambda}}Q(t)R_{-\omega t}\mathbf{U}_{\lambda t} \tag{77}$$

$$\approx \frac{\overline{\sigma}}{\sqrt{\lambda}}Q(t)R_{-\omega t}\mathbf{S}_{\lambda t} \quad \text{(using Baxendale and Greenwood (2011) theorem)} \tag{78}$$

Let $Q(t)$ be smooth with Fourier series coefficients decaying so rapidly that $Q(t) \sim Q_0 + Q_1 \cos \nu t + Q_2 \sin \nu t$, then

$$Q(t)R_{-\omega t} \sim Q_0 R_{-\omega t} + Q_1 R_{-\omega t} \cos \nu t + Q_2 R_{-\omega t} \sin \nu t.$$

By product-to-sum identities, we expect the emergence of frequencies ω and $\nu \pm \omega$.

5 COMPUTATIONAL STOCHASTIC APPROACHES

Earlier models were primarily described at the population level using system of differential equations, however, with ever increasing computational power, a drastic increase in the modeling complexity are being studied, most naturally modeled at the individual level. These types of individual or agent-based models capture a larger number of natural phenomena than typical population-level models. Though, there are still some challenges associated with handling of such agent-based models, where population, time and space are discrete and stochasticity related to each factors are distinct. The challenges with such models are: (i) numerical analysis is primarily possible rather than analytical results and (ii) data to thoroughly estimate or evaluate model parameters might be limited.

5.1 Agent-Based Models and Parameter Estimation as Emergent Behavior

The most common use of data to calibrate population dynamic models uses time-series data fitting in order to estimate values for model parameters which cannot be estimated directly from the literature (via biological field or lab studies)—typically contact rates between two populations (e.g., two different species, or susceptibles and infectives). However, in many cases involving wildlife, neglected tropical diseases, or other understudied phenomena, time-series data are not available, and direct observation in field or lab settings is impractical. In such instances, it may be possible to estimate such rates by using known information about the relevant populations' life cycles to build up a model which allows the resulting rates to be observed. This information includes, for wildlife and disease vectors, not only general demographic information such as birth and death rates and population density but more detailed day-to-day details such as nesting behavior, travel distances, feeding frequencies and habits, and ability to withstand biting or starvation. Of necessity these models are individual-based or agent-based, and due to the natural variation in behaviors from one individual to another, stochasticity is important on this scale.

Stochastic agent-based models (ABMs) have long been used to observe emergent behavior of a population or system: properties of the collective which cannot be deduced from the behavior of a single individual, but only from the behavior of many, like the way water pressure is formed by the forces of many water molecules, or the way every human's use of energy and materials (and influence on others' uses thereof) combines with that of others to have a net effect on the planet's resources. In the same way, the simulated actions of many individuals (agents) combine to produce a net activity rate in an agent-based model. Thus, the contact rates that drive the dynamics of classical compartmental population models can be estimated as the emergent result of noncompartmental models (i.e., ABMs).

This section presents a particular example as a case study. The protozoan parasite *Trypanosoma cruzi* causes Chagas disease throughout the Americas, with an estimated eight million people affected, primarily in rural Latin America. However, the parasite is maintained in sylvatic reservoir cycles involving hosts such as raccoons, opossums, coati, and woodrats, alongside the same type of insect vectors which then transmit the parasite to humans. Chagas disease is noted as a neglected tropical disease, and *T. cruzi* is even less-studied in sylvatic cycles; indeed, these cycles are difficult to study in the field, especially when financial resources are limited and thus dedicated to more pressing concerns. In addition, the few studies available of some *T. cruzi* vector species suggest that vector behavior in the lab and in the field may differ significantly, notably in response to climatic conditions. All these conditions make these sylvatic transmission cycles ideal candidates for study through ABMs.

The key process in sylvatic *T. cruzi* transmission cycles is the vector–host contact process which drives multiple infection events. Like mosquitoes and other insect vectors of infection, the triatomine bugs which transmit *T. cruzi* ingest the parasite into their guts through bloodmeals on hosts. Classical transmission to hosts, referred to as stercorarian infection, also follows from vector feeding events; however, in addition, many *T. cruzi* hosts are opportunistic feeders and sometimes eat the vectors, which live in the hosts' dens, thereby leading to oral transmission, which may be a dominant transmission route in some sylvatic cycles (Kribs-Zaleta, 2010). In compartmental population models, contact rates between two populations are typically described (following the law of mass action in chemistry) in terms of the product of the two population sizes, with a rate coefficient which may depend on the population sizes as well (often on one or the other). Only a very few studies have attempted to measure host–vector contact rates in a laboratory setting (e.g., Pippin, 1970), but even these have noted that the actual frequency of such contacts in the wild likely vary considerably from those obtained in a lab, where conditions typically facilitate contact and prevent either party (host or vector) from leaving or hiding.

In order to build up a model capable of describing relevant aspects of host and vector daily behaviors that lead to contacts, considerably more data are required than for a typical compartmental model. Yong et al. (2015) developed an agent-based model using more than forty parameters to simulate local interaction between hosts (either raccoons or woodrats) and vectors (either *Triatoma sanguisuga* or *Triatoma gerstaeckeri*), and thereby measure the distribution of contact rates. Agents in the model included not only the hosts and adult vectors, but also the host nests and dens, since the juvenile vectors, which far outnumber the adults, are not mobile enough to leave nests in search of new hosts. Parameters used to calibrate the model ranged from basic demographic information—birth and death rates, maturation rates for vectors, population densities per unit area of land as well as per nest or den—to the relation between feeding and travel for both host and vector, and the underlying ecological landscape(s) (which affects population densities).

The detailed information on host and vector feeding and travel used to construct the model of Yong et al. (2015) is a prime example of the kind of intricate detail (not present in classical compartmental models) that coordinates through ABMs to produce emergent properties, and can turn out to play a surprisingly important role in those properties. Hosts are assumed to change sleeping places (nests or dens) for one of two reasons: territorial foraging or so-called host irritability, when too many vectors try to bite it in one night. Because sleeping in a given place suggests that the host has been foraging there, temporarily depleting the food supply, each host has a given probability of changing sleeping places from one night to the next. In addition, when excessive vector feeding occurs, a host may get up in the middle of the night due to irritation and leave. When a host leaves, it selects another location from within those in a single day's (or night's) travel, which have gone unvisited the longest (since the food supply there has presumably had the most time to replenish). Vectors, meanwhile, feed and travel as a function of the time elapsed since their last feeding. For a given period after feeding, vectors lie quiescent, digesting. Past a first threshold, they are ready to feed again. If they have no opportunity to feed, then past a second threshold, adult vectors can leave the nest to seek a new host. When they disperse, they seek the nearest occupied nest within a day's travel (for them); otherwise they travel in a random direction each day seeking a new host. Past a third threshold, vectors of both stages can die from starvation. Vector dispersal is especially important because host-switching is the mechanism which spreads *T. cruzi* from one host to another, or from one transmission cycle to another.

Model simulations led to per-host and per-vector contact rate distributions, e.g., mean per-vector contact rates of once every 14 days for *T. sanguisuga* feeding on woodrats and once every 22 days for *T. gerstaeckeri* and woodrats, but they also led to some qualitative conclusions. Most notably, host irritability proved a key limiting factor in not only vector feeding rates but (consequently) vector densities. In addition, the mean feeding times fall between vector dispersal and vector starvation thresholds, so that vectors feed often enough to survive, but only by seeking hosts through dispersal. This result shows why these vectors spread the parasite to new hosts, and also contradicts classical compartmental models of vector–host infection like the Ross–MacDonald model which assume the contact rate proportional to (i.e., limited by) vector density—here the host density is the limiting factor. In short, sufficiently detailed agent-based models have the capacity to act as a virtual laboratory offering the same kinds of qualitative and quantitative insights as their real-life counterparts.

5.2 Algorithms for Simulating Stochastic Models

5.2.1 *Numerical Methods for Markov Chain Based Models*

5.2.1.1 Gillespie Method:

A popular (and probabilistically rigorous) method is called Gillespie's direct algorithm or stochastic simulation algorithm (SSA). In Gillespie's direct algorithm, simulation time scales with population size. The steps of the algorithm are

1. Label all possible events $E_1, E_2, \ldots, E_n$
2. Calculate rate at which each event occurs $R_1, R_2, \ldots, R_n$
3. The rate at which any event occurs is $R_{sum} = \sum_i R_i$
4. Calculate the time until next event (where U_1 is the uniform random number)

$$\delta t = \frac{-1}{R_{sum}} \log(U_1)$$

5. Generate uniform deviate U_2 and set $P = U_2 \times R_{sum}$
6. Event p occurs if

$$\sum_{m=1}^{p-1} R_m < P \leq \sum_{m=1}^{p} R_m$$

7. Update time $t = t + \delta t$ and return to Step 1.

5.2.1.2 First-Reaction Method:

In Gillespie's first-reaction method the absolute rates of events increase linearly with population size. This method is slower but more intuitive. For large population (hence, large per capita rates), time to next event decreases (need more iterations). The steps of the algorithm are:

1. Label all possible events $E_1, E_2, \ldots, E_n$
2. Calculate rate at which each event occurs $R_1, R_2, \ldots, R_n$
3. For each event, k, calculate the time until it next occurs (where U_k's are the k different uniform random numbers)

$$\delta t_k = \frac{-1}{R_k} \log(U_k)$$

4. Find the event, p, that happens first (that is, event that has the smallest δt)
5. Update time $t = t + \delta t_p$ and event p is performed. Return to Step 1.

5.2.1.3 Tau-Leaping Method:

In the Tau-leap method algorithm time is incremented as 0, τ, 2τ, 3τ, … for chosen τ. The number of events of each time within an interval of duration τ is approximately Poisson distributed with mean equal to the rate of the transition at the start of the interval divided by the duration of the interval. For example, in the epidemic susceptible-infected-recovered (SIR) model, between times t and $t + \tau$, the number of new infections is given by a Poisson random variate with mean $\beta\frac{IS}{N\tau}$. Note, the tau-leap method can occasionally give rise to a greater number of events than actually possible (e.g., a greater number of infections than there are susceptible individuals to be infected). This can be taken care of by setting the number of events possible at τ increment to be the maximum possible.

1. Choose small τ.
2. Choose increment $\Delta y_i = \min\{Y(t), q\}$, where $q \sim Poisson(R_i\tau)$ and R_i is the ith event.
3. Update state variable $Y(t+\tau) = Y(t) \pm \sum_{i=1}^{l} \Delta y_i$, where l events effect Y.

5.2.2 Numerical Methods for Stochastic Differential Equations Based Models

Suppose the stochastic differential equation is

$$dY(t) = \sum_{j=1}^{M} f_j(Y(t))dt + \sum_{j=1}^{M} g_j(y(t))dW_j(t).$$

The *Euler Maruyama method*:

$$Y(t+\tau) = Y(t) + \tau\sum_{j=1}^{M} f_j(Y(t)) + \sqrt{\tau}\sum_{j=1}^{M} g_j(y(t))Z_j$$

where Z_j being standard normal variables with mean 0 and variance 1 for all j. This scheme assumes that the drift and diffusion terms are constant between time steps, hence, the scheme improves as time steps becomes smaller.

The *Milstein method*:

$$Y(t+\tau) = Y(t) + \tau\sum_{j=1}^{M} f_j(Y(t)) + \sqrt{\tau}\sum_{j=1}^{M} g_j(y(t))Z_j$$
$$+\frac{1}{2}\left(\sum_{j=1}^{M} g_j'(y(t))\right)\left(\sum_{j=1}^{M} g_j(y(t))\right)\left(\tau Z_j^2 - \tau\right).$$

where g_j' is the derivative of g_j. The first three terms are similar to Euler Maruyama method but it also considers second-order term from Itô's formula (using and *Itô's lemma* and *stochastic Taylor expansions*).

6 CONCLUSION

Stochastic processes are mathematical approaches used to understand and analyze dynamical system with random component that changes over time. An equation that includes a random variable or a stochastic process is often referred as a *stochastic model*. Stochastic models can be analytically and computationally complex to analyze and may require in-depth probability and statistical theory and techniques. Some of the example of stochastic process are Poisson process, renewal process, branching process, semi-Markov process, time-reversible Markov chains, birth–death process, random walks, and Brownian motion. Stochastic models could be discrete and continuous in time and state space. In this chapter, certain types of stochastic population model are collected and some specific analytical approaches are described. The analytical approaches that are shown here are for computing approximations to persistence (for example, quasi-stationary distribution), extinction (for example, time to extinction), and periodicity (for example, sustained oscillations). In addition, numerical methods for simulating some stochastic models are also collected. A stochastic model can be used to compute measures such as (i) the amount of time the stochastic processes stays in state i before making a transition into a different state, (ii) time to extinction of the particular state (e.g., time for elimination of a epidemic), (iii) final state sizes (e.g., final epidemic size), (iv) time to reach peak of a population (e.g., epidemic peak), and (v) distribution of the states at any time.

Complexity in the stochastic model could involve increase in dimension. For example, stochastic population model (not covered in this chapter) may be developed to integrated a decision framework leading to design of a stochastic dynamic programming (SDP)-based model (Benjamin et al., 2009), which could be used to find the management strategy that maximizes future rewards. Such dynamic programming model gives optimal decisions over time in problems with complex state and decision variables (Chen, 2006; Niemi and Lehtonen, 2010). For example, a stochastic partial differential equation (SPDE)-based model could be derived for a population whose individuals randomly experience births, deaths, age, and size changes (Allen, 2009).

REFERENCES

Allen, E.J., 2009. Derivation of stochastic partial differential equations for size-and age-structured populations. J. Biol. Dyn. 3 (1), 73–86.

Altizer, S., Dobson, A., Hosseini, P., Hudson, P., Pascual, M., Rohani, P., 2006. Seasonality and the dynamics of infectious diseases. Ecol. Lett. 9 (4), 467–484.

Aparicio, J.P., Solari, H.G., 2001. Sustained oscillations in stochastic systems. Math. Biosci. 169 (1), 15–25.

Bani, R., Hameed, R., Szymanowski, S., Greenwood, P., Kribs-Zaleta, C., Mubayi, A., 2013. Influence of environmental factors on college alcohol drinking patterns. Math. Biosci. Eng. 10, 1281–1300.

Baxendale, P.H., Greenwood, P.E., 2011. Sustained oscillations for density dependent Markov processes. J. Math. Biol. 63 (3), 433–457.

Benjamin, L.R., Milne, A.E., Parsons, D.J., Cussans, J., Lutman, P.J.W., 2009. Using stochastic dynamic programming to support weed management decisions over a rotation. Weed Res. 49 (2), 207–216.

Black, A.J., McKane, A.J., 2012. Stochastic formulation of ecological models and their applications. Trends Ecol. Evol. 27 (6), 337–345.

Blackburn, T.M., Cassey, P., Lockwood, J.L., 2009. The role of species traits in the establishment success of exotic birds. Glob. Chang. Biol. 15 (12), 2852–2860.

Chen, F.H., 2006. A susceptible-infected epidemic model with voluntary vaccinations. J. Math. Biol. 53 (2), 253–272.

Corbin, J.D., D'Antonio, C.M., 2004. Competition between native perennial and exotic annual grasses: implications for an historical invasion. Ecology 85 (5), 1273–1283.

Darroch, J.N., Seneta, E., 1965. On quasi-stationary distributions in absorbing discrete-time finite Markov chains. J. Appl. Probab. 2 (1), 88–100.

Darroch, J.N., Seneta, E., 1967. On quasi-stationary distributions in absorbing continuous-time finite Markov chains. J. Appl. Probab. 4 (1), 192–196.

Davidson, A.D., Hamilton, M.J., Boyer, A.G., Brown, J.H., Ceballos, G., 2009. Multiple ecological pathways to extinction in mammals. Proc. Natl. Acad. Sci. 106 (26), 10702–10705.

Drake, J.M., Drury, K.L.S., Lodge, D.M., Blukacz, A., Yan, N.D., Dwyer, G., 2006. Demographic stochasticity, environmental variability, and windows of invasion risk for Bythotrephes longimanus in North America. Biol. Invasions 8 (4), 843–861.

Duncan, R.P., Blackburn, T.M., Rossinelli, S., Bacher, S., 2014. Quantifying invasion risk: the relationship between establishment probability and founding population size. Methods Ecol. Evol. 5 (11), 1255–1263.

Ebenman, B.O., Law, R., Borrvall, C., 2004. Community viability analysis: the response of ecological communities to species loss. Ecology 85 (9), 2591–2600.

Fieberg, J., Ellner, S.P., 2000. When is it meaningful to estimate an extinction probability? Ecology 81 (7), 2040–2047.

Fritz, S.A., Bininda-Emonds, O.R.P., Purvis, A., 2009. Geographical variation in predictors of mammalian extinction risk: big is bad, but only in the tropics. Ecol. Lett. 12 (6), 538–549.

Grassly, N.C., Fraser, C., 2006. Seasonal infectious disease epidemiology. Proc. R. Soc. London B. Biol. Sci. 273 (1600), 2541–2550.

Greenman, J., Kamo, M., Boots, M., 2004. External forcing of ecological and epidemiological systems: a resonance approach. Phys. D Nonlinear Phenom. 190 (1–2), 136–151.

Greenwood, P.E., Gordillo, L.F., 2009. Stochastic epidemic modeling. In: Chowell, G., Hayman, J.M., Bettencourt, L.M.A., Castillo-Chavez, P.C. (Eds.), Mathematical and Statistical Estimation Approaches in Epidemiology. Springer, Dordrecht, pp. 31–52.

Hasenauer, J., Schittler, D., Allgöwer, F., 2012. Analysis and simulation of division-and label-structured population models. Bull. Math. Biol. 74 (11), 2692–2732.

He, D., Earn, D.J.D., 2007. Epidemiological effects of seasonal oscillations in birth rates. Theor. Popul. Biol. 72 (2), 274–291.

Kenis, M., Auger-Rozenberg, M.-A., Roques, A., Timms, L., Péré, C., Cock, M.J.W., Settele, J., Augustin, S., Lopez-Vaamonde, C., 2009. Ecological effects of invasive alien insects. Biol. Invasions 11 (1), 21–45.

Kevorkian, J.K., Cole, J.D., 2012. Multiple Scale and Singular Perturbation Methods, vol. 114. Springer Science & Business Media.

Kloeden, P., Platen, E., 1991. Numerical methods for stochastic differential equations. Stoch. Hydrol. Hydraul. 5 (2), 172–172.

Kloeden, P.E., Platen, E., Schurz, H., 2012. Numerical Solution of SDE Through Computer Experiments. Springer Science & Business Media.

Kolar, C.S., Lodge, D.M., 2002. Ecological predictions and risk assessment for alien fishes in North America. Science 298 (5596), 1233–1236.

Kribs-Zaleta, C.M., 2010. Alternative transmission modes for Trypanosoma cruzi. Math. Biosci. Eng. 7 (3), 657–673.

Kuske, R., Gordillo, L.F., Greenwood, P., 2007. Sustained oscillations via coherence resonance in SIR. J. Theor. Biol. 245 (3), 459–469.

Lande, R., Engen, S., Saether, B.-E., 2003. Stochastic Population Dynamics in Ecology and Conservation. Oxford University Press.

Mata, M.A., Greenwood, P., Tyson, R., 2018. The relative contribution of direct and environmental transmission routes in stochastic avian flu epidemic recurrence: an approximate analysis. Bull. Math. Biol. 1–34. https://doi.org/10.1007/s11538-018-0414-6. [Epub ahead of print 2018-03-14].

Melbourne, B.A., Hastings, A., 2008. Extinction risk depends strongly on factors contributing to stochasticity. Nature 454 (7200), 100.

Méléard, S., Villemonais, D., et al., 2012. Quasi-stationary distributions and population processes. Probab. Surv. 9, 340–410.

Mubayi, A., Greenwood, P.E., 2013. Contextual interventions for controlling alcohol drinking. Math. Popul. Stud. 20 (1), 27–53.

Nåsell, I., 2001. Extinction and quasi-stationarity in the Verhulst logistic model. J. Theor. Biol. 211 (1), 11–27.

Nåsell, I., 2011. Extinction and Quasi-Stationarity in the Stochastic Logistic SIS Model. Springer.

Niemi, J.K., Lehtonen, H., 2010. Modelling pig sector dynamic adjustment to livestock epidemics with stochastic-duration trade disruptions. Eur. Rev. Agric. Econ. 38 (4), 529–551.

Palamara, G.M., Carrara, F., Smith, M.J., Petchey, O.L., 2016. The effects of demographic stochasticity and parameter uncertainty on predicting the establishment of introduced species. Ecol. Evol. 6 (23), 8440–8451.

Pippin, W.F., 1970. The biology and vector capability of Triatoma sanguisuga texana usinger and Triatoma gerstaeckeri (Stål) compared with Rhodnius prolixus (Stål) (Hemiptera: Triatominae). J. Med. Entomol. 7 (1), 30–45.

Pollard, J.H., 1975. Modelling human populations for projection purposes: some of the problems and challenges. Aust. J. Stat. 17 (2), 63–76.

Romanuk, T.N., Zhou, Y., Brose, U., Berlow, E.L., Williams, R.J., Martinez, N.D., 2009. Predicting invasion success in complex ecological networks. Philos. Trans. R. Soc. Lond. B Biol. Sci. 364 (1524), 1743–1754.

Ross, J.V., Pagendam, D.E., Pollett, P.K., 2009. On parameter estimation in population models II: multi-dimensional processes and transient dynamics. Theor. Popul. Biol. 75 (2-3), 123–132.

Sax, D.F., Stachowicz, J.J., Brown, J.H., Bruno, J.F., Dawson, M.N., Gaines, S.D., Grosberg, R.K., Hastings, A., Holt, R.D., Mayfield, M.M., et al., 2007. Ecological and evolutionary insights from species invasions. Trends Ecol. Evol. 22 (9), 465–471.

Smith, J.A., Martin, L., 1973. Do cells cycle? Proc. Natl. Acad. Sci. 70 (4), 1263–1267.

Vilà, M., Espinar, J.L., Hejda, M., Hulme, P.E., Jarošík, V., Maron, J.L., Pergl, J., Schaffner, U., Sun, Y., Pyšek, P., 2011. Ecological impacts of invasive alien plants: a meta-analysis of their effects on species, communities and ecosystems. Ecol. Lett. 14 (7), 702–708.

Yaglom, A.M., 1947. Certain limit theorems of the theory of branching random processes. Doklady Akad. Nauk SSSR (NS) 56, 795–798.

Yong, K.E., Mubayi, A., Kribs, C.M., 2015. Agent-based mathematical modeling as a tool for estimating *Trypanosoma cruzi* vector-host contact rates. Acta Trop. 151, 21–31.

Chapter 6

Analyzing Variety of Birth Intervals: A Stochastic Approach

Ram Chandra Yadava[1] **and Piyush Kant Rai**
Department of Statistics, Banaras Hindu University, Varanasi, India
[1]*Corresponding author: e-mail: rcyadava66@yahoo.co.in*

ABSTRACT

Usually, while studying human fertility through stochastic approach, mainly two dimensions of fertility are considered, viz., (i) how frequently females give births in a given interval of time and (ii) how the births are spaced. The second approach is normally called as study of fertility through birth intervals.

The present chapter mainly focuses its attention on the study of birth intervals. It is known that birth intervals are very much affected by sampling frame and are also affected by truncation of observations. Consequently, there are variety of birth intervals where intervals are ascertained under different conditions. This article attempts to derive probability distributions of different types of birth intervals, their uses, and limitations. It also discusses the problems of drawing valid inferences when population of females is heterogeneous with respect to fertility parameters.

Keywords: Sampling frame, Truncation effect, First birth interval, Closed birth interval, Most recent closed birth interval, Open birth interval, Forward birth interval, Interior birth interval

1 INTRODUCTION

Human fertility is a very complex biosociocultural phenomenon where no single disciplinary approach for its study can cognize the total reality. It is one of the most crucial factors in determining the demographic feature of a country. Precise estimates of human fertility are very important for formulation, implementation, and execution of various development plans as well as family welfare programs.

In order to have deeper understanding of the reproductive process, demographers have long felt need of quantification of different aspects of human fertility. Consequently, a large number of methods and scientific techniques

Handbook of Statistics, Vol. 40. https://doi.org/10.1016/bs.host.2018.09.003

have been developed to analyze various aspects of human fertility. The innovation of mathematical models for human fertility is such an attempt to explain empirical relationships of this complex interdisciplinary phenomenon.

A mathematical model is an abstraction of the real world in which the relevant relations between the real elements are replaced by similar relations between mathematical entities. A detailed discussion on mathematical, deterministic, and stochastic modeling is already presented in our earlier article (Yadava, 2018). Hence we do not present here further discussions on this issue.

It is pertinent to mention here that while studying human fertility through stochastic modeling usually two aspects of fertility phenomenon are studied, viz., (i) how frequently females give births and (ii) how these births are spaced.

These models have been referred as models for: (i) number of births/conceptions occurring to a female in a given period of time and (ii) models for study of distribution of random length of interval between births/conceptions. The second approach of the above is normally referred to as the study of "birth intervals."

In the present article we focus our attention only on the second approach [viz., (ii)] because a detailed description on the first approach [viz., (i)] is already given in our earlier article (Yadava, 2018). A brief description of various types of models on birth intervals is also presented in our earlier article (Yadava, 2018), but it does not include any discussion on some methodological issues such as sampling frame, truncation, and heterogeneity effects involved in the analysis of birth interval data. This article mainly focuses on these methodological issues. (A brief description of these methodological issues is given in the subsequent paragraphs.)

It is important to note that birth interval is essentially a "duration variable" and distribution of a duration variable is very much affected by its sampling frame apart from its own features. The sampling frame is defined as the method by which an individual is ascertained or identified as a member of the population to be sampled and the time reference for the duration variable to be measured (Menken and Sheps, 1972).

Theoretically, data on birth intervals may be obtained taking care of the following two conditions

1. Whether a cohort or a cross section of a population is examined and
2. The sampling frame, i.e., method of ascertainment and the type of interval measured (retrospective, prospective or straddling)

(A brief description of variety of birth intervals is given in Section 1.2.).

Intuitively, it is obvious that, unless a cohort is followed until every member has reached the end of the condition, the observed distribution of duration variable will be incomplete. Additionally, most surveys, irrespective of the method of data collection, produce only cross-sectional data, where the information pertains to the experience up to the date of the survey (or the end of the prospective study). In this situation the analyst is confronted with a set of incomplete information. In this context, it is important to distinguish three problems caused by the incomplete observation of birth intervals, viz., selectivity, censoring, and truncation. Distributions in a cohort are generally not identical with those in a cross section of a population at a specified time and hence separate methods are needed for their proper analysis.

Observed distributions of birth intervals depend in large part on the method by which the data are ascertained (see Sheps and Menken, 1973) and thus appropriate methodologies are needed for proper analysis of such data. In fact, many questions about the properties of birth intervals must be resolved before recommendations can be made for the data to be utilized for drawing proper inferences. It has been mentioned by many authors that unless proper care is taken, the analysis of such data may sometimes produce incorrect and misleading inferences (Henry, 1972; Pathak and Pandey, 1990; Sheps and Menken, 1973; Singh et al., 1979a,b; Srinivasan, 1980; Venkatacharya, 1969; Wolfers, 1968; and others).

The issue of proper analysis can be illustrated with the following examples. Suppose we are interested in analyzing ith order closed birth interval data. Then, ith order closed birth interval is defined for only those females who have given at least $(i+1)$ births prior to any investigation for ascertaining the birth interval. Thus all females who have not given at least $(i+1)$ births prior to the investigation do not become part of the study and thus do not play any role in shaping the distribution of ith order closed birth interval and this complicates the method of analysis.

The chance of not having at least $(i+1)$ births by a female depends mainly on two factors. First, the female may not be able to produce $(i+1)$ birth due to some voluntary or involuntary reasons such as practice of contraception or biological impairment of reproductive organ(s). Other reason may be non-availability of sufficient exposure period (marital duration) for having at least $(i+1)$ births. For example, it is biologically not possible for a female to have five births whose marital duration is only 3 years. This issue also complicates the problem of birth interval analysis which is known as truncation effect. This can be illustrated with the following example. Suppose there are females whose marital duration is 5 years. We want to study the distribution of time interval between marriage and first birth for such females. In this situation, the birth interval cannot exceed 5 years and consequently only such intervals can be part of the study which are less than 5 years. Thus larger intervals (which would have been included had the marital duration been large) will be excluded from the analysis resulting in inappropriate distribution of this birth interval. Thus any straightforward analysis of such data may provide vitiated results.

A pioneering work on truncation effect is given in detail in a classical paper by Sheps et al. (1970). The authors have demonstrated that unless and until one takes care of these facts, one may have incorrect and misleading results. Problems relating to truncation effect have been discussed in detail by other researchers also (see Pathak and Pandey, 1990; Poole, 1973; Sheps and Menken, 1972, 1973; Venkatacharya 1969).

It is known that data on birth intervals are normally collected using retrospective surveys. It is also a known fact that data on birth intervals are very much affected by memory bias and recall lapse in the sense that many of the respondents do not remember the timings of births especially of distant past. Consequently, many times data on closed birth intervals are obtained by considering only last and next to last birth prior to survey date. This may be called as "last closed birth interval" or "most recent closed birth interval." Data on this type of birth interval have been utilized by researchers to draw inferences regarding birth interval dynamics (Mode, 1985; Pandey et al., 1990; Poole, 1973; Sheps and Menken, 1973; Yadava and Sharma, 2004; and others). Although data relating to this interval may be more reliable, it is shown that their distribution will be different than usual closed birth interval and consequently it requires modification in the analysis of such data.

1.1 Heterogeneity and Selection

Besides sampling frame, heterogeneity with respect to parameters involved in the stochastic models may also alter the distribution. Hence, the selection effect within a heterogeneous population is also of our main concern. Heterogeneity in the underlying population places difficulties in the way of interpretation of all statistical data based on averages. If we take a sample of recently married females and follow the individuals in it month by month to establish the duration of first birth or consecutive births, we come face to face with the selection effect of heterogeneity. Those who are more fecund are likely to conceive sooner and consequently remaining will be on an average less fecund. Thus, statistics based on this more fecund group will exaggerate the rate of conception. If everyone had the same conception rate, then the selection effect would not exist.

1.2 Variety of Birth Intervals

For having more clarity on different types of birth intervals it is desirable to give a brief description of these which is presented below.

(i) The interval between marriage to first live birth is defined as *First Birth Interval*.
(ii) The interval between two successive live births is known as *Closed Birth Interval* or *Inter-live Birth Interval*.
(iii) *Last Closed Birth Interval* is the interval between the last and next to last live birth of a female. For example, if a female has given four births before the survey date, then her last closed birth interval is the time interval between her third and fourth births. This interval is also called as *Most Recent Closed Birth Interval*.

(iv) *Interior Birth Interval* is the closed birth interval that begins and ends in any segment of time (which may be the age-group or the marital duration or interval between two survey dates).

(v) The interval between the survey date and the date of last birth is referred to as *Open Birth Interval.*

(vi) The interval between survey date and the date of next live birth posterior to the survey date is known as *Forward Birth Interval.*

(vii) An interval is said to be straddling at a particular age or at a particular time point, if one birth occurs before that age or specified point of time and the other birth occurs later. Any closed birth interval that straddles the survey date is said to be *Straddling Birth Interval.*

The objective of the present article is to describe the distributions of variety of birth intervals under various sampling frames including truncation effect and related inferences. The issue of heterogeneity in fertility parameters on the distribution of birth intervals along with related inferences is also discussed in a separate section. The article also includes a separate section in which the issue of correlation between consecutive closed birth intervals is investigated.

It is pertinent to mention that some discussion on birth intervals is also presented in our earlier article (Yadava, 2018), but the above methodological issues are not discussed there. So, the present article may be considered to be a continuation of the previous article (Yadava, 2018).

An important feature of the present article is that it includes not only the theoretical distributions but also provides application of these with the help of real data which is likely to enhance its utility.

It is also important to mention that most of the results presented here are based on the researches done at the Centre of Population Studies, Department of Statistics, Banaras Hindu University, Varanasi, India, and the results should be seen in this perspective alone. Further it is worthwhile to mention that the symbols used here are mainly as used in the original papers published in different journals.

In this article, at many places, data of National Family Health Survey (NFHS) have been utilized to demonstrate the application of various theoretical results. Thus it seems desirable to give a brief description of NFHS for ready reference.

The National Family Health Survey (NFHS) is a large-scale, multiround survey conducted in a representative sample of households throughout India. Four rounds of the survey (NFHS-1, NFHS-2, NFHS-3, and NFHS-4) have been conducted since the first survey in 1992–93. The survey provides state and national information for India on fertility, infant, and child mortality, the practice of family planning, maternal and child health, reproductive health, nutrition, anemia, utilization, and quality of health and family planning services. The Ministry of Health and Family Welfare (MOHFW), Government of India, designated the International Institute for Population Sciences (IIPS), Mumbai, as the nodal agency, responsible for providing coordination and technical guidance for the survey. For more details, visit the website: http://rchiips.org/nfhs/.

2 IMPACT OF HETEROGENEITY ON DISTRIBUTIONS OF DURATION VARIABLES

In this section we try to study the impact of heterogeneity on the distributions of certain duration variables and related problems of inferences.

2.1 Selection Bias in Postpartum Amenorrhea Period From Follow-Up Studies and Its Adjustment

A paper entitled "An Adjustment of a Selection Bias in Postpartum Amenorrhea from Follow-up Studies" was published by Singh et al. (1979a,b) in *Journal of American Statistical Association, December 1979, Volume 74, Number 368, Theory and Methods Section.* Before discussing the issue of selection bias and its adjustment, it is desirable and interesting to describe the very genesis of the above paper.

A demographic survey of Varanasi (Rural), U.P., India, was conducted in 1969–70 under the auspices of Demographic Research Centre (presently known as Centre of Population Studies), Banaras Hindu University in which data on fertility, mortality, and migration along with other socioeconomic characteristics were collected from about 2500 households selected in the survey. In this survey, apart from other information, data on postpartum amenorrhea (PPA) period related to all births occurring during last 7 years from the date of survey were also collected.

After collection of data, it was thought that data on PPA may suffer from memory bias especially for births occurring 6–7 years before the survey date. To investigate this issue, all births occurring during last 7 years were classified according to year of birth and average PPA for births occurring in different years was obtained. After analysis, it was found that there was no variation (except from sampling fluctuations) in average PPA for births in different years indicating that there was no evidence of memory bias for births occurring in different years.

However, when data on PPA relating to last birth and next to last birth were analyzed, it was found that this mean was somewhat larger than the mean PPA for births occurring in a year.

However, when this issue was investigated on theory basis, no appropriate explanation could be given on the matter and the whole matter remained unexplained. One day, the following idea came in our mind.

Suppose there are two types of coins for which probability of success (head), i.e., $P(S)$ is a random variable which has the distribution as

$$P(S) = 0.5 \text{ with probability } 0.5 \text{ and } P(S) = 0.3 \text{ with probability } 0.5.$$

Suppose there are 100 coins of first type for which $P(S)=0.5$ and there are other 100 coins for which $P(S)=0.3$ so that the condition of

$$P(S)=0.5 \text{ with probability } 0.5 \text{ and } P(S) \\ =0.3 \text{ with probability } 0.5 \text{ is satisfied.}$$

Now these 200 coins are tossed together once, then, on an average, there will be 80 successes in all out of which 50 will be from coins of first type, while other 30 will be from second type. Thus the distribution of $P(S)$ will be

$$P(S)=0.5 \text{ with probability } 5/8 \text{ and } P(S)=0.3 \text{ with probability } 3/8.$$

Thus the proportion of two types of coins changes and coins for which $P(S)$ is larger are overrepresented. This gave the idea that the females who give births more frequently in a given time may be overrepresented resulting in a selection bias.

The whole issue is explained mathematically in the aforesaid paper. But the paper does not describe its technical and mathematical derivations in detail. Our aim is to provide a complete description of derivation of mathematical calculations which provide easiness to understand the facts for the readers of the paper.

In surveys employing the method of retrospective inquiries, information collected about events that occurred in distant past are subject to lapses of memory and other sources of biases that often depend on the recall period. Further, the responses on certain duration variables, that is, the interval between the occurrence of two defined events (accidents to a person, recurrence of a particular disease to a person, births to a female, appearance of menstruation after delivery, etc.) are relatively more affected by recall lapse than those relating to the number of such events occurring in a given period of time. Because of this phenomenon, prospective studies are preferred for obtaining data on duration variables, especially when the probability of occurrence of relatively longer intervals is small.

It should be noted, however, that apart from other factors the observed distributions can be affected by the choice of sampling frame (Sheps and Menken, 1972). Wolfers (1968), while analyzing the data on birth intervals, that is the time between successive births, pointed out that the significance of the observation often depends as much on the way in which data are collected as on the values themselves.

The duration of PPA, the interval from a delivery to the start of the first menstrual cycle, is an important component of the interval between two successive births. Hence, for modeling human fertility, it is desirable to ascertain PPA's distribution. Recently attempts have been made to obtain reliable data on the duration of PPA through follow-up studies. In such enquiries, females who give birth during a specified small interval of time (usually less than a year) are contacted regularly until the resumption of menstruation or the termination of survey period, whichever is earlier. Although such data are almost free from memory bias, a selection bias (overrepresentation of smaller PPA) is introduced. The selection bias arises mainly because of the fact that in follow-up studies the females with more frequent births have a relatively higher chance of inclusion in the study.

It should be emphasized here that most methodological work on PPA intervals has focused on reducing memory bias but has tended to ignore the selection bias. An exception is the article by Wolfers (1968) dealing with selection bias in birth intervals. He has distinguished between birth intervals measured from all births in a given period and birth intervals characteristic of women in a population. He has given a procedure to estimate the mean period of nonsusceptibility (the period of gestation + PPA interval), but no satisfactory procedure has yet been reported for computing the distribution of PPA in the population from data obtained through follow-up studies.

The objective of the present paper is to investigate the nature of the selection bias and to derive a procedure to obtain the distribution of PPA free from this bias. In view of the fact that the gestation period is almost constant, we have taken it as a constant and have considered the distribution of nonsusceptible period rather than the distribution of PPA because the two distributions will be similar except for a shift of origin. A detailed account of the bias is given in the following section, while the proposed procedure is outlined in the continued sections and an illustration of the technique is shown to observe data taken from Demographic Survey of Varanasi (Rural), India, 1969–70.

2.1.1 The Selection Bias

To examine the bias, consider a population of married females at time T, which is heterogeneous both with respect to period of nonsusceptibility g, which is the time that elapses between a conception and the first ovulation after delivery; and the conception rate φ implying that φdt is the probability of a conception in infinitely small interval dt.

The period of nonsusceptibility and the conception rate vary among the females and for any particular female may be age or parity dependent. For simplicity, however, we assume the period of nonsusceptibility and fecundability to be constant for each female. Such an assumption is not uncommon in modeling human fertility (Brass 1958; Dandekar 1955; and others). Thus, in the population every female can be characterized by a pair of values of φ and g ($0<\varphi<a$, $b<g<c$ where a, b, and c are positive and finite). Let the joint and marginal distributions of φ and g be denoted by $f(\varphi,g)$, $f_1(\varphi)$, and $f_2(g)$, respectively. Further, let the conditional distribution of φ for a given g be denoted by $f_3(\varphi|g)$.

For explaining the selection bias we have used the concept of asymptotic fertility rate. So it is desirable to explain the above rate.

Let $u(\varphi, g, t)dt$ be the expected number of births to a female in the interval $(t, t+dt)$ at a time t from the marriage of a female. Under certain conditions Henry (1972) has shown mathematically that, for given φ and g, $u(\varphi, g, t)$ is sum of a constant and an oscillating function that becomes damped with increasing t. Also for large t, $u(\varphi, g, t)$ can be approximated by a constant $\psi_{\phi,g} = \phi/(1+g\phi)$, which is designated as the asymptotic fertility rate for given ϕ and g (see eqn. (5) on p. 7 in Henry, 1972). Allowing for the heterogeneity in the population with respect to φ and g, the expected asymptotic fertility rate ψ is given by

$$\begin{aligned}\psi &= \int_b^c \int_0^a \lim_t u(\phi, g, t) d\phi dg \\ &= \iint \frac{\varphi}{1+g\varphi} f(\varphi, g) d\varphi dg\end{aligned} \tag{1}$$

For any $t > 0$, $u(\varphi, g, t)$ is nonnegative, finite, and $f(\varphi, g)$ is a probability distribution. Thus according to the dominated convergence theorem, the asymptotic expected fertility rate

$$\lim_t \int_b^c \int_0^a u(\varphi, g, t) d\varphi dg \tag{2}$$

is the same as the expected asymptotic fertility rate ψ. Hence we do not distinguish between the two rates and call them simply the asymptotic fertility rate. The asymptotic fertility rate ψ_g for given g is

$$\psi_g = \int \frac{\phi}{1+g\phi} f_3(\phi|g) d\phi \tag{3}$$

and

$$\psi = \int f_2(g) \psi_g dg \tag{4}$$

So

$$\begin{aligned}\psi &= \iint \frac{\phi}{1+g\phi} f_2(g) \frac{f(\phi, g)}{f_2(g)} d\phi dg \\ &= \iint \frac{\phi}{1+g\phi} f(\phi, g) d\phi dg\end{aligned} \tag{5}$$

Thus, in a follow-up study, in which the females giving birth in a small interval of time are followed for the ascertainment of PPA, the distribution of g becomes

$$f_4(g) = \left[\frac{\psi_g}{\psi}\right] f_2(g) = W_g f_2(g) \tag{6}$$

It is seen from above equation that $f_4(g)$ becomes the true distribution of g, i.e., $f_2(g)$ if and only if $W_g = 1$ for all g. This can happen when the asymptotic fertility of females in the population with different values of g is the same. If the value of W_g is not equal to 1 for all g, then we say that there is a selection bias.

It is worthwhile to mention that although the theory is based on asymptotic results, it does not pose a serious problem for practical applications. For example, Perrin and Sheps (1964) and Sheps and Menken (1971) have shown (using numerical examples) that in most of the cases the equilibrium (asymptotic result) is reached within a period of about 60 months from marriage. Thus, the procedure may be used for adjusting the PPA period of females having marriages of more than 5 years duration. The proportion of such females is quite large in the population of married females in the reproductive period.

2.1.1.1 We Examine the Behavior of W_g for the Following Three Cases

Case 1. If φ and g are independently distributed, ψ_g decreases as g increases and, in turn, W_g decreases monotonically. This shows that in follow-up studies smaller periods of nonsusceptibility are overrepresented.

Case 2. When φ and g are negatively associated, W_g generally decreases as g increases. This holds strictly when $F_3(\varphi|g) < F_3(\varphi|g^*)$ for $g < g^*$, where $F_3(\varphi|g)$ denotes the distribution function corresponding to $f_3(\varphi|g)$. Thus the pattern of bias is similar to that in Case 1 but its magnitude is more pronounced.

Case 3. When φ and g are positively associated, generally $F_3(\varphi|g) > F_3(\varphi|g^*)$ for $g < g^*$, hence, nothing definite can be said about the trend of W_g.

2.1.2 Procedure for Adjusting the Selection Bias

A procedure to find the true distribution of the period of nonsusceptibility from the data obtained through follow-up studies is outlined here when φ and g are independent and φ is constant.

Under this assumption

$$\frac{\psi_g}{\psi} = \frac{\left(\frac{\phi}{1+g\phi}\right)}{\int \left(\frac{\phi}{1+g\phi}\right) f_2(g)dg} \tag{7}$$

We know that

$$f_4(g) = \left[\frac{\psi_g}{\psi}\right] f_2(g) \tag{8}$$

So,

$$f_2(g) = \left[\frac{\psi}{\psi_g}\right] f_4(g) \tag{9}$$

We also know that $\psi_g = \phi/(1 + g\phi)$ (for φ constant).
So,

$$f_2(g) = \left[\frac{\psi}{\dfrac{\phi}{1+g\phi}}\right] f_4(g) = \psi\left[\frac{1+g\phi}{\phi}\right] f_4(g) \tag{10}$$

Since $f_2(g)$ is a probability density function, hence

$$\int f_2(g)dg = 1$$

Thus

$$\begin{aligned} &\frac{\psi}{\phi}\int(1+g\phi)f_4(g)dg = 1 \\ &\frac{\psi}{\phi}(1+\overline{g}'\phi) = 1 \\ &\psi = \frac{\phi}{(1+\overline{g}'\phi)} \end{aligned} \tag{11}$$

where $\overline{g}' = \int g f_4(g)dg$, i.e., $\overline{g}'$ is mean duration of nonsusceptibility for data from follow-up study.

Consequently,

$$f_2(g) = \left[\frac{1+g\varphi}{1+\overline{g}'\varphi}\right] f_4(g) \tag{12}$$

Using the above expression, the distribution $f_2(g)$ can be easily obtained from the distribution $f_4(g)$. It is to be noted that $f_4(g)$ is the distribution of g which is available from follow-up study. Obviously $\overline{g}'$ can be obtained from $f_4(g)$ as $\overline{g}' = \int g f_4(g)dg$.

Thus

$$\overline{g} = \int g f_2(g)dg, \tag{13}$$

which is the mean of nonsusceptible period for the females in the population can be obtained easily.

Alternatively, we can obtain the expression for $\overline{g}$ as follows:

$$\begin{aligned}
\overline{g} &= \int g f_2(g)dg \\
&= \int g\left[\frac{1+g\varphi}{1+\overline{g}'\varphi}\right] f_4(g)dg \\
&= \int g\left[\frac{1+\overline{g}'\varphi+g\varphi-\overline{g}'\varphi}{1+\overline{g}'\varphi}\right] f_4(g)dg \\
&= \int g\left[\frac{1+\overline{g}'\varphi}{1+\overline{g}'\varphi}+\frac{g\varphi-\overline{g}'\varphi}{1+\overline{g}'\varphi}\right] f_4(g)dg \\
&= \int g\left[\frac{1+\overline{g}'\varphi}{1+\overline{g}'\varphi}\right] f_4(g)dg + \int g\left[\frac{g\varphi-\overline{g}'\varphi}{1+\overline{g}'\varphi}\right] f_4(g)dg \\
&= \overline{g}' + \int g\left[\frac{g\varphi-\overline{g}'\varphi}{1+\overline{g}'\varphi}\right] f_4(g)dg \\
&= \overline{g}' + \frac{\phi}{1+\overline{g}'\phi}\int g(g-\overline{g}')f_4(g)dg \\
&= \overline{g}' + \frac{\phi}{1+\overline{g}'\phi}\left[\int g^2 f_4(g)dg - (\overline{g}')^2\right] \\
&= \overline{g}' + \frac{\phi}{1+\overline{g}'\phi}\sigma'^2 \\
&= \overline{g}' + \psi\sigma'^2
\end{aligned} \tag{14}$$

where $\psi = \phi/(1+\overline{g}'\phi)$ is the asymptotic fertility rate and σ'^2 is the variance of the probability density functions $f_4(g)$. Obviously $\overline{g} > \overline{g}'$. The equality holds if and only if σ'^2 is zero, that is, the population is homogeneous with respect to g also. Again, the difference between $\overline{g}$ and $\overline{g}'$ increases with φ, the level of fecundability.

The relationship between $f_2(g)$ and $f_4(g)$ provides the information on $f_2(g)$ which can be determined subject to the condition that φ is known. Usually φ is not known, but it may be estimated from the data on birth intervals or number of births in a given interval of time. This estimation procedure depends on the accuracy of the assumed (or estimated) value of φ. For examining the effect of difference in assumed (or estimated) and true value of φ on the estimated distribution $f_2(g)$, certain test calculations were performed which showed that for moderate differences in φ, the estimated distribution $f_2(g)$ does not deviate much from the true distribution. However for large differences in φ, the estimated distribution $f_2(g)$ may be quite different from the actual distribution. The detailed test calculations are given in the Table 1 of the paper by Singh et al. (1979a,b). The interested readers are advised to see the above table in the given paper.

We now demonstrate the application of adjustment of selection bias (assuming ϕ to be constant) with help of a real data set collected in the Demographic Survey of Varanasi (Rural) 1969–70.

TABLE 1 Distribution of the Duration of PPA

PPA in Months	Mid Value	Mid Value +9	Observed Distribution for Births Occurring Between 3 and 7 Years Before Reference Date		Computed Adjusted Proportions	Observed Frequency for Next-to-Last Birth	Expected Adjusted Frequency
			Frequency	*Proportion*			
0–4	2	11	361	0.3495	0.2725	123	117.4
5–9	7	16	105	0.1016	0.0920	40	39.70
10–14	12	21	263	0.2526	0.2626	104	113.20
15–19	17	26	138	0.1336	0.1546	68	66.60
20–24	22	31	105	0.1016	0.1303	57	56.20
25–29	27	36	36	0.0348	0.0490	22	21.10
30–34	32	41	10	0.0097	0.0149	17	16.80
35–39	37	46	15	0.0145	0.0241		
Total			1033	1.000	1.000	431	431.0
					$g'=19.7585$	$\overline{g}=21.5265$	$\chi^2(df=6)=1.10$

As already mentioned earlier, in the survey, data were obtained for PPA duration relating to births occurring in different years before the reference date of the survey.

While analyzing the pattern of PPA, Singh et al. (1971) used the observed distribution of PPA relating to births occurring in 1-year periods in different years before the date of survey. Hypothetically, such data may be treated as a result of a follow-up study, and consequently the procedure discussed above is applicable. Column 4 of Table 1 presents the observed (i.e., unadjusted) frequency distribution of the duration of PPA relating to births occurring between 3 and 7 years before the reference date of the survey and corresponding proportions are given in column 5 of the table. This distribution may be considered as observed distribution of $f_4(g)$. The distribution refers to the PPA following these births. The PPAs corresponding to births occurring within the last 3 years from the reference date have been excluded to avoid truncation bias. Using different models to the data of above survey the value of conception rate (i.e., ϕ) was estimated in the vicinity of 0.05 so we take the value of $\phi=0.05$. The adjusted distribution ($f_2(g)$) corresponding to the above observed distribution is presented in column 6 of the table. Column 7 presents the observed frequency distribution of PPA duration for next-to-last birth. To check the agreement between observed and expected distributions of $f_2(g)$, column 8 presents the expected frequencies related to frequencies given in column 7. The χ^2 test for testing the agreement between frequencies of column 7 and 8 was performed and the value of χ^2 is found to be 1.10 which is insignificant. So we can say that there is good agreement between the observed and theoretical distributions of $f_2(g)$ utilizing the proposed procedure of adjustment of selection bias.

It is to be pointed out that the above illustration assumes that ϕ is constant. However, Singh et al. (1979a,b) have also considered the case when ϕ varies among females.

2.2 Estimating Birth Interval Characteristic of Women

Wolfers (1968) has made the distinction between mean birth interval (births) and mean birth interval (women) and given certain examples to distinguish between the two mean birth intervals. However we still give a simple example, heuristic in nature, to demonstrate the difference between the two in a simpler way. Suppose, there are two females, one giving births on an average interval of 3 years, while the other gives births on an average interval of 2 years. So, for them the average birth interval would be $(2+3)/2=2.5$ years. Alternatively, let us consider a period of 6 years, and then the first female will give on an average two births in 6 years, while the other will be giving on an average three births in 6 years. So, in all there will be five birth intervals, two birth intervals of 3 years and three birth intervals of 2 years. So, for five births, total duration for births will be 12 years $(2\times 3 + 3\times 2)$, hence, average birth interval for one birth is $(12/5)=2.4$ years which is less than 2.5 years. This shows the basic difference between mean birth interval (births) and mean birth interval (women). The 2.5 years, mean birth interval relates to mean birth interval (women), while 2.4 years relates to mean birth interval (births).

We give a procedure to estimate mean birth interval (women) from data on mean birth interval (births). The present content may be considered to be an extension of the paper of Singh et al. (1979a,b). In Singh et al. (1979a,b) paper, the mean PPA (for women) was estimated from mean PPA for births. Here the procedure is proposed to estimate mean birth interval (women) from data on mean birth interval (births). We use here the same notations as given in Singh et al. (1979a,b).

This procedure may be considered to be an alternative procedure to the method suggested by Wolfers (1968) for obtaining the mean birth interval characteristic of women from the data on birth interval (births). It has been shown that Wolfers procedure involves some methodological problems. The proposed procedure is found to be relatively simple, and free from those methodological problems involved in Wolfers (1968).

2.2.1 *The Proposed Procedure and Its Applications*

Singh et al. (1979a,b) have developed a procedure for the adjustment of selection bias in the PPA. Here an alternative procedure for estimating the mean birth interval for women from the data on birth interval (births) is derived.

As discussed in the previous section a population of married women heterogeneous with respect to nonsusceptible period g and fecundability ϕ is considered. Let the joint and marginal distributions of φ and g be denoted by $f(\varphi, g)$, $f_1(\varphi)$, and $f_2(g)$, respectively, and let $\psi_{\phi,\ g} = \phi/(1+g\phi)$ be designated as the asymptotic fertility rate for given ϕ and g. Also, let the asymptotic fertility rate for given g is ψ_g and ψ_φ be the one when g is given. Further let ψ denote the asymptotic fertility rate in the population. In a population where the women giving birth in a small interval of time are included in a study, the distribution of φ and g will be

$$f_3(\phi) = \frac{\psi_\phi}{\psi} f_1(\phi) \text{ and } f_4(g) = \frac{\psi_g}{\psi} f_2(g) \tag{15}$$

and their joint density is given by

$$f_5(\phi, g) = \frac{\psi_{\phi,g}}{\psi} f(\phi, g) \tag{16}$$

It can be seen that $f_3(\varphi)$ and $f_4(g)$ are quite different, respectively, from $f_1(\varphi)$ and $f_2(g)$ (see Singh et al., 1979a,b).

Now, let X denote the length of the birth interval of the women giving birth in a small interval of time. In fact X is sum of two components, viz., g and Y where g is a nonsusceptible period and Y is a random variable representing menstruating interval and for given φ its pdf is

$$p(y) = \phi e^{-\phi y}, \quad y > 0 \tag{17}$$

Thus

$$E(X|\phi, g) = E(g+Y) = E(g) + E(Y) = g + \frac{1}{\phi} = \frac{1+g\phi}{\phi} \tag{18}$$

(for given ϕ and g)

Thus unconditional expectation of X is given as

$$\begin{aligned} E(X) &= \iint \frac{1+g\phi}{\phi} f_5(\phi, g)\, d\phi\, dg = \iint \frac{1+g\phi}{\phi} \frac{\phi}{1+g\phi} \frac{1}{\psi} f(\phi, g)\, d\phi\, dg \\ &= \frac{1}{\psi} \iint f(\phi, g)\, d\phi\, dg \\ &= \frac{1}{\psi} = M(\text{say}) \end{aligned} \tag{19}$$

For given ϕ and g

$$\begin{aligned} E(X^2) &= E(g^2) + E(Y^2) + 2E(gY) = g^2 + \frac{2}{\phi^2} + 2g\frac{1}{\phi} \left(\text{because } E(Y^2) = \frac{2}{\phi^2}\right) \\ &= \frac{g^2\phi^2 + 2 + 2g\phi}{\phi^2} \\ &= \frac{1 + (1+\phi g)^2}{\phi^2} \end{aligned} \tag{20}$$

(Note: There seems to be misprint in the expression for $E(X^2)$ of equation 3 of the paper Singh et al. (1987) written as $E(X^2) = \iint \frac{(1+g\phi)^2}{\phi^2} \frac{\phi}{1+g\phi} \frac{1}{\psi} f(\phi, g)\, d\phi\, dg$ which is rectified in this article.)

Thus, unconditional expectation of X^2 is

$$\begin{aligned} E(X^2) &= \iint \frac{(1+\phi g)^2 + 1}{\phi^2} f_5(\phi, g)\, d\phi\, dg \\ &= \iint \frac{(1+\phi g)^2 + 1}{\phi^2} \left(\frac{\phi}{1+\phi g}\right) \frac{1}{\psi} f(\phi, g)\, d\phi\, dg \\ &= \iint \frac{(1+\phi g)^2}{\phi^2} \left(\frac{\phi}{1+\phi g}\right) \frac{1}{\psi} f(\phi, g)\, d\phi\, dg + \iint \frac{1}{\phi^2} \left(\frac{\phi}{1+\phi g}\right) \frac{1}{\psi} f(\phi, g)\, d\phi\, dg \\ &= \iint \frac{(1+\phi g)}{\phi} \frac{1}{\psi} f(\phi, g)\, d\phi\, dg + \iint \frac{1}{\phi} \left(\frac{1}{1+\phi g}\right) \frac{1}{\psi} f(\phi, g)\, d\phi\, dg \\ &= M \iint \frac{(1+\phi g)}{\phi} f(\phi, g)\, d\phi\, dg + \iint \frac{1}{\phi} M \left(\frac{1}{1+\phi g}\right) f(\phi, g)\, d\phi\, dg \\ &= W \cdot M + \iint \frac{M}{\phi(1+\phi g)} f(\phi, g)\, d\phi\, dg \end{aligned} \tag{21}$$

Here, $W=\int\int\frac{1+g\phi}{\phi}f(\phi,g)d\phi dg$ is the mean birth interval for women. Thus,

$$\begin{aligned}
\frac{E(X^2)}{M} &= W+\int\int\frac{1}{\phi(1+\phi g)}f(\phi,g)d\phi dg\\
&= W+\int\int\frac{1}{\phi(1+\phi g)}f_1(\phi)f_2(g)d\phi dg\\
&= W+\int\frac{1}{\phi(1+\phi\overline{g}')}f_1(\phi)d\phi\left(\text{Because we know that }\frac{1}{1+\phi\overline{g}'}=\int\frac{1}{1+\phi g}f_2(g)dg\right)\\
&= W+\int\left(\frac{1}{\phi}-\frac{\overline{g}'}{(1+\phi\overline{g}')}\right)f_1(\phi)\ d\phi\left(\text{Here, we can write },\frac{1}{\phi(1+\phi\overline{g}')}=\frac{1}{\phi}-\frac{\overline{g}'}{1+\phi\overline{g}'}\right)\\
&= W+\int\frac{1}{\phi}f_1(\phi)d\phi-\int\frac{\overline{g}'}{1+\phi\overline{g}'}f_1(\phi)\ d\phi\\
&= W+\frac{1}{\phi_H}-\int\overline{g}'\left(1-\frac{\phi\overline{g}'}{1+\phi\overline{g}'}\right)f_1(\phi)\ d\phi\left(\text{Here, we can write, }\frac{1}{(1+\phi\overline{g}')}=1-\frac{\phi\overline{g}'}{1+\phi\overline{g}'}\right)\\
&= W+(W-\overline{g})-\left[\int\overline{g}'f_1(\phi)d\phi-\overline{g}'\int\overline{g}'\frac{\phi}{1+\phi\overline{g}'}f_1(\phi)d\phi\right]\\
&= 2W-\overline{g}-\left[\overline{g}'-\overline{g}'\frac{\overline{g}'}{M}\right]\quad\text{Here, }\int\frac{\phi}{1+\overline{g}'\phi}f_1(\phi)d\phi=\psi=\frac{1}{M}\\
&= 2W-\overline{g}-\overline{g}'\left[1-\frac{\overline{g}'}{M}\right]
\end{aligned}\tag{22}$$

Proceeding as discussed in Singh et al. (1979a,b), and denoting $E(X^2)/M=A$ (as given in Wolfers, 1968), it is obtained that

$$A=2W-\overline{g}-\overline{g}'\left(1-\frac{\overline{g}'}{M}\right)\tag{23}$$

where

$$\overline{g}=\int gf_2(g)dg\text{ and }\overline{g}'=\int gf_4(g)dg$$

It may be pointed out that the above expression is obtained by assuming $\overline{g}'$ to be constant with respect to φ. Thus the value of W can be estimated from above as

$$W=\frac{1}{2}\left[A+\overline{g}+\overline{g}'\left(1-\frac{\overline{g}'}{M}\right)\right]\tag{24}$$

with the knowledge of the values of A,M,$\overline{g}$, and $\overline{g}'$. In a study where data on birth intervals and the length of PPA are available from the women giving birth in a fixed period of time, the values of A,M, and $\overline{g}'$ are known, and the value of $\overline{g}$ can be estimated from the expression already proved as $\overline{g}'=\overline{g}-\left(\sigma'^2/M\right)$, where σ'^2 is the variance of the duration of PPA for women giving birth in a small interval of time.

The above procedure was applied to the real data collected in the Demographic Survey of Varanasi (Rural) 1969–70 and related results are given in the paper of Singh et al. (1987). They have also illustrated the application of the procedure considering some hypothetical distributions of φ and g. Interested readers may see the above paper for more details.

2.3 Impact of Heterogeneity on Time of First Conception

In this section, we study the impact of heterogeneity on the time of first conception by computing the conditional probability of conception in a given month.

Let X denote the time of first conception after marriage assuming time to be discrete (in months). Let p denote the fecundability then under the assumptions similar to Bernoulli trials, X follows a geometric distribution, i.e.,

$$\begin{aligned} P[X=x] &= q^{x-1}p \quad x=1,2,\ldots \quad 0<p<1 \quad q=1-p \\ &= 0 \quad \text{otherwise} \end{aligned} \tag{25}$$

In this context, the conditional probability of conception in xth month is given as

$$P[X=x|\text{ no conception till}(x-1)\text{th month}] = \frac{q^{x-1}p}{q^{x-1}} = p \tag{26}$$

because probability that no conception occurs till $(x-1)$th month is equal to q^{x-1}. This conditional probability is constant which is analogous to hazard rate in the context of life testing.

The above result is true if p is assumed to be constant. However, if a population of females is heterogeneous with respect to p, then the property that the conditional probability of conception in a month is constant is no more valid. We show this result below.

Let p have the pdf $f(p)$, then the probability of conception in first month is given as

$$P[x=1] = \int p \cdot f(p)dp = \overline{p}(\text{say}) \tag{27}$$

The conditional probability of conception in second month is

$$\begin{aligned} P[x=2|\text{ no conception in first month}] &= \frac{\int qp \cdot f(p)\,dp}{\int q \cdot f(p)\,dp} \\ &= \frac{\int p \cdot f(p)dp - \int p^2 \cdot f(p)dp}{1-\overline{p}} \\ &= \frac{\overline{p} - (\overline{p})^2 - \sigma_p^2}{1-\overline{p}} \end{aligned} \tag{28}$$

where σ_p^2 is variance of

$$\begin{aligned} p &= \frac{\overline{p}(1-\overline{p})-\sigma_p^2}{1-\overline{p}} \\ &= \overline{p} - \frac{\sigma_p^2}{1-\overline{p}} \end{aligned} \tag{29}$$

which is less than $\overline{p}$. Thus the conditional probability of conception in the second month is less than the probability of conception in first month.

A natural question arises: whether the conditional probability of conception in a month decreases over time (in months). The answer is "yes." We prove this result below. (The proof is already given in Sheps and Menken (1973); however, for ready reference, we present the result here also.)

Let c_x^* denote the probability that the female conceives in the xth month, then

$$\begin{aligned} c_x^* &= \int q^{x-1} p \cdot f(p)\, dp \\ &= \int q^{x-1} f(p) dp - \int q^x f(p) dp \\ &= c_x - c_{x+1} \text{ (say)} \end{aligned} \tag{30}$$

where

$$c_x = \int q^{x-1} f(p)\, dp \text{ and } c_{x+1} = \int q^x f(p)\, dp \tag{31}$$

Thus the conditional probability of conception in xth month is given as

$$P[X = x | \text{ no conception till}(x-1)\text{th month}] = \frac{c_x - c_{x-1}}{c_x} \tag{32}$$

Similarly

$$P[X = x+1 | \text{ no conception till } x\text{th month}] = \frac{c_{x+1} - c_{x+2}}{c_{x+1}} \tag{33}$$

Now we prove that

$$\begin{aligned} &\frac{c_x - c_{x+1}}{c_x} \geq \frac{c_{x+1} - c_{x+2}}{c_{x+1}} \\ &\text{i.e., } \frac{c_x - c_{x+1}}{c_x} - \frac{c_{x+1} - c_{x+2}}{c_{x+1}} \geq 0 \end{aligned} \tag{34}$$

We see that

$$\begin{aligned} \frac{c_x - c_{x+1}}{c_x} - \frac{c_{x+1} - c_{x+2}}{c_{x+1}} &= \frac{\left(c_x \cdot c_{x+1} - c_{x+1}^2\right) - \left(c_x \cdot c_{x+1} - c_x \cdot c_{x+2}\right)}{c_x \cdot c_{x+1}} \\ &= \frac{c_x \cdot c_{x+2} - c_{x+1}^2}{c_x \cdot c_{x+1}} \end{aligned} \tag{35}$$

We see that $c_x = \int q^{x-1} f(p)\,dp$ which may be considered to be $(x-1)$th moment of q (a function of p) while c_{x+1} and c_{x+2} may be considered to be xth and $(x+1)$th moments of q.

From Liapounoff's inequality for the moments, we know that

$$c_x \cdot c_{x+2} - c_{x+1}^2 \geq 0 \tag{36}$$

For ready reference we provide the Liapounoff's inequality:

Suppose X is a random variable which takes only nonnegative values, then

$$\left(\mu^{(b)}\right)^{a-c} \leq \left(\mu^{(c)}\right)^{a-b} \left(\mu^{(a)}\right)^{b-c}, \quad a \geq b \geq c \geq 0$$

where $\mu^{(a)}$, $\mu^{(b)}$, and $\mu^{(c)}$ are ath, bth, and cth order moment of X.

(In our case, $a = x+1$, $b = x$, and $c = x-1$).

Hence we see that conditional probability of conception in a month decreases with time (x).

Intuitionally, this result can be thought to be obvious because females with higher fecundability conceive early leaving less fecund females to conceive later.

Further

$$\frac{c_x - c_{x+1}}{c_x} = \frac{\int q^{x-1} p \cdot f(p)\,dp}{\int q^{x-1} f(p)\,dp} = \overline{p}_x\,(\text{say}) \tag{37}$$

may be interpreted as average fecundability of females, conceiving in xth month. Similarly, average fecundability of females conceiving in $(x+1)$th month will be

$$\overline{p}_{x+1} = \frac{c_{x+1} - c_{x+2}}{c_{x+1}} \tag{38}$$

Thus average fecundability of females conceiving in months $(x, x+1, x+2, \ldots, x+y)$ say $\overline{p}(x, x+y)$ will be

$$\begin{aligned}\overline{p}(x, x+y) &= \frac{c_x - c_{x+1}}{c_x} \cdot c_x + \frac{c_{x+1} - c_{x+2}}{c_{x+1}} \cdot c_{x+1} + \cdots + \frac{c_{x+y} - c_{x+y+1}}{c_{x+y}} \cdot c_{x+y} \\ &= \frac{c_x - c_{x+y+1}}{c_x + c_{x+1} + \cdots + c_{x+y}}\end{aligned} \tag{39}$$

Most probably this result is also derived in Sheps (1965).

From the above results we see that if the data on time of first conception show that the conditional probability of conception remains almost constant over time, then the population may be considered to be homogeneous with respect to the fecundability p. However if the data show that conditional probability of conception in a month decreases over time, then the population of females may be considered to be heterogeneous with respect to fecundability.

3 SAMPLING FRAME AS A DETERMINANT OF DURATION VARIABLE

In this section we present the impact of sampling frame on different types of birth intervals.

3.1 Usual Closed Birth Interval vs Most Recent Closed Birth Interval

The data on birth intervals can be obtained in quite different ways under different sampling frames, and hence appropriate techniques are needed for proper analysis of such data; otherwise, one may draw misleading and invalid conclusions (see Fisher, 1934; Rao, 1965; Sheps and Menken, 1972, 1973; Sheps et al., 1970; Singh et al., 1979a,b; Srinivasan, 1967a,b, 1980; Wolfers, 1968; Yadava and Sharma, 2004; and others). For example, a female having given three births from marriage to end of reproductive span or survey date has three birth intervals (i) interval from marriage to first birth, (ii) interval from first to second birth, and (iii) interval from second to third birth. In general, the interval between ith and $(i+1)$th births is normally called as closed birth interval (CBI) of order i or ith order CBI. In retrospective fertility surveys, the CBI of ith order is defined only for those females who have given at least $(i+1)$ births in the given marital duration. A CBI of any specific order is normally ascertained for all those women who ever experienced that interval, but for each female there will be a last CBI which is a parity-specific most recent CBI before the date of survey. To make a distinction between the two, the earlier is termed as usual CBI, while the latter as most recent CBI. Thus the ith order usual CBI is defined for those females who have given at least $(i+1)$ births, while ith order most recent CBI is defined for those females who have given exactly $(i+1)$ births. The most recent CBI is relatively easy to be ascertained and refers closely to the current fertility of women in the population under study. The distribution of most recent CBI is quite sensitive to small changes in fertility and robust with respect to nonfertility parameters (see Sehgal, 1971). The collection of birth interval data in underdeveloped countries suffers from many nonsampling errors, especially recall lapse on the part of respondents. These errors, which are sometimes serious, often vitiate the data if the duration of time elapsed since the occurrence of the event is large. Consequently, the data on most recent CBI seem to be less affected by such memory biases and, they are likely to be more reliable for the analysis of fertility changes among married women compared to usual CBI.

Usually, one may think that the distribution of CBI of any specific order may be the same as the distribution of most recent CBI of the same order. But it is not true. Sheps and Menken (1972) have remarked that for a given age (or marital duration), the mean of most recent CBI is somewhat larger than the other CBI based on the cohort approach, as this interval usually tends to select larger values more frequently resulting in higher mean. Sharma (2004) has also considered the distinction between the most recent CBI and usual CBI. He assumes a stationary population of females, where specific order of births is uniformly distributed over time, and demonstrated that if the population of females is homogeneous with respect to fertility parameters (conception rate and PPA period), then there is no difference between the distribution of usual CBI and most recent CBI, whereas if females are heterogeneous with respect to the fertility parameters, then the two distributions are distinct, and the mean of most recent CBI is somewhat larger than the mean of usual CBI of that order.

Some general distributions for the most recent CBI have been derived by Poole (1973) and Sheps and Menken (1972, 1973). Sheps and Menken (1973) have derived results regarding the most recent CBI considering general distributions between two events (births). Based on these results, they have given their remarks regarding most recent CBI, wherein one has to assume some specific forms of distribution functions of various components of a birth interval. Singh et al. (1988) derived probability models for the most recent CBI considering specific distributions for various components of CBI. Later, Pandey et al. (1998) extended the above model accounting for the nonexposure period in the beginning of the reproductive life caused due to adolescent sterility and temporary separation between the partners. Both have illustrated their models by applying them on an observed set of data. However, both the models do not attempt to study the distinction between the usual CBI and the most recent CBI. Abul-ata (1987) has derived expressions for the distributions of usual CBI and most recent CBI for specific parity with fixed marital duration by considering the interval between consecutive births as exponentially distributed. This assumption does not seem to be appropriate in the context of birth interval studies, because the timing between consecutive births not only includes the menstruating interval but also the gestation period and PPA period, which essentially cannot be ignored as gestation period itself is around 9 months and average PPA may be somewhat larger than 6 months.

Here the distinction between the distributions of various orders of usual CBI and most recent CBI under the condition of fixed marital duration is examined. In this context, first, the distinction between most recent CBI and usual CBI is demonstrated, empirically, by considering a real data set of NFHS-3. In Section 2, the results using the NFHS-3 (2005–06) data for Uttar Pradesh (U.P.) have been utilized, but the results for other states can also be demonstrated on similar lines. In the continuation, the distinction between distributions of usual CBI and most recent CBI for specific parity of fixed marital duration is theoretically demonstrated by taking certain hypothetical values of the fertility parameters.

3.1.1 Empirical Data on Closed Birth Intervals and Most Recent Closed Birth Interval for Specified Marital Duration

Before discussing theoretically the issues of truncation effect, effect of sampling frame, and usual closed birth interval vs most recent closed birth interval, we first demonstrate these issues with the help of real data collected in NFHS-3 for the state of U.P. We know that in NFHS-3 (as well as in other NFHS too), complete birth history of married females of age group 15–49 years was obtained from which birth intervals of different orders under different sampling frames can be easily generated.

After analysis of this data set, Tables 2–8 present means of closed birth intervals (including most recent closed birth interval) of different order for females of specified marital durations. The considered marital durations are 5–9, 10–14, 15–19, 20–24, 25–29, and 30+ years (Tables 2–7) and in the last, in Table 8 females of all marital durations have been considered.

3.1.1.1 A Critical Review of These Tables Reveals the Following Broader Facts

As for as the issue of usual closed birth interval vs most recent closed birth interval is concerned, for a given marital duration and fixed parity, the mean of the most recent closed birth interval is significantly larger than the means of other usual CBI. This particular result is common in every table (see Tables 2–8). The differences in the two means are somewhat between 3 and 6 months, which cannot be ignored on the basis of sampling or nonsampling errors in the data. For example, in Table 6 there are 136 females who have given exactly four births and among them the means of first and second order mean closed birth intervals are 34.39 and 33.74 months, while the mean closed birth interval of order 3 (most recent closed birth interval for these females) is 37.79 months. Obviously the mean of most recent CBI is significantly larger than the means of other two usual CBI. Similarly in the same table there are 138 females who have given exactly five births; for these females also the mean of most recent CBI is significantly larger than the means of other usual CBI. Similar results are also observed for females of other marital durations too including all marital durations (Table 8). This fact can be seen as the last figure in each row of each table.

It is pertinent to mention that usual CBI as well as most recent CBI relates to same females even though the most recent CBI becomes larger than the mean of usual CBI. We will try to give theoretical justification for observing such an unusual result. As far as the issue of truncation effect is concerned, it

TABLE 2 Various Order Mean Closed Birth Intervals (in Months) for Females Giving Different Number of Births in Marital Duration (5–9) Years (U.P. NFHS-3 Data)

	Mean Closed Birth Interval of Order									
Number of Births	*1st*	*2nd*	*3rd*	*4th*	*5th*	*6th*	*7th*	*8th*	*9th*	Frequency
1	—									—
2	34.24									679
3	26.32	29.62								494
4	22.57	23.11	26.75							218
5	19.07	20.65	21.82	25.03						71
6	—	—	—	—	—	—				81[a]
Total	29.04	26.93	25.52	25.53[b]	25.88[c]					1470

[a]*Very less number of observations.*
[b]*Mean based on very less number of observations.*
[c]*The calculated means are based on less number of observations and hence there reliability may be seen in this perspective.*

TABLE 3 Various Order Mean Closed Birth Intervals (in Months) for Females Giving Different Number of Births in Marital Duration (10–14) Years (U.P. NFHS-3 Data)

Number of Births	Mean Closed Birth Interval of Order									Frequency
	1st	*2nd*	*3rd*	*4th*	*5th*	*6th*	*7th*	*8th*	*9th*	
1	—									—
2	44.15									294
3	31.98	36.88								374
4	28.94	29.42	34.58							361
5	24.28	26.40	26.49	32.82						260
6	22.13	23.29	22.79	24.37	28.46					130
7	20.52	20.98	19.51	23.04	22.15	26.51				55
8	—	—	—	—	—	—	—	—		10[a]
9	—	—	—	—	—	—	—	—		2[b]
Total	30.93	29.91	28.91	28.92	25.93	25.68[c]	20.58[d]	17.50[d]		1486

[a]*Very less number of observations.*
[b]*The observations are very less hence no further analysis has been done.*
[c]*Mean based on very less number of observations*
[d]*The calculated means are based on less number of observations and hence there reliability may be seen in this perspective.*

TABLE 4 Various Order Mean Closed Birth Intervals (in Months) for Females Giving Different Number of Births in Marital Duration (15–19) Years (U.P. NFHS-3 Data)

	Mean Closed Birth Interval of Order									
Number of Births	*1st*	*2nd*	*3rd*	*4th*	*5th*	*6th*	*7th*	*8th*	*9th*	Frequency
1	—									—
2	46.08									175
3	35.29	43.40								271
4	30.13	32.95	38.44							279
5	28.32	31.01	29.26	36.94						268
6	29.22	25.79	27.89	28.65	32.40					223
7	26.93	26.36	23.35	23.18	26.83	30.19				111
8	23.24	22.97	24.8	22.55	23.74	24.23	33.24			74
9	—	—	—	—	—	—	—	—		18[a]
10	—	—	—	—	—	—	—	—	—	14[b]
Total	31.84	32.03	30.35	29.89	28.77	26.89	30.46[c]	23.78[b]	24.57[b]	1433

[a]*Very less number of observations.*
[b]*The observations are very less hence no further analysis has been done.*
[c]*Mean based on very less number of observations.*

TABLE 5 Various Order Mean Closed Birth Intervals (in Months) for Females Giving Different Number of Births in Marital Duration (20–24) Years (U.P. NFHS-3 Data)

	Mean Closed Birth Interval of Order									
Number of Births	*1st*	*2nd*	*3rd*	*4th*	*5th*	*6th*	*7th*	*8th*	*9th*	Frequency
1	—									—
2	45.20									138
3	37.03	48.47								223
4	33.43	31.31	39.41							196
5	30.25	30.52	33.45	37.72						206
6	30.75	32.11	30.78	30.84	37.74					183
7	28.78	26.96	27.22	25.56	28.73	36.37				145
8	25.33	26.21	26.08	24.98	26.47	25.49	35.45			87
9	23.46	24.53	22.61	22.76	23.82	23.70	25.29	36.74		66
10	24.42	25.06	21.55	23.74	21.97	25.81	26.97	23.94	28.39	31
Total	32.61	33.15	31.31	29.98	30.52	29.95	30.37	32.64	28.39	1275

TABLE 6 Various Order Mean Closed Birth Intervals (in Months) for Females Giving Different Number of Births in Marital Duration (25–29) Years (U.P. NFHS-3 Data)

	Mean Closed Birth Interval of Order									
Number of Births	*1st*	*2nd*	*3rd*	*4th*	*5th*	*6th*	*7th*	*8th*	*9th*	Frequency
1	—									—
2	52.82									65
3	38.16	46.87								116
4	34.39	33.74	37.79							136
5	32.16	35.67	33.76	36.09						138
6	32.08	34.57	30.71	32.35	36.81					150
7	28.79	30.17	29.01	33.10	30.60	33.02				105
8	29.12	30.10	28.60	29.29	28.76	29.94	34.93			101
9	29.91	29.27	27.73	27.39	24.64	25.25	27.06	30.75		64
10	27.25	25.76	26.12	33.18	28.82	26.71	24.00	24.76	29.94	17
Total	33.77	34.64	31.72	32.31	31.36	29.82	31.14	29.49	29.94	892

TABLE 7 Various Order Mean Closed Birth Intervals (in Months) for Females Giving Different Number of Births in Marital Duration (30+) Years (U.P. NFHS-3 Data)

Number of Births	Mean Closed Birth Interval of Order									Frequency
	1st	*2nd*	*3rd*	*4th*	*5th*	*6th*	*7th*	*8th*	*9th*	
1	—									—
2	46.56									18
3	39.40	51.45								49
4	35.38	41.56	39.45							78
5	42.79	41.59	37.59	41.76						63
6	32.40	29.59	37.93	32.33	39.00					58
7	27.81	31.8	35.76	31.43	30.54	33.43				54
8	29.06	29.17	30.11	34.17	35.89	27.89	44.22			36
9	28.19	25.16	30.53	25.16	30.13	30.19	30.09	31.19		32
10	24.81	24.86	22.29	26.33	27.00	29.95	26.38	27.86	34.19	21
Total	34.41	36.30	35.39	33.30	33.5	30.79	34.93	29.87	34.19	409

TABLE 8 Various Order Mean Closed Birth Intervals (in Months) for Females Giving Different Number of Births in All Marital Durations (U.P. NFHS-3 Data)

	Mean Closed Birth Interval of Order									
Number of Births	*1st*	*2nd*	*3rd*	*4th*	*5th*	*6th*	*7th*	*8th*	*9th*	Frequency
1	—									—
2	37.16									1621
3	31.93	38.22								1558
4	29.76	30.59	35.44							1270
5	28.46	30.29	30.02	35.38						1006
6	29.1	28.89	29.02	29.48	34.34					752
7	27.27	27.39	26.79	27.06	28.14	32.84				470
8	26.40	26.72	26.87	26.79	27.29	26.86	35.33			308
9	26.81	25.93	26.07	24.60	25.04	24.97	26.39	32.51		182
10	24.74	24.24	22.66	25.27	24.51	26.59	26.00	24.31	29.53	83
Total	31.19	31.69	30.58	30.45	30.10	29.20	31.13	29.94	29.53	7250

can be demonstrated from Table 2. We see that the first order CBI (i.e., interval between first and second birth) is 29.04 months (see first element of last row). This figure for Tables 3 and 4 is 30.93 and 31.84 months, respectively. The question is whether this (29.04) can be considered as true mean of first order CBI. The answer is no, because normally all these females do not have sufficient time to have at least two births in this marital duration of 5–9 years, and this CBI is defined only for those females who have given at least two births in this interval. Thus, larger birth intervals are excluded from the study resulting in lower observed mean. The issue is further illustrated from Table 2 itself by considering the mean CBI between second and third birth. This mean is found to be 26.93 months (see the last row and entry no. 2) which is lower than the interval between first and second births (29.04 months). In this case also larger proportion of longer birth intervals are excluded from the study. This is precisely the truncation effect as discussed in Sheps et al. (1970). Hence, the mean 29.04 months may be considered as an underestimate of mean for CBI of first order. However, this mean for the marital duration 10–14 years is 30.93 months, while for marital durations 15–19, 20–24, 25–29, and 30+ years, the means are 31.84, 32.61, 33.77, and 34.41 months, respectively. It is quite reasonable to assume that the probability of having at least two births in the marital duration 10–14 years and onward is almost one, and these means can be considered as reasonable estimates of the mean of the first order CBI.

3.1.2 Analytical Results on Usual Closed Birth Interval and Most Recent Closed Birth Interval

Till now we studied the difference between usual CBI and most recent CBI for specified parity based on data of NFHS-3. Now we try to study the whole issue analytically based on the concept of stochastic models. The results are published in the paper entitled "Usual Closed Birth Interval Versus Most Recent Closed Birth Interval" by Kumar and Yadava (2015). However for ready reference we present the results of above paper here also.

Here, certain models are derived to demonstrate the difference between usual CBI and most recent CBI for specified marital duration. Abul-ata (1987) has also derived such results. But in his approach, the distribution between consecutive births has been considered to be exponentially distributed implying that after a birth there is a possibility of another birth, just after it. This is totally unrealistic and can be viewed only as a theoretical result. A more realistic situation has been considered here, where after a birth the next conception may take place only after the nonsusceptible period (h), which is the sum of gestation period and PPA period. If one-to-one correspondence between conception and birth is assumed, then the number of conceptions is equivalent to number of births. Similarly, the duration of ith order CBI (duration between ith and $(i+1)$th births), $i \geq 1$, is same as the duration between ith and $(i+1)$th conceptions. So without loss of generality, we have derived the results for duration between consecutive conceptions called closed conception interval rather than between consecutive births. Before deriving the results, a brief description of various notations used (which are similar to the notations used in Abul-ata (1987)) is given below.

3.1.2.1 Notations

X_0 is the random length of the 0th order conception time, i.e., interval from marriage to first conception.

X_i is the random length of the interval from the (i)th conception to the $(i+1)$th conception when the potential duration of a fecund marriage is assumed to be infinite; i.e., every married women will experience the interval X_i with certainty. The density and distribution functions of variable X_i are $f_i(x)$ and $F_i(x)$, respectively. $i \geq 0$.

S_i is the total waiting time from marriage till the occurrence of $(i+1)$th conception, $S_i = \sum_{j=0}^{i} X_j$. The density and distribution function of variable S_i are $g_i(x)$ and $G_i(x)$, respectively.

T is the exposure time for conceptions measured at survey date. $T = T' - g$ where T' is the duration of marriage measured at survey date.

$P_i(T)$ is Pr [C (0, T) = i], where C (0, T) is the counting process that counts the number of conceptions in (0, T).

$X_i(T)$ is the random length of the interval from the (i)th conception to the $(i+1)$th conception, i.e., interval between (i)th conception to the $(i+1)$th conception conditional on length T. The density and distribution function of variable $X_i(T)$ are $h_i(x,T)$ and $H_i(x,T)$, respectively.

$X_i^l(T)$ is the random length of the last closed conception interval conditional on length T where conception happens to be the last conception in (0, T). The density and distribution function of variable $X_i^l(T)$ are $h_i^l(x,T)$ and $H_i^l(x,T)$, respectively.

3.1.2.2 Closed Conception Interval

$X_i(T)$ is a random variable denoting the length of the interval (if exists) between the (i)th conception and the $(i+1)$th conception, where the conception process is being truncated at T

$$X_i(T) \leq T.$$

Now, $H_i(x,T)$ is given as

$$\begin{aligned} H_i(x,T) &= \Pr(X_i(T) \leq x) \\ &= \Pr(X_i \leq x \mid S_{i+1} \leq T) \\ &= \frac{\Pr(X_i \leq x \cap S_{i+1} \leq T)}{G_{i+1}(T)} \end{aligned} \tag{40}$$

Now

$$\begin{aligned}\Pr(X_i \le x \cap S_{i+1} \le T) &= \int_0^x \Pr(S_{i+1} \le T \mid X_i \le y) f_i(y) dy \\ &= \int_0^x \Pr(S_{i+1} \le T - y) f_i(y) dy\end{aligned} \tag{41}$$

Thus, the distribution function for $X_i(T)$ is

$$H_i(x, T) = \frac{\int_0^x G_i(T-y) f_i(y) dy}{G_{i+1}(T)}, \quad x \le T \tag{42}$$

and the density function of $X_i(T)$ is

$$\begin{aligned}h_i(x, T) &= \frac{d}{dx} H_i(x, T) \\ &= \frac{G_i(T-x) f_i(x)}{G_{i+1}(T)}, \quad x \le T\end{aligned} \tag{43}$$

3.1.2.3 Retrospective Last Closed Conception Interval

The distribution function of $X_i^l(T)$ is defined as

$$\begin{aligned}H_i^l(x, T) &= \Pr\left(X_i^l(T) \le x\right) \\ &= \frac{\Pr(X_i \le x \cap C(0, T) = i+1)}{P_{(i+1)}(T)} \\ &= \frac{\int_0^x \Pr(C(0, T) = i+1 \mid X_i = y) f_i(y)\, dy}{P_{(i+1)}(T)} \\ &= \frac{\int_0^x \Pr(C(0, T-y) = i) f_i(y)\, dy}{P_{(i+1)}(T)}\end{aligned} \tag{44}$$

and

$$H_i^l(x, T) = \frac{\int_0^x P_i(T-y) f_i(y)\, dy}{P_{i+1}(T)} \tag{45}$$

Thus,

$$h_i^l(x,T) = \frac{P_i(T-x)f_i(x)}{P_{i+1}(T)}, \quad i \geq 1,\ x \leq T \tag{46}$$

Here, general stochastic models have been derived for the usual closed conception interval and the most recent conception interval. But for the analysis and comparison, some specific forms for $f_i(x)$ have been assumed. As it has been mentioned earlier that Abul-ata (1987) considered $f_i(x)$ to follow exponential distribution

$$f_i(x) = \lambda e^{-\lambda x}, \quad \lambda > 0,\ x > 0,\ i \geq 0 \tag{47}$$

and showed that

$$G_{i+1}(T) = \Pr(S_{i+1} \leq T) = \int_0^t g_{i+1}(s)ds = 1 - \sum_{j=0}^{i} e^{-\lambda T} \cdot \frac{(\lambda T)^j}{j!} \tag{48}$$

and the distribution for closed conception interval is given as

$$h_i(x,T) = \frac{\lambda e^{-\lambda x}\left(1 - \sum_{j=0}^{i-1} e^{-\lambda(T-x)} \cdot \frac{(\lambda(T-x))^j}{j!}\right)}{1 - \sum_{j=0}^{i} e^{-\lambda T} \cdot \frac{(\lambda T)^j}{j!}} \tag{49}$$

for $i \geq 1$, $x \leq T$, $\lambda \geq 0$ and expectation of $X_i(T)$

$$E(X_i(T)) = \frac{\lambda^{-1} - e^{-\lambda T} \cdot \left(T + \lambda^{-1} + \lambda^{-1} \sum_{j=0}^{i-1} \frac{(\lambda T)^{j+2}}{(j+2)!}\right)}{1 - \sum_{j=0}^{i} e^{-\lambda T} \cdot \frac{(\lambda T)^j}{j!}}, \quad \text{for } i \geq 1 \tag{50}$$

Similarly, the distribution of most recent closed conception interval is given as

$$\begin{aligned} h_i^l(x,T) &= \frac{\frac{(\lambda(T-x))^i e^{-\lambda(T-x)}}{i!}}{\frac{(\lambda T)^{i+1} e^{-\lambda T}}{(i+1)!}} \lambda e^{-\lambda x} \\ &= \frac{(i+1)(T-x)^i}{T^{i+1}}, \qquad i \geq 1,\ x \leq T \end{aligned} \tag{51}$$

The above distribution is independent of the parameter λ, i.e., conception rate.

The corresponding expression for mean is given as

$$E\left(X_i^l(T)\right) = \frac{T}{i+2} \quad i \geq 1 \tag{52}$$

The serious limitations of this model have already been mentioned. Now, the distributions of usual conception interval and most recent closed conception interval are derived based on the following assumptions:

- The female has led a married life throughout the period (0, *T*).
- The time interval of the first conception after marriage follows an exponential distribution with probability density function (pdf)

$$f_0(x) = \lambda e^{-\lambda x}, \quad \lambda \geq 0, \; x \geq 0$$

We call the parameter λ, the conception rate.

- The duration between *i*th and $(i+1)$th conceptions follows a displaced exponential distribution with pdf

$$f_i(x) = f(x) = \lambda e^{-\lambda(x-h)}, \quad x \geq 0, \; \lambda \geq 0, \; i \geq 1$$

This assumption implies that after a conception there is no possibility of another conception for a period *h*, where *h* is the sum of duration of pregnancy (*g*) and PPA associated with a birth (or conception).

- Every conception results in a live birth.

Note: Since the distribution of X_0 is different than the distribution of X_i, $i \geq 1$, hence the analysis is confined for closed conception interval of order $i \geq 1$.

These assumptions have also been used by many authors in deriving the models for either the number of births to a female or the birth intervals, and due to flexibility of the probability expressions, these models can easily be applied to data, relating to variety of situations (see Singh, 1968; Singh et al., 1971; Singh et al., 1974; and others).

Under the above assumptions

$$\begin{aligned} P(X_0 \leq x) &= 1 - e^{-\lambda x}, \quad x \geq 0 \\ P(X_i \leq x) &= 1 - e^{\{-\lambda(x-h)\}}, \quad x \geq h, \; i = 1,2,3\ldots \end{aligned} \tag{53}$$

and the sequence $\{X_0, X_1, X_2, \ldots\}$ are mutually independently distributed random variables. $G_{i+1}(T)$ is the distribution function of time for $(i+1)$th conception from marriage, which is the probability of at least $(i+1)$ conceptions in (0, *T*). Then the expression for $G_{i+1}(T) = P(X_0 + X_1 + X_2 + \cdots + X_i \leq T)$ is obtained (see Singh, 1968) as

$$G_{i+1}(T) = \begin{cases} 1 - e^{-\lambda(T-ih)} \displaystyle\sum_{j=0}^{i} \frac{\lambda^{j(T-ih)^j}}{j!}, & T \geq ih \\ 0 & \text{otherwise} \end{cases} \tag{54}$$

and

$$P_{i+1}(T)=\begin{cases} G_{i+1}(T)-G_{i+2}(T),\ \ T>(i-1)h \\ =e^{-\lambda(T-(i+1)h)}\displaystyle\sum_{j=0}^{i+1}\frac{\lambda^j(T-(i+1)h)^j}{j!}-e^{-\lambda(T-ih)}\sum_{j=0}^{i}\frac{\lambda^j(T-ih)^j}{j!} \\ G_{i+1}(T),\ \ \text{for } ih<T<(i-1)h \\ =1-e^{-\lambda(T-ih)}\displaystyle\sum_{j=0}^{i}\frac{\lambda^j(T-ih)^j}{j!} \\ 0\ ,\ \ \text{for} T<ih \end{cases} \tag{55}$$

Note: From the above, it is clear that for specified values of T and h, there will be a maximum possible value of i for which ith order closed conception interval is defined.

Now, substituting the values of $G_{i+1}(T)$ and $G_i(T-x)$ from the above, the density function for (i)th order usual closed conception interval for $i\geq 1$ is given as

$$h_i(x,T)=\begin{cases} \dfrac{G_i(T-x)}{G_{i+1}(T)}f(x),\ \ h<x<T-(i-1)h \\ 0,\qquad\qquad \text{otherwise} \end{cases} \tag{56}$$

The expected value of $X_i(T)$ is given as under

$$E(X_i(T))=\frac{1}{G_{i+1}(T)}\int_h^{T-(i-1)h} xG_i(T-x)f(x)dx$$

$$=\frac{1}{G_{i+1}(T)}\int_h^{T-(i-1)h}\lambda e^{-\lambda(x-h)}\left(1-e^{-\lambda(T-(i-1)h-x)}\sum_{j=0}^{i-1}\frac{\lambda^j(T-(i-1)h-x)^j}{j!}\right)xdx$$

$$=\frac{1}{G_{i+1}(T)}\left(\begin{array}{l}\dfrac{\lambda h+1}{\lambda}-\dfrac{\lambda T'+1}{\lambda}e^{-\lambda(T'-h)} \\ -e^{-\lambda(T'-h)}\left(\displaystyle\sum_{j=0}^{i-1}\left(h\frac{\lambda^i}{(j+1)!}(T'-h)^{j+1}+\frac{\lambda^i}{(j+2)!}(T'-h)^{j+2}\right)\right)\end{array}\right) \tag{57}$$

where $T'=T-(i-1)h$ and the expression for the density of most recent closed conception interval is given as under

$$h_i^l(x,T)=\begin{cases}\dfrac{P_i(T-x)}{P_{i+1}(T)}f(x), & h<x<T-ih\\[2ex] \dfrac{G_i(T-x)}{P_{i+1}(T)}f(x), & T-ih<x<T-(i-1)h\\[2ex] \text{because for } T-ih<x<T-(i-1)h,\ G_{(i+1)}(T-x)=0\\[2ex] 0, & \text{otherwise}\end{cases} \tag{58}$$

The corresponding expression for the mean of most recent closed conception interval is given as

$$E\left(X_i^l(T)\right)=\frac{1}{P_{i+1}(T)}\left(\int_h^{T''}xP_i(T-x)f(x)dx+\int_{T''}^{T'}xG_i(T-x)f(x)dx\right)$$

$$=\frac{1}{P_{i+1}(T)}\left[\begin{array}{l}\displaystyle\int_h^{T''}\left(e^{-\lambda(T-x-ih)}\sum_{j=0}^{i}\frac{\lambda^j(T-x-ih)^j}{j!}-e^{-\lambda(T-x-(i-1)h)}\right.\\[3ex] \displaystyle\qquad\left.\times\sum_{j=0}^{i-1}\frac{\lambda^j(T-x-(i-1)h)^j}{j!}\right)\lambda e^{-\lambda(x-h)}xdx\\[3ex] \displaystyle+\int_{T''}^{T'}\left(1-e^{-\lambda(T-x-(i-1)h)}\sum_{j=0}^{i-1}\frac{\lambda^j(T-x-(i-1)h)^j}{j!}\right)\lambda e^{-\lambda(x-h)}xdx\end{array}\right], \tag{59}$$

where $T'=T-(i-1)h$ and $T''=T-ih$.

The above expression can be written as

$$E\left[X_i^l(T)\right]=\frac{1}{P_{i+1}(T)}(\phi_1-\phi_2+\phi_3-\phi_4) \tag{60}$$

where

$$\phi_1=\int_h^{T''}\left(e^{-\lambda(T-x-ih)\sum_{j=0}^{i}\frac{\lambda^j(T-x-ih)^j}{j!}}\right)\lambda e^{-\lambda(x-h)}xdx \tag{61}$$

$$=\lambda e^{-\lambda(T''-h)}\left[\sum_{j=0}^{i}\frac{\lambda^j}{(j+2)!}\left\{h(j+2)(T''-h)^{(j+1)}+(T''-h)^{(j+2)}\right\}\right]$$

$$\phi_2 = \int_h^{T''} \left(e^{-\lambda(T-x-(i-1)h)\sum_{j=0}^{i-1} \frac{\lambda^j (T-x-(i-1)h)^j}{j!}} \right) \lambda e^{-\lambda(x-h)} x dx$$

$$= \lambda e^{-\lambda T''} \left[\sum_{j=0}^{i-1} \frac{\lambda^i}{(j+2)!} \left\{ h(j+2)T''^{(j+1)} - T''(j+2)h^{(j+1)} + T''^{(j+2)} - h^{(j+2)} \right\} \right] \tag{62}$$

$$\phi_3 = \int_{T''}^{T'} x\lambda e^{-\lambda(x-h)} dx$$
$$= e^{-\lambda(T''-h)}\left(T'' + \frac{1}{\lambda}\right) - e^{-\lambda(T'-h)}\left(T' + \frac{1}{\lambda}\right) \tag{63}$$

$$\phi_4 = \int_h^{T''} \left(e^{-\lambda(T-x-(i-1)h)} \sum_{j=0}^{i-1} \frac{\lambda^j (T-x-(i-1)h)^j}{j!} \right) \lambda e^{-\lambda(x-h)} x dx$$
$$= \lambda e^{-\lambda T''} \left[\sum_{j=0}^{i-1} \frac{\lambda^i}{(j+2)!} \left((j+2)T'' h^{(j+1)} + h^{(j+2)} \right) \right] \tag{64}$$

Substituting the values of ϕ_1, ϕ_2, ϕ_3, and ϕ_4 [from Expressions (61), (62), (63), and (64)] in Expression (60), the value of $E(X_i^l(T))$ can be obtained.

3.1.2.4 Conclusion

From the two derived Expressions (57) and (60) the values of means of usual closed conception interval and most recent closed conception interval for some hypothetical values of T, λ, h, and i are obtained. As both the expressions are functions of T, λ, h, and i, the behavior of $E(X_i(T))$ and $E(X_i^l(T))$ is analyzed by considering variations in all the above mentioned parameters. Table 9 presents the values of $E(X_i(T))$ and $E(X_i^l(T))$, for $T=60$ months, $h=12$ months, and for different values of $\lambda=0.04$, 0.06, and 0.08. Similar values are presented in Tables 10 and 11, for $T=90$ and 120 months, respectively. Similarly, Table 12, presents about the variation in $E(X_i(T))$ and $E(X_i^l(T))$ by changing h (i.e., $h=12$, 14, and 16 months) keeping $\lambda=0.06$ and $T=60$ months. Of course, similar tables can be prepared for other combinations of values of λ, T, and h also, but we restrict our investigation to these tables only. From these tables, the following points emerge conclusively

1. In all the cases, $E(X_i(T))$ is smaller than $E(X_i^l(T))$. However, for maximum possible value of i, both means are identically equal. This is because of the fact that in this situation, by definition, the usual closed conception interval is the same as the most recent closed conception interval.

TABLE 9 Values of $E(X_i(T))$ (in Months), $E(X_i^I(T))$ (in Months), and Difference Between $E(X_i(T))$ and $E(X_i^I(T))$ for $T = 60$ Months, $h = 12$ Months, and Different Values of λ ($\lambda = 0.04$, 0.06, and 0.08)

	$\lambda = 0.04$			$\lambda = 0.06$			$\lambda = 0.08$		
i	$E(X_i(T))$	$E(X_i(T))$	*Difference*	$E(X_i(T))$	$E(X_i(T))$	*Difference*	$E(X_i(T))$	$E(X_i(T))$	*Difference*
1	25.19	27.38	2.19	23.71	27.10	3.39	22.29	26.85	4.56
2	20.68	20.65	0.38	19.86	20.49	0.63	19.42	20.33	0.91
3	16.63	16.66	0.02	16.54	16.58	0.04	16.44	16.50	0.06
4	13.98	13.98	0.00	13.96	13.96	0.00	13.95	13.95	0.00

TABLE 10 Values of $E(X_i(T))$ (in Months), $E(X_i^I(T))$ (in Months), and Difference Between $E(X_i(T))$ and $E(X_i^I(T))$ for $T=90$ Months, $h=12$ Months, and Different Values of λ ($\lambda=0.04$, 0.06, and 0.08)

	$\lambda=0.04$			$\lambda=0.06$			$\lambda=0.08$		
i	$E(X_i(T))$	$E(X_i^I(T))$	*Difference*	$E(X_i(T))$	$E(X_i^I(T))$	*Difference*	$E(X_i(T))$	$E(X_i^I(T))$	*Difference*
1	30.43	37.30	6.87	26.88	37.00	10.12	24.02	36.74	12.72
2	25.87	28.04	2.17	24.32	27.83	3.52	22.76	27.64	4.89
3	21.87	22.50	0.63	21.29	22.35	1.06	20.65	22.22	1.57
4	18.68	18.82	0.14	18.48	18.72	0.24	18.26	18.63	0.37
5	16.18	16.20	0.01	16.13	16.15	0.02	16.06	16.10	0.04
6	14.23	14.23	0.00	14.21	14.21	0.00	14.20	14.20	0.00

TABLE 11 Values of $E(X_i(T))$ (in Months), $E(X_i^I(T))$ (in Months), and Difference Between $E(X_i(T))$ and $E(X_i^I(T))$ for $T=120$ Months, $h=12$ Months, and Different Values of λ ($\lambda=0.04$, 0.06, and 0.08)

	$\lambda=0.04$			$\lambda=0.06$			$\lambda=0.08$		
i	$E(X_i(T))$	$E(X_i(T))$	*Difference*	$E(X_i(T))$	$E(X_i(T))$	*Difference*	$E(X_i(T))$	$E(X_i(T))$	*Difference*
1	33.66	47.27	13.60	28.12	46.96	18.84	24.42	46.69	22.27
2	30.12	35.49	5.37	26.86	35.27	8.41	24.06	35.08	11.03
3	26.35	28.43	2.09	24.75	28.27	3.52	23.08	28.12	5.04
4	22.96	23.74	0.78	22.25	23.61	1.36	21.45	23.50	2.05
5	20.13	20.39	0.26	19.83	20.30	0.46	19.49	20.21	0.72
6	17.81	17.88	0.07	17.70	17.82	0.12	17.57	17.76	0.19
7	15.93	15.94	0.01	15.89	15.91	0.02	15.84	15.87	0.03
8	14.38	14.38	0.00	14.37	14.37	0.00	14.35	14.35	0.00
9	13.09	13.09	0.00	13.08	13.08	0.00	13.08	13.08	0.00

TABLE 12 Values of $E(X_i(T))$ (in Months), $E(X_i^I(T))$ (in Months), and Difference Between $E(X_i(T))$ and $E(X_i^I(T))$ for $T = 60$ Months, $\lambda = 0.06$ Months, and Different Values of h ($h = 12$, 14, and 16)

	$h = 12$			$h = 14$			$h = 16$		
i	$E(X_i(T))$	$E(X_i(T))$	*Difference*	$E(X_i(T))$	$E(X_i(T))$	*Difference*	$E(X_i(T))$	$E(X_i(T))$	*Difference*
1	23.71	27.10	3.39	25.39	28.20	2.81	27.07	29.31	2.24
2	19.86	20.49	0.63	21.11	21.43	0.32	22.33	22.44	0.11
3	16.54	16.58	0.04	17.46	17.46	0.00	18.34	18.33	0.00
4	13.96	13.96	0.00						

2. For fixed values of λ, h, and T, the difference between $E(X_i(T))$ and $E(X_i^l(T))$ decreases as i increases. Of course, as noted above, the difference is zero for maximum possible value of i.
3. For large values of T and smaller i, the mean of usual closed conception interval is almost the same as the mean of X_i (i.e., $(1/\lambda)+h$). This is because of the fact that for such values of i and T, $G_{i+1}(T)$ is almost equal to one and $E(X_i(T))$ becomes equal to $E(X_i)$.
4. As the value of λ is increased, $E(X_i(T))$ and $E(X_i^l(T))$ both decrease, but the decrease in $E(X_i^l(T))$ is relatively smaller in comparison to the decrease in $E(X_i(T))$.
5. As the value of T is increased, there is increment in the values of both the means, but $E(X_i^l(T))$ increases more rapidly.
6. As the value of h is increased and other parameters are fixed, there is increment in the value of both the means.

Till now, the results based on the computed values of $E(X_i(T))$ and $E(X_i^l(T))$ for different combinations of values of the parameters λ, T, and h are demonstrated. Now the issue is investigated by comparing the pdfs of the two distributions, viz., $h_i(x,T)$ and $h_i^l(x,T)$ (see Expressions (9) and (11)). The difference in the two distributions mainly occurs due to two ratios $R_1(x)$ and $R_2(x)$, which are explained below.

$$h_i(x,T)=\begin{cases}\dfrac{G_i(T-x)}{G_{\{i+1\}}(T)}f(x), & h<x<T-(i-1)h\\ 0, & \text{otherwise}\end{cases} \tag{65}$$

and

$$h_i^l(x,T)=\begin{cases}\dfrac{P_i(T-x)}{P_{i+1}(T)}f(x), & h<x<T-ih\\ \dfrac{G_i(T-x)}{P_{i+1}(T)}f(x), & T-ih<x<T-(i-1)h\\ 0, & \text{otherwise}\end{cases} \tag{66}$$

These can be written as

$$\begin{aligned}h_i(x,T)&=R_1(x)f(x)\\ h_i^l(x,T)&=R_2(x)f(x)\end{aligned} \tag{67}$$

where

$$R_1(x)=\begin{cases}\dfrac{G_i(T-x)}{G_{\{i+1\}}(T)}, & h<x<T-(i-1)h\\ 0, & \text{otherwise}\end{cases} \tag{68}$$

and

$$R_2(x) = \begin{cases} \dfrac{P_i(T-x)}{P_{i+1}(T)}, & h < x < T - ih \\ \dfrac{G_i(T-x)}{P_{i+1}(T)}, & T - ih < x < T - (i-1)h \\ 0, & \text{otherwise} \end{cases} \tag{69}$$

The range of x, for both the probability distributions or the ratios $R_1(x)$ and $R_2(x)$, is $(h, T-(i-1)h)$. The plots of $R_1(x)$ and $R_2(x)$ for some hypothetical values of λ, T, and h are drawn (as considered in the tables). These plots are presented in Figs. 1–3. $R_1(x)$ is of monotonically decreasing nature implying that apart from $f(x)$, it gives larger weights to smaller values of x and smaller weights to larger values of x. However, the behavior of $R_2(x)$ is not monotonic. For smaller values of i, the curve seems to be bell shaped, whereas for the larger values of i, the two ratios, i.e., $R_1(x)$ and $R_2(x)$, are almost identical. For smaller i, the behavior of $R_2(x)$ is such that it is giving, on an average, larger weights to the larger values of x in comparison to $R_1(x)$, implying that the $E(X_i^l(T))$ will be larger than the $E(X_i(T))$ for smaller values of i. But, for larger values of i, both the means will be almost equal. Thus, these figures give us the clue that why the $E(X_i^l(T))$ is larger than the $E(X_i(T))$ for smaller values of i and why for large values of i, both are almost equal. The figures also testify the remarks given by Sheps and Menken (1973).

That the distribution of usual closed conception interval is analogous to the distribution of usual CBI and similarly the distribution of most recent closed conception interval is analogous to the distribution of most recent CBI. The only difference is about the shift of origin in marital duration in the sense that conceptions in time $(0, T)$ are same as births in time $(g, g+T)$. Hence, both can be studied interchangeably by making suitable change in origin.

Above results are based on some simplifying assumptions such as closed conception intervals are identically distributed as displaced exponential distribution with parameters λ and h. It is not sure that what will be the behavior of $E(X_i(T))$ and $E(X_i^l(T))$, if other distributions for $f_i(x)$ are taken. However, these at least demonstrate that the most recent CBI is different from the usual CBI. Thus, more work is needed to study the relationship between the two under more realistic assumptions. It is also important to mention that the derived results are purely of theoretical nature, which do not account for any type of nonsampling error. However, it is a known fact that the birth interval data are prone to error of recall lapse and memory bias. In this context, the most recent CBI is likely to be more reliable than the other birth intervals. Thus, it is desirable to use the data of most recent CBI for studying fertility behavior of females between two births, but one should be cautious about accounting for the theoretical difference between the most recent CBI and the usual CBI.

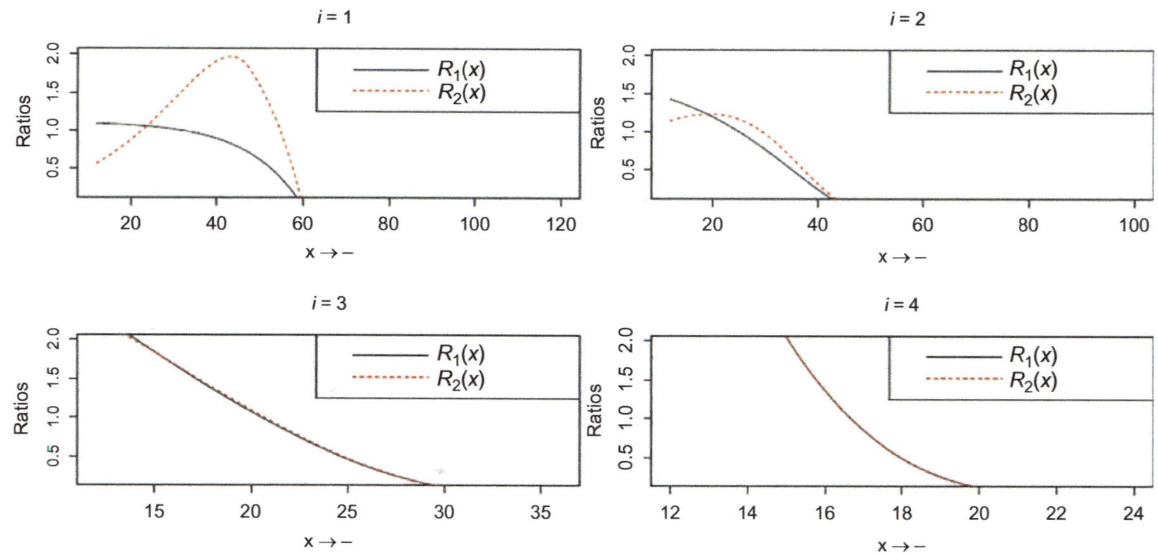

FIG. 1 Plots of $R_1(x)$ and $R_2(x)$ for $T=60$ months, $\lambda=0.04$, $h=12$ months, and $i=1$, 2, 3, and 4.

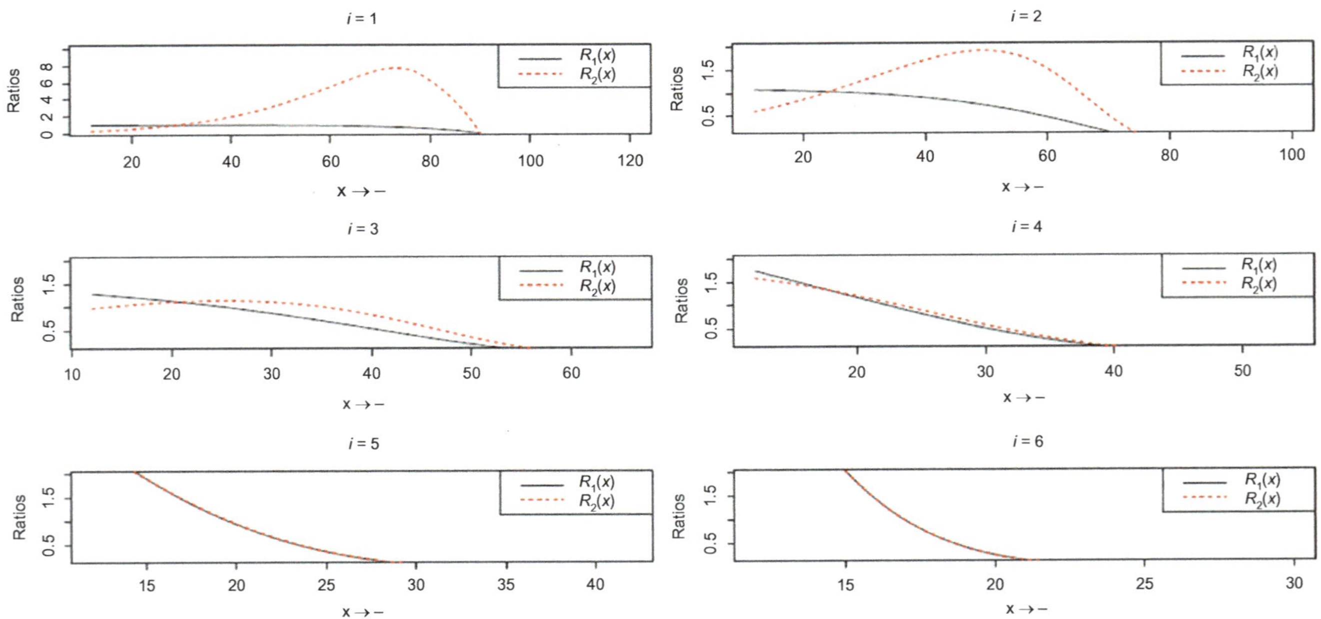

FIG. 2 Plots of $R_1(x)$ and $R_2(x)$ for $T=90$ months, $\lambda=0.06$, $h=12$ months, and different values of i.

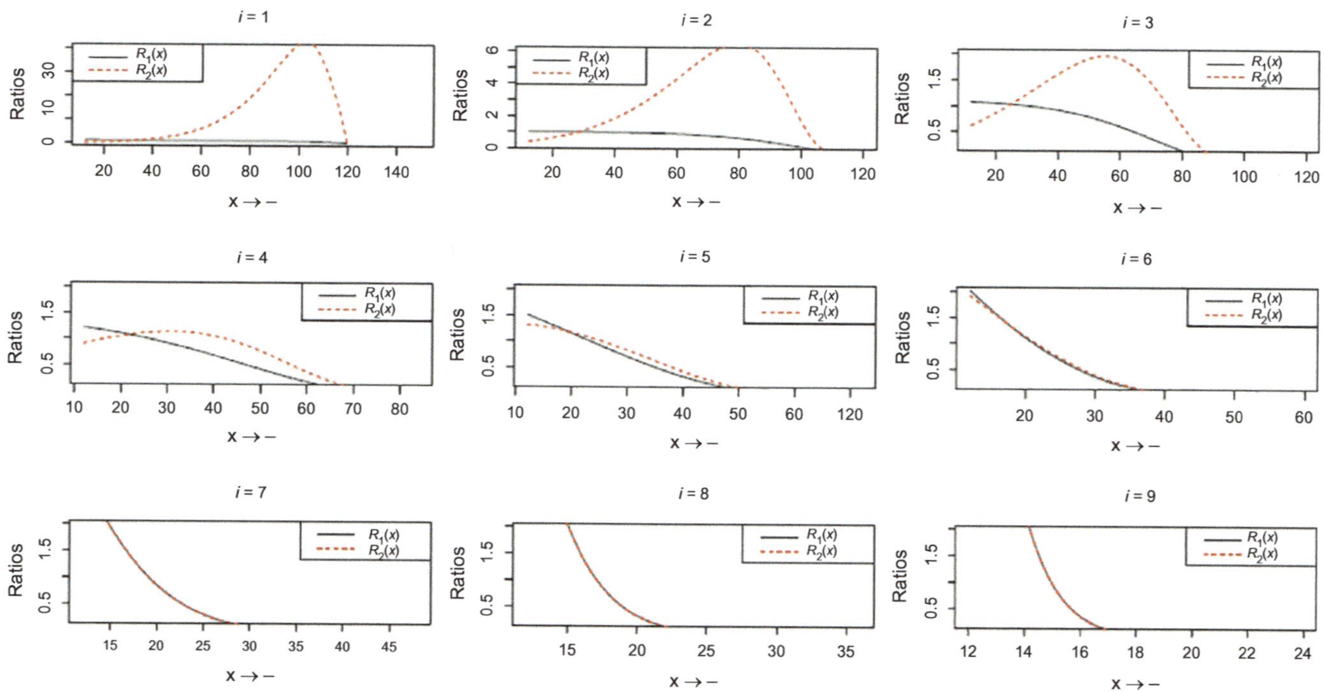

FIG. 3 Plots of $R_1(x)$ and $R_2(x)$ for $T=120$ months, $\lambda=0.08$, $h=12$ months, and different values of i.

3.2 Most Recent Closed Birth Interval Assuming *i*th Order Births to Be Uniformly Distributed Over Time

Till now we discussed the issue of closed birth interval vs most recent closed birth interval for fixed marital duration and parity. Now we discuss the above issue for the situation where ith order births (or births of any order) are uniformly distributed over time. The issue is first discussed when fertility parameters, viz., conception rate (λ) and nonsusceptible period (h) are constant. Subsequently we discuss the issue when the parameters are assumed to vary among females.

Suppose that the marriages in the population are uniformly distributed over time and there is no mortality among females after marriages till the end of the reproductive period and ith order births (or births of any order) are uniformly distributed over time. Further, first we assume that two major fertility parameters, viz., conception rate (λ) and nonsusceptible period associated with a birth (h) are same for all females. Later on, this assumption is relaxed to examine its effect on the distribution of most recent closed birth interval for a female.

Let X denote the interval between ith and $(i+1)$th births ($i \geq 1$), which is assumed to have pdf

$$f(x) = \lambda e^{-\lambda(x-h)} \quad x > h,\ \lambda > 0 \tag{70}$$

The above implies that conception rate (λ) remains constant between ith and $(i+1)$th births and if there is a birth at a time then there is no possibility of another birth in the subsequent h period.

Obviously,

$$E(X) = \frac{1}{\lambda} + h = \frac{(1+\lambda h)}{\lambda} \tag{71}$$

To derive the distribution of most recent closed birth interval we proceed as below.

For the situation where the ith birth has occurred y period before the survey date and $(i+1)$th birth has occurred at an interval of x from the ith birth so that x is the interval between ith and $(i+1)$th births then obviously the interval between ith and $(i+1)$th births will be the most recent birth interval of order i if no birth occurs after the occurrence of $(i+1)$th birth till the survey date. So for fixed y, the pdf of most recent closed birth interval say $f^*(x|y)$ will be

$$f^*(x|y) \propto f(x) Q(y-x) \tag{72}$$

where $Q(y-x)$ is the probability of no birth in an interval of length $(y-x)$.

Obviously

$$Q(y-x) = \begin{cases} 1 & \text{if } 0 < y-x < h \quad \text{i.e., } x < y < x+h \\ e^{-\lambda(y-x-h)} & \text{if } y-x > h \quad \text{i.e., } x+h < y < \infty \end{cases} \tag{73}$$

Since ith order births are assumed to be uniformly distributed over time, the unconditional distribution of the most recent closed birth interval say $f^*(x)$ is found to be

$$\begin{aligned} f^*(x) &\propto \int_{x}^{x+h} \lambda e^{-\lambda(x-h)} dy + \int_{x+h}^{\infty} \lambda\, e^{-\lambda(x-h)} e^{-\lambda(y-x-h)} dy \\ &\propto \lambda h e^{-\lambda(x-h)} + e^{-\lambda(x-h)} \\ &\propto (1+\lambda h) e^{-\lambda(x-h)} \quad h < x < \infty \end{aligned} \tag{74}$$

This is not a proper density, which can be made proper by multiplying by a constant k with the condition that

$$\int_0^\infty k f^*(x)\, dx = 1$$

$$\int_h^\infty k(1+\lambda h) e^{-\lambda(x-h)}\, dx = 1$$

i.e.,

$$k \cdot \frac{(1+\lambda h)}{\lambda} = 1 \text{ because, } \int_h^\infty \lambda e^{-\lambda(x-h)}\, dx = 1$$

so

$$k = \frac{\lambda}{1+\lambda h}$$

The above condition gives

$$f^*(x) = \begin{cases} \lambda e^{-\lambda(x-h)} & x > h \\ 0 & \text{otherwise} \end{cases} \tag{75}$$

This is the same as $f(x)$. Thus for such situation, the distribution of ith order most recent closed birth interval is the same as usual ith order closed birth interval based on cohort approach.

The next question arises whether the most recent closed birth interval will have the same distribution as the usual closed birth interval if the females are not having identical parameters of fertility, viz., conception rate and nonsusceptible period associated with a birth. In this context, we first consider the case when nonsusceptible period is variable while conception rate is assumed to be constant.

Let us consider the simplest case where the nonsusceptible period h is assumed to take two values h_1 and h_2, $h_1 < h_2$ with probabilities α and $1-\alpha$, respectively. For this case the random variable X can be considered as the mixture of two random variables X_1 and X_2 as $\alpha X_1 + (1-\alpha)X_2$, where X_1 has the pdf $f_1(x)$ as

$$\begin{aligned} f_1(x) &= \lambda e^{-\lambda(x-h_1)} \quad \lambda > 0,\ x > h_1 \\ &= 0 \qquad\qquad\quad \text{otherwise} \end{aligned} \tag{76}$$

while X_2 has the pdf,

$$\begin{aligned} f_2(x) &= \lambda e^{-\lambda(x-h_2)} \quad \lambda > 0,\ x > h_2 \\ &= 0 \qquad\qquad\quad \text{otherwise} \end{aligned} \tag{77}$$

and the pdf of X can be written as

$$f(x) = \alpha f_1(x) + (1-\alpha) f_2(x) \tag{78}$$

This can also be written as

$$\begin{aligned} f(x) &= 0 \ \text{ if } 0 < x < h_1 \\ &= \alpha\lambda\, e^{-\lambda(x-h_1)} \ \text{ if } h_1 < x < h_2 \\ &= \alpha\lambda\, e^{-\lambda(x-h_1)} + (1-\alpha)\lambda\, e^{-\lambda(x-h_2)} \ \text{ if } x > h_2 \\ &= 0 \ \text{ otherwise} \end{aligned} \tag{79}$$

As done earlier for fixed h, here also

$$f^*(x/y) \propto f(x) \cdot Q(y-x) \tag{80}$$

However in this case $f(x)$ is the mixture of $f_1(x)$ and $f_2(x)$, while $Q(y-x)$ will also be different for the two cases, i.e., $h = h_1$ and $h = h_2$. So for this case $f^*(x/y)$ can be written as

$$f^*(x/y) \propto [\alpha f_1(x) Q_1(y-x) + (1-\alpha) f_2(x) Q_2(y-x)] \tag{81}$$

where

$$\begin{aligned} Q_1(y-x) &= 1 \qquad\qquad\quad \text{for } 0 < (y-x) < h_1, \text{i.e., } x < y < x + h_1 \\ &= e^{-\lambda(y-x-h_1)} \ \text{ for } h_1 < (y-x) < \infty, \text{i.e., } x + h_1 < y < \infty \end{aligned} \tag{82}$$

and

$$\begin{aligned} Q_2(y-x) &= 1 \qquad\qquad\quad \text{for } 0 < (y-x) < h_2, \text{ i.e., } x < y < x + h_2 \\ &= e^{-\lambda(y-x-h_2)} \ \text{ for } h_2 < (y-x) < \infty, \text{ i.e., } x + h_2 < y < \infty \end{aligned} \tag{83}$$

Proceeding in the similar way as done for the case of fixed h, $f(x)$, the unconditional pdf of X say, $f^*(x)$, is obtained as

$$f^*(x) \propto \left[\begin{array}{l} \alpha\left(\int\limits_{x}^{x+h_1} \lambda e^{-\lambda(x-h_1)} dy + \int\limits_{x+h_1}^{\infty} \lambda e^{-\lambda(x-h_1)} e^{-\lambda(y-x-h_1)} dy \right) \\ \quad + (1-\alpha)\left(\int\limits_{x}^{x+h_2} \lambda e^{-\lambda(x-h_2)} dy + \int\limits_{x+h_2}^{\infty} \lambda e^{-\lambda(x-h_2)} e^{-\lambda(y-x-h_2)} dy \right) \end{array} \right] \tag{84}$$

$$f^*(x) \propto \left[\begin{array}{ll} \alpha(1+\lambda h_1) e^{-\lambda(x-h_1)} & \text{where } x > h_1 \\ +(1-\alpha)(1+\lambda h_2) e^{-\lambda(x-h_1)} & \text{where } x > h_2 \end{array} \right] \tag{85}$$

This can also be rewritten as

$$f^*(x) \propto \begin{cases} \alpha(1+\lambda h_1) e^{-\lambda(x-h_1)} & \text{for } h_1 < x < h_2 \\ \alpha(1+\lambda h_1) e^{-\lambda(x-h_1)} + (1-\alpha)(1+\lambda h_2) e^{-\lambda(x-h_1)} & \text{for } x > h_2 \end{cases} \tag{86}$$

Obviously this is not a proper density function. For making it proper density function, it is to be multiplied by a normalizing constant k such that $\int_0^\infty k f^*(x)\, dx = 1$.

So we get

$$\int\limits_0^\infty \left[\alpha(1+\lambda h_1) e^{-\lambda(x-h_1)} + (1-\alpha)(1+\lambda h_2) e^{-\lambda(x-h_2)} \right] k dx = 1$$

$$\text{i.e.,} \quad \frac{k[\alpha(1+\lambda h_1) + (1-\alpha)(1+\lambda h_2)]}{\lambda} = 1$$

$$\text{so } k = \frac{\lambda}{[\alpha(1+\lambda h_1) + (1-\alpha)(1+\lambda h_2)]}$$

So, the proper pdf is given as

$$\begin{aligned} f^*(x) &= \frac{\alpha(1+\lambda h_1) \lambda e^{-\lambda(x-h_1)}}{\alpha(1+\lambda h_1) + (1-\alpha)(1+\lambda h_2)} \quad \text{for} h_1 < x < h_2 \\ &= \frac{\alpha(1+\lambda h_1) \lambda e^{-\lambda(x-h_1)} + (1-\alpha)(1+\lambda h_2) \lambda e^{-\lambda(x-h_2)}}{\alpha(1+\lambda h_1) + (1-\alpha)(1+\lambda h_2)} \quad \text{for} x > h_2 \end{aligned} \tag{87}$$

This equation can be written as

$$\begin{aligned} f^*(x) &= \alpha^* \lambda e^{-\lambda(x-h_1)} \quad \text{for } h_1 < x < h_2 \\ &= \alpha^* \lambda e^{-\lambda(x-h_1)} + (1-\alpha^*) \lambda e^{-\lambda(x-h_2)} \quad \text{for } x > h_2 \end{aligned} \tag{88}$$

where

$$\alpha^* = \frac{\alpha(1+\lambda h_1)}{\alpha(1+\lambda h_1)+(1-\alpha)(1+\lambda h_2)} \quad \text{and}$$
$$(1-\alpha^*) = \frac{(1-\alpha)(1+\lambda h_2)}{\alpha(1+\lambda h_1)+(1-\alpha)(1+\lambda h_2)}$$

The above can also be rewritten as

$$\begin{aligned} f^*(x) =& \alpha^* \lambda e^{-\lambda(x-h_1)} \quad \text{where } x > h_1 \\ &+ (1-\alpha^*)\lambda e^{-\lambda(x-h_2)} \quad \text{where } x > h_2 \end{aligned} \tag{89}$$

This clearly indicates that $f^*(x)$ is different from $f(x)$ showing that the distribution of most recent closed birth interval is not same as the usual closed birth interval. However comparing the equations of both densities we find that, of course, the two distributions are not identical, but their forms are similar. In the case when nonsusceptible period is taken as variable and conception rate is taken as constant, the proportion of females with $h = h_1$ is α, while this proportion is α^* in the case of pdf of most recent closed birth interval. Further it is also seen that $\alpha^* < \alpha$ as shown below

$$\begin{aligned} \alpha^* &= \frac{\alpha(1+\lambda h_1)}{\alpha(1+\lambda h_1)+(1-\alpha)(1+\lambda h_2)} \\ &= \frac{\alpha(1+\lambda h_1)}{\alpha(1+\lambda h_1)+(1-\alpha)(1+\lambda h_1)+(1-\alpha)\lambda(h_2-h_1)} \\ &= \alpha\frac{1+\lambda h_1}{(1+\lambda h_1)+(1-\alpha)\lambda(h_2-h_1)} \\ &= \alpha \cdot C \end{aligned} \tag{90}$$

where

$$C = \frac{1+\lambda h_1}{(1+\lambda h_1)+(1-\alpha)\lambda(h_2-h_1)} \tag{91}$$
$$\text{obviously } C < 1$$

Thus $\alpha^* < \alpha$.

The above implies that in case of most recent closed birth interval, the proportion of females with smaller nonsusceptible period h_1 is smaller than this proportion in the population. This further indicates that mean of most recent closed birth interval should be larger than the mean of usual closed birth interval.

We next find the expression for the mean. The mean corresponding to pdf $f(x)$ is

$$E(X) = \alpha\left(\frac{1}{\lambda}+h_1\right)+(1-\alpha)\left(\frac{1}{\lambda}+h_2\right) = \frac{\alpha(1+\lambda h_1)+(1-\alpha)(1+\lambda h_2)}{\lambda} \tag{92}$$

Similarly the mean corresponding to $f^*(x)$ say $E^*(X)$ is

$$E^*(X) = \frac{\alpha(1+\lambda h_1)\left(\frac{1}{\lambda}+h_1\right)+(1-\alpha)(1+\lambda h_2)\left(\frac{1}{\lambda}+h_2\right)}{\alpha(1+\lambda h_1)+(1-\alpha)(1+\lambda h_2)} = \frac{\frac{\alpha(1+\lambda h_1)^2+(1-\alpha)(1+\lambda h_2)^2}{\lambda}}{\alpha(1+\lambda h_1)+(1-\alpha)(1+\lambda h_2)} \tag{93}$$

Now, we show that $E^*(X) > E(X)$.
This is shown as below

$$\begin{aligned}
&E^*(X)-E(X)\\
&=\left[\frac{\frac{\alpha(1+\lambda h_1)^2+(1-\alpha)(1+\lambda h_2)^2}{\lambda}}{\alpha(1+\lambda h_1)+(1-\alpha)(1+\lambda h_2)}\right]-\left[\frac{\alpha(1+\lambda h_1)+(1-\alpha)(1+\lambda h_2)}{\lambda}\right]\\
&=\frac{1}{\lambda}\left[\frac{\left\{\alpha(1+\lambda h_1)^2+(1-\alpha)(1+\lambda h_2)^2\right\}-\{\alpha(1+\lambda h_1)+(1-\alpha)(1+\lambda h_2)\}^2}{\alpha(1+\lambda h_1)+(1-\alpha)(1+\lambda h_2)}\right]\\
&=\frac{1}{\lambda}\left[\frac{\left\{\alpha(1+\lambda h_1)^2+(1-\alpha)(1+\lambda h_2)^2\right\}-\left\{\begin{array}{c}\alpha^2(1+\lambda h_1)^2+(1-\alpha)^2(1+\lambda h_2)^2\\+2\alpha(1-\alpha)(1+\lambda h_1)(1+\lambda h_2)\end{array}\right\}}{\alpha(1+\lambda h_1)+(1-\alpha)(1+\lambda h_2)}\right]\\
&=\frac{1}{\lambda}\left[\frac{\left\{\alpha(1+\lambda h_1)^2-\alpha^2(1+\lambda h_1)^2\right\}+\begin{array}{c}\left\{(1-\alpha)(1+\lambda h_2)^2-(1-\alpha)^2(1+\lambda h_2)^2\right\}\\-2\alpha(1-\alpha)(1+\lambda h_1)(1+\lambda h_2)\end{array}}{\alpha(1+\lambda h_1)+(1-\alpha)(1+\lambda h_2)}\right]\\
&=\frac{1}{\lambda}\left[\frac{\alpha(1-\alpha)(1+\lambda h_1)^2+\alpha(1-\alpha)(1+\lambda h_2)^2-2\alpha(1-\alpha)(1+\lambda h_1)(1+\lambda h_2)}{\alpha(1+\lambda h_1)+(1-\alpha)(1+\lambda h_2)}\right]\\
&=\frac{1}{\lambda}\left[\frac{\alpha(1-\alpha)\left\{(1+\lambda h_1)^2+(1+\lambda h_2)^2-2(1+\lambda h_1)(1+\lambda h_2)\right\}}{\alpha(1+\lambda h_1)+(1-\alpha)(1+\lambda h_2)}\right]\\
&=\frac{1}{\lambda}\left[\frac{\alpha(1-\alpha)\{(1+\lambda h_1)-(1+\lambda h_2)\}^2}{\alpha(1+\lambda h_1)+(1-\alpha)(1+\lambda h_2)}\right]\\
&=\frac{1}{\lambda}\frac{\alpha(1-\alpha)\lambda^2(h_2-h_1)^2}{\alpha(1+\lambda h_1)+(1-\alpha)(1+\lambda h_2)}\\
&=\frac{\lambda\alpha(1-\alpha)(h_2-h_1)^2}{\alpha(1+\lambda h_1)+(1-\alpha)(1+\lambda h_2)}
\end{aligned} \tag{94}$$

The above is positive. Hence, we see that $E^*(X) > E(X)$.

This shows that the mean corresponding to $f^*(x)$ is larger than the mean corresponding to$f(x)$.

We have shown above that $E^*(X) > E(X)$ assuming h takes two values, viz., h_1 and h_2 ($h_1 < h_2$) with particular probabilities α and $1-\alpha$, respectively. This result can be further generalized by assuming that h takes k values, viz., $h_1 < h_2 < \cdots < h_k$ with probabilities $\alpha_1, \alpha_2, \ldots\alpha_k$ such that $\alpha_1 + \alpha_2 + \cdots + \alpha_k = 1$.

In this case

$$E(X) = \frac{\alpha_1(1+\lambda h_1) + \alpha_2(1+\lambda h_2) + \cdots + \alpha_k(1+\lambda h_k)}{\lambda} = \frac{1}{\lambda}\sum_{i=1}^{k}\alpha_i(1+\lambda h_i) \quad (95)$$

and

$$E^*(X) = \frac{\dfrac{\alpha_1(1+\lambda h_1)^2 + \alpha_2(1+\lambda h_2)^2 + \cdots + \alpha_k(1+\lambda h_k)^2}{\lambda}}{\alpha_1(1+\lambda h_1) + \alpha_2(1+\lambda h_2) + \cdots + \alpha_k(1+\lambda h_k)} = \frac{\dfrac{1}{\lambda}\sum_{i=1}^{k}\alpha_i(1+\lambda h_i)^2}{\sum_{i=1}^{k}\alpha_i(1+\lambda h_i)} \quad (96)$$

Here also it can be shown that $E^*(X) > E(X)$ as

$$\begin{aligned} E^*(X) - E(X) &= \frac{\dfrac{1}{\lambda}\sum_{i=1}^{k}\alpha_i(1+\lambda h_i)^2}{\sum_{i=1}^{k}\alpha_i(1+\lambda h_i)} - \frac{\sum_{i=1}^{k}\alpha_i(1+\lambda h_i)}{\lambda} \\ &= \frac{1}{\lambda}\left[\frac{\sum_{i=1}^{k}\alpha_i(1+\lambda h_i)^2}{\sum_{i=1}^{k}\alpha_i(1+\lambda h_i)} - \sum_{i=1}^{k}\alpha_i(1+\lambda h_i)\right] \quad (97) \\ &= \lambda\sum_{i\neq j=1}^{k}\frac{\alpha_i(h_i - h_j)^2}{\alpha_i(1+\lambda h_i)} > 0 \end{aligned}$$

Thus, we see that $E^*(X) > E(X)$.

An Alternative Way to Prove: $E^*(X) > E(X)$.

This can also be proved in an alternative way. We see that

$$E^*(X) - E(X) = \frac{1}{\lambda}\left[\frac{\sum_{i=1}^{k}\alpha_i(1+\lambda h_i)^2}{\sum_{i=1}^{k}\alpha_i(1+\lambda h_i)} - \sum_{i=1}^{k}\alpha_i(1+\lambda h_i)\right]$$

If Y denotes a random variable taking values $(1+\lambda h_i)$, $i=1, 2, 3, \ldots k$ with probabilities α_i, $i=1, 2, 3, \ldots k$, then it can be easily seen that

$$E^*(X)-E(X)=\frac{1}{\lambda}\left[\frac{E(Y^2)}{E(Y)}-E(Y)\right]$$

$$=\frac{1}{\lambda}\left[\frac{E(Y^2)-\{E(Y)\}^2}{E(Y)}\right]$$

$$=\frac{1}{\lambda}\left[\frac{V(Y)}{E(Y)}\right]$$

where $V(Y)$ is variance of Y and $E(Y)$ is the mean of Y. Since Y takes only positive values, hence $E(Y)$ will be always positive and $V(Y)$ will be always positive. Thus $E^*(X)-E(X)$ will be always positive. This shows that the mean corresponding to $f^*(x)$ is larger than the mean corresponding to$f(x)$.

Now we attempt to obtain the expression for the distribution of the most recent closed birth interval when conception rate λ as well as nonsusceptible period h both are assumed to be variable. For the simplest case we assume that h takes two values h_1 and h_2 and λ also takes two values λ_1 and λ_2. The proportions for four combinations, viz., (h_1,λ_1), (h_1,λ_2), (h_1,λ_1), and (h_2,λ_1) are denoted as α_{11}, α_{12}, α_{21}, and α_{22}, respectively. Obviously sum of α_{11}, α_{12}, α_{21}, and α_{22} will be unity.

For this situation, the pdf of X will be

$$f(x)=\begin{cases}\alpha_{11}\lambda_1 e^{-\lambda_1(x-h_1)}+\alpha_{12}\lambda_2 e^{-\lambda_2(x-h_1)} & \text{if } h_1<x<h_2\\ \alpha_{11}\lambda_1 e^{-\lambda_1(x-h_1)}+\alpha_{12}\lambda_2 e^{-\lambda_2(x-h_1)}+\alpha_{21}\lambda_1 e^{-\lambda_1(x-h_1)} & \\ \quad +\alpha_{22}\lambda_2 e^{-\lambda_2(x-h2)} & \text{if } x>h_2\\ 0 & \text{otherwise}\end{cases} \tag{98}$$

Working on similar lines as derived earlier, the expression for $f^*(x)$ is obtained as

$$f^*(x)\propto\begin{cases}\alpha_{11}\lambda_1(1+\lambda_1 h_1)e^{-\lambda_1(x-h_1)}+\alpha_{12}\lambda_2(1+\lambda_2 h_1)e^{-\lambda_2(x-h_1)} & \text{if } h_1<x<h_2\\ \alpha_{11}\lambda_1(1+\lambda_1 h_1)e^{-\lambda_1(x-h_1)}+\alpha_{12}\lambda_2(1+\lambda_2 h_1)e^{-\lambda_2(x-h_1)} & \\ \quad +\alpha_{21}\lambda_1(1+\lambda_1 h_2)e^{-\lambda_1(x-h_1)}+\alpha_{22}\lambda_2(1+\lambda_2 h_2)e^{-\lambda_2(x-h_2)} & \text{if } x>h_2\end{cases} \tag{99}$$

Obviously this is also not a proper density function. So, for making it a proper density it should be multiplied by a normalizing constant k such that

$$\int_0^\infty kf^*(x)\,dx=1.$$

So, after solving we get

$$k = \frac{\lambda_1 \lambda_2}{[\alpha_{11}(1+\lambda_1 h_1)\lambda_2 + \alpha_{12}(1+\lambda_2 h_1)\lambda_1 + \alpha_{21}(1+\lambda_1 h_{21})\lambda_2 + \alpha_{22}(1+\lambda_2 h_{21})\lambda_1]} \tag{100}$$

This can be written as

$$k = \frac{\lambda_1 \lambda_2}{A} \tag{101}$$

where

$$A = [\alpha_{11}(1+\lambda_1 h_1)\lambda_2 + \alpha_{12}(1+\lambda_2 h_1)\lambda_1 + \alpha_{21}(1+\lambda_1 h_{21})\lambda_2 + \alpha_{22}(1+\lambda_2 h_{21})\lambda_1] \tag{102}$$

So, the proper pdf of $f^*(x)$ can be obtained incorporating the value of k in the above.

Also, the mean of most recent closed birth interval for this situation is obtained as

$$E^*(X) = \frac{1}{A}\left[\begin{array}{l} \alpha_{11}(1+\lambda_1 h_1)\left(\frac{1}{\lambda_1}+h_1\right)\lambda_2 + \alpha(1+\lambda_2 h_1)\left(\frac{1}{\lambda_2}+h_1\right)\lambda_1 + \alpha_{21}(1+\lambda_1 h_2)\left(\frac{1}{\lambda_1}+h_2\right)\lambda_2 \\ +\alpha_{22}(1+\lambda_2 h_2)\left(\frac{1}{\lambda_2}+h_2\right)\lambda_1 \end{array}\right] \tag{103}$$

However, the expression for the usual mean $E(X)$ is obtained as

$$E(X) = \left[\frac{\alpha_{11}(1+\lambda_1 h_1)}{\lambda_1} + \frac{\alpha_{12}(1+\lambda_2 h_1)}{\lambda_2} + \frac{\alpha_{21}(1+\lambda_1 h_2)}{\lambda_1} + \frac{\alpha_{22}(1+\lambda_2 h_2)}{\lambda_2}\right] \tag{104}$$

In this case also it is seen that $E^*(X)$ is greater than $E(X)$. The results can be further generalized if h takes values $h_1, h_2, \ldots h_k$ and λ takes values $\lambda_1, \lambda_2, \ldots \lambda_m$ with proper probability distribution.

3.3 Closed Birth Interval vs Straddling Birth Interval

Srinivasan (1967a,b) published a paper entitled "A Probability model applicable to the study of inter-live birth intervals and random segments of the same." He derived expressions for mean and variance of these inter-live birth intervals from which some fertility parameters may be estimated utilizing data on these intervals.

Suppose data are collected in a survey which gives information only about the last two births of any woman. Such a method of observation yields for each parity two kinds of separate data: complete intervals T_{1i}, i.e., interval between two successive births, i.e., ith and $(i+1)$th, and open interval T_{2i}, i.e., interval from the last live birth, i.e., ith to the survey point, implying that the following birth, i.e., $(i+1)$th having not occurred yet.

The above facts can be shown with the help of following diagram:

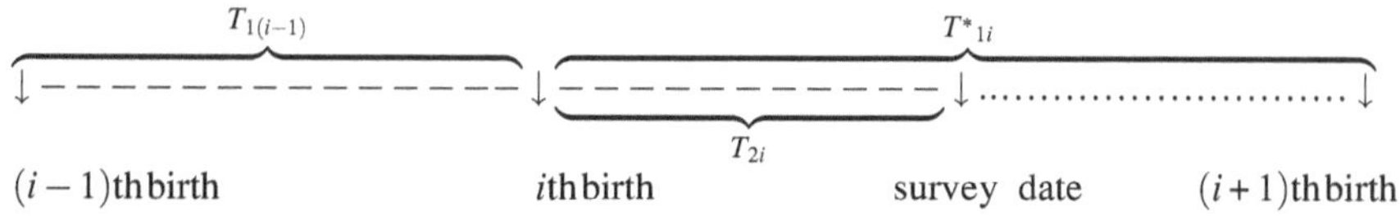

In this context it is assumed that the females giving ith birth will definitely move to the next birth (i.e., $(i+1)$th birth). For such females T_{2i} is a part of a complete interval T_{1i}.

Assuming that ith order births are uniformly distributed over time, Srinivasan assumed that for a given value of T_{1i}, T_{2i} is a random segment of the complete interval T_{1i} and the T_{2i} has a rectangular distribution such that

$$E(T_{2i}/T_{1i}) = \frac{1}{2}T_{1i} \tag{105}$$

and the nonconditional mean of T_{2i}

$$E(T_{2i}) = E \cdot E(T_{2i}/T_{1i}) = \frac{1}{2}E(T_{1i}) \tag{106}$$

The above gives a relation between means of T_{1i} and T_{2i} which implies that mean of T_{2i} is half of mean of T_{1i}. In fact T_{2i} is normally called as open interval or open birth interval.

All the above results are based on the implicit assumption that distributions of intervals T_{1i} including the fixed survey date is same as the one of all intervals T_{1i}.

Leridon (1969) has shown that the above implicit assumption, i.e., distributions of intervals T_{1i} including the survey date, is the same as the one of all intervals T_{1i} is not valid and consequently results derived by Srinivasan for T_{2i} are not valid.

The whole result is based on the following argument.

Let θ denote the survey date. As mentioned earlier, the ith order births are uniformly distributed over time. Let an ith birth occurs at time $\theta - u$, then this implies that this birth has occurred "u" time before the survey date. Suppose U_i denotes the random variable: time between the ith birth and the survey date. Then U_i has a uniform distribution and

$$P[u < U_i < u + du] = C_i du \tag{107}$$

where C_i is a constant not depending on u (because ith order births are uniformly distributed over time). Suppose there is a definite ith birth for which T_{1i} is x. This birth will be included in the study only if U_i, i.e., the interval between last birth and the survey date is less than x because if the value of U_i is more than x, then the $(i+1)$th birth will occur before the survey date and hence it cannot be a member of T_{1i} because $(i+1)$th birth must occur after the survey date. Thus, U_i must be less than x.

The above argument implies that only those T_{1i}'s will be included in the study for which $U_i<x$.

Now we see that

$$P[U_i<x]=\int_0^x C_i du = C\cdot x \tag{108}$$

This implies that the probability that an interval T_{1i} which includes the survey date is proportional to the length of T_{1i}. This further implies that larger intervals of T_{1i} have more chance of inclusion in the study than the smaller intervals. Consequently the distribution of such intervals says T_{1i}^* will be different from T_{1i}. This is precisely the comment of Leridon.

Thus if $f(x)$ denotes the pdf of T_{1i} and $\pi(x)$ denotes the pdf of T_{1i}^*, then

$$\pi(x)\propto xf(x) \tag{109}$$

and for $\pi(x)$ to be pdf, $\int \pi(x)dx=1$.

This gives

$$\pi(x)=\frac{xf(x)}{\mu} \tag{110}$$

where μ is the mean associated with pdf $f(x)$, i.e., mean of T_{1i}. This feature is also called as length biased sampling because the intervals of larger lengths have larger probability of inclusion in the study.

Obviously

$$\begin{aligned} E\left(T_{1i}^*\right) &= \int x\pi(x)dx \\ &= \int x^2 \frac{f(x)}{\mu} dx \end{aligned} \tag{111}$$

But

$$\int x^2 f(x)dx=\mu^2+\sigma^2 \tag{112}$$

where σ^2 is the variance associated with $f(x)$.

Thus

$$E\left(T_{1i}^*\right)=\mu+\frac{\sigma^2}{\mu} \tag{113}$$

Thus we see that mean of T_{1i}^* is larger than mean of T_{1i} and the difference between them increases with increase in the value of σ^2.

Although not mentioned by Leridon, the interval T_{1i}^* is called as straddling interval because it straddles the survey date.

In fact, T_{1i} and T_{1i}^* both are closed intervals between ith birth and $(i+1)$th birth, but the latter T_{1i}^* is different from T_{1i} mainly due to the condition that it includes the survey date.

In fact at first hand, the distinction between T_{1i} and T_{1i}^* is not visualized and perhaps because of this, Srinivasan perceived that the mean of open birth interval T_{2i} will be half of mean of closed birth interval T_{1i}.

In fact this issue is quite similar to the waiting time paradox discussed in Feller (1971). To understand the similarity between straddling birth interval and waiting time paradox we describe the paradox in brief below.

Suppose buses run randomly and the timing between consecutive buses follows an exponential distribution with pdf

$$f(x) = \alpha e^{-\alpha x} \quad \alpha > 0,\ x > 0 \tag{114}$$

which has mean $1/\alpha$.

Suppose a person reaches at the bus stop randomly then what will be his waiting time for the next bus?

There are two arguments:

Argument-1: Since exponential distribution has memory-less property, hence average waiting time for next bus will be $1/\alpha$.

Argument-2: Since buses run randomly and one reaches randomly at the bus stop, hence his average waiting time for next bus should be half of the average time between consecutive buses, i.e., his average waiting time should be $1/2\alpha$.

In fact in the above argument-2, it is implicitly assumed that the distribution of usual interval between consecutive buses will be the same as the distribution of interval between two buses in which the person reaches the bus stop, but this is not true. In fact the pdf of the interval for this situation will be

$$\begin{aligned} f'(x) &= \alpha^2 x e^{-\alpha x} \\ &= \frac{x f(x)}{\mu} \end{aligned} \tag{115}$$

where μ is the mean of time between two buses, i.e., $\mu = 1/\alpha$.

Obviously mean corresponding to $f'(x)$ is $2/\alpha$ and half of it is $1/\alpha$ which explains the paradox.

We clearly see that the distinction between T_{1i} and T_{1i}^* occurs only due to the condition that T_{1i}^* includes the survey date.

It is pertinent to mention that for getting data on straddling birth interval, one has to conduct a resurvey (follow-up) of the same females after a period of about 6–7 years so that they give their next $(i+1)$th birth. It is because of this limitation that straddling birth intervals are seldom used in the study of fertility; still these are studied because these enhance our knowledge about the difference between closed birth interval and straddling birth interval. We mentioned above that for getting data on straddling birth interval one has to conduct a prospective survey also, which can be considered to be a measure limitation of straddling birth interval. However, where a prospective study is conducted for some purpose then data on SBI may produce interesting results. In this context Sheps et al. (1970) have derived some general distributions for this interval and compared it with the distribution of usual CBI. Sehgal (1971) has also obtained certain interesting result through simulation. Singh (1989) have derived a generalized model for S.B.I. Yadava and Pandey (1989) have derived some explicit distributions of S.B.I.; they have also propounded a truncated version of distributions of S.B.I. by taking into account limited period of study. Yadava and Srivastava (1994) have also derived a model for S.B.I. by extending the model of Yadava and Pandey (1989) incorporating heterogeneity in the conception rate among females. Interested readers are advised to see the concern papers according to their requirements.

3.4 Open Birth Interval

In a retrospective study, the open birth interval for a female is defined as the interval between the last birth and the survey date. Sheps et al. (1970), while discussing the issue of truncation effect in closed and open birth interval data, have given expression for the pdf of open birth interval for a female who has a specified marital duration "t" and have given "n" births till time t such that her nth birth happens to be her last birth.

Let $U_n(x|t)$ be the pdf of the random variable designated as $U_n(X|t)$. Then $U_n(x|t) = \frac{g_{n-1}(t-x)\,Q_n(x)}{P_n(t)}$, where $g_{n-1}(x)$ is the pdf of $\sum_{i=0}^{n-1}\chi_i$ (where χ_i, $i \geq 0$ denotes the length of the ith order interval (the interval from the ith to the $(i+1)$th birth), given infinite time for reproduction), i.e., the pdf of interval from "0" to the occurrence of nth birth and $Q_n(x)$ is given as

$$Q_n(x) = \int_x^\infty f_n(y)dy, \quad \text{i.e.,} \quad P[\chi_n > x] \tag{116}$$

More simply, $Q_n(x)$ gives the probability that no birth occurs in an interval of length "x" after the nth birth. In fact $f_n(x)$ is the pdf of X_n, i.e., the length of interval of nth order interval, i.e., the interval from nth birth to the $(n+1)$th birth given infinite time of reproduction. $P_n(t)$ gives the probability of exactly n births in time $(0, t)$.

In a more simplified way, we can say that a female of marital duration "t" who has given exactly n births in time $(0, t)$ will have an open interval of length "x" if her nth birth occurs at time $(t-x)$ and no birth occurs in time $(t-x, t)$ of length "x" with the condition that n births have occurred in $(0, t)$, i.e.,

$$U_n(x|t) = \frac{g_{n-1}(t-x)\, Q_n(x)}{P_n(t)} \tag{117}$$

The above result is quite general and meaningful inferences can be drawn only when the form of $g_{n-1}(t-x)$, $Q_n(x)$, and $P_n(t)$ are known.

Now, we try to investigate the nature of $U_n(x|t)$ for given n and t. In this context first we give a very interesting and curious situation. Hypothetically, let us suppose that time between consecutive births follows an exponential distribution with pdf

$$f(x) = \lambda\, e^{-\lambda\, x}, \quad \lambda > 0, \quad x > 0, \tag{118}$$

and consecutive intervals are independent and identically distributed. Then, for a female, who has given n births in $(0, t)$, the pdf of open birth interval $U_n(x|t)$ will be

$$U_n(x|t) = \frac{g_{n-1}(t-x)\, Q_n(x)}{P_n(t)} \tag{119}$$

Under the given assumptions $g_{n-1}(t-x)$ is $\lambda\, e^{-\lambda(t-x)} \frac{[\lambda(t-x)]^{n-1}}{(n-1)!}$, $Q_n(x) = e^{-\lambda x}$ and

$$P_n(t) = \frac{e^{-\lambda\, t}(\lambda t)^n}{n!}. \tag{120}$$

Thus

$$\begin{aligned} U_n(x|t) &= \frac{\lambda e^{-\lambda(t-x)} \dfrac{[\lambda(t-x)]^{n-1}}{(n-1)!} e^{-\lambda\, x}}{\dfrac{e^{-\lambda\, t}(\lambda t)^n}{n!}} \\ &= \frac{\lambda^n\, e^{-\lambda t + \lambda\, x - \lambda\, x}(t-x)^{n-1} n!}{(n-1)! e^{-\lambda t} \lambda^n t^n} \\ &= \frac{n(t-x)^{n-1}}{t^n} \quad 0 < x < t \end{aligned} \tag{121}$$

It is verified that

$$\int_0^t U_n(x|t)dx = \int_0^t \frac{n(t-x)^{n-1}}{t^n} dx$$
$$= \left[\frac{-(t-x)^n}{n\ t^n} n\right]_0^t = \frac{t^n}{t^n} = 1 \tag{122}$$

It is interesting to note that the above density is totally independent of λ and hence no inference about λ can be drawn from data on open birth interval under the above condition. However the pdf $U_n(x|t)$ may not be totally independent of λ if the time between consecutive births follows some other distribution different from exponential distribution.

In this direction, Pathak (1971) proposed a probability model for open birth interval for fixed marital duration and parity based on some reasonable but simplifying assumptions.

Suppose a female has marital duration T units (years). The assumptions are similar to the assumptions made in Singh (1968). Under these assumptions time from marriage to first birth follows a displaced exponential distribution with pdf

$$f(t) = \lambda e^{-\lambda(t-g)} \quad \lambda > 0,\ t > g \tag{123}$$

where g is the gestation period associated with a live birth and time between consecutive births also follows a displaced exponential distribution with pdf

$$f^*(t) = \lambda e^{-\lambda(t-h)} \quad \lambda > 0,\ t > h \tag{124}$$

where h is nonsusceptible period associated with a birth (sum of gestation period and PPA).

Under the above assumptions, Pathak (1971) has shown that the pdf of O_i (open birth interval after ith birth at time T from marriage) is given as

$$\begin{aligned} u(o_i) &= f_1(t), \quad \text{for } h < t < T-(i-1)h-g \\ &= f_2(t), \quad \text{for } 0 < t < h \end{aligned} \tag{125}$$

where

$$f_1(t) = \frac{\lambda^i (T-g-(i-1)h)^{(i-1)} e^{-\lambda(T-g-ih)}}{(i-1)!} \tag{126}$$

$$f_2(t) = \frac{\lambda^i (T-g-(i-1)h)^{(i-1)} e^{-\lambda(T-t-g-(i-1)h)}}{(i-1)!} \tag{127}$$

[Note: Pathak (1971) has used the notation k for λ.]

Based on the above distribution of O_i, Pathak (1971) has computed mean open birth interval (in years) for specific marital duration (T) and various hypothetical values of λ and h (g is taken to be 0.75 years). These are given in Table 1 of Pathak (1971). However, for ready reference, we reproduce the table here also.

T (Years)	h (Years)	k (Annual)	All Parities	Last Parity					
				1	2	3	4	5	6
8	1.25	0.2	2.8974	3.6898	2.0743	1.2532	0.746	0.3961	0.1381
		0.3	2.5698	3.7192	2.1089	1.2852	0.7693	0.4056	0.1472
		0.4	2.2828	3.7467	2.1421	1.3164	0.7928	0.4164	0.1491
		0.6	1.8384	3.7965	2.2035	1.3765	0.8401	0.439	0.1525
	1.5	0.2	2.9418	3.7152	2.0154	1.1406	0.5887	0.2147	
		0.3	2.6293	3.7554	2.0616	1.1793	0.609	0.2179	
		0.4	2.3538	3.7927	2.1056	1.2175	0.6299	0.2213	
		0.6	1.9251	3.8593	2.187	1.2922	0.6733	0.2281	
5	1.25	0.2	1.9383	2.1846	1.0547	0.4605	0.1012		
		0.3	1.8413	2.2121	1.0814	0.4723	0.122		
		0.4	1.7454	2.2381	1.1077	0.4845	0.1027		
		0.6	1.5657	2.2859	1.1586	0.5096	0.1041		
	1.5	0.2	1.9774	2.2064	0.9801	0.3245			
		0.3	1.8949	2.2436	1.0117	0.3306			
		0.4	1.8112	2.2786	1.0431	0.3369			
		0.6	1.6501	2.3423	1.105	0.3498			

Based on the above table, Pathak (1971) has made the following observations.

> *From Table it is clear that for judging the sensitiveness of Open Birth Interval, one would have to know the mechanism underlying the variation in the parity progression. If we consider the observation period and the rest period (amenorrhea) as constants, we find that the pooled mean open birth interval (MOLIB) for all parities consistently decreases as the value of λ (i.e.,* k*) increases because for higher values of λ (i.e.,* k*) the more and more women would move to the higher orders of parity (see Table 6 of Pathak, 1971). If we break it up parity wise, we find that for all individual parities the trend has been reversed. This is because given the parities, the women with higher fecundability is expected to have higher MOLBI than the women with lower fecundability.*

Before making any comment on the observations made by Pathak (1971), we discuss first a methodological issue for expressions derived by Pathak.

We know that (see Sheps et al., 1970) the pdf of O_i, i.e., $u(O_i)$ should be

$$\frac{g_{i-1}(T-t)Q_i(t)}{P_i(T)} \tag{128}$$

where $g_{i-1}(T-t)dt$ is the probability that ith birth occurs between time $(T-t, T-t+dt)$, $Q_i(t)$ is the probability that no birth occurs between $(T-t,\ T)$ (of length t) and $P_i(T)$ is probability of exactly i births in $(0, T)$.

[In Sheps et al. (1970) notation, here "n" is replaced by i, t is replaced by T, and "x" is replaced by "t."]

If we compare this pdf with that of Pathak (1971) we see that the numerator is the same, but Pathak has ignored the denominator, i.e., $P_i(T)$.

We have shown earlier that if timing between consecutive births follows exponential distribution with pdf

$$f(t)=\lambda e^{-\lambda t} \quad \lambda>0,\ t>0 \tag{129}$$

then the pdf of O_i is

$$\frac{g_{i-1}(T-t)Q_i(t)}{P_i(T)}=\frac{\dfrac{\lambda e^{-\lambda(T-t)}\left[\lambda(T-t)\right]^{i-1}}{\underline{|i-1}}\cdot e^{-\lambda t}}{\dfrac{e^{-\lambda T}(\lambda T)^i}{\underline{|i}}}=\frac{i\cdot(T-t)^{i-1}}{T^i} \quad 0<t<T \tag{130}$$

which is independent of λ.

In fact, the above pdf becomes independent of λ only because it is to be divided by $P_i(T)$ so that all terms containing λ are canceled out by each other and the pdf becomes free from λ.

In the same way, if we divide the pdf mentioned in Pathak (1971) by $P_i(T)$ we may see that many of the terms containing λ may be canceled out (at least approximately). For example, we know that

$$P_i(T) = e^{-\lambda(T-g-ih)}\left(\sum_{m=0}^{i}\frac{\left[\lambda(T-g-ih)\right]^m}{\lfloor m}\right) - e^{-\lambda(T-g-(i-1)h)}\left(\sum_{m=0}^{i-1}\frac{\left[\lambda(T-g-(i-1)h)\right]^m}{\lfloor m}\right) \tag{131}$$

(see Singh, 1968)

If we approximate $P_i(T)$ by

$$e^{-\lambda(T-g-ih)}\frac{\left[\lambda(T-g-ih)\right]^i}{\lfloor i} \tag{132}$$

then $f_1(t)$ becomes free from λ as shown below

$$f_1(t) = \frac{\dfrac{\lambda^i\left(T-t-g-(i-1)h\right)^i}{\lfloor i-1}e^{-\lambda(T-g-ih)}}{\dfrac{\left[\lambda\left(T-t-g-(i-1)h\right)\right]^i}{\lfloor i}e^{-\lambda(T-g-ih)}} = \frac{\left(T-t-g-(i-1)h\right)^{i-1}\cdot i}{\left(T-g-ih\right)^i} \tag{133}$$

which is totally independent of λ.

Of course the term $(f_2(t))/(P_i(T))$ may involve a few terms which include λ, but its impact on MOLBI may be small.

Perhaps it is because of this fact that by changing the value of λ substantially from 0.2 to 0.6, the mean open birth interval for parity 2 for $T=8$ years changes very little (i.e., from 2.0743 to 2.2035, i.e., a difference of only 0.13 year). Similar small differences are also seen for other parities too. So, it seems that for given marital duration and parity, the mean open birth interval is totally insensitive to the level of fecundability (λ) and hence it is undesirable to draw any meaningful inference about λ from data on open birth interval for fixed parity and marital duration.

Of course, the mean open birth interval regardless of parity is quite sensitive to λ and hence meaningful inferences can be drawn from such data utilizing appropriate techniques.

Based on similar approach Singh et al. (1982) have derived a parity-dependent model for open birth interval. They have also found similar results as obtained above and hence we do not discuss it here. Interested readers may see the above paper for further details.

It is pertinent to mention here that the above results have been derived on the assumption that all females remain fecund at least up to the end of observational period (T). Thus, if some females become sterile in between $(0, T)$, then the methodology needs further modifications.

The above results are given for finite value of T. Sheps et al. (1970) have given the pdf of open birth interval for large value of T (i.e., $T\to\infty$)

regardless of parity assuming timing between consecutive births having identical distributions. If we denote this pdf by $u^*(t)$, then

$$u^*(t) \rightarrow \frac{Q(t)}{E(\chi)} \tag{134}$$

where $E(\chi)$ is the mean of closed birth interval when the marital duration is very large (i.e., tending to infinity). The above results have been obtained by the use of renewal theory where open birth interval may be treated as backward recurrence time in a renewal process.

It is pertinent to mention that $Q(t)$ is nothing but $1-F(t)$, where $F(t)$ is distribution function of the closed birth interval.

Similar result has also been obtained by Srinivasan (1968) assuming ith order births to be uniformly distributed over time. In his notation, if T_i represents the closed birth interval between ith and $(i+1)$th births $(i \geq 1)$ and U_i represents the open birth interval after ith birth, then the pdf of U_i say $u_i(t)$ is given as

$$u_i(t) = \frac{1-F_i(t)}{\mu} \tag{135}$$

where $F_i(t)$ is the distribution function associated with T_i. From above, obviously

$$E(U_i) = \frac{E(T_i^2)}{2E(T_i)} \tag{136}$$

It is to be noted that above result is true only for those women who continue to reproduce at least up to $(i+1)$th birth. Further, if some females stop child bearing after ith birth, then mean open birth interval of such females will be different. Using this concept Srinivasan (1968) proposed to estimate IPPR values for different parities. In a series of papers, Yadava and his associates (Yadava and Bhattacharya, 1985; Yadava and Saxena, 1989; Yadava et al., 1992; Yadava et al., 2013) have proposed methods to estimate PPR utilizing data on open and closed birth intervals. These have been discussed in detail in Yadava (2018). Hence we do not make any further discussion on these papers here.

3.4.1 Estimation of Proportion of Fecund Females Among Females With a Specified Open Birth Interval Group

We have mentioned above that Yadava et al. (1992) proposed a method to estimate parity progression ratio (PPR) "α_i" for parity "i" for females having open birth intervals between $(0, C)$. In this, they have computed $\alpha_i^*(0,C) = \dfrac{\int_0^C \alpha_i[1-F_i(t)]\,dt}{\int_0^C \alpha_i[1-F_i(t)]\,dt + (1-\alpha_i)C}$ where $F_i(t)$ is the distribution function of open birth interval of order "i." This α^*_i has been called as instantaneous parity progression ratio (IPPR) for females having open birth intervals $(0,C)$ at the time of survey. This α^*_i may also be interpreted as the proportion of fecund females among the females having open birth intervals $(0,C)$. Taking this concept into account, Yadava and Srivastava (1994) have proposed a procedure to estimate the proportion of fecund females among females having open birth interval between C_1 to C_2 $(C_1 < C_2)$. This is based on the following facts.

Above derivations are based on the assumptions that ith order births are uniformly distributed over time. Let $B_i dt$ be the number of ith order births between time t to $t+dt$ where B_i is a constant. In fact, the total number of females having open birth interval between $(0,C)$ will be $B_i\alpha_i\int_0^C[1-F_i(t)]dt+(1-\alpha_i)C$ and number of fecund females with open birth intervals between $(0,C)$ will be $B_i\alpha_i\int_0^C[1-F_i(t)]dt$. Thus, the total number of females having open birth intervals $(0,C_1)$ will be $B_i\alpha_i\int_0^{C_1}[1-F_i(t)]dt+(1-\alpha_i)C_1$ and number of fecund females among these will be $B_i\alpha_i\int_0^{C_1}[1-F_i(t)]dt$. Similarly, the total number of females with open birth intervals $(0,C_2)$ will be $B_i\alpha_i\int_0^{C_2}[1-F_i(t)]dt+(1-\alpha_i)C_2$ and number of fecund females among these will be $B_i\alpha_i\int_0^{C_2}[1-F_i(t)]dt$. So, the total number of females having open birth intervals between (C_1,C_2) will be $B_i\alpha_i\int_0^{C_2}[1-F_i(t)]dt+(1-\alpha_i)C_2-B_i\alpha_i\int_0^{C_1}[1-F_i(t)]dt+(1-\alpha_i)C_1=A-B$, here A is the total number of females having open birth intervals $(0,C_2)$, and B is the total number of females with open birth intervals $(0,C_1)$. Also, the number of fecund females with open birth intervals between (C_1,C_2) will be $B_i\alpha_i\int_0^{C_2}[1-F_i(t)]dt-B_i\alpha_i\int_0^{C_1}[1-F_i(t)]dt=D-E$, here D is the number of fecund females having open birth intervals $(0,C_2)$, and E is the number of fecund females with open birth intervals $(0,C_1)$. Thus, the proportion of fecund females having birth intervals between (C_1,C_2) is given by $\frac{D-E}{A-B}$.

This concept was mainly thought that proportion of fecund females among females with larger open birth intervals may be quite small, implying that any attempt to persuade such females for any family planning program may not be effective in reducing the level of fertility. Yadava and Srivastava (1994) have also applied above procedure for estimation of proportion of fecund females having different open birth intervals data collected in the survey "Evaluation of Impact of Development Activities and Fertility Regulation

Programmes on Population Growth Rates in Rural Areas" conducted under the auspices of Centre of Population Studies, Banaras Hindu University during 1978. They found that in most of the cases the females with open birth intervals between (C_1, C_2) where $C_1 = 60$ months are quite small in comparison to their PPR values as shown in the given table:

Estimates of proportion of fecund females from observed distribution of closed birth intervals for different open birth interval groups

Parity	PPR of Order i, α_i $(i = 1, 2, 3)$	Interval (C_1, C_2)	$\alpha_i^*(C_1, C_2)$
1	0.996	60–84	0.9364
		60–120	0.8948
2	0.950	60–84	0.4563
		60–120	0.2852
3	0.945	60–84	0.3176
		60–120	0.1787

For more details, the readers are advised to see the concerned paper.

3.4.2 Open Birth Interval as Measure of Fertility

Till now we discussed about the distributions of closed and open birth intervals and related means. We have also mentioned earlier that closed birth intervals are only useful for studying spacing pattern between births and are not capable of providing any useful measure of fertility simply because of the fact that closed birth intervals are defined only for those females who still continue to reproduce at various orders of births.

However, open birth interval is defined for all married females and thus it can be thought to provide some meaningful idea about the level of fertility of a given population.

In this context, Srinivasan (1980) proposed a simple regression model between total marital fertility rate (TMFR) and average open birth interval. Utilizing certain data sets he found that the fitted regression line was providing reasonably good estimates of TMFR for different situations.

Based on similar approach Yadava and Kumar (2002) also proposed a regression line where the dependent variable was TFR, while independent variable was proportion of females with open birth interval more than 5 years in population of married females in the age group 15–49 years. When the methodology was applied to data of NFHS-2, it gave reasonably satisfactory estimates of TFR for different states of India and different groups of the population (R^2 values more than 0.9).

Thus, we see that along with data on open and closed birth intervals one can get good estimates of PPR, while data on only open birth interval may provide good estimate of TFR. Of course, it is desirable that the proposed procedure may be applied to other sets of data to check the adequacy of these procedures which may provide further confidence on the application of these procedures to draw meaningful inferences.

3.4.3 Computation of $\frac{E(T_i^2)}{2E(T_i)}$ for Usual Closed Birth Interval and Most Recent Closed Birth Interval

Now we discuss another methodological issue in respect of finding the value of $\frac{E(T_i^2)}{2E(T_i)}$.

We have shown earlier that the distribution of usual closed birth interval of order "*i*" is different from the most recent closed birth interval if the population of females is heterogeneous with respect to fertility parameters, viz., conception rate and nonsusceptible period assuming that *i*th order births are uniformly distributed over time. While applying the data on closed and open birth intervals to estimate IPPR and PPR, Srinivasan (1968) as well as Yadava et al. (1992) have utilized the data on most recent closed birth interval for computation of $\frac{E(T_i^2)}{2E(T_i)}$. A natural question arises whether for the computation of $\frac{E(T_i^2)}{2E(T_i)}$, data on most recent closed birth interval should be used or the data on usual closed birth interval based on cohort approach should be used?

Theoretically it seems that the data on usual closed birth interval should be used for the computation of $\frac{E(T_i^2)}{2E(T_i)}$. Looking into this issue Sharma (2004) has computed the values of $\frac{E(T_i^2)}{2E(T_i)}$ for usual closed birth interval as well as most recent closed birth interval utilizing the data of NFHS-2, for parities 1, 2, 3, and 4. To avoid the truncation effect, for computation of $\frac{E(T_i^2)}{2E(T_i)}$ in context of usual closed birth interval, Sharma (2004) has considered only those females whose marital duration was

1. More than 15 years
2. More than 20 years

The computed values of $E(T_i)$, $E(T_i^2)$, and $\frac{E(T_i^2)}{2E(T_i)}$ for closed birth intervals based on cohort approach as well as for the most recent closed birth interval are given in Tables 13 and 14.

From these tables it is evident that computed values of $E(T_i)$ as well as $E(T_i^2)$ are distinctly different for the two types of closed birth intervals and computed values for most recent closed birth intervals are higher than that of cohort based birth intervals. However interestingly enough the computed ratios $\frac{E(T_i^2)}{2E(T_i)}$ are almost the same for the two cases. It is worthwhile to mention that the formulae developed by Srinivasan (1968) as well as Yadava et al.

TABLE 13 Computed Values of $E(T_i)$, $E(T_i^2)$, and $\frac{E(T_i^2)}{2E(T_i)}$ for the Closed Birth Intervals Based on Cohort Approach

Parity	$E(T_i)$	$E(T_i^2)$	$\frac{E(T_i^2)}{2E(T_i)}$
For marital duration 15+ years			
1	32.71	1472.32	22.51
2	32.51	3091.80	21.38
3	32.22	1342.27	20.48
4	31.38	1293.15	20.60
For marital duration 20+ years			
1	33.20	1542.52	22.94
2	33.05	1454.45	22.00
3	31.77	1342.09	21.12
4	31.62	1328.47	21.01

TABLE 14 Computed Values of $E(T_i)$, $E(T_i^2)$, and $\frac{E(T_i^2)}{2E(T_i)}$ for the Most Recent Closed Birth Interval

Parity	$E(T_i)$	$E(T_i^2)$	$\frac{E(T_i^2)}{2E(T_i)}$
1	36.21	1649.46	22.77
2	35.80	1591.12	22.22
3	35.57	1579.98	22.21
4	35.89	1609.77	22.43

(1992) for the estimation of IPPR and PPR involve expression $\frac{E(T_i^2)}{2E(T_i)}$ rather than $E(T_i)$ and $E(T_i^2)$ separately. Consequently estimates of PPR computed utilizing data on closed birth interval either based on cohort approach or most recent closed birth interval are likely to give almost identical estimates of PPR.

In fact Sharma (2004) has also obtained the estimates of PPRs based on the data of above two types which are given in Table 15.

From the table it is clearly seen whether we use data on most recent closed birth interval or cohort-based closed birth interval, the estimates of PPR are not going to be affected to any appreciable degree. However, one point needs

TABLE 15 Computed Values of Parity Progression Ratios Under Two Cases

Parity	Based on Most Recent Closed Birth Interval (Last Closed Birth Interval)	Based on Usual Closed Birth Interval (Cohort Approach)	
		15+ Years Marital Duration	*20+ Years Marital Duration*
1	0.9785	0.9785	0.9839
2	0.9037	0.9049	0.9087
3	0.8620	0.8630	0.8660
4	0.7641	0.7718	0.7738

special attention that the most recent closed birth interval for any specific parity is based on fertility experience of females in recent past, while closed birth intervals based on cohort approach are on an average based on the fertility experience of females of earlier periods. Further if fertility has changed over time, especially in the context of spacing pattern, then the estimates of $E(T_i)$ and $E\left(T_i^2\right)$ may not be comparable for the two cases. However, there are evidences that during last few decades closed birth intervals have not changed much, although fertility has shown appreciable change over time. In the light of the above facts, it is preferable to take values of $E(T_i)$ and $E\left(T_i^2\right)$ for most recent closed birth interval rather than closed birth interval based on cohort approach.

3.5 Forward Birth Interval

We have defined open birth interval as the time between last birth and the survey date. On the same line a forward birth interval may be defined as the time between survey date and the next birth posterior to the survey date. It is similar to forward recurrence time in renewal process. Open birth interval may be treated as backward recurrence time. It is obvious that in retrospective studies data on open birth interval can be obtained easily which may be utilized for drawing inferences about fertility behavior of females. However, for getting data on forward birth interval a prospective study is needed which may be considered to be a serious limitation of forward birth interval. Still in some situations, forward birth interval may be useful in fertility studies.

In the renewal theory, it is well known that the limiting forms of the probability density functions of backward recurrence time and forward recurrence time are identical on the assumption that renewal densities do not change over time. However, if renewal density changes after some time (say survey point), then the distribution of open birth interval and forward birth interval will not be identical. Such a situation may exist if certain family planning program is introduced at the time of survey so that the fertility parameter, viz., conception rate, may change after survey. It is desirable to measure this change so that adequacy of the family planning program may be assessed by comparing the distributions of the open birth interval and forward birth interval.

In this context, Singh et al. (1998) derived a probability model for forward birth interval under the assumption that a family planning program has been introduced in the population at the survey point. They have also obtained expressions for mean and variance under different situations. Although the paper is published quite long back but for ready reference, we describe the model in detail here also.

Singh et al. (1998) have derived the probability distribution of forward birth interval as follows:

Suppose a female is married at time zero (0) and the successive births to her occur at times $T_1+T_2+T_3+\cdots+T_r$ for $r=1, 2, \ldots$ where T_1 is the time of first birth since marriage and T_r $(r>1)$ is the time between the $(r-1)$th and rth births. Further, let us assume that the interval $T_1, T_2, T_3, \ldots T_r$ are mutually independent random variables and $T_2, T_3, \ldots T_r$ have a common distribution $f(t)$. However T_1 has a different probability density function, because the starting point of T_1 is not a birth. From Cox and Miller (1965), it can be seen that the above birth process is a modified renewal process.

Suppose that a survey is conducted at time T^*, where T^* is a distant point since marriage and let U_t and V_t, respectively, be the open and forward birth intervals, respectively, then the limiting probability density function of U_t and V_t (for simplicity we write it as T) will be the same and is given by $f_1(t)=\frac{1-F(t)}{\mu}$, where $\mu=E(T_r)$; $r>1$ and $F(t)=\int_0^t f(x)\,dx$. This is on the assumption that the renewal densities do not change over time. If we assume that a female has (i) a constant risk of conception "m," (ii) a constant period of nonsusceptibility "h" associated with each delivery, and (iii) there is one-to-one correspondence between a conception and a live birth, then we have $f(t)=me^{-m(t-h)}$, $t>h$.

Suppose $f_1(t)$ denotes the pdf of V_t then

$$\begin{aligned} f_1(t) &= \frac{m}{1+mh} \quad \text{for } 0<t<h \\ &= \frac{m}{1+mh}e^{-m(t-h)} \quad \text{for } h<t<\infty \end{aligned} \tag{137}$$

Then

$$\begin{aligned} E(T) &= \int_0^h t\frac{m}{1+mh}dt + \int_h^\infty t\frac{m}{1+mh}e^{-m(t-h)} \\ &= \frac{mh^2}{2(1+mh)} + \int_0^\infty \frac{m}{1+mh}e^{-mu}(u+h)\,du \end{aligned} \tag{138}$$

To solve the second term we make the transformation $u=t-h$.

We may note that the random variable $u=t-h$ has an exponential distribution with pdf me^{-mu}, $m>0$, $u>0$ and mean $1/m$.

Thus, $\int_0^\infty \frac{m}{1+mh}e^{-mu}(u+h)du = \frac{\frac{1}{m}+h}{1+mh} = \frac{1+mh}{m(1+mh)} = \frac{1}{m}$.

Thus,

$$E(T) = \frac{mh^2}{2(1+mh)} + \frac{1}{m} \tag{139}$$

Now,

$$\begin{aligned} E(T^2) &= \int_0^h t^2 \frac{m}{(1+mh)}dt + \int_h^\infty t^2 \frac{m}{1+mh}e^{-m(t-h)}du \\ &= \frac{mh^3}{3(1+mh)} + \int_0^\infty (u+h)^2 \frac{m}{1+mh}e^{-mu}du \end{aligned} \tag{140}$$

Now,

$$\begin{aligned} \int_0^\infty (u+h)^2 \frac{m}{1+mh}e^{-mu}du &= \int_0^\infty \left(u^2+h^2+2uh\right)\frac{m}{1+mh}e^{-mu}du \\ &= \frac{\int_0^\infty u^2 e^{-mu}du + \int_0^\infty h^2 e^{-mu}du + 2h\int_0^\infty u\, e^{-mu}du}{1+mh} \\ &= \frac{E(U^2)+2hE(U)+h^2}{1+mh} \\ &= \frac{\frac{2}{m^2}+\frac{2h}{m}+h^2}{1+mh} \\ &= \frac{2+2mh+h^2m^2}{m^2(1+mh)} \\ &= \frac{1+2m\,h+h^2m^2}{m^2(1+mh)} + \frac{1}{m^2(1+mh)} \\ &= \frac{(1+mh)^2}{m^2(1+mh)} + \frac{1}{m^2(1+mh)} \\ &= \frac{(1+mh)}{m^2} + \frac{1}{m^2(1+mh)} \end{aligned} \tag{141}$$

Thus,

$$E(T^2) = \frac{mh^3}{3(1+mh)} + \frac{(1+mh)}{m^2} + \frac{1}{m^2(1+mh)} \tag{142}$$

Now,

$$V(T) = E(T^2) - [E(T)]^2 = \frac{mh^3}{3(1+mh)} + \frac{(1+mh)}{m^2} + \frac{1}{m^2(1+mh)} - \left[\frac{mh^2}{2(1+mh)} + \frac{1}{m}\right]^2$$

$$= \frac{mh^3}{3(1+mh)} + \frac{(1+mh)}{m^2} + \frac{1}{m^2(1+mh)} - \left[\frac{m^2h^4}{4(1+mh)^2} + \frac{1}{m^2} + 2\frac{mh^2}{2(1+mh)}\frac{1}{m}\right]$$

$$= \frac{mh^3}{3(1+mh)} - \frac{m^2h^4}{4(1+mh)^2} + \frac{(1+mh)}{m^2} - \frac{1}{m^2} + \frac{1}{m^2(1+mh)} - \frac{h^2}{(1+mh)}$$

$$= \frac{mh^3}{12(1+mh)}\left[4 - \frac{3mh}{1+mh}\right] + \frac{1}{m^2}(1+mh-1) + \frac{1}{(1+mh)}\left[\frac{1}{m^2} - h^2\right]$$

$$= \frac{mh^3}{12(1+mh)}\left[\frac{4+4mh-3mh}{1+mh}\right] + \frac{h}{m} + \frac{1}{(1+mh)}\frac{(1+mh)(1-mh)}{m^2}$$

$$= \frac{mh^3}{12(1+mh)}\frac{(4+mh)}{(1+mh)} + \frac{h}{m} + \frac{(1-mh)}{m^2}$$

$$= \frac{mh^3}{12(1+mh)}\frac{[3+(1+mh)]}{(1+mh)} + \frac{mh+1-mh}{m^2}$$

$$= \frac{mh^3}{12(1+mh)}\frac{[3+(1+mh)]}{(1+mh)} + \frac{1}{m^2}$$

$$= \frac{3mh^3}{12(1+mh)^2} + mh^3\frac{(1+mh)}{12(1+mh)^2} + \frac{1}{m^2}$$

$$= \frac{3m^3h^3 + m^3h^3(1+mh) + 12(1+mh)^2}{12m^2(1+mh)^2}$$

$$= \frac{m^3h^3(4+mh) + 12(1+mh)^2}{12m^2(1+mh)^2} \tag{143}$$

Singh et al. (1998) have obtained the following expression for the variance of T

$$V(T) = \frac{6+(1+mh)^2}{12m^2} + \frac{2}{3m^2(1+mh)} - \frac{1}{4m^2(1+mh)^2} \tag{144}$$

Now we demonstrate that value of $V(T)$ obtained in Eqs. (143) and (144) is same.

From Eq. (144), we have

$$
\begin{aligned}
V(T) &= \frac{6+(1+mh)^2}{12m^2} + \frac{2}{3m^2(1+mh)} - \frac{1}{4m^2(1+mh)^2} \\
&= \frac{(1+mh)^4 + 6(1+mh)^2 + 8(1+mh) - 3}{12m^2(1+mh)^2} \\
&= \frac{(1+mh)^4 + 8(1+mh) + 6(1+mh)^2 - 3}{12m^2(1+mh)^2} \\
&= \frac{(1+mh)\left[8+(1+mh)^3\right] + 6(1+mh)^2 - 3}{12m^2(1+mh)^2} \\
&= \frac{(1+mh)\ \left[8+(1+mh)^3\right] + 6(1+mh)^2 - 3}{12m^2(1+mh)^2} \\
&= \frac{(1+mh)[8+1+m^3h^3+3mh+3m^2h^2] + 6(1+mh)^2 - 3}{12m^2(1+mh)^2} \\
&= \frac{(1+mh)[9+m^3h^3+3mh(1+mh)] + 6(1+mh)^2 - 3}{12m^2(1+mh)^2} \\
&= \frac{(9+m^3h^3)(1+mh) + 3mh(1+mh)^2 + 6(1+mh)^2 - 3}{12m^2(1+mh)^2} \\
&= \frac{9+9mh+m^3h^3+m^4h^4+3mh+3m^3h^3+ + 6m^2h^2 + 6(1+mh)^2 - 3}{12m^2(1+mh)^2} \\
&= \frac{6+12mh+4m^3h^3+m^4h^4+6m^2h^2+6(1+mh)^2}{12m^2(1+mh)^2} \\
&= \frac{6+12mh+6m^2h^2+4m^3h^3+m^4h^4+6(1+mh)^2}{12m^2(1+mh)^2} \\
&= \frac{6(1+mh)^2+4m^3h^3+m^4h^4+6(1+mh)^2}{12m^2(1+mh)^2} \\
&= \frac{12(1+mh)^2+4m^3h^3+m^4h^4}{12m^2(1+mh)^2} \\
&= \frac{12(1+mh)^2+m^3h^3(4+mh)}{12m^2(1+mh)^2} \\
&= \frac{m^3h^3(4+mh)+12(1+mh)^2}{12m^2(1+mh)^2}
\end{aligned}
\tag{145}
$$

This is same as Eq. (143) mentioned above.

We can also obtained the values of $E(T)$ and $E(T^2)$ in an alternative way which is explained below.

The random variable T may be treated as a random variable having the pdf $\frac{1-F(x)}{\mu}$, where $F(x)$ is the distribution function of random variable X, where it denotes the time between consecutive births and μ represents its mean, i.e., $E(X)$. Let $f(x)$ denote the pdf of X, then $f(x)=me^{-m(x-h)}$ $m>0,\ \ x>h$.

$$\begin{aligned} E(X) &= \int_h^\infty x\, me^{-m(x-h)}dx \\ &= \int_0^\infty (u+h)me^{-m\,u}du \ \text{ where } u=x-h \\ &= \frac{1}{m}+h \\ &= \frac{1+mh}{m} \end{aligned} \tag{146}$$

Now,

$$\begin{aligned} E(X^2) &= \int_h^\infty x^2\, me^{-m(x-h)}dx \\ &= \int_0^\infty (u+h)^2 me^{-m\,u}du \ \text{ where} u=x-h \\ &= \int_0^\infty u^2 me^{-m\,u}du + 2h\int_0^\infty ume^{-m\,u}du + \int_0^\infty h^2 me^{-m\,u}du \\ &= \frac{2}{m^2}+\frac{2h}{m}+h^2 \\ &= \frac{2+2mh+h^2m^2}{m^2} = \frac{(1+mh)^2}{m^2}+\frac{1}{m^2} \end{aligned} \tag{147}$$

We have already shown that the mean corresponding to pdf of $\frac{1-F(x)}{\mu}$ is $\frac{E(X^2)}{2E(X)}$.

So, $E(T)=\frac{E(X^2)}{2E(X)}$.

Thus,

$$E(T) = \frac{\left(\frac{1+mh}{m}\right)^2 + \frac{1}{m^2}}{2\left(\frac{1+mh}{m}\right)} = \frac{\left(\frac{1+mh}{m}\right)^2 \left(\frac{m}{1+mh}\right)}{2} + \frac{\frac{1}{m^2}\frac{m}{1+mh}}{2}$$
$$= \frac{1+mh}{2m} + \frac{1}{2m(1+mh)} \tag{148}$$

Also, it is known that $E(T^2) = \frac{E(X^3)}{3\,E(X)}$.

Now,

$$E(X^3) = \int_0^\infty m e^{-m\,u}\left(u^3 + 3u^2 h + 3u\,h^2 + h^3\right) du$$
$$= \frac{6}{m^3} + \frac{6h}{m^2} + \frac{3h^2}{m} + h^3$$
$$= \frac{6 + 6hm + 3h^2 m^2 + m^3 h^3}{m^3} \tag{149}$$
$$= \frac{1 + 3mh + 3\,(mh)^2 + m^3 h^3 + (5 + 3mh)}{m^3}$$
$$= \frac{(1+mh)^3}{m^3} + \frac{(5+3mh)}{m^3}$$

Thus,

$$E(T^2) = \frac{E(X^3)}{3\,E(X)} = \frac{\left[\frac{(1+mh)^3}{m^3} + \frac{(5+3mh)}{m^3}\right]}{\left(\frac{1+mh}{m}\right)} \tag{150}$$

Now, it can be checked that the expressions of $E(T)$ and $E(T^2)$ obtained from Eqs. (139) and (142) will yield the same results as to those obtained from Eqs. (148) and (150), respectively. This is shown from a hypothetical example given below:

Suppose $m = \frac{1}{2}$ and $h = 2$, then $E(T)$ and $E(T^2)$ obtained from Eqs. (139) and (142) are $E(T) = \frac{2}{4} + 2 = \frac{5}{2}$

$$E(T^2) = \frac{4}{6} + 8 + 2 = \frac{4 + 48 + 12}{6} = \frac{64}{6} = \frac{32}{3}$$

The values of $E(T)$ and $E(T^2)$ obtained Eqs. (148) and (150) are

$$E(T) = \frac{2}{1} + \frac{1}{2} = \frac{5}{2}$$

and

$$E(T^2) = \frac{E(X^3)}{3E(X)} = \frac{64+64}{3\times 4} = \frac{32}{3}$$

Thus we see that the values of $E(T)$ and $E(T^2)$ obtained from both ways are the same.

Singh et al. (1998) also obtain mean and variance of the forward birth interval if after time T^*, α proportion of the females use some contraceptives of a given effectiveness so that the risk parameter m becomes m_1 and pdf will also be changed. It will be given as

$$\begin{aligned} f_1^*(t) &= \frac{m}{1+mh} \quad \text{for } 0<t<g \\ &= \frac{\alpha m_1}{1+mh} e^{-m_1(t-g)} + \frac{m}{1+mh} - \frac{\alpha m}{1+mh} e^{-m_1(t-g)} \quad \text{for } g<t<h \\ &= \frac{\alpha m_1}{1+mh} e^{-m_1(t-g)} + \frac{\alpha m}{1+mh} e^{-m_1(t-h)} + \frac{(1-\alpha)m}{1+mh} e^{-m(t-h)} \\ &\quad - \frac{\alpha m}{1+mh} e^{-m_1(t-g)}, \quad \text{for } h<t<\infty \end{aligned} \tag{151}$$

They have also computed the values of mean and variance associated with pdf of $f_1^*(t)$. The interested readers may see the above paper for further details. The model involves four parameters α, h, m, and m_1. If h and m are known, the remaining parameters α and m_1 can be obtained by method of moments. The parameter "h" can be estimated from the data collected in the survey, whereas "m" can be estimated from the data on open birth interval.

3.6 Interior Birth Interval

We have already mentioned earlier that the interior birth interval is the closed birth interval that begins and ends in any segment of time (which may be the age group or marital duration or interval between two survey dates). The above interval is not only dependent on the fertility parameters involved in the model but also dependent upon the condition of the occurrences of birth in a given age interval/marital duration. All these make the distribution quite complex and perhaps due to this reason not much work has been done on this birth interval. Still some researchers have derived probability distributions of interior birth intervals under varying sets of assumptions [see Bhattacharya and Singh (1986), Pandey et al. (1987), and Singh (1989)]. Interested readers are advised to see these papers for more details.

4 ANALYZING CONSECUTIVE CLOSED BIRTH INTERVALS: A CORRELATION ANALYSIS

Most of the researches relating to closed birth intervals are the studies of distributions of closed birth intervals under different sampling frames. The observed distributions exhibit the variations in length of closed birth intervals for specific parities. These variations show the "between" variation among females. These "between" variations may occur due to two major causes, viz., (i) variation occurring due to random nature of the phenomenon and (ii) variation occurring due to variation in certain fertility parameters among females.

There is large number of studies, both of theoretical as well as applied nature, available in literature which deals with the study of variation in length of closed birth interval among females for specified parity. However, there is another dimension of the variation which is almost untouched. This variation may be called as "within" variation that is what is the nature of variation in closed birth intervals for a female over parities?

(1). Are the distributions identical over parities?
(2). Are the distributions over parities independent?
(3). If the distributions are not independent, then what is the nature of relationship among them?

In this article these issues are investigated theoretically as well as with the help of data. In this context the data are from NFHS-2. The issues are investigated applying the correlation analysis of consecutive closed birth intervals. The related paper is published as Yadava and Sharma (2007) but we explain the whole procedure in detail in this article also.

4.1 Consecutive Closed Birth Intervals

The closed birth interval between ith and $(i+1)$th births $(i \geq 1)$ is composed of the following three components (if fetal losses between these are ignored).

(i) PPA period following ith birth
(ii) Menstruating interval between ith and $(i+1)$th births
(iii) Gestation period associated with $(i+1)$th birth

Though there are little "within" as well as "between" variations in the gestation period associated with a live birth, however, for all practical purposes this period may be taken as constant for all births. Thus practically variation in closed birth interval can occur only due to variations in PPA period and menstruating interval. Thus if we denote PPA after ith birth by X_i, menstruating interval between ith and $(i+1)$th births by Y_i, and gestation period associated with $(i+1)$th birth by G, then variation in ith order closed birth interval can occur only through $X_i + Y_i$.

On the similar lines, the closed birth interval between $(i+1)$th and $(i+2)$th births may be considered as $X_{i+1}+Y_{i+1}+G$ and variation in this closed birth interval can occur only through $X_{i+1}+Y_{i+1}$.

If we assume that PPA is same for all females and fecundability is constant for all females as well as over parities, then the distributions of closed birth intervals over parities become identical as well as independent. However, if there are variations in either PPA or fecundability or in both, then the issue needs further investigation.

There are evidences that there is variation in PPA period among females. However, there are also evidences that the PPA period for a female generally does not vary over parities especially in the case where there is no infant death. Similarly, if it is assumed that fecundability also does not change over parity but PPA varies among females, then, under the above assumptions, the two closed birth intervals, viz., $X_i+Y_i+\mathrm{G}$ and $X_{i+1}+Y_{i+1}+G$ are identical but may not be independent. Thus a correlation analysis of the two closed birth intervals, viz., X_i+Y_i+G and $X_{i+1}+Y_{i+1}+G$ is likely to provide some insight on the nature of the phenomenon.

It is important to mention that there is a wide scope in doing the correlation analysis of two consecutive closed birth intervals if data on X_i and X_{i+1} as well as on Y_i and Y_{i+1} are available separately. Unfortunately such separate data are not available. However data on closed birth intervals classified according to marital duration and parity can easily be obtained by analyzing the NFHS-2 raw data on birth histories.

However, data on PPA are available only related to those births which occurred during last 3 years from the date of survey. Even for such cases, many of the observations are "continuing" and thus no separate complete data are available on X_i and X_{i+1}. Similarly separate data on Y_i and Y_{i+1} are also not available. Thus there is no scope of doing any correlation analysis on the two components, viz., X and Y separately utilizing NFHS-2 data. Thus one has no option than to make the analysis taking $X+Y$ together rather than taking them separately.

Now we attempt to derive the theoretical expression for correlation coefficient between two closed birth intervals under certain assumptions. As discussed above, if X_{ij} denotes the duration of PPA for jth female after ith birth and $X_{i+1,j}$ denotes the same after the $(i+1)$th birth and if we further assume that $X_{ij}=X_{i+1,j}$ for all j, then X_i and X_{i+1} are identical having correlation coefficient one. If we further assume that fecundability is same for all females, then Y_i and Y_{i+1} are independent and identical and then correlation coefficient between $\mathrm{X}_i+\mathrm{Y}_i=\mathrm{Z}_i$ and $\mathrm{X}_{i+1}+\mathrm{Y}_{i+1}=\mathrm{Z}_{i+1}$ is obtained as

$$r=\frac{Cov(Z_i, Z_{i+1})}{SD(Z_i)SD(Z_{i+1})} \tag{152}$$

Since under the above assumptions, distributions of Z_i and Z_{i+1} are identical (though not independent), the denominator of above expression becomes $\mathrm{Var}(\mathrm{Z}_i)$ or $\mathrm{Var}(\mathrm{Z}_{i+1})$.

Now $\text{Cov}(Z_i, Z_{i+1})$ is $\text{Cov}(X_i+Y_i, X_{i+1}+Y_{i+1})$ i.e.,

$$E\left[\left\{(X_i+Y_i)-\left(\overline{X}_i+\overline{Y}_i\right)\right\}\left\{(X_{i+1}+Y_{i+1})-\left(\overline{X}_{i+1}+\overline{Y}_{i+1}\right)\right\}\right] \tag{153}$$

Since we have assumed that $X_i = X_{i+1}$, hence

$$\begin{aligned}
&\text{Cov}(Z_i, Z_{i+1}) = E\left[\left\{(X_i+Y_i)-\left(\overline{X}_i+\overline{Y}_i\right)\right\}\left\{(X_i+Y_{i+1})-\left(\overline{X}_i+\overline{Y}_{i+1}\right)\right\}\right]\\
&\text{i.e., } E\left[\left\{\left(X_i-\overline{X}_i\right)+\left(Y_i-\overline{Y}_i\right)\right\}\left\{\left(X_i-\overline{X}_i\right)+\left(Y_{i+1}-\overline{Y}_{i+1}\right)\right\}\right]\\
&\text{i.e., } E\left[\left\{\left(X_i-\overline{X}_i\right)^2+\left(X_i-\overline{X}_i\right)\left(Y_{i+1}-\overline{Y}_{i+1}\right)+\left(Y_i-\overline{Y}_i\right)\left(X_i-\overline{X}_i\right)+\left(Y_i-\overline{Y}_i\right)\left(Y_{i+1}-\overline{Y}_{i+1}\right)\right\}\right]\\
&\text{i.e., } [\text{Var}(X_i)+\text{Cov}(X_i, Y_{i+1})+\text{Cov}(X_i, Y_i)+\text{Cov}(Y_i, Y_{i+1})]
\end{aligned} \tag{154}$$

Since Y_i and Y_{i+1} are assumed to be independent and identical, $\text{Cov}(Y_i, Y_{i+1})$ is zero and $\text{Cov}(X_i, Y_i)$ and $\text{Cov}(X_i, Y_{i+1})$ are equal.

Thus

$$\text{Cov}(Z_i, Z_{i+1}) = \text{Var}(X_i) + 2\ \text{Cov}(X_i, Y_i). \tag{155}$$

Thus under the assumptions, the expression for correlation coefficient (say "r") between the two closed birth intervals is found to be

$$r = \frac{\text{Var}(X_i) + 2\,\text{Cov}\,(X_i, Y_i)}{\text{Var}(Z_i)} \tag{156}$$

or

$$r = \frac{\text{Var}(X_{i+1}) + 2\ \text{Cov}\,(X_{i+1}, Y_{i+1})}{\text{Var}(Z_{i+1})} \tag{157}$$

If it is further assumed that X_i and Y_i are also independent, then correlation coefficient between Z_i and Z_{i+1} is

$$r = \frac{\text{Var}(X_i)}{\text{Var}(Z_i)} \tag{158}$$

or

$$r = \frac{\text{Var}(X_{i+1})}{\text{Var}(Z_{i+1})} \tag{159}$$

(Since X_i and X_{i+1} as well as Z_i and Z_{i+1} are identical under the assumptions.)

From above equation it is evident that if X_i and Y_i are independent, then the correlation coefficient r will be zero if $\text{Var}(X_i)=0$ which implies that the PPA duration after ith birth is constant. Further if PPA is not constant, i.e., the population is heterogeneous with respect to PPA, then the correlation coefficients r will be always positive.

On the other hand if X_i and Y_i are positively correlated, then also the correlation coefficient is positive, but if these are negatively correlated, then nothing definite can be said about the value of r (correlation coefficient).

4.2 Application of the Analysis on NFHS-2 Data

We now present the above correlation analysis along with data on consecutive closed birth intervals. Table 16 presents the observed bivariate distribution of closed birth intervals between first and second births and between second and third births. Here only those females have been considered whose marital duration is more than 15 years. Further it is a known fact that PPA period is usually reduced in case of infant deaths; hence to overcome this problem, only those females have been included whose children are still alive. Obviously the table includes only those females who have given at least three births. The marginal distributions of two closed birth intervals are presented in Table 17. The means and variances of the two closed birth intervals are also shown in the table. Similar results are also presented for second and third births and third and fourth births in Tables 18 and 19.

A critical review of Table 17 shows that means as well as variances of the two closed birth intervals are almost equal. This clearly indicates that fertility behavior of females between first and second births and between second and third births is almost similar and thus there is no change in spacing pattern at least up to third birth. The χ^2 (Chi-square) test also shows that the two distributions do not vary significantly. Almost similar results are also obtained in case of closed birth intervals between second and third births and between third and fourth births which are presented in Tables 18 and 19, respectively, although the mean and variance in case of interval between third and fourth births are found to be slightly lower than the corresponding values for birth

TABLE 16 Bivariate Distribution of Closed Birth Intervals (in Months) Between First and Second and Between Second and Third Births for Females Having Marital Duration At Least 15 Years Under Condition That Births Are Alive

First to Second Birth	Second to Third Birth						
	0–11	*12–23*	*24–35*	*36–47*	*48–59*	*60+*	*Total*
0–11	9	15	15	8	6	1	54
12–23	23	161	243	93	40	55	615
24–35	16	247	319	132	58	51	823
36–47	6	110	158	100	38	43	455
48–59	2	25	54	36	23	21	161
60+	9	47	61	47	23	24	211
Total	65	605	850	416	188	195	2319

TABLE 17 Marginal Distributions of Closed Birth Intervals Between First and Second Births and Between Second and Third Births Under Condition That Births Are Alive

Closed Birth Interval (in Months)	First to Second Birth	Second to Third Birth
0–11	54	65
12–23	615	605
24–35	823	850
36–47	455	416
48–59	161	188
60+	211	195
Total	2319	2319
Mean (months)	34.06	34.19
Variance $(\text{months})^2$	361.34	354.10
Correlation coefficient $(r)=0.113$		
$\chi^2_{0.05}(5)=6.00$		

TABLE 18 Bivariate Distribution of Closed Birth Intervals (in Months) Between Second and Third and Between Third and Fourth Births for Females Having Marital Duration At Least 15 Years Under Condition That Births Are Alive

Second to Third Birth	Third to Fourth Birth						
	0–11	*12–23*	*24–35*	*36–47*	*48–59*	*60+*	*Total*
0–11	7	11	11	12	3	3	47
12–23	13	130	214	68	28	21	474
24–35	15	186	257	109	45	45	657
36–47	11	73	111	74	22	20	311
48–59	3	31	34	27	21	16	132
60+	8	32	29	15	10	15	109
Total	57	463	656	305	129	120	1730

TABLE 19 Marginal Distributions of Closed Birth Intervals Between Second and Third and Between Third and Fourth Births for Females Having At Least 15 Years Marital Duration Under Condition That Births Are Alive

Closed Birth Interval (in Months)	Second to Third Birth	Third to Fourth Birth
0–11	47	57
12–23	474	463
24–35	657	656
36–47	311	305
48–59	132	129
60+	109	120
Total	1730	1730
Mean (years)	34.74	32.76
Variance $(\text{years})^2$	377.47	304.91
Correlation coefficient $(r)=0.112$		
$\chi^2_{0.05}(5)=1.71$		

interval between second and third births. This may be mainly due to the truncation effect. It should be mentioned here that in case of Table 18, only those females have been considered who have given at least four births so that closed birth interval between third and fourth births is also defined for such females.

As far as the correlation analysis is concerned, the observed correlation coefficient between lengths of two closed birth intervals (interval between first and second births and interval between second and third births) is found to be 0.113. Similarly for the case of second to third and third to fourth births, the correlation is observed to be 0.112. Although the values are significantly different from zero, the level of correlation coefficients is definitely low. We have given theoretical expression for correlation coefficient under certain assumptions. The expression involves values of variances of closed birth interval as well as PPA period and covariance between PPA period and menstruating interval.

The observed mean and variance for two closed birth intervals in Table 17 are 34.06 months and 361.34 $(\text{months})^2$ and 34.19 months and 354.10 $(\text{months})^2$, respectively, which are almost same. However, as far as variance of PPA period is concerned, no direct data are available on PPA period from which variance could be computed directly. However, Sharma (2004) has suggested an indirect procedure to estimate the distribution of PPA period utilizing data on current status of females regarding their PPA period utilizing data for births occurring during last 3 years for NFHS-2 (for more details for finding the distribution of PPA, readers are advised to see Sharma et al., 2007). In fact he has estimated the distribution of PPA for females along with its mean and variance. The value of variance has been counted to be 89.0 $(\text{months})^2$. This estimated distribution of PPA might not be the true distribution of PPA in its strict sense on cohort approach, but in the absence of any further knowledge, we take the variance of this distribution as variance of PPA as an approximation.

We have discussed above that under the assumption of independence between PPA and menstruating interval, correlation coefficient between two closed birth intervals is the ratio of variances of PPA and closed birth interval. Thus under this assumption, the theoretical value of correlation coefficient between first and second order closed birth intervals is found to be 0.246 if we consider Eq. (158) and it will be 0.251 if we consider Eq. (159). Similarly the theoretical value of correlation coefficient between birth intervals of order two and three is found to 0.236 if we consider the variance of closed birth interval of second order (use of Eq. 158). However, if we consider the variance of birth interval of third order, the theoretical value of correlation coefficient is found to be 0.292 (use of Eq. 159), although this value might be more affected by the truncation effect.

In any case, the theoretical values of correlation coefficient under the assumption of independence between PPA and menstruating interval are significantly larger (more than two times) than the observed correlation coefficients. This clearly indicates that the assumption of independence between PPA and menstruating interval may not be valid for data under consideration. The values suggest a negative association between PPA and menstruating interval indicating that in case of smaller duration of PPA, couples might be practicing some kind of abstinence for some period even after the expiry of PPA period. This situation might be responsible for negative association between PPA period and menstruating interval. Such abstinence may be almost absent in case of larger duration of PPA period. Similar situations for negative association between PPA period and menstruating interval are reported in other studies also where effects of sociocultural factors on fertility have been discussed in detail (Mandelbaum, 1974; Pandey, 2002; and others) (Tables 20 and 21).

It is pertinent to mention here that although the consecutive closed birth intervals are independent for a female, a positive correlation is observed between them mainly due to heterogeneity in PPA duration among females.

Here it has been shown that consecutive closed birth intervals are positively correlated if females are heterogeneous with respect to their PPA period. Recently, Anoop Kumar and R.C. Yadava (paper yet to be published)

TABLE 20 Observed Variances of Closed Birth Intervals and Postpartum Amenorrhea Period and Theoretical Values of Correlation Coefficients Between Birth Intervals of Consecutive Parities Calculated by Eqs. (158) and (159)

Order of Closed Birth Intervals	Marital Duration 15+ Years	
	Variance of Closed Birth Interval $(Months)^2$	*Variance of PPA* $(Months)^2$
1	361.10[a]	89.0
2	377.47[a]	89.0
2	354.10[b]	89.0
3	304.91[b]	89.0

Note: The variances of PPA period for all orders have been taken to be same.
[a]*Considering those females who have given at least three births.*
[b]*Considering those females who have given at least four births.*

TABLE 21 Theoretical Value of Correlation Coefficients

(1) Between first and second order closed birth interval (using Eq. 158)	0.246
(2) Between first and second order closed birth interval (using Eq. 159)	0.236
(3) Between second and third order closed birth interval (using Eq. 158)	0.251
(4) Between second and third order closed birth interval (using Eq. 159)	0.292

have derived similar result considering females to be heterogeneous with respect to their fecundability level. (Since the paper is yet to be published, we do not give any further detail of this paper).

Most of the results obtained here are based on some simplifying assumptions thus; further researches can be done making more realistic assumptions of the fertility process.

REFERENCES

Abul-ata, M.F., 1987. Stochastic Models of Birth Intervals According to Data Ascertainment Method and Relevant Fertility Indices. A PhD thesis, Department of Biostatistics, University of North Carolina at Chapel Hill, Institute of Statistics.

Bhattacharya, B.N., Singh, K.K., 1986. A probability model for interior birth interval and its applications. Can. Stud. Popul. 13 (2), 167–180.

Brass, W., 1958. The distribution of births in human population in rural Taiwan. Popul. Stud. 12, 51–52.

Cox, D.R., Miller, H.D., 1965. The Theory of Stochastic Processes. Methuen and Co. Ltd, London.

Dandekar, V.M., 1955. Certain modified forms of binomial and Poisson distributions. Sankhya 15, 237–251.

Feller, W., 1971. An Introduction to Probability Theory and Its Applications. vol. 2. John Wiley and Sons, New York.

Fisher, R.A., 1934. The effect of the methods of ascertainment upon the estimation of frequencies. Ann. Eugen. 6, 13–25.

Henry, L., 1972. On the Measurement of Human Fertility. Elsevier Publishing, Amsterdam.

Kumar, A., Yadava, R.C., 2015. Usual closed birth interval versus most recent closed birth interval. J. Data Sci. 13 (1), 73–94.

Leridon, H., 1969. Some comments on an article by K. Srinivasan. Popul. Stud. 23, 102.

Mandelbaum, D.G., 1974. Human Fertility in India: Social Components and Policy Perspectives. University of California Press, Berkeley, p. 144.

Menken, J.A., Sheps, M.C., 1972. The sampling frame as a determinant of observed distributions of duration variables. In: Greville, T.N.E. (Ed.), Population Dynamics. Academic, New York, pp. 57–87.

Mode, C.J., 1985. Stochastic Processes in Demography and Their Computer Implementation. Springer-Verlag, Berlin.

Pandey, R., 2002. Bio Social Determinants of Human Fertility. An unpublished PhD thesis. Banaras Hindu University, India.

Pandey, A., Singh, K.K., Dwivedi, S.N., 1987. On the distribution of interior birth interval. Demogr. India 15 (2), 283–287.

Pandey, A., Dwivedi, S.N., Mishra, R.N., 1990. Distribution of closed birth interval with some biosocial components: a stochastic model and its applications. J. Math. Sociol. 16, 89–106.

Pandey, A., Dwivedi, S.N., Mishra, R.N., 1998. A stochastic model for the study of last closed birth interval with some biosocial components. Math. Popul. Stud. 7 (1), 1–27.

Pathak, K.B., 1971. A stochastic model for the study of open birth interval—a cohort approach. Sankhya B 33, 305–314.

Pathak, K.B., Pandey, A., 1990. Application of a stochastic model to a certain truncation bias and systematic component of variation in closed birth intervals. Sankhya B 52 (3), 261–270.

Perrin, E.B., Sheps, M.C., 1964. Human reproduction: a stochastic process. Biometrics 20, 28–45.

Poole, K., 1973. Fertility measures based on birth interval data. Theor. Popul. Biol. 4, 357–387.

Rao, C.R., 1965. On Discrete Distributions Arising Out of Method of Ascertainment. In: Classical and Contagious Distributions. Statistical Publishing Society, Calcutta, pp. 320–332. Chapter.

Sehgal, J.M., 1971. Indices of Fertility Derived from Data on the Length of Birth Intervals, Using Different Ascertainment Plans. A PhD thesis, Department of Biostatistics, University of North Carolina at Chapel Hill, Institute of Statistics, Mimco, p. 768.

Sharma, S.S., 2004. A Study on Some Mathematical Models for Birth Intervals. A PhD thesis, Banaras Hindu University.

Sharma, S.S., Shukla, A.K., Tiwari, A.K., Yadava, R.C., 2007. Duration of postpartum amenorrhea period. J. Fam. Welf. 53 (1), 39–51.

Sheps, M.C., 1965. Application of probability models to the study of patterns of human reproduction. In: Sheps, M.C., Ridley, J.C. (Eds.), Public Health and Population Change. University of Pittsburgh Press, Pittsburgh, pp. 307–332.

Sheps, M.C., Menken, J.A., 1971. A model for studying birth rates given time dependent changes in reproductive parameters. Biometrics 27, 325–343.

Sheps, M.C., Menken, J.A., 1972. Distribution of birth intervals according to the sampling frame. Theor. Popul. Biol. 3, 1–26.

Sheps, M.C., Menken, J.A., 1973. Mathematical Models of Conception and Birth. University of Chicago Press, Chicago.

Sheps, M.C., Menken, J.A., Ridley, J.C., Linger, J.W., 1970. Truncation effect in closed and open birth interval data. J. Am. Stat. Assoc. 65, 678–693.

Singh, S.N., 1968. A chance mechanism of variation in the number of births per couple. J. Am. Stat. Assoc. 63 (321), 209–213.

Singh, V.K., 1989. Estimation of fecundability and risk of foetal wastage from data on straddling birth intervals. In: Singh, S.N., Premi, M.K., Bhatia, P.S., Ashish Bose, (Eds.), Population Transition in India. B.R. Publishers, New Delhi, pp. 401–409.

Singh, S.N., Bhattacharya, B.N., Yadava, R.C., 1971. On some distributions for couple fertilty and their applications. In: Singh, S.N. (Ed.), Proceeding of All-India Seminar on Demography and Statistics. Banaras Hindu University, Varanasi, India.

Singh, S.N., Bhattacharya, B.N., Yadava, R.C., 1974. A parity dependent model for number of births and its applications. Sankhya B 36 (1), 93–102.

Singh, U.P., Singh, V.K., Singh, O.P., 1988. A parity dependent model for the most recent birth interval, Can. Stud. in Popul., 15(1), 25–36.

Singh, S.N., Yadava, R.C., Pandey, A., 1979a. A probability model for closed birth interval. Health Popul. Perspect. Issues 2, 224–230.

Singh, S.N., Yadava, R.C., Pandey, A., 1979b. On a generalized distribution of open birth interval regardless of parity. J. Sci. Res. 29 (1), 167–170.

Singh, S.N., Yadava, R.C., Chakrabarty, K.C., 1982. A parity dependent model for open birth interval. Sankhya B, 212–218.

Singh, S.N., Yadava, R.C., Pandey, A., 1987. On estimating the mean birth interval characteristic of women. Can. Stud. Popul. 14 (1), 1–8.

Singh, S.N., Yadava, R.C., Pandey, A., 1998. A probabilty model for forward birth interval. Health Popul. Perspect. Issues 1 (4), 295–301.

Srinivasan, K., 1967a. Analytical Models for Two Types of Birth Intervals With Applications to Indian Population. A PhD thesis, University of Kerala, Trivandrum, India.

Srinivasan, K., 1967b. A probability model applicable to the study of inter-live birth intervals and random segments of the same. Popul. Stud. 21, 63–70.

Srinivasan, K., 1968. A set of analytical models for the study of open birth intervals. Demography 5, 34–44.

Srinivasan, K., 1980. Birth interval analysis in fertility surveys: an illustrative application to data from Fiji fertility survey. Scientific Report No. 7. International Statistical Institute, Netherlands.

Venkatacharya, K., 1969. Certain implications of short marital durations in the analysis of live birth intervals. Sankhya B 31, 53–68.

Wolfers, D., 1968. Determinants of birth intervals and their means. Popul. Stud. 22, 253–262.

Yadava, R.C., 2018. Stochastic modelling of some natural phenomena: a special reference to human fertility. In: Srinivasa Rao, A.S.R., Rao, C.R. (Eds.), Handbook of Statistics: Integrated Population Biology and Modelling Part A, vol. 39, pp. 187–274.

Yadava, R.C., Bhattacharya, M., 1985. Estimation of Parity Progression Ratios From Closed and Open Birth Interval Data. Mimeo, Centre of Population Studies, Banaras Hindu University, Varanasi, India.

Yadava, R.C., Kumar, A., 2002. On an indirect estimation of total fertility rate from open birth interval data. Demogr. India 31, 211–222.

Yadava, R.C., Pandey, A., 1989. On the distribution of straddling birth intervals. Biom. J. 31 (7), 855–863.

Yadava, R.C., Saxena, N.C., 1989. On the estimation of parity progression and instantaneous parity progression ratios. In: Singh, S.N., Premi, M.K., Bhatia, P.S., Ashish Bose, (Eds.), Population Transition in India, vol. 1. B.R. Publishing, Delhi, India, pp. 357–362.

Yadava, R.C., Sharma, S.S., 2004. Closed birth intervals versus most recent closed birth interval. Demogr. India 33 (2), 249–263.

Yadava, R.C., Sharma, S.S., 2007. The distribution of consecutive closed birth intervals in females in Uttar Pradesh. J. Biosci. 39, 185–199.

Yadava, R.C., Srivastava, M., 1994. On estimation of a probability distribution for straddling birth interval. Janasamkhya XII, 43–56. Number IX 2.

Yadava, R.C., Pandey, A., Saxena, N.C., 1992. Estimation of parity progression ratios from the truncated distribution of closed and open birth intervals. Math. Biosci. 110 (2), 181–190.

Yadava, R.C., Kumar, A., Pratap, M., 2013. Estimation of parity progression ratios from open and closed birth interval data. J. Data Sci. 11 (3), 607–621.

Section VIII

GWAS, Species Divergence and Bayesian Item Response Theory

Chapter 7

Detection of Quantitative Trait Loci From Genome-Wide Association Studies

David A. Spade[1]

Department of Mathematical Sciences, University of Wisconsin–Milwaukee, Milwaukee, WI, United States

[1]Corresponding author: e-mail: spade@uwm.edu

ABSTRACT

This chapter presents a class of methods for detecting single-nucleotide polymorphisms (SNPs) that impact the expression of quantitative traits. First, techniques that work for both phased and unphased SNPs and do not require knowledge of the evolutionary history among individuals are presented. Second, a set of approaches that incorporate the marginal evolutionary history among individuals to estimate a covariance structure among the trait values are given. The chapter closes with a method that applies to unphased SNP detection while still incorporating phylogenetic information among the trait values.

Keywords: Statistical phylogenetics, Statistical genomics, Genetic marker detection, Single-nucleotide polymorphism, SNP phasing

1 INTRODUCTION

This chapter provides a description of common methods of detecting and localizing variations in the genome that are associated with the expression of disease in humans. A genome-wide association study (GWAS) is an example of such an examination of all the variants along the genome. Genetic variations can come in several forms. Structural variations commonly come in the form of *copy-number variations* (CNV), which arise when the number of copies of a gene varies from individual to individual. GWAS have been used to locate CNV that are associated with gastric cancer (Roukos, 2009), colorectal cancer (Fernandez-Rozadilla et al., 2013), and opioid dependence (Li et al., 2015).

The other common class of genetic variation is composed of genetic markers. One type of genetic marker is a *minisatellite*, which is a stretch of DNA

Handbook of Statistics, Vol. 40. https://doi.org/10.1016/bs.host.2018.09.004

that is typically 10–60 base pairs (bp) long and that is repeated several (5–500) times along the genome. The other common type of genetic marker is a *single-nucleotide polymorphism* (SNP). An SNP is a variation in which a single nucleotide changes to another in a certain stretch of DNA. GWAS have been used to link SNPs to an array of human diseases, including asthma (Immervoll et al., 2001), chronic migraines (McCarthy et al., 2001), systemic lupus erythematosus (SLE) (Stevens et al., 2001), and type II diabetes (Tsunoda et al., 2001).

With the variety of diseases that have a genetic component and the decreasing time and expense with which it is possible to extract genetic information at the molecular level, it is becoming increasingly important to understand how to use these genetic data to describe associations between genetic variations and disease. The ability to do this has clear implications in targeted gene therapy and personalized medicine. This chapter provides descriptions of a few of the many methods that are available to make such determinations.

The methods described in this chapter rely on sequences of SNPs as the genetic data input. These data often require a preprocessing step known as *phasing* of genotypes into haplotypes before analysis may begin. Section 2 provides a description of some methods of doing this. Since the description of example problems in the GWAS setting can be very involved, Section 3 provides a small working example based on five individuals and ten phased, or five unphased SNPs. The methods of SNP detection and localization can be grouped into two broad classes: nontree-based techniques and tree-based techniques. Section 4 provides a description of nontree-based methods for both phased and unphased SNP data, and Section 5 provides a description of some tree-based techniques for phased SNP data. Section 6 describes a tree-based approach for unphased SNP data. The chapter concludes with a brief discussion of all of the information presented in the first six sections.

2 SNP DATA AND DATA PREPARATION

This section presents a brief description of some methods of data preparation that typically need to be carried out before analysis of the genomic data may begin. Specifically, a description of how SNP data appear is given, along with descriptions of a pair of techniques for phasing SNP data into haplotypes.

2.1 Description of the Data

When SNP sequences are received, they are typically in the form of genotype sequences, which are sequences with allele pairs coded by 0, H, and 1. The possible alleles are 0, which represents the major allele, or the allele that occurs most frequently in the population, and 1, which corresponds to the minor allele. Since each individual in a diploid population inherits two copies of a gene, one from each parent, two alleles exist at every SNP site. What are unknown are the

TABLE 1 Example of SNP Data Before and After Phasing

Individual	Sequence Before Phasing	Haplotypes After Phasing
1	0H10H11	0010111
		0110011
2	0H0001H	0100011
		0000010
3	HH011H1	1001101
		0101111
4	10H0110	1000110
		1010110
5	01H1H1H	0111011
		0101110

allele pairs, or haplotypes, at each SNP site. The genotype 0 corresponds to the 00 allele pair. The 1 genotype corresponds to the 11 allele pair. These two genotypes are unambiguous. The H genotype is ambiguous, as it may represent either the 01 or the 10 haplotype. For most methods of SNP detection, these haplotypes need to be inferred before analysis can begin. This is done through a process known as phasing. Table 1 provides an example of how the genetic data look before phasing and after phasing. Several approaches are available to phase genotypes into haplotypes. Section 2.2 describes some of them.

2.2 Methods of Phasing

The methods of phasing can be grouped into three classes: computational imputation, expectation-maximization (EM)-based methods, and methods based on the coalescent model (Kingman, 1982) of evolution. Sections 2.2.1–2.2.3 describe these classes, and examples are provided of a computational imputation approach and two coalescent-based methods.

2.2.1 Computational Imputation

Computational techniques make use of information from several individuals to infer haplotype phase. A common approach is to phase unrelated individuals by examining sets of common haplotypes that can explain the observed genotype data (Clark, 1990; Gusfield, 2003; Wang and Xu, 2003). The number of unrelated individuals is a key determinant in the accuracy of phase estimation. The accuracy improves with more individuals. Related individuals can be phased by examining haplotypes that are shared identically by descent (Kong et al., 2008) between individuals within families. Here, the method of Clark (1990) will be described in detail, and other techniques will be mentioned.

2.2.1.1 Clark (1990) Method

The method of Clark (1990) uses unambiguous haplotypes that come from individuals with at most one ambiguous, or *heterozygous* marker and parsimony, which depends on finding solutions that minimize the number of unique haplotypes. This approach was one of the first methods of haplotype inference for three or more markers in unrelated individuals. Suppose that a series of SNP sequences from diploid individuals with many heterozygous sites is obtained. A sufficient number of individuals in the sample all but guarantees that at least one homozygous individual or an individual with at most one heterozygous site is present. To recognize a homozygote, look for a lack of ambiguous sites. If an individual is homozygous, then one unambiguous haplotype has been identified. If an individual has one heterozygous site, then two unambiguous haplotypes have been identified. The algorithm begins by finding all homozygotes and single-site heterozygotes and then tabulating the resulting known haplotypes.

For each known haplotype, the remaining unresolved sequences are examined and a determination is made as to whether the known haplotype can be made from some combination of the ambiguous sites. When such a haplotype is found, the complement of the haplotype is recovered as another potential haplotype. This chain continues until either all haplotypes have been recovered or until no new haplotypes can be recovered. Fig. 1 illustrates the chain of inference.

Example 1. Suppose that the genotype sequence 0100110 is observed. This is unambiguously H_1H_1, where $H_1 = 0100110$. Suppose that the genotype sequence 0H0H110 is also observed. This sequence has two ambiguous sites, resulting in the following four possible haplotypes: $H_{21} = 0000110$, $H_{22} = 0001110$, $H_{23} = 0100110$, and $H_{24} = 0101110$. One of these, H_{23} is the

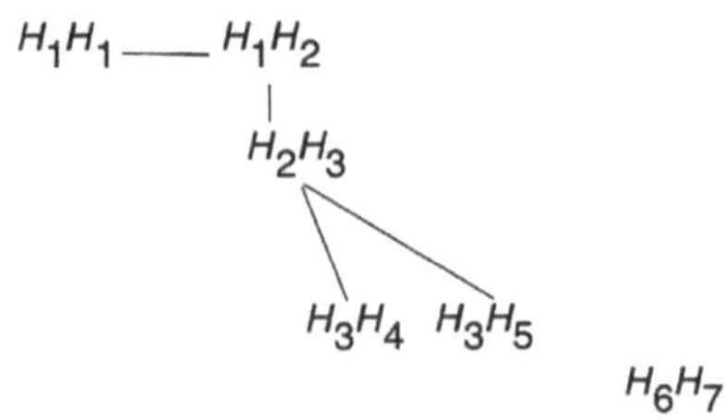

FIG. 1 Begin with H_1H_1 as a homozygote for haplotype H_1. If an ambiguous genotype is also observed, then it can arise from haplotypes H_1 and H_2. A similar process yields H_2H_3, H_3H_4, and H_3H_5. The haplotype pair H_6H_7 cannot be derived from any of the known haplotypes.

original haplotype H_1. Of the remaining three haplotypes, only one, H_{22}, could have given the ambiguous genotype. We thus infer that $H_2 =$ 0001110, and that the genotype sequence is composed of the haplotype pair H_1H_2. If the genotype sequence 01HH110 is then observed, then it is necessary to consider which possible configurations with both H_1 and H_2 can explain this genotype. There are four possible haplotypes here. $H_{31} =$ 0100110, $H_{32} = 0101110$, $H_{33} = 0110110$, and $H_{34} = 0111110$. Observe that because in H_2, the second site has a 0, there is no haplotype of these four that is going to give the 01HH110 genotype if H_2 is the allele that is inherited. Examining H_1, it can be seen that the H_{34} haplotype 0111110 is the only one of these four haplotypes that can be paired with H_1 to obtain the genotype. Thus, $H_3 = H_{34}$, and the haplotype pair is H_1H_3. Finally, suppose the genotype 00HH100 is observed. The possible haplotypes that can arise from this genotype are $H_{41} = 0000100$, $H_{42} = 0001100$, $H_{43} = 0010100$, and $H_{44} =$ 0011100. To get the genotype 00HH100, one of these four haplotypes must be paired with either H_1 or H_3. However, both H_1 and H_3 have a 1 in site 2, while the genotype has a 0 in site 2. Therefore, none of the four possible haplotypes combined with the ancestral H_1 and H_3 haplotypes explains the genotype. This genotype is unresolved. Call its components H_4 and H_5.

This algorithm suffers from three important limitations. First, if it is not possible to recover any homozygotes or single-site heterozygotes, the process cannot get started. Second, the process may result in unresolved haplotypes at the end. Avoiding this issue requires that alleles already on one chromosome be *tightly linked*, i.e., the alleles on one chromosome will be inherited as a unit more frequently than not. The tightness of allele linkage also plays a role in the third limitation, where genotypes may be phased incorrectly if a crossover product of two haplotypes is identical to another true haplotype. These limitations have resulted in the advent of other computational phasing techniques, including parsimony methods (Gusfield, 2003; Wang and Xu, 2003) and long-range phasing (Kong et al., 2008).

2.2.2 EM-Based Methods

The phasing methods that arose after the Clark (1990) method are primarily based on the EM algorithm. These methods use no prior information on the haplotype frequency distribution, but rely on a multinomial model for haplotype frequencies. The goal is to maximize the multinomial likelihood. Such methods were developed by Excoffier and Slatkin (1995) and Long et al. (1995), as well as by Hawley and Kidd (1995) in the HAPLO procedure. The EM algorithm has also been used in more complex methods, such as fastPHASE (Scheet and Stephens, 2006) and HaploRec (Eronen et al., 2006), and it has been extended to partition-ligation EM algorithms (Qin et al., 2002) for larger numbers of markers. Here, the Excoffier and Slatkin (1995) method is described in detail. The Excoffier and Slatkin (1995) method begins with likelihood estimation, which comprises the expectation (E) step of the EM algorithm. Suppose a sample of n phenotypes, or sequences of genotypes, is collected, and let m be the number of distinct phenotypes in the sample. Let $P_1, \ldots, P_m$ denote the population phenotype frequencies. Then the probability of the sample data is given by

$$\mathrm{P}(\text{sample}|P_1,\ldots,P_m) = \frac{n!}{n_1!\ldots n_m!} P_1^{n_1} \ldots P_m^{n_m},$$

where n_i is the count of phenotype i in the sample. The likelihood, however, needs to be in a form that can be used to estimate haplotype frequencies. To this end, let c_j denote the number of haplotype pairs that lead to phenotype j. This quantity is a function of s_j, the number of heterozygous loci. The relationship is as follows:

$$c_j = \begin{cases} 2^{s_j-1} & \text{if } s_j > 0 \\ 0 & \text{if } s_j = 0. \end{cases}$$

Then

$$\begin{aligned} P_j &= \sum_{i=1}^{n_i} \mathrm{P}(\text{genotype } i) \\ &= \sum_{i=1}^{c_j} \mathrm{P}(h_k, h_l), \end{aligned}$$

where $\mathrm{P}(h_k, h_l)$ denotes the probability that the ith genotype is made of haplotypes h_k and h_l. Then

$$\mathrm{P}(h_k, h_l) = \begin{cases} p_k^2 & \text{if } k = l \\ 2p_k p_l & \text{if } k \neq l, \end{cases}$$

where p_k and p_l are the population frequencies of haplotypes k and l. This gives the likelihood

$$\mathcal{L}(P_1,\ldots,P_m) = a \prod_{j=1}^{m} \left(\sum_{i=1}^{c_j} \mathrm{P}(h_{ik}, h_{il}) \right)^{n_i},$$

where a is the multinomial coefficient.

The algorithm begins with an initial set $p_1^{(0)}, \ldots, p_h^{(0)}$ of haplotype frequencies. These initial values are used to estimate haplotype frequencies in the next step. At iteration g, $\mathrm{P}(h_k, h_l)^{(g)}$ is given by

$$\mathrm{P}(h_k, h_l)^{(g)} = \frac{n_j}{n} \frac{\mathrm{P}_j(h_k, h_l)^{(g)}}{P_j^{(g)}},$$

where $P_j(h_k, h_l) = \Pr(\text{genotype } h_k h_l \text{ is in phenotype } j)$. The likelihood is then maximized through a gene-counting method (Ceppelini et al., 1955; Smith, 1957) to construct estimates

$$\hat{p}_t^{(g+1)} = \frac{1}{2}\sum_{j=1}^{m}\sum_{i=1}^{c_j}\delta_{it}P_j(h_k,h_l)^{(g)},$$

where δ_{it} denotes the number of times haplotype t is present in genotype i, so $\delta_{it} = 0$, 1, or 2.

EM-based methods suffer from two key limitations. First, for large numbers of markers, they can be inaccurate. Furthermore, the variances of the haplotype frequencies are unknown, and estimating these variances can be quite difficult to do.

2.2.3 Coalescent-Based Methods

In recent years, the trend in haplotype phasing technology has been toward approximate coalescent models (McVean and Cardin, 2005). These types of models account for the fact that new haplotypes arise from old ones through the processes of mutation and recombination. Over small genetic regions, mutations and recombinations are rare. As a result, haplotypes tend to look similar to each other. As an example, if the haplotypes 1100 and 0001 are present in a sample, it is also likely that the haplotype 1101 is also in the sample. This haplotype arises through a recombination event. It may also be likely that the haplotype 0011, which is formed through a mutation, is observed. However, observing haplotype 1111 is less likely because it requires both a recombination event and a mutation, two mutations in 1100, or three mutations in 0001. Many statistical phasing methods, such as PHASE (Stephens and Scheet, 2005; Stephens et al., 2001), fastPHASE (Scheet and Stephens, 2006), BEAGLE (Browning and Browning, 2007), IMPUTE2 (Howie et al., 2009), and MACH (Li et al., 2010) use the approximate coalescent model. Here, PHASE (Stephens and Scheet, 2005; Stephens et al., 2001) and BEAGLE (Browning and Browning, 2007) are described in detail. Many of these approaches use the approximate coalescent model to construct a hidden Markov model (HMM), and the parameters of the HMM are estimated through iterative procedures.

2.2.3.1 PHASE (Stephens and Scheet, 2005; Stephens et al., 2001)

Suppose a sample of n diploid individuals is drawn from a population. Let $\mathbf{G}$ denote the vector $(G_1, \ldots, G_n)$, which represents the observed genotypes for the individuals, and let $\mathbf{H} = (H_1, \ldots, H_n)$ represent the unknown corresponding haplotype pairs. Denote the unknown population haplotype frequencies by $\mathbf{F} = (F_1, \ldots, F_M)$, and let $\mathbf{f} = (f_1, \ldots, f_M)$ denote the unknown sample haplotype frequencies, where M is the number of possible haplotypes.

PHASE (Stephens and Scheet, 2005; Stephens et al., 2001) views the unknown haplotypes as random variables and estimates their conditional distribution given the observed genotypes. PHASE (Stephens and Scheet, 2005; Stephens et al., 2001) uses a Gibbs sampler (Gil, 1996) to obtain an approximate sample from the distribution P(**H**|**G**). This results in the construction of a Markov chain $\mathbf{H}^{(0)}$, $\mathbf{H}^{(1)}$, ... with stationary distribution P(**H**|**G**) in the following way. For $t = 0, 1, 2, \ldots$, obtain $\mathbf{H}^{(t+1)}$ using three steps.

1. Choose an individual i uniformly at random from all heterozygous individuals.
2. Sample $H_i^{(t+1)}$ from $P(H_i|\mathbf{G},\mathbf{H}_{-i}^{(t)})$, where $\mathbf{H}_{-i}$ is the set of haplotypes excluding individual i.
3. Set $H_j^{(t+1)} = H_j^{(t)}$ for $j = 1, \ldots, n$, $j \neq i$.

The difficult part of the process is step 2. In order to make it easier to sample from $P(H_i|\mathbf{G}, \mathbf{H}_{-i})$, let $H_i = (h_{i1}, h_{i2})$ denote a haplotype pair consistent with genotype G_i. Then

$$\begin{aligned} P(H_i|\mathbf{G},\mathbf{H}_{-i}) &\propto P(H_i|\mathbf{H}_{-i}) \\ &\propto p(h_{1i}|\mathbf{H}_{-i})p(h_{2i}|\mathbf{H}_{-i},h_{1i}). \end{aligned}$$

In order for $p(\cdot, \mathbf{H})$ to be known, it is necessary to assume that the type of mutant offspring is independent of the type of parent. While this assumption is biologically unrealistic, the resulting model works reasonably well for SNP data. The conditional distribution $p(\cdot|\mathbf{H})$ is approximated as suggested by Stephens and Donnelly (2000). For a general mutation model with types in the set E and reversible mutation matrix P,

$$p(h|\mathbf{H}) \approx \sum_{\alpha \in E} \sum_{s=0}^{\infty} \frac{r_\alpha}{r} \left(\frac{\theta}{r+\theta}\right)^s \frac{r}{r+\theta} P_{\alpha h}^s,$$

where r_α is the number of haplotypes of type α in the set $\mathbf{H}$, r is the total number of haplotypes in $\mathbf{H}$, and θ is a scaled mutation rate. At this point, it is necessary to describe the process of sampling from $p(H_i|\mathbf{H}_{-i})$. First, θ can be estimated using $\hat{\theta} = \frac{S}{\log(2n)}$, where S is the number of polymorphic sites. To sample from $p(H_i|\mathbf{H}_{-i})$, begin with some initial haplotype reconstruction $\mathbf{H}^{(0)}$. The Gibbs sampler is adjusted in the following way. For $t = 0, 1, 2, \ldots$, $\mathbf{H}^{(t+1)}$ is obtained from $\mathbf{H}^{(t)}$ through the following three steps.

1. Choose an individual i uniformly at random from all heterozygous individuals.
2. Select a subset R of ambiguous sites in individual i to update. Let $\mathbf{H}(R)$ denote the haplotype information for individual i at the loci in R, and let $\mathbf{H}(-R)$ denote the complement of $\mathbf{H}(R)$ in $\mathbf{H}$. Sample $\mathbf{H}^{(t+1)}(R)$ from $P(\mathbf{H}(R)|\mathbf{G}, \mathbf{H}^{(t)}(-R))$.
3. Set $\mathbf{H}^{(t+1)}(-R) = \mathbf{H}^{(t)}(-R)$.

Once the haplotypes are sampled, the sample haplotype frequencies are estimated from the empirical distribution of the haplotypes. A single point estimate $\hat{\mathbf{H}}^{SSD}$ of $\mathbf{H}$ along with estimates of $\mathbf{Q} = (q_{ij})$, where q_{ij} is the probability that the phase call for individual i at locus j is correct, is computed. These estimates are obtained by minimizing the posterior expected loss, where the loss function $L(\hat{\mathbf{H}}^{SSD}, \mathbf{Q}|\mathbf{H})$ is equal to the quantity

$$-\max\left\{\sum_{(i,j)\in C}\log(q_{ij})+\sum_{(i,j)\notin C}\log(1-q_{ij}),\sum_{(i,j)\notin C}\log(1-q_{ij})+\sum_{(i,j)\in C}\log(q_{ij})\right\},$$

where $(i, j) \in C$ if the phase call for individual i at locus j is correct. The following example partially carries out this process.

Example 2. Table 2 gives the genotype data for this example. Here, $n = 5$ and sites 1, 2, and 3 are polymorphic. Thus, $\hat{\theta}=\frac{3}{\log(10)}=1.443$. We now enumerate the haplotype pairs that are consistent with each genotype. These pairs are in Table 3. Suppose that we want the probability that for $G_5, H_5=\begin{matrix}01100\\10001\end{matrix}$. Here, $E=\{0,1\}^5$, and $|E| = 32$. We also have $r =$ the number of haplotypes in $\mathbf{H}_{-5} = 28$. Let $\alpha_1 = 00100$, $\alpha_2 = 01100$, $\alpha_3 = 01101$, $\alpha_4 = 00101$, $\alpha_5 = 10100$, $\alpha_6 = 11101$, $\alpha_7 = 11100$, $\alpha_8 = 10101$, $\alpha_9 = 00000$, $\alpha_{10} = 00001$, $\alpha_{11} = 10000$, and $\alpha_{12} = 10001$. Then $r_{\alpha_6}=r_{\alpha_7}=1$, $r_{\alpha_2}=r_{\alpha_3}=r_{\alpha_9}=r_{\alpha_{10}}=r_{\alpha_{11}}=r_{\alpha_{12}}=2$, $r_{\alpha_5}=r_{\alpha_8}=3$, and $r_{\alpha_1}=r_{\alpha_4}=4$.

$$\begin{aligned}\mathrm{P}(H_5|\mathbf{H}_{-5})&=\mathrm{P}(01100|H_5)\mathrm{P}(10001|\mathbf{H}_{-5},01100)\\&=\mathrm{P}(01100|\mathbf{H}_{-5})\mathrm{P}(10001|\mathbf{H}_{-5}).\end{aligned}$$

2.2.3.2 BEAGLE (Browning and Browning, 2007)

The HMM for haplotype frequencies in BEAGLE (Browning and Browning, 2007) is developed from a graph of haplotypes. This graph is called the localized haplotype cluster model. This model clusters haplotypes at each marker

TABLE 2 Genotype Data for Example

Individual	Genotype
1	G_1 = 0H10H
2	G_2 = HH10H
3	G_3 = H0H0H
4	G_4 = H0H0H
5	G_5 = HHH0H

TABLE 3 Haplotype Pairs

Individual	Haplotype Pairs			
1	00100	01100		
	01101	00101		
2	00100	00101	01100	10100
	11101	11100	10101	01101
3	00000	00001	00100	10000
	10101	10100	10001	00101
4	00000	00001	00100	10000
	10101	10100	10001	00101
5	00000	00001	00100	00101
	11101	11100	11001	11000
5	01000	01001	01101	01100
	10101	10100	10000	10001

to improve the prediction of alleles at markers to the right of a given marker given the alleles at the markers to the left. Suppose that a sample of haplotypes for M markers is obtained, and that the haplotypes have no missing alleles. A localized haplotype cluster model for the sample is a directed, acyclic graph with four properties:

1. The graph has a root node with no incoming edges and a terminal node with no outgoing edges.
2. The graph is leveled with $M + 1$ levels. Each node A has a level, m. All incoming edges to A have the parent node at level $m - 1$ and all outgoing edges have the child node at level $m + 1$. The root has level 0, and the terminal node has level M.
3. For each $m = 1, 2, \ldots, M$, each edge with the child node at level m is labeled with an allele for the mth marker. Two edges from the same parent node cannot be labeled with the same allele.
4. For each sampled haplotype, there is a path from the root to the terminal node such that the mth allele of the haplotype is the label of the mth edge of the path. Each edge of the graph has at least one haplotype in the sample whose path traverses the edge.

For each edge e, let $n(e)$ be the number of sampled haplotypes whose path traverses the edge, and let $n_P(e)$ be the number of haplotypes whose path traverses the parent node of e.

The localized haplotype cluster model gives rise to an HMM in which the states are the edges of the graph. Each state emits with probability 1 the allele that labels the edge. Initial state probabilities are given by $P(e) = n(e)/n_P(e)$ if the parent node of edge e is the root and $P(e)$ is zero otherwise. The transition probabilities are $P(e_1|e_2) = n(e_1)/n_P(e_1)$ if the parent node of e_1 is the child node of e_2, and $P(e_1|e_2) = 0$ otherwise. This describes a haploid HMM, but a diploid HMM is required since diploid data are observed. The diploid HMM arises from ordered pairs of edges in each level of the graph. Since the graph is leveled, it can be partitioned into classes L_m, where $m = 1, 2, \ldots, M$ is the level of the edge's child node in the graph. The state space for the HMM is the union over m of $(L_m \times L_m)$, and the symbol that is emitted for each state is the unordered pair of alleles that label the state's ordered pair of edges. If (e_1, e_2) is a state of the HMM, then the unordered genotype that is determined by the two alleles that label the edges is emitted with probability 1. Under the assumption of Hardy–Weinberg Equilibrium, $P(e_1, e_2) = P(e_1)P(e_2)$, and $P((e_1, e_2)|(e_3, e_4)) = P(e_1|e_3)P(e_2|e_4)$.

To sample from the HMM, let g_m be the genotype at marker m, and let $s_m = (e_1, e_2)$ be an ordered pair of edges in $L_m \times L_m$. For any s_m in the diploid HMM and with the individual genotypes $\{g_1, \ldots, g_M\}$, define

$$\alpha_m(s_m) = P(g_1, \ldots, g_m, s_m).$$

For $m = 1, 2, \ldots, M$, $\alpha_m(s_m)$ is computed as follows. To initialize the computation, set

$$\alpha_1(s_1) = P(g_1, s_1) = P(s_1)P(g_1|s_1).$$

For $m + 1$,

$$\begin{aligned}
\alpha_{m+1}(s_{m+1}) &= P(g_1, \ldots, g_{m+1}, s_{m+1}) \\
&= \sum_{s_m} P(g_1, \ldots, g_{m+1}, s_m, s_{m+1}) \\
&= \sum_{s_m} P(g_1, \ldots, g_m, s_m) P(g_{m+1}|s_{m+1}) P(s_{m+1}|s_m) \\
&= P(g_{m+1}|s_{m+1}) \sum_{s_m} \alpha_m(s_m) P(s_{m+1}|s_m).
\end{aligned}$$

The probabilities $P(g_{m+1}|s_{m+1})$ are always 0 or 1 depending on whether the genotype is consistent with the edges in s_{m+1}. In the case of missing alleles, if both alleles of g_m are missing, $P(g_m|s_m) = 1$. If one allele is missing, $P(g_m|s_m) = 1$ if the non-missing allele labels one of the ordered edges of s_m, and $P(g_m|s_m) = 0$ otherwise. Sampling the hidden states conditionally on the genotypes proceeds backward by induction. To begin, randomly choose the state s_M with probability proportional to $\alpha_M(s_M)$. Given $s_{m+1}, \ldots, s_M$, choose s_m with probability

$$P(s_m|s_{m+1}, \ldots, s_M, g_1, \ldots, g_M) = \frac{P(g_{m+1}|s_{m+1}) P(s_{m+1}|s_m) \alpha_m(s_m)}{\alpha_{m+1}(s_{m+1})}.$$

By the use of a Viterbi algorithm (Rabiner, 1989), the most likely haplotype pairs conditional on the genotype data can be obtained. The following example illustrates the $\alpha_m(s_m)$ computation and the calculation of sampling probabilities.

Example 3. Consider the haplotypes and their counts in Table 4. These haplotypes can be represented by the localized haplotype cluster model in Fig. 2. In this figure, a solid line denotes an edge terminating at a 0, and a dashed line denotes an edge terminating at a 1.

Table 5 gives $n(e_i)$ and $n_P(e_i)$ for $i = A, B, \ldots, L$ to be used in probability calculation.

Suppose the genotypes are $g_1 = ?/?$, $g_2 = 0/0$, $g_3 = 0/1$, and $g_4 = 0/1$, where? denotes a missing allele. We begin with probability calculation. We need to find $\alpha_1(e_A, e_A)$, $\alpha_1(e_A, e_B)$, $\alpha_1(e_B, e_A)$, and $\alpha_1(e_B, e_B)$.

$$\begin{aligned}\alpha_1(e_A,e_A) &= P(s_1=(e_A,e_A))P(?/?|s_1=(e_A,e_A)) \\ &= P(e_A)P(e_A)(1) \\ &= \left(\frac{319}{585}\right)^2.\end{aligned}$$

Similarly,

$$\begin{aligned}\alpha_1(e_A,e_B) &= \frac{319(266)}{585^2}, \\ \alpha_1(e_B,e_A) &= \frac{319(266)}{585^2}, \text{ and} \\ \alpha_1(e_B,e_B) &= \left(\frac{266}{585}\right)^2.\end{aligned}$$

TABLE 4 Haplotypes and Counts

Haplotype	Count
0000	97
0001	38
0011	59
0110	125
1000	86
1001	75
1011	105

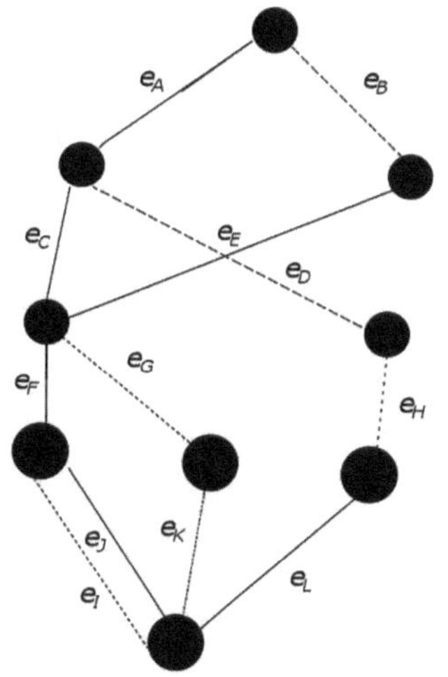

FIG. 2 Localized haplotype cluster model for BEAGLE example. All edges are directed from *top* to *bottom*, so *arrows* indicating this are omitted.

TABLE 5 Values of $n(e)$ and $n_P(e)$ for Beagle Example

e	$n(e)$	$n_P(e)$
A	97 + 38 + 59 + 125 = 319	585
B	86 + 75 + 105 = 266	585
C	97 + 38 + 59 = 194	319
D	125	319
E	86 + 75 + 105 = 266	266
F	97 + 38 + 86 + 75 = 296	194 + 266 = 460
G	59 + 105 = 164	460
H	125	125
I	38 + 75 = 113	296
J	97 + 86 = 183	296
K	59 + 105 = 164	164
L	125	125

For step 2, the genotype is $g_2 = 0/0$, and we need to find $\alpha_2(e_C, e_C)$, $\alpha_2(e_C, e_D)$, $\alpha_2(e_C, e_E)$, $\alpha_2(e_D, e_C)$, $\alpha_2(e_D, e_D)$, $\alpha_2(e_D, e_E)$, $\alpha_2(e_E, e_C)$, $\alpha_2(e_E, e_D)$, and $\alpha_2(e_E, e_E)$. The pairs (e_C, e_D), (e_D, e_C), (e_D, e_D), (e_D, e_E), and (e_E, e_D) are inconsistent with g_2, so $P(0/0|s_2) = 0$ for s_2 equal to any of these five pairs.

$$\begin{aligned}\alpha_2(e_C,e_C) = {} & \mathrm{P}(0/0|s_2=(e_C,e_C))[\alpha_1(e_A,e_A)\mathrm{P}(s_2=(e_C,e_C)|s_1=(e_A,e_A))] \\ & +\mathrm{P}(0/0|s_2=(e_C,e_C))[\alpha_1(e_A,e_B)\mathrm{P}(s_2=(e_C,e_C)|s_1=(e_A,e_B))] \\ & +\mathrm{P}(0/0|s_2=(e_C,e_C))[\alpha_1(e_B,e_A)\mathrm{P}(s_2=(e_C,e_C)|s_1=(e_B,e_A))] \\ & +\mathrm{P}(0/0|s_2=(e_C,e_C))[\alpha_1(e_B,e_B)\mathrm{P}(s_2=(e_C,e_C)|s_1=(e_B,e_B))] \\ = {} & 1\left(\frac{319}{585}\right)^2\left(\frac{194}{319}\right)^2+0+0+0=\left(\frac{194}{585}\right)^2.\end{aligned}$$

Similarly,

$$\begin{aligned}\alpha_2(e_C,e_E) &= \frac{266(194)}{585^2}, \\ \alpha_2(e_E,e_C) &= \frac{266(194)}{585^2}, \text{ and} \\ \alpha_2(e_E,e_E) &= \left(\frac{266}{585}\right)^2.\end{aligned}$$

For site 3, the genotype is $g_3 = 0/1$, and the possible pairs are (e_F, e_F), (e_F, e_G), (e_F, e_H), (e_G, e_F), (e_G, e_G), (e_G, e_H), (e_H, e_F), (e_H, e_G), and (e_H, e_H). Note that the genotype g_3 is inconsistent with (e_F, e_F), (e_G, e_G), (e_G, e_H), (e_H, e_G), and (e_H, e_H), so $\mathrm{P}(0/1|s_3) = 0$ for s_3 equal to any of these pairs. Furthermore, $\mathrm{P}(s_3 = (e_F, e_H)|s_2) = \mathrm{P}(s_3 = (e_H, e_F)|s_2) = 0$ for $s_2 = (e_C, e_C)$, (e_C, e_E), (e_E, e_C), and (e_E, e_E). Therefore, $\alpha_3(e_F, e_H) = \alpha_3(e_H, e_F) = 0$.

$$\begin{aligned}\alpha_3(e_F,e_G) = {} & \mathrm{P}(0/1|s_3=(e_F,e_G))\alpha_2(e_C,e_C)\mathrm{P}(s_3=(e_F,e_G)|s_2=(e_C,e_C)) \\ & +\mathrm{P}(0/1|s_3=(e_F,e_G))\alpha_2(e_C,e_E)\mathrm{P}(s_3=(e_F,e_G)|s_2=(e_C,e_E)) \\ & +\mathrm{P}(0/1|s_3=(e_F,e_G))\alpha_2(e_E,e_C)\mathrm{P}(s_3=(e_F,e_G)|s_2=(e_E,e_C)) \\ & +\mathrm{P}(0/1|s_3=(e_F,e_G))\alpha_2(e_E,e_E)\mathrm{P}(s_3=(e_F,e_G)|s_2=(e_E,e_E)) \\ = {} & \left(\frac{194}{585}\right)^2\frac{296}{460}+\frac{194(266)}{585^2}\frac{296}{460}\frac{164}{460} \\ & +\frac{194(266)}{585^2}\frac{296}{460}\frac{164}{460}+\left(\frac{266}{585}\right)^2\frac{296(164)}{460^2} \\ = {} & \frac{296(194)}{585^2}.\end{aligned}$$

Similarly, $\alpha_3(e_G,e_F)=\frac{296(194)}{585^2}$. For site 4, the genotype is $g_4 = 0/1$, and it is straightforward to show that for all possible pairs (e_1, e_2) except for (e_I, e_K) and (e_K, e_I), $\alpha_4(e_1, e_2) = 0$. We can find that

$$\alpha_4(e_I,e_K)=\alpha_4(e_K,e_I)=\frac{113(164)}{585^2}.$$

To initialize the sampling probability portion, set $s_4 = (e_I, e_K)$ with probability

$$\alpha_4(e_I,e_K)/(\alpha_4(e_I,e_K)+\alpha_4(e_K,e_I))=\frac{1}{2}.$$

Next, sample $s_3 = (e_F, e_G)$ given $s_4 = (e_I, e_K)$ with probability

$$\frac{P(g_4 = 0/1 | s_4 = (e_I, e_K))P(s_4 = (e_I, e_K) | s_3 = (e_F, e_G))\alpha_3(e_F, e_G)}{\alpha_4(e_I, e_K)} = 1.$$

Next, $s_2 = (e_C, e_E)$ given $s_3 = (e_F, e_G)$ is sampled with probability

$$\frac{P(g_3 = 0/1 | s_3 = (e_F, e_G))P(s_3 = (e_F, e_G) | s_2 = (e_C, e_E))\alpha_2(e_C, e_E)}{\alpha_3(e_F, e_G)} = 0.2439$$

Finally, sample $s_1 = (e_A, e_B)$ given $s_2 = (e_C, e_E)$ with probability

$$\frac{P(g_2 = 0/0 | s_2 = (e_C, e_E))P(s_2 = (e_C, e_E) | s_1 = (e_A, e_B))\alpha_1(e_A, e_B)}{\alpha_2(e_C, e_E)} = 1.$$

The path $s_1 = (e_A, e_B)$, $s_2 = (e_C, e_E)$, $s_3 = (e_F, e_G)$, $s_4 = (e_I, e_K)$ is sampled with probability 0.5(1)(0.2439)(1) = 0.12195. This path results in the haplotype pair 0000 1011. The reverse path, resulting in the haplotype pair 1011 0000, is also sampled with probability 0.12195.

3 WORKING EXAMPLE

This short section presents the genetic and trait data that will be used in illustrations of several methods of SNP detection that will be described in Sections 4 and 5. Table 6 presents the phased genetic data, as well as the unphased genetic data. The fourth column gives the trait data.

4 NONTREE-BASED METHODS OF GWAS ANALYSIS

Several methods exist to analyze GWAS and to detect quantitative trait loci (QTL) without incorporation of marginal evolutionary structure for each site. The primary approaches are single-marker association (SMA), haplotype association mapping (HAM), and regression-based techniques. In Sections 4.1–4.3, these procedures are described.

TABLE 6 Genetic and Trait Data for Later Examples

Individual	Phased Sequence	Unphased Sequence	Trait Value
1	0100110111	H01H1	96.3112
2	0010100111	0HHH1	77.8221
3	0111110011	H1101	99.5216
4	0111011011	H1HH1	97.1203
5	0100001011	H00H1	94.9828

4.1 Single-Marker Association

A common approach to nontree-based GWAS analysis for QTL detection is to examine each marker independently and determine which sites show a difference in mean trait values for different characters at the site. This approach can be carried out for both phased and unphased SNPs.

4.1.1 Phased SNP Data

When SNP data are phased, there are only two possible characters, 0 and 1, at each site. Consequently, significance of the marker at a given site is assessed through the use of a two-sample t-test. To do this, assume that for trait values $y_1, \ldots, y_n$, where n is the number of individuals in the study, if in site i, $i = 1, \ldots, M$, the individual has a 0 at the given site, its trait value comes from a normal distribution with mean μ_{0i} and variance σ_i^2, and for individuals with a 1 at the site, the trait values are normally distributed with mean μ_{1i} and variance σ_i^2. All trait values are assumed to be independent. Note that the variances are assumed equal for individuals with a 0 and those with a 1 at the given site. Let $\overline{y}_{0i}$ denote the sample mean trait value for individuals with a 0 at site i, and let s_{0i}^2 denote the sample variance. Let $\overline{y}_{1i}$ and s_{1i}^2 be defined similarly for individuals with a 1 at site i. To begin, letting n_{0i} and n_{1i} denote the number of individuals with a 0 and a 1 at site i, respectively, compute the pooled sample variance

$$s_{pi}^2 = \frac{(n_{0i}-1)s_{0i}^2 + (n_{1i}-1)s_{1i}^2}{n_{0i}+n_{1i}-2}.$$

The null hypothesis for this test is that there is no difference between the mean trait values for individuals with a 0 at site i compared to those with a 1 at site i. Mathematically, we write

$$H_{0i}: \mu_{0i} = \mu_{1i} \text{ or } \mu_{0i} - \mu_{1i} = 0$$
$$H_{1i}: \mu_{0i} - \mu_{1i} \neq 0.$$

Under H_{0i}, the statistic

$$t_i = \frac{\overline{y}_{0i} - \overline{y}_{1i}}{s_{pi}\sqrt{\dfrac{1}{n_{0i}} + \dfrac{1}{n_{1i}}}}$$

follows a t distribution with $n_{0i} + n_{1i} - 2$ degrees of freedom. This distribution is used to compute the p-value $\mathrm{P}(|T_{n_{0i}+n_{1i}-2}| \geq |t_i|)$, and the ith marker is deemed significant if the p-value is less than some predetermined value α. This procedure works if the goal is to determine the significance of only one marker. In practice, however, we carry out tests of significance for all

M markers. In order to do this while controlling the probability that at least one marker is deemed significant when it is not, an adjustment is made to α such that H_{0i} is rejected if the p-value is less than α/M. This is known as the Bonferroni procedure, and is the procedure used in the example to follow this subsection.

Example 4. Using the data from Section 3, we aim to make determinations as to which markers are significant to the expression of the trait. The tests are carried out at overall level $\alpha = 0.05$, so each null hypothesis is rejected if the p-value for the individual test is less than $0.05/7 = 0.0071$. The process is carried out in detail for site 2, and the summary statistics and t-statistics, as well as the p-values, are given in Table 7. Significant markers are noted with an asterisk (*). Observe that the significance of sites 1, 9, and 10 cannot be assessed since only one allele is present at each of these sites. For site 2, $\overline{y}_{02} = 96.9840$, and $\overline{y}_{12} = 77.8221$. We have $n_{02} = 1$ and $n_{12} = 4$. Furthermore, $s_{02}^2 = 3.6385$, and $s_{12}^2 = 0$, so $s_{p2}^2 = 3.6365$. Our t-statistic is

$$t_2 = \frac{77.8221 - 96.9840}{\sqrt{3.6385\left(\frac{1}{4} + \frac{1}{1}\right)}} = -8.985.$$

The p-value for this test is $P(|T_3| \geq 8.985) = 0.003 < 0.0071$, so we reject H_{02} and conclude that marker 2 has a significant impact on trait expression. Thus, we conclude that the marker at site 2 is significant to the expression of the trait, but that none of the others have a significant impact on trait values.

4.1.2 Unphased SNP Data

For unphased SNP data, the principle behind SMA is the same as it is for phased data, but now there are up to three groups at each site i. There is a group of individuals with a 0 at site i, a group of individuals with an H, and a group of individuals with a 1. Let μ_{0i}, μ_{Hi}, and μ_{1i} denote the corresponding mean trait values. Assessing significance here requires an analysis of variance (ANOVA). The ANOVA assumptions for trait values are as follows, where Y_{kji} is the trait value for individual j in group k at site i.

$$Y_{kji} = \mu_{kj} + \varepsilon_{kji},$$

where the ε_{kji} are independent and identically distributed according to a normal distribution with mean 0 and variance σ_i^2. The null hypothesis here is that all of the group means are equal. In other words,

$$H_{0i} : \mu_{0i} = \mu_{Hi} = \mu_{1i}$$
$$H_{1i} : \text{At least one } \mu_{ki} \text{ differs from the others.}$$

TABLE 7 Summary Statistics and Test Results for Phased Example

Marker	(n_0, n_1)	$(\overline{y}_0, \overline{y}_1)$	(s_0^2, s_1^2)	s_p^2	t	p-Value
2	(1,4)	(77.822, 96.984)	(0,3.638)	3.638	−8.985	0.003*
3	(2,3)	(95.647, 91.488)	(0.882, 141.509)	94.634	0.468	0.6715
4	(3,2)	(89.705, 98.321)	(106.350, 2.883)	71.861	−1.113	0.3468
5	(2,3)	(96.052, 91.218)	(2.284,137.170)	92.208	0.551	0.6198
6	(2,3)	(86.402, 97.651)	(147.245, 2.788)	50.940	−1.726	0.1828
7	(3,2)	(91.218, 96.052)	(137.170, 2.284)	92.208	−0.551	0.6198
8	(3,2)	(97.208, 87.067)	(5.156, 170.923)	60.412	1.429	0.2482

In order to test this hypothesis, compute two statistics. The mean square due to the SNP (MSS), where K_i is the number of distinct characters present at site i, is given by

$$\mathrm{MSS}_i = \frac{1}{K_i - 1}\sum_{k=1}^{K_i} n_{ki}(\overline{y}_{k\cdot i} - \overline{y}_{\cdot\cdot})^2,$$

where $\overline{y}_{k\cdot i}$ mean trait value for group k at site i and $\overline{y}_{\cdot\cdot}$ is the overall mean trait value. The mean square due to error (MSE) is given by

$$\mathrm{MSE}_i = \frac{1}{n - K_i}\sum_{k=1}^{K_i}\sum_{j=1}^{n_{ki}} (y_{kji} - \overline{y}_{k\cdot i})^2.$$

Under H_{0i}, the quantity $F_i = \frac{\mathrm{MSS}_i}{\mathrm{MSE}_i}$ follows an F distribution with $K_i - 1$ and $n - K_i$ degrees of freedom. For testing M sites, H_{0i} is rejected if the p-value $\mathrm{P}(F_{K_i-1,n-K_i} \geq F_i)$ is less than α/M.

Example 5. In the example data from Section 3, note that site 5 of the unphased data set cannot be tested in this way because there is only one group of characters at this site. For site 2, there are three groups, so $K_2 = 3$. We have $n_{02} = 2$, $n_{H2} = 1$, and $n_{12} = 2$. Furthermore, $\overline{y}_{0\cdot 2} = 95.647$, $\overline{y}_{H\cdot 2} = 77.8221$, $\overline{y}_{1\cdot 2} = 98.321$, and $\overline{y}_{\cdot\cdot} = 93.152$. This gives

$$\begin{aligned}\mathrm{MSS}_2 &= \frac{1}{2}\left[2(95.647 - 93.152)^2 + 1(77.8221 - 93.152)^2 + 2(98.321 - 93.152)^2\right]\\ &= 150.446.\end{aligned}$$

$$\begin{aligned}\mathrm{MSE}_2 &= \frac{1}{2}\left[(96.311 - 95.647)^2 + (94.9828 - 95.647)^2 + (77.8221 - 77.8221)^2\right]\\ &\quad + \frac{1}{2}\left[(99.522 - 98.321)^2 + (97.120 - 98.321)^2\right]\\ &= 1.883.\end{aligned}$$

$$F_2 = \frac{150.446}{1.883} = 79.909.$$

The p-value is $\mathrm{P}(F_{2,2} \geq 79.897) = 0.0123$. Since we are testing four sites, we reject H_{02} if the p-value is less than $0.05/4 = 0.0125$. We conclude that the marker at site 2 is significant to the expression of the trait. The remainder of the summary statistics, F statistics, and p-values is in Table 8. Significant markers are noted with an asterisk. When the SNP data are unphased, the markers at sites 1 and 2 are deemed significant to the expression of the trait. An interesting point here is that, while with the phased data, neither site 3 nor site 4 was significant, when these two sites are joined into an unphased genotype, the unphased SNP at site 2 is significant to the expression of the trait.

TABLE 8 Summary Statistics and Test Results for Unphased Data

Marker	$(\overline{y}_{0\cdot}, \overline{y}_{H\cdot}, \overline{y}_{1\cdot})$	MSS	MSE	*F*	*p*-Value
1	(77.822, 96.959, N/A)	293.742	3.638	80.7319	0.003*
2	(95.647, 77.8221, 98.321)	150.446	1.883	79.909	0.0124*
3	(94.983, 87.471, 97.916)	56.647	95.682	0.595	0.6281
4	(99.522, 91.559, N/A)	50.721	84.645	0.599	0.495

4.1.3 Multiple Comparison Procedures

The Bonferroni procedure of multiple comparisons can be problematic in this setting. Suppose we want to control familywise error rate (FWER) at level $\alpha = 0.05$, and we are examining 10^6 markers in the genome. Then our *p*-value threshold will be 5×10^{-8}, which is overly conservative. A Sidak correction alleviates some, albeit not very much, of this conservatism by showing that if M tests are being carried out, then we reject any given null hypothesis if the *p*-value for the corresponding test is less than $1-(1-\alpha)^{\frac{1}{M}}$, then FWER is controlled at level α. If $\alpha = 0.05$ and $M = 10^6$, then we reject a given null hypothesis for a *p*-value less than 5.129×10^{-8}. This is slightly less conservative than the Bonferroni method. Neither of these procedures, however, controls the false discovery rate (FDR), which is defined as the probability that at least one null hypothesis is falsely rejected. In other words, if V is the number of rejected null hypotheses that are true, then

$$\mathrm{FDR} = \mathrm{P}(V \geq 1).$$

A common technique for controlling FDR is provided by Benjamini and Hochberg (1995). Let M_0 be the number of null hypotheses that are true, and let α again be the overall significance level. Define the false discovery proportion (FDP) as

$$\mathrm{FDP} = \frac{B}{K},$$

where B is the number of Type I errors and K is a number of tests that could result in a false discovery. Then

$$\mathrm{FDR} = \mathbb{E}[\mathrm{FDP}] = \mathbb{E}\left[\frac{B}{K}|K>0\right]\mathrm{P}(K>0).$$

We compute a *p*-value $P_{(i)}$ for each of tests $i = 1, \ldots, m$, and we order them so that

$$P_{(1)} < P_{(2)} < \ldots < P_{(M)}.$$

Under H_0, these p-values follow a Uniform(0,1) distribution. Furthermore, all tests are independent. Let $l_i = \frac{i\alpha}{M}$, and let

$$R = \max\{i : P_{(i)} < l_i\}.$$

Our p-value threshold is to reject H_0 if $P_{(i)} < P_{(R)}$. For example, suppose we again have $M = 10^6$ tests. If $P_{(10)} = 4.5 \times 10^{-7}$ and $P_{(11)} = 6 \times 10^{-7}$, then

$$l_{10} = \frac{10(0.05)}{10^6} = 5 \times 10^{-7},$$

so we reject the null hypothesis corresponding to this p-value. We also have

$$l_{11} = \frac{11(0.05)}{10^6} = 5.5 \times 10^{-7},$$

and we would not reject the null hypothesis corresponding to this p-value. Thus, $R = 10$, and our p-value threshold is $P_{(10)}$. Since

$$\text{FDR} \leq \frac{M_0}{M} \leq \alpha,$$

this process controls FDR at level α. Storey (2003) presents a q-statistic that can be used to control FDR as well, but the Benjamini and Hochberg (1995) method is the more common of the two approaches.

4.2 Haplotype Association Mapping

SMA can be successful in identifying known QTL, but the structure of phased SNP data allows only two genetic groups to be modeled at one time. Furthermore, it is common to have phased SNP sites that contain only one of the two possible alleles. In these cases, it is not possible to assess significance of the SNP to the trait values. HAM addresses this issue by examining windows of r SNPs at a time. The SNPs in this window must be contiguous. In other words, if the r SNPs in the window are listed in order of their site number, there may not be any gaps. A common choice of r is 3. See Pletcher et al. (2004) for more information on this choice. Trait values are grouped according to the phased haplotype in the given window. To assess significance, a modified ANOVA F statistic is used. The modified F statistic differs from the typical F statistic by a weighting factor that accounts for overall population structure among the individuals and the genetic diversity among the individuals in a given haplotype group. Let G_i denote the number of haplotype groups for window i, $i = 1, 2, \ldots, n - r + 1$, and let μ_{gi} denote the mean trait value in group g for window i.

Let j index the trait values in group g, so that $j = 1, 2, \ldots, n_{gi}$. The ANOVA model is given by

$$Y_{gji} = \mu_{gi} + \varepsilon_{gji},$$

where the ε_{gji} are independent and identically distributed with mean 0 and variance σ_i^2. The test statistic is given by

$$F_i = \frac{\frac{\text{SSG}_i}{df_{G_i}}}{\frac{\text{SSE}_i}{df_{E_i}}},$$

where SSG_i is the sum of squares due to the groups in window i and SSE_i is the sum of squares due to error in window i. Mathematically,

$$\text{SSG}_i = \sum_{g=1}^{G_i} w_{gi} n_{gi} (\bar{y}_{g\cdot i} - \bar{y}_{\cdot\cdot})^2,$$

$$df_{G_i} = G_i - 1$$

$$\text{SSE}_i = \sum_{g=1}^{G_i} \sum_{j=1}^{n_{gi}} w_{gi} (y_{gji} - \bar{y}_{g\cdot i})^2, \text{ and}$$

$$df_{E_i} = n - G_i.$$

The weight w_{gi} is a factor that represents the genetic diversity in haplotype group g in window i. This weight is given by Pletcher et al. (2004) as

$$w_{gi} = \frac{\text{number of sites where individuals in the group disagree}}{\text{total number of sites that contain genetic information}}.$$

The statistic F_i is compared to an F distribution with $G_i - 1$ and $n - G_i$ degrees of freedom, and if M groups are tested, the null hypothesis that all group means are equal is rejected if the p-value $P(F_{G_i-1,n-G_i} \geq F_i)$ is less than α/M for some specified value of α. Next, an example is presented where the calculations are done in detail for one 3-SNP window, and the results of significance tests for the remaining 3-SNP windows are presented in a table.

Example 6. The data for this example are presented in Table 9. The specified value of α is 0.05. Begin with the 3-SNP window containing sites 1, 2, and 3. There are three haplotype groups here. Individuals 1, 2, 3, and 4 are in group 1 with haplotype 001. Thus, $n_{11} = 4$. Individuals 5, 6, and 10 are in group 2 with haplotype 000, and individuals 7, 8, and 9 are in group 3 with haplotype 110. Therefore, $n_{21} = n_{31} = 3$. Next, compute the weights. We have $w_{11} = \frac{1}{6}$,

TABLE 9 Data for HAM Example

Individual	Phased Sequence (6 sites)	Trait Value
1	001100	92.372
2	001100	93.419
3	001100	92.996
4	001110	94.019
5	000100	78.278
6	000100	76.932
7	110100	89.145
8	110100	91.913
9	110110	88.816
10	000110	81.473

$w_{21} = \frac{1}{6}$, and $w_{31} = \frac{1}{6}$. The group means and overall mean are $\bar{y}_{1\cdot1} = 93.2015$, $\bar{y}_{2\cdot1} = 78.8943$, $\bar{y}_{3\cdot1} = 89.958$, and $\bar{y}_{\cdot\cdot} = 87.9363$. This gives

$$\begin{aligned}
\text{SSG}_1 &= \frac{4}{6}(93.2015 - 87.9363)^2 + \frac{3}{6}(78.8943 - 87.9363)^2 + \frac{3}{6}(89.958 - 87.9363)^2 \\
&= 61.4047. \\
\text{SSE}_1 &= 2.588. \\
F_1 &= \frac{\frac{61.4047}{3-1}}{\frac{2.588}{10-3}} \\
&= 71.1903.
\end{aligned}$$

We reject $H_{01} : \mu_{11} = \mu_{21} = \mu_{31}$ if $\mathrm{P}(F_{2,\,7} \geq 71.1903) < 0.05/4 = 0.0125$. This probability is 2.23×10^{-5}, so we reject the hypothesis of equal group means and conclude that the (1,2,3) window is significant to the expression of the trait. The remainder of the results are presented in Table 10.

The windows (1,2,3) and (2,3,4) are deemed to have a significant impact on trait expression. Here the sample size was fairly small. In these cases, it is common to use a Kruskal–Wallis test, as in Wang and Paigen (2002). Bootstrap methods have also been used to obtain a distribution of F statistics, and this distribution is used to assess significance. Information on estimating the FDR of these approaches can be found in Benjamini and Yekutieli (2001)

TABLE 10 Results of HAM for 6-Site Example

Window	Group Means	SSG	SSE	F	p-Value
(1,2,3)	(93.202, 78.894, 89.958)	61.404	2.588	71.190	2.23×10^{-5}
(2,3,4)	(93.202, 78.894, 89.958)	61.404	2.588	71.190	2.23×10^{-5}
(3,4,5)	(92.929, 94.019, 84.067, 85.1445)	25.158	66.242	0.760	0.5564
(4,5,6)	(87.865, 88.103)	0.0455	193.2086	0.0019	0.9665

and Storey and Tibshirani (2003). Estimation of the FWER and the generalized FWER can be done via methods presented in Westfall and Young (1993), Churchill and Doerge (1994), and Dudoit et al. (2004). Since haplotype association mapping relies on inferred haplotypes, it is not possible to carry out this procedure for unphased genotypes without some extra considerations. It is not the case that one should simply choose a window size of three and use it do carry out HAM. In fact, in haplotype-based methods of QTL detection, a major concern is the determination of haplotype block size. There are many methods of doing this, many of which are based on recombination or linkage disequilibrium (LD), which is a nonrandom association of alleles at different loci in the population. Here, two methods of block size determination, 4gamete (Wang and Paigen, 2002) and HapBlock (Zhang et al., 2005), will be described in detail, while several others will be mentioned at the end of this section.

4.2.1 4gamete (Wang and Paigen, 2002)

Wang and Paigen (2002) describe a technique that makes use of observed recombination crossovers between loci using an extension of the four-gamete test (FGT). Previous approaches in this vein define a test statistic and use this as a way to identify haplotype blocks provided that this test statistic falls above some predetermined threshold. 4gamete (Wang and Paigen, 2002) does not have such a requirement, so there is less inherent subjectivity in this approach. Suppose we have a set of SNPs. To begin, perform the FGT between each pair of SNPs. A pair of SNPs passes the FGT if the pair fails to contain one of the pairs 00, 01, 10, or 11. A block is defined as a set of contiguous, ordered markers for which there is no evidence of recombination. Naturally, the infinite sites model is assumed here. The infinite sites model assumes that there are an infinite number of sites at which mutations may occur, that no more than one mutation occurs at any given site, and that there are no recombination events. The process of block selection under this procedure is effectively the same as finding contiguous sets of compatible SNPs. The process ends when a pair of SNPs is found that fails the FGT.
Example 7. Consider the set of six individuals and six SNP sites given in Table 11. We can see that there is no pair of SNPs that fail the FGT, so our haplotype block contains all six SNPs.

4.2.2 HapBlock

HapBlock (Zhang et al., 2005) is a linkage disequilibrium (LD)-based suite of programs that sets out to minimize the number of maximally informative SNPs (i.e., tag SNPs) within blocks. LD refers to a nonrandom association between trait values at different sites. HapBlock (Zhang et al., 2005) also aims to minimize the total number of blocks among these partitions with the minimum number of tag SNPs. This set of programs allows three different definitions of blocks.

TABLE 11 SNPs for 4gamete Example

Site	1	2	3	4	5	6
	0	0	1	1	0	1
	0	0	0	1	1	1
	1	0	1	1	0	0
	0	0	0	1	1	1
	1	1	1	1	0	1
	0	0	1	1	0	1

1. Coverage of common haplotypes (Patil et al., 2001) where a certain percentage or more of haplotypes must be common haplotypes.
2. LD-based blocks (Gabriel et al., 2001; Zhang and Jin, 2003) rely on LD statistic D'. Let p_0 and p_1 denote the proportion of gametes 0 and 1 at a given locus, and let $D = p_{01} - p_0p_1$, where p_{01} is the proportion of the time that 0 and 1 occur in the same gamete, and p_0 and p_1 denote the proportions of 0 and 1 alleles, respectively. Assume p_0 and p_1 are both positive. If $D \neq 0$, then there is no illusion that the association between 0 and 1 alleles is random, so LD must be present. Let $D_{\max}$ be defined as

$$D_{\max} = \begin{cases} \max\{-p_0p_1, -(1-p_0)(1-p_1)\} & \text{when } D < 0 \\ \min\{p_0(1-p_1), p_1(1-p_0)\} & \text{when } D > 0. \end{cases}$$

   The D' statistic is defined as $D' = D/D_{\max}$. In effect, the test for LD is very much akin to carrying out a test of independence, where in each block, we require at least a certain percentage of SNP pairs having strong LD. In other words, $|D'| > K$, where K is some prespecified threshold.
3. No historical evidence of recombination (Wang and Paigen, 2002; Zhang and Jin, 2003). This is based on the FGT.

The set of programs allows for five definitions of tag SNPs:

1. As described by Patil et al. (2001) and implemented by Zhang et al. (2002), in a block, the minimum set of SNPs that can uniquely distinguish a certain percentage of all the haplotypes is deemed a set of tag SNPs.
2. Johnson et al. (2001) use the minimum set of SNPs that can account for a certain percentage of overall haplotype diversity as a set of tag SNPs.
3. Nothnagel et al. (2002) use entropy for haplotype block partitioning. If there are n haplotypes and haplotype i occurs with frequency p_i, the entropy of the haplotypes is given by

$$S = -\sum_{i=1}^{n} p_i \log p_i.$$

The tag SNP set is the smallest set of SNPs that can account for a certain percentage of overall haplotype entropy.

4. The measure r^2 of LD has been proposed by Zhang and Jin (2003) and Carlson et al. (2004) for selecting tag SNPs due to the proportionality of r^2 to statistical power in association studies. The value of r^2 is given by

$$r^2 = \frac{D^2}{p_0(1-p_0)p_1(1-p_1)}.$$

All pairwise r^2 values between SNPs in a block and the SNPs that are not in the block are computed. For a given SNP outside of the block, the maximum value of r^2 is used to quantify the predictive power of the block. Furthermore, the minimum value of r^2 over all SNPs outside the block is taken as the overall prediction power. As with the other measures, the smallest set of SNPs with prediction power above a certain threshold is used as a set of tag SNPs.

5. Stram et al. (2003) give a formal measure, R_h^2, to describe the uncertainty in predicting haplotype from genotypes, i.e., phasing, and uses it to select tag SNPs. This quantity is computed for each common haplotype based on a subset of SNPs over all SNPs in a block. The smallest R_h^2 among all common haplotypes is used as a measure of overall prediction strength. The smallest set of SNPs with an R_h^2 value above a given threshold is used as a set of tag SNPs. The value of R_h^2 is given as follows. Let $\delta_h(H)$ denote the estimated number of h haplotypes contained in the unknown pair of haplotypes H, so that $\delta_h(H)$ is 0, 1, or 2. Estimates are computed based on the genotype G. The expected value of $\delta_h(H_i)$ from individual i with genotype G_i is

$$\mathbb{E}[\delta_h(H_i)] = \frac{\sum_{H\sim G_i} \delta_h(H) p_{h_1} p_{h_2}}{\sum_{H\sim G_i} p_{h_1} p_{h2}},$$

where $\sum_{H\sim G_i}$ is a sum over the ordered haplotype pairs $H = (h_1, h_2)$ with respective frequencies p_{h_1} p_{h_2} that are compatible with genotype G_i. HapBlock (Zhang et al., 2005) uses a dynamic programming approach to search all possible block partitions, and it uses an enumeration method to search for tag SNPs in each block. While slower than many other methods, it guarantees that a block partition with the minimum number of tag SNPs is found.

4.2.3 Other Approaches

Many other techniques of determining block size can be found in the literature. Some depend on sliding, yet fixed blocks (Guo et al., 2009; Hao et al., 2012; Lorenz et al., 2010; Lu et al., 2011; Zhao et al., 2007). Combining SNPs in a fixed window size (Lu et al., 2012; Yan et al., 2009; Yomenaru et al.,

2012), some techniques use overlapping, fixed-size sliding windows (Lorenz et al., 2010), and still others depend on variable-length sliding windows of 2–10 SNPs. Still others (Bafna et al., 2003) question the very need for block selection as a precursor to choosing sets of tag SNPs. They give a formal definition of SNP informativeness and use a minimum set of informative SNPs to construct three distinct tag SNP selection algorithms. The underlying point here is that whether or not the researcher believes in a need for block selection, it is important to have some idea of how to select informative SNPs before analysis and detection of QTL via haplotype-based methods can begin.

4.3 Regression-Based Methods of QTL Detection

Another common approach to detection of QTL is to construct a linear regression model to assess significance of a given marker. The advantages of this approach over SMA and HAM are that environmental and marker covariates can be incorporated into the model. Here, the models will be discussed briefly. The three most common models are the interval-mapping model (Lander and Botstein, 1989, 1994), the composite interval-mapping model (Jansen, 1993; Zeng, 1993, 1994), and the extended composite interval-mapping model.

Interval-mapping (Lander and Botstein, 1989, 1994) models the QTL effect through use of a mixture model where the component distributions are normal with common error variance and that has a mean phenotype based on the regression model

$$y_i = \mu + \mathbf{Q}_q \alpha + \varepsilon_i,$$

where y_i is the phenotype for individual i, $i = 1, 2, \ldots, n$, μ is an overall mean effect, $\mathbf{Q}_q$ is a coded value for one of the n_Q genotypes q, $q = 1, \ldots, n_Q$, α is the effect of an allele substitution, and ε_i is a random error term that is normally distributed with mean 0 and variance σ^2. This model does not incorporate marker or environmental covariates. Composite interval mapping (Jansen, 1993; Zeng, 1993, 1994) incorporates marker covariates, such as confounding associations with both the gene and the trait. This model thus controls for additional QTL in the genome. Here, the mean phenotype is represented by the linear relationship

$$y_i = \mu + Q_q \alpha + \sum_{k=1}^{M} M_{ik} \beta_k + \varepsilon_i,$$

where M_{ik} is a coded value for covariate marker $k = 1, \ldots, K$ and β_k is a regression parameter for marker covariate k, and all other terms are as defined before. This model, while controlling for marker covariates, does not control for environmental covariates. The extended composite interval model does this. Here, the phenotypic values are represented by

$$y_i = \mu + Q_q \alpha + \mathbf{X}_i^T \boldsymbol{\beta} + \varepsilon_i,$$

where $\mathbf{X}_i$ is the ith row of the design matrix for fixed effects, including the marker covariates, and $\boldsymbol{\beta}$ is a vector of regression parameters. In both the composite interval and the extended composite interval models, the covariate parameters are effectively treated as nuisance parameters in order to assess the significance of the regression on the genotypic values. This analysis is carried out for each marker along the genome using standard ANOVA-style techniques to assess whether there is a difference among the mean phenotypic values for each value of the genotype.

5 TREE-BASED METHODS FOR PHASED DATA

Several tree-based methods exist for phased data. All of the techniques described here require the construction of local perfect, or near-perfect, phylogenetic trees. This concept is taken up in Section 5.1.

5.1 Perfect Phylogeny Construction

The construction of a perfect phylogeny relies on the concept of compatible SNP intervals.

5.1.1 *Compatibility and the Four-Gamete Test*

For a given site, the trait values are related by a true phylogenetic tree. This tree, in most settings, can only be partially inferred. In order to infer this tree, it is necessary to make determinations about which contiguous SNPs are *compatible* with the site of interest. Two sites are compatible if their configuration could have arisen without a recombination event. The assessment of compatibility relies on the use of the FGT. Two sites that pass the FGT are said to be compatible. It is important to note that, while the FGT is the standard approach to detecting recombination events, it is limited by low statistical power in the detection of such events. The fact that a configuration passes the FGT does not imply that no recombination events occurred. It simply means that recombination is not necessary for the configuration to arise. Consequently, many recombination events can go undetected by the FGT.

The idea for inference of local phylogenies presented by Mailund et al. (2006) is to construct local compatible SNP intervals. At site j, construct an interval $[a_j, b_j]$ of compatible SNPs. This interval must contain j, and there can be no gaps in the interval. In other words, if $j-2$ is in the interval, then $j-1$ must also be in the interval. Furthermore, all SNP sites that are in the interval must be compatible with each other. Using the data from Section 3, we will construct a local compatible SNP interval for site 3.

Example 8. For site 3, the genetic data are $\{0,1,1,1,0\}$. Examine site 2. The pairs made with site 3 are $\{10,01,11,11,10\}$. This group does not contain 00, so sites 2 and 3 are compatible. Next, examine the next closest site to 3, which is site 4. The pairs made with site 3 are $\{00,10,11,11,00\}$. The 01 pair is missing, so sites 3 and 4 are compatible. Recall that in order to be a part of the interval, site 4 must also be compatible with site 2. The pairs made with sites 2 and 4 are $\{10, 00, 11, 11, 10\}$. Since the 01 pair is missing, sites 2 and 4 are compatible, and the interval is now [2, 4]. Moving on to site 1, the pattern is $\{0, 0, 0, 0, 0\}$, so site 1 is clearly compatible with sites 2, 3, and 4 since any pair that begins with a 1 is not possible. The pairings with sites 1–4 and 5 are as follows: $\{01, 01, 01, 00, 00\}$ for site 1, $\{11, 01, 11, 10, 10\}$ for site 2, $\{01, 11, 11, 10, 00\}$ for site 3, and $\{01, 01, 11, 10, 00\}$ for site 4. Since site 5 is incompatible with sites 3 and 4, the compatible SNP interval is [1, 4].

5.1.2 Construction of Phylogenetic Tree

Mailund et al. (2006) go about constructing a local phylogenetic tree for a given site by adding splits in the tree based on where 0s and 1s appear in the compatible SNP interval. Splits are added by beginning at the site of interest and moving in a left-right-left-right alternating pattern until the tree is resolved. The tree is resolved when no node has more than two descendants emanating from it. This is illustrated using the compatible SNP interval constructed in the previous section. The initial tree is shown in Fig. 3.

At site 3, individuals 1 and 5 have a 0, and individuals 2, 3, and 4 have a 1. Adding the split accordingly gives the tree in Fig. 4.

Here, the (2,3,4) clade remains unresolved. In order to rectify this, move to site 2. Since the (1,5) clade is resolved, it is necessary only to examine individuals 2, 3, and 4. At site 2, individual 2 has a 0, while individuals 3 and 4 have a 1. Adding the required split results in the tree shown in Fig. 5.

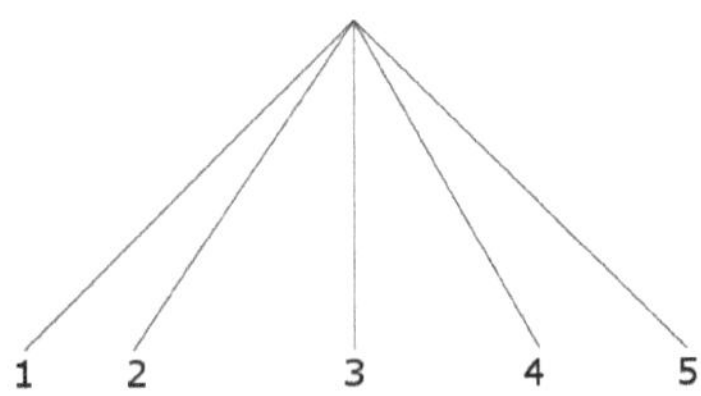

FIG. 3 The initial tree for perfect phylogeny construction.

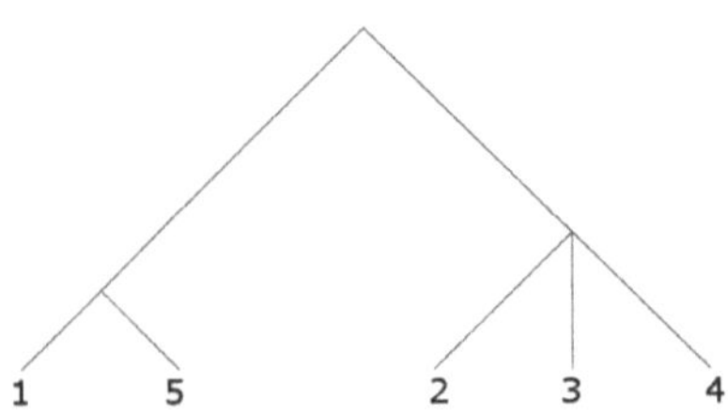

FIG. 4 The resulting tree after adding the split based on site 3.

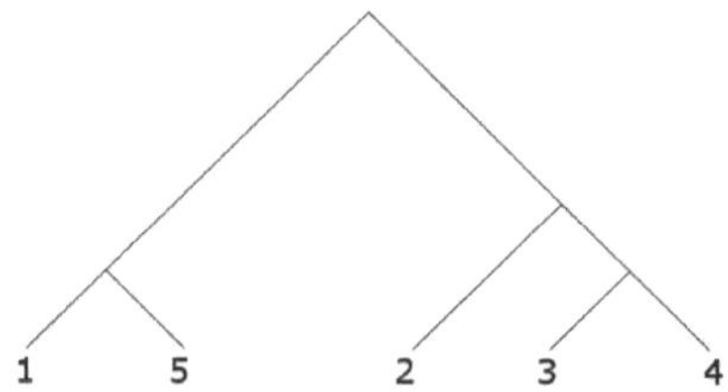

FIG. 5 The resulting tree after adding the split based on site 2.

The tree is now fully resolved. This is the inferred marginal phylogeny at site 3. Here, the SNP interval is larger than what is required to resolve the tree. This is more the exception rather than the rule, however, as in real-data studies with hundreds or even thousands of individuals, it is extremely rare to have compatible SNP intervals consisting of more than one or two sites. In these cases, Mailund et al. (2006) prescribe making use of neighboring incompatible SNPs to construct an interval that is large enough to resolve the local phylogeny. This is in contrast to Bardel et al. (2005) and Templeton et al. (2005), who restrict tree construction to compatible SNPs only. They argue correctly that restricting tree construction only to compatible sites results in a natural clustering of the tree that can be used for SNP detection and localization. However, in situations where the interval results in only one site being used to estimate the tree, the structure of the tree is such that it is only inferred which individuals are on each side of the root. While this may be sufficient for determination of which SNPs are significant to the expression of discrete traits, such as in case versus control studies, for quantitative traits, a more granular clustering of the tree is usually required. Consequently, Mailund et al. (2006) infer the full local tree and then cluster it later on. Furthermore, incompatible SNPs add splits at lower levels of the tree than do compatible SNPs, so the overall structure of the inferred tree is not severely impacted by the addition of incompatible SNPs. For example, the use of incompatible SNPs will not result in individuals being placed on the incorrect side of the root. Next, an example of the construction of a tree using incompatibilities is presented.

Example 9. The goal here is to construct the marginal phylogenetic tree at site 6. First, construct the compatible SNP interval. Examining site 5, the pairs are $\{11, 01, 11, 10, 00\}$ between sites 5 and 6, so these two sites are incompatible. Next, examine site 7. The pairs between sites 6 and 7 are $\{11, 00, 10, 11, 01\}$, so the compatible interval consists only of site 6. Using this to add a split to the tree puts individuals 2 and 5 on the left side of the root and individuals 1, 3, and 4 on the right, as seen in Fig. 6.

The (1,3,4) clade is not resolved, so examine the incompatible marker at site 5. Individuals 1 and 3 have a 1 at site 5, while individual 4 has a 0. Adding the split resolves the tree, as seen in Fig. 7

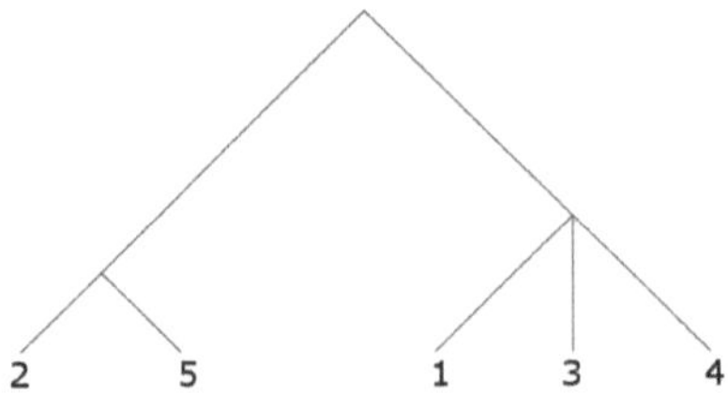

FIG. 6 Tree after adding a split based on site 6.

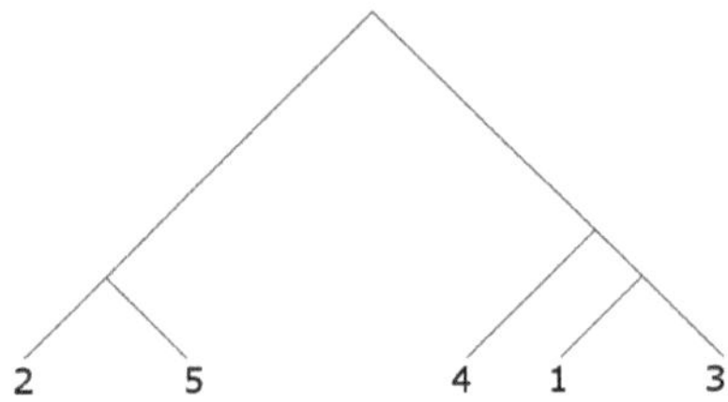

FIG. 7 Resolved tree after adding a split based on site 5.

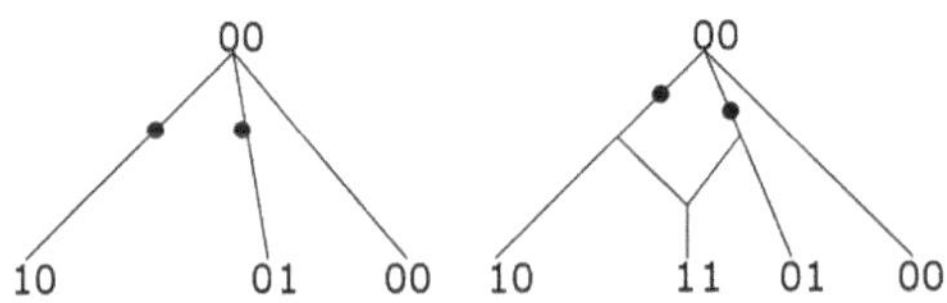

FIG. 8 Illustration of mutation (*left*) and recombination (*right*) events.

5.1.3 Why Use Marginal Phylogenetic Trees?

One of the limitations of perfect phylogenetic trees is that such trees do not exist that model incompatibilities on both sides of a recombination point at the same time. In this setting, the genealogical history of a chromosome is represented by an ancestral recombination graph (ARG). An ARG is a phylogenetic network that represents both coalescent and recombination events simultaneously. They account for the fact that local phylogenetic trees exist at each site, but that neighboring trees can be incompatible due to recombination events. Fig. 8 shows an ARG.

In this figure, suppose that the combination 00 is present at the root. On the left, the result of a mutation from 00 to 10 is given in the leftmost branch, and the center branch is the result of a mutation from 00 to 01. The right diagram shows a recombination, where the individuals in the leftmost two branches have exchanged their second alleles. This gives rise to the 11 individual. Wu (2011) uses the approach of inferring ARGs to estimate marginal phylogenetic trees based on SNP data. However, from SNP data, ARGs can be very difficult to estimate accurately, since SNP sequences represent a very small fraction of the genome. When dealing with SNP data, it is much easier to estimate marginal phylogenies directly.

5.2 QBlossoc

Besenbacher et al. (2009) develop a tree-based approach to SNP detection and localization for quantitative traits known as quantitative trait block association mapping (QBlossoc). In order to carry out this procedure for a given site, the local tree is inferred by the method outlined in Section 5.1. The tree is then clustered and scored. This procedure is carried out in one of two ways. Sites can be scored independently or in pairs. Sites are typically scored independently in haploid individuals. To score sites independently, observe that each tree T defines a set of models for the trait data. Any edge e of the tree splits the leaves of the tree into two disjoint sets S_i and S_j. The trait values for the leaves in S_i have a different distribution than do the trait values for the leaves in S_j. In a biological sense, this means that a mutation that affects the distribution of the trait values has occurred somewhere along the edge e. A second edge, say e', will split one of S_i or S_j into two new disjoint sets. In general, a set of k edges will split the leaves of T into $k + 1$ disjoint sets. For each subset S_i of leaves on the tree, assign a mean μ_i to the trait values corresponding to the leaves in S_i, so that the trait values for the leaves in S_i are independent and identically distributed according to a normal distribution with mean μ_i and variance σ^2. Given a tree with n leaves numbered 1, …, n with corresponding trait values $\mathbf{Y} = y_1, \ldots, y_n$ and a model Θ that divides the leaves into k disjoint subsets $S_1, \ldots, S_k$ with means $\mu_1, \ldots, \mu_k$, the conditional likelihood is given by

$$\mathcal{L}(\mathbf{y}|\Theta,\boldsymbol{\mu},\sigma^2)=\prod_{i=1}^{k}\prod_{j\in S_i}\frac{1}{\sqrt{2\pi\sigma^2}}e^{-\frac{1}{2\sigma^2}(y_j-\mu_i)^2}.$$

To compute the score for the tree, first estimate $\boldsymbol{\mu}$ and σ^2 via maximum likelihood. Then, for each model Θ, compute the Bayesian information criterion (BIC). The score for the tree is given by

$$\text{Score}(T)=\max_{\Theta,\boldsymbol{\mu},\sigma^2}(-2\ \ln\mathcal{L}(\mathbf{y}|\Theta,\boldsymbol{\mu},\sigma^2)+k\ \ln(n)).$$

In the following example, we present the sets of clusters that are consistent with the tree in site 3 (recall Fig. 5), and we compute scores for each of them. This will lead to the overall score for the tree. In clustering the tree, consider the two edges incident to the root as one edge. The clusters are presented in Table 12.

TABLE 12 Clusters for QBlossoc Score Computation

Number of Edges	Clusters
$k = 1$	$S_1 = (1, 5)$, $S_2 = (2, 3, 4)$
$k = 2$	$S_1 = (1)$, $S_2 = (5)$, $S_3 = (2, 3, 4)$
$k = 2$	$S_1 = (2)$, $S_2 = (1, 5)$, $S_3 = (3, 4)$
$k = 3$	$S_1 = (2)$, $S_2 = (3)$, $S_3 = (4)$, $S_4 = (1, 5)$

TABLE 13 Scores for QBlossoc at Site 3

Model	BIC
$S_1 = (1, 5)$, $S_2 = (2, 3, 4)$	33.0095
$S_1 = (1)$, $S_2 = (5)$, $S_3 = (2, 3, 4)$	34.6034
$S_1 = (2)$, $S_2 = (1, 5)$, $S_3 = (3, 4)$	13.0052
$S_1 = (2)$, $S_2 = (3)$, $S_3 = (4)$, $S_4 = (1, 5)$	7.3593

Example

We explore a subset of possible clusterings of the tree. For a given cluster S_i, the maximum likelihood estimator of μ_i is $\overline{y}_{i\cdot}$, where $\overline{y}_{i\cdot}$ denotes the sample mean trait value for the leaves in the set S_i. The maximum likelihood estimator of σ^2 for a model Θ is

$$\hat{\sigma}^2 = \frac{1}{n}\sum_{i=1}^{k+1}\sum_{j=1}^{n_i}(y_{ij} - \overline{y}_{i\cdot})^2,$$

where n_i is the number of leaves in S_i. With this in mind, there are BICs to compute for four models. The results for site 3 are in Table 13. If these were the only possible sets of clusters of the tree, the score for the tree at site 3 would be the maximum BIC, which is 34.6034. This technique works well if the goal is to analyze each site independently, but often, it is desired to analyze pairs of sites since diploid individuals have two alleles apiece. QBlossoc (Besenbacher et al., 2009) also presents a way to do this. In this setting, each individual has two leaves. Consequently, a model that splits the leaves into k subsets results in $k(k + 1)/2$ different combinations of sets in which two leaves from an individual can reside. The trait values in this setting are modeled using a normal distribution with mean μ_{ij} for individuals with one leaf in S_i and the other in S_j, and variance σ^2. Given n individuals, denote by h_{11}, h_{12},

..., h_{n1}, h_{n2} the leaves, with h_{i1} and h_{i2} coming from the same individual. The model Θ again divides the leaves into k subsets S_1, ..., S_k. The likelihood here is

$$\mathcal{L}(\mathbf{y}|\Theta,\boldsymbol{\mu},\sigma^2)=\prod_{i=1}^{K}\prod_{j=1}^{i}\prod_{\substack{h_{x1}\in S_i \text{ and } H_{x2}\in S_j \text{ or}\\ h_{x2}\in S_i \text{ and } h_{x1}\in S_j}}\frac{1}{\sqrt{2\pi\sigma^2}}e^{-\frac{1}{2\sigma^2}(y_{xj}-\mu_{ij})^2}.$$

The score for the tree is another BIC, given by

$$\text{Score}(T)=\max_{\Theta,\boldsymbol{\mu},\sigma^2}\left(-2\ \ln\mathcal{L}(\mathbf{y}|\Theta,\boldsymbol{\mu},\sigma^2)+K\ \ln(n)\right),$$

where K is the number of nonempty pairs of classes. In the following example, we construct the tree based on this alternate view of genetic data and compute the BIC for one model.

Example 10. Here, our genetic data appear as in Table 14. Suppose the goal is to build a tree for the new version of site 3. Site 3 is not compatible with either of sites 2 or 4, but we will use them to resolve the tree. The inferred tree is in Fig. 9.

Partition the tree so that the haplotypes with 0 in site 3 are on one side of the root and those with 1 in site 3 are on the other side of the root. Then $S_1=\{22,41,51,52\}$ and $S_2=\{11,12,21,31,32,42\}$. For individual 1, we have h_{11} and h_{12} both being in S_2. Furthermore, h_{21} is in S_2 and h_{22} is in S_1, h_{31} and h_{32} are in S_2, h_{41} is in S_1 and h_{42} is in S_2, and h_{51} and h_{52} are both in S_1. We thus

TABLE 14 Genetic Data for Pairs Analysis

Individual	Haplotype
1	00101
	10111
2	01101
	00011
3	01101
	11101
4	01011
	11101
5	00011
	00001

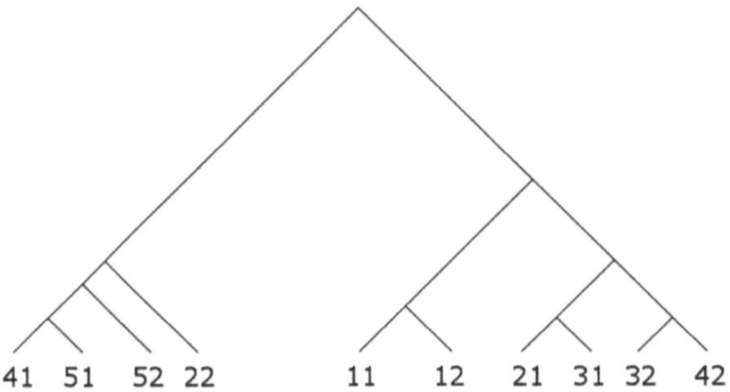

FIG. 9 Tree for paired site analysis example.

need to estimate μ_{11}, μ_{12}, and μ_{22}. There are $K = 3$ nonempty pairs of classes, and n is still equal to 5. The BIC under the specified model is 38.2770. This comes from the estimates $\hat{\mu}_{11} = 94.9828$, $\hat{\mu}_{12} = 87.471$, $\hat{\mu}_{22} = 97.196$, and $\hat{\sigma}^2 = 38.2728$. To complete QBlossoc (Besenbacher et al., 2009), a specific number, say R, of markers is chosen, and the sites with the highest R scores are then chosen as significant to the expression of the trait.

5.3 Models of Allele Substitution

The next two methods of tree-based GWAS analysis require the use of probability models for allele substitution. The two models that are described here are common choices for such settings. To begin, let $\mathbf{Q}$ be a matrix that specifies the probability that a 0 changes to a 1 or a 1 changes to a 0 in an infinitesimal amount of time. Let π_0 and π_1 denote the equilibrium frequencies of the 0 and 1 alleles, respectively. The transition probability matrix after time t is given by

$$\mathbf{P}(t) = e^{\mathbf{Q}t}.$$

In the process of matrix exponentiation, it is necessary to find the spectral decomposition $\mathbf{Q} = \mathbf{U}\boldsymbol{\Lambda}\mathbf{U}^T$, where $\mathbf{U}$ is a matrix whose columns are the normalized eigenvectors of $\mathbf{Q}$, and $\boldsymbol{\Lambda}$ is the diagonal matrix of eigenvalues of $\mathbf{Q}$. Then

$$\mathbf{P}(t) = \mathbf{U}e^{\boldsymbol{\Lambda}t}\mathbf{U}^T.$$

The first model described here is the M2 model of Lewis (2001). For this model, the genetic distances are also computed as they are needed to carry out the method described in Section 5.4. For the M2 model, assume that $\pi_0 = \pi_1 = \frac{1}{2}$, and that α is the instantaneous rate of change from a 0 to a 1 or a 1 to a 0. The rate matrix then is

$$\mathbf{Q} = \begin{bmatrix} -\frac{1}{2}\alpha & \frac{1}{2}\alpha \\ \frac{1}{2}\alpha & -\frac{1}{2}\alpha \end{bmatrix}.$$

This is a symmetric matrix, so eigenvalues and eigenvectors can be found directly. The eigenvalue matrix is

$$\mathbf{\Lambda} = \begin{bmatrix} 0 & 0 \\ 0 & -\alpha \end{bmatrix}.$$

The eigenvector matrix is

$$\mathbf{U} = \begin{bmatrix} \frac{1}{\sqrt{2}} & \frac{1}{\sqrt{2}} \\ \frac{1}{\sqrt{2}} & -\frac{1}{\sqrt{2}} \end{bmatrix}.$$

This gives the transition matrix

$$\begin{aligned} \mathbf{P}_{M2}(t) &= \begin{bmatrix} \frac{1}{\sqrt{2}} & \frac{1}{\sqrt{2}} \\ \frac{1}{\sqrt{2}} & -\frac{1}{\sqrt{2}} \end{bmatrix} \begin{bmatrix} 1 & 0 \\ 0 & e^{-\alpha t} \end{bmatrix} \begin{bmatrix} \frac{1}{\sqrt{2}} & \frac{1}{\sqrt{2}} \\ \frac{1}{\sqrt{2}} & -\frac{1}{\sqrt{2}} \end{bmatrix} \\ &= \begin{bmatrix} \frac{1}{2}+\frac{1}{2}e^{-\alpha t} & \frac{1}{2}-\frac{1}{2}e^{-\alpha t} \\ \frac{1}{2}-\frac{1}{2}e^{-\alpha t} & \frac{1}{2}+\frac{1}{2}e^{-\alpha t} \end{bmatrix}. \end{aligned}$$

From these probabilities, model distances can be computed. Let S denote the proportion of sites that show allele differences between two sequences. Then the expected value of S is

$$\begin{aligned} \mathbb{E}[S] &= \pi_0 \mathbf{P}_{M201}(t) + \pi_1 \mathbf{P}_{M210}(t) \\ &= \frac{1}{2}\left(1 - e^{-\alpha t}\right). \end{aligned}$$

The goal is to estimate αt. Setting $S = \mathbb{E}[S]$, we have

$$\begin{aligned} S &= \frac{1}{2}\left(1 - e^{-\alpha t}\right), \text{ so that} \\ \hat{\alpha t} &= -\ln(1 - 2S). \end{aligned}$$

Model distances (Kimura, 1980) are given by

$$\begin{aligned} d &= -2t \sum_{i=0,1} \pi_i \mathbf{Q}_{ii} \\ &= 2\alpha t. \end{aligned}$$

Therefore,

$$\hat{d} = -2 \ln(1 - 2S).$$

The key limitation of this model is that it assumes that the equilibrium frequencies of alleles 0 and 1 are equal. This is a biologically unrealistic assumption, and the approach described in Section 5.5 depends on the use of a model that does not rely on this assumption. The rate matrix for this model is

$$\mathbf{Q}=\begin{bmatrix}-\alpha\pi_1 & \alpha\pi_1\\ \alpha\pi_0 & -\alpha\pi_0\end{bmatrix}.$$

This matrix is not symmetric, but it can be written as the product of a symmetric matrix $\mathbf{B}$ and a diagonal matrix $\mathbf{D}$. Then construct the symmetric matrix $\mathbf{M}=\mathbf{D}^{\frac{1}{2}}\mathbf{B}\mathbf{D}^{\frac{1}{2}}$. Here, the diagonal matrix is

$$\mathbf{D}=\begin{bmatrix}\pi_1 & 0\\ 0 & \pi_0\end{bmatrix},$$

and the symmetric matrix $\mathbf{B}$ is

$$\mathbf{B}=\begin{bmatrix}-\alpha & \alpha\\ \alpha & -\alpha\end{bmatrix}.$$

This gives the matrix

$$\mathbf{M}=\begin{bmatrix}-\alpha\pi_1 & \alpha\sqrt{\pi_0\pi_1}\\ \alpha\sqrt{\pi_0\pi_1} & -\alpha\pi_0\end{bmatrix}.$$

Now we obtain the spectral decomposition of $\mathbf{M}$. We find that

$$\mathbf{\Lambda}=\begin{bmatrix}0 & 0\\ 0 & -\alpha\end{bmatrix},$$

and

$$\mathbf{U}=\begin{bmatrix}\sqrt{\pi_0} & -\sqrt{\pi_1}\\ \sqrt{\pi_1} & \sqrt{\pi_0}\end{bmatrix}.$$

The transition probabilities are given by

$$\begin{aligned}\mathbf{P}(t)&=\mathbf{D}^{-\frac{1}{2}}\mathbf{U}e^{\mathbf{\Lambda}t}\mathbf{U}^T\mathbf{D}^{\frac{1}{2}}\\ &=\begin{bmatrix}\pi_0+\pi_1e^{-\alpha t} & \pi_1-\pi_1e^{-\alpha t}\\ \pi_0-\pi_0e^{-\alpha t} & \pi_1+\pi_0e^{-\alpha t}\end{bmatrix}.\end{aligned}$$

One desirable property of the transition matrix is that it is time-reversible. This means that for all (i, j) pairs, $\mathbf{P}_{ij}(t)\pi_i = \mathbf{P}_{ji}(t)\pi_j$. In other words, the Markov chain $(\mathrm{X}_t)_{t\geq 0}$ that is specified by the transition matrix $\mathbf{P}$ looks the same forward and backward. Another way to express this is that for all (i, j) pairs,

$$\mathrm{P}(\mathrm{X}_0=i,\mathrm{X}_t=j)=\mathrm{P}(\mathrm{X}_0=j,\mathrm{X}_t=i).$$

This property can be used as a way to show that $[\pi_0 \pi_1]$ is the unique stationary, or equilibrium distribution for the model. Here, it is shown that both the M2 (Lewis, 2001) model and the model with unequal equilibrium frequencies, from here on referred to as the general time-reversible model of allele substitution (GTRA), are time-reversible. Observe that since this result is trivial for $i = j$, it is only necessary to show that $\mathbf{P}_{01}(t)\pi_0 = \mathbf{P}_{10}(t)\pi_1$ for each model. For the M2 (Lewis, 2001) model,

$$\mathbf{P}_{01}(t)\pi_0 = \frac{1}{2}\left(\frac{1}{2} - \frac{1}{2}e^{-\alpha t}\right) = \frac{1}{4}(1 - e^{-\alpha t}).$$

$$\mathbf{P}_{10}(t)\pi_1 = \frac{1}{2}\left(\frac{1}{2} - \frac{1}{2}e^{-\alpha t}\right) = \frac{1}{4}(1 - e^{-\alpha t}).$$

For the GTRA model,

$$\mathbf{P}_{01}(t)\pi_0 = \pi_0(\pi_1 - \pi_1 e^{-\alpha t}) = \pi_0\pi_1(1 - e^{-\alpha t})$$
$$\mathbf{P}_{10}(t)\pi_1 = \pi_1(\pi_0 - \pi_0 e^{-\alpha t}) = \pi_0\pi_1(1 - e^{-\alpha t}).$$

Therefore, both the M2 (Lewis, 2001) and the GTRA model are time-reversible. These models both have the property that they are *time-homogeneous*, which means that for all (i, j),

$$\mathrm{P}(\mathrm{X}_{t+s} = i | \mathrm{X}_t = j) = \mathrm{P}(\mathrm{X}_t = i | \mathrm{X}_0 = j)$$

for all $s \in \mathbb{R}$. Time-homogeneity guarantees that the Chapman–Kolmogorov equation holds. The Chapman–Kolmogorov equation is as follows.
Theorem 1 *If* $\mathbf{P}$ *is a time-homogeneous transition matrix, then for any* $s, t \geq 0$,

$$\mathbf{P}(t+s) = \mathbf{P}(t)\mathbf{P}(s).$$

Theorem 1 enables the process of phylogenetic likelihood calculation that is described in Section 5.5.

5.4 Likelihood Score Approach

The Thompson and Kubatko (2013) method, unlike QBlossoc (Besenbacher et al., 2009), incorporates a local dependence structure among the trait values, and they do this by using estimates of the branch lengths of the marginal phylogeny to estimate the marginal covariance between trait values. To assess significance, the likelihood score approach (LSA) method (Thompson and Kubatko, 2013) uses permutation testing. To begin, the marginal phylogenetic tree at each site is estimated using the same approach described by Mailund et al. (2006). Next, the internal node sequences of the tree are estimated using a fast heuristic method introduced by Rogers and Swofford (1998).

In order to understand this technique, it is important to understand the concept of *most parsimonious reconstruction* (MPR).

5.4.1 Most Parsimonious Reconstruction

Several methods of MPR exist in the literature, and for information on methods other than the one described here, one can consult Farris (1970), Adams (1972), Hartigan (1973), Archie (1989), and Swofford and Maddison (1992), among others. Here, we describe the Fitch (1971) method of MPR for internal node sequence reconstruction. The approach assumes that a set of descendant sequences that evolved from a common ancestral sequence is available, that the ancestral relationships among these sequences are known, and that any allele may replace any other allele at any site without regard to the nature of prior placements in the site or elsewhere. The reconstruction of ancestral allele sequences follows two phases. In the preliminary phase, alleles are placed on all ancestral nodes, and in the final phase, corrections to the preliminary assignments are made. In the preliminary phase, the preliminary nodal set consists of all alleles that are present in both immediately descendant sets. If neither allele is present in both immediately descendant nodes at the given site, then the union of the sets of alleles that are present at the given site is assigned to the ancestral node. Here, we present an example of the preliminary phase for the tree in Fig. 10.

Example 11. The data for this example are in Table 15. The data consist of 5 individuals with 50 site SNP sequences. In order to demonstrate the

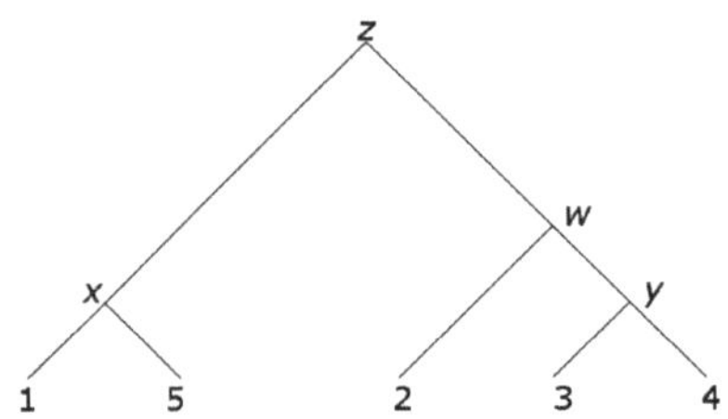

FIG. 10 The phylogenetic tree with nodes labeled w, x, y, and z.

TABLE 15 Data at the Leaves for the MPR Example

Leaf	1–5	6–10	11–50
1	00011	00000	0
2	01101	11100	0
3	01110	01011	0
4	01100	00001	0
5	10011	10000	0

preliminary phase, we examine site 4. At site 4, leaves 1 and 5 both have allele 1. Since the 1 allele is present in both immediate descendants of x, the preliminary nodal set at x is 1. Leaf 3 has allele 1 at site 4 and leaf 4 has allele 0 at site 4. Since neither allele is present in both immediate descendants of y, the preliminary nodal set at y is 0/1. Leaf 2 contains allele 0 at site 4, and the preliminary nodal set for y is 0/1. Since 0 is present in the preliminary nodal sets of both immediate descendants of w, the preliminary nodal set at w in site 4 is 0. With the preliminary nodal set being 1 for node x and 0 for node w, neither allele is present in both immediate descendants of z, the preliminary nodal set at z is 0/1, and the preliminary phase at site 4 is complete. The full preliminary nodal sets for each node and site are given in Table 16. The final phase consists of a six-step algorithm that involves going the other direction down the tree and correcting inferences that were made in the preliminary phase. First, set the sequence at the root node z as the final sequence at node z. This initializes the final phase. The remainder of the algorithm is as follows.

1. If the preliminary nodal set contains all alleles present in the final nodal set of its immediate ancestor, go to step b. Otherwise, proceed to step c.
2. Eliminate all alleles from the preliminary nodal set that are not present in the final nodal set of its immediate ancestor and go to step d.
3. If the preliminary nodal set was formed by a union of its descendant sets, go to step d. Otherwise, go to step e.
4. Add to the preliminary nodal set any alleles in the final nodal set of the immediate ancestor that are not already present in the preliminary nodal set and move to step f.
5. Add to the preliminary nodal set any alleles that are not already present provided that they are present in both the final nodal set of the immediate ancestor and in at least one of the two immediately descendant preliminary nodal sets and go to step f.
6. The preliminary nodal set being examined is now final. Descend one node as long as any nodes still have the preliminary nodal set and return to step 1. If all nodes have gone through the final phase, the algorithm is complete.

The next example shows the final phase.

TABLE 16 Preliminary MPR Reconstruction

Node	1–5	6–10	11–50
x	0/10011	0/10000	0
y	0110/10	00/100/11	0
w	01100/1	0/110/100/1	0
z	00/10/10/11	0/10/1000	0

Example 12. To illustrate the final phase of the MPR for internal node sequences, we again examine site 4. The preliminary nodal set at the root node *z* is 0/1, so to begin, set 0/1 as the final nodal set at site 4 for node *z*. Now examine node *x*. The preliminary nodal set for *x* at site 4 is 1. Since this set does not contain all of the alleles present in the final nodal set of *z*, move to step c. Since the preliminary nodal set was not formed through a union of descendant sets of *x*, move to step e. At site 4, allele 0 is present in the final nodal set at *z*, but not in either of the immediately descendant sets of *x*, so 0 is not added to the preliminary nodal set of *x*. The final nodal set at *x* is therefore 1. The preliminary nodal set at *w* is 0, so we move to step c as this does not contain all alleles present in the final nodal set of the immediate ancestor *z*. The preliminary nodal set at *w* was not formed through a union of its descendant sets, so we move to step e. At site 4, the immediate descendants have preliminary nodal sets 0 for leaf 2 and 0/1 for node *y*. Since allele 1 is present in the final nodal set of *z* and in the preliminary nodal set of one of the immediate descendant sets of *w*, 1 is added to the preliminary nodal set at *w*. The final nodal set of *w* is 0/1. The preliminary nodal set at *y* is 0/1. This set contains all alleles that are present in the final nodal set of its immediate ancestor *w*, so we move to step b. This set contains no alleles that are not present in the final nodal set of *w*, so we move to step d. The final nodal set of *w* does not contain any alleles that are not present in the preliminary nodal set at *y*. The final nodal set at *y* is therefore 0/1, and the MPR for the internal nodes at site 4 is complete. The full reconstruction is shown in Table 17.

5.4.2 The Rogers and Swofford (1998) Method of Ancestral Sequence Reconstruction

The goal of the Rogers and Swofford (1998) process is to choose a reconstruction as the most parsimonious reconstruction of the internal node sequences. The set of reconstructions is called an MPR set, and the Rogers and Swofford (1998) procedure uses it to find one MPR that makes necessarily short branches as short as possible, thus forcing as many changes as possible onto longer branches. The algorithm proceeds in five steps.

TABLE 17 Final MPR Reconstruction

Node	1–5	6–10	11–50
x	0/10011	0/10000	0
y	0110/10	00/100/11	0
w	01100/1	0/110/100/1	0
z	00/10/10/11	0/10/1000	0

1. For each ancestral sequence and each site, find all MPR sets, holding fixed all sequences and sites that were fixed in earlier cycles of the algorithm. If all sequences and sites are now fixed, stop. Otherwise, move to step 2.
2. For each branch of the tree, count the number of sites at which the two connected sites are fixed for the same allele (S) or potentially have the same allele (PS). That is, at least one of the two sequences has an ambiguous MPR and their MPR sets intersect.
3. Rank all branches in decreasing order of $S + PS$. If two branches have the same value for this sum, rank the one with the larger PS value ahead of the other.
4. For each site, proceed through the branches in rank order until a branch is found for which one of the two connected sequences is unambiguous, the other sequence is ambiguous, and the MPR sets of the two intersect. Set the ambiguous sequence equal to the fixed sequence at that site. When one such branch has been found, or none are found, go to the next site. No more than one branch may be fixed at any site in a single cycle of the algorithm. Ambiguous sites in terminal sequences, such as gaps or polymorphisms, are also fixed by this step.
5. If all sequences at all sites have now been fixed, stop. Otherwise, go to step 1.

Illustrating this is difficult to do with only two sites. In the following example, we examine a data set with 50 sites.

Example 13. The genetic data at the leaves for this example is in Table 15, and the final MPR reconstruction for the internal nodes is given in Table 17. To illustrate the use of the Rogers and Swofford (1998) method, we begin by ranking the branches of the tree in the manner previously described. These rankings are given in Table 18. The tree with labeled branches can be seen in Fig. 11.

TABLE 18 First Step of Rogers and Swofford (1998) for Site 1, Node x

Branch	S	PS	$S + PS$	Rank
b_1	48	2	50	7
b_2	48	2	50	7
b_3	46	4	50	4
b_4	47	3	50	5
b_5	47	3	50	5
b_6	43	7	50	2
b_7	42	8	50	1
b_8	44	6	50	3

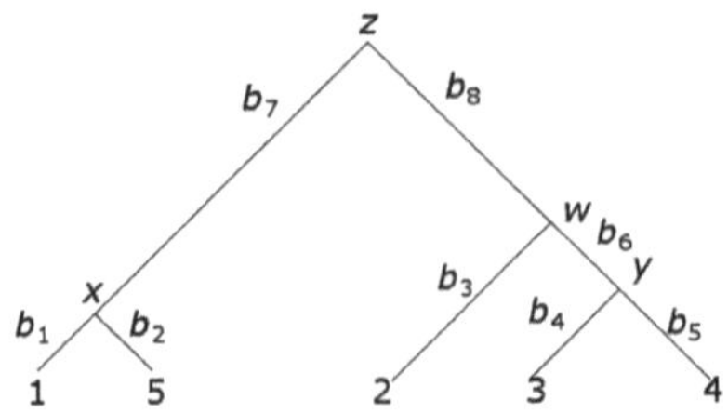

FIG. 11 The tree to be used to find the Rogers and Swofford (1998) internal node sequence reconstruction.

TABLE 19 Rogers and Swofford (1998) Reconstruction of Internal Node Sequences

Node	1–5	6–10	11–50
x	00011	00000	0
y	01100	01001	0
w	01101	01000	0
z	00011	00000	0

All branches have an $S + PS$ value of 50, but branch b_7 has a PS value of 8, which is the largest of any branch. Thus, we aim to infer the allele in site 1 for node x. We choose this because at this site, x has an ambiguity, while at the other end of b_7, node z is fixed for 0. Thus, it is inferred that x has a 0 at site 1. Per the description of the algorithm, we move to site 2, keeping in mind that in this cycle of the algorithm, b_7 cannot be fixed for site 2 since it has already been fixed for another site. The final inferred internal node sequences can be seen in Table 19. The method of Thompson and Kubatko (2013) continues with estimation of the branch lengths of the marginal phylogenetic tree. Using the number of SNPs used to resolve the tree at site i, they use the proportion $\hat{p}$ = the proportion of SNPs that differ between the ancestral and terminal nodes to compute the distance

$$\hat{d} = -2\ \ln(1 - 2\hat{p})$$

between the sequences at the ancestral and terminal nodes. This distance is then used as the estimate of the length of the connecting branch. If $\hat{p} = 0$, then the distance is estimated using $\hat{d} = \frac{1}{\text{number of SNPs}}$. This distance arises from the M2 model of Lewis (2001). For each marginal phylogeny, the estimated branch lengths are used to estimate the covariance matrix among the trait values based on the inferred tree. The covariance between two quantitative trait values is proportional to the length of their shared evolutionary history. The technique only accounts for the overall structure of the tree, so at each SNP site, the tree is partitioned into k clusters based on the earliest $k - 1$ edges of the tree. The matrix $\mathbf{D}$ is defined by

$$\mathbf{D}_{ij} = \begin{cases} 1 & \text{if observation } i \text{ fallis in cluster } j \\ 0 & \text{otherwise} \end{cases}$$

for $i = 1, \ldots, n$ and $j = 1, \ldots, k$, where n is the number of observations. Each cluster has its own mean, so consider the mean vector $\boldsymbol{\mu} = (\mu_1, \ldots, \mu_K)^T$. Then the trait values $\mathbf{Y} = (Y_1, \ldots, Y_n)$ are assumed to follow a normal distribution with mean $\mathbf{D}\boldsymbol{\mu}$ and covariance matrix $\sigma^2\mathbf{V}$. The quantities $\boldsymbol{\mu}$ and σ^2 are estimated via maximum likelihood, so that

$$\hat{\boldsymbol{\mu}} = (\mathbf{D}^T\mathbf{V}^{-1}\mathbf{D})^{-1}\mathbf{D}^T\mathbf{V}^{-1}\mathbf{Y}, \text{ and}$$

$$\hat{\sigma}^2 = \frac{1}{n}(\mathbf{Y} - \mathbf{D}\hat{\boldsymbol{\mu}})^T\mathbf{V}^{-1}(\mathbf{Y} - \mathbf{D}\hat{\boldsymbol{\mu}}).$$

Hypothesis testing is carried out based on the likelihood score statistic, which is given by

$$LSS = \max_k \left\{2 \ln \mathcal{L}(\hat{\boldsymbol{\mu}}, \hat{\sigma}^2 | \mathbf{Y}, \mathbf{V}) - k \ln(n)\right\}.$$

At each site, a tree is scored according to the LSS for varying numbers of clusters, and the tree score is the maximum score over the number of clusters. In order to detect significant SNPs, permutation testing is carried out. The trait values are permuted among the tips of the estimated phylogenetic tree. The local tree is scored for each of these permuted data sets. These permutations and their corresponding scores give an estimated distribution of the LSS for the tree at the given site, so the p-value is estimated by computing the proportion of data sets scoring higher than the observed data at each locus. Naturally, sites with low p-values are deemed significant. While this method shows an incremental improvement in statistical detection power over QBlossoc (Besenbacher et al., 2009), it does not control the FDR in any explicit way. Furthermore, Bayesian statisticians will object to the use of permutation testing on the grounds that it relies on hypothetical data. In other words, permutations of the same values are not what sets of random samples from a population typically look like.

Example 14. Consider the observed trait data $\mathbf{Y} = (97.835, 96.209, 98.279, 96.498, 95.479, 97.425)^T$. Furthermore, assume that the inferred tree is the one given in Fig. 12.

If the tree is not clustered, it has to be treated as though each leaf is in a different cluster. Therefore, $\mathbf{D}_1$ here is given by

$$\mathbf{D}_1 = \begin{bmatrix} 1 & 0 & 0 & 0 & 0 & 0 \\ 0 & 0 & 0 & 0 & 1 & 0 \\ 0 & 1 & 0 & 0 & 0 & 0 \\ 0 & 0 & 1 & 0 & 0 & 0 \\ 0 & 0 & 0 & 0 & 0 & 1 \\ 0 & 0 & 0 & 1 & 0 & 0 \end{bmatrix}.$$

The $\mathbf{V}_1$ matrix is

$$\mathbf{V}_1 = \begin{bmatrix} 0.7 & 0 & 0.65 & 0.41 & 0 & 0 \\ 0 & 0.7 & 0 & 0 & 0.62 & 0.3 \\ 0.65 & 0 & 0.7 & 0.41 & 0 & 0 \\ 0.41 & 0 & 0.41 & 0.7 & 0 & 0 \\ 0 & 0.62 & 0 & 0 & 0.7 & 0.3 \\ 0 & 0.3 & 0 & 0 & 0.3 & 0.7 \end{bmatrix}.$$

If the tree is partitioned into two clusters, then the clustered version of the tree is as seen in Fig. 13.

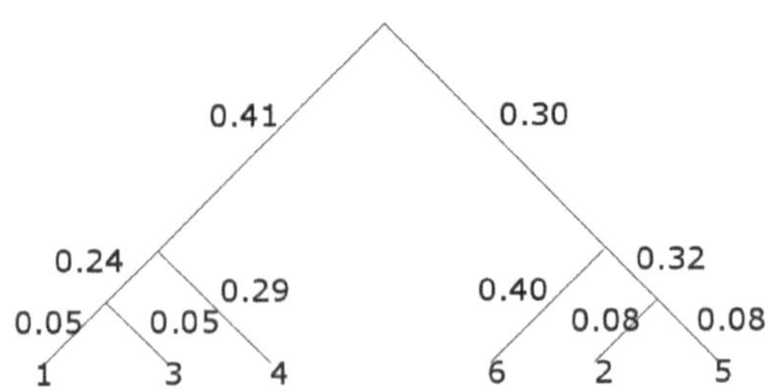

FIG. 12 The inferred tree for the permutation test.

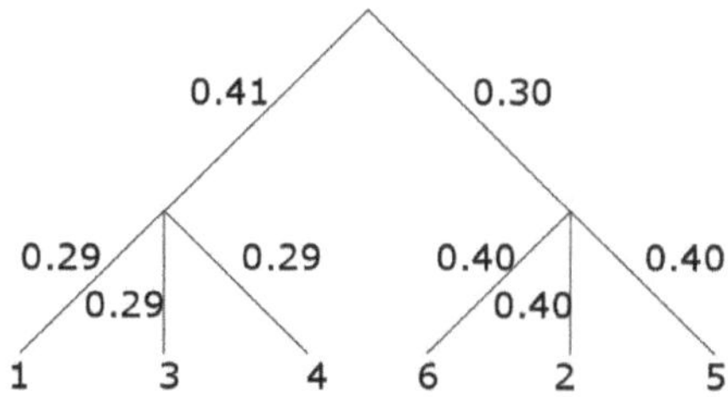

FIG. 13 The inferred tree partitioned into two clusters.

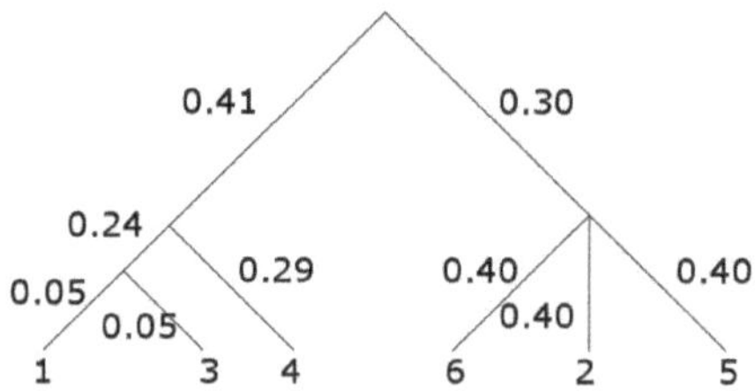

FIG. 14 One of two possible partitions of the tree into three clusters.

The $\mathbf{D}_2$ matrix is a 6 × 2 matrix given by

$$\mathbf{D}_2 = \begin{bmatrix} 1 & 0 \\ 0 & 1 \\ 1 & 0 \\ 1 & 0 \\ 0 & 1 \\ 0 & 1 \end{bmatrix}.$$

The $\mathbf{V}_2$ matrix is then

$$\mathbf{V}_2 = \begin{bmatrix} 0.7 & 0 & 0.41 & 0.41 & 0 & 0 \\ 0 & 0.7 & 0 & 0 & 0.3 & 0.3 \\ 0.41 & 0 & 0.7 & 0.41 & 0 & 0 \\ 0.41 & 0 & 0.41 & 0.7 & 0 & 0 \\ 0 & 0.3 & 0 & 0 & 0.7 & 0.3 \\ 0 & 0.3 & 0 & 0 & 0.3 & 0.7 \end{bmatrix}.$$

There are two possible partitions that result in three clusters. In one, the inferred tree is clustered as in Fig. 14, and the corresponding $\mathbf{D}_3$ and $\mathbf{V}_3$ matrices are given by

$$\mathbf{D}_3 = \begin{bmatrix} 1 & 0 & 0 \\ 0 & 0 & 1 \\ 1 & 0 & 0 \\ 0 & 1 & 0 \\ 0 & 0 & 1 \\ 0 & 0 & 1 \end{bmatrix}, \text{ and}$$

$$\mathbf{V}_3 = \begin{bmatrix} 0.7 & 0 & 0.65 & 0.41 & 0 & 0 \\ 0 & 0.7 & 0 & 0 & 0.3 & 0.3 \\ 0.65 & 0 & 0.7 & 0.41 & 0 & 0 \\ 0.41 & 0 & 0.41 & 0.7 & 0 & 0 \\ 0 & 0.3 & 0 & 0 & 0.7 & 0.3 \\ 0 & 0.3 & 0 & 0 & 0.3 & 0.7 \end{bmatrix}.$$

The other partition that can result in three clusters is shown in Fig. 15. The corresponding $\mathbf{D}_4$ and $\mathbf{V}_4$ matrices are given by

$$\mathbf{D}_4 = \begin{bmatrix} 1 & 0 & 0 \\ 0 & 0 & 1 \\ 1 & 0 & 0 \\ 1 & 0 & 0 \\ 0 & 0 & 1 \\ 0 & 1 & 0 \end{bmatrix} \text{ and}$$

$$\mathbf{V}_4 = \begin{bmatrix} 0.7 & 0 & 0.41 & 0.41 & 0 & 0 \\ 0 & 0.7 & 0 & 0 & 0.62 & 0.3 \\ 0.41 & 0 & 0.7 & 0.41 & 0 & 0 \\ 0.41 & 0 & 0.41 & 0.7 & 0 & 0 \\ 0 & 0.62 & 0 & 0 & 0.7 & 0.3 \\ 0 & 0.3 & 0 & 0 & 0.3 & 0.7 \end{bmatrix}.$$

If we partition the tree into four clusters, then there are three possible partitions. One such partition is (1,3,4,(6,2,5)), and the other is ((1,3,4),6,2,5). A third

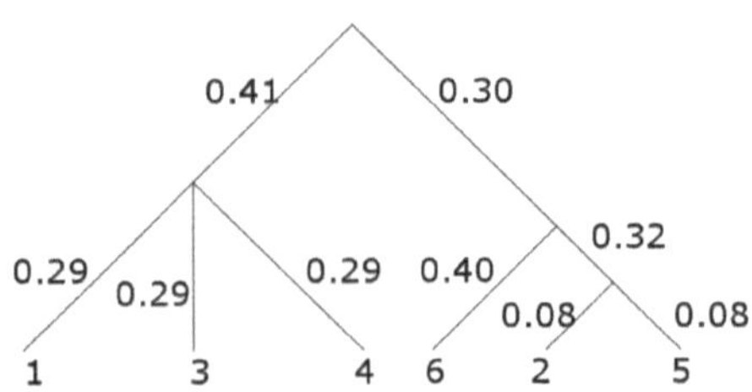

FIG. 15 The other possible partition of the tree into three clusters.

partition gives the clustering ((1,3),4,6,(2,5)). The corresponding $\mathbf{D}_5$, $\mathbf{V}_5$, $\mathbf{D}_6$, $\mathbf{V}_6$, and $\mathbf{D}_7$ matrices are given below. It is straightforward to see that $\mathbf{V}_7 = \mathbf{V}_1$.

$$\mathbf{D}_5 = \begin{bmatrix} 1 & 0 & 0 & 0 \\ 0 & 0 & 0 & 1 \\ 0 & 1 & 0 & 0 \\ 0 & 0 & 1 & 0 \\ 0 & 0 & 0 & 1 \\ 0 & 0 & 0 & 1 \end{bmatrix} \quad \mathbf{V}_5 = \begin{bmatrix} 0.7 & 0 & 0.65 & 0.41 & 0 & 0 \\ 0 & 0.7 & 0 & 0 & 0.3 & 0.3 \\ 0.65 & 0 & 0.7 & 0.41 & 0 & 0 \\ 0.41 & 0 & 0.41 & 0.7 & 0 & 0 \\ 0 & 0.3 & 0 & 0 & 0.7 & 0.3 \\ 0 & 0.3 & 0 & 0 & 0.3 & 0.7 \end{bmatrix}$$

$$\mathbf{D}_6 = \begin{bmatrix} 1 & 0 & 0 & 0 \\ 0 & 0 & 1 & 0 \\ 1 & 0 & 0 & 0 \\ 1 & 0 & 0 & 0 \\ 0 & 0 & 0 & 1 \\ 0 & 1 & 0 & 0 \end{bmatrix} \quad \mathbf{V}_6 = \begin{bmatrix} 0.7 & 0 & 0.41 & 0.41 & 0 & 0 \\ 0 & 0.7 & 0 & 0 & 0.62 & 0.3 \\ 0.41 & 0 & 0.7 & 0.41 & 0 & 0 \\ 0.41 & 0 & 0.41 & 0.7 & 0 & 0 \\ 0 & 0.62 & 0 & 0 & 0.7 & 0.3 \\ 0 & 0.3 & 0 & 0 & 0.3 & 0.7 \end{bmatrix}$$

$$\mathbf{D}_7 = \begin{bmatrix} 1 & 0 & 0 & 0 \\ 0 & 0 & 0 & 1 \\ 1 & 0 & 0 & 0 \\ 0 & 1 & 0 & 0 \\ 0 & 0 & 0 & 1 \\ 0 & 0 & 1 & 0 \end{bmatrix}.$$

Finally, for a set of five clusters, the two possible clusterings are (1,3,4,6,(2,5)) and ((1,3),4,6,2,5). We have that $\mathbf{V}_8 = \mathbf{V}_9 = \mathbf{V}_1$. The $\mathbf{D}_8$ and $\mathbf{D}_9$ matrices are given by

$$\mathbf{D}_8 = \begin{bmatrix} 1 & 0 & 0 & 0 & 0 \\ 0 & 0 & 0 & 0 & 1 \\ 0 & 1 & 0 & 0 & 0 \\ 0 & 0 & 1 & 0 & 0 \\ 0 & 0 & 0 & 0 & 1 \\ 0 & 0 & 0 & 1 & 0 \end{bmatrix} \quad \mathbf{D}_9 = \begin{bmatrix} 1 & 0 & 0 & 0 & 0 \\ 0 & 0 & 0 & 1 & 0 \\ 1 & 0 & 0 & 0 & 0 \\ 0 & 1 & 0 & 0 & 0 \\ 0 & 0 & 0 & 0 & 1 \\ 0 & 0 & 1 & 0 & 0 \end{bmatrix}.$$

At this point, nine sets of maximum likelihood estimates need to be computed. These estimates, along with the BIC for each cluster, are presented in Table 20. For partition i, the trait values $\mathbf{Y}$ are assumed to follow a normal distribution with mean $\mathbf{D}_i\hat{\boldsymbol{\mu}}_i$ and covariance matrix $\hat{\sigma}_i^2\mathbf{V}_i$. Observe that in order for this method to work, the tree must be clustered. Without clustering, in situations where each individual has only one observation, the maximum likelihood estimate of σ^2 is always 0. Thus, we begin with the two-cluster setups. Since the maximum LSS here is -13.933, the overall score for the tree at this hypothetical site is -13.933. In order to assess significance, we carry out a toy example of a permutation test. Consider the following five permutations of the trait data: $\mathbf{Y}_{p1}$ = (97.425, 96.498, 97.835, 95.479, 96.209, 98.279), $\mathbf{Y}_{p2}$ = (96.498, 98.279, 95.479, 97.425, 97.835, 96.209), $\mathbf{Y}_{p3}$ = (98.279, 96.209, 95.479,

TABLE 20 Maximum Likelihood Estimates and BIC for Each Cluster Based on Observed Data

Cluster	$\hat{\mu}_i$	$\hat{\sigma}_i^2$	BIC
2	(97.313, 96.595)	1.2162	−25.236
3	(98.057, 96.498, 96.371)	0.8620	−19.720
4	(97.312, 97.425, 95.844)	1.170	−21.550
5	(97.835, 98.279, 96.498, 96.593)	1.103	−21.260
6	(97.313, 97.425, 96.209, 95.479)	1.058	−20.952
7	(98.057, 96.498, 97.425, 95.844)	0.884	−19.869
8	(97.835, 98.279, 96.498, 97.425, 95.844)	0.555	−17.079
9	(98.057, 96.498, 97.425, 96.209, 95.479)	0.329	−13.933

97.425, 96.498, 97.835), $\mathbf{Y}_{p4} = (96.498, 96.209, 95.479, 97.425, 97.835, 98.279)$, and $\mathbf{Y}_{p5} = (96.209, 95.479, 97.425, 96.498, 97.835, 98.279)$. For each of these, we cluster the tree in the same ways as we did for the observed data, and we find the LSS for the tree given each permutation. The LSS values for each permutation, in order from 1 to 5, are −12.977, −11.113, −5.960, −22.130, and −25.640. Three of the five permutations have higher LSS scores than the data, so the estimated p-value is $3/5 = 0.6$. This site would not be declared significant to the expression of the trait.

5.5 Bayesian SNP Detection

Spade developed a Bayesian method of QTL detection from phased SNP data that incorporates the population covariance structure among the trait values and does not rely on permutation testing. The approach begins with the inference of the marginal phylogenetic tree topology at each SNP site in the manner prescribed by Mailund et al. (2006). Once the tree topology is inferred, the branch lengths are estimated via maximum likelihood. In order to understand phylogenetic maximum likelihood estimation, we need to take an excursion into phylogenetic likelihood calculation.

5.5.1 *Phylogenetic Likelihood Calculation*

Phylogenetic likelihood calculation relies on the Markov property, which means that if an individual has a particular SNP sequence, then the SNP sequences that evolve from it over the next v units of time depend only on the current sequence and the genetic changes that occur over the next v units of time. The Markov property enables the use of the peeling algorithm described by Felsenstein (1981). The peeling algorithm relies on the conditional likelihood $\mathcal{L}_s^{(i)}(k)$, which represents the probability of everything that occurs from node k down the tree at site i, given that the allele s is in site i at node k. The computation of phylogenetic likelihoods is elegantly described recursively. Suppose node k has immediate descendants l and m at the bottom ends of branches of lengths t_l and t_m, respectively. Then

$$\mathcal{L}_s^{(i)}(k) = \left(\sum_z \Pr(z|s,t_l)\mathcal{L}_z^{(i)}(l)\right)\left(\sum_y \Pr(y|s,t_m)\mathcal{L}_y^{(i)}(m)\right). \tag{1}$$

The intuition behind (1) is that, under an assumption of independent evolution between lineages, the probability of everything at or below a node k given that k has allele s in site i is equal to the product of the probabilities of the corresponding events in the two descendant lineages of k.

The peeling algorithm begins with the assignment of conditional likelihoods to the tips of the tree. Since data are observed at the tips of the tree, the conditional likelihood of an allele s at a particular tip is assigned a value of 1 if s is observed at the given tip and 0 otherwise. Once conditional likelihoods are assigned at the tips, the calculation of the value in (1) is performed for all nodes that have only tips as their immediate descendants. This calculation is then done for each node of the tree in a postorder traversal. This succession continues all the way to the root node, say 0, to obtain $\mathcal{L}_s^{(i)}(0)$. The computation of the site likelihood concludes by summing $\pi_s\mathcal{L}_s^{(i)}(0)$ over all alleles that can occur at the root in site i. If nodes 1 and 2 are the immediate descendants of the root, then the ith site likelihood for a genetic data set $\mathbf{D}$ is written as

$$\mathcal{L}^{(i)}(\mathbf{D}|\mathbf{t}) = \sum_{s_0} \pi_{s_0}\left(\sum_{s_1} \mathrm{P}_{s_0 s_1}(t_1)\mathcal{L}_{s_1}^{(i)}(1)\right)\left(\sum_{s_2} \mathrm{P}_{s_0 s_2}(t_2)\mathcal{L}_{s_2}^{(i)}(2)\right),$$

where $\mathrm{P}_{ab}(t)$ is the probability that allele a is substituted by allele b over time t. These probabilities come from the transition matrix for the substitution model being used. For a full data set with M sites, we assume independent evolution between sites. Consequently, the full likelihood of $\mathbf{D}$ is given by

$$\mathcal{L}(\mathbf{D}|\mathbf{t}) = \prod_{i=1}^{M} \mathcal{L}^{(i)}(\mathbf{D}|\mathbf{t}).$$

If no assumptions are made about the placement of the root in the tree, it is possible to make use of the Pulley Principle (Felsenstein, 1981), which states that the likelihood of a rooted tree is equal to the likelihood of the unrooted tree that results from removing the root and connecting end-to-end the two branches that are incident to the root. Let t_1 and t_2 denote the lengths of these branches. Then the likelihood of the data at site k given the tree is written as

$$\mathcal{L}^{(i)}(\mathbf{D}|\mathbf{t}) = \sum_{s_1}\sum_{s_2} \mathrm{P}_{s_1 s_2}(t_1+t_2)\mathcal{L}_{s_1}^{(k)}(1)\mathcal{L}_{s_2}^{(k)}(2).$$

In the following example, we compute the site likelihood for a six-tip tree under the GTRA model.

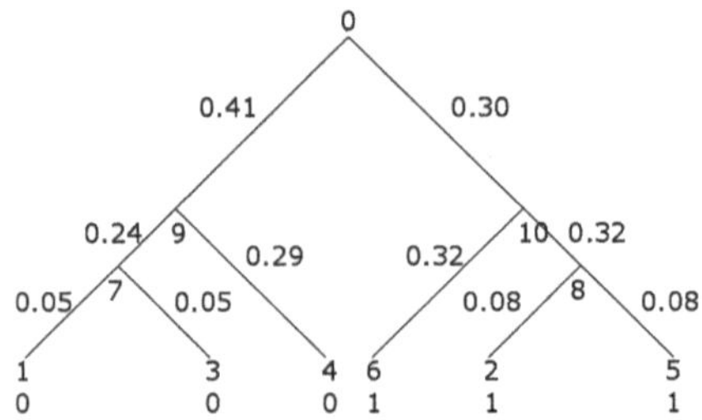

FIG. 16 The tree for the example of likelihood calculation. The root is labeled with a 0. The internal nodes are labeled as 7–10, while the tips are labeled 1–6. Each branch is labeled with its length, and the data at the tips is shown below the tip labels.

Example 15. The tree in this example is in Fig. 16. Assume that in the GTRA model, $\alpha = 1$, $\pi_0 = 0.8$, and $\pi_1 = 0.2$. Then the transition matrix is given by

$$\mathrm{P}(t) = \begin{bmatrix} 0.8+0.2e^{-t} & 0.2-0.2e^{-t} \\ 0.8-0.8e^{-t} & 0.2+0.8e^{-t} \end{bmatrix}.$$

We compute the likelihood of the branch lengths of the tree given the data at the leaves and the tree topology. To begin, find the conditional likelihood at the tips of the tree. We have

$$\begin{aligned} \mathcal{L}_0(1) &= \mathcal{L}_0(3) = \mathcal{L}_0(4) = 1 \\ \mathcal{L}_1(1) &= \mathcal{L}_1(3) = \mathcal{L}_1(4) = 0 \\ \mathcal{L}_0(2) &= \mathcal{L}_0(5) = \mathcal{L}_0(6) = 0 \\ \mathcal{L}_1(2) &= \mathcal{L}_1(5) = \mathcal{L}_1(6) = 1. \end{aligned}$$

These conditional likelihoods enable computation of $\mathcal{L}_0(7)$, $\mathcal{L}_1(7)$, $\mathcal{L}_0(8)$, and $\mathcal{L}_1(8)$. We have

$$\begin{aligned} \mathcal{L}_0(7) &= (\mathrm{P}_{00}(0.05)\mathcal{L}_0(1) + \mathrm{P}_{01}(0.05)\mathcal{L}_1(1)) \\ &\quad \times (\mathrm{P}_{00}(0.05)\mathcal{L}_0(3) + \mathrm{P}_{01}(0.05)\mathcal{L}_1(3)) \\ &= (0.8+0.2e^{-0.05})^2 = 0.9806. \\ \mathcal{L}_1(7) &= (\mathrm{P}_{10}(0.05)\mathcal{L}_0(1) + \mathrm{P}_{11}(0.05)\mathcal{L}_1(1)) \\ &\quad \times (\mathrm{P}_{10}(0.05)\mathcal{L}_0(3) + \mathrm{P}_{11}(0.05)\mathcal{L}_1(3)) \\ &= (0.8-0.8e^{-0.05})^2 = 0.0015. \\ \mathcal{L}_0(8) &= (\mathrm{P}_{00}(0.08)\mathcal{L}_0(2) + \mathrm{P}_{01}(0.08)\mathcal{L}_1(2)) \\ &\quad \times (\mathrm{P}_{00}(0.08)\mathcal{L}_0(5) + \mathrm{P}_{01}(0.08)\mathcal{L}_1(2)) \\ &= (0.2-0.2e^{-0.08})^2 = 0.0002 \\ \mathcal{L}_1(8) &= (\mathrm{P}_{10}(0.08)\mathcal{L}_0(2) + \mathrm{P}_{11}(0.08)\mathcal{L}_1(2)) \\ &\quad \times (\mathrm{P}_{10}(0.08)\mathcal{L}_0(5) + \mathrm{P}_{11}(0.08)\mathcal{L}_1(5)) \\ &= (0.2+0.8e^{-0.08})^2 = 0.8808. \end{aligned}$$

From here, we move up one level of the tree and compute $\mathcal{L}_0(9)$, $\mathcal{L}_1(9)$, $\mathcal{L}_0(10)$ and $\mathcal{L}_1(10)$. Here, we have

$$\begin{aligned}
\mathcal{L}_0(9) &= (\mathrm{P}_{00}(0.24)\mathcal{L}_0(7) + \mathrm{P}_{01}(0.24)\mathcal{L}_1(7)) \\
&\quad \times (\mathrm{P}_{00}(0.29)\mathcal{L}_0(4) + \mathrm{P}_{01}(0.29)\mathcal{L}_1(4)) \\
&= 0.8916. \\
\mathcal{L}_1(9) &= (\mathrm{P}_{10}(0.24)\mathcal{L}_0(7) + \mathrm{P}_{11}(0.24)\mathcal{L}_1(7)) \\
&\quad \times (\mathrm{P}_{10}(0.29)\mathcal{L}_0(4) + \mathrm{P}_{11}(0.29)\mathcal{L}_1(4)) \\
&= 0.0340. \\
\mathcal{L}_0(10) &= (\mathrm{P}_{00}(0.32)\mathcal{L}_0(8) + \mathrm{P}_{01}(0.32)\mathcal{L}_1(8)) \\
&\quad \times (\mathrm{P}_{00}(0.4)\mathcal{L}_0(6) + \mathrm{P}_{01}(0.4)\mathcal{L}_1(6)) \\
&= 0.0032. \\
\mathcal{L}_1(10) &= (\mathrm{P}_{10}(0.32)\mathcal{L}_0(8) + \mathrm{P}_{11}(0.32)\mathcal{L}_1(8)) \\
&\quad \times (\mathrm{P}_{10}(0.4)\mathcal{L}_0(6) + \mathrm{P}_{11}(0.4)\mathcal{L}_1(6)) \\
&= 0.5065.
\end{aligned}$$

At this point, compute $\mathcal{L}_0(0)$ and $\mathcal{L}_1(0)$.

$$\begin{aligned}
\mathcal{L}_0(0) &= (\mathrm{P}_{00}(0.41)\mathcal{L}_0(9) + \mathrm{P}_{01}(0.41)\mathcal{L}_1(9)) \\
&\quad \times (\mathrm{P}_{00}(0.3)\mathcal{L}_0(10) + \mathrm{P}_{01}(0.32)\mathcal{L}_1(10)) \\
&= 0.0245. \\
\mathcal{L}_1(0) &= (\mathrm{P}_{10}(0.41)\mathcal{L}_0(9) + \mathrm{P}_{11}(0.41)\mathcal{L}_1(9)) \\
&\quad \times (\mathrm{P}_{10}(0.3)\mathcal{L}_0(10) + \mathrm{P}_{11}(0.3)\mathcal{L}_1(10)) \\
&= 0.1065.
\end{aligned}$$

This means that the likelihood is given by

$$\begin{aligned}
\mathcal{L}((0,0,0,1,1,1)|\mathbf{T},\mathbf{t}) &= \pi_0\mathcal{L}_0(0) + \pi_1\mathcal{L}_1(0) \\
&= 0.8(0.0245) + 0.2(0.1065) \\
&= 0.0409.
\end{aligned}$$

Once the branch lengths of the tree are estimated the estimates are used in the same way as described by Thompson and Kubatko (2013) to give a matrix that represents a dependence structure among the trait values.

5.5.2 Bayesian Detection of QTL

The trait values $\mathbf{Y}$ for site i are assumed to follow a normal distribution with mean vector $\boldsymbol{\mu}$ and covariance matrix $\sigma^2\mathbf{V}_i$. Instead of estimating $\boldsymbol{\mu}$ and σ^2 using maximum likelihood, this procedure relies on Bayes estimation. This technique does not require clustering of the tree, so this procedure does not give up any of the information contained in the branch lengths. Incorporating all of the information in the tree should increase detection power. There are two natural ways to assign prior distributions to $\boldsymbol{\mu}$ and σ^2. One way is to use conjugate prior distributions so that $\boldsymbol{\mu}$ is assumed normal with mean vector $\boldsymbol{\mu}_0$ and covariance matrix $\tau^2\mathbf{I}$, and σ^2 is assumed to follow an inverse gamma distribution with shape parameter α and rate parameter β. Furthermore, an improper noninformative prior distribution is assigned to $\mathbf{V}_i$, so that $p(\mathbf{V}_i) \propto 1$. The posterior density is thus given by

$$p(\boldsymbol{\mu},\sigma^2,\mathbf{V}_i|\mathbf{Y}) \propto p(\mathbf{y}|\boldsymbol{\mu},\sigma^2,\mathbf{V})p(\boldsymbol{\mu})p(\sigma^2)p(\mathbf{V}_i)$$
$$= (\sigma^2)^{-\frac{n}{2}-\alpha-1} e^{-\frac{1}{2\sigma^2}(\mathbf{y}-\boldsymbol{\mu})^T\mathbf{V}_i^{-1}(\mathbf{y}-\boldsymbol{\mu})-\frac{1}{2\tau^2}(\boldsymbol{\mu}-\boldsymbol{\mu}_0)^T(\boldsymbol{\mu}-\boldsymbol{\mu}_0)-\frac{\beta}{\sigma^2}}.$$

The parameters $\boldsymbol{\mu}$ and σ^2 are sampled directly by first sampling $\boldsymbol{\mu}$ from the $N(\boldsymbol{\mu}^*, \boldsymbol{\Sigma}^*)$ distribution, where

$$\boldsymbol{\mu}|\mathbf{V}_i,\sigma^2,\mathbf{y} \sim N(\boldsymbol{\mu}^*,\boldsymbol{\Sigma}^*), \text{ where}$$
$$\boldsymbol{\mu}^* = \left(\frac{1}{\sigma^2}\mathbf{V}_i^{-1} + \frac{1}{\tau^2}\mathbf{I}\right)^{-1}\left(\frac{1}{\sigma^2}\mathbf{V}_i^{-1}\mathbf{y} + \frac{1}{\tau^2}\boldsymbol{\mu}_0\right) \text{ and}$$
$$\boldsymbol{\Sigma}^* = \left(\frac{1}{\sigma^2}\mathbf{V}_i^{-1} + \frac{1}{\tau^2}\mathbf{I}\right)^{-1}.$$

Next, σ^2 is sampled from an inverse gamma distribution with shape parameter α^* and scale parameter β^*, where

$$\alpha^* = \frac{n}{2} + \alpha \text{ and}$$
$$\beta^* = \frac{1}{2}(\mathbf{y}-\boldsymbol{\mu})^T\mathbf{V}_i^{-1}(\mathbf{y}-\boldsymbol{\mu}).$$

The trouble with assigning a normal prior distribution to $\boldsymbol{\mu}$ is that we lack information about $\boldsymbol{\mu}$. Naturally, we would want to assign different means to trait values that correspond to individuals with a 0 at the trait locus than to individuals with a 1. Without knowledge of which loci are associated with the trait, it is difficult to know what mean to assign to individuals with a 0 and what mean to assign to individuals with a 1. Instead, it is natural to assign an improper, noninformative prior distribution to $\boldsymbol{\mu}$ so that $p(\boldsymbol{\mu}) \propto 1$. The posterior density is then given by

$$p(\boldsymbol{\mu},\sigma^2,\mathbf{V}_i|\mathbf{Y}) \propto (\sigma^2)^{-\frac{n}{2}-\alpha-1} e^{-\frac{\beta}{\sigma^2}-\frac{1}{2\sigma^2}(\mathbf{y}-\boldsymbol{\mu})^T\mathbf{V}_i^{-1}(\mathbf{y}-\boldsymbol{\mu})}.$$

Under this posterior density, the sampling scheme is as follows. The mean vector $\boldsymbol{\mu}$ is sampled from a $N(\mathbf{y}, \sigma^2\mathbf{V}_i)$ distribution and σ^2 is sampled from an inverse gamma distribution with shape parameter $\tilde{\alpha} = \frac{n}{2} + \alpha$ and scale parameter

$$\tilde{\beta} = \frac{1}{2}(\mathbf{y}-\boldsymbol{\mu})^T\mathbf{V}_i^{-1}(\mathbf{y}-\boldsymbol{\mu}).$$

Some number K of samples are drawn from the posterior density. Once these samples are drawn, compute mean sampled values $\hat{\boldsymbol{\mu}}_i$ of $\boldsymbol{\mu}$ and $\hat{\sigma}_i^2$ of σ^2. Next, compute the posterior density $p(\hat{\boldsymbol{\mu}}_i, \hat{\sigma}_i^2, \mathbf{V}_i|\mathbf{Y})$ of these values. The idea is that values of $\hat{\boldsymbol{\mu}}_i$ and $\hat{\sigma}_i^2$ in regions of high probability will have higher posterior densities than estimates that are not in these areas. With that in mind, compute estimates of $\boldsymbol{\mu}$ and σ^2 for each site. To compare sites i and j, compute the posterior density ratio

$$\frac{p(\hat{\boldsymbol{\mu}}_i, \hat{\sigma}_i^2, \mathbf{V}_i|\mathbf{Y})}{p(\hat{\boldsymbol{\mu}}_j, \hat{\sigma}_j^2, \mathbf{V}_j|\mathbf{Y})}$$

for each pair (i, j) of sites. Specify a threshold A, and for each site, compute p_i, the proportion of sites j for which the posterior density ratio exceeds A. Rank p_i in descending order. For some specified quantity α, select the sites with the $100\alpha\%$ highest values of p_i as sites that are significant to the expression of the trait. In lieu of an example here, we present the results of a simulation study and how the Bayesian method performs in comparison to the method of Thompson and Kubatko (2013) and QBlossoc (Besenbacher et al., 2009).

5.5.3 Simulation Study

In this study, genetic data were generated for 1000 haploid individuals. Haploid individuals were chosen to avoid the need for phasing of bi-allelic genotypes into haplotypes. For each individual, a 100,000 site SNP sequence was generated using equilibrium frequencies $\pi_0 = 0.8$ and $\pi_1 = 0.2$. To simulate trait data, a disease locus is chosen at random, and the local tree and its branch lengths are estimated. The trait data are generated according to a generalized Hansen process (Hansen, 1997), which is represented as follows. Let $X_i(t)$ denote the state (0 or 1) of individual i at time t. Then

$$\frac{dY_i(t)}{dt} = \gamma(\theta_i(t) - Y_i(t)) + \sigma_Y \frac{dB_i(t)}{dt}, \text{ where}$$

$$\theta_i(t) = \begin{cases} 80 & \text{if } X_i(t) = 0 \\ 100 & \text{if } X_i(t) = 1, \end{cases}$$

$B_i(t)$ is a standard Brownian motion process, γ is a parameter that represents the strength of selection, and σ_Y is the standard deviation of the process per unit of time. In this study, γ varies among 5, 7.5, and 10, while σ_Y varies among 10, 20, 30, and 40. A noninformative prior distribution is placed on the mean vector $\boldsymbol{\mu}$, and σ^2 is assumed to follow an inverse gamma distribution with shape and rate parameters equal to 1. In determining significance for a given site i, the proportion of sites j for which

$$\frac{p(\hat{\boldsymbol{\mu}}_i, \hat{\sigma}_i^2, \mathbf{V}_i|\mathbf{Y})}{p(\hat{\boldsymbol{\mu}}_j, \hat{\sigma}_j^2, \mathbf{V}_j|\mathbf{Y})} > 3$$

TABLE 21 Detection Probabilities for All Three Methods

Parameter Values	QBlossoc	LSS	Posterior Density Ratio
$\gamma = 5$, $\sigma_Y = 10$	0.84	0.81	0.94
$\gamma = 5$, $\sigma_Y = 20$	0.81	0.82	0.95
$\gamma = 5$, $\sigma_Y = 30$	0.78	0.78	0.92
$\gamma = 5$, $\sigma_Y = 40$	0.75	0.76	0.91
$\gamma = 7.5$, $\sigma_Y = 10$	0.89	0.91	0.94
$\gamma = 7.5$, $\sigma_Y = 20$	0.82	0.84	0.93
$\gamma = 7.5$, $\sigma_Y = 30$	0.80	0.81	0.97
$\gamma = 7.5$, $\sigma_Y = 40$	0.77	0.79	0.91
$\gamma = 10$, $\sigma_Y = 10$	0.98	0.97	0.90
$\gamma = 10$, $\sigma_Y = 20$	0.89	0.91	0.96
$\gamma = 10$, $\sigma_Y = 30$	0.83	0.81	0.95
$\gamma = 10$, $\sigma_Y = 40$	0.81	0.78	0.97

is computed. A sample of size 50 is used to compute $\hat{\boldsymbol{\mu}}_i$ and $\hat{\sigma}_i^2$. This process is carried out 100 times for each of the 12 possible combinations of γ and σ_Y, and for each combination, a locus i is deemed significant if p_i is in the top 0.5% of p quantities. The detection power for QBlossoc (Besenbacher et al., 2009), the Thompson and Kubatko (2013) method, and the Bayesian method are given in Table 21 for each combination of γ and σ_Y. A plot of the detection probabilities for each method at each value of γ for varying values of σ_Y is provided in Fig. 17.

The Bayesian approach appears to increase detection power substantially over both QBlossoc (Besenbacher et al., 2009) and the LSA (Thompson and Kubatko, 2013) method in almost all cases. Furthermore, as σ_Y increases, both QBlossoc (Besenbacher et al., 2009) and the LSA (Thompson and Kubatko, 2013) approach demonstrate a steady decrease in power as σ_Y increases. The Bayesian method does not appear to suffer from this limitation, which suggests that it is fairly robust to changes in the variability of the trait data.

6 TREE-BASED GWAS ANALYSIS FROM UNPHASED GENETIC DATA

Unphased SNP data can be problematic for tree-based analysis of GWAS because no natural method exists to represent the evolutionary relationships among individuals due to the presence of heterozygous genotypes. Zhang et al. (2012) developed an approach to detection of QTLs known as HTreeQA

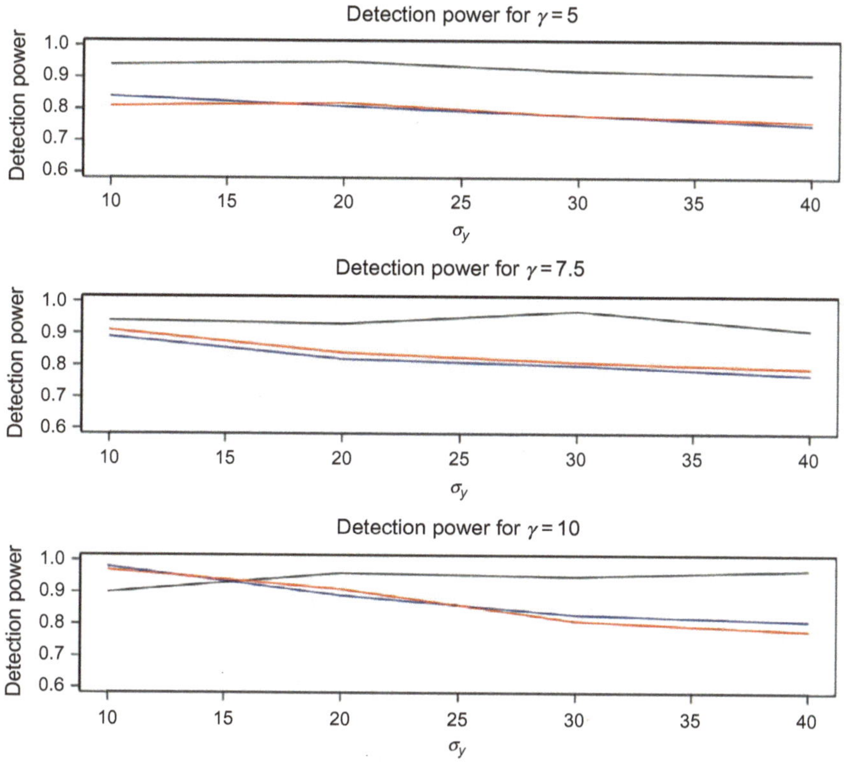

FIG. 17 Plots of the detection proportions for $\gamma = 5$ (*top*), $\gamma = 7.5$ (*middle*), and $\gamma = 10$ (*bottom*) as σ_Y varies among 10, 20, 30, and 40. The power curve for QBlossoc (Besenbacher et al., 2009) is in *blue*, the power curve for the LSA (Thompson and Kubatko, 2013) approach is in *red*, and the power curve for the posterior density ratio approach is in *black*.

that involves two steps. The first step is construction of a tristate semiperfect phylogeny (TSSP), and the second step is data analysis. Sections 6.1 and 6.2 describe these processes.

6.1 Local Semiperfect Phylogeny Construction

To begin, it is necessary to define some notation. Let $\mathcal{M}$ be a matrix of genotypes, and let $\mathcal{C}_{u,v}(\mathcal{M})$ denote an interval in $\mathcal{M}$. Let $S_1, \ldots, S_m$ denote the unphased SNPs. Suppose that there are m SNPs and n individuals in the study. Then unphased SNPs are transformed in the following way. Any 0 allele is written as the triplet (0,0,0). Any H allele is written as (1,0,1), and any 1 allele is written as (1,1,0). Two SNPs S_i and S_j are compatible if any two pairs of columns in the transformed SNPs S_i' and S_j' pass the FGT.

The TSSP is an extension of the multistate perfect phylogenetic tree (MPPT) (Gusfield, 2010). The purpose of the MPPT was to relax the assumption of the infinite sites model and handle rare events involving multiple mutations. The issue with a perfect phylogeny is that it does not handle heterozygous sites well. The TSSP treats the heterozygous allele as a third state, and consequently, is a relaxed version of the perfect phylogenetic tree. Zhang et al. (2012) define the TSSP in the following way.

Definition 1 Given an interval $\mathcal{C}_{u,v}(\mathcal{M})$ of n genotypes and m SNPs, a tristate semiperfect phylogenetic tree is a tree in which the genotype sequences are the tips and the SNPs are the edges. An SNP corresponds to an edge if only two of the three possible alleles are observed, and an SNP corresponds to two edges if all three alleles are observed. Given an allele of any SNP, the subgraph induced by all nodes that carry the allele is still a connected subgraph.

This definition is best understood through an example. We will examine each SNP here and construct the TSSP step by step, with clarifying comments after each step.

Example 16. The untransformed data are given in Table 22, and the transformed data are given in Table 23. All pairs of columns in S_1' and S_2' pass the four-gamete test except for columns 2 in S_1' and 3 in S_2', so S_1 and S_2 are compatible. All pairs of columns among S_2', S_3', S_4', and S_5' pass the

TABLE 22 Untransformed SNP Data

Genotype	S_1	S_2	S_3	S_4	S_5
1	0	0	H	H	0
2	0	0	H	0	1
3	1	H	0	0	0
4	1	1	0	0	0
5	H	H	0	0	0

TABLE 23 Transformed Unphased SNP Data

Genotype	S_1'	S_2'	S_3'	S_4'	S_5'
1	000	000	101	101	000
2	000	000	101	000	110
3	110	101	000	000	000
4	110	110	000	000	000
5	101	101	000	000	000

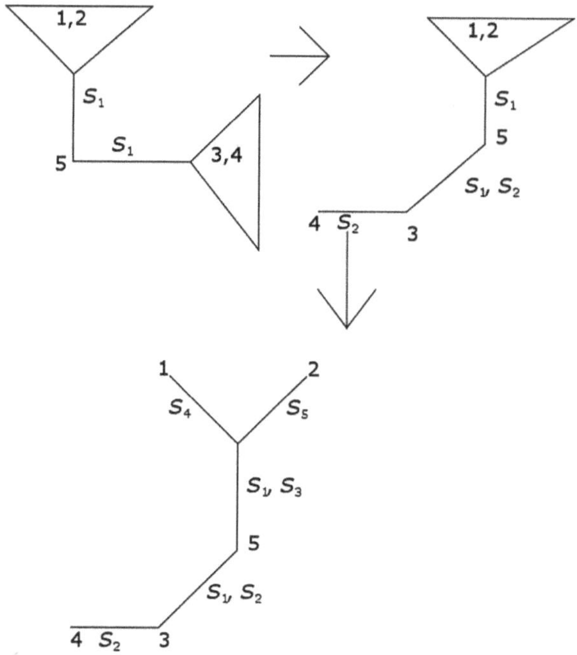

FIG. 18 Illustration of the construction of the TSSP. *Top left*: The resulting tree topology after using site 1. *Top right*: The tree that results from adding site 2. *Bottom*: The final tree after using the entire compatible SNP interval.

four-gamete test, and all of these sites are compatible with S_1', so our compatible SNP interval is [1,5]. This means that a TSSP exists for this interval.

At S_1, we have all three alleles present, so S_1 labels two edges. These edges will separate the groups (1,2), (3,4), and 5. Using this site gives the tree in the top left panel of Fig. 18. Moving to S_2, all three alleles are present, so S_2 labels two edges. These edges separate groups (1,2), 4, and (3,5). This split is shown in the top right panel of Fig. 18. The S_3 arrangement contains only two alleles, so it labels one edge. This edge splits groups (1,2) and (3,4,5). Similarly, S_4 contains only two alleles and labels the edge that splits groups 1 and (2,3,4,5). Finally, S_5 contains only two alleles, and it labels the edge that splits groups 2 and (1,3,4,5). The final TSSP is shown in the bottom panel of Fig. 18.

6.2 Determination of Significance

The HTreeQA (Zhang et al., 2012) method operates in a fundamentally different way than do any of the tree-based methods for phased SNP data. HTreeQA (Zhang et al., 2012) looks for the significance of compatible SNP intervals rather than detecting individual SNPs. However, with large data sets, the lack of compatibilities enables detection of either very small intervals or of individual SNPs themselves. HTreeQA (Zhang et al., 2012) begins by considering all possible partitions of the tree that result from removing edges. Like the Thompson and Kubatko (2013) method, this technique will not work if all edges are removed from the tree. Removing $p - 1$ edges from the tree partitions the tree into p groups with sizes t_i, $i = 1, \ldots, p$. The test statistic here is given by

$$F = \frac{\sum_{i=1}^{p} t_i(\overline{Y}_i - \overline{Y})^2/(p-1)}{\sum_{i=1}^{p}\sum_{j=1}^{t_i}(Y_{ij} - \overline{Y}_i)^2/(n-p)},$$

which follows an F distribution with $p - 1$ and $n - p$ degrees of freedom for normally distributed trait values. In this situation, the p-value based on the $F_{p-1,\ n-p}$ distribution is reported. Otherwise, permutation testing is used, and the p-value of the test statistic for the given partition is computed in the same way as Thompson and Kubatko (2013) describe. All partitions are examined and the corresponding p-value is computed. The smallest p-value is reported as the p-value for the tree. In the following example, assume that the trait values are normally distributed.

Example 17. We examine all but one of the partitions of the tree that result from removing one edge, one partition that results from removing two edges, and one partition that results from removing three edges. The partitions are listed in Table 24. There are 18 other partitions that result from removing two or three edges. In order to report a p-value, it would be necessary to compute an overall p-value for all of them. For the sake of illustration, suppose that the listed partitions are the only partitions of the tree. The results for these partitions are listed in Table 25. We compute the F-statistic and the p-value for the

TABLE 24 Partitions for Data Analysis

Edge Label	Groups
S_2	4, (1,2,3,5)
(S_1, S_2)	(1,2,5), (3,4)
(S_1, S_3)	(1,2), (3,4,5)
S_4	1, (2,3,4,5)
(S_1, S_2), (S_1, S_3)	(1,2), (3,4), 5
(S_1, S_2), (S_1, S_3), S_4	(1,2), 3,4,5

TABLE 25 Results for HTreeQA (Zhang et al., 2012)

Groups	Group Means	DF	*F*	*p*-Value
(1,2,3,5),4	(92.1594, 97.1203)	1,3	0.207	0.680
(1,2,5), (3,4)	(98.321, 89.754)	1,3	1.244	0.347
(1,2), (3,4,5)	(87.0667, 97.2082)	1,3	2.043	0.248
1, (2,3,4,5)	(96.3112, 92.3617)	1,3	0.128	0.744
(1,2), (3,4), 5	(87.0665, 98.321, 94.9828)	2,2	0.753	0.571
(1,2),3,4,5	(87.0665, 99.5216,97.1205,94.9828)	3,1	0.261	0.855

(S_1', S_3') partition, which results in groups (1, 2) and (3, 4, 5). Name these groups 1 and 2, respectively. We have $t_1 = 2$, $t_2 = 3$, $p = 2$, $\overline{Y}_1 = 87.0667$, $\overline{Y}_2 = 97.2082$, and $\overline{Y} = 93.1516$. This gives

$$F = \frac{123.42/(2-1)}{181.2354/(5-2)} = 2.043.$$

The p-value is $\mathrm{P}(F_{1,3} \geq 2.043) = 0.248$. If the partitions in Table 24 were the only partitions of the tree, then the p-value for the tree would be reported as 0.248. This interval would not be deemed significant to the expression of the trait at any reasonable level of significance.

7 CONCLUSION

This chapter aims to provide the reader with an understanding of some common techniques of GWAS analysis. We examine four broad situations: nontree-based analysis for phased and unphased SNPs and tree-based analysis for each type of SNP data. It is clear that nontree-based methods can be carried out much more quickly than can tree-based methods, but that computational gain comes with a cost. That cost is the loss of detection power that accompanies the unrealistic assumption of independence among the trait values.

The possibilities for analyzing GWAS for unphased data while incorporating local population dependence structure are still not well-understood. For example, it remains unclear how such a dependence structure would be estimated. While tree-based analysis of phased data leads to a natural way of doing this, the TSSP does not provide information about the order of evolution or ancestral patterns. It is thus impossible to use the amount of shared evolutionary history as a proxy for the covariance. One idea is to estimate a rooted tree that allows up to three descendants from each internal node.

However, while there does appear to be some promise in this technique, common phylogenetics software is equipped to handle only binary trees. Since the type of tree proposed may have either two or three descendants from each internal node, putting tree construction and likelihood estimation aside, it is a nontrivial task even to determine how many internal nodes such a tree will have. The number of internal nodes is highly dependent on the structure of the genetic data. Even covariance estimation directly from the trait data is difficult to do well because we only observe one trait value per individual. However, it is not in dispute that a covariance structure exists among the trait values, and it needs to be taken into account in order to analyze GWAS accurately.

While this chapter presents several techniques of GWAS analysis, the presentation is far from exhaustive. For example, there are other methods of nontree-based GWAS analysis that the chapter did not have the space to cover. Among these methods are quantitative haplotype pattern mining (QHPM) (Onkamo et al., 2002) and HapMiner (Li and Jiang, 2005). Still other tree-based methods are available that differ from the ones described here to model evolutionary histories among species, subspecies, or strains (Yang et al., 2011). These methods work by inferring an accurate global ancestral recombination graph (Larribe et al., 2002; Minichiello and Durbin, 2006; Morris et al., 2002) instead of inferring a local tree at each marker. Other techniques, including Genomic Control (Devlin and Roeder, 1999), EIGENSTRAT (Price et al., 2006), and EMMA (Kang et al., 2008) build global statistical models that are used to account for genetic effects. It is clear that the literature contains many other methods of GWAS analysis, and the interested reader is encouraged to examine some of these other methods.

REFERENCES

Adams, E.N., 1972. Consensus techniques and the comparison of taxonomic trees. Syst. Zool. 21 (4), 390–397.

Archie, J.W., 1989. Homoplasy excess ratios: new indices for measuring levels of homoplasy in phylogenetic systematics and a critique of the consistency index. Syst. Zool. 38 (3), 253–269.

Bafna, V., Halldorsson, B.V., Schwartz, R., Clark, G., Istrail, S., 2003. Haplotypes and informative SNP selection algorithms: don't block out information. In: Proceedings of the Seventh International Conference in Computational Molecular Biology (RECOMB2003). ACM Press, pp. 19–27.

Bardel, C., Danjean, V., Hugot, J.P., Darlu, P., Génin, E., 2005. On the use of haplotype phylogeny to detect disease susceptibility loci. BMC Genet. 6, 24.

Benjamini, Y., Hochberg, Y., 1995. Controlling the false discovery rate: a practical and powerful approach to multiple testing. J. R. Stat. Soc. Ser. B 57 (1), 289–300.

Benjamini, Y., Yekutieli, D., 2001. The control of the false discovery rate in multiple testing under dependency. Ann. Stat. 29 (4), 1165–1188.

Besenbacher, S., Mailund, T., Schierup, M.H., 2009. Local phylogeny mapping of quantitative traits: higher accuracy and better ranking than single marker association in genomewide scans. Genetics 181 (2), 747–753.

Browning, S.R., Browning, B.L., 2007. Rapid and accurate haplotype phasing and missing-data inference for whole-genome association studies by use of localized haplotype clustering. Am. J. Hum. Genet. 81, 1084–1097.

Carlson, C.S., Eberle, M.A., Rieder, M.J., Yi, Q., Kruglyak, L., Nickerson, D.A., 2004. Selecting a maximally informative set of single-nucleotide polymorphisms for association analysis using linkage disequilibrium. Am. J. Human Genet. 74, 106–120.

Ceppelini, R., Siniscalco, M., Smith, C.A.B., 1955. The estimation of gene frequencies in a random mating population. Ann. Hum. Genet. 20 (2), 97–115.

Churchill, G.A., Doerge, R.W., 1994. Empirical threshold values for quantitative trait mapping. Genetics 138 (3), 963–971.

Clark, A.G., 1990. Inference of haplotypes from PCR-amplified samples of diploid populations. Mol. Biol. Evol. 7 (2), 111–122.

Devlin, B., Roeder, K., 1999. Genomic control for association studies. Biometrics 55 (4), 997–1004.

Dudoit, S., van der Laan, M.J., Pollard, K.S., 2004. Multiple testing. Part I. Single-step procedures for control of general type I error rates. Stat. Appl. Genet. Mol. Biol. 3. Article 13.

Eronen, L., Geerts, F., Toivonen, H., 2006. Haplorec: efficient and accurate large-scale reconstruction of haplotypes. BMC Bioinformatics 7, 542.

Excoffier, L., Slatkin, M., 1995. Maximum-likelihood estimation of molecular haplotype frequencies in a diploid population. Mol. Biol. Evol. 12, 921–927.

Farris, J.S., 1970. Methods for computing Wagner trees. Syst. Zool. 19 (1), 83–92.

Felsenstein, J., 1981. Evolutionary trees from DNA sequences: a maximum likelihood approach. Mol. Phylogenet. Evol. 17 (6), 368–376.

Fernandez-Rozadilla, C., Brea-Fernandez, A., Bessa, X., Alvarez-Urturi, C., Abulí, A., Clofent, J., Payá, A., EPICOLON Consortium, Jover, R., Xicola, R., Llor, X., Andreu, M., Castells, A., Carracedo, A., Castellvi-Bel, S., Ruiz-Ponte, C., 2013. BMPR1A mutations in early-onset colorectal cancer with mismatch repair proficiency. Clin. Genet. 84 (1), 94–96.

Fitch, W.M., 1971. Toward defining the course of evolution: minimum change for a specific tree topology. Syst. Zool. 20 (4), 406–416.

Gabriel, S.B., Schaffner, S.F., Nguyen, H., Moore, J.M., Roy, J., Blumenstiel, B., Higgins, J., DeFeliece, M., Lochner, A., Faggart, M., et al., 2001. The structure of haplotype blocks in the human genome. Science 296, 2225–2229.

Gilks, W.R., Richardson, S., Spiegelhalter, D.J. (Eds.), 1996. Markov Chain Monte Carlo in Practice. Chapman and Hall, London.

Guo, Y., Li, J., Bonham, A.J., Wang, Y., Deng, H., 2009. Gains in power for exhaustive analyses of haplotypes using variable-sized sliding window strategy: a comparison of association-mapping strategies. Eur. J. Hum. Genet. 17, 785–792. https://doi.org/10.1534/genetics.104.031799. PMID: 15489534.

Gusfield, D., 2003. Haplotype inference by pure parsimony. In: Combinatorial Pattern Matching, Proceedings, vol. 2676, pp. 144–155.

Gusfield, D., 2010. The multi-state perfect phylogeny problem with missing and removable data: solutions via integer-programming and chordal graph theory. J. Comput. Biol. 17 (3), 383–399.

Hansen, T.F., 1997. Stabilizing selection and the comparative analysis of adaptation. Evolution 51 (5), 1341–1351.

Hao, C., Wang, Y., Hou, J., Feuillet, C., Balfourier, F., Zhang, X., 2012. Association mapping and haplotype analysis of a 3.1-Mb genomic region in fusarium head blight resistance on wheat chromosome 3BS. PLoS One 7(10). https://doi.org/10.1371/journal.pone.004644. E46444.

Hartigan, J.A., 1973. Minimum mutation fits to a given tree. Biometrics 29 (1), 53–65.

Hawley, M.E., Kidd, K.K., 1995. HAPLO: a program using the E-M algorithm to estimate the frequencies of multi-site haplotypes. J. Hered. 86, 409–411.

Howie, B.N., Donnelly, P., Marchini, J., 2009. A flexible and accurate genotype imputation method for the next generation of genome-wide association studies. PLoS Genet. 5, e1000529.

Immervoll, T., Loesgen, S., Dütsch, G., Gohlke, H., Herbon, N., Klugbauer, S., Dempfle, A., Bickeböller, H., Becker-Follmann, J., Rüschendorf, F., Saar, K., Reis, A., Wichmann, H.E., Wjst, M., 2001. Fine mapping and single-nucleotide polymorphism association results of candidate genes for asthma and related phenotypes. Hum. Mutat. 18 (4), 327–336.

Jansen, R.C., 1993. Interval mapping of multiple quantitative trait loci. Genetics 135, 205–211.

Johnson, G.C.L., Esposito, L., Barratt, B.J., Smith, A.N., Heward, J., Di Genova, G., Ueda, H., Cordell, H.J., Eaves, I.A., Dudbridge, F., et al., 2001. Haplotype tagging for identification of common disease genes. Nat. Genet. 29, 233–237.

Kang, H.M., Zaitlen, N.A., Wade, C.M., Kirby, A., Heckerman, D., Daly, M.T., Eskin, E., 2008. Efficient control of population structure in model organism association mapping. Genetics 178 (3), 1709–1723.

Kimura, M., 1980. A simple method for estimating evolutionary rates of base substitution through comparative studies of nucleotide sequences. J. Mol. Evol. 16 (2), 111–120.

Kingman, J.F.C., 1982. The coalescent. Stoch. Process. Appl. 13 (3), 235–248.

Kong, A., Masson, G., Frigge, M.L., Gylfason, A., Zusmanovich, P., Thorleifsson, G., Olason, P.I., Ingason, A., Steinberg, S., Rafnar, T., Sulem, P., Mouy, M., Jonsson, F., Thorsteinsdottir, U., Gudbjartsson, D.F., Stefansson, H., Stefansson, K., 2008. Detection of sharing by descent, long-range phasing, and haplotype imputation. Nat. Genet. 40, 1068–1075.

Lander, E.S., Botstein, D., 1989. Mapping mendelian factors underlying quantitative traits. Genetics 121, 185–199.

Lander, E.S., Botstein, D., 1994. Corrigendum. Genetics 136, 705.

Larribe, F., Lessard, S., Schork, N.J., 2002. Gene mapping via the ancestral recombination graph. Theor. Popul. Biol. 62 (2), 215–229.

Lewis, P.O., 2001. A likelihood approach to estimating phylogeny from discrete morphological character data. Syst. Biol. 50 (6), 913–925.

Li, J., Jiang, T., 2005. Haplotype-based linkage disequilibrium mapping via direct data mining. Bioinformatics 21 (24), 4384–4393.

Li, Y., Willer, C.J., Ding, J., Scheet, P., Abecasis, G.R., 2010. MaCH: using sequence and genotype data to estimate haplotypes and unobserved genotypes. Genet. Epidemiol. 34 (8), 816–834.

Li, D., Zhao, H., Kranzler, H.R., Li, M.D., Jensen, K.P., Zayats, T., Farrer, L.A., Gelernter, J., 2015. Genome-wide association study of copy number variations (CNVS) and opioid dependence. Neuropsychopharmacology 40 (4), 1016–1026.

Long, J.C., Williams, R.C., Urbanek, M., 1995. An E-M algorithm and testing strategy for multiple-locus haplotypes. Am. J. Hum. Genet. 56, 799–810.

Lorenz, A.J., Hamblin, M.T., Jannink, J.L., 2010. Performance of single nucleotide polymorphisms versus haplotypes for genome-wide association analysis in barley. PLoS One 5(11). https://doi.org/10.1371/journal.pone.0014079. PMID:21124933.

Lu, Y., Shah, T., Hao, Z., Taba, S., Zhang, S., Gao, S., et al., 2011. Comparative SNP and haplotype analysis reveals a higher genetic diversity and rapider LD decay in tropical than temperate germplasm in maize. PLoS One 6(9). https://doi.org/10.1371/journal.pone.0024861. PMID: 21949770.

Lu, Y., Xu, J., Yuan, Z., Hao, Z., Xie, C., Li, X., 2012. Comparative LD mapping using single SNPs and haplotypes identifies QTL for plant height and biomass as secondary traits of drought tolerance in maize. Mol. Breed. 30 (1), 407–418.

Mailund, T., Besenbacher, S., Schierup, M.H., 2006. Whole-genome association mapping by incompatibilities and local perfect phylogenies. BMC Bioinformatics 7, 454.

McCarthy, L.C., Hosford, D.A., Riley, J.H., Bird, M.I., White, N.J., Hewett, D.R., Peroutka, S.J., Griffiths, L.R., Boyd, P.R., Lea, R.A., Bhatti, S.M., Hosking, L.K., Hood, C.M., Jones, K.W., Handley, A.R., Rallan, R., Lewis, K.F., Yeo, A.J., Williams, P.M., Priest, R.C., Khan, P., Donnelly, C., Lumsden, S.M., O'Sullivan, J., See, C.G., Smart, D.H., Shaw-Hawkins, S., Patel, J., Langrish, T.C., Feniuk, W., Knowles, R.G., Thomas, M., Libri, V., Montgomery, D.S., Manasco, P.K., Xu, C.F., Dykes, C., Humphrey, P.P., Roses, A.D., Purvis, I.J., 2001. Single-nucleotide polymorphism alleles in the insulin receptor gene are associated with typical migraine. Genomics 78 (3), 135–149.

McVean, G.A., Cardin, N.J., 2005. Approximating the coalescent with recombination. Philos. Trans. R. Soc. Lond. B Biol. Sci. 360 (1459), 1387–1393.

Minichiello, M.J., Durbin, R., 2006. Mapping trait loci by use of inferred ancestral recombination graphs. Am. J. Hum. Genet. 79 (5), 910–922.

Morris, A.P., Whittaker, J.C., Balding, D.J., 2002. Fine-scale mapping of disease loci via shattered coalescent modeling of genealogies. Am. J. Hum. Genet. 70 (3), 686–707.

Nothnagel, M., Furst, R., Rohde, K., 2002. Entropy as a measure for linkage disequilibrium over multilocus haplotype blocks. Hum. Hered. 54, 186–198.

Onkamo, P., Ollikainen, V., Sevon, P., Toivonen, H., Mannila, H.T., Kere, J., 2002. Association analysis for quantitative traits by data mining: QHPM. Ann. Hum. Genet. 66 (5–6), 419–429.

Patil, N., Berno, A.J., Hinds, D.A., Barrett, W.A., Doshi, J.M., Hacker, C.R., Kautzer, C.R., Lee, D.H., Marjoribancks, C., McDonough, D.P., et al., 2001. Blocks of limited haplotype diversity revealed by high-resolution scanning of human chromosome 21. Science 294, 1719–1723.

Pletcher, M.T., McClurg, P., Batalov, S., Su, A.I., Barnes, S.W., Lagler, E., Korstanje, R., Wang, X., Nusskern, D., Bogue, M.A., Mural, R.J., Paigen, B., Wiltshire, T., 2004. Use of a dense single nucleotide polymorphism map for in silico mapping in the mouse. PLoS Biol. 2, e393. 101371/journal.pbio.0020393, 12.

Price, A.L., Patterson, N.J., Plenge, R.M., Weinblatt, M.E., Shadick, N.A., Reich, D., 2006. Principal components analysis corrects for stratification in genome-wide association studies. Nat. Genet. 38 (8), 904–909.

Qin, Z.S., Niu, T., Liu, J.S., 2002. Partition-ligation-expectation-maximization algorithm for haplotype inference with single-nucleotide polymorphisms. Am. J. Hum. Genet. 71 (1), 1242–1247.

Rabiner, L.R., 1989. A tutorial on hidden Markov models and selected applications in speech recognition. Proc. IEEE 77 (2), 257–286.

Rogers, J.S., Swofford, D.L., 1998. A fast method for approximating maximum likelihoods of phylogenetic trees from nucleotide sequences. Syst. Biol. 47 (1), 77–89.

Roukos, D.H., 2009. Genome-wide association studies and aggressive surgery toward individual prevention, and improved local control and overall survival for gastric cancer. Ann. Surg. Oncol. 16 (4), 795–798.

Scheet, P., Stephens, M., 2006. A fast and flexible statistical model for large-scale population genotype data: applications to inferring missing genotypes and haplotypic phase. Am. J. Human Genet. 78 (4), 629–644.

Smith, C.A.B., 1957. Counting methods in genetical statistics. Ann. Hum. Genet. 21 (3), 254–276.

Stephens, M., Donnelly, P., 2000. Inference in molecular population genetics. J. R. Stat. Soc. Ser. B 62, 605–655.

Stephens, M., Scheet, P., 2005. Accounting for decay of linkage disequilibrium in haplotype inference and missing-data imputation. Am. J. Hum. Genet. 76, 449–462.

Stephens, M., Smith, N.J., Donnelly, P., 2001. A new statistical method for haplotype reconstruction from population data. Am. J. Hum. Genet. 68, 978–989.

Stevens, A., Ray, D., Alansari, A., Hajeer, A., Thomson, W., Donn, R., Ollier, W.E., Worthington, J., Davis, J.R., 2001. Characterization of prolactin gene polymorphism and its association with systemic lupus erythematous. Arthritis Rheumatol. 44 (10), 2358–2366.

Storey, J.D., 2003. The positive false discovery rate: a Bayesian interpretation and the q-value. Ann. Stat. 31, 2013–2035.

Storey, J.D., Tibshirani, R., 2003. Statistical significance for genomewide studies. Proc. Natl. Acad. Sci. U.S.A. 100 (16), 9440–9445.

Stram, D.O., Haiman, C.A., Hirschhorn, J.N., Altshuler, D., Kolonel, L.N., Henderson, B.E., Pike, M.C., 2003. Choosing haplotype-tagging SNPs based on unphased genotype data using preliminary sample of unrelated subjects with an example from the multiethic cohort study. Hum. Hered. 55, 27–36.

Swofford, D.L., Maddison, W.P., 1992. Parsimony, character-state reconstructions, and evolutionary inferences. In: Mayden, R.L. (Ed.), Systematics, Historical Ecology and North American Freshwater Fishes. Stanford University Press, Palo Alto, CA, pp. 186–223.

Templeton, A.R., Maxwell, T., Posada, D., Stangård, J.H., Boerwinkle, E., Sing, C.F., 2005. Tree-scanning: a method for using haplotype trees in phenotype/genotype studies. Genetics 169 (1), 441–453.

Thompson, K.L., Kubatko, L.S., 2013. Using ancestral information to detect and localize quantitative trait loci in genome-wide association studies. BMC Bioinformatics 14, 200–209.

Tsunoda, K., Sanke, T., Nakagawa, T., Furuta, H., Nanjo, K., 2001. Single-nucleotide polymorphism (D68D, T to C) in the syntaxin 1A gene correlates to age at onset and insulin requirement in type II diabetes patients. Diabetologia 44 (11), 2092–2097.

Wang, X., Paigen, B., 2002. Quantitative trait loci and candidate genes regulating HDL cholesterol: a murine chromosome map. Arterioscler. Thromb. Vasc. Biol. 22 (9), 1390–1401.

Wang, L., Xu, Y., 2003. Haplotype inference by maximum parsimony. Bioinformatics 19 (14), 1773–1780.

Westfall, P.H., Young, S.S., 1993. Re-Sampling-Based Multiple Testing: Examples and Methods for p-Value Adjustment. Wiley Interscience.

Wu, Y., 2011. New methods for inference of local tree topologies with recombinant SNP sequences in populations. IEEE/ACM Trans. Comput. Biol. Bioinform. 8 (1), 182–193.

Yan, J., Shah, T., Warburton, M.L., Buckler, E.S., McMullen, M.D., Crouch, J., 2009. Genetic characterization and linkage disequilibrium estimation of a global maize collection using SNP markers. PLoS One 4(12). https://doi.org/10.1371/journal.pone.0008451. PMID: 20041112.

Yang, H., Wang, J.R., Didion, J.P., Buus, R.J., Bell, T.A., Welsh, C.E., Bonhomme, F., Yu, A.H., Nachman, M.W., Pialek, J., Tucker, P., Boursot, P., McMillan, L., Churchill, G., de Villena, F.P., 2011. Subspecific origin and haplotype diversity in the laboratory mouse. Nat. Genet. 43 (7), 648–655.

Yomenaru, J.I., Yamamoto, T., Ebana, K., Yamarnoto, E., Nagasaki, H., Shibaya, T., et al., 2012. Genome-wide haplotype changes produced by artificial selection during modern rice breeding in Japan. PLoS One 7(3). https://doi.org/10.1371/journal.pone.0032982. PMID: 17339218.

Zeng, Z.B., 1993. Theoretical basis for separation of multiple linked gene effects in mapping quantitative trait loci. Proc. Natl. Acad. Sci. U.S.A. 90 (23), 10972–10976.

Zeng, Z.B., 1994. Precision mapping of quantitative trait loci. Genetics 136, 1457–1468.

Zhang, K., Jin, L., 2003. Haploblockfinder: haplotype block analysis. Bioinformatics 19, 1300–1301.

Zhang, K., Deng, M., Chen, T., Waterman, M.S., Sun, F., 2002. A dynamic programming algorithm for haplotype partitioning. Proc. Natl. Acad. Sci. U.S.A. 99, 7335–7339.

Zhang, K., Qin, Z., Chen, T., Liu, J.S., Waterman, M.S., Sun, F., 2005. HapBlock: haplotype block partitioning and tag SNP selection software using a set of dynamic programming algorithms. Bioinformatics 21 (1), 131–134.

Zhang, Z., Zhang, X., Wang, W., 2012. HTreeQA: using semi-perfect phylogeny trees in quantitative trait loci study on genotype data. G3: Genes, Genomes, Genetics 2 (2), 175–189.

Zhao, L.P., Li, S.S., Shen, F., 2007. A haplotype-linkage analysis method for estimating recombination rates using dense SNP trio data. Genet. Epidemiol. 31 (2), 154–172. https://doi.org/10.1002/gepi.20198. PMID: 17219374.

FURTHER READING

Wang, N., Akey, J.M., Zhang, K., Chakraborty, K., Jin, L., 2002. Distribution of recombination crossovers and the origin of haplotype blocks: the interplay of population history, recombination, and mutation. Am. J. Human Genetl. 71, 1227–1234.

Chapter 8

Bayesian Item Response Theory for Cancer Biomarker Discovery

Katabathula Ramachandra Murthy*[,1], Salendra Singh*, David Tuck[†] and Vinay Varadan*[,1]

*Case Comprehensive Cancer Center, Case Western Reserve University, Cleveland, OH, United States

[†]Curis, Inc., Lexington, MA, United States

[1]Corresponding authors: e-mail: rmk146@case.edu; vxv89@case.edu

ABSTRACT

Item response theory (IRT), which originated from psychometric theory, relates to latent variable models for psychological assessment . IRT's contributions have been less extensively studied in cancer research even though IRT has become prevalent in the measurement of ability and item parameters with respect to latent traits. Benefits of IRT include comprehensive analyses and reduction of measurement error, meaningful scaling of latent variables, evaluation of test and item bias, greater accuracy in the assessment of change due to therapeutic intervention, and evaluation of model and person fit. Advances in cancer biology and human genetics have resulted in the identification of driver events and critical biologic dependencies. These advances, together with the development of drugs for specific biologic targets hold promise for an era of personalized oncology treatment. Therefore, robust discovery of biomarkers associated with the biologic mechanisms that drive cancers could provide clinicians with the ability to tailor therapeutic strategies for individual patients, thus improving clinical outcomes. Incorporating biomarkers in clinical trials during drug development may also allow for determining the optimal patient population for a given drug, thus maximizing its efficacy. Thus, this chapter explores the application of different IRT models in the discovery of biomarkers and altered biologic processes in cancer. The utility and limitations of the IRT models in the discovery of biomarkers or biological mechanisms are highlighted through simulations and application to genome-scale molecular profiling studies of cancers.

Keywords: Clinical trials, IRT, 1PL, 2PL, 3PL, Rasch, Graded response, Partial credit, Rating scale, WAIC, Biomarker information function

Handbook of Statistics, Vol. 40. https://doi.org/10.1016/bs.host.2018.09.005

1 INTRODUCTION

The pathobiology of complex diseases such as cancer typically involves coordinated dysregulation of multiple genes interacting in intricate ways that are not yet fully understood. Characterizing the complex interplay of cellular processes in cancer using in-depth molecular measurements would enable uncovering the key mechanisms underlying its development and progression. Significant advances in molecular measurement technologies have resulted in a wide range of tumor profiling capabilities ranging from measuring individual molecular characteristics per tumor cell in a given tissue sample, such as genomic copy number alterations using fluorescent in situ hybridization (FISH) and protein expression levels by immunohistochemistry (IHC); to tissue-level expression levels of a few genes using quantitative polymerase chain reaction (qPCR); and ultimately to genome-scale measurements of mutations, copy number alterations, and gene expression levels using next generation sequencing (NGS) of DNA or RNA derived from tissues or cells. Each of these types of measurement technologies are useful in the diagnosis, prognosis, and treatment of cancer, with genome-scale measurement technologies quickly becoming more reliable and cost-effective, thus likely replacing the more focused measurement technologies in both research and clinical settings (Kamalakaran et al., 2013).

These technological advances in genome-scale measurement technologies have enabled large-scale molecular profiling of cancer tissues by translational scientists, an effort exemplified by The Cancer Genome Atlas (TCGA) consortium that was initiated and funded by the National Cancer Institute and National Human Genome Research Institute. Over the past decade the TCGA has generated genome-scale molecular profiles of over 11, 000 cancer samples, spanning 33 cancer types, totaling over 2.5 petabytes of data. Such large-scale profiling efforts have resulted in a paradigm-shift in our understanding of cancer, posing both new challenges and opportunities for the treatment of this complex and aggressive set of diseases. On the one hand, large-scale genomic profiling of cancers has revealed that tumors typically harbor irregularities across genomic (mutations and structural variations), transcriptomic (differential expression of coding and noncoding RNAs), and epigenetic (DNA methylation and histone modifications) modalities. These multiomic abnormalities have revealed substantial molecular heterogeneity resulting in multiple molecularly-defined subtypes even within a single cancer type such as breast cancer, each of which exhibit different clinical outcomes and response rates to frontline therapies (Lehmann et al., 2011, 2016; Morganella et al., 2016; Nik-Zainal et al., 2016; Perou et al., 2010; Sanchez-Garcia et al., 2014; Varadan et al., 2016a, b; Yates et al., 2017). Thus, one of the principal challenges of this postgenomic era of cancer research is deciphering how diverse sets of molecular alterations jointly contribute to an individual tumors phenotype in order to enable development of personalized strategies in the clinic. This challenge has spawned an entire research field in bioinformatics focused on developing methodologies for integrative analysis of multiomics cancer profiles (Creixell et al., 2015; Dimitrova et al., 2017; Greenblum et al., 2011; Kristensen et al., 2012; Osmanbeyoglu et al., 2014; Razi et al., 2015; Sedgewick et al., 2013; Tarca et al., 2008; Vandin et al., 2011; Varadan et al., 2012; Vaske et al., 2010).

On the other hand, pan-cancer integrative analyses, which involves the analysis of molecular profiles across multiple cancer types, have identified commonalities in oncogenic processes in cancers arising in completely different organs (Ciriello et al., 2013; Ding et al., 2018). The recent Pan-Cancer Atlas set of publications provides new opportunities to identify molecularly targeted treatment strategies that can be applied to different cancer types, irrespective of their tissue of origin (Kruger, 2018). Indeed, the clinical implication of this paradigm-shifting view of cancer is being explored in two large clinical trials at the National Cancer Institute, known as NCI-MATCH and NCI-MPACT. Patients in the NCI-MATCH trial are assigned to receive a specific targeted treatment based on molecular profiles of their individual tumor samples, irrespective of the tissue of origin of the cancer (Coyne et al., 2017; Do et al., 2015). If successful, these trials will usher in a new paradigm of precision cancer medicine, where molecular biomarkers will be employed in the clinic to assign patients to specific interventions.

The above challenge and opportunity thus strongly motivate the development of new computational approaches to identify cancer biomarkers from molecular profiling data of patient tumors. Biomarker discovery specific to cancer has been an area of significant interest in the statistical and informatics communities, given the unique challenges of the relatively small numbers of patients ($<10^2$) when compared to number of diverse molecular features ($\gg 10^4$) measured per patient. This chapter is motivated by this dual-challenge of integrating diverse molecular data (multiomics measurements) in relatively small patient populations, to discover the subset of biologically or clinically relevant molecular features (biomarkers). In machine learning or data science literature, mining informative features is an active area of research. Most of the literature is focused on either feature transformation to generated new set of features or selecting feature subsets from the original features. The true meaning of features get modified while transforming them and selection of features requires exhaustive search of feature space due to the high correlations (Chandrashekar and Sahin, 2014; Cunningham and Ghahramani, 2015; Sreevani and Murthy, 2017; Sreevani et al., 2018). Whereas, latent theory provides a flexibility of independence to observable feature by modeling an unobservably (latent) variables. Thus, we focus on the application of a well-developed area of statistical analysis, item response theory (IRT), also known as latent trait theory, to this challenge of multiomics biomarker discovery in cancer.

Goal of the chapter is to model different biological aspects of cancers using biomarker information of patients and IRT models on the latent trait. Moreover, identifying informative biomarkers for cancers in a clinical trail is the primary goal. Firstly, different item response theory models are reviewed in perspective of clinical study. And, then a criterion function called biomarker (item) information function has been discussed which describes properties of biomarkers on latent scale. Subsequently followed by Bayesian framework for IRT models and simulated experimental study for biomarker selection.

2 ITEM RESPONSE THEORY

2.1 Introduction

IRT has been introduced by Lord and Novick (1968) to improve the precision in psychological measurement, along with Rasch's (Andrich, 1981) treatment of probabilistic models in cognitive testing. IRT can be consider as an extension of "classical factor analysis" and is a useful tool in test construction because it provides a theoretical framework for modeling underlying properties of examinees and test items (questions) on a uniform metric. The majority of IRT models are stochastic in nature; that is, examinee's responses (e.g., true (1) vs false (0)) are assumed to be probabilistic. Also, IRT models are predicated on the existence of latent variables: constructs that cannot be directly measured, yet are inferred to exist based on peripheral associations among measurable qualities. The concept of a latent variable is well established in the minds of many psychologists and statisticians, as factor analysis which is also assumes the existence of latent variables. Fundamentally, it must be assumed that latent variables can account for all observed covariation between test items. In other words, the observed relations between items should disappear given their association with one or more latent variables (i.e., local independence) (Bock, 1997).

IRT models assume that the probability of an examinee passing (responding correctly or affirmatively) a test item is a function of two sets of parameters: (i) their position on the latent trait, the examinee's parameter; and (ii) the characteristics of the test item, the item parameters. The function is a theoretical proposition for how variables in a system are related as similar to a curve fitting (regression) models. More interestingly, the relation is a nonlinear in nature, and thus it is not possible to employ the simple linear models. However, the relation does tend to take on two interesting forms, the normal ogive (the integral of the Gaussian function) and the logistic ogive (a sigmoid-shaped function) (Thomas, 2011; van der Linden and Hambleton, 2013).

This book chapter discusses models rested on the assumption that a single latent variable accounts for the observed intercorrelations among test items—unidimensional scales. However, the importance of multidimensional models in psychological assessment has long been recognized (van der Linden and Hambleton, 2013). The well-known IRT models for dichotomous responses, for instance, adjust response data for such properties of test items as their difficulty, discriminating power, or liability to guessing. The emphasis in this book chapter, however, is not only on IRT models for handling dichotomously scored data, in fact more on the polytomous models (Birnbaum, 1968). That is, if responses are made to the individual answer choices of a test item (e.g., multiple-choice item), they clearly depend on the properties of the answer choices, which have to be parametrized in addition to other properties of the items. Various models for this case, generally known as polytomous IRT models, are the tunnel vision of this chapter.

Novel biomarker selection in cancer clinical studies is one of the important application in the field of computational biology. Biomarker (short for biological marker) (Mayeux, 2004) is a biological measurement of a biological state. By definition, a biomarker is "a characteristic that is objectively measured and evaluated as an indicator of normal biological processes, pathogenic processes or pharmacological responses to a therapeutic intervention." At molecular level, it is defined as a biological molecule found in blood, other body fluids, or tissues that is a sign of a normal or abnormal process, or of a condition or disease. A biomarker may be used to see how well the patient responds to a treatment for a disease or condition (Strimbu and Tavel, 2010). Clinical trials and studies are research studies for digging new ways to improve treatments and the quality of life of diseased people (Aggarwal et al., 2004). These studies help researchers to design cancer clinical trials to test new ways to treat cancer, find and diagnose cancer, prevent cancer, manage symptoms of cancer and side effects from its treatment, etc. (Lewis et al., 2003). The goal of this chapter is to analyze the utility of IRT models in selection of novel biomarkers for a clinical study.

In perspective of IRT and cancer clinical studies, each biomarker is treated as a test item and patient is considered as an examinee. With the help of IRT models and corresponding item information functions (see Section 3), different properties, like difficulty, discrimination, etc., of biomarkers can be modeled, and accordingly informative biomarkers can be selected. Firstly, dichotomous IRT models have been discussed, subsequently polytomous models will be explained and, finally item (biomarker) information functions for both dichotomous and polytomous models will be discussed for biomarker selection.

2.2 Dichotomous Models

The dichotomous IRT models assume data with dichotomous (0/1) responses of each patient to a biomarker. For example, in cancer genomics, gene expression microarrays, microRNA arrays, methylation arrays, array CGH, SNP arrays and sequencing, etc. can be dichotomized to interrogate the genomes of several types of cancers (Cancer Genome Atlas Research Network et al., 2008).

2.2.1 One-Parameter Logistic (1PL or Rasch) Model

The Rasch model (Rasch, 1960), the one-parameter logistic model, is one of the simplest and the most widely used item response model. In the 1PL model, the probability of a correct response is defined as,

$$\pi_{ik} = \pi(r_{ik} = 1|\theta_i, \beta_k) = \frac{e^{(\theta_i - \beta_k)}}{1 + e^{(\theta_i - \beta_k)}}. \quad (1)$$

Here θ_i denotes respond ability of a patient i to the therapy/clinical trial and the biomarker difficulty parameter β_k; r_{ik} is the random variable that represents i-th patient's response to k-th biomarker; $\forall i \in \{1, 2, \ldots, N\}$, $\forall k \in \{1, 2, \ldots, M\}$; N is the number of patients and M is the number of biomarkers.

In Fig. 1A, five biomarker characteristic curves (BCCs), i.e., π_{ik}, corresponding to Eq. (1) are plotted with different biomarker difficulty (β_k) values. Each BCC describes the biomarker-specific relationship between the patient's ability level to respond (θ) and the probability (π) of a correct response (i.e., 1) to the biomarker. The difficulty parameter β_k is the point on the ability or latent scale (θ) that corresponds to a probability of a correct response of 1/2. A biomarker is said to be easier when the probability of success is higher in comparison with another biomarker given the same ability level. In Fig. 1A, the plotted BCCs from the left to the right have increasing biomarker difficulty. It can be seen that, to maintain a probability of success of 1/2 on each biomarker, one should increase its ability level from − 2 to 2, starting from the left BCC to the rightmost BCC. An important feature of BCCs corresponding to the Rasch model is that the BCCs are parallel to one another. That means, for these biomarkers an increase in ability leads to the same increase in the probability of success. It is said that the biomarkers discriminate in the same way between success probabilities for related ability levels.

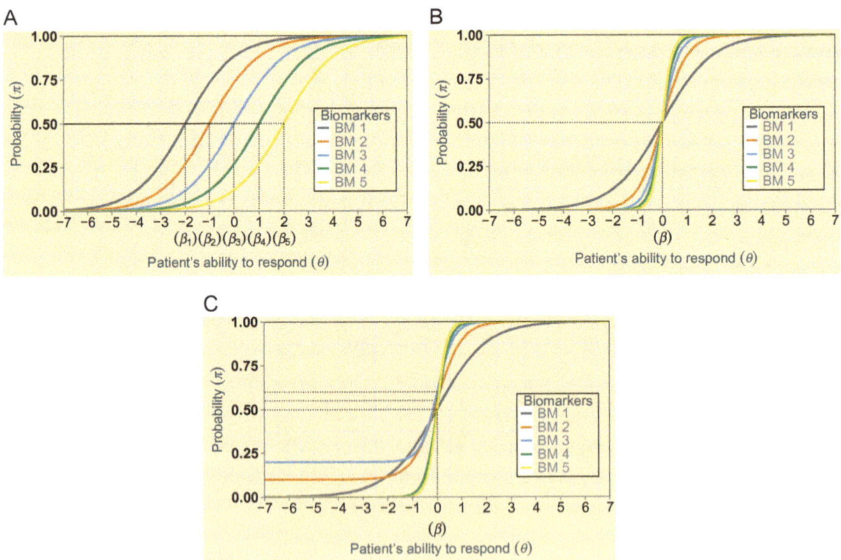

FIG. 1 Interpretation of parameters for dichotomous models. (A) Rasch. (B) 2PL. (C) 3PL.

2.2.2 Two-Parameter Logistic Model

In the two-parameter logistic model (2PL), a biomarker discrimination parameter (α) is added to the 1PL model, which leads to

$$\pi_{ik} = \pi(r_{ik} = 1 | \theta_i, \beta_k, \alpha_k) = \frac{e^{\alpha_k(\theta_i - \beta_k)}}{1 + e^{\alpha_k(\theta_i - \beta_k)}}. \tag{2}$$

As a result, the BCC has a slope parameter α_k and the biomarkers are no longer equally related to the ability parameter. In Fig. 1B, five BCCs for the two-parameter IRT model with the same difficulty parameter ($\beta_k = \beta = 0; \forall k$) are plotted. The slope of each BCC is characterized by the discrimination parameter α_k. The five BCCs have discrimination parameter values of 1, 2, 3, 4, and 5, respectively. The BCC with $\alpha_5 = 5$ has the steepest slope. The higher (lower) the discrimination parameter, the better (poorer) the biomarker is capable of discriminating between low and high ability levels of response to the therapy. Note that the biomarker's discrimination value may strongly related to the biomarker's difficulty value (van der Linden and Hambleton, 2013). A biomarker of high discrimination is only useful in the area of the biomarker's difficulty level that corresponds to a certain region of the ability scale. In Fig. 1B, it can be seen that the biomarker (BM_5) with the steepest slope is useful in the region between −1 and 1 of the ability scale, whereas the biomarker with the flattest BCC (BM_1) is useful in the region between −4 and 4. Note that the slope parameter $\alpha \in (0, \infty)$.

2.2.3 Three-Parameter Logistic Model

The two-parameter model can be extended to allow for guessing by introducing a nonzero lower asymptote (γ) for the BCC; that is,

$$\pi_{ik} = \pi(r_{ik} = 1 | \theta_i, \beta_k, \alpha_k, \gamma_k) = \gamma_k + (1 - \gamma_k)\frac{e^{\alpha_k(\theta_i - \beta_k)}}{1 + e^{\alpha_k(\theta_i - \beta_k)}}; \tag{3}$$

where γ_k is known as the guessing parameter of biomarker k. The probability of a correct response is given by a guessing parameter plus a second term representing the probability of a correct response depending on biomarker parameter values and the ability level of patient i. When $\gamma_k = 0$, the three-parameter model resembles the two-parameter model. For $\gamma_k > 0$, the interpretation of β_k is changed. In the three-parameter model, the proportion responding correctly at β_k equals $(1 + \gamma_k)/2$, and in the two-parameter model β_k is the value of θ_i at which a patient has a probability of 1/2 of responding correctly (Fig. 1B and C) to the biomarkers. The parameter γ can be accountable for noise in the biomarker responses.

In Fig. 1C, three BCCs of the three-parameter model are plotted with the same difficulty level but with five different levels of discrimination same as in the case of 2PL and guessing, low(0), medium (0.1), high (0.2), low (0), and low (0), respectively. The height of the lower asymptote is the guessing level of the biomarker and corresponds to the probability of success when guessing the response. It can be seen that for high-ability respondents, the effect of guessing on the success probability is very small since the five BCCs are almost identical at the higher end of the ability scale.

2.3 Polytomous Models

Occasionally, responses to clinical study and biomarkers are recorded in several mutually exclusive categories intended to represent increasing levels or amounts of the latent trait being measured. In this case, the observable categories (e.g., A, B, C, D) have an intended order $A < B < C < D$. These categories may be defined by response alternatives such as strongly not responded, not responded, responded, strongly responded, or they may be defined as observed stages toward the completion of a biomarker. For convenience, ordered categories of this type will be labeled 0, 1, 2, ..., m; where category 0 indicates the lowest level of response of a patient to a biomarker, and category m represents the highest.

The purpose in defining more than two levels of responses to a biomarker is to seek more information about a patient's location θ_i on the latent trait being measured than would be available from the dichotomous scoring of that biomarker. In estimating θ_i from observations based on a set of ordered response categories, it is essential that the intended order of these categories be taken into account. The first issue in developing a measurement model for ordered response categories for a biomarker of how this intended order ($0 < 1 < 2, \ldots, < m$) is to be incorporated into the model. In particular, what are the implications of this intended order for the model probabilities.

2.3.1 Graded Response Model

Samejima (1968) has proposed graded-response model that is built on the cumulative boundary measurement approach and is one of the major methodology in polytomous IRT models that is distinct from the Rasch-type models. Initially, two models have been developed to estimate patient's response ability to the therapy on a latent trait from their responses to a set of biomarkers with an ordered polytomous response format. The two models have a similar framework and differed only in that one employed a normal ogive to model responses and the second one used a logistic function to construct a BCC. In keeping other polytomous models, this chapter discusses the graded response model (GRM) that refers to the logistic function of Samejima's (1968) original two models. The GRM (logistic) is built on the 2PL model because the dichotomous model is used as the function to obtain the cumulative boundary functions defined as,

$$\pi^*_{ikl} = \pi^*(r_{ik} = l > 0 | \theta_i, \beta_{kl}, \alpha_k) = \frac{e^{\alpha_k(\theta_i - \beta_{kl})}}{1 + e^{\alpha_k(\theta_i - \beta_{kl})}}, \tag{4}$$

where α_k is the discrimination and β_{kl} is the boundary location parameter of cumulative probabilities of lth category of kth biomarker. Note that, in GRM, the β_{kl} are always in increasing order with respect to ordered categories. π^*_{ikl} is the probability of ith patient responding in or above category response l for kth biomarker. Note that, $\pi^*_{ik0} = 1$ and $\pi^*_{ikm_k} = \pi_{ikm_k}$; m_k is simultaneously the maximum score for kth biomarker. And, the probability of ith patient responding in the category l of kth biomarker is

$$\pi_{ikl} = \pi^*_{ikl} - \pi^*_{ik(l+1)}; \tag{5}$$

and $\pi_{ik0} = 1 - \pi^*_{ik1}$. The boundary locations (β) are defined at the inflection point of the Category Boundary Response Function (CBRF), that is, the point on the latent trait continuum where $\pi^* = 0.5$. This contrast with polytomous Rasch Models where boundary (difficulty) locations are defined at the crossing points of adjacent BCCs. In GRM, these BCC crossing points need not be where the CBRF has $\pi^* = 0.5$. Fig. 2 shows an example of CBRFs and thecorresponding BCCs of a biomarker for the parameters $\beta_{k\bullet} = \{-3, -1, 1, 3\}$ and $\alpha_k = 1$. One can observe that the coefficients of $\overrightarrow{\beta_{i\bullet}}$ positioned on latent trait exactly where $\pi^* = 0.5$ in CBRF (Fig. 2A), whereas these coefficients are not the crossing points of adjacent BCCs (for $\beta_{k1} = -3$ and $\beta_{k4} = 3$ in Fig. 2B).

2.3.2 *Partial Credit Model*

The partial credit and rating scale models (RSMs) are due to Rasch. These models are built on the requirement of model parameter separability, with the associated issues of sufficiency, specific objectivity, and fundamental measurement. Master (Eiji, 1992; Masters, 1982) formulated the partial credit model (PCM) by utilizing the Rasch dichotomous models. The PCM is developed by assuming that the probability of choosing the lth response over the $(l - 1)$th response is motivated by the dichotomous models. The probability of choosing the lth response from m_k possible responses of biomarker k boil downs to,

$$\begin{aligned}
\pi_{ikl} = \pi\left(r_{ik} = l > 0 | \theta_i, \overrightarrow{\beta_k}, \alpha_k\right) &= \frac{e^{\sum_{j=1}^{l}\left(\alpha_k\theta_i - \beta_{k(j-1)j}\right)}}{1 + \sum_{t=1}^{m_k} e^{\sum_{j=1}^{t}\left(\alpha_k\theta_i - \beta_{k(j-1)j}\right)}} \\
\pi\left(r_{ik} = 0 | \theta_i, \overrightarrow{\beta_k}, \alpha_k\right) &= \frac{1}{1 + \sum_{t=1}^{m_k} e^{\sum_{j=1}^{t}\left(\alpha_k\theta_i - \beta_{k(j-1)j}\right)}}.
\end{aligned} \tag{6}$$

Here r_{ik} represents the response of ith patient to kth biomarker, m_k is simultaneously the maximum score category and number of step difficulty

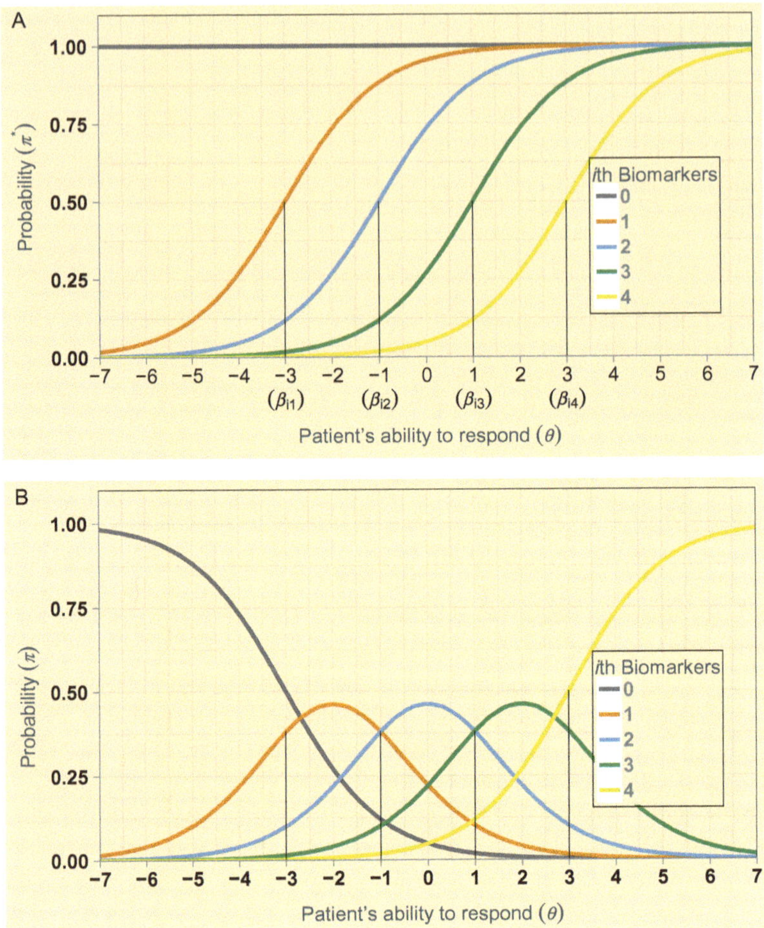

FIG. 2 Interpretation of graded response model. (A) CBRFs of GRM for a biomarker. (B) BCCs of GRM for a biomarker.

parameters for kth biomarker, $\beta_{k(j-1)j}$ is the difficulty parameter from (j − 1)th to jth category of kth biomarker; $\vec{\beta_k} = \left(\beta_{k01}, \beta_{k12}, \ldots, \beta_{k(m_k-1)m_k}\right)$; α_k, which is a discrimination parameter for kth biomarker; θ_i is the ith patient's ability to respond to the therapy; $\forall i = 1, 2, \ldots, N$; $\forall k = 1, 2, \ldots, M$.

According to Masters (1982), $\beta_{k(j-1)j}$ in Eq. (6) is called as biomarker's step parameter as $\pi_{ikl} = \pi_{ik(l-1)} \Longleftrightarrow \theta_i = \beta_{k(l-1)l}$. Note that, the difficulty parameters $\beta_{k(j-1)j}$ are not sequentially ordered within biomarker k as the parameters represent the relative magnitude of the adjacent probabilities π_{ikl} and $\pi_{ik(l-1)}$. Although the intersection points of BCCs of the PCM are easily interpretable, the peak points of these BCCs for the middle categories are not, because $\frac{\partial \pi_{ikl}}{\partial \theta} = 0 \Rightarrow \sum_{l=1}^{m_k} j.\pi_{ikl} = l$. Thus, the peak of BCCs (π_{ikl}) is depends on all other probabilities of the kth biomaker, $j = 1, 2, \ldots, m_k$.

For the dichotomous IRT models, the discriminating power is only depends on the biomarker discriminating parameters, i.e., $\alpha's$. However, for the polytomous item models, the discriminating power of each BCC depends on the combination of the slope and step difficulty parameters. Andrich (1978) distinguished these two kinds of a discriminating power and retained only the threshold discrimination in his Rasch family of the RSMs. The slope parameter (α) indicates the degree to which categorical response varies among biomarkers as ability level changes. In Fig. 3A and B, one can observe that the degree of steepness change from left to right when discrimination (α) changes from 1 to 2 with same difficulty steps. The concept of the biomarker discriminating power is closely related to the item reliability index in classical test theory.

Fig. 3A narrates the BCC for a biomarker with $\alpha_k = l$ and $\beta_{k01} = -3$, $\beta_{k12} = -1$, $\beta_{k23} = 1$, $\beta_{k34} = 3$. If β_{k12} and β_{k23} are brought closer together by changing to -0.5 and 0.5, respectively, then the probability of responding to the second category decreases, as illustrated in Fig. 3C. In other words, the range of the θ of patients who are more likely to respond to the second category than the other categories decreases from $(-1, 1)$ to $(-0.5, 0.5)$. If the slope parameter is changed from 1 to 2, as shown in Fig. 3B, the intersection points of all trace lines are left unchanged and the curves become steeper. The item discriminating power of these BCCs increases for all categorical responses.

In Fig. 3D, the second biomarker step parameter is made larger than the third biomarker step parameter ($\beta_{k12} = 0.5 > \beta_{k23} = -0.5$) then BCC of π_{ik2} drops. For the whole range of θ, the probabilities of categories except second are higher than the probability of the second categorical response. If one trace each of the BCCs, one can find that every patient who is more likely to respond to the second category than other categories is most likely to respond to the other category. Consequently, the marginal frequency of the second categorical response becomes quite small compared to the other response frequencies.

2.3.3 Rating Scale Model

The RSM is a special case of PCM (Andrich, 1978; Muraki, 1990). In RSM, the parameter $\beta_{k(j-1)j}$ is additively decomposed as $\beta_k + \tau_{(j-1)j}$ and the probability of responses is defined as,

$$
\begin{aligned}
\pi_{ikl} = \pi\left(r_{ik} = l > 0 \mid \theta_i, \beta_k, \vec{\tau}, \alpha_k\right) &= \frac{e^{\sum_{j=1}^{l}\left(\alpha_k\theta_i - \beta_k - \tau_{(j-1)j}\right)}}{1 + \sum_{t=1}^{m} e^{\sum_{j=1}^{t}\left(\alpha_k\theta_i - \beta_k - \tau_{(j-1)j}\right)}} \\
\pi\left(r_{ik} = 0 \mid \theta_i, \beta_k, \vec{\tau}, \alpha_k\right) &= \frac{1}{1 + \sum_{t=1}^{m} e^{\sum_{j=1}^{t}\left(\alpha_k\theta_i - \beta_k - \tau_{(j-1)j}\right)}}.
\end{aligned}
\tag{7}
$$

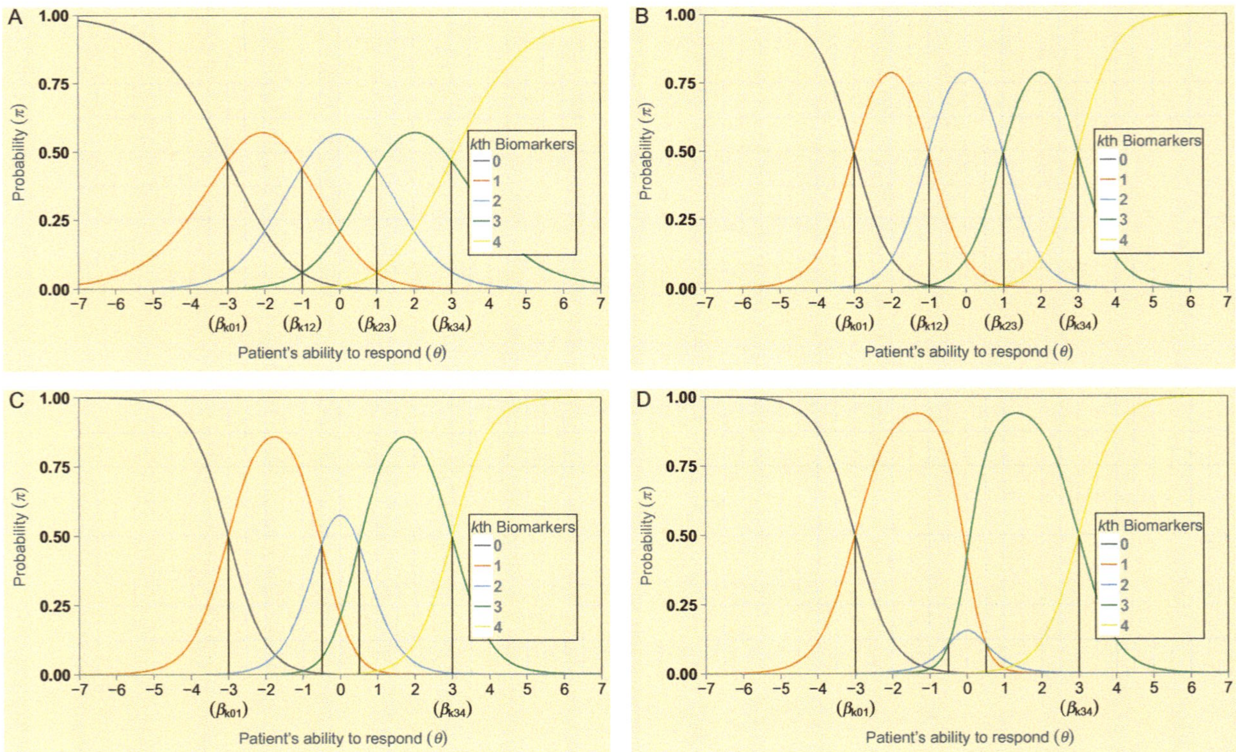

FIG. 3 Interpretation of partial credit model. (A) BCC of PCM when $\alpha = 1$. (B) BCC of PCM when $\alpha = 2$. (C) BCC of PCM with different difficulties. (D) BCC of PCM with different difficulties.

Here, m is simultaneously the maximum score and number of step difficulty parameters per biomarker, β_k is the location of biomarker k; $\tau_{(j-1)j}$ is the jth step rating scale to reach j from $(j - 1)$, constant across biomarkers and $\vec{\tau} = \left(\tau_{01}, \tau_{12}, \ldots, \tau_{(m-1)m}\right)$ is rating scale vector, which common to all biomarkers. The RSM can be applied to any situations where the PCM is fitted. Since the values of the biomarker step parameters $\beta_{k(j-1)j}$ are not necessarily ordered within item k, the rating parameters τ_j are not sequentially ordered for $j = 1$, 2, …, m. The parameter τ_j is interpreted as the relative difficulty of step j in comparing other steps within a biomarker. Note that, RSM forces equal discrimination around the location of biomarkers. Fig. 4A and B represents two BCCs with different location parameter values (-1 and 1 from left to right, respectively) and different discrimination. Also, note that the rating

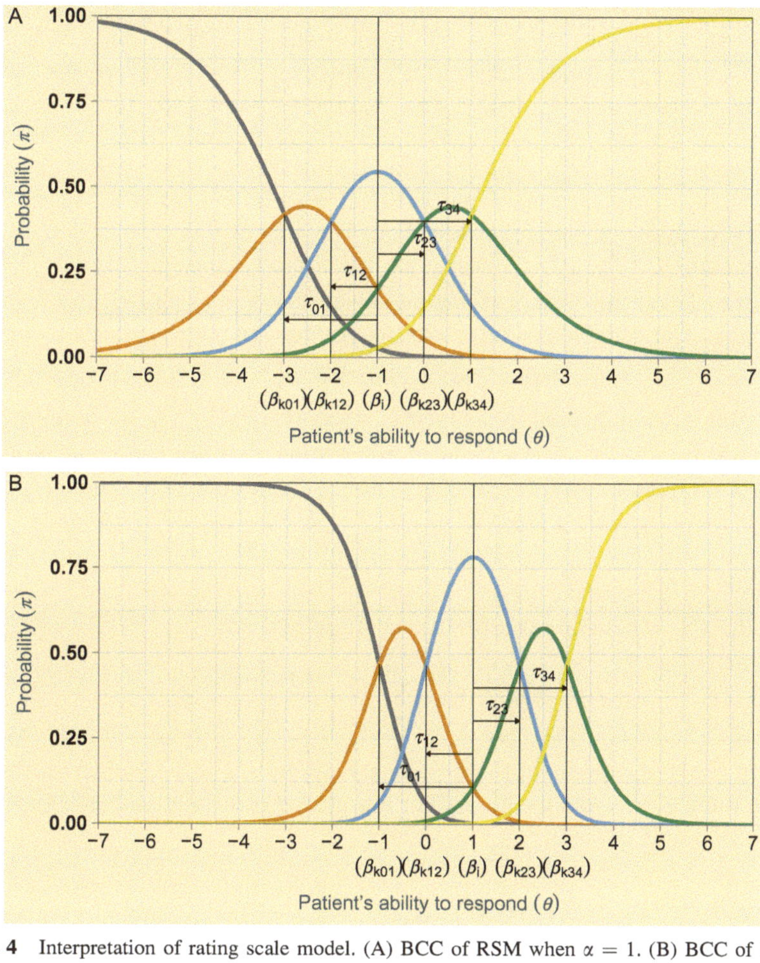

FIG. 4 Interpretation of rating scale model. (A) BCC of RSM when $\alpha = 1$. (B) BCC of RSM when $\alpha = 2$.

scale assumes equal discrimination for the two biomarkers in the figure. RSM's BCCs follow similar properties as of PCM's as it is a special case of it.

3 BIOMARKER INFORMATION FUNCTION

The information within each biomarker plays an important role in the IRT as it is possible to use this information to evaluate precisely how individual biomarkers included in the clinical study measure the level of a given latent trait θ (patient's ability to respond to the therapy). In case of biomarker selection, biomarker parameters represent the difficulty of biomarkers while patients parameters represent the response ability level to the therapy. The higher the patient's response ability relative to the difficulty of a biomarker (parameter β describes the degree of biomarker difficulty, the level of influence of the biomarker on the patient), the higher the probability of the correct response to that biomarker. The information about a specific value θ_i depends on the number and the qualities of the biomarkers used to evaluate that parameter. In the case of IRT models, each biomarker has its specific information function which provides information about how precisely, and what level of the analyzed latent trait, a given biomarker measures. This function is called "biomarker information function (BIF)" as each biomarker is an item in this specific problem of biomarker selection. The goal of this chapter is to utilize the properties of the BIFs to select the novel biomarkers.

Let us denote the random vector of biomarkers as $\vec{r} = (r_{\bullet 1}, r_{\bullet 2}, \ldots, r_{\bullet M})$; where $r_{\bullet j}$ is a random variable that represents the response of all the patients to jth biomarker. By the local independence assumption, $\pi\left(\vec{r}\,|\theta\right) = \prod_{k=1}^{M} \pi(r_{\bullet k}|\theta)$. Thus, the likelihood function of patients response ability θ is nothing but $L(\theta) = \prod_{k=1}^{M} \pi(r_{\bullet k}|\theta)$. The amount of information, $I(\theta)$ (total information function), is given by a set of M biomarkers for the fixed value of θ, is define as,

$$I(\theta) = E\left\{\frac{\partial logL}{\partial\theta}\right\}^2$$

And the above equation can be further simplified into

$$I(\theta) = \sum_{k=1}^{M} I_k(\theta) = \sum_{k=1}^{M}\left[-E\left(\frac{\partial^2 log\pi_{\bullet k\bullet}}{\partial\theta^2}\right)\right].$$

Here $I_k(\theta)$ be the amount of information given by a specific biomarker for the fixed value of θ, when the biomarker is a graded one (polytomous). And BIF, $I_k(\theta)$, is simplified to,

$$I_k(\theta) = \sum_{l=1}^{m_k} I_{\bullet kl}(\theta) = \sum_{l=1}^{m_k} \frac{\left(\pi'_{\bullet kl}\right)^2}{\pi_{\bullet kl}}. \tag{8}$$

The above definition of BIF contains first-order derivative of any IRT model's BCC ($\pi_{\bullet kl}$) and $\bullet$ denotes the continuous scale of θ. In case, where biomarkers are dichotomous in nature, then the BIF is

$$I_k(\theta) = \frac{\left(\pi'_{\bullet k}\right)^2}{\pi_{\bullet k}\pi^c_{\bullet k}}, \tag{9}$$

where $\pi_{\bullet k}$ is a BCC of any dichotomous model and $\pi^c_{\bullet k} = 1 - \pi_{\bullet k}$. With help of Eqs. (8) and (9), the BIFs of dichotomous and polytomous models are simplified as follows:

3.1 One Parameter Logistic

$$I_k(\theta) = \frac{e^{(\theta-\beta_k)}}{\left(1+e^{(\theta-\beta_k)}\right)^2}.$$

The maximum of information for all biomarkers in the 1PL model is 0.25 and, the BIF is unimodal and symmetric about β_k, where the maximum is attained.

3.2 Two Parameter Logistic

$$I_k(\theta) = \frac{\alpha_k^2 e^{\alpha_k(\theta-\beta_k)}}{\left(1+e^{\alpha_k(\theta-\beta_k)}\right)^2}.$$

Because when $\pi_{ik} = (1 - \pi_{ik}) = 0.5$, the product $\pi_{ik}(1 - \pi_{ik})$ is at its maximum and maximum of biomarker information is $\alpha_k^2 * 0.25$. When a biomarker is calibrated with the two-parameter model and $\alpha_k > 1.0$, then the biomarker contributes more information for estimating θ than when the 1PL model is used for calibration. Moreover, the area under the curve (AUC) $I_k(\theta)$ is equal to α_k (with the D scaling constant), i.e., $\int_{-\infty}^{+\infty} I_k(\theta)d\theta = D\alpha_k$ (Lord and Novick, 1968). Therefore, BIF provides "how quickly the BCC is changing" over the latent trait.

3.3 Three Parameter Logistic

$$I_k(\theta) = \alpha_k^2\left(\frac{\pi_{ik}-\gamma_k}{1-\gamma_k}\right)^2\left(\frac{1-\pi_{ik}}{\pi_{ik}}\right);$$

here π_{ik} is BCC of three-parameter logistic (3PL) model. Because guessing behavior reflects "noise," it may be intuited that effect of a nonzero γ_k will reduce the amount information available for locating patient on the θ latent continuum. For a given α_k and β_k, a biomarker provides more information for patient estimation when $\gamma_k = 0$ than when it is nonzero. Therefore, for the 3PL model the upper limit of $I_k(\theta)$ is given by the more restrictive 2PL model. In contrast to the 1PL and 2PL models with their maximum item information at β_k, the peak of 3PL biomarker information does not occur at β_k, but slightly above it. This offset from β_k is given by $\frac{ln\left(1/2+\sqrt{1+8\gamma_k/2}\right)}{\alpha_k}$. At this location the maximum biomarker information value (Lord, 2012) is $\frac{\alpha_k^2}{8(1-\gamma_k)^2}\left[1-20\gamma_k-8\gamma_k^2+(1+8\gamma_k)^{1.5}\right]$. The area under the BIF (Lord and Novick, 1968), in this case, is $\int_{-\infty}^{+\infty} I_k(\theta)d\theta = D\alpha_k \frac{\gamma_k ln(\gamma_k)+1-\gamma_k}{1-\gamma_k}$.

3.4 Graded Response Model

In polytomous models, it is possible to determine the amount of information provided by each response category with in a biomarker. From Eqs. (5) and (8),

$$I_k(\theta) = \sum_{l=1}^{m_k} \frac{\left(\pi'_{\bullet kl}\right)^2}{\pi_{\bullet kl}} = \sum_{l=1}^{m_k} \frac{\left[\left(\pi^*_{\bullet kl}\right)' - \left(\pi^*_{\bullet k(l+1)}\right)'\right]^2}{\pi^*_{\bullet kl} - \pi^*_{\bullet k(l+1)}}$$

As discussed before, $\pi^*_{\bullet kl}$ is, in effect, the 2PL model and the first-order derivative of $\pi^*_{\bullet kl}$ is $\alpha_k \pi^*_{\bullet kl}(1-\pi^*_{\bullet kl})$. It is discussed earlier that, for each biomarker, the boundary locations should be in increasing order with the ordered categories. BIF with more number of categories will have more information. In GRM, BIF's maximum location depends upon how the boundary locations are distributed, i.e., symmetrically or asymmetrically, around its mean (of boundary locations) and number of boundary locations, i.e., odd or even. If the number of categories is odd and symmetrical, then maximum location of BIF attains at the mean of boundary locations. For asymmetrical boundary locations, the maximum of BIF shifts toward the direction of dominant sign of the boundary locations. Fig. 5A contains two biomarkers with three categories and with biomarker locations (−2.0, 0.0, 2.0) and (−2.0, −1.0, 2.0). Clearly, the maximum of BIF of *BM* 1 attains at 0.0 (mean of boundary locations), whereas the maximum of *BM* 2's BIF is at −1.5 (shifted toward the direction of dominant sign). The maximum of BIFs occurs at different positions in case of even and symmetrical biomarker locations depends upon the distance between middle two locations. For example, in Fig. 5B, the maximum of BIF for *BM* 1 (with location parameters (−1.0, −0.5, 0.5, 1.0)) is at 0.0 since the distance between the middle two locations is small (i.e., 1.0); whereas for *BM* 2 (with location parameters

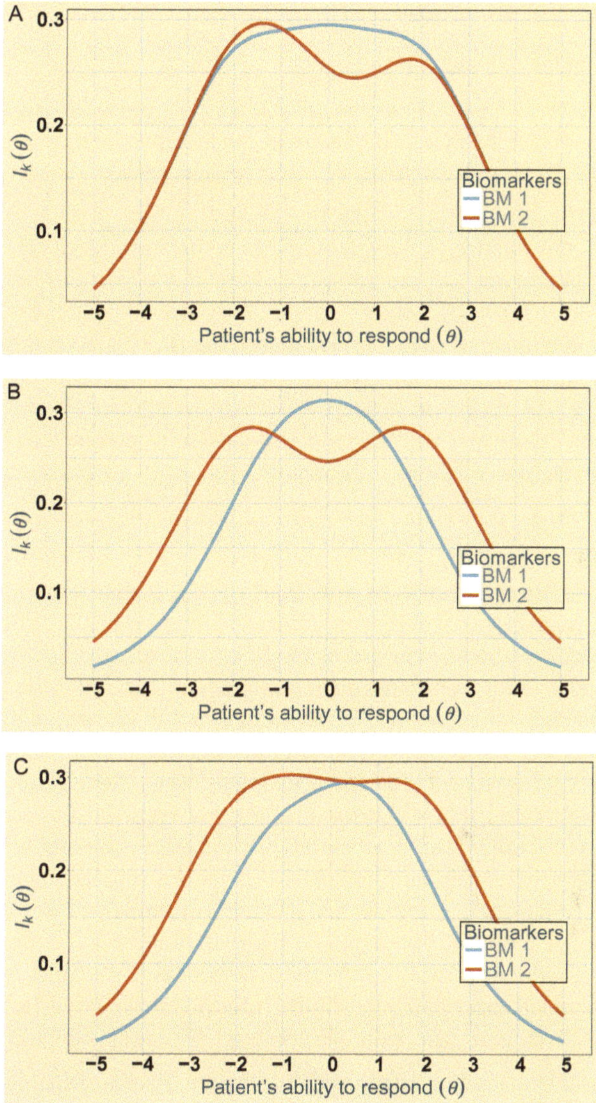

FIG. 5 BIFs of graded response model with $\alpha_k = 1; \forall k = 1, 2$. (A) Asymmetrical and odd number of boundary locations. (B) Symmetrical and even number of boundary locations. (C) Asymmetrical and even number of boundary locations.

(−2.0, −1.5, 1.5, 2.0)), the distance is larger (i.e., 3.0) and the BIF is bimodal with two different maximum locations. In case of asymmetrical and even biomarker locations, the BIFs are bimodal as shown in Fig. 5C; where the biomarker locations are (−2.0, −1.75, 0.5, 1.0) and (−2.0, −1, 1.5, 2.0) respectively.

3.5 Partial Credit Model

The BIF for PCM is simplified to,

$$I_k(\theta) = \sum_{l=1}^{m_k} l^2 * \pi_{\bullet kl} - \left(\sum_{l=1}^{m_k} l * \pi_{\bullet kl} \right)^2.$$

In case of PCM, biomarkers with the same number of categories provide the same total information, although biomarkers that have more categories yield more information across latent trait θ than do biomarkers with fewer categories (Dodd and Koch, 1987). Moreover, the maximum biomarker information occurs within the range of transition locations (Fig 6). However, the distribution of biomarker information is affected by the range of step difficulty parameters (β), the number of reversals of step difficulty parameters on the latent trait, and the distance between the reversed step difficulty parameters (Eiji, 1992). Fig. 6A depicts four different biomarkers with four graded categories with difficulty vectors $\vec{\beta_1} = (2.0, 1.0, -1.0, -2.0)$, $\vec{\beta_2} = (2.0, -1.0, 1.0, -2.0)$, $\vec{\beta_3} = (-2.0, 1.0, -1.0, 2.0)$, $\vec{\beta_4} = (-2.0, -1.0, 1.0, 2.0)$, respectively. It is clear that, the flattest information function was for *BM* 4, which had the step difficulty estimates in sequential order from easiest to most difficult and the most peaked information function was for *BM* 1 which had step difficulty parameters in reverse order from most difficult to easiest. For *BM* 2 the first and last step were reversed, while for *BM* 3 the second and third step parameters were reversed. This leaded to *BM*2 with more information than *BM*3 because the step that were reversed represented nonadjacent steps. Thus, these information function analyses revealed that when the distance between the first and last step is held constant, the more that the step parameters are out of sequential order and the greater the number of the steps are displaced, the more the information function will be. When both the number of steps that were not out of sequence and the degree of displacement of the steps were held constant, biomarkers with a small distance between the first and last step parameters yielded more information than biomarkers with a larger distance between the first and last step parameters. This relationship is demonstrated with *BM* 1 and 2 (step difficulties are $\vec{\beta_1} = (-2.0, 2.0, -1.0, 1.0)$, $\vec{\beta_2} = (-2.0, 1.0, 2.0, -1.0)$, respectively) in Fig. 6B. Both *BM* 1 and 2 had a single parameter that was out of order by two steps. *BM* 2, which had a distance between the first and last step parameters of 1.0, yielded a slightly more peaked information function than *BM* 2, which had a distance between the first and last step parameters of 3.0. Thus, the closer the proximity of the first and last step the more was the information function. It was also discovered that the size of the distance between the step parameters that were out of sequential order can override the effects of the proximity of the first and last step parameters on the peakedness of the information

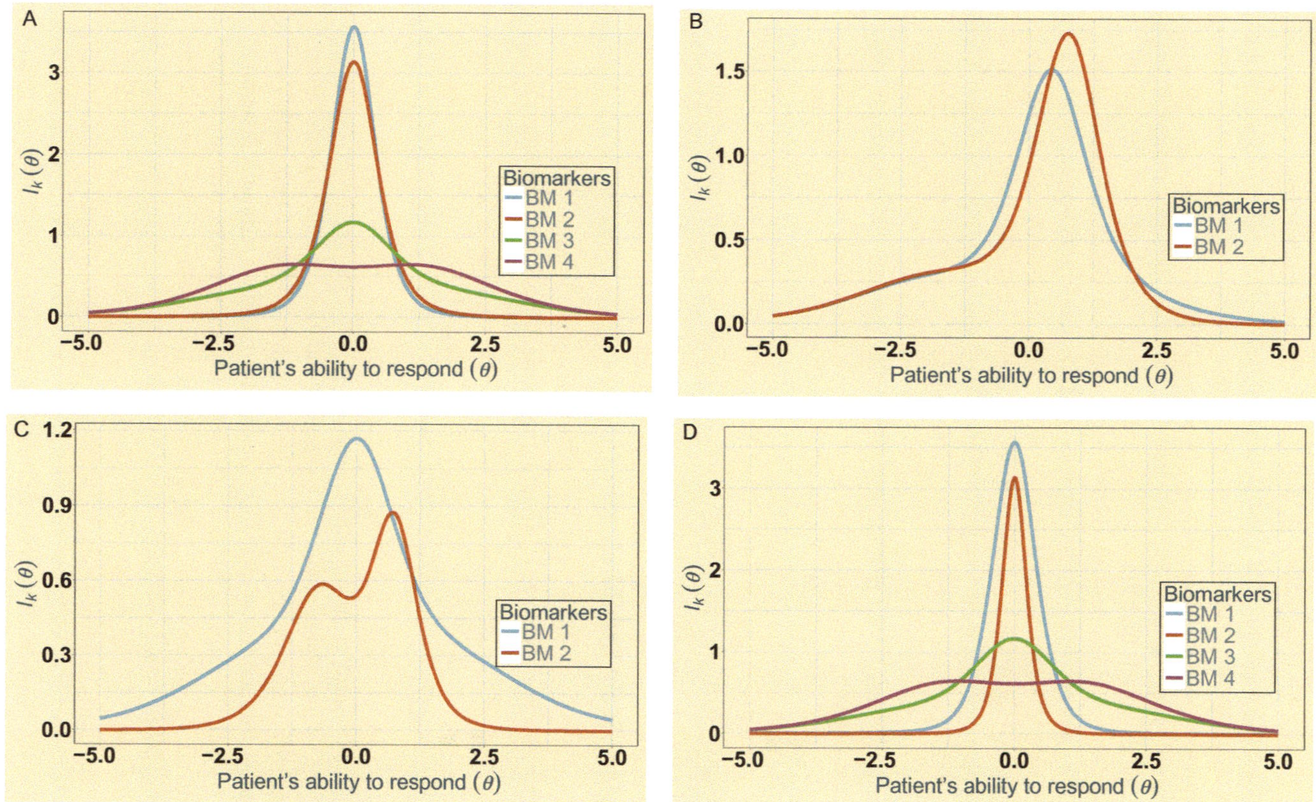

FIG. 6 BIF of partial credit model. (A) Sequential and out of sequential ordered step parameters with $\alpha_k = 1$. (B) Out of sequential order with different distance between the first and last step parameters. (C) Different distances between the first and last parameters and different magnitude of the distance between the reversed step parameters. (D) Sequential and out of sequential ordered step parameters with $\alpha_2 = 2$, $\alpha_k = 1$; $k \neq 2$.

function. Fig. 6C shows two biomarkers with step difficulty parameters $\vec{\beta_1} = (-2.0, 1.0, -1.0, 2.0)$, $\vec{\beta_2} = (-2.0, -1.0, 2.0, 1.0)$ and each biomarker having a reversal of adjacent step difficulty parameters. For *BM* 1, the second and third steps were reversed, and the distance between the first and last step parameters was 4.0. For *BM* 2, the third and fourth step parameters were reversed, and the distance between the first and last step parameters was 3.0. *BM* 1, which had the distance between the first and last step parameters, yielded a more peaked information function because the distance between the two reversed (i.e., 2.0) was greater than the distance between the two reversed steps for *BM* 2 (i.e., 3.0). The importance of α can be visualize from Fig. 6D where *BM* 2 is more steeper than *BM* 1 and discriminate between patients with smaller difference of ability between -1 and 1. The effect of the proximity of the first and last step parameters on the peakedness of the information function was dominated by the distance between the reversed step parameters. Thus, the greater the magnitude of the distance between the steps that were out of sequential order, the more peaked was the information function. In general, for a fixed distance between the first and last difficulty parameters, a biomarker that has more difficulty parameters that are in sequential order, and the greater the distance between difficulty parameters that are out of order, the more peaked is the BIF.

3.6 Rating Scale Model

The BIF for RSM is similar to that of PCM except m_k (number categories in biomarker k) is changed to m (similar rating scale for all the biomarkers),

$$I_k(\theta) = \sum_{l=1}^{m} l^2 * \pi_{\bullet kl} - \left(\sum_{l=1}^{m} l * \pi_{\bullet kl} \right)^2.$$

With respect to the RSM's information functions, the location of the maximum of the biomarker information function is affected by the symmetric nature of the rating scales about the biomarker location, the number of rating scales, and the range of the rating scales. In general, biomarkers with four rating scales produce more total information across the continuum than biomarkers with three rating scales. Therefore, only biomarkers with the same number of rating scales produce the same total information across the latent continuum (Dodd et al., 1995).

The location of the biomarker's maximum information is affected by whether the rating scales are symmetrically or asymmetrically distributed about the biomarker's location and whether there is an odd or even number of ratings scales. In case of an odd number of asymmetric thresholds, the location of the maximum information shifts away from the biomarker location in the direction of the dominant sign of the scales. The BIF of *BM* 1 in Fig. 7A

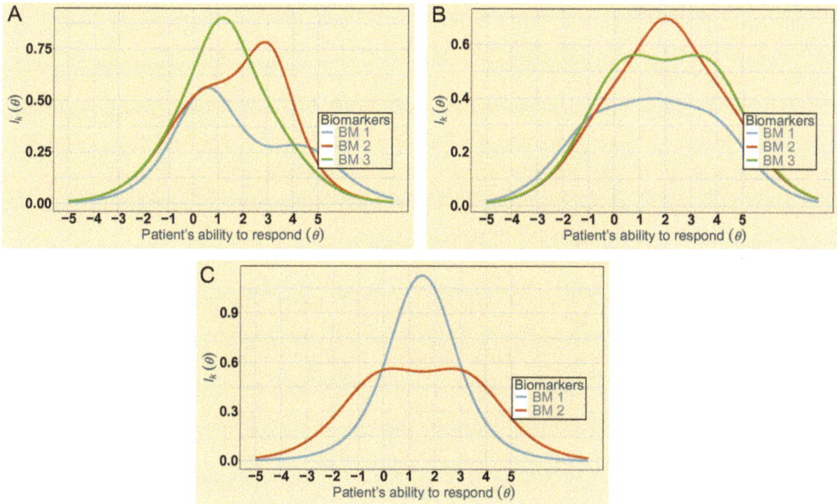

FIG. 7 BIF of rating scale model. (A) Asymmetrical rating scales with $\alpha_k = 1; \forall k$. (B) Symmetrical rating scales with $\alpha_k = 1; \forall k$. (C) Different span of rating scales.

is with location 2.0 and asymmetric rating scales (−2.0, − 1, 2.5). From the figure, the maximum information of the BIF is attained at 0.5 which is shift to left, i.e., toward dominant sign of the scales, of the 2.0. With an even number of scales, the degree of shift in the location of the biomarker's maximum information is directly related to the distance between adjacent thresholds. Specifically, the greater the distance between the two middle rating scales, the greater the shift (Dodd and Koch, 1994). The *BM* 2 and *BM* 3 in Fig. 7A with location 1.5 and asymmetric scales (−2.0, −0.5, 1.5, 1.75) and (−2.0, −0.5, 0, 1.75), respectively, depicts this kind of behavior. The shift is large for *BM* 2 as distance between the middle rating scales is 2 compare to that of in *BM* 3 with the distance 0.5. A different phenomena occurs with symmetric thresholds. With an odd number of rating scales that are symmetric about the biomarker's location, the peak of the biomarker information function occurs at β_k. For example in, Fig. 7B, the BIF of *BM* 1 with symmetric rating scales (−2.5, 0, 2.5) peaks at 1.5, which is also its location. For biomarkers with even rating scales that are symmetric about the biomarker's location, the distance between the middle two rating scales affects the shape as well as the location of the information function's maximum. If the range between the middle two rating scales is less than 2, then the peak of the information function occurs at the biomarker's location (*BM* 2 in Fig. 7B with location 2 and rating scales (−2.5, −0.5, 0.5, 2.5)). However, if the middle two rating scales are greater than (or equal) 2 apart, then the biomarker information may be asymmetrically bimodal (*BM* 3 in Fig. 7B with location 2 and rating scales (−2.5, −1, 1, 2.5)). In general, rating scales with thresholds that span a small θ range produce a more peaked information function than when the rating scales have a wider θ range (Ayala, 2013) (the BIF of two BMs with location 1.5 and rating scales (−1, −0.5, 0.5, 1) and (−2.5, −1, 1, 2.5), respectively, in Fig. 7C).

Obviously, one can ask, whether dichotomously or polytomously scored biomarkers have more information? Samejima (1968) has proved that more information will be given by a biomarker if a response category is added between two adjacent categories. That means, more information will be obtained, if more response categories are set up.

4 BAYESIAN FRAME WORK FOR IRT MODELS

As discussed in Section 2.1, IRT models make two key assumptions: (i) a change in the latent variable leading to a change in the probability of a specified patient response is completely described by the BCC (ii) responses to a pair of biomarkers are statistically independent when the underlying latent variable (the biomarkers measure a unidimensional latent variable) is held constant. The parameters of IRT model are treated as random variables in the Bayesian framework, and can provide with prior distributions that reflect the uncertainty about the true parameters values before observing the data. So far, the IRT models are discussed for the observed data those describe the data-generating process as a function of unknown parameters and are known as likelihood models. The likelihood models are the density of the data conditional on the model parameters. Thus, the Bayesian requires prior and likelihood specifications. The posterior distribution is constructed by combining prior knowledge with the information from the data. Bayesian inference conditioned on the data and inferences about parameters can be made directly from their posterior densities. The Bayesian inference is made with the help of Bayes' theorem, i.e., $\pi(\theta|r) \propto \pi(r|\theta)p(\theta)$.

In IRT Bayesian modeling, the patient and biomarker parameters are often of interest, and the objective of inferences is their posterior distributions. The posterior information is most often summarized by reporting the posterior mean and standard deviation. The prior densities of biomarker parameters play an important role in modeling the patient's responses. The joint posterior density of these parameters, with the help of Bayes' theorem, can be defined as,

$$\pi\left(\theta, \vec{\xi}\,|r\right) = \frac{\pi\left(r|\theta, \vec{\xi}\right)\pi\left(\theta, \vec{\xi}\right)}{\pi(r)}$$

$$= \frac{\pi\left(r|\theta, \vec{\xi}\right)\pi(\theta)\pi\left(\vec{\xi}\right)}{\pi(r)};$$

where $\vec{\xi}$ is vector of all the required biomarker parameters specific to a model and the prior densities of θ and $\vec{\xi}$ are assumed to be independent from each other. But, the biomarker parameters, $\vec{\xi}$, may not be independent from each other. For example, the difficulty and discrimination parameters are related as one increases other one may increase or decrease. In order to compute the posterior of the patient's parameter θ, the other parameters are eliminated by marginalizing as,

$$\pi(\theta|r) = \int \pi\left(\theta, \vec{\xi}\,|r\right) d\,\vec{\xi}$$
$$= \int \frac{\pi\left(r|\theta, \vec{\xi}\right)\pi(\theta)\pi\left(\vec{\xi}\right)}{\pi(r)} d\,\vec{\xi}$$

The marginal posterior of parameter θ is obtained by integrating out the biomarker parameters. Similarly, the marginal posterior of the biomarker's parameters is obtained by integrating out the patient's parameters. Multiple integrals need to be evaluated when the marginal posterior of a single component of, for example, the vector of patient's parameters is required. Evaluating the marginal posterior integrals remains a challenging task. Simulation-based methods, i.e., MCMC, can be used to generate samples from the marginal posteriors.

4.1 Markov Chain Monte Carlo Methods (MCMC)

Simulation of posterior distribution values (sampling) directly is often difficult and challenging. Usually, most of the problems have intractable marginal posteriors and huge dimensionality that lead to different challenges in obtaining directly simulated values from the posteriors. Markov chain Monte Carlo (MCMC) simulation methods build sequences that converge to the posterior distribution and sample averages are computed to estimate posterior expectations. In MCMC, the samples are drawn via a constructed Markov chain that has the target distribution as its stationary distribution. Gibbs sampling is one of the most popular MCMC model which is introduced by Chen et al. (2012) named after the physicist Josiah Willard Gibbs. Gibbs sampling is a random walk through parameter space. The walk starts at some arbitrary point, and at each point in the walk, the next step depends only on the current position, and on no previous positions. The Metropolis–Hastings (M-H) algorithm was developed by Metropolis et al. (1953) and generalized by Hastings (1970). The M-H algorithm generalizes the Gibbs sampler as it is sampling from a conditional distribution, from which it is difficult to sample directly. M-H is a two-step model, in which the first step, a candidate is drawn from a proposal density that may be chosen to approximate the desired posterior density and in second, this draw is accepted or rejected based on a specific acceptance criterion. The efficient model for generating Monte Carlo steps is Hamiltonian Monte Carlo (HMC) (Hoffman and Gelman, 2014). HMC uses a proposal distribution that changes depending upon the current position and the direction in which the posterior distribution increases depend upon gradient and wraps the proposal distribution toward the gradient. In recent years, the number of MCMC methods is growing rapidly, and for interesting and complete overview one can refer to Chen et al. (2012), Gilks et al. (1995), and Robert (2004). This chapter uses "RStan" to simulate Bayesian IRT models, which is the R interface to Stan for simulating MCMC using HMC. One can use Gibbs and H-M models with help of JAGS and BUGS (Kruschke, 2014).

4.2 Model Selection

Different Bayesian models can be evaluated and compared in several ways. The fit of Bayesian model to data can be assessed using posterior and prior predictive checks (when evaluating potential replications involving new parameter values), or, more generally, mixed checks for hierarchical models. The challenge is to estimate predictive model accuracy, correcting for the bias inherent in evaluating a model's predictions of the data that were used to fit it. A simple way to estimate out-of-sample prediction error is cross-validation, but researchers have always sought alternative measures, as cross-validation requires repeated model fits and can run into trouble with sparse data. For practical reasons alone, there remains a place for simple bias corrections such as Akaike information criterion, deviance information criterion, Bayesian information criterion, and WAIC (Watanabe-Akaike information criterion), and all these can be viewed as approximations to different versions of cross-validation (Gelman et al., 2014). This chapter utilizes WAIC for model comparison, which is a fully Bayesian approach for estimating the out-of-sample expectation, starting with the computed log pointwise posterior predictive density and then adding a correction for effective number of parameters to adjust for over-fitting. WAIC uses the variance of individual terms in the log predictive density summed over the N patients,

$$p_{waic} = \sum_{i=1}^{N} V_{posterior}\left(log\, \pi\left(r_{ik}|\theta, \vec{\xi}\right)\right).$$

It is more stable because it computes the variance separately for each data point and then averages; the averaging yields stability. For bias correction, WAIC uses,

$$\widehat{elppd} = lppd - p_{waic}.$$

Here *lppd* is log pointwise predictive density and is defined as $log\prod_{i=1}^{n}\pi_{posterior}(r_{ik})$; where $\pi_{posterior}$ is the posterior distribution, r_{ik} is the patient's response data, and N is number of patients. Compared to AIC and DIC, WAIC has the desirable property of averaging over the posterior distribution rather than conditioning on a point estimate.

5 EXPERIMENTAL STUDY

5.1 Model Selection

Let us assume that there are handful of dichotomous and polytomous Bayesian models that have been fit for a given clinical trial data. The WAIC is one of the best measure to select the best fit to the dataset. To analyze the power of WAIC, the discussed dichotomous and polytomous models have been fitted on simulated data from the models itself. Figs. 8 and 9 contain WAIC on y-axis and, different N and M on x-axis using different simulated data and fitted model. Even the difference between WAIC values look small (pictorially), but the y-axis scale is larger. So, smaller pictorial difference represent sufficient larger difference in WAIC values. Note that, smaller WAIC represents best fit. Fig. 8 describes the nature of the dichotomous models to different data where the data is generated from the models themselves. Clearly, one can observe that any given model is the best for the data generated by that model, e.g., Rasch is the best fit for Rasch generated data. As the number of patients and biomarkers are increasing, the differences of WAIC between different models also increasing. That means, as the likelihood is getting informative, the fit is getting refined. In case of polytomous models (Fig. 8), the nature of the three models are quite distinguishable over all the simulated data. The WAIC measures of GRM and PCM are of similar kind as the can model intracategory discrimination differently for each biomarker. Also, GRM's and PCM's WAIC values are smaller on GRM and PCM simulated data, respectively. Also, RSM is a poor fit for GRM and PCM as it can't model varying intradiscrimination of the biomarkers. Whereas, RSM outperforms others on RSM simulated data. Thus, WAIC is the best measure to select the model for the given data. Once best fitted model has been selected then biomarkers can selected using this model properties.

5.2 Biomarker Selection

5.2.1 Dichotomous Models

In this section, a simulated study for biomarker selection has been presented. For this purpose, simulated data ($N = 500$, $M = 10$) from 2PL and 3PL have been generated and all three dichotomous models are analyzed for biomarker selection. The difficulty and discrimination values for 2PL data are fixed to $\beta = \{-0.6, -0.3202, -0.3033, 0.0923, 0.3602, 0.6816, 0.6844, 0.8973, 0.9826, -2.4749\}$, and $\alpha = \{0.5, 1.0, 1.5, 2.0, 0.5, 1.0, 1.5, 2.0, 0.5, 1.0\}$, respectively. Similarly for 3PL model data, the difficulty, discrimination, and guessing parameter values are $\beta = \{-0.9008, -0.685, -0.5007, -0.4837, -0.4208, -0.0729, -0.0047, 0.6004, 0.6785, 1.7897\}$, $\alpha = \{0.5, 1.0, 1.5, 2.0, 0.5, 1.0, 1.5, 2.0, 0.5, 1.0\}$, and $\gamma = \{0.0, 0.3, 0.5, 0.7, 0.9, 0.0, 0.3, 0.5, 0.7, 0.9\}$, respectively. The AUC and the maximum (peak) of

A

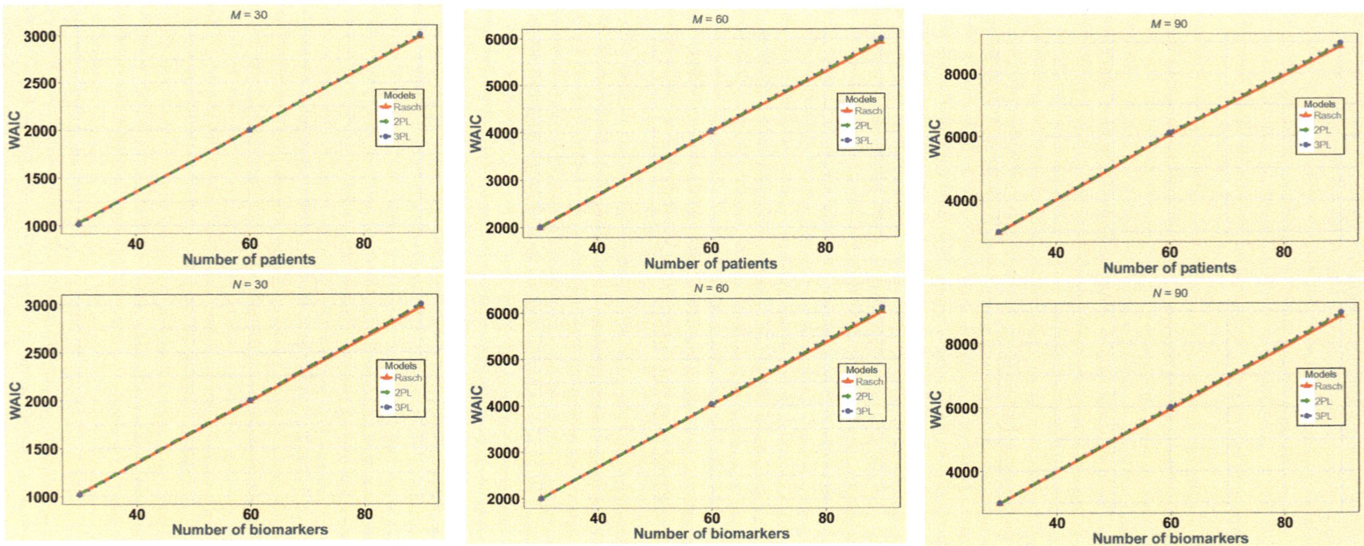

FIG. 8 WAIC measures for dichotomous models. (A) Rasch data. (B) 2PL data. (C) 3PL data.

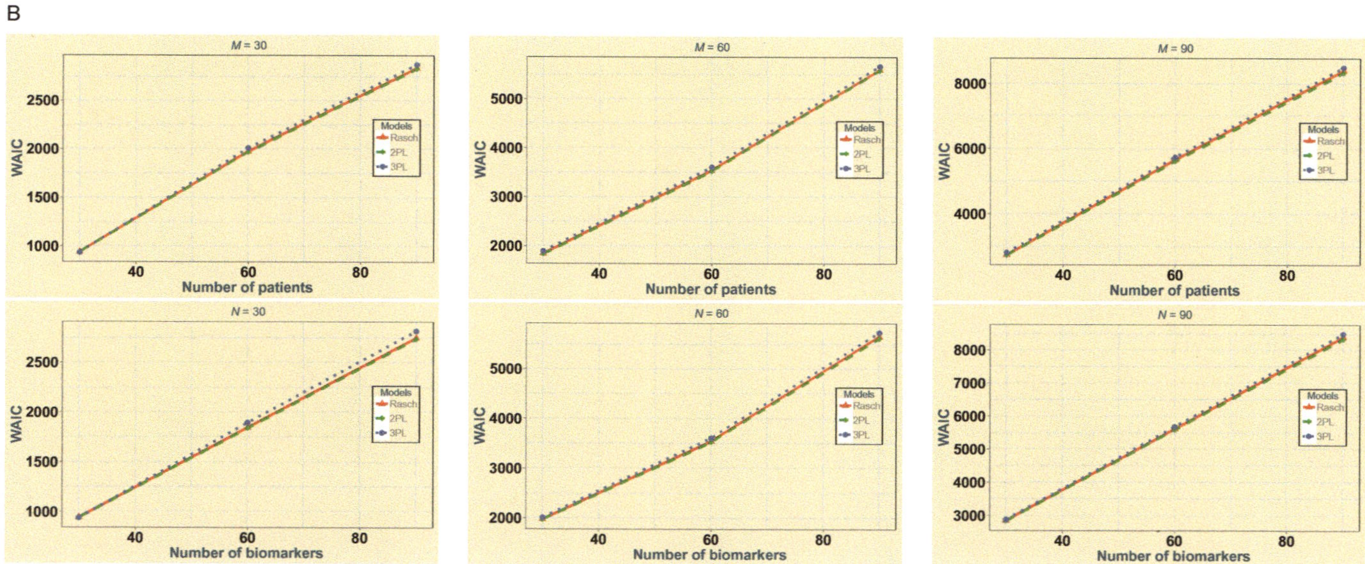
B
M = 30
M = 60
M = 90
N = 30
N = 60
N = 90
WAIC
Number of patients
Number of biomarkers
Models
Rasch
2PL
3PL

FIG. 8—Cont'd

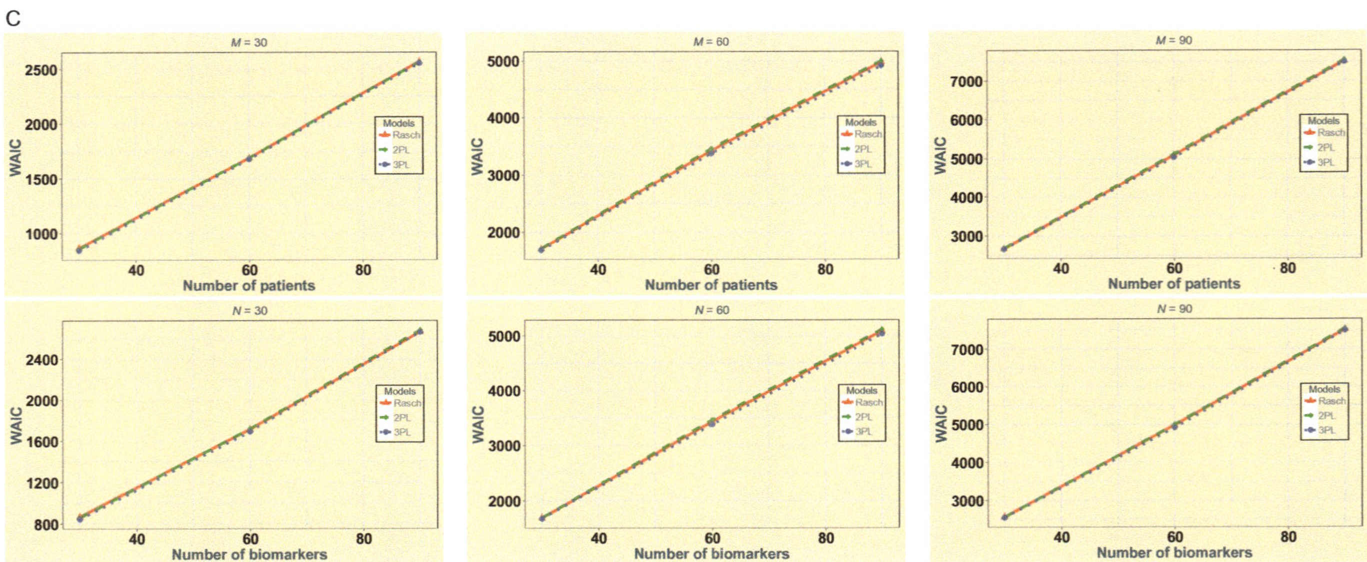
C
M = 30
M = 60
M = 90
N = 30
N = 60
N = 90
WAIC
Number of patients
Number of biomarkers
Models
Rasch
2PL
3PL
2500
2000
1500
1000
2400
1600
1200
800
5000
4000
3000
7000
6000
40
60
80

FIG. 8—Cont'd

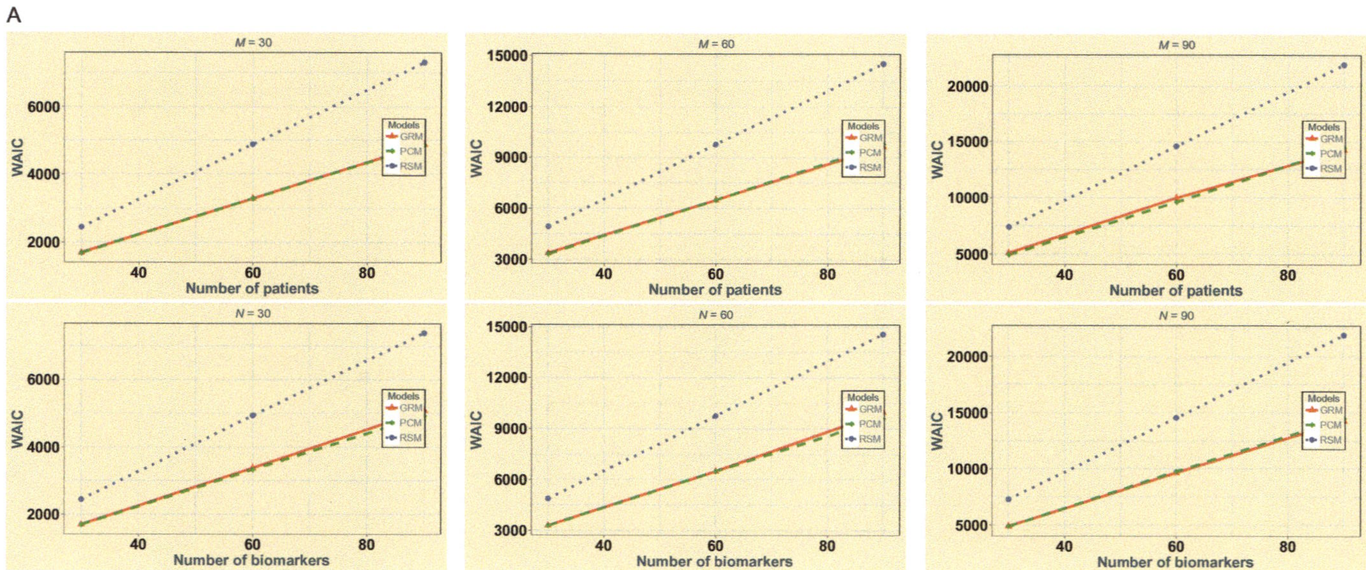

FIG. 9 WAIC measures for polytomous models. (A) GRM data. (B) PCM data. (C) RSM data.

B

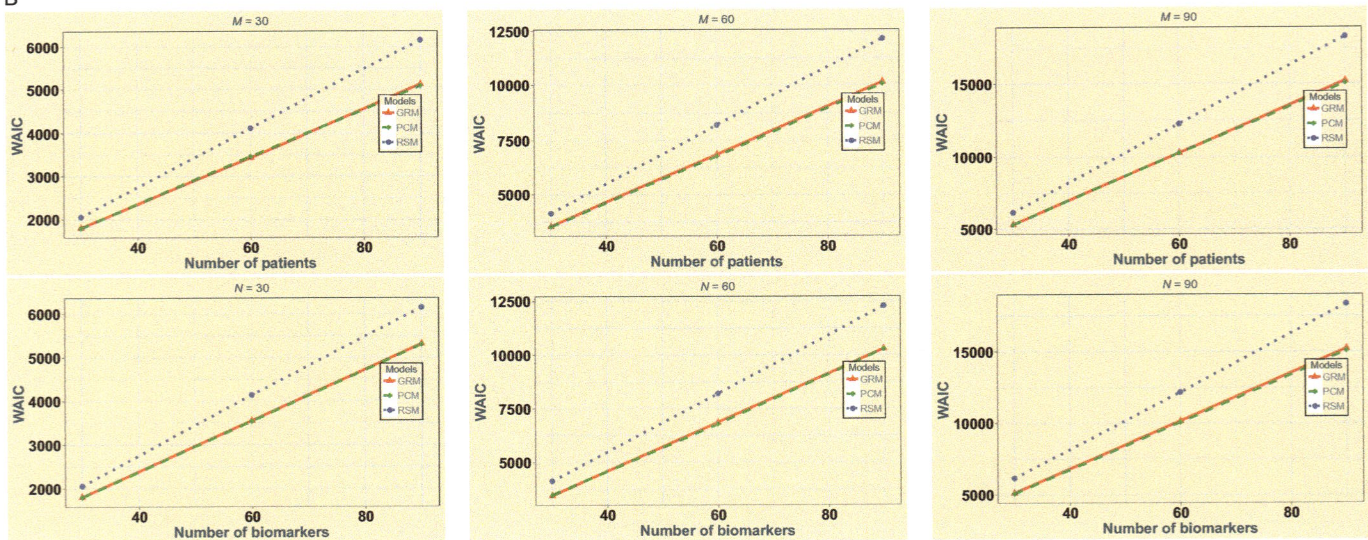

M = 30
6000
5000
4000
3000
2000
WAIC
40
60
80
Number of patients
Models
GRM
PCM
RSM
M = 60
12500
10000
7500
5000
M = 90
15000
10000
5000
N = 30
Number of biomarkers
N = 60
N = 90

FIG. 9—Cont'd

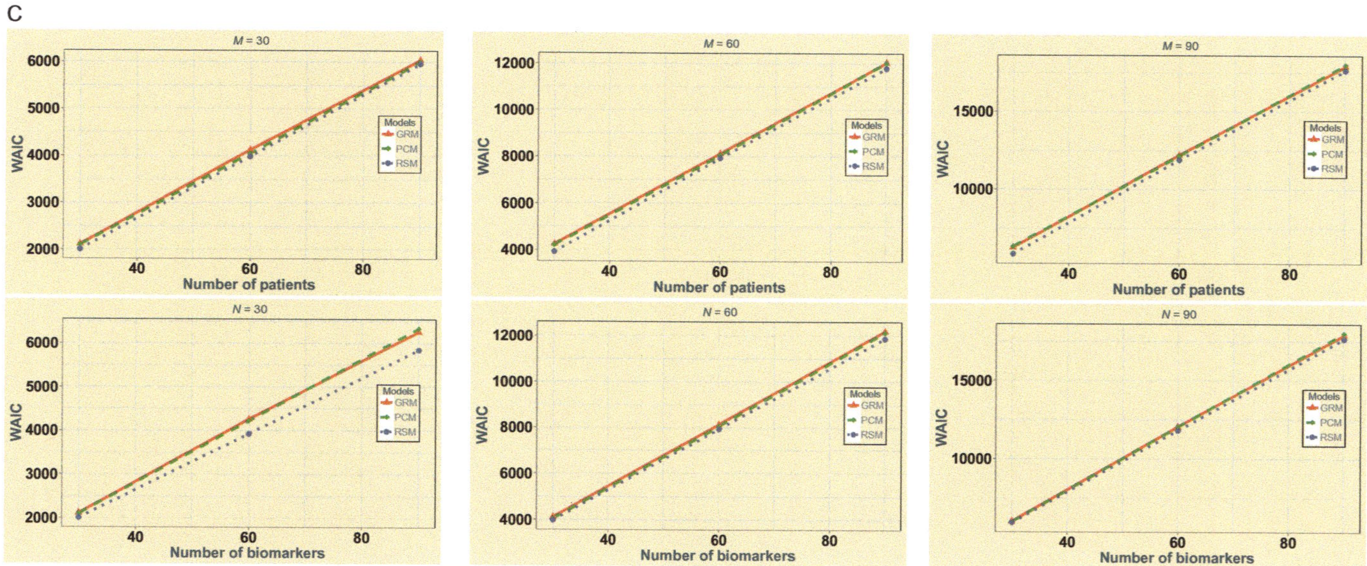
C
M = 30
M = 60
M = 90
N = 30
N = 60
N = 90
WAIC
Number of patients
Number of biomarkers
Models
GRM
PCM
RSM

FIG. 9—Cont'd

BIF are the measures that are considered to rank the biomarkers because AUC predicts the span of BIF and peak informs about the discrimination capability of different patients over the latent trait (θ). Note that, through the study all models are with hierarchical Bayesian framework. Clearly, as discussed earlier, 1PL model's BIF maximum values (peak) is 0.25 and AUC is 1.0 for both the 2PL and 3PL simulated data sets (Tables 1 and 2), respectively. Thus, 1PL model's BIF may not be useful to select informative biomarkers as all the BIFs depict similar information. One may rank the biomarkers according to the Bayesian estimated mean difficulty values, but here only maximum (peak) and AUC are used for ranking. Whereas, BIFs of 2PL and 3PL produce different information (in terms of AUC and peak) for different biomarkers depending on their properties. For 2PL simulated data (Table 1), both *2PL* and *3PL* models perform more or less similarly. One can observe that, in this example, the ranking of biomarkers matches with that of the discrimination of 2PL data as BIF involves the first-order derivative. This phenomena much more clear from Fig. 10B and C. Whereas, Fig. 10A proves that 1PL BIFs are noninformative in nature. According to 2PL and 3PL and using the results of Table 1 and Fig. 10, biomarker set {3, 4, 7, 8} enough informative to model patient's responses. Similar kind of observation can be made for 3PL simulated data (Table 2 and Fig. 11) and the biomarker set {3, 6, 7} may be informative. Note that, the 3PL BIF also considers guessing parameter and it leads to different rankings. The 3PL ranking is more useful in real life applications as the data may contains noise. One should note that, as mentioned before, first use WAIC for picking best model and then select informative biomarkers.

5.2.2 Polytomous Models

In this simulation study, GRM simulated polytomous data ($N = 500, M = 10$) has been used for biomarker selection. Unlike dichotomous, BIFs of polytomous models are difficulty to interpret because of their nature of multimodality. The boundary locations of biomarkers for GRM simulated data are presented in Table 3 and discrimination parameter values are similar to that of dichotomous in the previous section. Table 4 presents the ranking of the 10 biomarkers using AUC and maximum peak value of each BIF, and Table 5 contains WAIC measures of these models. From the latter table, it is clear that both GRM and PCM are the best fits for this data. Whereas, RSM is not a suitable fir for this data and is evident from the WAIC and BIF's (Fig. 12C). As discussed earlier, more categories bring more information. Therefore, the markers $BM2$, $BM4$, and $BM9$ should be more informative compare to the others. According the ranking of GRM (Table 4), these three biomarkers are ranked in top four for both AUC and peak ranking. Even, the biomarkers $BM1$ and BM_3 (that are having only two categories) are ranked poorly in GRM rankings. In case of PCM, these biomarkers are not consistent

TABLE 1 Rankings of Biomarkers for 2PL Simulated Data

Biomarker	Rasch				2PL				3PL			
	AUC	*Ranking*	*Peak*	*Ranking*	*AUC*	*Ranking*	*Peak*	*Ranking*	*AUC*	*Ranking*	*Peak*	*Ranking*
BM1	1	5	0.25	10	1.677	4	0.7031	4	1.151	4	0.3794	4
BM2	1	4	0.25	6	1.3121	8	0.4306	8	0.8256	8	0.1884	8
BM3	1	9	0.25	9	1.207	3	0.3645	3	0.7489	3	0.1744	3
BM4	1	6	0.25	3	1.0848	7	0.2946	7	0.6894	7	0.136	7
BM5	1	7	0.25	7	0.6801	6	0.1184	6	0.4122	6	0.0565	6
BM6	1	2	0.25	2	0.6339	2	0.115	10	0.3582	2	0.0474	2
BM7	1	3	0.25	8	0.6269	10	0.1039	2	0.1908	1	0.0352	10
BM8	1	8	0.25	1	0.3611	1	0.0406	1	0.1902	10	0.0201	1
BM9	1	1	0.25	4	0.2432	9	0.0227	9	0.1444	5	0.0138	5
BM10	0.99	10	0.25	5	0.2251	5	0.0203	5	0.1344	9	0.0118	9

TABLE 2 Rankings of Biomarkers for 3PL Simulated Data

Biomarker	Rasch				2PL				3PL			
	AUC	*Ranking*	*Peak*	*Ranking*	*AUC*	*Ranking*	*Peak*	*Ranking*	*AUC*	*Ranking*	*Peak*	*Ranking*
BM1	1	7	0.25	9	2.3	3	1.3225	3	0.3959	6	0.0839	7
BM2	1	8	0.25	2	1.7532	7	0.7685	7	0.3908	7	0.0826	3
BM3	1	1	0.25	7	1.6889	6	0.7132	6	0.3372	3	0.0741	6
BM4	1	2	0.25	8	1.4934	4	0.5581	4	0.3116	8	0.0636	8
BM5	1	6	0.25	5	1.4367	2	0.5163	2	0.2831	2	0.0595	2
BM6	0.99	3	0.25	4	1.3654	8	0.4665	8	0.1844	1	0.0258	1
BM7	0.99	9	0.25	10	1.0261	5	0.273	5	0.1265	4	0.0192	4
BM8	0.99	4	0.25	1	0.9916	1	0.2479	1	0.0318	9	0.0039	9
BM9	0.98	10	0.25	3	0.7602	10	0.1594	10	0.0189	10	0.0027	5
BM10	0.97	5	0.25	6	0.6095	9	0.1058	9	0.0188	5	0.0024	10

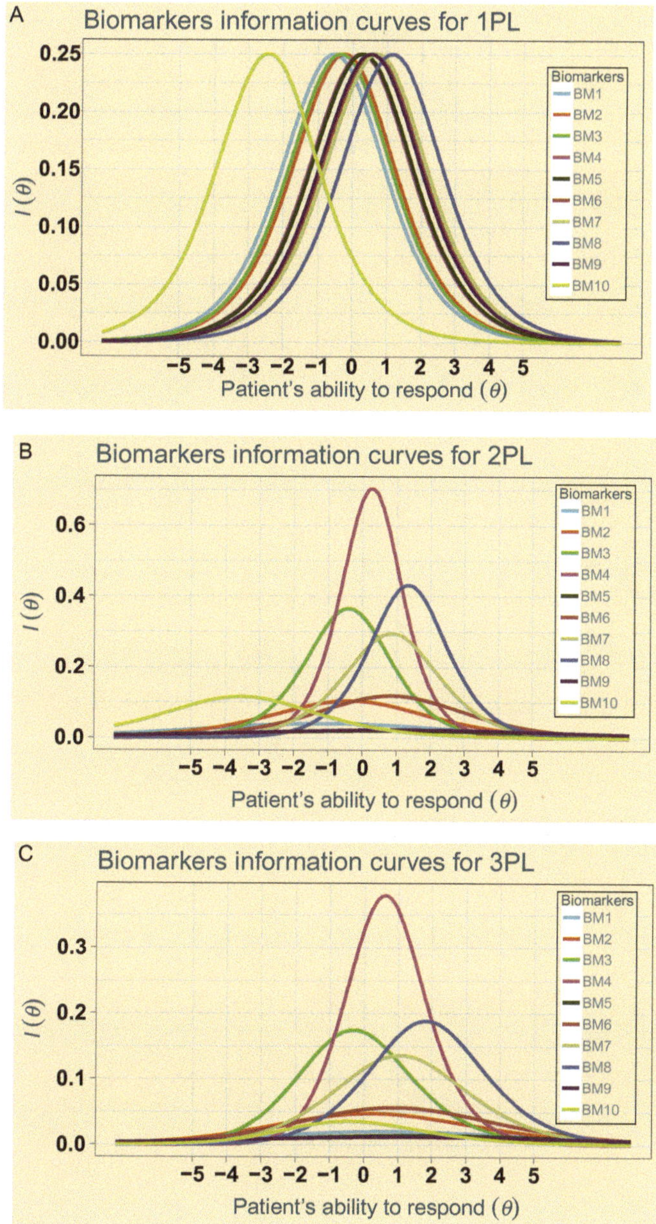

FIG. 10 BIF of dichotomous models for simulated 2PL data. (A) Rasch model's BIF. (B) 2PL model's BIF. (C) 3PL model's BIF.

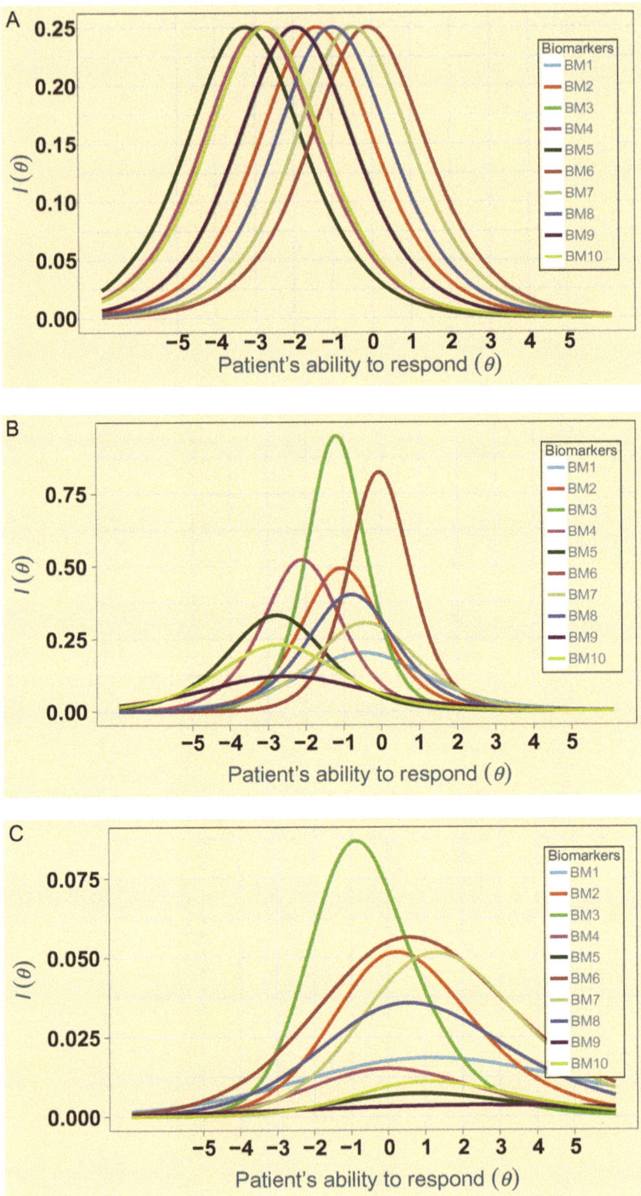

FIG. 11 BIF of dichotomous models for simulated 3PL data. (A) Rasch model's BIF. (B) 2PL model's BIF. (C) 3PL model's BIF.

TABLE 3 Boundary Location (BL) Values for GRM Simulated Data

BL	Categories			
	1	*2*	*3*	*4*
$\vec{\beta}_1$	−0.5316	0.7278		
$\vec{\beta}_2$	−0.6999	−0.4834	−0.1384	0.6462
$\vec{\beta}_3$	−0.2490	0.6637		
$\vec{\beta}_4$	−0.6614	−0.0826	0.2994	0.3077
$\vec{\beta}_5$	−0.4205	−0.3937	0.3379	
$\vec{\beta}_6$	−0.9752	−0.1347	0.0090	
$\vec{\beta}_7$	−0.4938	0.2592	0.4144	
$\vec{\beta}_8$	−0.6133	0.6426	0.8092	
$\vec{\beta}_9$	−0.6728	−0.2390	0.9281	0.9668
$\vec{\beta}_{10}$	−0.9959	−0.2863	−0.2002	

with number of categories and this is expected as data is not suitable for PCM. So, GRM rankings are more useful to select the informative biomarkers in this case.

5.3 Applications of IRT in Cancer Research

So far, importance and application of IRT models have been demonstrated with the help of simulated data. Most of the cases, this kind of simulations help to understand the properties of the models and their applicability to real life data. This section discusses interesting applications of IRT models to decode the biological underpinnings of cancer.

Li and Hong (2001) studied the relationship between tissue phenotypes (different types of cancers, responses of cell lines to different chemical compounds, time to tumor recurrence after treatment, etc.) and gene expression profiles on the whole-genome scale. They developed two different formulations of the polytomous Rasch model for analysis of microarray gene expression data. In the first variation, they treated samples as "students" and genes as "questions (items)." Their goal was to infer a single latent trait of a given patient sample using the expression level of a cluster of genes. Thus multiple clusters of genes are used to estimate different latent traits for a given sample.

TABLE 4 Rankings of Biomarkers for GRM Simulated Data

	GRM				PCM				RSM			
	Peak	*Ranking*	*AUC*	*Ranking*	*Peak*	*Ranking*	*AUC*	*Ranking*	*Peak*	*Ranking*	*AUC*	*Ranking*
BM1	3.1170	4	8.4073	2	2.0737	4	2.0086	7	0.8629	1	4.3832	1
BM2	2.8673	2	7.2818	9	1.2121	5	1.4748	5	0.8629	3	3.4648	3
BM3	2.4075	6	6.2936	4	1.1038	6	1.2679	9	0.8629	8	2.5075	7
BM4	2.2842	9	5.5883	6	1.0897	7	1.1655	4	0.8629	9	2.0229	10
BM5	1.8998	10	5.3690	10	0.9810	9	1.1583	1	0.8629	7	2.0100	5
BM6	1.3301	8	4.5842	8	0.9654	2	1.1139	8	0.8629	10	1.9343	8
BM7	1.0163	5	3.3457	5	0.8670	10	1.0448	6	0.8629	6	1.7687	6
BM8	0.6257	3	2.5982	7	0.7727	8	0.9908	2	0.8629	5	0.9886	9
BM9	0.5712	7	2.3002	3	0.5964	1	0.9907	3	0.8628	2	0.7701	2
BM10	0.4648	1	1.9583	1	0.5798	3	0.9045	10	0.8626	4	0.4109	4

TABLE 5 WAIC Results for GRM Simulated Data

	GRM	PCM	RSM
WAIC	2080.82	2083.33	2268.86

Subsequently, the inferred latent traits of a given sample are used as input to a regression framework to predict a sample's phenotype, such as response to a drug. The study revealed the usefulness of IRT in estimating the latent factors of patient samples using the clusters of genes with similar expression profiles followed by relating the inferred latent traits to observed phenotypes. In the second variation, the authors explore identification of differentially expressed genes between cancer subtypes. In this application, they treated genes as "student" and samples as 'questions.' Their method predicted different classes of acute leukemia and identified genes that are either over or underexpressed within cancers or cell lines of the same type as compared to cancers or cell lines of other types. This study reveals the flexibility of the IRT framework in identifying genes related to cancer subtypes while also enabling the prediction of cancer phenotypes by generating summaries of a large number of genes as a latent trait. There are few other applications of IRT in order to study epigenetic alterations in bladder cancer (Houseman et al., 2007), the methylation silencing of gene expression in bladder tumors (Scott and Ip, 2002), the relationship between latent variables and phenotypes (Kastrin, 2009).

Most of the above models considered single assays to decipher the biological phenotypes of diseased tissues. The estimates may be more powerful if multiple molecular measurements are integrated together. Tong and Coombes (2012) developed a framework that uses item response theory (integIRTy) to recognize altered genes through the integrative analysis of multiple high-throughput measurements including gene expression, DNA copy number and DNA methylation. In this study, the authors considered genes as students and patient samples as questions (items). They used IRT to addressed different challenges of integrative analysis: (i) the heterogeneity between multiomics measurements on the same gene can be resolved by the autoadjustable item parameters, making the estimation of an "overall latent trait of alteration" possible and meaningful, (ii) integration of different molecular measurements can be easily achieved by assembling a larger test with different multiomics supplying the items, (iii) the estimated latent trait of gene alteration as well as the item difficulty parameters are intuitive and can be used for further inference. integIRTy's main aim is to model the latent trait (ability) of each gene to be altered in patient samples using information from across all assay types and samples with the help of 2PL IRT model. If patient's responses got altered in large number of genes (low-item difficulty) then they have less-useful information than

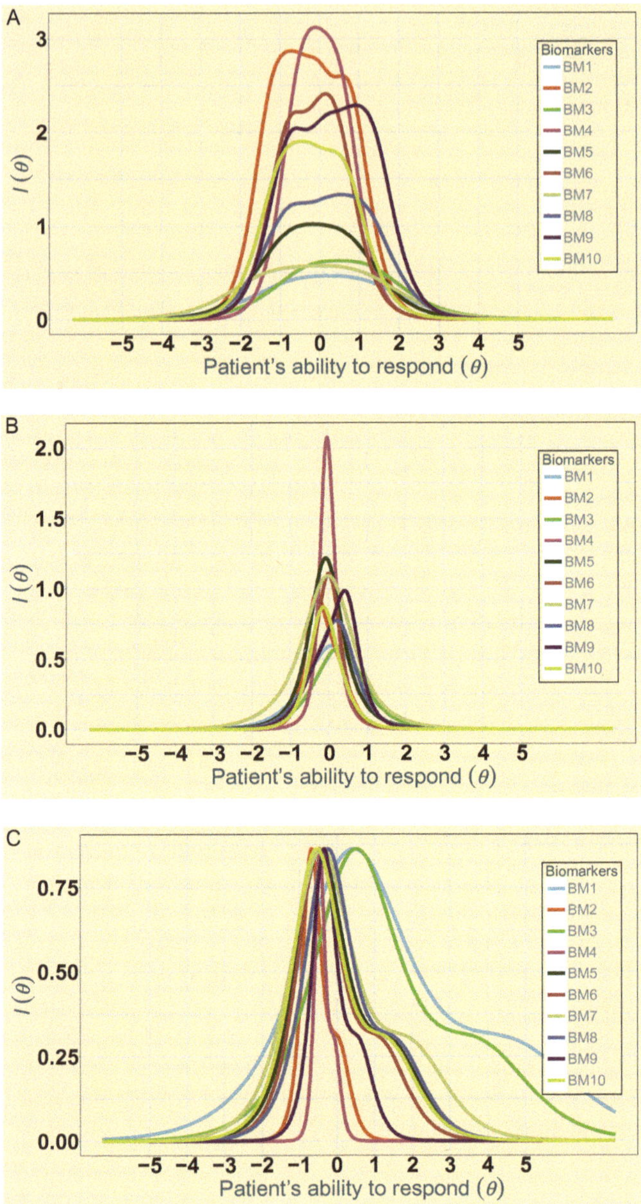

FIG. 12 BIFs of polytomous models for simulated PCM data. (A) GRM model's BIF. (B) PCM model's BIF. (C) RSM model's BIF.

patient's responses with fewer number of altered genes (high-item difficulty). Patients group with an idiosyncratic set of altered genes have low-item discrimination and weighted lesser than the patients group with similar patterns of altered genes tend to have a high-item discrimination. Novel genes are identified by estimating the latent ability of gene alteration from each assay and after integrating the multiple assays. The novelty of the integIRTy has been established by applying on two TCGA datasets, ovarian carcinoma (OV), glioblastoma (GBM) (Tong and Coombes, 2012), and one breast cancer (BRCA) dataset (Pollack et al., 2002). These data contains multiple high-throughput assays interrogating the same set of samples. The integIRTy has been compared with Students *t*-test and Wilcoxon rank-sum test and to the CNAmet approach for integrating different assays.

An important aspect of this model is their use of dichotomization of the molecular measurements in order to apply 2PL model. For gene expression data, they used the log ratio of intensities from tumor versus healthy reference tissue, followed by selection of a cutoff value for dichotomization. Similarly, the DNA methylation data was dichotomized in a biologically meaningful way by discretizing the beta values obtained from the arrays in to three categories denoted as unmethylated ($\beta < 0.25$), partly methylated ($0.25 \leq \beta \leq 0.75$), and highly methylated ($\beta > 0.75$). Following that, a gene's methylation status was coded as altered if the group membership for the given tumor sample differed from the normal reference. Finally, in case of DNA copy number, if the absolute adjusted *log*2 ratio of a gene in a particular sample was larger than a fixed threshold, it is assigned to 1 or else to 0.

The alteration patterns of genes across multiple assays are then evaluated using IntegIRTy. The comparison of alteration from integrated OV, BRCA and GBM datasets and individual assays is presented in Fig. 13. One can observe that there are only few genes with severe alterations in all assays. Secondly, there is not much correlation between assays in any of the datasets. Notably, the individual assays contributed differently to the integrated data. In case of both OV and GBM, the DNA methylation data appears to be most influenced by the integrated latent trait (fourth row and second column panel of Fig. 13A and C), followed by CN (fourth row and third column panel of Fig. 13A and C) and expression (fourth row and first column panel of Fig. 13A and C), respectively. In the BRCA dataset, gene expression has a larger impact on the integrated latent trait than copy number (third row and first column panel of Fig. 13B). Therefore, it is clear that the correlation between the estimated latent traits derived from integrated assays and individual assays is well behaved. Thus, the integrated data represents the overall propensity of a gene to be altered in at least one of the assays. And, finally, the novel altered genes in different cancers are discovered only when all the assays (expression, methylation, and copy number) are integrated to estimate the latent trait (Tong and Coombes, 2012). Thus, the ability to identify altered genes that cannot be identified by individual assays proved the advantages of data integration as well as the effectiveness of IRT (2PL) model.

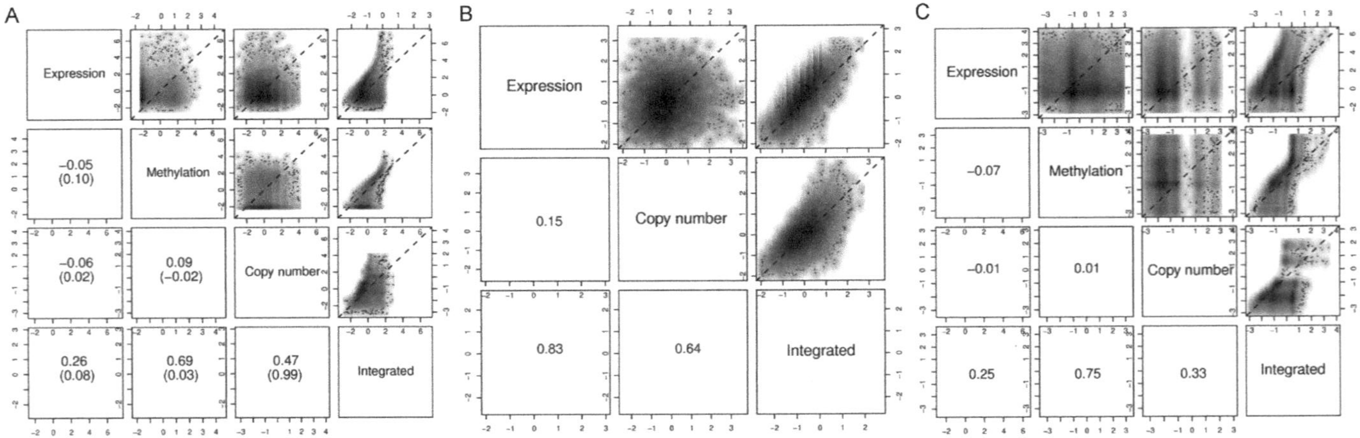

FIG. 13 Upper panels represent pair-wise smoothed density of estimated latent trait for alteration and lower panels represent Spearman rank correlations among different individual and integrated upper panels. *Reproduced from Tong, P., Coombes, K. R., 2012. integIRTy: a method to identify genes altered in cancer by accounting for multiple mechanisms of regulation using item response theory. Bioinformatics 28 (22), 2861–2869.*

We now review another novel application of IRT in oncology clinical trials where patient reported outcomes (PROs), such as symptom scales or more broad-based health-related quality-of-life measures, play an important role. PROs are helpful to evaluate cancer treatments, as well as for supportive and palliative oncology care. Garcia et al. (2007) have developed the Patient Reported Outcomes Measurement Information System (PROMIS) network, part of the National Institutes of Health Roadmap Initiative, which aims to improve appreciably how PROs are selected and assessed in clinical research, including clinical trials. They developed item banks to assess PROs and an item bank is a group of questions, the measurement properties to carefully calibrate an operational definition of a concept (e.g., pain, fatigue, anxiety) and accurately assess the entire continuum (e.g., severity or frequency) of that concept. IRT modeling has been adopted by PROMIS for formation and application of item banks for use in clinical oncology research. Moreover, IRT modeling has been used to specify how each item should be used to measure the concept of interest (e.g., pain, fatigue, anxiety). As a part of PROMIS, Quinn et al. (2014) used IRT to incorporate new potential items (22) in the study of pediatric patients with cancer, chronic kidney disease, obesity, rehabilitation needs, rheumatic disease, and sickle cell disease. For anger, anxiety, and depressive symptoms, they successfully added new items (e.g., "I was angry when things didn't go my way," "It was hard for me to be away from home because I had pain," "I could pick up coins from a table," etc.) that may render those item banks more robust and flexible. And they demonstrated the usefulness of IRT for continually enhancing health measurement while maintaining a consistent underlying measurement system. Therefore, IRT can be used to add items to a bank (without altering the original item parameters), using concurrent calibration of the (potential) new items with the original item response data that was used as the basis of scale construction.

6 FUTURE DEVELOPMENTS OF IRT IN BIOLOGICAL MODELLING

In literature, the utility of IRT is limited to dichotomous models in cancer research. As discussed earlier, polytomous models are more powerful as compared to dichotomous ones, to model different biological properties of biomarkers and patient's respond ability to the therapy. Further, multiomics profiles of biological samples are more naturally modeled as polytomous variables. There are multiple challenges in discretization such as number of categories for each biomarkers, proper threshold selection, biologically meaningful ordered categories, etc. Furthermore, in many clinical studies/trials, the patient's final responses to drugs are themselves polytomous. For example, in breast cancer, the extent of a patient's response to neoadjuvant chemotherapy is characterized using a measure called residual cancer burden (RCB), which can take one of four categories, RCB 0–3 (Symmans et al., 2007). Therefore, IRT models that can combine a polytomous patient response to therapy with polytomous biomarkers measurements would be very helpful in biomarker identification using clinical trial datasets. Current, IRT literature did not explore this direction to develop models and BIFs that fit this kind of data. Also, in IRT applications to clinical trails, the BIFs of polytomous models are multimodal (this is assessed from the experimental study also) in nature and they may be hard to interpret. And, the measures such as AUC and maximum peak of BIFs may not be the suitable ones to select informative biomarkers. Therefore, addressing these challenges would broaden the applicability and utility of IRT frameworks in cancer biomarker identification.

7 STAN CODES FOR IRT MODELS

This section provides stan models for the discussed IRT models. These models help to simulate MCMC for posteriors. Most of the stan models can be found in https://education-stan.github.io/ and that those not available are presented here.

Three-Parameter Logistic

```
// Stan Model for "Three Parameter Logistic Model"
data {
   int<lower=1> M;                          // Number of Biomarkers
   int<lower=1> N;                          // Number of Patients
   int<lower=1> T;                          // Total Number of
                                               observations (=I*J)
   int<lower=1, upper=M> Bio_Index[T];      // Indexing for
                                               Biomarkers
   int<lower=1, upper=N> Pat_Index[T];      // Indexing for Patients
   int<lower=0, upper=1> r[T];              // Patients Responses
}
parameters {
   vector[M] beta;                          // Difficulty Parameters
   vector[N] theta;                         // Patient's Parameters
   real<lower=0> sigma;                     // Standard Deviation of
                                               Patient's Parameter
   vector<lower = 0>[M] alpha;              // Discrimination of
                                               each Biomarkers
   vector<lower = 0, upper=1>[M] gamma;     // Guessing Parameter
}
```

```
model{
   alpha ~ lognormal(1,1);                 // Prior Distribution
                                              for 'Discrimination'
   gamma ~ normal(0.5,1);                  // Prior Distribution
                                              for 'Guessing'
   target += normal_lpdf(beta | 0, 3);     // Prior Distribution for
                                              'Difficulty'
   theta ~ normal(0, sigma);               // Prior Distribution
                                              for Patient's
                                              Parameter
   sigma ~ exponential(.1);                // Hyper Prior
                                              Distribution for
                                              Sigma
   r ~ bernoulli( gamma[Bio_Index] + (1 - gamma[Bio_Index]) .*
      inv_logit(alpha[Bio_Index].*theta[Pat_Index] - beta
      [Bio_Index]) );
   // Three Parameter Logistic Sampling
}
// Generating response estimations to compute "WAIC"
generated quantities{
   vector[T] r_generation;
    for (i in 1:T)
     r_generation[i] = bernoulli_lpmf( r[i] | gamma[Bio_Index[i]] +
        (1-gamma[Bio_Index[i]]) * inv_logit(alpha[Bio_Index[i]] *
        theta[Pat_Index[i]] - beta[Bio_Index[i]]) );
}
```

Graded Response

```
// Stan Model for "Graded Response Model"
functions {
   // Function to Simulate Graded Response Model Categories
   real grm(int y, real theta, vector beta) {
     vector[rows(beta) + 1] cummilative_prob;
     vector[rows(beta) + 1] probs;
     cummilative_prob = append_row(rep_vector(1.0, 1), inv_logit
         (theta - beta) );
     for(i in 1:rows(beta))
       probs[i] = cummilative_prob[i] - cummilative_prob[i+1];
     probs[rows(beta)+1] = cummilative_prob[rows(beta)+1];
     return categorical_lpmf(y + 1 | probs);
   }
```

```
    // Function to Simulate Graded Response Model BCC
    vector lgrm(real theta, vector beta) {
      vector[rows(beta) + 1] cummilative_prob;
      vector[rows(beta) + 1] probs;
      cummilative_prob = append_row(rep_vector(1.0, 1), inv_logit
          (theta - beta));
      for(i in 1:rows(beta))
         probs[i] = cummilative_prob[i] - cummilative_prob[i+1];
      probs[rows(beta)+1] = cummilative_prob[rows(beta)+1];
      return probs;
    }
}
data {
   int<lower=1> M;                               // Number of Biomarkers
   int<lower=1> N;                               // Number of Patients
   int<lower=1> T;                               // Total Number of
                                                    observations (=I*J)
   int<lower=1, upper=M> Bio_Index[T];  // Indexing for Biomarkers
   int<lower=1, upper=N> Pat_Index[T];  // Indexing for Patients
   int<lower=0> r[T];                            // Patients Responses
}
transformed data {
   int m[M];  // parameters per item
   int pos[M];  // first position in beta vector for item
   m = rep_array(0, M);
   for(i in 1:T)
      if(r[i] > m[Bio_Index[i]])
         m[Bio_Index[i]] = r[i];
   pos[1] = 1;
   for(i in 2:M)
      pos[i] = m[i-1] + pos[i-1];
}
parameters {
   vector<lower=0>[M] alpha;                     // Discrimination
                                                    Parameter
   vector[sum(m)-1] beta_free;                   // Difficulty
                                                    Parameter
   vector[N] theta;                              // Patient's Parameter
   real<lower=0>    sigma;                       // Hyper    Parameter    for
    Patient's
       Parameter
}
transformed parameters {
   vector[sum(m)] beta;
   vector[sum(m)] beta_sorted;
   beta[1:(sum(m)-1)] = beta_free;
   beta[sum(m)] = -1*sum(beta_free);
```

```
    // Sorted Order Condition for Boundary Locations
    for(i in 1:M)
      beta_sorted[pos[i]:(pos[i]+m[i]-1)] = sort_asc(segment
          (beta, pos[i], m[i]));
}
model {
    alpha ~ lognormal(1, 1);               // Prior Distribution for
                                              Discrimination
    target += normal_lpdf(beta | 0, 3);    // Prior Distribution for
                                              Difficulty
    theta ~ normal(0, sigma);              // Prior Distribution for
                                              Patient's

        Parameter
    sigma ~ exponential(.1);               // Hyper Prior
                                              Distribution for Sigma
    // GRM Sampling
    for (i in 1:T)
      target += grm(r[i], theta[Pat_Index[i]]*alpha[Bio_Index[i]],
        segment(beta_sorted, pos[Bio_Index[i]], m[Bio_Index[i]]));
}
// Generating response estimations to compute "WAIC"
generated quantities {
    vector[T] r_generation;
    for (i in 1:T)
      r_generation[i] = categorical_lpmf(r[i]+1|lgrm(theta
          [Pat_Index[i]]*alpha[Bio_Index[i]],segment(beta_sorted,
          pos[Bio_Index[i]], m[Bio_Index[i]])));
}
```

REFERENCES

Aggarwal, B.B., Bhardwaj, A., Aggarwal, R.S., Seeram, N.P., Shishodia, S., Takada, Y., 2004. Role of resveratrol in prevention and therapy of cancer: preclinical and clinical studies. Anticancer Res. 24 (5A), 2783–2840.

Andrich, D., 1978. A rating formulation for ordered response categories. Psychometrika 43 (4), 561–573.

Andrich, D., 1981. Book review: probabilistic models for some intelligence and attainment tests (expanded edition: Georg Rasch Chicago: The University of Chicago Press, 1980, 199 pp. $15 hardcover, $7 paperback). Appl. Psychol. Meas. 5 (4), 545–550.

Ayala, R.J.D., 2013. The Theory and Practice of Item Response Theory. Guilford Publications.

Birnbaum, A., 1968. Some Latent Trait Models and Their Use in Inferring an Examinee's Ability. Addison-Wesley, Reading, pp. 397–479.

Bock, R.D., 1997. A brief history of item theory response. Educ. Meas. Issues Pract. 16 (4), 21–33.

Cancer Genome Atlas Research Network, 2008. Comprehensive genomic characterization defines human glioblastoma genes and core pathways. Nature 455 (7216), 1061.

Chandrashekar, G., Sahin, F., 2014. A survey on feature selection methods. Comput. Electr. Eng. 40 (1), 16–28.

Chen, M.-H., Shao, Q.-M., Ibrahim, J.G., 2012. Monte Carlo Methods in Bayesian Computation. Springer Science & Business Media.

Ciriello, G., Miller, M.L., Aksoy, B.A., Senbabaoglu, Y., Schultz, N., Sander, C., 2013. Emerging landscape of oncogenic signatures across human cancers. Nat. Genet. 45 (10), 1127.

Coyne, G.O., Takebe, N., Chen, A.P., 2017. Defining precision: the precision medicine initiative trials NCI-MPACT and NCI-MATCH. Curr. Probl. Cancer 41 (3), 182–193.

Creixell, P., Reimand, J., Haider, S., Wu, G., Shibata, T., et al., 2015. Pathway and network analysis of cancer genomes. Nat. Methods 12 (7), 615.

Cunningham, J.P., Ghahramani, Z., 2015. Linear dimensionality reduction: survey, insights, and generalizations. J. Mach. Learn. Res. 16 (1), 2859–2900.

Dimitrova, N., Nagaraj, A.B., Razi, A., Singh, S., Kamalakaran, S., Banerjee, N., Joseph, P., et al., 2017. InFlo: a novel systems biology framework identifies cAMP-CREB1 axis as a key modulator of platinum resistance in ovarian cancer. Oncogene 36 (17), 2472.

Ding, L., Bailey, M.H., Porta-Pardo, E., Thorsson, V., et al., 2018. Perspective on oncogenic processes at the end of the beginning of cancer genomics. Cell 173 (2), 305–320.

Do, K., O'Sullivan, G.C., Chen, A.P., 2015. An overview of the NCI precision medicine trials-NCI MATCH and MPACT. Chin. Clin. Oncol. 4 (3), 31.

Dodd, B.G., Koch, W.R., 1987. Effects of variations in item step values on item and test information in the partial credit model. Appl. Psychol. Meas. 11 (4), 371–384.

Dodd, B.G., Koch, W.R., 1994. Item and scale information functions for the successive intervals Rasch model. Educ. Psychol. Meas. 54 (4), 873–885.

Dodd, B.G., Ayala, R. J. D., Koch, W.R., 1995. Computerized adaptive testing with polytomous items. Appl. Psychol. Meas. 19 (1), 5–22.

Eiji, M., 1992. A generalized partial credit model: application of an EM algorithm. ETS Res. Rep. Ser. (1), i-30.

Garcia, S.F., Cella, D., Clauser, S.B., Flynn, K.E., Lad, T., Lai, J.-S., Reeve, B.B., Smith, A.W., Stone, A.A., Weinfurt, K., 2007. Standardizing patient-reported outcomes assessment in cancer clinical trials: a patient-reported outcomes measurement information system initiative. J. Clin. Oncol. 25 (32), 5106–5112.

Gelman, A., Hwang, J., Vehtari, A., 2014. Understanding predictive information criteria for Bayesian models. Stat. Comput. 24 (6), 997–1016.

Gilks, W.R., Richardson, S., Spiegelhalter, D., 1995. Markov Chain Monte Carlo in Practice. CRC Press.

Greenblum, S.I., Efroni, S., Schaefer, C.F., Buetow, K.H., 2011. The pathologist: an automated tool for pathway-centric analysis. BMC Bioinformatics 12 (1), 133.

Hastings, W.K., 1970. Monte Carlo sampling methods using Markov chains and their applications. Biometrika 57 (1), 97–109.

Hoffman, M.D., Gelman, A., 2014. The No-U-turn sampler: adaptively setting path lengths in Hamiltonian Monte Carlo. J. Mach. Learn. Res. 15 (1), 1593–1623.

Houseman, E.A., Marsit, C., Karagas, M., Ryan, L.M., 2007. Penalized item response theory models: application to epigenetic alterations in bladder cancer. Biometrics 63 (4), 1269–1277.

Kamalakaran, S., Varadan, V., Janevski, A., Banerjee, N., Tuck, D., McCombie, W.R., Dimitrova, N., Harris, L.N., 2013. Translating next generation sequencing to practice: opportunities and necessary steps. Mol. Oncol. 7 (4), 743–755.

Kastrin, A., 2009. Item response theory modeling for microarray gene expression data. Metodoloski Zvezki 6 (1), 51.

Kristensen, V.N., Vaske, C.J., Ursini-Siegel, J., Loo, P.V., Nordgard, S.H., et al., 2012. Integrated molecular profiles of invasive breast tumors and ductal carcinoma in situ (DCIS) reveal differential vascular and interleukin signaling. Proc. Natl. Acad. Sci. 109 (8), 2802–2807.

Kruger, R., 2018. Charting a course to a cure. Cell 172 (2), 277.

Kruschke, J., 2014. Doing Bayesian Data Analysis: A Tutorial With R, JAGS, and Stan. Academic Press.

Lehmann, B.D., Bauer, J.A., Chen, X., Sanders, M.E., Chakravarthy, A.B., Shyr, Y., Pietenpol, J.A., 2011. Identification of human triple-negative breast cancer subtypes and preclinical models for selection of targeted therapies. J. Clin. Invest. 121 (7), 2750–2767.

Lehmann, B.D., Jovanović, B., Chen, X., Estrada, M.V., Johnson, K.N., Shyr, Y., Moses, H.L., Sanders, M.E., Pietenpol, J.A., 2016. Refinement of triple-negative breast cancer molecular subtypes: implications for neoadjuvant chemotherapy selection. PLoS One 11 (6), e0157368.

Lewis, J.H., Kilgore, M.L., Goldman, D.P., Trimble, E.L., Kaplan, R., Montello, M.J., Housman, M.G., Escarce, J.J., 2003. Participation of patients 65 years of age or older in cancer clinical trials. J. Clin. Oncol. 21 (7), 1383–1389.

Li, H., Hong, F., 2001. Cluster-Rasch models for microarray gene expression data. Genome Biol. 2(8). research0031.1.

Lord, F.M., 2012. Applications of Item Response Theory to Practical Testing Problems. Routledge.

Lord, F.M., Novick, M.R., 1968. Statistical Theories of Mental Test Scores. Addison-Wesley Pub. Co., Reading, MA

Masters, G.N., 1982. A rasch model for partial credit scoring. Psychometrika 47 (2), 149–174.

Mayeux, R., 2004. Biomarkers: potential uses and limitations. NeuroRX 1 (2), 182–188.

Metropolis, N., Rosenbluth, A.W., Rosenbluth, M.N., Teller, A.H., Teller, E., 1953. Equation of state calculations by fast computing machines. J. Chem. Phys. 21 (6), 1087–1092.

Morganella, S., Alexandrov, L.B., Glodzik, D., Zou, X., Davies, H., et al., 2016. The topography of mutational processes in breast cancer genomes. Nat. Commun. 7, 11383.

Muraki, E., 1990. Fitting a polytomous item response model to Likert-type data. Appl. Psychol. Meas. 14 (1), 59–71.

Nik-Zainal, S., Davies, H., Staaf, J., Ramakrishna, M., Glodzik, D., et al., 2016. Landscape of somatic mutations in 560 breast cancer whole-genome sequences. Nature 534 (7605), 47.

Osmanbeyoglu, H.U., Raphael, P., Bromberg, J.F., Leslie, C.S., 2014. Linking signaling pathways to transcriptional programs in breast cancer. Genome Res. 24 (11), 1869–1880.

Perou, C.M., Parker, J.S., Prat, A., Ellis, M.J., Bernard, P.S., 2010. Clinical implementation of the intrinsic subtypes of breast cancer. Lancet Oncol. 11 (8), 718–719.

Pollack, J.R., Sørlie, T., Perou, C.M., Rees, C.A., Jeffrey, S.S., Lonning, P.E., Tibshirani, R., Botstein, D., Børresen-Dale, A.-L., Brown, P.O., 2002. Microarray analysis reveals a major direct role of DNA copy number alteration in the transcriptional program of human breast tumors. Proc. Natl. Acad. Sci. 99 (20), 12963–12968.

Quinn, H., Thissen, D., Liu, Y., Magnus, B., Lai, J.-S., Amtmann, D., Varni, J.W., Gross, H.E., DeWalt, D.A., 2014. Using item response theory to enrich and expand the PROMIS® pediatric self report banks. Health Qual. Life Outcomes 12 (1), 160.

Rasch, G., 1960. Studies in Mathematical Psychology: I. Probabilistic Models for Some Intelligence and Attainment Tests. Nielsen & Lydiche.

Razi, A., Banerjee, N., Dimitrova, N., Varadan, V., 2015. Non-linear Bayesian framework to determine the transcriptional effects of cancer-associated genomic aberrations. In: 37th Annual International Conference of the IEEE Engineering in Medicine and Biology Society (EMBC), pp. 6514–6518.

Robert, C.P., 2004. Monte Carlo Methods. Wiley Online Library.

Samejima, F., 1968. Estimation of latent ability using a response pattern of graded scores. ETS Res. Rep. Ser. 1968 (1), i-169.

Sanchez-Garcia, F., Villagrasa, P., Matsui, J., Kotliar, D., Castro, V., et al., 2014. Integration of genomic data enables selective discovery of breast cancer drivers. Cell 159 (6), 1461–1475.

Scott, S.L., Ip, E.H., 2002. Empirical Bayes and item-clustering effects in a latent variable hierarchical model: a case study from the national assessment of educational progress. J. Am. Stat. Assoc. 97 (458), 409–419.

Sedgewick, A.J., Benz, S.C., Rabizadeh, S., Soon-Shiong, P., Vaske, C.J., 2013. Learning subgroup-specific regulatory interactions and regulator independence with PARADIGM. Bioinformatics 29 (13), i62–i70.

Sreevani, Murthy, C.A., 2017. Bridging feature selection and extraction: compound feature generation. IEEE Trans. Knowl. Data Eng. 29 (4), 757–770.

Sreevani, Murthy, C.A., Chanda, B., 2018. Generation of compound features based on feature interaction for classification. Expert. Syst. Appl. 108, 61–73.

Strimbu, K., Tavel, J.A., 2010. What are biomarkers? Curr. Opin. HIV AIDS 5 (6), 463.

Symmans, W.F., Peintinger, F., Hatzis, C., Rajan, R., Kuerer, H., Valero, V., et al., 2007. Measurement of residual breast cancer burden to predict survival after neoadjuvant chemotherapy. J. Clin. Oncol. 25 (28), 4414–4422.

Tarca, A.L., Draghici, S., Khatri, P., Hassan, S.S., Mittal, P., Kim, J.S., Kim, C.J., Kusanovic, J.P., Romero, R., 2008. A novel signaling pathway impact analysis. Bioinformatics 25 (1), 75–82.

Thomas, M.L., 2011. The value of item response theory in clinical assessment: a review. Assessment 18 (3), 291–307.

Tong, P., Coombes, K.R., 2012. integIRTy: a method to identify genes altered in cancer by accounting for multiple mechanisms of regulation using item response theory. Bioinformatics 28 (22), 2861–2869.

van der Linden, W. J., Hambleton, R.K., 2013. Handbook of Modern Item Response Theory. Springer Science & Business Media.

Vandin, F., Upfal, E., Raphael, B.J., 2011. Algorithms for detecting significantly mutated pathways in cancer. J. Comput. Biol. 18 (3), 507–522.

Varadan, V., Mittal, P., Vaske, C.J., Benz, S.C., 2012. The integration of biological pathway knowledge in cancer genomics: a review of existing computational approaches. IEEE Signal Process. Mag. 29 (1), 35–50.

Varadan, V., Gilmore, H., Miskimen, K. L. S., Tuck, D., Parsai, S., et al., 2016a. Immune signatures following single dose trastuzumab predict pathologic response to preoperativetrastuzumab and chemotherapy in HER2-positive early breast cancer. Clin. Cancer Res. 22 (13), 3249–3259.

Varadan, V., Kamalakaran, S., Gilmore, H., Banerjee, N., Janevski, A., Miskimen, K. L. S., 2016b. Brief-exposure to preoperative bevacizumab reveals a TGF-β signature predictive of response in HER2-negative breast cancers. Int. J. Cancer 138 (3), 747–757.

Vaske, C.J., Benz, S.C., Sanborn, J.Z., Earl, D., Szeto, C., Zhu, J., Haussler, D., Stuart, J.M., 2010. Inference of patient-specific pathway activities from multi-dimensional cancer genomics data using PARADIGM. Bioinformatics 26 (12), i237–i245.

Yates, L.R., Knappskog, S., Wedge, D., Farmery, J. H. R., Gonzalez, S., et al., 2017. Genomic evolution of breast cancer metastasis and relapse. Cancer Cell 32 (2), 169–184.

Chapter 9

Effects of Phenotypic Plasticity and Unpredictability of Selection Environment on Niche Separation and Species Divergence

Narayan Behera[1]
Department of Physics, Dayananda Sagar University, Bengaluru, India
[1]Corresponding author: e-mail: chairman-phy@dsu.edu.in

ABSTRACT

Evolution of a phenotypically plastic trait is examined under weak optimizing selection in a heterogeneous environment. Selection operates toward optimum phenotype that depends upon the environment. Density-dependent number regulation is caused by non-uniform food supply. The juveniles do not disperse but develop in the parental patch. An adult has a certain probability to migrate to a different patch of selection. Density-dependent number regulation operates after selection. There is one Gaussian food distribution over the patches for a haploid population. The model is individual based. A mutation model has been developed to find the evolutionary stable strategies genotypic values. In the absence of phenotypic plasticity, multiple populations are observed at evolutionary equilibrium with clear niche separation. The genotypes are specialists. Splitting of the population occurs beyond a critical value of food width (defined as the food distance between patches where food supply is half of the maximum value) at fixed migration width. This critical value is correlated with migration distribution. The genotypes develop generalist tendency with the increase of the level of plasticity. Eventually genetic differentiation is removed and one population is found at equilibrium beyond a critical value of plasticity. The genotypes become generalists. Phenotypic plasticity increases niche width and tempers niche separation. The critical value of food width increases with higher migration width. Phenotypic plasticity allows niche separation in the presence of two Gaussian food distributions if the food width happens to be below a critical value and the food distance between the peaks of the two Gaussian food distributions lies above a threshold range. In the presence of additive allelic effect of a diploid population, polymorphic state is observed in the presence of

Handbook of Statistics, Vol. 40. https://doi.org/10.1016/bs.host.2018.09.002

phenotypic plasticity. An initially monomorphic population evolves into two genotypically diverging lineages. The local abundance of food causes local genotypic specialization leading to polymorphism. Niche separation becomes easier with higher selection intensity.

Keywords: Spatial effect, Evolutionary ecology, Niche separation, Phenotypic plasticity, Food width, Migration distribution, Polymorphism, Evolutionary equilibrium

1 INTRODUCTION

Many organisms live in heterogeneous environments. They display phenotypic plasticity in order to cope with this heterogeneity. It is believed that natural selection molds the level of phenotypic plasticity in a given trait. Phenotypic plasticity is the ability of a single genotype to produce multiple phenotypes in response to variation in the environment. Yet its evolutionary significance of how it impacts diversification and speciation is not fully understood. Historically, research has concentrated on genetic and ecological causes of diversification. It is increasingly clear that the responsiveness to changes in the environment commonly seen in development might play a key role in diversification and speciation. It is an important aspect of global environmental change that the modern world faces.

It is an important problem to study the evolution of phenotypic plasticity in a heterogeneous environment where unequal food supplies in different environments cause density-dependent number regulation. The interrelationship between migration and population persistence was studied in demographic models (Okubo, 1980). They came to the conclusion that a population cannot survive in a suitable patch if individuals disperse too quickly into neighboring habitats that are unsuitable. Mayr (1963) has concluded that adaptation of a population at the periphery might be limited by gene flow from central populations. The interactions between gene flow and local selection pressure can limit the geographical range of a species (Kirkpatrick and Barton, 1997). This model keeps track of the evolutionary and demographic changes in a continuously distributed population. A quantitative trait evolves under the influence of stabilizing selection and gene flow. The important factor in the model is the inclusion of density-dependent population regulation so that limited-range equilibrium is possible. The basic tendency for gene flow from better-adapted regions to delimit local adaptation elsewhere may be compensated by phenotypic plasticity and by increase in genetic variation caused by the admixture (Grant and Grant, 1994). The peripheral populations act as demographic sinks due to bad adaptation caused by the gene flow. These regions are maintained by migration from better-adapted populations. Gene flow acts as a mechanism that limits the geographical range of a species. In a series of computer simulation models within the genetic algorithm framework, it was shown that phenotypic plasticity plays a strong role in evolution of species, adaptations, and diversifications (Behera and Nanjundiah, 1995, 1996, 1997, 2004). Spatial effects are important in the evolution of mutualism of species (Yamamura et al., 2004). Phenotypic diversity also leads to more stability of ecosystems (Loreau and Behera, 1999).

Christiansen and Loeschcke (1980) have studied the evolution of a species with respect to its environment by considering the evolution of characters that determine its niche. They have used a model of exploitative competition with a Gaussian resource spectrum. Evolution is considered at a gene locus that influences the niche position. The allelic contributions to the genotypic values are additive. Intraspecific competition does not play an important role when the resource abundance spectrum imposes strong directional selection. So the genotypes that utilize the more abundant resources will have a higher selective value. But intraspecific competition becomes important when evolution due to this directional selection approaches the goal (i.e., the niche is at the resource optimum). Evolution makes the mode of the niche converge to the resource optimum when allele contributions are small in comparison to the distance between the mode of the niche and the resource optimum. The globally stable equilibrium will maintain at most two alleles in the population when the allele contributions are of the same order of magnitude as the above distance. Intraspecific exploitative competition can be a dominant force for the maintenance of two-allele polymorphism.

The limitations on the evolution of adaptive plasticity and niche width have been examined (Whitlock, 1996). It is shown that species with narrower niche width fix beneficial alleles with high probability in less number of generations, develop lighter drift load, and have a smaller number of deleterious alleles at mutation–selection equilibrium. The rate of evolutionary response is slow for species with larger niche width. The evolution of niche width and plasticity does not depend only on fitness in different environments but also related to the evolutionary process per se. The effective population size increases in proportion to niche width due to the availability of more resources to a species.

The profile of phenotypes produced by a genotype across environments is called as the reaction norm. The optimum reaction norm evolves when selection is totally predictable in a heterogeneous environment (Via and Lande, 1985). The optimal reaction norm is the collection of optimal phenotypes that would evolve if separate evolution occurred in each patch of a heterogeneous environment. When adults migrate to a different patch of environment, the selection environment will be different than the environment of development. Scheiner (1998) considered the effect of genetic architecture and spatial structure in individual-based model to study the evolution of phenotypic plasticity. He considered the presence of different types of loci such as nonplastic loci (insensitive to environment) and plastic loci (sensitive to environment in linear fashion and quadratic fashion). He took into migration pattern to consider the spatial structure. For the phenotypic plasticity to be selected two conditions were necessary: genotypic expression must be very strongly correlated with environments, and the changes in environment must be large in-between generations. He has found evolved differences between genotypes in local habitats due to limited adult migration. Here, unpredictability of selection causes the population structure to influence the evolution of reaction norms. Plasticity becomes more favored with high migration patterns. In all locally differentiated reaction norms, phenotypic plasticity was lower than the phenotypic plasticity selected for. Curvature appeared in evolved reaction norms even if the optimal reaction norm was linear in the presence of all three types of loci. Higher migration rates are correlated with steeper reaction norms. In addition to unpredictability of selection, edge effect due to finite size of the multipatch system also plays a role in creating local genotypic differentiation. As selection becomes unpredictable due to habitat change between development and selection, the slope of the evolved reaction norm becomes shallower than the slope of the optimum reaction norm (Gavrilets and Scheiner, 1993; Sasaki and de Jong, 1999). It leads to evolutionarily stable bet-hedging reaction norm constituting a compromise between the phenotypic optima in different environments. Density regulation after selection has a much stronger influence on the evolution of reaction norm than density regulation before selection. Due to differential productivity of patches, a source–sink structure of the environment is created. It strongly affects how the bet-hedging strategy weighs the different phenotypic optima. If the variability of density regulation after selection among patches is very large, no single reaction norm is evolutionarily stable. As a result polymorphic reaction norm develops in the evolutionary stable population.

Sasaki and de Jong (1999) have simulated a two-patch model with the result that unpredictability of selection leads to evolutionarily stable bet-hedging reaction norm constituting a compromise between the phenotypic optima in the different patches. Polymorphism is a discontinuous genetic variation that gives rise to several different types of individuals among the members of a single species. A discontinuous genetic variation divides the individuals of a population into many distinct forms. The polymorphic response in the evolutionarily stable population occurs in the presence of very strong selection. This model is basically a zygote pool model. de Jong and Behera (2002) have extended the above analysis by considering a zygote dispersal stage where there exists a correlation between the patch of selection and the patch of development. The life cycle includes three stages in which density-dependent number regulation can take place: in the environment of development, and in the environment of selection before and after selection. In this model, the influence of density-dependent number regulation strongly depends on the exact life cycle. Differences in productivity between patches cause major changes in phenotypic evolution.

Two-allele polymorphism is observed in the presence of Gaussian food supply and intraspecific competition (Christiansen and Loeschcke, 1980). Similar type of limitation for the species range is observed in the presence of gene flow, local selection pressure, and density-dependent population regulation

(Kirkpatrick and Barton, 1997). The ecological interactions such as adult migration and nonuniform food supply resulting in differential density regulation are major evolutionary forces leading to niche separation with phenotypic divergence. This kind of evolutionary branching under certain ecological conditions has been investigated where the initial monomorphic population can split into diverging phenotypes after attaining the fitness minimum (Doebeli and Dieckmann, 2000).

An initially monomorphic population can evolve and split into two phenotypically diverging lineages through evolutionary branching. A genetic polymorphism of different alleles can be generated in an initially monomorphic diploid population by means of frequency-dependent evolution when the effect of mutations on the phenotype is small (Kisdi and Geritz, 1999). The branching on the evolutionary tree can be imagined as the emergence of two distinct strategies by short evolutionary steps. Geritz and Kisdi (2000) have investigated the evolutionary outcome of sexually reproducing organisms. Evolutionary branching occurs in same ecological situations in clonal as well as sexual populations. The evolution of assortative mating with respect to the ecological trait is favored. In adaptive speciation approach the evolutionary driving force is mediated by the ecological interactions (Doebeli and Dieckmann, 2000). The population may be taken toward a state of fitness minimum in phenotypic space due to directional selection. Then it split into two diverging phenotypic clusters after attaining the fitness minimum. With assortative mating for the desired ecological characters in sexual population, evolutionary branching can lead to speciation.

Adaptive dynamics is a framework to study evolutionary branching, where ecological interactions induce disruptive selection and a consequent split into two distinct phenotypes. Sympatric speciation is a probable outcome of competition for resources (Dieckmann and Dobeli, 1999). They have used multilocus genetics to describe sexual reproduction in an individual-based model and have considered the evolution of assortative mating depending on an ecological character affecting the resource use. Reproductive isolation between ecologically diverging subpopulations can be generated by the evolution of assortative mating. Rundle and Whitlock (2001) have examined an experimental design to detect ecologically dependent isolation. When hybrids develop reduced fitness as a result of interaction between their phenotype and environment, ecologically dependent isolation can happen. Ecological speciation occurs when reproductive isolation evolves due to divergent natural selection. The intermediate phenotypes suffer reduced fitness through ecological mechanisms. A model has been developed to detect the reasons for hybrid unfitness. It considers the expected phenotype of an individual under the influence of outcrossing by taking into account two environments. Ecological speciation demands that the fitness of hybrids be reduced relative to the parental forms in each habitat and this reduction should be the result of ecological mechanisms.

The inclusion of additive allelic effect and genetic recombination during reproduction of sexual population does not change this basic conclusion. Basically, local abundance of food causes local genotypic specialization leading to niche separation and polymorphism. The differential density regulation in the patches caused by nonuniform food supply makes the genotypes to adopt specialist tendency to survive in the local environments. Here, the basic evolutionary driving forces are the ecological interactions due to the nonuniform food supply leading to differential density-dependent regulation and unpredictability in selection. Similar type of evolutionary branching under certain ecological conditions occurs so that the population splits into two diverging lineages after attaining the fitness minimum (Doebeli and Dieckmann, 2000). Evolutionary branching is also a probable outcome of competition for resources in an individual-based model where sexual reproduction is described by multilocus genetics (Dieckmann and Dobeli, 1999). Geritz and Kisdi (2000) have also found related type of polymorphic consequences in sexually reproducing populations. Another important area of application of the present result is ecological speciation which occurs due to ecological mechanisms. Ecological speciation requires that the fitness of hybrids be reduced relative to the parental forms. A model was designed to detect the hybrid unfitness and hence ecologically dependent isolation (Rundle and Whitlock, 2001). The problems of evolution of phenotypic plasticity, niche separation, ecological speciation, and environmental adaptation were investigated in the recent past (Cloyed and Eason, 2017; Hollander et al., 2015; Pfenning et al., 2010; Plante et al., 2011; Reed et al., 2011). Furthermore, the relationships between phenotypic plasticity, migration distribution, niche width, and species divergence were investigated (Behera and de Jong, 2018). Unrestricted phenotypic plasticity removes niche separation and genetic variation. The unequal food supply in different patches gives rise to differential density regulation in the patches. The adult migration makes the genotypes concentrated near the patches where there is local abundance of food supply. Hence the genotypic differentiation arises at evolutionary equilibrium. The genotypic values in the groups of populations are separated by gaps. The local adaptability causes the genotypes to be specialists. They can adapt well only in the local environment. The branching at evolutionary equilibrium was observed where two distinct genotypes evolve from a single monomorphic population (Behera et al., 2018).

2 MODEL

2.1 Life Cycle

In the Scheiner (1998) computer simulation, adults migrate but in the models of Via and Lande (1985) and Gavrilets and Scheiner (1993) zygotes disperse. Adult migration generally disconnects the habitats of development and selection.

It leads to imperfect predictability of future selection. In de Jong (1990, 1995) work, a zygote pool exists from which zygotes disperse at random to a series of habitats. In these habitats, individuals are developed and selected. Adults join a mating pool after leaving their habitat. Random mating results in a zygote pool of the next generation. Here, selection and development are in the same direction. As a result the optimum reaction norm evolves. The unpredictability of selection comes into picture when zygotes disperse at random to a series of habitats where they become adults and are selected in another habitat after migration. Surviving adults leave this selection habitat and join a mating pool. Here, unpredictability of selection leads to the evolved reaction norm that is shallower than the slope optimal reaction norm.

Development occurs in discrete habitat characterized by an environmental variable x with a distribution given by f_x. In the same way selection is characterized by an environmental variable y with a distribution f_y. These distributions depend on the habitats, not on the organism. The mean environmental value is given as $\bar{x}$ and the variance is $\sigma^2\{x\}$. The conditional probability of being selected in any patch y if developed in patch x is given by $f_{y|x}$. The probability distribution implies covariance $\text{cov}\{x,y\}$ between the environmental values x and y. If the range of values of x and y are similar, $|\text{cov}\{x,y\}| \leq \sigma^2\{x\}$. The genotypic reaction norm is determined by the environmental variable x in the habitat. The genotypic reaction norm is a function of the environment defined by $g = g_0 + g_1x + g_2x_2 + \cdots$. Here, g_0 equals the reaction norm height at $x = 0$, g_1 represents the slope, and g_2 the curvature of the reaction norm, all defined at $x = 0$. The genotypic value of an individual that develops in patch with environmental value x is given by the function $g(x) = g_0 + g_1x + g_2x_2 + \cdots$. The coefficients g_i are genetically variable and differ between genotypes. A mean genotypic value of each coefficient and a genetic variance covariance matrix of the coefficients exist in a population. The function connecting phenotypic optima in all habitats y is given as $c = c_0 + c_1y + c_2y^2 + \cdots$. It is the optimum reaction norm. Quadratic optimizing selection is the simplest case.

The evolution of the reaction norm of a quantitative trait and the consequent evolution of niche width of a haploid asexual population is considered. The phenotype of an individual is determined by its genotype and an environmental variable which depends on the environment of development. The environment consists of several different patches so that the phenotype of an individual differs in each patch. That is why the environment is heterogeneous. The population life cycle is described in Fig. 1.

Zygotes arrive in a patch in the development environment. After the phenotype is determined in the patch of development, surviving adult individuals migrate to another patch. In the patch of selection, adults are subject to viability selection and then density-dependent number regulation after selection. The density-dependent number regulation is caused by nonuniform food supply which is taken to be Gaussian. Viability selection is optimizing.

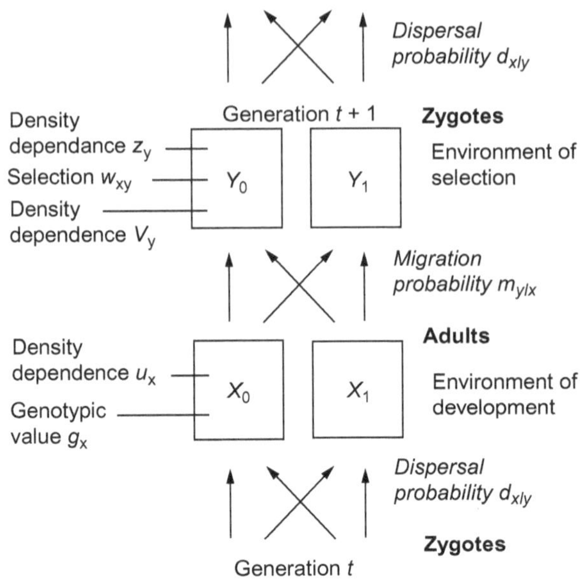

FIG. 1 The population cycle.

The optimum phenotype varies from patch to patch depending on the environmental variable. Surviving individuals reproduce and die. Then the offspring disperse to the patch of development in the next generation. Hence the population cycle is completed. The cycle iterations continue until the evolutionary equilibrium is reached.

2.2 Model Specification

There are n different values of an environmental variable corresponding to different patches in the environment of development and the environment of selection. A juvenile develops in patch x and obtains the genotypic value $g(x)$ according to the wild-type reaction norm g. In generation t, the $N_x(t)$ juveniles arrive in patch x. In generation t, the $N_x(t)$ juveniles arrive in patch x and are subject to density-dependent number regulation. The probability of survival is $u_x = \exp[-\alpha_x N_x]$ where α_x is the sensitivity to density regulation in patch x. A surviving individual migrates from its development patch x to a selection patch y with a conditional probability $m_{y|x}$ where $\sum_{y=1}^{n} m_{y|x} = 1$ for each x. The number of individuals in patch y after migration and before selection equals

$$N_{1,y}(t) = \sum_{x=1}^{n} N_x \cdot m_{y|x}. \tag{1}$$

In patch y, Gaussian optimizing selection on the phenotypic trait $g(x)$ occurs. The viability of an individual that developed in patch x (with phenotype $g(x)$) and moved to selection patch y is given by

$$w_{xy}^{g} \equiv W_y[g(x)] = \exp\left\{-s(\theta(y) - g(x))^2\right\} \tag{2}$$

where $\theta(y)$ is the optimum trait value in patch y and $s = 1/(2\sigma^2)$ measures the intensity of optimizing selection in Gaussian selection of width σ^2. After selection the number of individuals in selection patch y is given by

$$N_{2,y}(t) = N_{1,y} w_{xy}^{g}. \tag{3}$$

Density-dependent number regulation after selection leads to the probability of survival as

$$z_y \equiv Z_y\left[N_{2,y}(t)\right] = \exp\left\{-\gamma_y N_{2,y}\right\} \tag{4}$$

where γ_y is the sensitivity in patch y to density regulation after selection. The number of individuals in patch y after selection equals

$$N_{3,y} = N_{2,y} z_y. \tag{5}$$

Let $d_{x|y}$ be the dispersal probability of a zygote from selection patch y to development patch x. Each surviving individual reproduces with fecundity F. The number of zygotes arriving in development patch x in generation $t+1$ equals

$$N_x(t+1) = \sum_{y=1}^{n} N_{3,y} d_{x|y} F. \tag{6}$$

2.2.1 *Fitness*

In the zygote pool model, the population of zygotes is fully mixed before dispersing to the development habitats x. The dispersal probability to a patch x is equal for each patch y and equal to the frequency of occurrence of the patches x. The population is fully mixed in the zygote pool. So the recurrence equations of the numbers of each genotype can be written down from zygote pool to zygote pool.

At evolutionary equilibrium, the expectation of fitness W_g for a particular genotype g can be found from the recurrence equations (Sasaki and de Jong, 1999) as

$$E\left[W_g\right] = \sum\nolimits_x \sum\nolimits_y d_x u_x^* m_{y|x} v_y^* w_{xy}^{g^*} z_y^* d_{x|y} F, \tag{7}$$

where the symbol * refers to the individual number at evolutionary equilibrium. Weak selection at stable equilibrium number leads to a compromise ESS reaction norm defined by

$$g(x)^* = \sum\nolimits_y m_{y|x} v_y^* w_{xy}^{g^*} z_y^* d_{x|y} \theta(y) \Big/ \sum\nolimits_y m_{y|x} v_y^* w_{xy}^{g^*} z_y^* d_{x|y} \tag{8}$$

The evolved compromise reaction norm is weighted toward the optimum of the selection environment with least stringent density regulation, highest migration, and dispersal. Adults leaving their patch of development are counted. The total number of each genotype after a one-generation interval is tracked. Fitness is defined as the expected number of adult offspring of adult individuals leaving their patch of development for each genotype. We use the population growth rate of each population of one genotype. As these are haploid populations, this is an adequate measure of fitness.

The total number leaving any development patch x is $N_x(t)$ in generation t; in the next generation $t+1$ the total number leaving any development patch k and descendant of an individual leaving patch x in generation t equals

$$N_x(t+1) = N_x(t)\sum_k \sum_y \left(m_{y|x} v_y w^g_{x,y} z_y F\right) d_{k|y}. \tag{9}$$

At stable equilibrium, the total number becomes

$$N^* = \sum_x N^*_x \cdot \sum_k \sum_y \left(m_{y|x} v^*_y w^g_{x,y} z^*_y F d_{k|y}\right). \tag{10}$$

The condition giving the ESS reaction norm is that no mutant can invade: the growth rate of any mutant is less than 1. Genotypic fitness is defined as the expected contribution to the next generation after averaging over all individuals of a genotype in the present generation. The number of individuals of the genotype g leaving any development patch x in generation t is denoted as $N^g_x(t)$. The expected contribution to the next generation of individuals of genotype g equals the growth rate of the population of genotype g.

The expected fitness $W^g = N^g(t+1)/N^g(t)$. Averaging over all patch types, the fitness of the population becomes

$$W_g = \sum_x \sum_y \sum_k \left(N^*_x m_{y|x} v^*_y w^{g^*}_{xy} z^*_y F d_{k|y}\right) / \sum_x N^*_x \tag{11}$$

The condition for ESS reaction norm is that no rare mutant can invade in a population. The fitness of the resident genotype is at maximum and equal to one. All individuals have equal probability to undergo mutation at a certain very low rate. The population size is not affected by invasion, so also the density-dependent viabilities. The condition for the resident strategy of genotypic value g^r to be a globally stable ESS is that for all mutant strategy g^m, the mutant fitness is less than the fitness of one of the resident strategy g^r. The ESS genotypic value involves a compromise between different optima.

The fitness is at a maximum at $g(x)^*$. The implicit solution for the evolved reaction norm is given by

$$g(x)^* = \sum_y \sum_k m_{y|x} w^{g^*}_{xy} z^*_y d_{k|y} \theta(y) / \sum_y \sum_k m_{y|x} w^{g^*}_{xy} z^*_y d_{k|y} \tag{12}$$

With weak Gaussian selection, fitness has a maximum at $g(x)^*$. $g(x)^*$ is an ESS under weak selection.

2.2.2 Density-Dependent Number Regulation

A "source" environment is a patch where food supply is high so that an individual survives with high probability. Conversely, food supply is low in a "sink" environment so that density regulation is large leading to lower survival probability of a population. The presence of "source" and "sink" environments highly influences the evolved reaction norm. In order to appreciate the effect of source and sink environments, it is useful to define the "flow" of individuals through the patches of environments of development and of selection. The flow summarizes the relative frequency of organisms meeting each sequence of patches y and x, according to

$$f(y|\,x) = m_{y|x} v_y^* w_{xy}^{g^*} z_y^* d_{x|y} / \left(\sum\nolimits_y m_{y|x} v_y^* w_{xy}^{g^*} z_y^* d_{x|y} \right) \tag{13}$$

The flow is counted from just before migration of adults in generation t, to just before migration of adults in generation $t+1$. Flow acts as if it is the biological frequency of the environment combinations.

Given the definition of flow

$$g(x)^* = \sum\nolimits_y f(y|\,x)\,\theta_y = E\left[\theta_y|\,x\right] \tag{14}$$

in all models with equal selection intensity s in all environments y, the difference is in the definition of the flow. In the above E denotes the expected value.

Equal viability due to density-dependent number regulation before and after selection extinguishes the effect of density-dependence on selection. If $z_y^* = z^*$ and $v_y^* = v_y$ in all the patches of the environment of selection, the flow becomes, for Gaussian selection:

$$f(y|\,x) = m_{y|x} w_{xy}^{g^*} d_{x|y} / \left(\sum\nolimits_y m_{y|x} w_{xy}^{g^*} d_{x|y} \right). \tag{15}$$

The evolved phenotype $g(x)$ is the same as for a pure genotypic frequency model, given we are using Gaussian optimizing selection.

2.2.3 Evolved Reaction Norm

We define the linear reaction norm as $g(x) = g_0 + g_1\,x$ where g_0 is the reaction norm height, g_1 is the reaction norm slope, and x refers to the environment of development. The optimal reaction norm is defined as $c(y) = c_0 + c_1\,y$ where c_0 is the optimal reaction norm height, c_1 is the reaction norm slope, and y denotes the environment of selection. Let $r(x,y)$ denote the migration correlation between the patch of development x and the patch of selection y. If selection is predictable, $r(x,y) = 1$, then the evolved reaction norm is given by $g_0 = c_0$ and $g_1 = c_1$. But, as the selection becomes unpredictable, $0 \leq r(x,y) \leq 1$, the evolved reaction norm might differ from the optimal reaction norm. Weak selection leads to the bet-hedging of the reaction norm, while strong selection gives polymorphism. The slope $\overline{g}_1$ of the evolved mean reaction norm $\overline{g} = \overline{g}_0 + \overline{g}_1\, x$ differs, depending upon the unpredictability of selection and quadratic or Gaussian optimizing fitness function. When genotypic reaction norms are linear and optimum reaction norm is linear (meaning only the first two coefficients exist in the genetic variation), the evolved mean genotypic reaction norm coefficients are given by (de Jong, 1999)

$$g_0 = c_0 + c_1\overline{y} - \overline{g}_1\overline{x} \tag{16a}$$

and

$$g_1 = c_1\,(\mathrm{Cov}(x, y)/\mathrm{Var}\,(x)) \tag{16b}$$

But now the density dependence influences the distribution of individuals over the patches. The optimal reaction norm is always obtained when the food supply is uniform over the patches and migration is symmetric. The deviation from the optimal reaction norm occurs due to the asymmetry in the effective migration distribution. This is caused by two factors. One is the mirroring of migration due to the edges of the environment. This happens when the number of patches is finite. It can be better illustrated by a simple example. Suppose the migration width is three, and there is equal a priori probability of migration. If the straight migration occurs at the right boundary, there is no possibility of migration to one patch right of the boundary. This migration probability will be mirrored to the left patch of the boundary so that the probability of migration to this patch will be 0.666. This can cause asymmetry in the migration distribution. The other factor is the density-dependent number regulation due to nonuniform food supply. The nonuniform food supply will cause differential density regulation over the patches. This can create asymmetry in the migration distribution.

The predicted ESS g_0 and g_1 values depend on the number of individuals at evolutionary equilibrium. But this number cannot be found analytically. Furthermore, the small deviations of the evolved g_0 and g_1 values from the optimal c_0 and c_1 values are difficult to find as selection is weak and the model is individual-based. So it seems best to develop a mutation model where deviations in g_0 and g_1 are introduced with low probability.

2.3 Computer Simulation

The numerical iterations are carried out in a 15-patch version of the model in order to bring out clearly the influence of density-dependent number regulation caused by nonuniform food supply. The one food distribution is considered as Gaussian with a peak occurring in the middle patch. The food width is defined as the food distance between patches where food supply is half of the maximum value. The amount of food remains the same, thus food width only changes food distribution over the patches. The migration distribution is characterized by the width and peakedness. When the migration probabilities to one of the neighboring patches as determined by the migration width are equal, the peakedness of migration becomes one. In general, the migration probability to successive neighboring patches decreases geometrically by a factor whose value is given by the peakedness. When the peakedness of adult migration is high the probability that the individual will migrate to one of the neighboring patches (determined by the migration width) is small. For example, when peakedness = 1 and migration width = 7, the individual has equal a priori probability (equal to 1/7 = 0.142) to stay in the parental patch or migrate to one of the three neighboring patches in the left and three in the right. But, as the peakedness becomes five, the probability of the individual to stay in the parental patch is equal to 0.668, while migration probability to one of the three left neighboring patches decreases geometrically by a factor of 5. The same situation holds for migration to the right neighboring patches. The juveniles do not disperse but stay in the same environment. After reproduction, a mutational deviation is introduced to the genotypic values of the offspring. More genetic variation is created by mutation.

The program is individual-based and object-oriented. The individuals are considered as objects with associated properties. In every stage of the life cycle, each individual is tracked whether dead or alive. The optimum phenotypes of patch 0 is zero, patch 1 is one and so on. Initially the g_0 values of the individuals are chosen randomly within the interval of -1 and 15. Let $g_0[j]$ and $g_1[j]$ are the genotypes of the jth individual, and x be a uniformly distributed pseudorandom number in the interval of 0 and 1. The following expressions are used to find the initial g_0 and g_1 values of an individual.

$$g_0[j] = (p+1)x - 1 \tag{17a}$$

and

$$g_1[j] = \{1/(p-1)\}\{(p+1)x - 1\} - \{1/(p-1)\}\, g_0[j] \tag{17b}$$

where p denotes the total number of patches in the model. One can see that the values of $g_1[j]$ lie approximately within -1 and 1. The starting population is 150. The normal selection intensity is taken as 0.07. The reproduction occurs asexually with a fecundity equal to two. The mutation probability for an individual's genotype is 0.001. The simulation is iterated for 10,000 generations. Then the ESS genotypic values are obtained, and the distribution of phenotypes over the patches is found. The computer simulation data were analyzed using SPSS statistical package.

We have analyzed three different models: (A) haploid population with single Gaussian food supply, (B) haploid population with double Gaussian food supply, (C) diploid population having additive allelic effect of gene interaction and double Gaussian food supply.

3 RESULTS

3.1 Haploid Population With Single Gaussian Food Supply

The most important conclusion is that genotypes at evolutionary equilibrium become specialists with clear genotypic differentiation and niche separation in the absence of phenotypic plasticity. But they gradually develop generalist tendency as the level of plasticity is increased and eventually become fully generalists at unrestricted plasticity.

3.1.1 Population Differentiation Without Plasticity

There is clear niche separation without phenotypic plasticity. There is a threshold value of food width beyond which many populations are observed at evolutionary equilibrium (for fixed migration width). This threshold value increases for low peakedness (Fig. 2). The threshold value rises with high migration width. Niche separation is always possible if the migration width is greater than the food width. For fairly restricted food width, niche separation occurs below a critical value of migration width. For instance, when the food width = 1.9 and peakedness = 1, the critical migration width = 9 (Fig. 2).

Fig. 3A shows that a monomorphic population splits into five populations at evolutionary equilibrium with distinct niches. The niche width becomes narrower. When a large number of samples are analyzed, five and six populations are observed at equilibrium with probabilities of 0.6 and 0.4, respectively. The

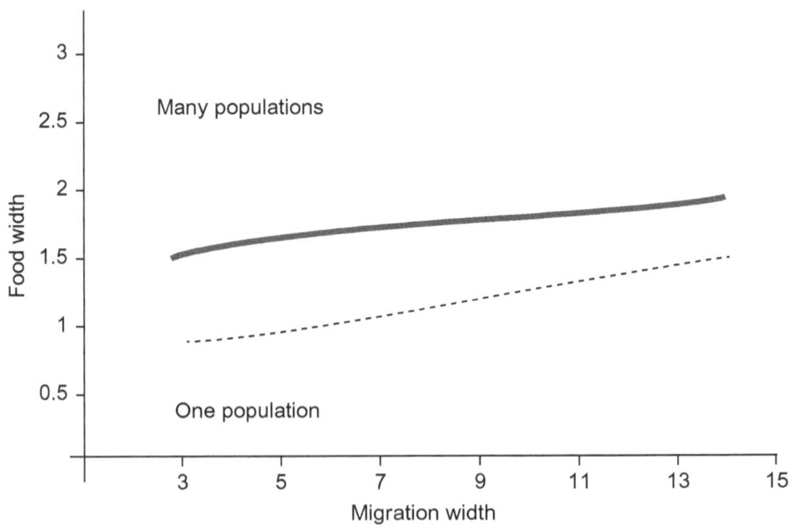

FIG. 2 The critical value of food width beyond which many populations (specialist genotypes) are observed at evolutionary equilibrium. No plasticity, peakedness = 1 (*solid line*), peakedness = 5 (*dashed line*). The results are averaged over six simulations.

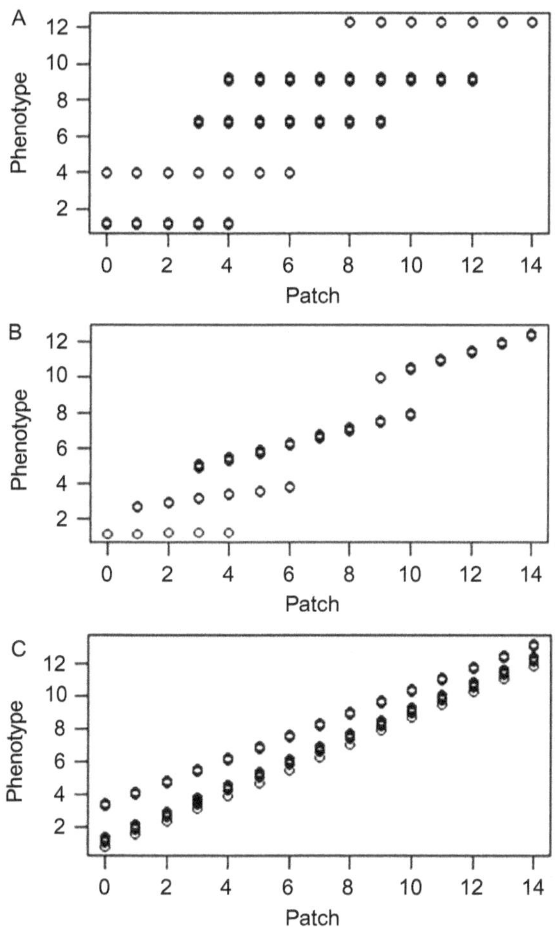

FIG. 3 The phenotypes of the population at evolutionary equilibrium. Model parameters: migration width = 3, migration peakedness = 1, food width = 4, and juveniles stay at home environment. Single simulation. (A) No plasticity, five distinct populations. (B) Limited plasticity (g_1 varies from −0.3 to 0.3), four populations. (C) Limited plasticity (g_1 varies from −0.8 to 0.8), two populations.

clear gap in the phenotypic value between two neighboring populations implies niche separation. The phenotypic gaps are equally placed.

When the peakedness is increased to five in the model corresponding to Fig. 3A, seven distinct populations are obtained at evolutionary equilibrium (Behera and de Jong, 2018). Without plasticity, low migration width and high peakedness give many populations, while the reverse situation generates two populations. When the food width is larger, the differential density regulation among patches is relatively smaller. This causes higher survival probability in the parental patch. Thus, there is less necessity for an individual to migrate to a neighboring patch in order to survive. Hence we find more specialist genotypes and many populations as the food width increases. The higher migration width increases the propensity of an individual to migrate to a neighboring patch. So we need a greater food width with less differential density regulation so that individuals will be restricted in the same patches producing more specialist genotypes. Similarly, when peakedness is lower, an individual can migrate to other patches with greater probability. So the food width has to be higher in order to find many populations at equilibrium. The above argument also works for the fact that the splitting of the population at equilibrium occurs for a higher value of food width when the level of plasticity increases.

3.1.2 Allowing Increasing Level of Plasticity

As the degree of plasticity is increased the number of distinct populations at equilibrium decreases leading to larger niche width. The number of distinct populations at evolutionary equilibrium is reduced to four with lower value of the degree of plasticity (Fig. 3B) and to two with high limited plasticity (Fig. 3C). When g_1 varies from −0.5 to 0.5, the statistical probabilities of finding 1, 2, 3, or 4 populations at equilibrium are 0.12, 0.68, 0.13, and 0.07, respectively. When the degree of plasticity increases, the splitting of the population at equilibrium occurs for a higher value of the food width. Some limited plasticity is required to remove genetic differentiation and get one population at equilibrium. Genotypes develop generalist tendency. Fig. 4 shows the genotypic values at evolutionary equilibrium with limited plasticity.

3.1.3 Phenotypic Plasticity Reduces Genetic Variation

With unrestricted plasticity only one population is found at equilibrium. The niche width is the largest. The statistical probability of finding one population is always one when a large number of simulations are investigated. It should be mentioned that one population is already observed with very high value of the degree of plasticity (i.e., when g_1 varies from −0.9 to 0.9).

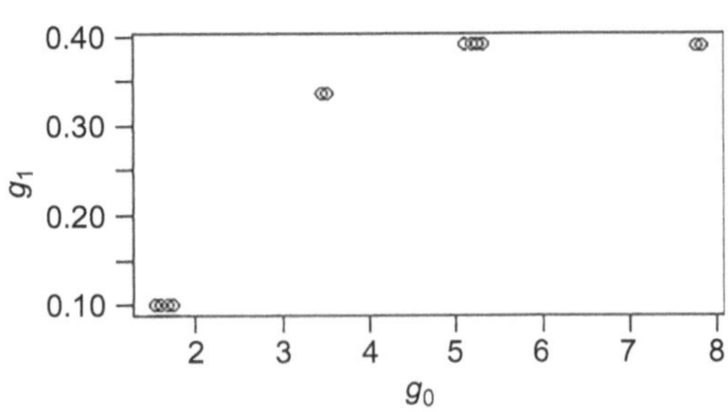

FIG. 4 The genotypes of the population at evolutionary equilibrium. Model parameters: migration width = 3, migration peakedness = 1, food width = 4, and juveniles stay at home environment. Single simulation. Limited plasticity (g_1 varies from −0.3 to 0.3), genotypes (g_0 and g_1).

One can never get niche separation with unrestricted plasticity even if the food width is very high. Genotypes are always generalists.

3.1.4 Population Number at Evolutionary Equilibrium

The population number at evolutionary equilibrium varies in the range of 1000 and 1500. This population number increases with the level of plasticity and migration peakedness but decreases with migration width. With higher degree of plasticity, the individuals can adapt better in a variable environment. So the death of an individual will be less likely. The population number at equilibrium will increase. When the peakedness is greater, an individual is confined to the same patch with a higher probability and less likely to go extinct due to a hostile environment. The same argument applies to the case when the migration width is lower. So, more number of populations will survive at evolutionary equilibrium.

3.2 Haploid Population With Double Gaussian Food Supply

We have investigated the evolution of phenotypic plasticity and the problem of niche separation in the presence of two overlapping Gaussian food distributions in a heterogeneous environment. The migration pattern and food width can influence the probability of niche separation at evolutionary equilibrium. We have examined the necessary conditions of niche separation as a function of food width and food distance.

Niche separation happens when a single monomorphic population splits into two distinct populations at evolutionary equilibrium. Niche separation occurs with and without phenotypic plasticity below certain critical value of food width and above a threshold value for the food distance. The critical and threshold values are different in systems without and with plasticity. These values also depend on other model parameters. A typical food distribution for which niche separation occurs is given as food distance = 0.4 and food width = 2.25.

3.2.1 Niche Separation Without Plasticity

A monomorphic population splits into two distinct populations with a clear gap in the phenotypic value (Fig. 5A). This gap in the phenotypic value between neighboring populations implies niche separation. This phenotypic gap size increases with food distance but becomes narrower as food width increases. As the food distance decreases, the niche separation is apparently lost giving rise to a single population at evolutionary equilibrium (data not shown). One of the necessary conditions for niche separation is that the food distance should be above 1.25. When the food width increases, more number of populations are observed at equilibrium. The phenotypes of the three such populations are depicted in Fig. 5B.

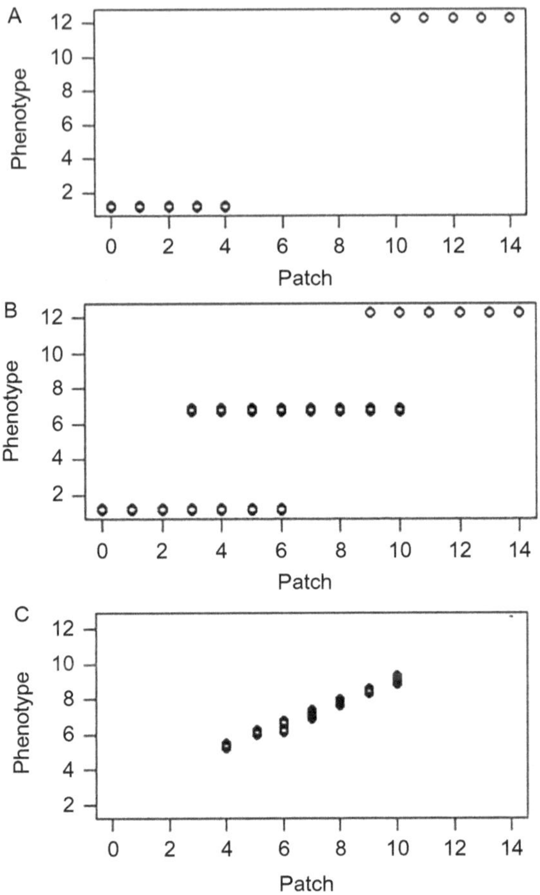

FIG. 5 The phenotypes and genotypes of the population at evolutionary equilibrium. Model parameters: migration width = 7, migration peakedness = 1, and juveniles stay at home environment. Single simulation. (A) No plasticity. Food distance = 5, food width = 1, two populations. (B) Food distance = 5, food width = 4, three populations. (C) Unrestricted plasticity, food distance = 1, food width = 1, one population, genotype is $g_0 = 5.8$ and $g_1 = 0.25$).

3.2.2 Niche Separation With Plasticity

The situation changes with the presence of unlimited plasticity. The phenotypes of the two populations are less shallow than their counterpart without plasticity. In some simulations, the phenotype can slightly decrease with the increase in patch number. This can happen purely due to random effect. As the food distance between the two Gaussian distributions decreases, a single population is obtained at equilibrium (Fig. 5C). One can see that the phenotypic values increase with patch number due to the presence of unrestricted plasticity. The niche separation is eliminated when the food width increases so that one monomorphic population is found at evolutionary equilibrium (Behera and de Jong, 2018). Compare the corresponding case in Fig. 5B where three populations are observed without plasticity.

3.2.3 Limiting Conditions for Niche Separation

There is a critical value of food width below which niche separation occurs at evolutionary equilibrium (for fixed migration width). This critical value increases for low migration peakedness. The critical value of food width increases with migration width. The necessary but not sufficient conditions for niche separation (in the presence of plasticity) are that the food width should be below a critical value (2.75) and the food distance must be above a threshold value (1.25). These limiting values are found after averaging over six simulations. In the absence of plasticity, the necessary but not sufficient criterion for niche separation to happen is different. In this case the food width should be above a threshold value (see Section 4). In general, niche separation is somewhat easier without plasticity. The population number at equilibrium increases with food width. Phenotypic plasticity allows niche separation so that an initially monomorphic population evolves to a point where it splits into two genotypically diverging lineages. It should be noted that phenotypic plasticity never allows niche separation for a single Gaussian food distribution.

3.3 Diploid Population Having Additive Allelic Effect Gene Interaction and Double Food Supply

The problem of sympatric speciation in a heterogeneous environment in diploid sexual populations is explored (Behera et al., 2018). The genetic interaction is introduced through additive allelic effect in the determination of the fitness of an individual. The ESS genotypic values will be determined through a mutation model in the limit of weak selection. The simulation is individual-based as before. Two types of models: (i) female gametes stay at parental patch but male gametes disperse (plant-like reproduction) and (ii) both male and female gametes stay at parental patches (animal-like reproduction) are investigated. An important question is if sympatric speciation can occur in a sexual population in the presence of phenotypic plasticity. For this purpose, the fitness of hybrid offspring is examined. The effect of selection strength on the probability of niche separation is examined. Finally phenotypic plasticity's importance in niche separation is studied.

The additive effect of alleles is included in the genotypes of an individual in the following manner (Gavrilets and de Jong, 1993). The new genotype of the *j*th individual becomes.

$$Ng_0[j] = g_0[j]/(2\mathrm{m}) \tag{18a}$$

$$Ng_1[j] = g_1[j]/(2\mathrm{m}) \tag{18b}$$

where m is the number of loci, $Ng_0[j]$ and $Ng_1[j]$ are the new genotypic values of the jth individual. The selection intensity is taken as 0.07. The male and female gametes are produced with fecundities of $4F$ and F, respectively. The mutation probability is 0.0001 per locus for an individual. The mutation is introduced after reproduction. The starting population is 150. The simulation is iterated for 7000 generations. The number of genetic loci is taken to be 25. The ESS genotypic values are obtained, and the distribution of phenotypes over the patches is found.

Niche separation and polymorphic behavior are observed in diploid sexual populations at evolutionary equilibrium. Two types of mating situations are explored: (i) plant like where female gametes sedentary but male gametes and zygotes disperse and (ii) animal like where both male and female gametes stay in the parental patches, while zygotes disperse.

3.3.1 Existence of Polymorphism

Polymorphism is found at equilibrium without and with phenotypic plasticity when (i) female gametes stay at parental patch but male gametes and zygotes disperse (Fig. 6A) and (ii) male and female gametes stay in parental patches, while zygotes disperse (Fig. 6B).

A monomorphic population splits into two distinct populations at evolutionary equilibrium. In order to check if real speciation occurs, the fitness of the hybrid population corresponding to the situation in Fig. 6 is examined in 7001st generation. The male and female gametes were chosen at random and united. The hybrid population refers to the cases when male and female gametes come from two different groups of populations. There is no reduced fitness for hybrid population (Behera et al., 2018). This implies that there is no speciation but only polymorphism exists.

3.3.2 Effect of Phenotypic Plasticity

Niche separation is more difficult in the presence of plasticity as is generally the case with haploid population. The phenotypic gap at the boundary separating two populations is smaller with phenotypic plasticity. It can also be seen that the phenotypes increase with the patch number in the presence of plasticity, while they remain nearly constant without plasticity.

3.3.3 Effect of Selection Strength

Niche separation becomes easier with higher selection strength. This can be seen clearly by the fact that there is a large gap in the phenotypic value separating the two populations at evolutionary equilibrium (Fig. 6A and B).

3.3.4 Dependence of Polymorphism on Dispersal Pattern

In addition to the above two situations corresponding to only female gametes or both male as well as female gametes staying in the parental patches,

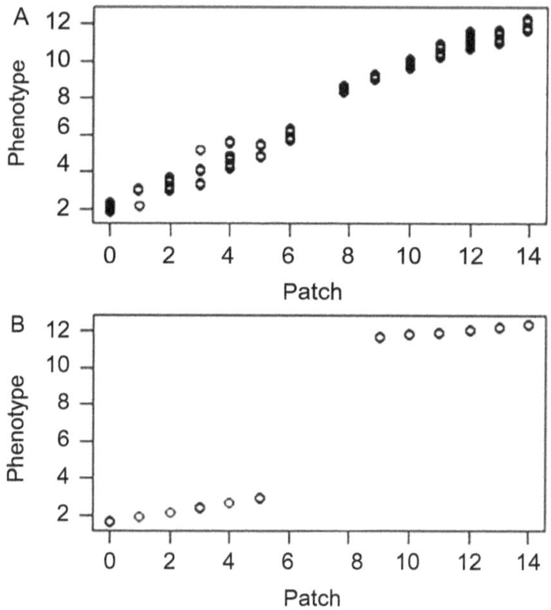

FIG. 6 The phenotypes and genotypes of the population at evolutionary equilibrium. Model parameters: migration width = 7, migration peakedness = 1, zygote dispersal width = 7, zygote dispersal peakedness = 1, food distance = 5 and food width = 1, loci number = 25. Single simulation. (A) Female gametes stay at parental patch, male gametes dispersal width = 7 and male gametes dispersal peakedness = 1. Full plasticity, selection intensity = 0.1. Genotypes: two sets are $(g_0, g_1) = (3.8, 0.4)$ and $(5.5, 0.43)$. (B) Both male and female gametes stay at parental patches. Full plasticity, selection intensity = 0.7. Genotypes: two sets are $(g_0, g_1) = (2.5, 0.4)$ and $(7.2, 0.43)$.

polymorphism also occurs in the following cases for all values of dispersal width and peakedness: (a) female gametes and zygotes stay at parental patches but male gametes disperse, (b) zygotes stay in parental patch but male and female gametes disperse, and (c) zygotes, male, and female gametes disperse (data not shown). However, in the following situations, polymorphism at evolutionary equilibrium occurs only for low dispersal width and high peakedness when (a) male gametes and zygotes stay in parental patch but female gametes disperse, and (b) male gametes stay in parental patch but zygotes and female gametes disperse.

4 DISCUSSION

The present model attempts to investigate the relationships among gene flow, phenotypic plasticity, niche width, and adaptation. Unrestricted phenotypic plasticity removes niche separation and genetic variation. The unequal food supply in different patches gives rise to differential density regulation in the patches. The adult migration makes the genotypes concentrated near the patches where there is local abundance of food supply. Hence the genotypic differentiation arises at evolutionary equilibrium. The genotypic values in the groups of populations are separated by gaps. The local adaptability causes the genotypes to be specialists. They can adapt well only in the local environment.

Without plasticity, the niche widths at evolutionary equilibrium become narrower with the persistence of specialist genotypes. The genotypes specialize to local patches due to differential density regulation and adult migration. Related type of niche limitation occurs due to gene flow and local adaptation which constrain the species to a limited range (Kirkpatrick and Barton, 1997). The larger the number of specialist genotypes, the shorter the niche widths. This is advantageous to the species. Because species with narrower niche widths can have a higher rate of evolutionary response, can fix beneficial alleles with high probability, can have lower frequency of deleterious alleles at mutation–selection equilibrium, and can persist longer as species in the sense that they can increase mean fitness faster (Whitlock, 1996). The Gaussian food distribution plays an important role in the evolution of specialist genotypes which is a form of polymorphism. Similar type of two-allele polymorphism is observed in the presence of Gaussian food supply and intraspecific competition (Christiansen and Loeschcke, 1980).

Without phenotypic plasticity, low migration width and high peakedness give many specialist populations at equilibrium. This situation reduces the probability of migration of an individual to neighboring patches. So the individuals are more likely to be confined in the local patches, thus making them more specialists. When the food width is larger, the differential density regulation among patches is relatively smaller. This causes higher survival probability in the parental patch. Thus, there is less necessity for an individual to migrate to a neighboring patch in order to survive. Hence we find more specialist genotypes and many populations as the food width increases.

When phenotypic plasticity is present, the genotypes can adapt in a wider range of environments without undergoing extinction. So the genotypes develop generalist tendency as the degree of phenotypic plasticity is increased. Finally there remain only generalist genotypes of a single population beyond a critical value of the level of plasticity. Here, phenotypic plasticity prevails over genetic variation. With unlimited plasticity, evolved generalist genotype gives rise to broader niche width for the species. Species with larger niche width have slower rate of evolutionary response (Whitlock, 1996). The splitting of the population at equilibrium occurs for a higher value of food width when the level of plasticity increases but remains below the critical value. The population never splits when plasticity is unlimited even for very large value of the food width. The population number at evolutionary equilibrium increases with the level of plasticity as plastic individuals adapt better in a variable environment. Plasticity leads to the increase of niche width. As niche width increases, the effective population size will increase proportionately (Whitlock, 1996).

When juveniles disperse with certain frequency, the specialist tendency of the genotypes is reduced. So there is less probability of niche separation. Here, we have seen that phenotypic plasticity increases niche width and tempers niche separation. It is obviously an interesting problem to investigate the situation where phenotypic plasticity allows niche separation. This can happen in a heterogeneous environment when two Gaussian food distributions are clearly separated (Behera and de Jong, 2018).

There has been considerable controversy whether phenotypic plasticity can allow separation and speciation. The present model shows that phenotypic plasticity allows niche separation in asexual population. This asexual model can act as a precursor to understand the problem of niche separation, reproductive isolation, and speciation in sexual populations.

The larger the food distance, the greater is the probability of niche separation. The monomorphic population cannot split into two or more populations at evolutionary equilibrium if this food distance is below a critical value. The food supply becomes less in more number of patches when the peaks of two Gaussian food distributions are separated by large distance. This causes the density regulation to be higher in these patches making the survival probability of the individuals lower. As evolution proceeds, the populations get more concentrated in the patches where food supply is abundant. Hence local genotypic differentiation takes place. The existence of phenotypic plasticity makes the phenotype differ in different patches. The threshold value of the food distance beyond which niche separation occurs is relatively higher in the presence of phenotypic plasticity.

Niche separation is observed for higher food width when there is no plasticity. It is lost with unrestricted plasticity for the corresponding value of food width. In the absence of plasticity when the food width is larger, there is higher survival probability of an individual in the parental patch as the differential density regulation among patches is smaller. Thus, there is less need for the individual to migrate to a neighboring patch to survive. So the genotypes develop specialist tendency and niche separation is inevitable. But unrestricted plasticity makes the individuals suitably adapted in patches other than the parental patch. So there is no gap in the phenotypic value over the patches. So, a single population at evolutionary equilibrium is seen.

There is a critical value of food width below which niche separation occurs in the presence of unlimited plasticity. This happens when the distance between the peaks of the two distributions is sufficiently large. As the food widths of the Gaussian distributions become narrower, there is more differences in the food supply in the patches causing more differential density regulation. So the genotypes cannot survive in the patches where density regulation is very high. Hence the niche separation in genotypic space becomes easier. As the migration width increases, more number of individuals will undergo strong density regulation where food supply is low. This will make local genetic differentiation easier. So instead of one continuous population, we can find it split into two populations. The critical food width has to increase. However, with larger peakedness of migration, the individuals are more confined to the parental patch. This produces more specialistic tendency. So the critical food width below which niche separation can happen decreases. With larger food width, there is less differential density regulation and more number of individuals are likely to survive at evolutionary equilibrium.

Here, it is seen that local genotypic differentiation and niche limitation occur due to differential density regulation and adult migration when the food distance of the two Gaussian food distributions is sufficiently large. Similar type of limitation for the species range is observed in the presence of gene flow, local selection pressure, and density-dependent population regulation (Kirkpatrick and Barton, 1997). The ecological interactions such as adult migration and nonuniform food supply resulting in differential density regulation are major evolutionary forces leading to niche separation with phenotypic divergence. This kind of evolutionary branching under certain ecological conditions has been investigated where the initial monomorphic population can split into diverging phenotypes after attaining the fitness minimum (Doebeli and Dieckmann, 2000). The branching at evolutionary equilibrium is observed where two distinct genotypes evolve from a single monomorphic population. Related type of evolutionary branching is found in asexual populations where a monomorphic population evolves into a polymorphic state of different strategies (Kisdi, 1999). Same class of genetic polymorphism and evolutionary branching occur in diploid sexual organisms as the evolutionary outcome of an initially monomorphic population (Geritz and Kisdi, 2000; Kisdi and Geritz, 1999). In this model, phenotypic plasticity can allow niche separation under certain situations. An interesting question is to examine whether niche separation and polymorphism can happen in a sexual population in the presence of phenotypic plasticity. This occurs in a diploid sexual population with the additive allelic effect when the two Gaussian food distributions are separated by sufficient distance (Behera et al., 2018).

It was examined if niche separation and speciation can occur in diploid sexual populations without or with plasticity. Speciation is not found as there is no reduction of the fitness of the hybrid population. The first conclusion is that polymorphism can occur without and with phenotypic plasticity. A typical example is the situation when female gametes stay in the parental patch but male gametes disperse (which can be found in plants). An initially monomorphic population evolves and splits into two diverging phenotypes at evolutionary equilibrium. This happens when the peaks of the two Gaussian food distributions are separated beyond a threshold value and the food widths lie below a critical value. These are the basic conditions for niche separation in the evolution of asexual populations (de Jong and Behera, 2002). The inclusion of additive allelic effect and genetic recombination during reproduction of sexual population does not change this basic conclusion. Basically, local abundance of food causes local genotypic specialization leading to niche separation and polymorphism. The differential density regulation in the patches caused by nonuniform food supply makes the genotypes to adopt specialist tendency to survive in the local environments.

Here, the basic evolutionary driving forces are the ecological interactions due to nonuniform food supply leading to differential density-dependent regulation and unpredictability in selection. Similar type of evolutionary branching under certain ecological conditions occurs so that the population splits into two diverging lineages after attaining the fitness minimum (Doebeli and Dieckmann, 2000). Evolutionary branching is also a probable outcome of competition for resources in an individual-based model where sexual reproduction is described by multilocus genetics (Dieckmann and Dobeli, 1999). Geritz and Kisdi (2000) have also found related type of polymorphic consequences in sexually reproducing populations. Another important area of application of the present result is ecological speciation which occurs due to ecological mechanisms. Ecological speciation requires that the fitness of hybrids be reduced relative to the parental forms. A model was designed to detect the hybrid unfitness and hence ecologically dependent isolation (Rundle and Whitlock, 2001).

Phenotypic plasticity allows niche separation in sexual populations although the probability of niche separation is easier without plasticity. Phenotypic plasticity provides a mechanism for survival of a genotype in a habitat even when the food supply is only minimum. The population is well adapted in a wide range of environments. The presence of phenotypic plasticity does not cause niche separation so difficult in asexual populations under similar situations (de Jong and Behera, 2002). In sexual populations, the additive effect of alleles provides a probable mechanism for better adaptation of the population at the boundary so that the probability of niche separation is less difficult.

The higher selection intensity makes the niche separation easier. As the selection intensity increases, the probability of survival of an individual at the boundary (where food supply is minimum) becomes more difficult. So the genotypes at more number of patches near the boundary cannot survive.

REFERENCES

Behera, N., de Jong, G., 2018. Evolution of Phenotypic Plasticity and Niche Separation in Haploid Population. unpublished.

Behera, N., Nanjundiah, V., 1995. An investigation into the role of phenotypic plasticity in evolution. J. Theor. Biol. 172 (3), 225–234.

Behera, N., Nanjundiah, V., 1996. The consequence of phenotypic plasticity in cyclically varying environments: a genetic algorithm study. J. Theor. Biol. 178 (2), 135–144.

Behera, N., Nanjundiah, V., 1997. *Trans*-gene regulation in adaptive evolution: a genetic algorithm model. J. Theor. Biol. 188, 153–162.

Behera, N., Nanjundiah, V., 2004. Phenotypic plasticity can potentiate rapid evolutionary change. J. Theor. Biol. 226, 177–184.

Behera, N., de Jong, G., Yamauchi, A., 2018. Polymorphic State at Evolutionary Equilibrium of a Diploid Population with Additive Allelic Effect. unpublished.

Christiansen, F.B., Loeschcke, V., 1980. Evolution and intraspecific exploitative competition I. One-locus theory for small additive gene effects. Theor. Popul. Biol. 18, 297–313.

Cloyed, C.S., Eason, P.K., 2017. Niche partitioning and the role of intraspecific niche variation in structuring a guild of generalist anurans. R. Soc. Open Sci. 4, 170060. https://doi.org/10.1098/rsos.170060. Published 15 March 2017.

de Jong, G., 1990. Quantitative genetics of reaction norms. J. Evol. Biol. 3, 447–468.

de Jong, G., 1995. Phenotypic plasticity as a product of selection in a variable environment. Am. Nat. 145, 493–512.

de Jong, G., 1999. Unpredictable selection in a structured population leads to local genetic differentiation in evolved reaction norms. J. Evol. Biol. 12, 839–851.

de Jong, G., Behera, N., 2002. The influence of life history differences on the evolution of reaction norms. Evol. Ecol. Res. 4, 1–25.

Dieckmann, U., Dobeli, M., 1999. On the origin of species by sympatric speciation. Nature 400, 354–357.

Doebeli, M., Dieckmann, U., 2000. Evolutionary branching and sympatric speciation caused by different type of ecological interactions. Am. Nat. 156 (S4), S77–S101.

Gavrilets, S.Y., de Jong, G., 1993. Pleiotropic models of polygenic variation, stabilizing selection and epistasis. Genetics 134, 609–625.

Gavrilets, S.Y., Scheiner, S.M., 1993. The genetics of phenotypic plasticity. V. Evolution of reaction norm shape. J. Evol. Biol. 6, 31–48.

Geritz, A.H.S., Kisdi, E., 2000. Adaptive dynamics in diploid, sexual populations and the evolution of reproductive isolation. Proc. R. Soc. Lond. 267, 1671–1678.

Grant, P.R., Grant, B.R., 1994. Phenotypic and genetic effects of hybridization in Darwin's finches. Evolution 48, 297–316.

Hollander, J., et al., 2015. New frontiers in phenotypic plasticity and evolution. Heredity 115, 273–275.

Kirkpatrick, M., Barton, N.H., 1997a. Evolution of a species' range. Am. Nat. 150, 1–23.

Kisdi, E., 1999. Evolutionary branching under asymmetric competition. J. Theor. Biol. 97, 149–162.

Kisdi, E., Geritz, S.A.H., 1999. Adaptive dynamics in allele space: evolution of genetic polymorphism by small mutations in a heterogeneous environment. Evolution 53, 993–1008.

Loreau, M., Behera, N., 1999. Phenotypic diversity and stability in ecosystem processes. Theor. Popul. Biol. 56, 29–47.

Mayr, E., 1963. Animal Species and Evolution. Harvard University Press, Belknap, Cambridge, MA.

Okubo, A., 1980. Diffusions and Ecological Problems: Mathematical Models. Springer, Berlin.

Pfenning, D.W., et al., 2010. Phenotypic plasticity's impacts on diversification and speciation. Trends Ecol. Evol. 25, 429.

Plante, A.F., Fernandez, J.M., Haddix, M.L., Steinweg, J.M., Conant, R.C., 2011. Biological, chemical and thermal indices of soil organic matter stability in four grassland soils. Soil Biol. Biochem. 43, 1051–1058.

Reed, T.E., et al., 2011. Interacting effects of phenotypic plasticity and evolution on population persistence in a changing climate. Conserv. Biol. 25, 56–63.

Rundle, H.D., Whitlock, M.C., 2001. A genetic interpretation of ecologically dependent isolation. Evolution 55, 198–201.

Sasaki, A., de Jong, G., 1999. Density-dependence and unpredictable selection in a heterogeneous environment: compromise and polymorphism in the ESS reaction norm. Evolution 53, 1329–1342.

Scheiner, S.M., 1998a. The genetics of phenotypic plasticity. VII. Evolution in a spatially structured environment. J. Evol. Biol. 11, 303–320.

Via, S., Lande, R., 1985. Genotype-environmental interaction and the evolution of phenotypic plasticity. Evolution 39, 505–522.

Whitlock, M.C., 1996. The red queen beats the jack-of-all-traders: the limitations on the evolution of phenotypic plasticity and niche breadth. Am. Nat. 148, S65–S77.

Yamamura, Y., Higashi, M., Behera, N., Wakano, J.Y., 2004. Evolution of mutualism through spatial effect. J. Theor. Biol. 226, 421–428.

FURTHER READING

Barton, N.H., Charlesworth, B., 1984. Genetic revolutions, founder effect and speciation. Annu. Rev. Ecol. Syst. 15, 133–164.

Barton, N.H., Hewitt, G.M., 1985. Analysis of hybrid zones. Annu. Rev. Ecol. Syst. 16, 113–148.

Behera, N., 1994. Phenotypic plasticity and genetic assimilation in development and evolution. Bionature 14, 1–22.

Behera, N., 1997. Effect of phenotypic plasticity on adaptation and evolution: a genetic algorithm analysis. Curr. Sci. 73, 968–976.

Bush, G.L., 1994. Sympatric speciation in animals: new wine in old bottles. Trends Ecol. Evol. 9, 285–288.

Doebeli, M., 1996. A quantitative genetic competition model for sympatric speciation. J. Evol. Biol. 9, 893–909.

Glazier, D.S., 1987. Toward a predictive theory of speciation: the ecology of isolate selection. J. Theor. Biol. 126, 323–333.

Johannesson, K., 2001. Parallel speciation: a key to sympatric divergence. Trends Ecol. Evol. 16, 148–152.

Metz, J.A.J., Nisbet, R., Geritz, S.A.H., 1992. How should we define 'fitness' for general ecological scenarios? Trends Ecol. Evol. 7, 198–202.

Rice, W.R., Hostert, E.E., 1993. Laboratory experiments on speciation: what have we learned in 40 years? Evolution 47, 1637–1653.

Smith, M.J., 1966. Sympatric speciation. Am. Nat. 100, 637–650.

Tregenza, T., Butlin, R.K., 1999. Speciation without isolation. Nature 400, 311–312.

West-Eberhard, M.J., 1983. Sexual selection, social competition, and speciation. Q. Rev. Biol. 58, 155–183.

Section IX

Aging and Age-Structured Population Dynamics

Chapter 10

Theory and Applications of Backward Probabilities and Prevalences in Cross-Longitudinal Surveys

Nicolas Brouard[1]
National Institute for Demographic Studies (INED), Paris, France
[1]*Corresponding author: e-mail: brouard@ined.fr*

ABSTRACT

In this chapter we introduce the backward probability and the backward prevalence. Both measures are of minor interest compared to forward probability and forward prevalence; however, they can bring a better understanding of the past population dynamic. Moreover, we show that through the calculation of backward probabilities, one can reconstruct prevalences at any age for older generations. Here, the demographic interest lies in the comparison of the three prevalences: cross-sectional, forward, and backward. In order to accomplish our task, we first review theories of Markov chains with (i) an age-independent transition matrix and, (ii) when transitions vary with age. The second theory leads to an interesting property called "weak ergodicity" that allows us to predict future prevalences for younger generations. It is important to mention here that backward probability was rapidly defined in 1980 (Brouard, 1980) using longitudinal information of French women's participation in economic activity between 1977 and 1978, and most results presented in the publication of 1980 are reviewed for this study.

This chapter also shows that in a stationary multistate population, cross-sectional, forward, and backward prevalences are identical at each age. If they are not, as in the case of the economic activity of French women which changed after the 1968s revolution (women's liberation, contraception, and abortion laws), our approach enables a clearer, faster, and synthetic analysis of these changes without the need to wait 20 or 30 years until these women meave the labor market.

Then, we review demographic tools, now widely used in mortality analysis that compare "cross-sectional prevalence of survival" and period mortality table. We extend them to multistates methods, particularly to methods developed in the mid-1990s to estimate disability-free life expectancies.

Handbook of Statistics, Vol. 40. https://doi.org/10.1016/bs.host.2018.11.009

But in current cross-sectional surveys, as in the case of health surveys, people do not respond to the survey at every wave, so the health status may be unknown or known with exposure times unequal between individuals, making multinomial logistic regression useless. A more complex model like the interpolated Markov chain model must be applied. Since the software IMaCh has been developed in the late 1990s, the use of the interpolated Markov chain model greatly increased among health and disability researchers and more accurate estimates of healthy life expectancy can be calculated.

It is also important to mention here that the latest experimental version 0.99 of IMaCh integrates the calculation of the backward prevalences with its confidence intervals. In this chapter, we apply this latest IMaCh version to show the relevance of backward and forward prevalences to understand the aging of societies resorting recent data from cross-longitudinal surveys in Italy and in the United States.

Keywords: Demography, Statistics, Multistate, Mortality, Labor force, Cross-longitudinal survey, Life expectancy, Markov chain, Ergodicity

MSC: 00A71, 47A35

A backward probability can be defined as the probability to act today while being conditioned by an event which will occur in the future.

A classical probability or "forward" probability is, for example, the probability of having a child 2 years after marriage; or more generally since the time elapsed since marriage, but a demographer might be interested in studying fertility according to the time remaining before the divorce, because a potential separation could explain a lower fertility. A backward probability can also be considered as a measure of indirect variables and latent variables that are unknown in the present but nevertheless influence the future.

Forward probabilities in demography, economics, and statistics are common and powerful. Backward probabilities are clearly less relevant because literature on this concept seems to be rare or even nonexistent. But are they completely useless? We will explore their power in a better understanding of our past history using an already published example (Brouard, 1980) as well as to propose a new software, IMaCh 0.99 (2019; http://euroves.ined.fr/imach) which estimates backward probabilities (as well as their companion the backward prevalences) from adequate data. These data are provided by a particular type of statistical survey that are becoming more common nowadays and are now called longitudinal cross-sectional surveys or cross-longitudinal surveys (Lièvre et al., 2003).

In simple longitudinal surveys, such as labor force surveys, where mortality before retirement age is negligible, backward probabilities are as natural as forward probabilities, but they are different and can be combined to produce backward prevalence. Forward probabilities can be applied to a fictitious young cohort to produce forward age-specific prevalences. Then, they can be compared with current cross-sectional prevalences and even with the forward prevalences of younger cohorts.

If the concept of forward probability and forward prevalences is well known in the case of a period life table, the concept of backward probabilities and backward prevalences makes sense only in multistate models.

1 BACKWARD PROBABILITIES ESTIMATED FROM CHAINED LABOR FORCE SURVEYS

Let us take the example of a labor force survey where women had their first interview in 1977 describing their economic activity status and interviewed again 1 year later in 1978 to measure changes. From this follow-up, age-specific input/output matrices can be constructed similar to Table 1 describing the repartition of the $N_{..}$ women into $N_{1.}$ active and $N_{2.}$ inactive women at age x in 1977 as well as the various flows which occurred in one year up to age $x + 1$ in 1978.

1.1 Probability or Forward Probability

From this table, we can calculate the age-specific probability (also called forward probability) of leaving the labor force in one year by $\hat{c}_x = \frac{N_{12}}{N_{1.}}$, as well as the probability of entering or reentering the labor force by $\hat{a}_x = \frac{N_{21}}{N_{2.}}$.

We can construct age-specific (forward) transition matrices P_x

$$P_x = \begin{pmatrix} 1-\hat{c}_x & \hat{c}_x \\ \hat{a}_x & 1-\hat{a}_x \end{pmatrix}$$

and by right multiplying them, we can calculate the probability of being out of the labor force after n years for someone in the labor force at age x as well as the probability of being in the labor force after n years for someone outside the labor force at age x, etc.

TABLE 1 Matrix of Input/Output by Age: Two Linked Labor Surveys

Age x + 1 1978	Active	Inactive	
Age x 1977			
Actives	N_{11}	N_{12}	$N_{1.}$
Inactives	N_{21}	N_{22}	$N_{2.}$
	$N_{.1}$	$N_{.2}$	$N_{..}$

A step by step description can be done in order to better understand the figures and the ergodic properties of multiplying stochastic matrices:

- If the probability for a woman inactive at age 30 in 1977 to be active at age 31 in 1978 is 12%;
- if the probability for a woman inactive at age 31 in 1977 to be active at age 32 in 1978 is 12.04%;
- if the probability for a woman active at age 30 in 1977 to be still active at age 31 in 1978 is 94%;
- if the probability for a woman active at age 31 in 1977 to be still active at age 32 in 1978 is 95%;

then we can write, using the Markov assumption that the probabilities depend only from the current state, that the probability for a woman active at age 30 to be out of the labor at 32 is

$$\begin{array}{ccccc} {}_2P_{30} & = & P_{30} & & P_{31} \\ \begin{pmatrix} 90\% & 10\% \\ 22\% & 78\% \end{pmatrix} & = & \begin{pmatrix} \mathbf{94\%} & 6\% \\ \mathbf{12\%} & 88\% \end{pmatrix} & & \begin{pmatrix} \mathbf{95\%} & 5\% \\ \mathbf{12.04\%} & 87.96\% \end{pmatrix} \end{array}$$

And for a projection in k years ahead:

$${}_kP_x = P_x P_{x+1} \cdots P_{x+k-1}$$

The product matrix benefits, when k is increasing, of weak ergodicity properties which can be visually appreciated in Fig. 1, providing a stable curve, which is also the limit of the projections (see Fig. 2).

It has be called the "period" activity curve, because it is computed only from age-specific incidences estimated during the 1977–1978 period of these two first chained labor force surveys (Brouard, 1980). Now, it will be also called the forward prevalence of activity.

Comparing the cross-sectional curves of activity by age (in 1977 and 1978) with projected curves and more interestingly with the ultimate stable curve, we can draw some interesting sociological conclusions concerning this period in France.

Because of the development of massive contraception in the mid-1970s (contraception law in 1967, abortion law in 1975), women changed their employment and maternity behavior. It has become easier to postpone the formation of a family or to supplement one's family while retaining a job and enter or reenter the labor market even if the children were already born.

The period (forward) prevalence is induced by the probabilities a_x and c_x, which are represented on the graph in Fig. 3.

The cross-sectional prevalence observed in 1977 or 1978 is not stable. It is analogous to an age pyramid from a census which differs from the stable population induced by the age specific fertility and mortality rates.

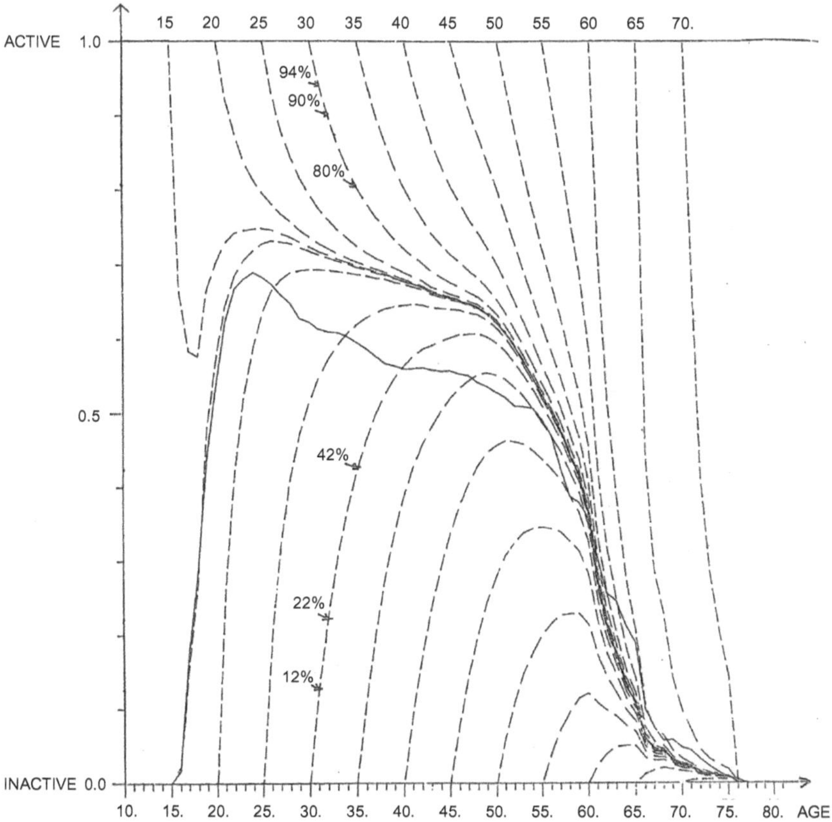

FIG. 1 Probability for a woman inactive at age 30 in 1977 to be active at age 31 (12%) in 1978, 32 (22%) in 1979 as well as (on top), probability for a woman active at age 30 to be still active at 31 (94%), 32 (90%), etc. *From Brouard, N., 1980. Espérance de vie active, reprises d'activité féminine: un modèle. Revue Econ. 31 (6), 1260–1287. https://doi.org/10.2307/3501190 (special issue on Economy and Demography).*

1.2 Backward Probability

According to Table 1 we can compute a different probability $b^{12} = \frac{N_{12}}{N_{.2}}$ defined as the probability to be inactive at age x knowing that we will be active at age $x + 1$. It can be called a *backward* probability, as well as for $b^{21} = \frac{N_{21}}{N_{.1}}$ which is the probability to be active at age x knowing that will be inactive at age $x + 1$.

Let us remark that extending the classical notation in demography for a probability of death ${}_hq_x$, the full notation should be ${}_1p_x^{ij}$ and ${}_{-1}b_{x+1}^{ij}$. We will keep the left index notation only for matrices.

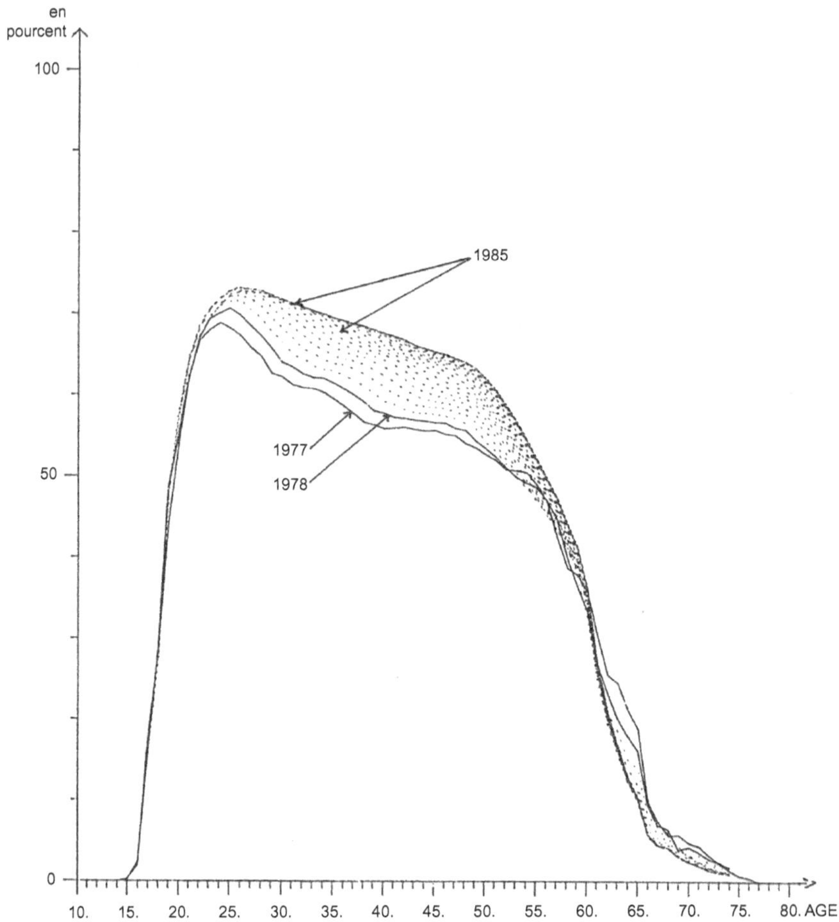

FIG. 2 Observed activity ratios by age in 1977 and 1978 and projected in the future until convergence is reached. Projections after year 1985 can be distinguished from the limit only between ages 40 and 50. *From Brouard, N., 1980. Espérance de vie active, reprises d'activité féminine: un modèle. Revue Econ. 31 (6), 1260–1287. https://doi.org/10.2307/3501190 (special issue on Economy and Demography).*

Then we can make backward projections

$$B_{x+1} = \begin{pmatrix} 1-b^{21} & b^{12} \\ b^{21} & 1-b^{12} \end{pmatrix}$$

$$_{-n}B'_{x+1} = B'_{x+1}B'_{x}\cdots B'_{x-n}$$

and because of the weak ergodicity property of stochastic matrices again we get convergence in the past in Fig. 4.

But what is very interesting in this backward prevalence is that it corresponded to the age-specific activity ratios that prevailed in France before 1968,

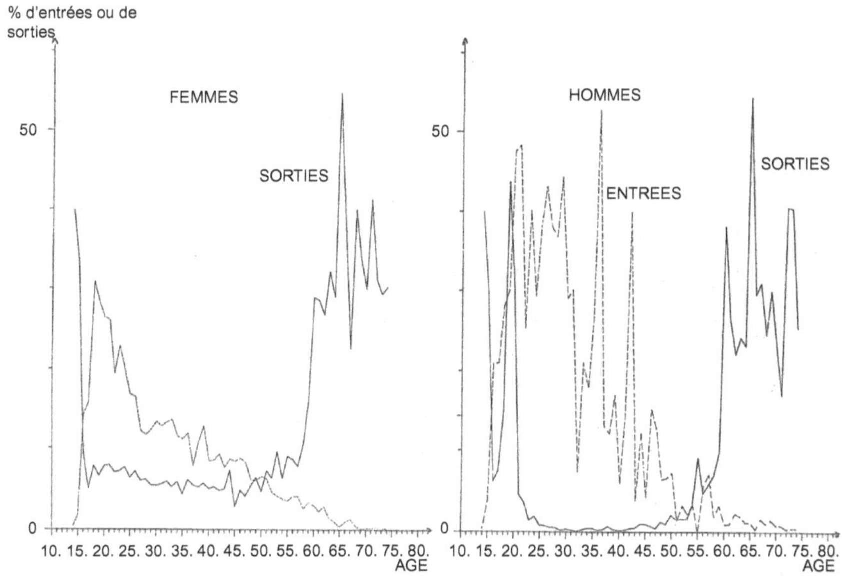

FIG. 3 Probability of entering a_x or leaving c_c the labor force in France (1977–1978) by sex. *From Brouard, N., 1980. Espérance de vie active, reprises d'activité féminine: un modèle. Revue Econ. 31 (6), 1260–1287. https://doi.org/10.2307/3501190 (special issue on Economy and Demography).*

with the well-known "M" shape: the women were at work young but left their jobs to raise their families and some of them did return to the labor force when the children were already grownup.

We can appreciate the quality of both projections in Fig. 5 by comparing with the activity ratios observed during the censuses of 1968, 1982, and 1990 in France.

We remark that the forward prevalence better fits the prevalence of 1982 at younger ages and that of 1990 at older ages: we can see in Fig. 2 that the horizon of 1982 is only 4 years after 1978 and that the horizon of 1990 is 5 years after 1985 and cannot be distinguished from the limit.

If the backward prevalence reveals the "M" shape and if this backward projection is correct at older ages proving that the retirement age was higher in 1968 than in 1977–1978, the hollow of the wave is not low enough and reflecting also the fact that the limit of the backward projection cannot be estimated earlier than in 1972 which is close to stable backward prevalence as shown in Fig. 6.

From a demographic point of view, the analogy with life table analysis is simple. A life table is nothing more than a projection of cohort survival according to the age-specific mortality rates observed over a period of time. And projections applied to complete cohorts of newborns are less reliable than those applied to middle-aged cohorts.

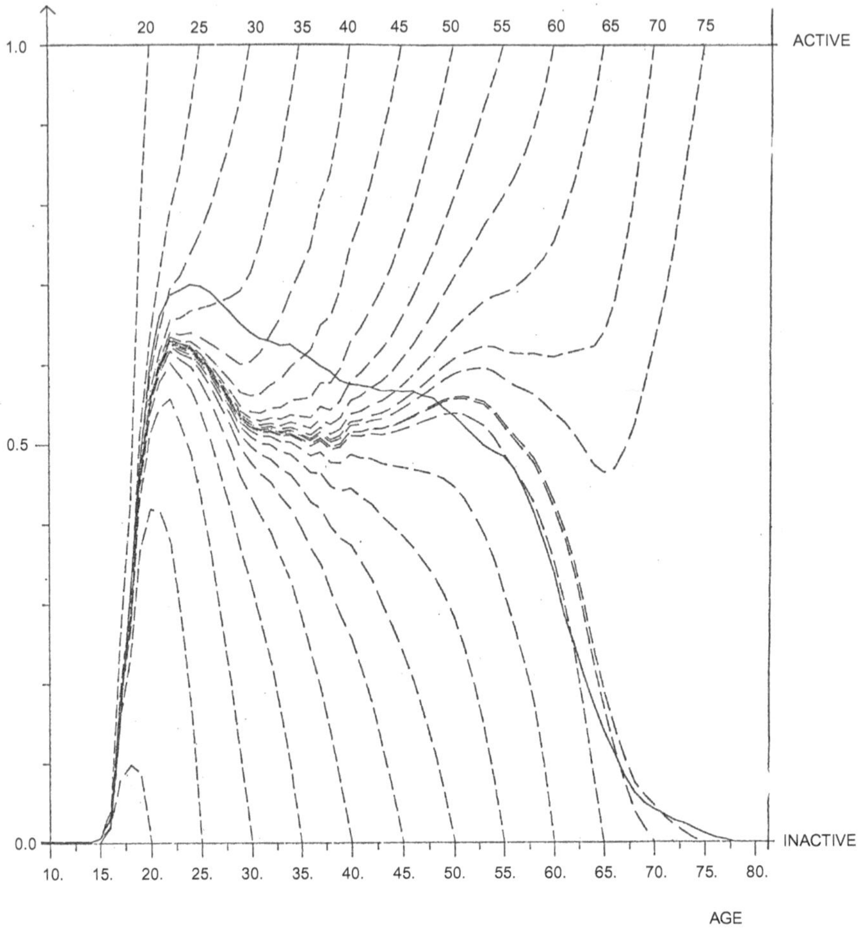

FIG. 4 Backward probability to be active at $x - h$ in a cohort, active or inactive at age x. *From Brouard, N., 1980. Espérance de vie active, reprises d'activité féminine: un modèle. Revue Econ. 31 (6), 1260–1287. https://doi.org/10.2307/3501190 (special issue on Economy and Demography).*

After this intuitive presentation of the forward and the backward probabilities describing their ability to predict changes in populations by amplifying apparently minor age-related changes, we return to theoretical parts. We will also discuss the divergence observed in Fig. 6 with the retroprojections into the future.

1.3 Markov Chains and Strong Ergodicity

In this section, we will suppose that the transition matrix P_x is independent of age, which means that $c_x = c$ and $a_x = a$. We will review the properties of a simple Markov chain, the convergence into the future, but also the divergence

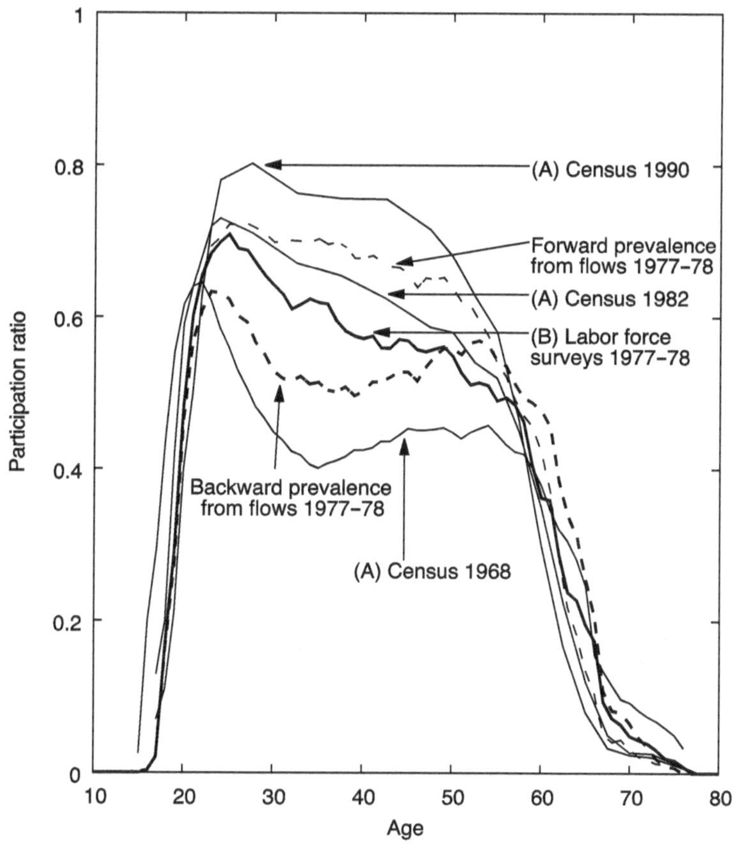

FIG. 5 Female labor participation ratios by age from (A) Censuses in 1968, 1982, and 1990. (B) From the labor force surveys in 1977–1978 (raw data). Comparison with the forward prevalence and backward prevalence computed from the chained labor forces in 1977 and 1978.

into the past (retroprojection). Then, we will introduce the back probabilities which will induce a different Markov chain, which will converge into the past and subsequently diverge into the future.

Let us remember some properties of a Markov chain with n states. The transition matrix P of a Markov chain is a stochastic matrix which means that the sum of the coefficients of each row is one ($\sum_{j=1,n} p_{ij} = 1, \forall i$). Let us suppose for simplicity that the n roots, real or complex, $\lambda_1, \lambda_2, \ldots, \lambda_n$ of $\det(P - \lambda I)$ are distinct and not equal to zero. Then the space engendered by the n eigenvectors is also of dimension n. The matrix P can be diagonalized in the space of the n latent vectors.

Let U be the matrix of the n right eigenvectors in column (U_i, $i = 1, n$), $V^\top$ the transposed matrix of the left eigenvectors in column (${V_i}^\top, i = 1, n$) and Λ

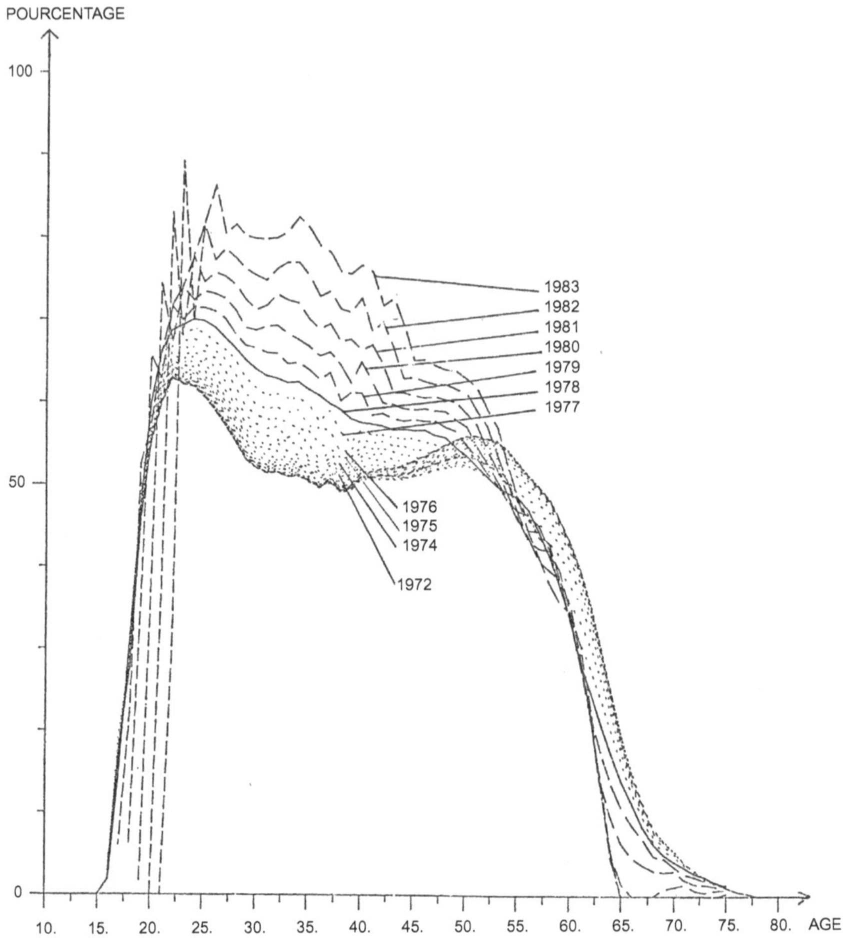

FIG. 6 Backward projections of female participation rates (France 1977–1978). Backward convergence to the situation before 1968. Years before 1972 are hard to be distinguished from the limit. Retroprojections into the future are divergent. *From Brouard, N., 1980. Espérance de vie active, reprises d'activité féminine: un modèle. Revue Econ. 31 (6), 1260–1287. https://doi.org/10.2307/3501190 (special issue on Economy and Demography).*

the diagonal matrix of the n eigen values ordered by decreasing magnitude $|\lambda_1| > |\lambda_2| > \cdots > |\lambda_n|$. Then, we can write

$$PU_i = \lambda_i U_i$$
$$V_j^\top P = \lambda_j V_j^\top$$
$$V_j^\top P U_i = \lambda_j V_j^\top U_i = V_j^\top \lambda_i U_i$$
$$\text{or } (\lambda_j - \lambda_i) V_j^\top U_i = 0$$
$$\text{and } V_j^\top U_i = \begin{cases} 0 \text{ if } i \neq j \\ k_i \text{ if } j = i \end{cases}$$

It is easy to normalize the eigenvectors such that $k_i = 1$ and $V_j^{\top} U_i = \delta_{ij}$ (where δ_{ij} is equal to 1 if $i = j$ and 1 if $i \neq j$), $V^{\top}U = I$, $U^{\top}V = I$ so that $U^{-1} = V^{\top}$.

Then we can write

$$PU = U\Lambda$$
$$V^{\top}P = \Lambda V^{\top}$$
$$\Lambda = \begin{pmatrix} \lambda_1 & & \\ & \ddots & \\ & & \lambda_n \end{pmatrix}$$
$$V^{\top}PU = \Lambda V^{\top}U = \Lambda$$

and by right multiplying first line by V^t

$$PUV^{\top} = P = U\Lambda V^{\top} = \sum_{i=1,n} \lambda_i U_i V_i^{\top} = U_1 V_1^{\top} + \sum_{i=2,n} \lambda_i U_i V_i^{\top}$$

Also,

$$P^2 = U\Lambda V^{\top} U\Lambda V^{\top} = U\Lambda^2 V^{\top} \tag{1}$$

$$P^k = U\Lambda^k V^{\top} = \sum_{i=1,n} \lambda_i^k U_i V_i^{\top} \tag{2}$$

Because of the stochasticity of the matrix P, $\lambda = 1$ is a latent root and $e^{\top} = (1, 1, \ldots, 1)^{\top}$ is a (right) latent vector, $Pe = e$. Let $V_1^{\top}$ be the corresponding left latent vector with latent value to 1, it is also a latent vector for any matrix P^k and is the stationary distribution of probabilities

$$V_1^{\top}P = V_1^{\top} = V_1^{\top}P^k \quad \forall k$$
$$V_{1h} = \sum_j V_{1j} P_{jh} \quad \forall h \tag{3}$$

If a left latent vector W of P is associated to a latent value λ, it is also the left latent vector of the iterated matrix P^k with latent value λ^k

$$\lambda^k W_h = \sum_j W_j P_{jh}^k \quad \forall h \text{ and } k \tag{4}$$

$$|\lambda^k||W_h| \leq \sum_j |W_j| P_{jh}^k \quad \forall h \text{ and } k \tag{5}$$

Let us notate h_{max} the index of the highest module $|W_h|$, $h = 1, n$

$$|\lambda^k||W_{h_{\max}}| \leq |W_{h_{\max}}| \sum_j P_{jh}^k \leq |W_{h_{\max}}| \sum_j \max_j P_{jh}^k \leq |W_{h_{\max}}| n$$

thus $|\lambda^k| \leq n$ for any k which implies that $|\lambda| \leq 1$. The diagonal matrix Λ can then be ordered by decreasing value of the magnitude of the latent values $1 < |\lambda_2| < \ldots < |\lambda_n|$.

Because of the stochasticity of the transition of a Markov chain, Eq. (2) can be written

$$P^k = U\Lambda^k V^\top = U_1 V_1{}^\top + \sum_{i=2,n} \lambda_i^k U_i V_i{}^\top \tag{6}$$

The transition probabilities P_{ij}^k reach a limit V_{1j} ($j = 1, n$) in the sense that $\lim_{k\to\infty} P_{ij}^k = V_{1j}, \forall i$. In order to prove it, it is sufficient to right multiply by P_{jh}, to sum over j and to use Eq. (3)

$$\lim_{k\to\infty} \sum_{j=1}^{n} P_{ij}^k P_{jh} = \sum_{j=1}^{n} V_{1j} P_{jh} = V_{1h}$$

We can also see from Eq. (6) that the speed of the convergence depends essentially on the module of second highest latent value $|\lambda_2|$.

1.3.1 Two States: Activity and Inactivity by Age

In the simple case of two states, active and inactive, the transition matrix P can be written

$$P = \begin{pmatrix} 1-c & c \\ a & 1-a \end{pmatrix} = U\Lambda V^\top \tag{7}$$

$$\Lambda = \begin{pmatrix} 1 & 0 \\ 0 & 1-c-a \end{pmatrix} \tag{8}$$

$$U = \frac{1}{\sqrt{c+a}} \begin{pmatrix} 1 & -c \\ 1 & a \end{pmatrix} \tag{9}$$

$$V^\top = \frac{1}{\sqrt{c+a}} \begin{pmatrix} a & c \\ -1 & 1 \end{pmatrix} = U^{-1} \tag{10}$$

$$(c+a)P = \begin{pmatrix} a & c \\ a & c \end{pmatrix} + (1-c-a)\begin{pmatrix} c & c \\ -a & a \end{pmatrix} \tag{11}$$

$$(c+a)P^k = (c+a)U\Lambda^k V^\top = \begin{pmatrix} a & c \\ a & c \end{pmatrix} + (1-c-a)^k \begin{pmatrix} c & c \\ -a & a \end{pmatrix} \tag{12}$$

We suppose that $0 < a < 1$ and $0 < c < 1$. We will study the case with absorbing states in a next section.

Let A_x and I_x be the number of people, respectively, active and inactive at age x, N be the total population of age x or $x + 1$, $N = A_x + I_x$, y_x the prevalence of activity $y_x = A_x/N$. Then, the numbers of people active and inactive at age $x + 1$ and $x + k$ are given by the formula

$$(A_{x+1} \;\; I_{x+1}) = (A_x \;\; I_x)P \tag{13}$$

$$(A_{x+k} \;\; I_{x+k}) = (A_x \;\; I_x)P^k \tag{14}$$

In terms of prevalence of activity the formula is

$$(y_{x+1} \;\; 1-y_{x+1}) = (y_x \;\; 1-y_x)P \tag{15}$$

$$(y_{x+k} \;\; 1-y_{x+k}) = (y_x \;\; 1-y_x)P^k \tag{16}$$

and we can write

$$\lim_{k\to\infty} P^k = \begin{pmatrix} \frac{a}{a+c} & \frac{c}{a+c} \\ \frac{a}{a+c} & \frac{c}{a+c} \end{pmatrix} \tag{17}$$

or

$$(A_{x+k} \;\; I_{x+k}) \simeq (A_x \;\; I_x) \begin{pmatrix} \frac{a}{a+c} & \frac{c}{a+c} \\ \frac{a}{a+c} & \frac{c}{a+c} \end{pmatrix} \tag{18}$$

$$\lim_{k\to\infty} A_{x+k} = \frac{a}{a+c}(A_x + I_x) = \frac{a}{a+c}N + \tag{19}$$

$$+ \frac{(1-c-a)^k}{a+c}(cA_x - aI_x) \tag{20}$$

$$y_{x+k} = \frac{a}{a+c} + \frac{(1-c-a)^k}{a+c}\frac{cA_x - aI_x}{N} \tag{21}$$

$$y_{x+k} = \frac{a}{a+c} + (1-c-a)^k\left(y_x - \frac{a}{a+c}\right) \tag{22}$$

When k is big and because $|\lambda_2| = |1 - c - a| < 1$, the prevalence of activity converges to a limit independent of the initial prevalence

$$\lim_{k\to\infty} y_{x+k} = y_L = \frac{a}{a+c} \tag{23}$$

We can also remark that when the initial prevalence at age x is the stationary prevalence $y_x = y_L$, the prevalence at any age $x + k$ is constant and also equal to the stationary prevalence.

When $1 - c - a = 0$ the limit $y_L = a = 1 - c$ is reached in one step. When $1 - c - a \leq 0$ the convergence is alternated. When $a + c$ is close to the unreachable (by definition) maximum of 2, the convergence to $\frac{1}{2}$ is slow and alternated.

Let us now review how we classically define a retroprojection that is unrelated to what we call a backward prevalence.

1.3.2 Two States, Retroprojection

As none of the n eigenvalues of P is equal to zero, we can calculate the number of people active and inactive at age x by reversing the recurrence and using the inverse P^{-1} of the transition matrix P

$$\begin{pmatrix} y_x & 1-y_x \end{pmatrix} = P^{-1}\begin{pmatrix} y_{x+1} & 1-y_{x+1} \end{pmatrix} \tag{24}$$

$$P^{-1} = \frac{1}{1-a-c}\begin{pmatrix} 1-a & -c \\ -a & 1-c \end{pmatrix} \tag{25}$$

$$P^{-1} = U\Lambda^{-1}U^{-1} \tag{26}$$

or, with numbers instead of prevalence and after k iterations

$$\begin{pmatrix} A_x & I_x \end{pmatrix} = P^{-1}\begin{pmatrix} A_{x+1} & I_{x+1} \end{pmatrix} \tag{27}$$

$$\begin{pmatrix} A_{x-k} & I_{x-k} \end{pmatrix} = (P^{-1})^k\begin{pmatrix} A_{x+1} & I_{x+1} \end{pmatrix} \tag{28}$$

$$(c+a)P^{-k} = (c+a)U\Lambda^{-k}V^{\top} = \begin{pmatrix} a & c \\ a & c \end{pmatrix} + (1-c-a)^{-k}\begin{pmatrix} c & c \\ -a & a \end{pmatrix} \tag{29}$$

It is now obvious that the process is divergent and that the new prevalence is very sensitive to the prevalence value at age $x + 1$

$$\begin{pmatrix} A_{x-k} & I_{x-k} \end{pmatrix} \simeq \frac{1}{(1-c-a)^k}\begin{pmatrix} c & c \\ -a & a \end{pmatrix}\begin{pmatrix} A_{x+1} & I_{x+1} \end{pmatrix} \tag{30}$$

$$y_{x-k} = \frac{a}{a+c} + (1-c-a)^{-k}\left(y_{x+1} - \frac{a}{a+c}\right) \tag{31}$$

1.3.3 Two States, Back Probability

In order to reconstruct an initial table, M, with numbers, similar to Table 1, we consider that the population is of size N (we simplify the notation $N_{..}$) for all ages x.

The population at age x is divided into two subpopulations, the active people yN and the inactive people $(1 - y)N$. Let c and a be the constant parameters of the forward matrix of transition P_x, then we can write

$$M(c,a,y,N) = \begin{pmatrix} (1-c)yN & cyN \\ a(1-y)N & (1-a)(1-y)N \end{pmatrix} \tag{32}$$

The active population at age $x + 1$ and year $t + 1$ is then $(1 - c)yN + a(1 - y)N$ out of which $a(1 - y)N$ people were inactive at age x and year t. The backward probability for a woman to be inactive at age x knowing that she will be active at age $x + 1$ is then

$$\gamma = \frac{a(1-y)}{(1-c)y + a(1-y)} \tag{33}$$

Also the backward probability for a woman to have been active at age x knowing that she is inactive at age $x + 1$ is

$$\alpha = \frac{cy}{cy + (1-a)(1-y)} \tag{34}$$

The prevalence of activity at age $x + 1$ is

$$y^* = (1-c)y + a(1-y) = a + y(1-c-a) \tag{35}$$

is constant but differs from the prevalence y unless $y = \frac{a}{a+c}$.

With the notations α, γ and y^*, we can write M as

$$M(\alpha,\gamma,y^*,N) = \begin{pmatrix} (1-\gamma)y^*N & \alpha(1-y^*)N \\ \gamma y^*N & (1-\alpha)(1-y^*)N \end{pmatrix} \tag{36}$$

so that the transposed matrix $M^\top(\alpha, \gamma, y^*, N)$ has an analogous structure to $M(c, a, y, N)$.

The backward matrix of probabilities is thus dependent of the initial prevalence y (at age x) and can be written

$$B_{x+1}(y) = \begin{pmatrix} 1-\gamma(y) & \alpha(y) \\ \gamma(y) & 1-\alpha(y) \end{pmatrix} \tag{37}$$

Its transposed matrix

$$B_{x+1}{}^\top(y) = \begin{pmatrix} 1-\gamma(y) & \gamma(y) \\ \alpha(y) & 1-\alpha(y) \end{pmatrix} \tag{38}$$

is a stochastic matrix of transition from age $x + 1$ to x which corresponds to a Markov chain converging into the past.

We can also write

$$M = N\begin{pmatrix} 1-\gamma & \alpha \\ \gamma & 1-\alpha \end{pmatrix}\begin{pmatrix} y^* & 0 \\ 0 & 1-y^* \end{pmatrix} = N\begin{pmatrix} y & 0 \\ & 1-y \end{pmatrix}\begin{pmatrix} 1-c & c \\ a & 1-a \end{pmatrix} \tag{39}$$

We can use the notation w_x^i for the prevalence in state i at age x, and here with only two states, activity and nonactivity, we have $w_x^1 = y_x$, $w_x^2 = 1 - y_x$. Also using the notation W_x^d for the diagonal matrix of prevalences in each state, we have

$$W_x^d = \begin{pmatrix} w_x^1 & 0 \\ 0 & w_x^2 \end{pmatrix} = \begin{pmatrix} y_x & 0 \\ 0 & 1-y_x \end{pmatrix}. \tag{40}$$

Using matrices, M can be written

$$M = NB_{x+1}W_{x+1}^d = NW_x^dP_x \tag{41}$$

and the duality can be expressed in this form hereafter

$$B_{x+1}W^d_{x+1} = W^d_x P_x. \tag{42}$$

The matrix transition of the backward process, or backward Markov chain, is

$$B_{x+1}{}^\top = (W^d_{x+1})^{-1} P_x{}^\top W^d_x \tag{43}$$

or

$$\begin{pmatrix} 1-\gamma & \gamma \\ \alpha & 1-\alpha \end{pmatrix} = \begin{pmatrix} \frac{1}{y_{x+1}} & 0 \\ 0 & \frac{1}{1-y_{x+1}} \end{pmatrix} \begin{pmatrix} 1-c & a \\ c & 1-a \end{pmatrix} \begin{pmatrix} y_x & 0 \\ 0 & 1-y_x \end{pmatrix} \tag{44}$$

from which the values of γ (Eq. 33) and α (Eq. 34) can be easily verified.

Introducing the logit function defined by $\operatorname{logit} x = \log \frac{x}{1-x}$, we can write

$$\frac{1}{\gamma} - 1 = \frac{1-c}{a} \frac{y}{1-y}$$

$$\frac{1}{\alpha} - 1 = \frac{1-a}{c} \frac{1-y}{y} \text{ and by log, we get}$$

$$\operatorname{logit} \gamma = -\log\left(\frac{1-c}{a}\right) - \operatorname{logit} y$$

$$\operatorname{logit} \alpha = -\log\left(\frac{1-a}{c}\right) + \operatorname{logit} y$$

From last two equations we get

$$\operatorname{logit} y = \frac{1}{2}\left(\operatorname{logit} \alpha - \operatorname{logit} \gamma + \log \frac{(1-a)a}{(1-c)c}\right)$$

$$\operatorname{logit} \gamma + \operatorname{logit} \alpha = \operatorname{logit} c + \operatorname{logit} a$$

Also, in last equation, the sum of the two logits is independent of the initial prevalence y.

The left eigenvectors of the backward transition matrix $B_{x+1}{}^\top$ must satisfy $B_{x+1}{}^\top U^* = U^* \Lambda^*$ and the right eigenvector $V^* \top B_{x+1}{}^\top = \Lambda^* V^{*\top}$. Their values can be deduced from former equations involving c, a, and y (Eq. (7) and following).

$$\Lambda^* = \begin{pmatrix} 1 & 0 \\ 0 & 1-\gamma-\alpha \end{pmatrix} \tag{45}$$

$$U^* = \frac{1}{\sqrt{\gamma+\alpha}} \begin{pmatrix} 1 & -\gamma \\ 1 & \alpha \end{pmatrix} \tag{46}$$

$$V^{*\top} = \frac{1}{\sqrt{\gamma+\alpha}} \begin{pmatrix} \alpha & \gamma \\ -1 & 1 \end{pmatrix} = U^{*-1} \tag{47}$$

$$B^{\top} = U^{*}\Lambda^{*}V^{\top} \tag{48}$$

$$(\gamma+\alpha)(B^{\top}) = \begin{pmatrix} \alpha & \gamma \\ \alpha & \gamma \end{pmatrix} + (1-\gamma-\alpha)\begin{pmatrix} \gamma & \gamma \\ -\alpha & \alpha \end{pmatrix} \tag{49}$$

$$(\gamma+\alpha)(B^{\top})^{k} = (\gamma+\alpha)U\Lambda^{k}V^{\top} = \begin{pmatrix} \alpha & \gamma \\ \alpha & \gamma \end{pmatrix} + (1-\gamma-\alpha)^{k}\begin{pmatrix} \gamma & \gamma \\ -\alpha & \alpha \end{pmatrix} \tag{50}$$

Using the decomposition of Eq. (7) we can directly write the duality equation (Eq. 43)

$$P^{\top} = U^{-1^{\top}}\Lambda U^{\top} = V\Lambda V^{-1} \tag{51}$$

$$B_{x+1}{}^{\top} = (W^{d}_{x+1})^{-1}V\Lambda V^{-1}W^{d}_{x} \tag{52}$$

$$B_{x+1}{}^{\top} = ((W^{d}_{x+1})^{-1}V)\Lambda((W^{d}_{x})^{-1}V)^{-1} \tag{53}$$

but the latter decomposition does not correspond to the diagonalization. Also we can understand that the eigenvalues of $B_{x+1}{}^{\top}$ will not be very different from those of P_x because in the real life the prevalences at age $x + 1$, W^{d}_{x+1}, will be very close to the prevalences at age x, W^{d}_{x}. Using our example of two states, the second eigenvalue of matrix P_x, $1 - c - a$ is different from the second eigenvalue of $B_{x+1}{}^{\top}$, $1-\gamma-\alpha$, because $c + a \neq \gamma + \alpha$ but according to Eq. (45) the difference is slim and the equality is exact via the logit transformation.

All the results of the previous subsection are still valid: the prevalence at younger ages converges to a limit y^{*}_{L} verifying:

$$y^{*}_{L} = \frac{\alpha}{\alpha+\gamma}. \tag{54}$$

Property 1. *If the initial prevalence of the forward process corresponds to the stationary state, it corresponds also to the stationary prevalence of the backward process.*

Proof. Replacing y by $y_L = \frac{a}{a+c}$ in Eqs. (33) and (34) we find that $\gamma(y_L) = c$ and $\alpha(y_L) = a$ so that $y^{*}_{L} = \frac{\alpha}{\alpha+\gamma} = \frac{a}{a+c} = y_L$. □

We have also an obvious property

Property 2. *If the convergence to the forward stationary prevalence is alternated, the convergence to the backward stationary prevalence is also alternated.*

Proof. The convergence of the forward process changes to alternated as soon as the second eigenvalue $1 - a - c$ of the forward transition matrix is negative. When $1 - a + c = 0$ in Eqs. (33) and (34) we found that $\gamma = 1 - y$ and $\alpha = y$ so that the second eigen value of the backward transition $1 - \alpha - \gamma$ is also equal to zero when the convergence starts to be alternated too. □

1.4 Weak Ergodicity

In the previous section, we discussed the simple case when a_x and c_x were independent of the age x, but we have seen in the first part of this chapter that the main sociological interest resides in the fact that the probability to enter or reenter the labor force as well as the probability to leave the labor force are varying by age as shown in Fig. 3.

Despite the ups and downs that can be observed in Fig. 3 (and which are due to the fact that the numbers of people observed by single age are too few), the variations of the parameters a_x and c_x between to successive ages are significative but small so that the product of a high number of transition matrices will slightly vary with age, and not converge to a fixed matrix with equal rows.

But an important phenomena will remain. This phenomena is called the weak ergodicity and its property is that the product matrix ${}_kP_x = P_xP_{x+1}\cdots P_{x+k-1}$ tends to a matrix with equal rows. Rows are changing with k, but tend to be equal. It has been demonstrated in many different ways (Cohen, 1979; Lopez, 1961) but H. Lebras's proof is particularly simple (Le Bras, 1971).

In terms of prevalence the formula (16) is now

$$(y_{x+1} \quad 1-y_{x+1}) = (y_x \quad 1-y_x)P_x \tag{55}$$

$$\text{or } (y_{x+k} \quad 1-y_{x+k}) = (y_x \quad 1-y_x)P_xP_{x+1}\cdots P_{x+k-1}. \tag{56}$$

It is useful to calculate the forward prevalence of activity starting from a cohort of only active women $(1 \quad 0)$ at a quinquennial age (as in Fig. 4) or inactive women $(0 \quad 1)$ in order to visualize how the stationary prevalence is bound by these two extreme upper (y^U) and lower curves (y^L) up to convergence

$$\left(y^U_{x+k} \quad 1-y^U_{x+k}\right) = (1 \quad 0)_kP_x \tag{57}$$

$$\left(y^L_{x+k} \quad 1-y^L_{x+k}\right) = (0 \quad 1)_kP_x \tag{58}$$

We can see that once the convergence is reached for a product k matrices, the convergence holds for any further projection. At the limit, we can write

$${}_kP_x = P_xP_{x+1}\cdots P_{x+k-1} \simeq \begin{pmatrix} {}_ky_{x+k} & 1-{}_ky_{x+k} \\ {}_ky_{x+k} & 1-{}_ky_{x+k} \end{pmatrix} \text{ or} \tag{59}$$

$$\simeq \begin{pmatrix} 1 & 1 \\ 1 & 1 \end{pmatrix} \begin{pmatrix} {}_ky_{x+k} \\ 1-{}_ky_{x+k} \end{pmatrix} \tag{60}$$

where ${}_ky_{x+k}$ is the stationary prevalence of activity at age $x+k$ computed with the product of k matrices since age x.

It is probably easier to use the notation y^{∞}_x for the stationary prevalence of activity at age x (and $1-y^{\infty}_x$ for the prevalence of inactivity)

$$\begin{pmatrix} y^{\infty}_x & 1-y^{\infty}_x \\ y^{\infty}_x & 1-y^{\infty}_x \end{pmatrix} = \lim_{k\to\infty} \begin{pmatrix} 1 & 0 \\ 0 & 1 \end{pmatrix} {}_kP_{x-k} = \lim_{k\to\infty} \begin{pmatrix} 1 & 0 \\ 0 & 1 \end{pmatrix} P_{x-k}P_{x-k+1}\cdots P_{x-1}. \tag{61}$$

And as

$$\begin{pmatrix} y_x^\infty & 1-y_x^\infty \\ y_x^\infty & 1-y_x^\infty \end{pmatrix} = \begin{pmatrix} 1 & 1 \\ 1 & 1 \end{pmatrix} \begin{pmatrix} y_x^\infty \\ 1-y_x^\infty \end{pmatrix},$$

we get

$$\begin{pmatrix} 1 & 1 \\ 1 & 1 \end{pmatrix} \begin{pmatrix} y_x^\infty \\ 1-y_x^\infty \end{pmatrix} = \lim_{k\to\infty} {}_kP_{x-k} = \lim_{k\to\infty} P_{x-k}P_{x-k+1}\cdots P_{x-1}. \tag{62}$$

If the convergence is not reached with enough precision after a product of k matrices, the rows are not identical and we will keep both upper and lower bounds

$$\begin{pmatrix} {}_ky_x^U & 1-{}_ky_x^U \\ {}_ky_x^L & 1-{}_ky_x^L \end{pmatrix} = \begin{pmatrix} 1 & 0 \\ 0 & 1 \end{pmatrix} {}_kP_{x-k} = {}_kP_{x-k}. \tag{63}$$

If the stationary prevalence is known up to age x, the stationary prevalence at age $x + 1$ is simply given by the recurrent Eq. (16)

$$\left(y_{x+1}^\infty \quad 1-y_{x+1}^\infty\right) = \left(y_x^\infty \quad 1-y_x^\infty\right) P_x \tag{64}$$

$$\begin{pmatrix} 1 & 1 \\ 1 & 1 \end{pmatrix} \begin{pmatrix} y_{x+1}^\infty \\ 1-y_{x+1}^\infty \end{pmatrix} = \begin{pmatrix} 1 & 1 \\ 1 & 1 \end{pmatrix} \begin{pmatrix} y_x^\infty \\ 1-y_x^\infty \end{pmatrix} P_x. \tag{65}$$

We must also consider a difficulty for young people. Indeed, in order to achieve convergence at a young age, say age 15, we need to start calculating the matrix product from a much younger age, something like age 10 or 5. It means that if data on flows were only available from age 15 to 74, we need to extrapolate the function c_x and a_x of Fig. 3 from ages 5 or 10 to age 15 in order to get the convergent prevalence at age 15 (see Fig. 1). In this example, it was easy to extrapolate the probabilities to enter the labor force before age 15 as something close to zero and the probabilities to leave the labor force as something very high. But if these assumptions were wrong, the stationary forward prevalence could not be accurately estimated. Therefore the recurrent Eq. (64) is useful if and only if the stationary prevalence is known at the first age x_0 where the parameters of the transition matrix P_{x_0} (here c_{x_0} and a_{x_0}) start to be known.

The series of left eigenvectors of each age specific transition matrix play an important role in the calculation of the stable prevalence but no easy formula could be found yet. Therefore, the stable prevalence has been calculated with the use of a computer in order to get the product of matrices.

The same conclusion applies to the age specific backward prevalence which cannot be calculated by easy formulas involving the latent left vectors of the age specific matrices of back probabilities.

In the case of the back prevalence of activity at old ages, and in order to get convergence, we supposed that after age 74 it was not possible to enter the labor force ($a_x = 0$) and that the probability to leave the labor force (c_x) was close to 1. γ_{x+1} and α_{x+1} have been extrapolated in the same way on both extremities.

The backward equation of recurrence is simply

$$\begin{pmatrix} y_x^{*\infty} & 1-y_x^{*\infty} \end{pmatrix} = \begin{pmatrix} y_{x+1}^{*\infty} & 1-y_{x+1}^{*\infty} \end{pmatrix} B_{x+1}{}^{\top} \tag{66}$$

$$\begin{pmatrix} 1 & 1 \\ 1 & 1 \end{pmatrix}\begin{pmatrix} y_x^{*\infty} & 0 \\ 0 & 1-y_x^{*\infty} \end{pmatrix} = \begin{pmatrix} 1 & 1 \\ 1 & 1 \end{pmatrix}\begin{pmatrix} y_{x+1}^{*\infty} & 0 \\ 0 & 1-y_{x+1}^{*\infty} \end{pmatrix} B_{x+1}{}^{\top} \tag{67}$$

and when the convergence is not reachable, we have to deal with the upper and lower back prevalences, starting at age $x + k$ to get prevalences at age x implying k backward transitions matrices

$$\begin{pmatrix} {}_k y_x^{*U} & 1-{}_k y_x^{*U} \\ {}_k y_x^{*L} & 1-{}_k y_x^{*L} \end{pmatrix} = \begin{pmatrix} 1 & 0 \\ 0 & 1 \end{pmatrix} {}_{-k}B_{x+k}{}^{\top} = {}_{-k}B_{x+k}{}^{\top} = B_{x+k}{}^{\top} B_{x+k-1}{}^{\top} \cdots B_{x+1}{}^{\top}. \tag{68}$$

We can write an equation similar to Eq. (42)

$$M = N^{\cdot\cdot} B_{x+1}^{O} W_{x+1,t+1}^{dO} = N^{\cdot\cdot} W_x^{dO} P_x^{O} \tag{69}$$

$$B_{x+1}^{O} W_{x+1,t+1}^{dO} = W_{x,t}^{dO} P_x^{O} \tag{70}$$

where $W_{x+1,t+1}^{dO} = \mathrm{diag}(w_{x+1,t+1}^{1O}, w_{x+1,t+1}^{2O}, \ldots, w_{x+1,t+1}^{iO}, \ldots)$ is the diagonal matrix of the prevalence in state i at age $x + 1$ and date $t + 1$. We use the superscript "^{O}" when the data are observed from the survey.

1.5 Backward Prevalence of a Specific Cohort

We represented in Fig. 7 a detail of some trajectories:

- the observed prevalences (in activity and in inactivity) of the cohort (x, t) observed at the first round at t and age x, $W_{x,t}^{O}$, and at the second round at $t + 1$ and age $x + 1$, $W_{x+1,t+1}^{O}$;
- the same observed trajectory of the younger cohort ($x - 1$, t), $W_{x-1,t}^{O}$ and $W_{x,t+1}^{O}$.

From the observed prevalences of the (x, t) cohort at t and $t + 1$ we deduced the forward matrix of transition P_x^{O} and the backward matrix B_{x+1}^{O}, and from the observed prevalences of the ($x - 1$, t) cohort at t and $t + 1$ we deduced the forward matrix of transition P_{x-1}^{O} and the backward matrix B_x^{O}.

Eq. (70) can written under the elementary form

$$b_{x+1}^{ijO} w_{x+1}^{jO} = w_x^{iO} p_x^{ijO} \quad \forall i, j, \tag{71}$$

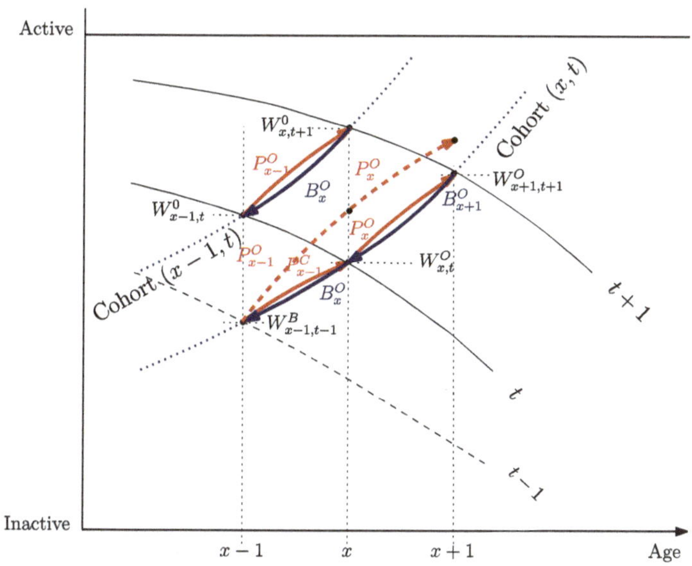

FIG. 7 Zoom on trajectories of two cohorts (x, t) and $(x - 1, t)$ and their observed prevalence in activity at first survey in t and second survey in $t + 1$.

and thus, the Markov forward projection of the observed prevalence at (x, t) by P_x^O

$$W_{x+1,t+1}{}^{\top} = \left(w_{x+1,t+1}^1 \;\; w_{x+1,t+1}^2\right)^{\top} = \left(w_{x,t}^{1O} \;\; w_{x,t}^{2O}\right)^{\top} P_x^O \tag{72}$$

which also can be written under the elementary form

$$w_{x+1}^j = \sum_i w_{x,t}^{iO} p_x^{ijO} \quad \forall j, \tag{73}$$

is equal the observed prevalence at age $x + 1$. The proof is obtained by simple replacement

$$w_{x+1}^j = \sum_i b_{x+1}^{ijO} w_{x+1}^{jO} = w_{x+1}^{jO} \quad \forall j. \tag{74}$$

and because the transposed matrix $B_{x+1}^O{}^{\top}$ is stochastic and $\sum_i b_{x+1}^{ijO} = 1, \forall j$.

Under the assumption that the age specific observed backward prevalences, B_x^O, are constant over time, we can deduce the backward prevalence of the cohort (x, t) at age $x - 1$ by applying the Markov backward projection B_x^O to the observed prevalence at age x ($W_{x,t}^O$).

But the trick here is that the process is no more reversible. In fact, if the pair (B_x^O, P_{x-1}^O) applied to $W_{x,t+1}^O$, moved to $W_{x-1,t}^O$ and $W_{x,t+1}^O$ again, the same pair (B_x^O, P_{x-1}^O) applied to $W_{x,t}^O$ moves to the so-called backward prevalence at age $x - 1$, $W_{x-1,t-1}^B$, but then to a prevalence at age x which differs from $W_{x,t}^O$. The reason comes from the fact that

$$B_x^O W_{x,t}^{dO} \neq W_{x-1,t-1}^{dB} P_{x-1}^O \tag{75}$$

even if

$$B_x^O W_{x,t+1}^{dO} = W_{x-1,t}^{dO} P_{x-1}^O. \tag{76}$$

It reminds us that by changing the prevalence level at age x and keeping the backward matrix unchanged, the forward matrix at age x cannot remain constant and *has to be changed* to a matrix P_{x-1}^C, satisfying equation

$$B_x^O W_{x,t}^{dO} = W_{x-1,t-1}^{dB} P_{x-1}^C \tag{77}$$

in order to move back to the same cohort.

In Fig. 7, we applied this constant forward matrix P_{x-1}^O but also applied P_x^O to the result. It can be seen that this trajectory (red dashed) is different from the trajectory of the cohort (x, t).

Having this remark in mind, we can now calculate the backward prevalences of the same cohort (x, t) at any age $x - k$ by chaining the constant backward matrices B_x^{OT} into a matrix $({}_{-k}B_x^O)^\top$ and we get $\left(w_{x-k,t-k}^{1B} \;\; w_{x-k,t-k}^{2B}\right)^\top$

$$(W_{x-k,t-k}^B)^\top = (W_{x+1,t+1}^O)^\top ({}_{-(k+1)}B_{x+1}^O)^\top \tag{78}$$

$$= (W_{x+1,t+1}^O)^\top B_{x+1}^{O\ \top} B_x^{O\top} B_{x-1}^{O\ \top} \cdots B_{x-k+1}^{O\ \top} \tag{79}$$

$$= (W_{x,t}^O)^\top B_x^{O\ \top} B_{x-1}^{O\ \top} \cdots B_{x-k+1}^{O\ \top} \tag{80}$$

$$= (W_{x,t}^O)^\top ({}_{-k}B_x^O)^\top \tag{81}$$

with

$$({}_{-k}B_x^B)^\top = B_x^{O\ \top} B_{x-1}^{O\ \top} \cdots B_{x-k+1}^{O\ \top}. \tag{82}$$

In the case of J states, the above equations are still valid with

$$W_{x-k,t-k}^B = \left(w_{x-k}^{11} \;\; w_{x-k}^{12} \;\; \cdots \;\; w_{x-k}^{1J}\right) \tag{83}$$

$$W_{x,t}^O = \left(w_x^{11O} \;\; w_x^{12O} \;\; \cdots \;\; w_x^{1JO}\right) \tag{84}$$

1.6 Forward Prevalence of a Specific Cohort

Similarly, assuming that the observed forward transition matrices at any age x, P_x^O are constant over time, we can deduce the forward prevalences of the cohort (x, t) in any state i and age $x + k$

$$W^F_{x+k,t+k} = W^O_{x,t} \cdot {}_kP^F_x = W^O_{x+1,t+1} \cdot {}_{k-1}P^F_{x+1} \tag{85}$$

with

$${}_kP^F_x = P^O_x P^O_{x+1} \cdots P^O_{x+k-1} \tag{86}$$

2 BACKWARD PROBABILITIES WITH TRANSIENT STATES AND AN ABSORBING STATE

As in the previous section, we will take an easy example in order to present the methodology before entering the theoretical part.

Using data from the Italian SILC survey (Giudici et al., 2017) which is a follow-up study with two interviews on the health status (active or inactive), we get the following table

		Act.(1)	*In.*(2)	*Dead*
Active(1)	9983	7998	1932	53
Inactive	6607	1828	4525	254
Total	16590	9826	6457	307

This table can also be represented on the Lexis diagrams of Fig. 8. The classical approach using forward probabilities is on the right. From a cohort of

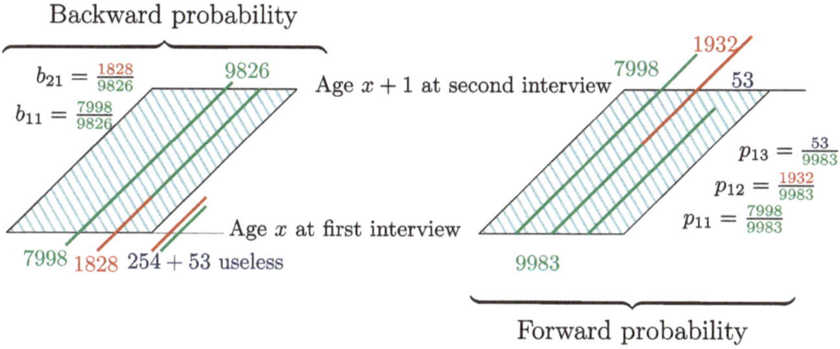

FIG. 8 Lexis diagram of forward and backward approaches with *differential mortality*. *Green* for active (state 1), *red* for inactive (state 2), and *blue* for death (state 3).

9983 people active at age x, only 7998 were still active (state 1) at age $x + 1$, 1932 were inactive at the date of the second interview (state 2), and 4 died (state 3) before the second interview. The same process is applied to the 6457 inactive people at the first interview. If we consider a Markov chain model with three states, active, inactive, and dead, the various probabilities are represented on the figure.

On the left side of the same Fig. 8 we defined the backward probabilities. Starting from the 9826 people who survived and were active at the second interview, we know that 7998 were active (1) and 1828 were inactive (2) at the first interview. Thus, $b_{21} = \frac{1828}{9826}$ is the probability to be inactive (1) a year before, knowing the current status (2) at age x and $b_{11} = \frac{7998}{9826}$, etc.

In the forward approach, we computed $p_{13} = \frac{53}{9983}$ as the probability for an active person to die between the two interviews. In the backward approach, even if we could define $b_{13} = \frac{53}{307}$ as the probability, knowing that we will die between the two interviews, to be active at the first interview, we are only interested in past history of a cohort of survivors,

It could be argued that backward prevalences involve the past history of a very selected population who by chance survived to old ages but it is not completely true because this backward cohort is starting with few people at old ages but its size increases at younger ages and is very numerous at very young ages.

Also the same argument could apply to forward prevalences of a classical life table where probability of death at old ages is estimated from a highly selected population and is not reflecting the mortality of the total population if that population was reaching these old ages.

Let us now move to the theoretical part using the case with only three alive states (1, 2, 3) but with an absorbing state (4). But it is not difficult to generalize to an abitrary number of nonabsorbing states. This section involves only matrices and is presented here because our IMaCh software which is able to treat an abitrary number of alive states, uses such kind of matrix products. We will add the superscript "e" to distinguish the full extended matrices from the matrices reduced to the transient states.

Let us first consider the M matrix which describes how $N^{\cdot\cdot}$ people in various alive states i at age x are found at age $x + 1$ in state j at the next wave.

Sometimes we will remove the x index for better readability. $N^{\cdot 4}_{x+1} = N^{14} + N^{24} + N^{34}$ people died between the two waves.

$$
\begin{array}{cc}
N^{1\cdot}_x & \\
N^{2\cdot}_x & \\
N^{3\cdot}_x & \\
N^{\cdot\cdot} &
\end{array}
\begin{array}{c}
\begin{pmatrix}
N^{11} & N^{12} & N^{13} & N^{14} \\
N^{21} & N^{22} & N^{23} & N^{24} \\
N^{31} & N^{32} & N^{33} & N^{34}
\end{pmatrix} \\
\begin{matrix} N^{\cdot 1}_{x+1} & N^{\cdot 2}_{x+1} & N^{\cdot 3}_{x+1} & N^{\cdot 4}_{x+1} \end{matrix}
\end{array}
$$

Using an arbitrary last row, we can rewrite M as a squared matrix

$$M=\begin{pmatrix} N^{11} & N^{12} & N^{13} & N^{14} \\ N^{21} & N^{22} & N^{23} & N^{24} \\ N^{31} & N^{32} & N^{33} & N^{34} \\ 0 & 0 & 0 & N^{\cdot\cdot} \end{pmatrix}$$

$$=\begin{pmatrix} N_x^{1\cdot} & 0 & 0 & 0 \\ 0 & N_x^{2\cdot} & 0 & 0 \\ 0 & 0 & N_x^{3\cdot} & 0 \\ 0 & 0 & 0 & N^{\cdot\cdot} \end{pmatrix}\begin{pmatrix} p^{11} & p^{12} & p^{13} & p^{14} \\ p^{21} & p^{22} & p^{23} & p^{24} \\ p^{31} & p^{32} & p^{33} & p^{34} \\ 0 & 0 & 0 & 1 \end{pmatrix}$$

$$M=\begin{pmatrix} b^{11} & b^{12} & b^{13} & b^{24} \\ b^{21} & b^{22} & b^{23} & b^{24} \\ b^{31} & b^{32} & b^{33} & b^{34} \\ 0 & 0 & 0 & 1 \end{pmatrix}\begin{pmatrix} N_{x+1}^{\cdot 1} & 0 & 0 & 0 \\ 0 & N_{x+1}^{\cdot 2} & 0 & 0 \\ 0 & 0 & N_{x+1}^{\cdot 3} & 0 \\ 0 & 0 & 0 & N^{\cdot\cdot} \end{pmatrix} \tag{87}$$

$$=B_{x+1}^{e}\,\mathrm{diag}(N_{x+1}^{\cdot 1},N_{x+1}^{\cdot 2},N_{x+1}^{\cdot 3},N^{\cdot\cdot}) \tag{88}$$

$$M=N^{\cdot\cdot}\begin{pmatrix} b^{11} & b^{12} & b^{13} & b^{24} \\ b^{21} & b^{22} & b^{23} & b^{24} \\ b^{31} & b^{32} & b^{33} & b^{34} \\ 0 & 0 & 0 & 1 \end{pmatrix}\begin{pmatrix} w_{x+1}^{1eO} & 0 & 0 & 0 \\ 0 & w_{x+1}^{2eO} & 0 & 0 \\ 0 & 0 & w_{x+1}^{3eO} & 0 \\ 0 & 0 & 0 & 1 \end{pmatrix} \tag{89}$$

$$=N^{\cdot\cdot}B_{x+1}^{e}\,\mathrm{diag}(w_{x+1}^{1eO},w_{x+1}^{2eO},w_{x+1}^{3eO},1) \tag{90}$$

$$M=N^{\cdot\cdot}\begin{pmatrix} w_x^{1O} & 0 & 0 & 0 \\ 0 & w_x^{2O} & 0 & 0 \\ 0 & 0 & w_x^{3O} & 0 \\ 0 & 0 & 0 & 1 \end{pmatrix}\begin{pmatrix} p^{11} & p^{12} & p^{13} & p^{14} \\ p^{21} & p^{22} & p^{23} & p^{24} \\ p^{31} & p^{32} & p^{33} & p^{34} \\ 0 & 0 & 0 & 1 \end{pmatrix} \tag{91}$$

$$=N^{\cdot\cdot}\cdot\mathrm{diag}(w_x^{1O},w_x^{2O},w_x^{3O},1)\cdot P_x \tag{92}$$

where $w_x^{iO}=\frac{N_x^{i\cdot}}{N^{\cdot\cdot}}$ is the observed prevalence in state i at age x and $p^{ij}=\frac{N^{ij}}{N_x^{i\cdot}}$ is the forward probability to be in state j (j = 1, …, 4) at x + 1, being in state i, (i = 1, …, 3), at age x. $w_{x+1}^{jeO}=\frac{N_{x+1}^{\cdot j}}{N^{\cdot\cdot}}$ is the extended observed prevalence at age x + 1 in state j including death, while $w_{x+1}^{jO}=\frac{N_{x+1}^{\cdot j}}{N_{x+1}^{\cdot\cdot}-N_{x+1}^{\cdot 4}}$ is the observed prevalence among the survivors at age x + 1 in state j.

The link between the extended matrices B_{x+1}^{e} and P_x^{e} with the extended prevalences at x + 1 and the observed prevalances at x is given by the following equation, similar to Eq. (70)

$$B_{x+1}^{e}\,\mathrm{diag}(w_{x+1}^{1eO},w_{x+1}^{2eO},\ldots,1)=\mathrm{diag}(w_x^{1O},w_x^{2O},\ldots,1)\cdot P_x^{e}. \tag{93}$$

2.1 Chaining Forward for a Specific Cohort

As the set (P_x^{eO}, $\forall x$) is a set of stochastic matrices, we can suppose that they are constant over time and chain them in the same manner that with did in Section 1.6 and using Eq. (86) in order get the forward extended prevalences at age $x + 1$ and year $t + 1$ from the observed prevalences at age x and year t

$$\begin{pmatrix} w_{x+1,t+1}^{1eF} & w_{x+1,t+1}^{2eF} & w_{x+1,t+1}^{3eF} & 0 \end{pmatrix} = \begin{pmatrix} w_{x,t}^{1O} & w_{x,t}^{2O} & w_{x,t}^{3O} & 0 \end{pmatrix} \cdot P_x^{e} \tag{94}$$

$$\left(W_{x+1,t+1}^{eF}\right)^{\top} = \left(W_{x,t}^{O}\right)^{\top} \cdot P_x^{eO} \tag{95}$$

And we can chain forward up to age $x + k$

$$\left(W_{x+2,t+2}^{eF}\right)^{\top} = \left(W_{x,t}^{O}\right)^{\top} P_x^{eO} P_{x+1}^{eO} = \left(W_{x,t}^{O}\right)^{\top} \cdot {}_2P_x^{eO} \tag{96}$$

$$\left(W_{x+k,t+k}^{eF}\right)^{\top} = \left(W_{x,t}^{O}\right)^{\top} P_x^{eO} P_{x+1}^{eO} \cdots P_{x+k-1}^{eO} \tag{97}$$

$$\left(W_{x+k,t+k}^{eF}\right)^{\top} = \left(W_{x,t}^{O}\right)^{\top} \cdot {}_kP_x^{eO}. \tag{98}$$

In the case of three alive transient states and death, it can be written

$$\begin{pmatrix} w_{x+k}^{1eF} & w_{x+k}^{2eF} & w_{x+k}^{3eF} & w_{x+k}^{4eF} \end{pmatrix} = \begin{pmatrix} w_{x,t}^{1O} & w_{x,t}^{2O} & w_{x,t}^{3O} & 0 \end{pmatrix} \begin{pmatrix} {}_kp_x^{11} & {}_kp_x^{12} & {}_kp_x^{13} & {}_kp_x^{14} \\ {}_kp_x^{21} & {}_kp_x^{22} & {}_kp_x^{23} & {}_kp_x^{24} \\ {}_kp_x^{31} & {}_kp_x^{32} & {}_kp_x^{33} & {}_kp_x^{34} \\ 0 & 0 & 0 & 1 \end{pmatrix} \tag{99}$$

We can also rewrite the full extended matrix ${}_kP_x^{eO}$ in order to display the submatrix of the transient state only

$${}_kP_x^{eO} = \begin{pmatrix} {}_kP_x^{O} & e - {}_kP_x^{O} \\ 0 & 1 \end{pmatrix} \quad e = \begin{pmatrix} 1 \\ 1 \\ 1 \end{pmatrix} \tag{100}$$

$${}_kP_x^{O} = \begin{pmatrix} {}_kp_x^{11} & {}_kp_x^{12} & {}_kp_x^{13} \\ {}_kp_x^{21} & {}_kp_x^{22} & {}_kp_x^{23} \\ {}_kp_x^{31} & {}_kp_x^{32} & {}_kp_x^{33} \end{pmatrix} \tag{101}$$

Even if the matrix ${}_kP_x^{O}$ is not stochastic, the property of ergodicity applies to this submatrix and when k increases, its rows tend to be identical. The matrix changes with the horizon k but is unique and depends only on the flows (p_y^{ij} $\forall y \geq x$) and no more on the prevalences at age x.

The reader may refer to a complete proof, similar to Lebras's proof, in the appendix of our article of 2003 (Lièvre et al., 2003).

We can remark that the vector of the observed prevalence at age x is a sum of three matrices

$$\left(w_{x,t}^{1O}\ \ w_{x,t}^{2O}\ \ w_{x,t}^{3O}\ \ 0\right) = w_{x,t}^{1O}(1\ \ 0\ \ 0\ \ 0) + w_{x,t}^{2O}(0\ \ 1\ \ 0\ \ 0) + w_{x,t}^{3O}(0\ \ 0\ \ 1\ \ 0). \tag{102}$$

Therefore we can focus on the behavior of each of them

$$\left(w_{x+k}^{11eF}\ \ w_{x+k}^{12eF}\ \ w_{x+k}^{13eF}\ \ w_{x+k}^{14eF}\right) = (1\ \ 0\ \ 0\ \ 0)\,{}_kP_x^{eO} \tag{103}$$

$$= (1\ \ 0\ \ 0\ \ 0)\begin{pmatrix} {}_kp_x^{11} & {}_kp_x^{12} & {}_kp_x^{13} & {}_kp_x^{14} \\ {}_kp_x^{21} & {}_kp_x^{22} & {}_kp_x^{23} & {}_kp_x^{24} \\ {}_kp_x^{31} & {}_kp_x^{32} & {}_kp_x^{33} & {}_kp_x^{34} \\ 0 & 0 & 0 & 1 \end{pmatrix} \tag{104}$$

$$= \left({}_kp_x^{11}\ \ {}_kp_x^{12}\ \ {}_kp_x^{13}\ \ {}_kp_x^{14}\right) \tag{105}$$

$$\left(w_{x+k}^{21eF}\ \ w_{x+k}^{22eF}\ \ w_{x+k}^{23eF}\ \ w_{x+k}^{24eF}\right) = (0\ \ 1\ \ 0\ \ 0)\,{}_kP_x^{eO} = \left({}_kp_x^{21}\ \ {}_kp_x^{22}\ \ {}_kp_x^{23}\ \ {}_kp_x^{24}\right) \tag{106}$$

$$\left(w_{x+k}^{31eF}\ \ w_{x+k}^{32eF}\ \ w_{x+k}^{33eF}\ \ w_{x+k}^{34eF}\right) = (0\ \ 0\ \ 1\ \ 0)\,{}_kP_x^{eO} = \left({}_kp_x^{11}\ \ {}_kp_x^{32}\ \ {}_kp_x^{33}\ \ {}_kp_x^{34}\right) \tag{107}$$

or of its summary

$$\begin{pmatrix} w_{x+k}^{11eF} & w_{x+k}^{12eF} & w_{x+k}^{13eF} & w_{x+k}^{14eF} \\ w_{x+k}^{21eF} & w_{x+k}^{22eF} & w_{x+k}^{23eF} & w_{x+k}^{24eF} \\ w_{x+k}^{31eF} & w_{x+k}^{32eF} & w_{x+k}^{33eF} & w_{x+k}^{34eF} \end{pmatrix} = \begin{pmatrix} 1 & 0 & 0 & 0 \\ 0 & 1 & 0 & 0 \\ 0 & 0 & 1 & 0 \\ 0 & 0 & 0 & 1 \end{pmatrix}\begin{pmatrix} {}_kp_x^{11} & {}_kp_x^{12} & {}_kp_x^{13} & {}_kp_x^{14} \\ {}_kp_x^{21} & {}_kp_x^{22} & {}_kp_x^{23} & {}_kp_x^{24} \\ {}_kp_x^{31} & {}_kp_x^{32} & {}_kp_x^{33} & {}_kp_x^{34} \\ 0 & 0 & 0 & 1 \end{pmatrix} \tag{108}$$

and we draw in Fig. 9 the nine elements of matrix ${}_kP_x$, starting at age $x = 55$, up to an increasing age $x = 55 + k$. In addition, we visualized the survival curves in each of the three states, $s^i = 1 - p^{i4} = p^{i1} + p^{i2} + p^{i3}$.

We can see that the mortality of a person living in an institution is much more important than the mortality of a person living in coresidence and is similar to that of a person living alone.

But because of the high mortality at these old ages, all the 9 elements are collapsing to zero and the ergodicity is not obvious.

Also for each of three starting values i, we got the result that the forward extended prevalences w_{x+k}^{ijeF} in state j at age $x + k$ are simply the element (i, j) of the extended matrix ${}_kP_x^{eO}$, ${}_kp_x^{ij}$.

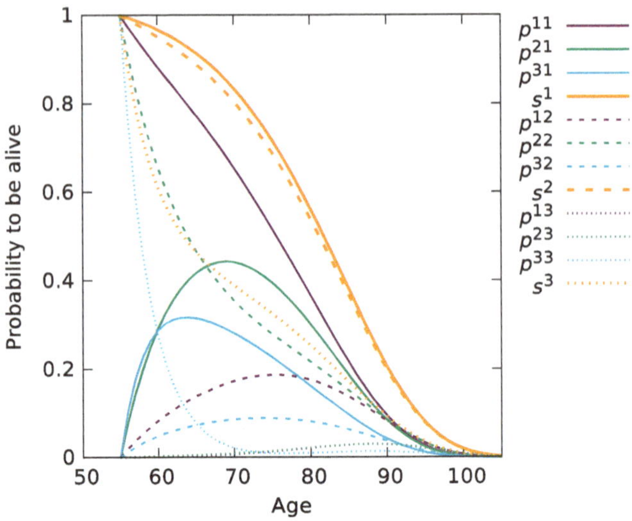

FIG. 9 Survival functions for a person living initially at age 55 either in coresidence ($i = 1$), alone (2) or in an institution (3). Probability ${}_{x-55}p_{55}^{ij}$ to be at age x alive in state $j = 1, 2,$ or 3 and in any state (s_i), being in state i at age 55 (HRS 1998–2014).

The forward prevalences, w_{x+k}^{ijF}, are simply obtained by dividing by the probabilities of survival in k years being initially in state i at age x. This probability being $1 - {}_kp_x^{i4}$, the forward prevalences are

$$w_{x+k}^{ijF} = \frac{{}_kp_x^{ij}}{{}_kp_x^{i1} + {}_kp_x^{i2} + {}_kp_x^{i3}} \tag{109}$$

In Fig. 10, we divided each row of the matrix by the corresponding probability to survive and now, the nine elements are clearly converging to three different curves which are the unique forward prevalences in each of the three states, coresidence, alone and in institutions.

The extended prevalences of a cohort are then a weighted mean of the three state-specific prevalences and the forward prevalences are the results of their division by the overall probability to survive.

From Eq. (99), we can write w_{x+k}^{4e} as the overall probability to die within k years when the population at age x is distributed in state i according to w_x^{iO}

$$w_{x+k}^{4e} = w_x^{1O}\,{}_kp_x^{14} + w_x^{2O}\,{}_kp_x^{24} + w_x^{3O}\,{}_kp_x^{34} \tag{110}$$

$$= 1 - \left(w_{x+k}^{1eF} + w_{x+k}^{2eF} + w_{x+k}^{3eF}\right) \tag{111}$$

and its opposite, the probability to survive is a weighted mean of the various probabilities of survive when originally in a state

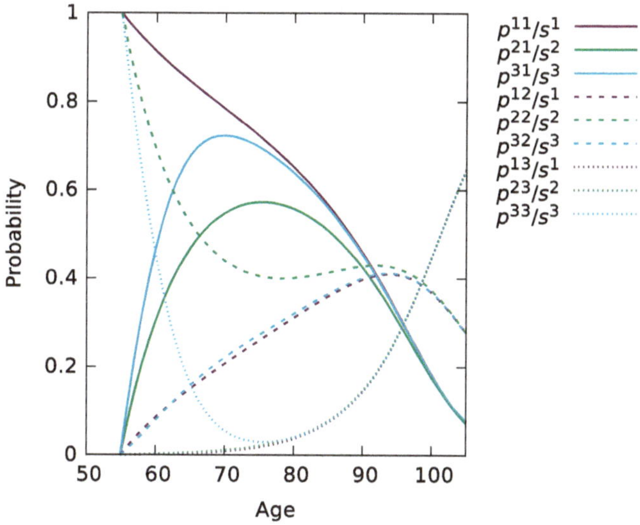

FIG. 10 Identical to Fig. 9 but divided by probability to survive ${}_{x-55}p_{55}^{ij}/{}_{x-55}s_{55}^{i}$. Convergence to the forward prevalences.

$$
\begin{aligned}
{}_k s_x^{wO} &= 1 - w_{x+k}^{4eF} && (112)\\
&= \sum_i w_x^{iO}\, {}_k s_x^i && (113)\\
&= w_x^{1O}\left({}_kp_x^{11} + {}_kp_x^{12} + {}_kp_x^{13}\right) + w_x^{2O}\left({}_kp_x^{21} + {}_kp_x^{22} + {}_kp_x^{23}\right) \\
&\quad + w_x^{3O}\left({}_kp_x^{31} + {}_kp_x^{32} + {}_kp_x^{33}\right). && (114)
\end{aligned}
$$

The prevalences at age $x+k$ are simply the proportions of people in each state among the survivors at age $x+k$

$$
w_{x+k}^{jF} = \frac{w_{x+k}^{jeF}}{{}_ks_x^w} = w_x^{1O} \cdot \frac{{}_kp_x^{1j}}{{}_ks_x^w} + w_x^{2O} \cdot \frac{{}_kp_x^{2j}}{{}_ks_x^w} + w_x^{3O} \cdot \frac{{}_kp_x^{3j}}{{}_ks_x^w}. \tag{115}
$$

Using matrices, we can write

$$
(W_{x+k,t+k}^F)^\top = (W_{x+k,t+k}^{eF})^\top \frac{1}{{}_ks_x^{wO}} = (W_{x,t}^O)^\top \cdot \frac{1}{{}_ks_x^{wO}} \cdot {}_kP_x^{eO} = (W_{x,t}^O)^\top \overline{{}_kP_x^{eO}} \tag{116}
$$

with

$$
\overline{{}_kP_x^{eO}} = \frac{1}{{}_ks_x^{wO}} \cdot {}_kP_x^{eO}, \tag{117}
$$

We also have the following recurrences

$$
\overline{{}_kP_x^{eO}} = \overline{{}_{k-1}P_x^{eO}} \cdot \overline{P_{x+k-1}^{eO}} \tag{118}
$$

$$(W^F_{x+k,t+k})^\top = (W^F_{x+k-1,t+k-1})^\top \overline{P^{eO}_{x+k-1}} \tag{119}$$

with

$$\overline{P^{eO}_{x+k-1}} = \frac{P^{eO}_{x+k-1}}{{}_{k-1}\sigma^{wO}_x} \tag{120}$$

$$\overline{P^{eO}_x} = \frac{P^{eO}_x}{\sigma^{wO}_x} \tag{121}$$

and

$${}_k s^i_x = {}_{k-1} s^i_x \cdot s^i_{x+k-1} \tag{122}$$

$${}_{k-1}\sigma^{wO}_x = \frac{{}_k s^{wO}_x}{{}_{k-1}s^{wO}_x} = \sum_i w^{iO}_x \cdot s^i_{x+k-1} \frac{{}_{k-1}s^i_x}{\sum_j w^{jO}_x \cdot {}_{k-1}s^j_x}. \tag{123}$$

With Eq. (119), we are back to an inhomogenous Markov chain without absorbing state but this Eq. (119) differs from Eq. (56) because the coefficient ${}_{k-1}\sigma^{wO}_x = \frac{{}_k s^{wO}_x}{{}_{k-1}s^{wO}_x}$ depends on the level of the prevalences at age x and not only on the elements of the last matrix P^{eO}_{x+k-1}. We could have suspected than the forward prevalences were not unique but we saw that they are not dependent on both observed and state-specific initial prevalences.

If we had used forward Markov matrices that would have been resized into a stochastic matrix at each age

$$\widetilde{P^{eO}_{x+k-1}} = \frac{P^{eO}_{x+k-1}}{s^{wO}_{x+k}} \tag{124}$$

$${}_k\widetilde{P^{eO}_x} = \widetilde{P^{eO}_x} \cdot \widetilde{P^{eO}_{x+1}} \cdots \widetilde{P^{eO}_{x+k-1}} \tag{125}$$

we would have not found a unique forward prevalence unless the mortality from any state was identical at any age. And we just saw in the above example that it is usually false. For disability versus disability-free, the differential is about 4 at age 70 and only 1.5 at age 95.

In order to calculate the prevalence at age x, we could start from any observed prevalence w^{iO}_{x-k} in state i at age $x-k$ or from a set of prevalences $(W^O_{x-k})^\top$.

Hence transposing Eq. (115) and using the notation ${}_{-k}w^{jF}_x$, the observed prevalence in state j at age x is given by the formula

$${}_{-k}w^{jF}_x = \frac{{}_{-k}w^{jeF}_x}{{}_k s^w_{x-k}} = w^{1O}_{x-k} \cdot \frac{{}_k p^{1j}_{x-k}}{{}_k s^w_{x-k}} + w^{2O}_{x-k} \cdot \frac{{}_k p^{2j}_{x-k}}{{}_k s^w_{x-k}} + w^{3O}_{x-k} \cdot \frac{{}_k p^{3j}_{x-k}}{{}_k s^w_{x-k}}. \tag{126}$$

Using matrices, it follows

$$\left({}_{-k}W^{F}_{x,t}\right)^{\top}=\left({}_{-k}W^{eF}_{x,t}\right)^{\top}\frac{1}{{}_{k}s^{wO}_{x-k}}=\left(W^{O}_{x-k,t-k}\right)^{\top}\cdot\frac{1}{{}_{k}s^{wO}_{x-k}}\cdot{}_{k}P^{eO}_{x-k}=\left(W^{O}_{x-k,t-k}\right)^{\top}\overline{{}_{k}P^{eO}_{x-k}} \tag{127}$$

with

$$\overline{{}_{k}P^{eO}_{x-k}}=\frac{1}{{}_{k}s^{wO}_{x-k}}\cdot{}_{k}P^{eO}_{x-k}, \tag{128}$$

Thus, the unique forward prevalence at age x is the limit of ${}_{-k}w^{jF}_{x}$ when k increases to infinity. Using Eq. (108) we can simply remark the unique forward prevalences at age x in any state j are given by any column of the matrix $\overline{{}_{k}P^{eO}_{x-k}}$ when k increases to infinity.

In practice, we compute the matrix ${}_{k}P^{eO}_{x-k}$ which is the product of the k matrices $P^{eO}_{x-k}P^{eO}_{x-k+1}\cdots P^{eO}_{x-2}P^{eO}_{x-1}$. Remembering Eq. (108), each row i corresponds to the extended forward prevalences by states. Thus a simple division of each element (i, j) of the row i by the sum of the first elements (which is probability ${}_{k}s^{i}_{x-k}$ to survive until age x when in state i at age $x-k$, $1-{}_{k}p^{i4}_{x-k}$), will output the forward prevalences in each state. Any row should output the same results.

In fact, the software checks that for each column, the relative difference between the maximum and the minimum is less than a given tolerance. When this tolerance is reached, the forward prevalence is obtained.

And thus we can write the forward prevalences using an expression which is only dependent of the ${}_{k}p^{ij}_{x-k}$ and not of any observed prevalence

$$\begin{aligned}{}_{k}S^{d}_{x-k}=\operatorname{diag}\big(&{}_{k}p^{11}_{x-k}+{}_{k}p^{12}_{x-k}+{}_{k}p^{13}_{x-k},\\&{}_{k}p^{21}_{x-k}+{}_{k}p^{22}_{x-k}+{}_{k}p^{23}_{x-k},\\&{}_{k}p^{31}_{x-k}+{}_{k}p^{32}_{x-k}+{}_{k}p^{33}_{x-k}\big)\end{aligned} \tag{129}$$

$$=\begin{pmatrix}{}_{k}s^{1}_{x-k}&0&0\\0&{}_{k}s^{2}_{x-k}&0\\0&0&{}_{k}s^{3}_{x-k}\end{pmatrix} \tag{130}$$

$$W^{F\infty\top}_{x}=\begin{pmatrix}w^{1F}_{x}&w^{2F}_{x}&w^{3F}_{x}\\w^{1F}_{x}&w^{2F}_{x}&w^{3F}_{x}\\w^{1F}_{x}&w^{2F}_{x}&w^{3F}_{x}\end{pmatrix}=\lim_{k\to\infty}\left({}_{k}S^{d}_{x-k}\right)^{-1}\cdot{}_{k}P_{x-k} \tag{131}$$

$$W^{F\infty\top}_{x}=\lim_{k\to\infty}\left({}_{k}S^{d}_{x-k}\right)^{-1}{}_{k}P_{x-k}=\lim_{k\to\infty}\left({}_{k}S^{d}_{x-k}\right)^{-1}P_{x-k}P_{x-k+1}\cdots P_{x-1} \tag{132}$$

Also, applying Eq. (119) with $k=0$ we get the recurrent equation

$$\left(W^{F}_{x,t}\right)^{\top}=\left(W^{F}_{x-1,t-1}\right)^{\top}\overline{P^{eO}_{x-1}}=\left(W^{F}_{x-1,t-1}\right)^{\top}\frac{P^{eO}_{x-1}}{\sigma^{wO}_{x-1}} \tag{133}$$

2.2 Chaining Backward for a Specific Cohort

Let us now review similar results for the backward prevalence.

The backward probability b^{ij}_{x+1} at age $x+1$ is defined as below

$$b^{ij}_{x+1} = \frac{N^{ij}}{N^{\cdot j}_{x+1}} \quad i=1,3 \;\; j=1,4 \tag{134}$$

with

$$N^{\cdot j}_{x+1} = \sum_{i=1}^{3} N^{ij} = N^{\cdot\cdot} \sum_{i=1}^{3} w^{i}_{x} p^{ij} = w^{je}_{x+1} N^{\cdot\cdot}, \;\; j=1,4 \tag{135}$$

$$\text{or } \; w^{je}_{x+1} = \sum_{i=1}^{3} w^{i}_{x} p^{ij} = w^{1}_{x} p^{1j} + w^{2}_{x} p^{2j} + w^{3}_{x} p^{3j}. \tag{136}$$

According to Eq. (93) and using a matrix notation, we can write

$$B^{e}_{x+1} = \mathrm{diag}(w^{1}_{x}, w^{2}_{x}, w^{3}_{x}, 1) P^{e}_{x} (\mathrm{diag}(w^{1e}_{x+1}, w^{2e}_{x+1}, w^{3e}_{x+1}, 1))^{-1} \tag{137}$$

or

$$B^{e}_{x+1} = \begin{pmatrix} w^{1}_{x} & 0 & 0 & 0 \\ 0 & w^{2}_{x} & 0 & 0 \\ 0 & 0 & w^{3}_{x} & 0 \\ 0 & 0 & 0 & 1 \end{pmatrix} \begin{pmatrix} p^{11} & p^{12} & p^{13} & p^{14} \\ p^{21} & p^{22} & p^{23} & p^{24} \\ p^{31} & p^{32} & p^{33} & p^{34} \\ 0 & 0 & 0 & 1 \end{pmatrix} \cdot \tag{138}$$

$$\begin{pmatrix} \frac{1}{w^{1}_{x}p^{11} + w^{2}_{x}p^{21} + w^{3}_{x}p^{31}} & 0 & 0 & 0 \\ 0 & \frac{1}{w^{1}_{x}p^{12} + w^{2}_{x}p^{22} + w^{3}_{x}p^{32}} & 0 & 0 \\ 0 & 0 & \frac{1}{w^{1}_{x}p^{13} + w^{2}_{x}p^{23} + w^{3}_{x}p^{33}} & 0 \\ 0 & 0 & 0 & 1 \end{pmatrix} \tag{139}$$

$$B^{e}_{x+1} = \begin{pmatrix} \frac{w^{1}_{x}p^{11}}{w^{1}_{x}p^{11} + w^{2}_{x}p^{21} + w^{3}_{x}p^{31}} & \frac{w^{1}_{x}p^{12}}{w^{1}_{x}p^{12} + w^{2}_{x}p^{22} + w^{3}_{x}p^{32}} & \frac{w^{1}_{x}p^{13}}{w^{1}_{x}p^{13} + w^{2}_{x}p^{23} + w^{3}_{x}p^{33}} & \frac{w^{1}_{x}p^{14}}{w^{1}_{x}p^{14} + w^{2}_{x}p^{24} + w^{3}_{x}p^{34}} \\ \frac{w^{2}_{x}p^{21}}{w^{1}_{x}p^{11} + w^{2}_{x}p^{21} + w^{3}_{x}p^{31}} & \frac{w^{2}_{x}p^{22}}{w^{1}_{x}p^{12} + w^{2}_{x}p^{22} + w^{3}_{x}p^{32}} & \frac{w^{2}_{x}p^{23}}{w^{1}_{x}p^{13} + w^{2}_{x}p^{23} + w^{3}_{x}p^{33}} & \frac{w^{2}_{x}p^{24}}{w^{1}_{x}p^{14} + w^{2}_{x}p^{24} + w^{3}_{x}p^{34}} \\ \frac{w^{3}_{x}p^{31}}{w^{1}_{x}p^{11} + w^{2}_{x}p^{21} + w^{3}_{x}p^{31}} & \frac{w^{3}_{x}p^{32}}{w^{1}_{x}p^{12} + w^{2}_{x}p^{22} + w^{3}_{x}p^{32}} & \frac{w^{3}_{x}p^{33}}{w^{1}_{x}p^{13} + w^{2}_{x}p^{23} + w^{3}_{x}p^{33}} & \frac{w^{3}_{x}p^{34}}{w^{1}_{x}p^{14} + w^{2}_{x}p^{24} + w^{3}_{x}p^{34}} \\ 0 & 0 & 0 & 1 \end{pmatrix}. \tag{140}$$

The arbitrary last row is used so that the product of matrices used in our IMaCh software, concerns only full square matrices of the same size. But we are only interested in the 3 by three submatrix of the alive states

$$B_{x+1} = \begin{pmatrix} w_x^1 & 0 & 0 \\ 0 & w_x^2 & 0 \\ 0 & 0 & w_x^3 \end{pmatrix} \begin{pmatrix} p^{11} & p^{12} & p^{13} \\ p^{21} & p^{22} & p^{23} \\ p^{31} & p^{32} & p^{33} \end{pmatrix} \cdot \tag{141}$$

$$\begin{pmatrix} \frac{1}{w_x^1 p^{11} + w_x^2 p^{21} + w_x^3 p^{31}} & 0 & 0 \\ 0 & \frac{1}{w_x^1 p^{12} + w_x^2 p^{22} + w_x^3 p^{32}} & 0 \\ 0 & 0 & \frac{1}{w_x^1 p^{13} + w_x^2 p^{23} + w_x^3 p^{33}} \end{pmatrix} \tag{142}$$

$$B_{x+1} = \begin{pmatrix} \frac{w_x^1 p^{11}}{w_x^1 p^{11} + w_x^2 p^{21} + w_x^3 p^{31}} & \frac{w_x^1 p^{12}}{w_x^1 p^{12} + w_x^2 p^{22} + w_x^3 p^{32}} & \frac{w_x^1 p^{13}}{w_x^1 p^{13} + w_x^2 p^{23} + w_x^3 p^{33}} \\ \frac{w_x^2 p^{21}}{w_x^1 p^{11} + w_x^2 p^{21} + w_x^3 p^{31}} & \frac{w_x^2 p^{22}}{w_x^1 p^{12} + w_x^2 p^{22} + w_x^3 p^{32}} & \frac{w_x^2 p^{23}}{w_x^1 p^{13} + w_x^2 p^{23} + w_x^3 p^{33}} \\ \frac{w_x^3 p^{31}}{w_x^1 p^{11} + w_x^2 p^{21} + w_x^3 p^{31}} & \frac{w_x^3 p^{32}}{w_x^1 p^{12} + w_x^2 p^{22} + w_x^3 p^{32}} & \frac{w_x^3 p^{33}}{w_x^1 p^{13} + w_x^2 p^{23} + w_x^3 p^{33}} \end{pmatrix} . \tag{143}$$

which is equivalent to Eq. (42)

$$B_{x+1} = \text{diag}(w_x^i) P_x \, \text{diag}\left(\frac{1}{\sum_i w_x^i p_x^{ij}} \right) \tag{144}$$

The transposed matrix $(B_{x+1})^\top$ (in fact, only its 2 by 2 submatrix of alive states)

$$(B_{x+1})^\top = \text{diag}\left(\frac{1}{\sum_i w_x^i p_x^{ij}} \right) P_x{}^\top \text{diag}(w_x^i) \tag{145}$$

is stochastic and corresponds to the transition matrix of the backward Markov chain within the alive states.

We can then right multiply the matrices to get the limit of the product of backward matrices

$$({}_{-k}B_{x+k})^\top = (B_{x+k})^\top (B_{x+k-1})^\top \cdots (B_{x+2})^\top (B_{x+1})^\top . \tag{146}$$

In order to highlight the backward prevalences w_x^{iB} and forward prevalences w_x^{iF} at exact age x as a limit when $k \to \infty$, and using the notations generalizing to J states

$$(w_x^B)^\top = \begin{pmatrix} w_x^{1B} & w_x^{2B} & \cdots & w_x^{JB} \end{pmatrix} \tag{147}$$

$$(W_x^B)^\top = \begin{pmatrix} w_x^{1B} & w_x^{2B} & \cdots & w_x^{JB} \\ \vdots & \vdots & & \vdots \\ w_x^{1B} & w_x^{2B} & \cdots & w_x^{JB} \end{pmatrix} \tag{148}$$

we can write

$$(W_x^{B\infty})^\top = \lim_{k\to\infty} ({}_{-k}B_{x+k})^\top = \lim_{k\to\infty} (B_{x+k})^\top (B_{x+k-1})^\top \cdots (B_{x+2})^\top (B_{x+1})^\top \tag{149}$$

so that with Eq. (133) we get the generalized recurrent equations

$$(W_x^B)^\top = (W_{x+1}^B)^\top (B_{x+1})^\top \tag{150}$$

$$(W_x^F)^\top = (W_{x-1}^F)^\top \overline{P_{x-1}^{eO}} = (W_{x-1}^F)^\top \frac{P_{x-1}^{eO}}{\sigma_{x-1}^{wO}}. \tag{151}$$

2.2.1 Stationary Population, Forward and Backward Prevalences

Let us study again the simple example of the labor force and the activity ratio of women in France. We suppose that the mortality is neglectable or, and this is similar, that there is no differential of mortality between active and inactive women.

Let us consider a cohort, named $\mathcal{F}$, of N young women of age 14 born at year t and a set of matrices M similar to Table 1 for each age between 14 and 74 years. In order to simplify, we suppose that the size of the cohort at each age is constant and equal to $N^{\cdot\cdot} = N$.

We can set the age-specific transition matrices P_x using the series of c_x, a_x observed in France in 1977–1978 from $x = 14$ to 74 years, calculate the number of active and inactive women as well as the stationary forward prevalence y_x^∞ of this cohort $\mathcal{F}$ at each age x.

Let us suppose now that the population is stationary in the sense that each year, the number of young women reaching the age of 14 during any year t is constant and equal to N. We also suppose that the age-specific transition matrices P_x are constant over time. It is easy to understand that the cross-sectional prevalence of this stationary population at any time t is constant and equal to the forward prevalence at each age. Also, this prevalence, y_x^∞ satisfies the recurrent Eq. (64).

On the other way, we can consider a cohort, named $\mathcal{B}$, of N old women of age 75 and compute the backward transition matrices B_{x+1}

(series of γ_x, α_x for $x = 15$ to 75 years) at each age to get the number of active and inactive women at each younger age as well as the stationary backward prevalence $y_x^{*\infty}$ which must statisfy Eq. (66).

Are there conditions that must satisfy c_x, a_x, γ_x, and α_x in order to produce equal forward and backward stable prevalences?

From Eqs. (64) and (66) or, better, from generalized Eqs. (150) and (151) at age x, we get

$$\begin{cases} (W_{x+1}^F)^\top = (W_x^F)^\top \overline{P_x} \\ (W_x^B)^\top = (W_{x+1}^B)^\top (B_{x+1})^\top. \end{cases} \tag{152}$$

If we assume the equality at age $x + 1$, $(W_{x+1}^F)^\top = (W_{x+1}^B)^\top$ we get

$$(W_x^B)^\top = (W_x^F)^\top \overline{P_x} (B_{x+1})^\top \tag{153}$$

so that the equality at any age x imposes the necessary following condition

$$(W_x^B)^\top = (W_x^B)^\top \overline{P_x} (B_{x+1})^\top \tag{154}$$

which does not seem to make any sense.

Let us now investigate a different approach which will result in the following theorem.

Theorem 1. *If the observed prevalence in an cohort of survivors is equal, for any state and age, to the stationary backward prevalence, the stationary forward prevalence is also equal to both of them.*

Proof. We can consider a cohort of old people, $\mathcal{B}$, whose transitions between states and two adjacent ages $x + 1$ and x are known. Thus, we can calculate the age specific observed prevalences in any state as well as the stationary backward prevalences, $(W_x^B)^\top$ constructed for example, by recurrence using the backward generalized Eq. (150) and starting from a very old age. Let us suppose that the observed backward prevalence is identical at any age and state i to the corresponding stationary backward prevalence. We could say that this old cohort $\mathcal{B}$ is "backward stationary."

Let us now calculate the forward transitions, $\overline{P_x(\mathcal{B})}$, as well as the forward prevalences $(W_x^F(\mathcal{B}))^\top$ of this backward stationary cohort $\mathcal{B}$ in any state i using the generalized forward recurrent Eq. (151) while starting yet from a very young age onward.

We will use a mathematical induction to prove that this forward prevalence $(W_x^F(\mathcal{B}))^\top$ is equal to the backward prevalence $(W_x^B)^\top$.

Making the hypothesis that the theorem is true up to age x, we will prove that it is also true at age $x + 1$. More precisely let us suppose that this forward prevalence is equal to the backward prevalence starting from this very young age up to age x (Fig. 11)

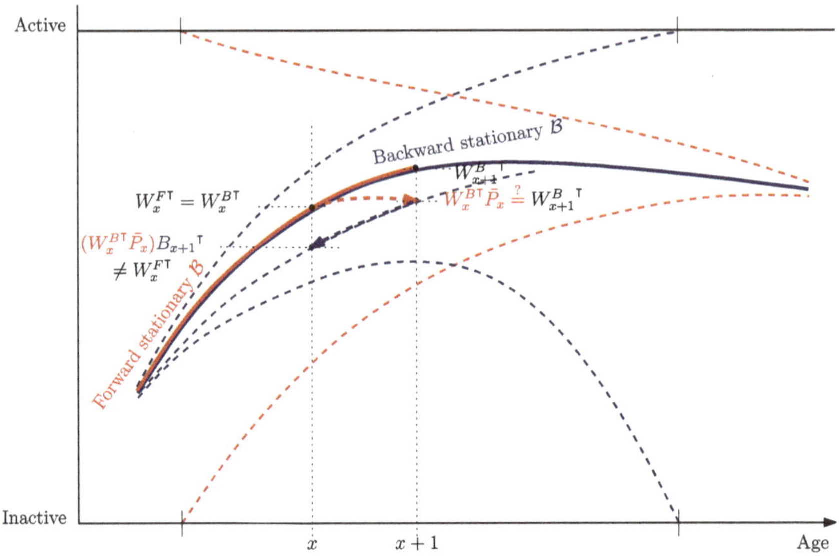

FIG. 11 Help for the proof of Theorem 1. Considering a backward stationary cohort $\mathcal{B}$, its associated forward prevalence should be identical to the backward prevalence. *Blue* is for backward projections, *red* for forward projections. *Dotted lines* are examples of trajectories either from activity or inactivity.

$$(W_y^F)^\top = (W_y^B)^\top \quad \forall y \leq x. \tag{155}$$

We can then use Eq. (151) and above Eq. (155) at age $x-1$ to calculate the forward prevalence at age $x+1$, $(W_{x+1}^F)^\top = (W_x^F)^\top \overline{P_x} = (W_x^B)^\top \overline{P_x}$.

And finally, we can certify using a proof by contradiction that this calculated forward prevalence $(W_x^B)^\top \overline{P_x}$ is identical to the backward stationary prevalence $(W_{x+1}^B)^\top$: In fact, if $(W_x^B)^\top \overline{P_x}$ was not equal to $(W_{x+1}^B)^\top$, the resulting prevalence calculated from the backward recurrent Eq. (150) at age x, $((W_x^B)^\top \overline{P_x})(B_{x+1})^\top$, would differ from the prevalence $(W_x^B)^\top$ and that is in contradiction with our former hypothesis of Eq. (155).

Also, because each of the two directions used, first forward and then backward, is directions which are convergins, we will not have precision issues.

This completes the mathematical induction and achieves the proof of the theorem.

A similar theorem applies to a cohort whose observed prevalences are equal to the stationary forward prevalences: the backward prevalences are identical to the observed and forward prevalences.

This theorem is important in practice because it means that if the cross-sectional prevalences observed during a first pass are identical to the prevalences deduced from the matrix of the forward transitions measured at a second pass, the prevalences deduced from the backward transitions must also be identical.

It can be said differently: if the forward prevalences are close to the cross-sectional prevalences, the backward prevalences are also close.

To be convinced and highlight the previous theorem, we drew in Figs. 12 and 13 the forward (violet) and backward (green) probabilities from the chain survey on the French female labor force (in bold lines) and their associated stationary series (in fine lines). A series and its associated will produce identical forward and backward prevalences.

As the situation in France was very far from stationnarity, the forward and backward probabilities were very different. But their associated probabilities, either forward or backward are closer from each other.

In demography, it is a good practice to use a fictitious cohort in order to emphasize the changes in mortality or fertility rates observed during a short period of time. But in this chapter, we are introducing the concept of a fictitious cohort of old people whose characteristics were measured when they were younger.

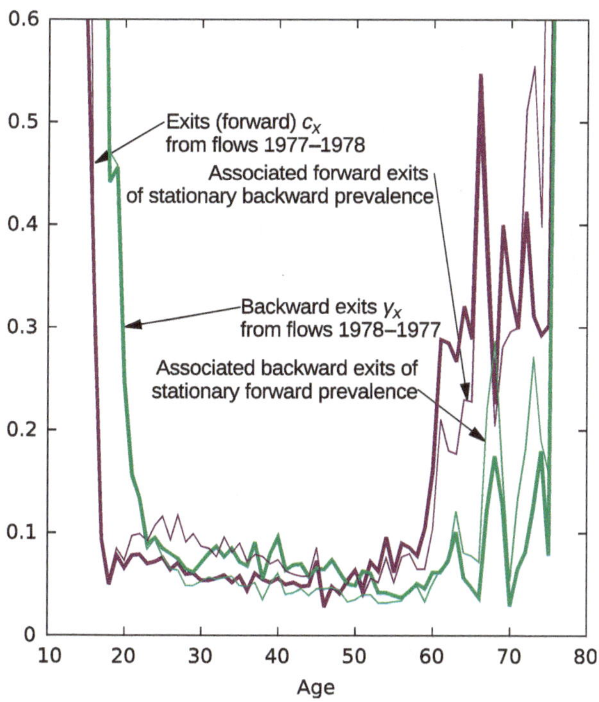

FIG. 12 Exits: initial forward c_x and backward γ_x probabilities as well as their associated $c_x(\mathcal{B})$ and $\gamma_x(\mathcal{F})$.

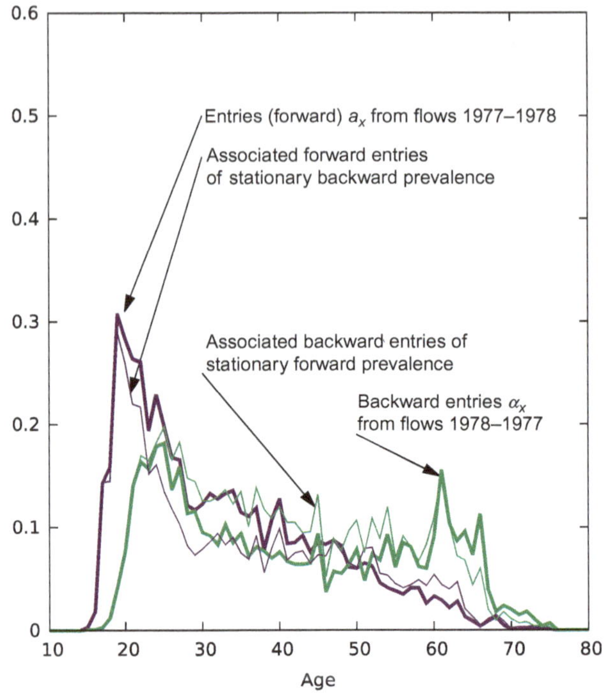

FIG. 13 Entries: initial forward a_x and backward α_x probabilities as well as their associated $a_x(\mathcal{B})$ and $\alpha_x(\mathcal{F})$.

The backward prevalences will reflect the behavior of the population in the past and the forward prevalences the behavior of the population in the future. The cross-sectional prevalences reflect the current situation mixing old and young people.

These three concepts are represented on the Lexis diagram of Fig. 14.

On this figure, we can see a "census" vertical line at time t and a period of time between $t-\frac{1}{2}$ and $t+\frac{1}{2}$. Usually, age specific rates or transitions are estimated in a square between two exact ages, but there values are similar to those estimated in a "perspective" diamond centered at the same mean date and mean age. These forward probabilities or rates belonging to a diamond can be moved to the right to correspond to what a "young" cohort could expect in its future if the rates remained constant. If we use mortality rates, we can compute the survival function that this young cohort could expect as well as its life expectancy.

The introduction of backward probabilities or backward rates estimated in the same diamonds is new. These diamonds and their corresponding backward values can be moved to the left in order to belong to a same cohort of "old" people. Unfortunately the backward probability of mortality is useless but for multistate models estimated from cross-longitudinal surveys, the concept of backward probability is meaningful, and, for example, backward prevalence in a health state can be estimated in an old cohort at each age.

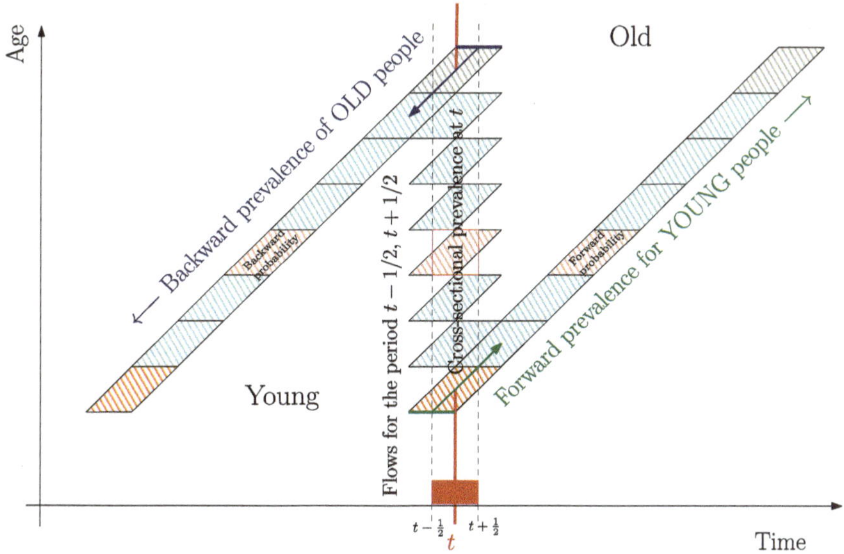

FIG. 14 Age-specific forward probabilities and age-specific backward probabilities linked for young, respectively, old cohorts.

The calculation of backward prevalences is therefore an important contribution to highlight the behavioral changes of the human populations: if a significant change in behavior is initiated, the backward prevalence will generally be in the opposite direction of the forward prevalence relative to cross-sectional prevalence.

Before the presentation of some results concerning these newly backward prevalences, it seems important to explain, using the analysis of mortality, how big the difference between a cross-sectional index and a forward or period index can be.

Let us imagine a stationary population, i.e., a closed population with a constant number of births per unit of time and constant age specific mortality rates. In each cohort, the survival function is constant and its area under the curve or life expectancy is constant too.

Let us now assume that the number of births is still constant and equal to 1 per unit of time but that the mortality corresponded to the historical observed mortality by age since more than one and half century (see Fig. 15).

By analogy with the cross-sectional prevalence of disability or of any characteristic which can be measured during a census or a cross-sectional survey, we will name this "observed" proportion of survivors since birth, the "cross-sectional prevalence of life."

Fig. 16 shows the cross-sectional prevalences of survival by gender in France on the mid-1982 when this concept has been first published (Brouard, 1986). They can be compared with the corresponding period life tables deduced from the mortality rates observed in France during 1982.

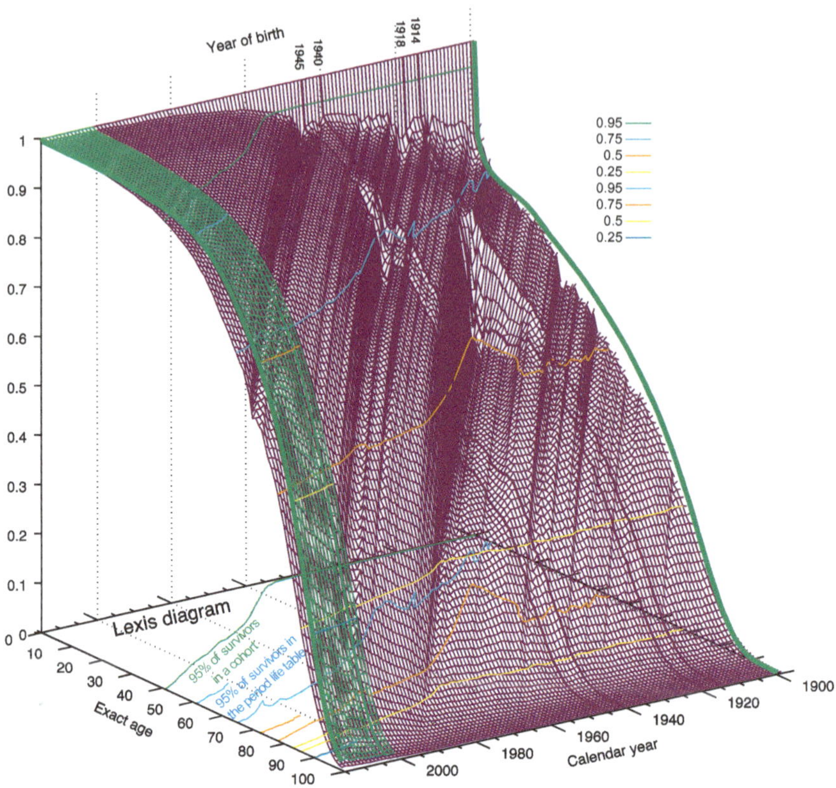

FIG. 15 Proportion of male survivors from birth in France observed between 1899 and 2015, by exact age and calendar year. Depletions due to the World Wars I and II. Comparison with the annual period life tables since year 2000 and in 1899 (*green surfaces*). Contour plots at 95% (5% loss) of cohorts surface (*green line*) and period surface (*blue line*) are also reported on the Lexis diagram of the basis.

If the area under the period life table is the classical life expectancy at birth e_0, the area under the cross-sectional life prevalence, which I sometimes named d_0 (for duration), is a "crosse-sectional" index (Brouard, 1986), now the so-called C.A.L. index, for cross-sectional average length of life (Canudas-Romo and Guillot, 2015; Goldstein and Wachter, 2006; Guillot, 2003; Luy, 2006).

In France, the difference between both indices which is visualized in Fig. 17 is still high (about 6 years for men and 5 years for women in 2015), reflecting the fact that the current age structure of the French (male) population is still much younger than what the low current mortality levels will impose. These days, the difference is getting smaller because the larger gap was due to the historic decline in infant mortality and young adults, which could be seen until 1950 (Fig. 15). After 1950, infant mortality was already

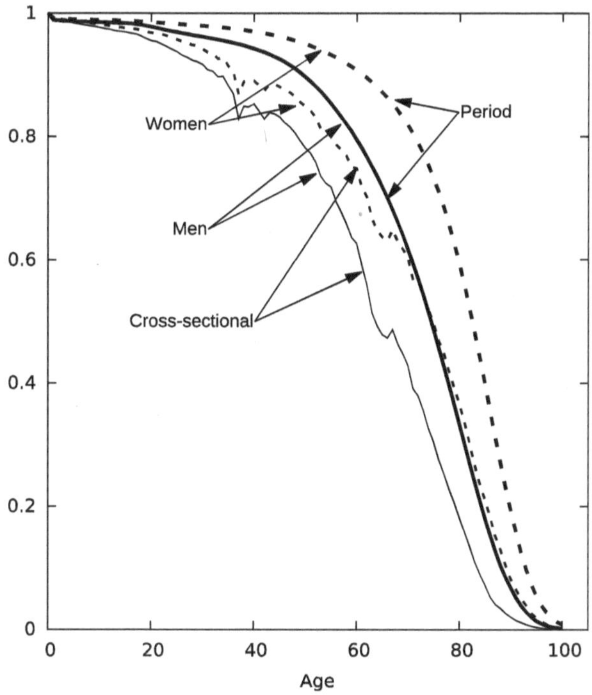

FIG. 16 Cross-sectional prevalences of survival by gender in France in 1982. Comparison with the period life tables in 1982.

low, so even though it continued to decline in relative variation, absolute variations were small, minimizing the tempo effect. We can see in this figure that according to the 2015 period life table, 95% of French men could reach the age of 50, but in reality within any male cohort ever born in France, 5% already died before the age of 40 (95% of the most recent cohort was only 40 years old).

The tempo effect was aggravated by world wars, particularly because of the tragic mortality caused by the First World War of 1914–1918 among men (Fig. 15).

In 1982, assuming that the mortality rates observed in 1982 would remain constant in the future, it would have been possible to predict the cross-sectional prevalence of survival, each year to come, by projecting survival of each cohort under this hypothesis of mortality constancy.

Such predictions for 1992, 2002, 2012, 2022, and 2032 are represented in Fig. 18. The convergence to be the period prevalence, or period life table of 1982, is almost achieved by 2022. The last comparison, which is of great interest, concerns cross-sectional prevalence of survival in 2015. The 1982 mortality table was a good predictor of survivors but, obviously, mortality did not stop decreasing after 1982 and especially after 70 years where the

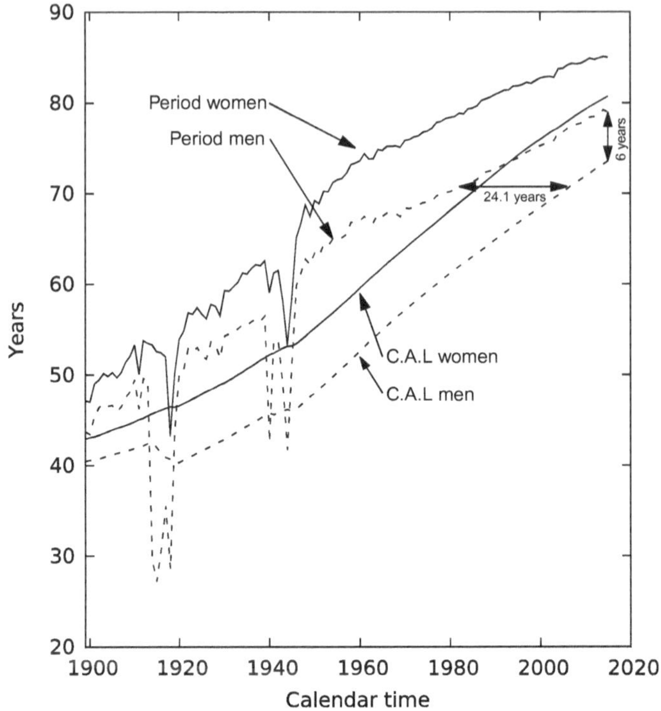

FIG. 17 Cross-sectional duration of survival or C.A.L (d_0) and life expectancy (e_0) in France by gender from 1899 to 2015.

cross-sectional is much greater than the period. For men, the "expected" life in 1982, $e_0(1982)$, is reached by the effective duration d_0 only after 24.1 years in February 2006.

This is very similar to Figs. 2 and 5 when the economic activity of 1978 was projected in the future using the transitions measured between 1977 and 1978 up to the limit (or forward prevalence) and finally compared with the cross-sectional prevalence observed in the next censuses.

We can also say that life expectancy at birth is an advanced indicator of mortality. This advance is currently around 24 years but it overestimates the actual mortality in France of about 6 years.

Some economists, demographers, and epidemiologists mixed age-specific cross-sectional prevalences of activity or disability within a period survival curve in order to summarize the resulting curve by its area and naming it the "working life expectancy" (Wolfbein, 1949) or "health expectancy" (Sullivan, 1971). These indices are hybrid, mixing a potential future of mortality decline with a observed prevalence which cannot be constant because its components the in and out flows are not constant over time in particular because the mortality is not constant and its life expectancy is different from the C.A.L.

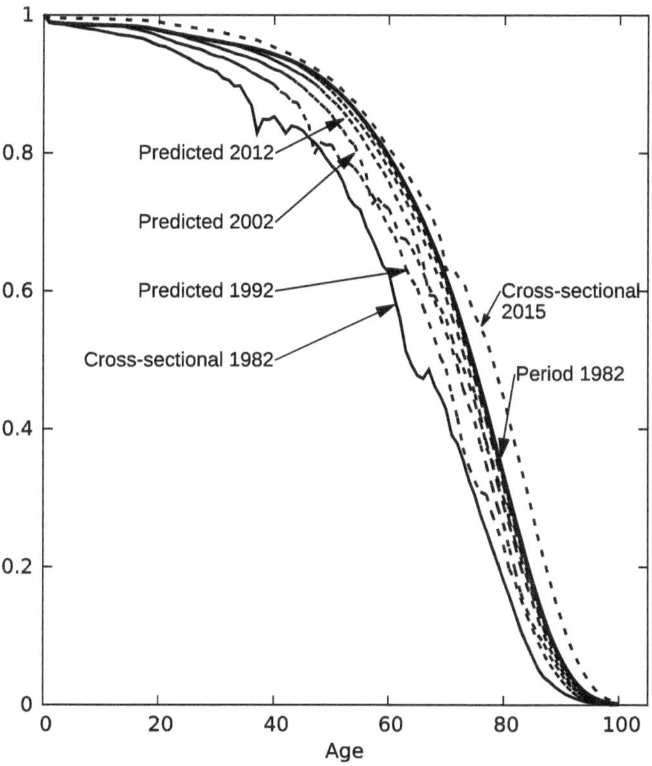

FIG. 18 Forecast of the male 1982 cross-sectional prevalence of survival under the hypothesis of a constant mortality (1982) in 1992, 2002, 2012, 2022, and 2032. Comparison with the period life table of 1982 and the observed cross-sectional life prevalence of 2015.

We proposed in 1992 (Bonneuil et al., 1992) to calculate two indices, (a) a purely cross-sectional, calculated as the cross-sectional prevalence of activity or disability at each age multiplied by the cross-sectional prevalence of survival at the same age and (b) a second, a purely period index, that is to say computed only from the age-specific flows between states (healthy, disabled, dead) observed during a given period, so that their comparison will highlight the tempo effect of mortality but also of disability (Luy et al., 2018).

In this end of the chapter, we simply transposed the two demographic analysis techniques (economic status change models of 1980 and the mortality models of 1986) to the analysis of the more recent phenomenon that is the evolution of disability and dependence.

With the availability of more and more cross-longitudinal surveys, such as the Longitudinal Studies Of Aging (LSOA I and LSOA II) or the Health and Retirement Study (HRS), the concept of health expectancy developed by Sullivan has evolved into a more valuable indicator-based solely on age-specific transition rates between health states and death (Lièvre et al., 2003).

The international REVES network (https://reves.site.ined.fr/en/) has been working since the mid-1980s for the development of such health indicators.

As a founding member of this network in the late 1980s, we developed with the help of Agnès Lièvre and Christopher Heathcote, a computer program called IMaCh (http://euroves.ined.fr/imach) that estimates healthy life expectancy and age-specific forward prevalences (Lièvre et al., 2003). In today's transversal-longitudinal surveys, people are unfortunately not interviewed at each of the passages, so that transitions between states have different exposure times and must be estimated by a statistical model like the interpolated Markov chain model (Laditka and Wolf, 1998). The IMaCh software has already been used by specialized researchers in health and disability.

The software evolved since the early versions of 1999 through a first main publication in 2003. It is currently a software of 12,800 lines of C code. One of the main constraint is the time for the likelihood of the sample to be maximized and IMaCh users are invited to start with time intervals close to the mean interval between waves before decreasing the interval down to a single month which dramatically increases the number of matrix products and the time to maximum likelihood. Therefore, we derived our own method of optimization (Brouard and Heathcote, 2015) to speed up the convergence and got grants from Intel software in order to produce IMaCh Windows 64 bits executables compiled with their fast Intel compiler.

But it is only recently that we have developed the concept of backward prevalence and that the experimental version 0.99 of IMaCh (2019) makes it possible to calculate backward prevalences by age. These backward prevalences are plotted with the cross-sectional prevalences as well as with the forward prevalences for useful comparisons.

With the general decline of mortality in the developed countries, we are usually observing a lower period prevalence of disability than the cross-sectional prevalence. This indicates that disability is currently declining or postponed. But by using the estimation provided by IMaCh on two almost identically designed cross-longitudinal surveys (LSOA I and LSOA II), we have been able to measure the changes in the transitions rates over a decade and found that the decline in the mortality of dependent people fell rapidly and contravened the general decline in dependency in the United States (Crimmins et al., 2009).

2.3 Some Estimations of Backward Prevalences

We will use two examples in order to highlight the various advantages of the calculation of backward probabilities.

2.3.1 Did the 2008 Economic Crisis in Italy Impact the Health Expectancies?

Results from the 2013 longitudinal release of the Italian "Statistics on Income and Living Condition" survey (EU-SILC) (Giudici et al., 2017) are shown

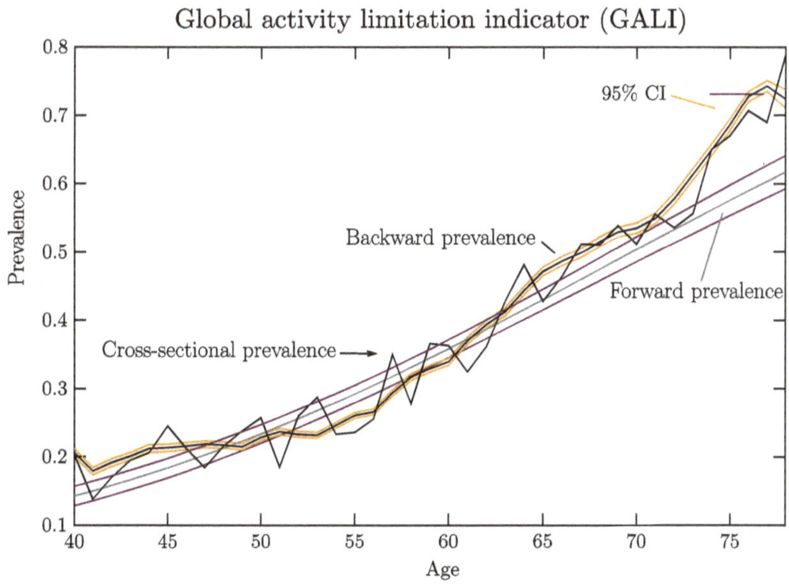

FIG. 19 Italian SILC survey: backward, forward, and cross-sectional prevalences are similar. 95% confidence intervals are represented. No influence of the 2008 economic crisis on health expectancy.

in Fig. 19 that the cross-sectional, forward, as well as backward prevalences of disability measured by the Gali index are almost identical. It is a proof that the 2008 economic crisis in Italy has not had a major impact on health expectancies. This stationarity is also seen when changing the Gali index with the self-rated health index or even the chronic diseases index.

Also we can remark that younger generations will be less disabled at very old ages.

2.3.2 Health and Retirement Study: Changes in Living Arrangements

In this second example, we explored the properties of the backward prevalences using data from the recent (2017) US Health and Retirement Study with its nine waves (1998–2014). Using an earlier version of HRS (with eight waves), we were not only interested in health changes, but also in how living conditions evolved according to the disability status of a person during his or her life cycle (Shih, 2016).

In this chapter, we are limiting our investigation to three different living arrangement statuses, "coresidence," "alone," and "in institution" and how people move between the three states and death which is an increasingly competing risk after age 50 (see Fig. 20).

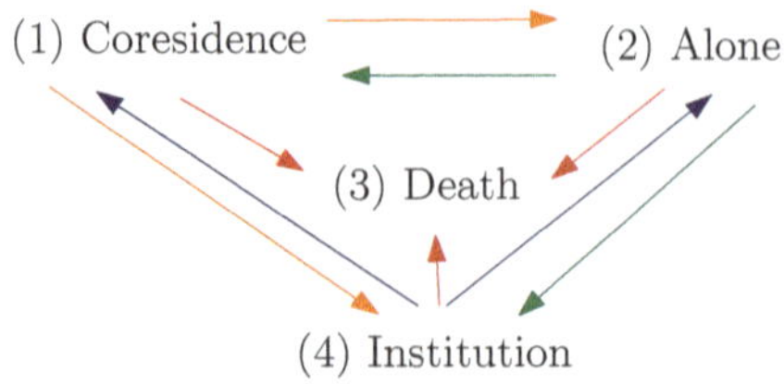

FIG. 20 Diagram of the transitions between the four states.

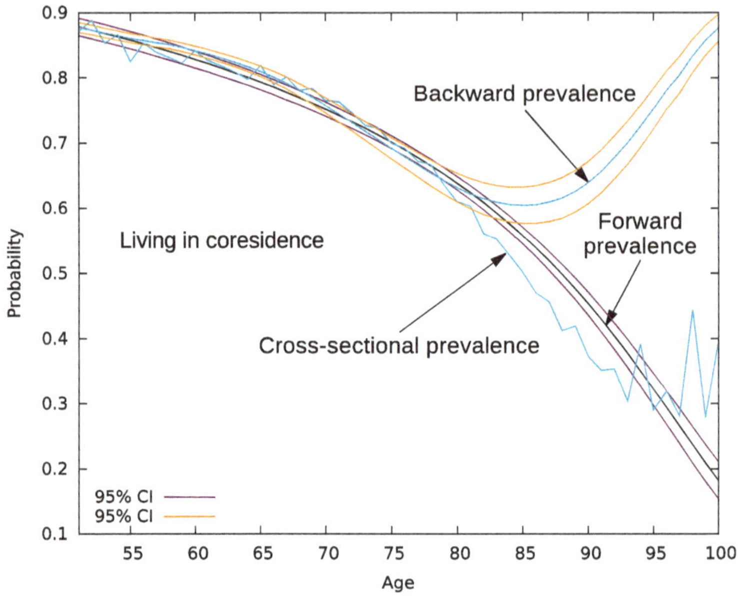

FIG. 21 Cross-sectional, forward, and backward prevalences with 95% confidence intervals of living in coresidence (HRS 1988–2014). Living in coresidence at old ages was more frequent is the past.

- Results shown in Fig. 21 describe how living in coresidence will slightly increase between age 80 and 90 (significative difference between the cross-sectional and forward prevalences at those ages). The backward prevalence also shows that for old cohorts, living in coresidence after age 85 concerned a majority of families and increased with age.
- Results in Fig. 22 indicate that "living alone" is shifting to older ages: the peak was at age 85 (backward prevalence), is around 90 (cross-sectional), and will reach the age of 95 (forward prevalence).
- Results shown in Fig. 23 indicate that "living in an institution" will not change in the near future. It was very occasional in the past, because the backward prevalence was very low.

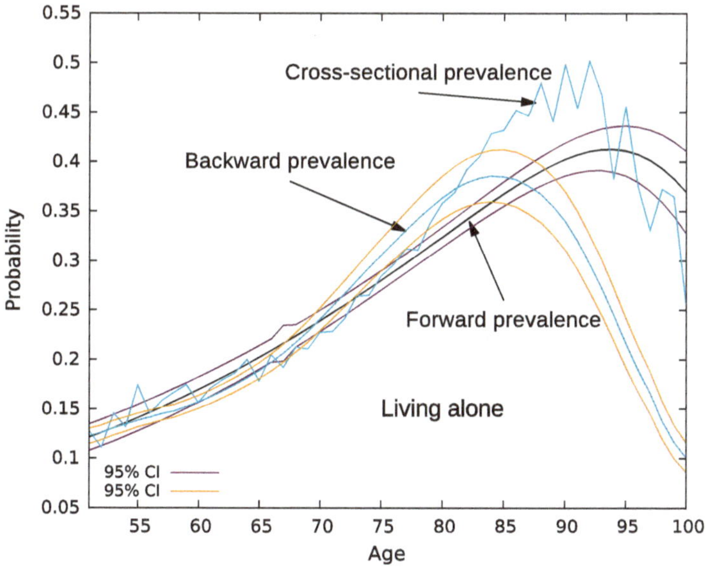

FIG. 22 "Living alone": shift to older ages.

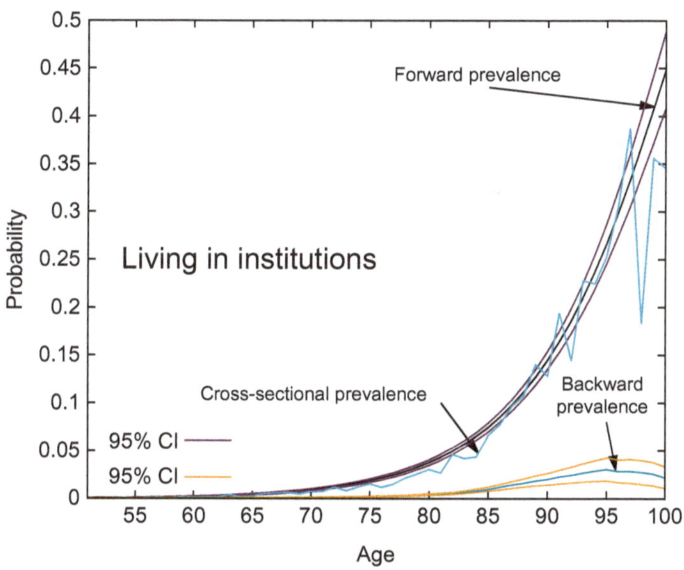

FIG. 23 "Living in an institution" will not increase in the future but was very rare among the old cohorts.

The most interessing change concerns the aging of the persons living alone and we can even indicate, by using projections starting from the mid period of the nine waves, i.e., August 2003, that this modal age of 94 will be reached around 2024 (Fig. 24).

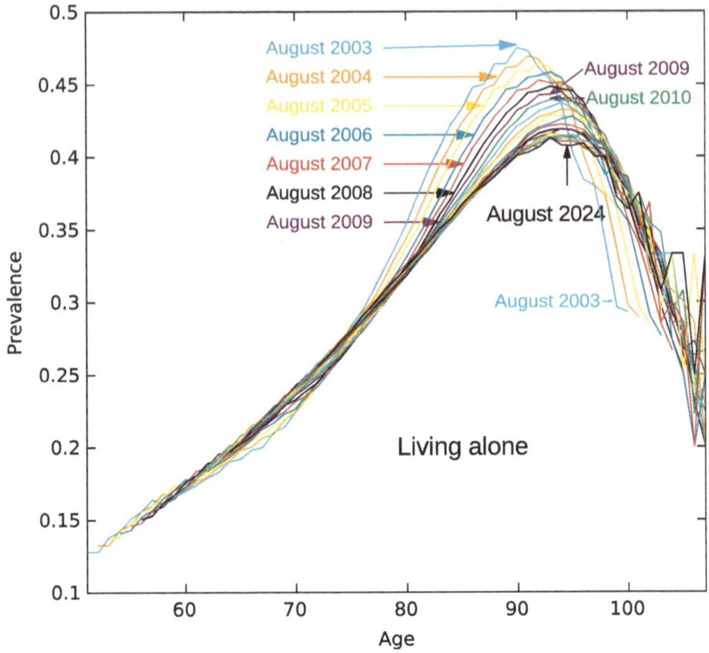

FIG. 24 (Forward) Projection of prevalence of living alone from the observed in August 2003 to August 2024 (HRS 2017). The modal age will reach 94 in 2024. Results from IMaCh 0.99.

2.4 Limitations Concerning the Estimation of Forward and Backward Prevalences

We cannot conclude this chapter without discussing the major limitations of these forward and backward prevalences. Concerning the forward prevalences, they are synthesized in Fig. 25. It can be seen that from people living alone at age 50 (state 2), only half of them are still living alone 10 years after at age 60 (prev(2,2)). From people living in coresidence at age 50, only 16% will be alone at age 60 (prev(1,2)). And among those living in an institution at age 50, 18% will return home alone at 60 (prev(3.2)). If these results seem plausible and interesting, we can see that those probabilities curves are not converging rapidly but at a late age, around age 90 or 95.

It is therefore difficult to obtain the estimate of the forward stationary prevalence of "living alone" unless assumptions are made about transition forces before the age of 50. It has already been discussed in the case of the activity ratio before age 15 and was easily solved (see Section 1.4).

But here, there are no obvious assumptions about these transition rates before the age of 50, so that before age 90, the exact prevalence could only be framed by higher and lower prevalences. Fortunately, in the IMaCh software, the probabilities are estimated using a multinomial logistic

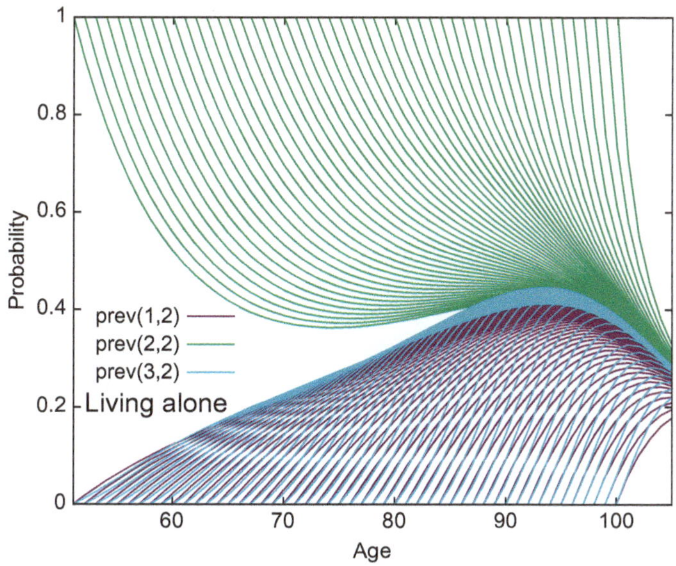

FIG. 25 Slow convergence to "living alone" (state 2): Forward probabilities to stay in the "living alone" state (prev(2,2)) or to exit from a coresidence (state 1) (prev(1,2)) or from an institution (prev(3,2)).

regression model, so that it is possible, at the cost of a wider confidence interval, to extrapolate the transition probabilities outside the age range. But we can see that the 95% confidence intervals of Fig. 23 are not huge.

This discussion on the limitations does also concern the backward prevalences. It can be seen in Fig. 26 that the backward convergence is more rapid giving more credits to its estimation in this example.

2.5 Perspectives Concerning the Estimation of Backward Prevalences

Let us review first the computation of the forward prevalences in practice. IMaCh optimizes the likelihood of the sample defined by the product of the *forward* probabilities p_x^{ij}, of being observed in state j at $x + 1$ (or next wave) being observed in state i at the first wave (at age x) in order to (1) obtain the maximum likelihood parameters.

Then IMaCh (2) recalculates the individual transition matrices at each age, (3) builds the product of matrices using Eq. (132) until convergence is reached.

Concerning the backward prevalences, the current version 0.99 of IMaCh is simply calculating the backward matrices at each age x, B_{x+1} using Eq. (141) which requires the estimated matrix P_x and the *observed* cross-sectional prevalences w_x^i in any state i and age x during a wave (or the average

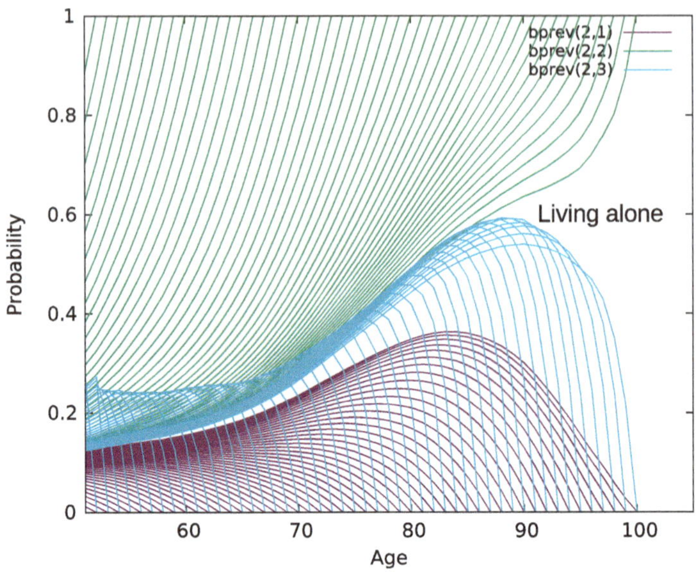

FIG. 26 Convergence to the stationary backward prevalence of "living alone" from each of the three states.

of the cross-sectional prevalences observed during different waves). We also have to ask the user to agree with a smoothing process applied to the age-specific cross-sectional prevalences. Then, IMaCh uses Eq. (149) in order to get the convergence at age x and the corresponding backward prevalences.

But an interesting perspective could be to build a kind of "backward" likelihood made of observed back probabilities in order to directly estimate the backward parameters and finally the backward prevalences by age.

3 CONCLUSION

Even if the concept of back probability and backward stationary prevalence were already described in our old article of 1980 (Brouard, 1980), the mathematical theory behind this concept was never developed.

We revisited the strong and weak ergodicity properties of Markov chains and applied them to backward transitions even in the presence of an absorbing state.

Therefore we introduced a new theorem concerning the equivalence of the backward prevalences with the cross-sectional as well as with the forward prevalences in the stationary case.

In the first and in the last sections, we tried to persuade the reader that both concepts of forward and backward prevalences could be of great interest to analyze sociological changes in our modern societies.

We also know that the estimation of such prevalences requires a lot of data and computing. Data provided by cross-longitudinal surveys are more and more numerous in the United States, in Asia and in Europe. And most of the time, they are made available to researchers free of charge.

Similarly, tedious estimates of these age-transition forces as well as both backward and forward prevalences can only be made by appropriate software, such as IMaCh. Let us mention two other softwares offering multistate life tables analyses, currently limited to the calculation of health expectancies but which could be extended to the calculation of backward and forward prevalences: the SPACE (Stochastic Population Analysis for Complex Events) program, developed by Cai et al. (2010) and ELECT (Estimation of life expectancies using continuous-time multistate survival models) from Ardo van den Hout which utilizes the 'msm' package for R developed by Jackson (2011).

Fortunately, these two softwares, like IMaCh, are GPL licensed softwares that can be used and modified freely. Hopefully they will be used and improved in the future by new generations of economists, demographers, and epidemiologists.

ACKNOWLEDGMENTS

The author would like to thank a few persons: (a) Yao-Chi Shih, from the National Chung Cheng University of Taiwan for his idea to study living arrangements in the United States as a multistate model; (b) Cristina Giudici and Eleanora Trappolini from the La Sapienza University of Rome in Italy for their idea to measure if the Italian economic crisis of 2008 did affect the health expectancies by using the Italian SILC survey; (c) Saito Yasuhiko from the Nihon University of Tokyo in Japan for his continuous support to this research. Also the IMaCh software got a grant from the Japan Society for the Promotion of Science (Grant-in-Aid for Scientific Research # 25293121 and Intel software provided free licenses for the Intel C compiler, mostly efficient on Windows 64 bit platforms. The author would like to thank Marília Nepomuceno, postdoctoral researcher at the Max Planck Institute for Demographic Research in Rostock, Germany, for her help to improve the readability of, otherwise, more obscure sections and an anonymous referee.

REFERENCES

Bonneuil, N., Brouard, N., Robine, J.-M., 1992. Methods of calculation of health expectancy: application to the LSOA surveys (1984-86-88). In: 5th Meeting of the International Network on Health Expectancy (REVES-5): Future Uses of Health Expectancy Indices, Ottawa, February 19-20-21, p. 10. http://euroreves.ined.fr/imach/Bonneuil-Brouard-Robine-Ottawa-1992.pdf.

Brouard, N., 1980. Espérance de vie active, reprises d'activité féminine: un modèle. Revue Econo. 31 (6), 1260–1287. https://doi.org/10.2307/3501190 (special issue on Economy and Demography).

Brouard, N., 1986. Structure et dynamique des populations. la pyramide des années à vivre, aspects nationaux et exemples régionaux. Espace Pop. Soc. (2), 157–168. https://doi.org/10.17605/OSF.IO/2UCYH. (T. Riffe in 2018 on SocArxiv, Transl.).

Brouard, N., Heathcote, C., 2015. IMaCh is proposing to measure health expectancy with one index. In: Health Expectancy: Is It Possible to Measure Population Health With One Index? http://euroreves.ined.fr/imach/singapore-Brouard-Heathcote-2015.pdf.

Cai, L., Hayward, M., Saito, Y., Lubitz, J., Hagedorn, A., Crimmins, E., 2010. Estimation of multi-state life table functions and their variability from complex survey data using the SPACE program. Demogr. Res. 22 (6), 129–158. https://doi.org/10.4054/DemRes.2010.22.6.

Canudas-Romo, V., Guillot, M., 2015. Truncated cross-sectional average length of life: a measure for comparing the mortality history of cohorts. Popul. Stud. 69 (2), 147–159. https://doi.org/10.1080/00324728.2015.1019955. PMID: 25871610.

Cohen, J.E., 1979. Ergodic theorems in demography. Bull. Am. Math. Soc. (N.S.) 1 (2), 275–295. https://projecteuclid.org:443/euclid.bams/1183544080.

Crimmins, E., Hayward, M.D., Hagedorn, A., Saito, Y., Brouard, N., 2009. Change in disability-free life expectancy for Americans 70 years old and older. Demography 46 (3), 627–646. https://doi.org/10.1353/dem.0.0070.

Giudici, C., Brouard, N., Coppola, L., Trappolini, E., 2017. Disability trends among Italian adult population. A multistate approach. In: Squaring the Curve in Less Developed Countries: Thinking Public Policies for Healthy Life Expectancies. XXVIII International Population Conference, Cape-Town (South Africa), 29 Oct–4 Nov 2017, Poster.

Goldstein, J.R., Wachter, K.W., 2006. Relationships between period and cohort life expectancy: gaps and lags. Popul. Stud. 60 (3), 257–269. https://doi.org/10.1080/00324720600895876 PMID: 17060053.

Guillot, M., 2003. The cross-sectional average length of life (CAL): a cross-sectional mortality measure that reflects the experience of cohorts. Popul. Stud. 57 (1), 41–54. https://doi.org/10.1080/0032472032000061712.

Jackson, C.H., 2011. Multi-state models for panel data: the msm package for R. J. Stat. Softw. 38 (8), 1–29. http://www.jstatsoft.org/v38/i08/.

Laditka, S.B., Wolf, D.A., 1998. New methods for analyzing active life expectancy. J. Aging Health 10 (2), 214–241. https://doi.org/10.1177/089826439801000206.

Le Bras, H., 1971. Éléments pour une théorie des populations instables. Population 26 (3), 525–572. https://doi.org/10.2307/1529142.

Lièvre, A., Brouard, N., Heathcote, C., 2003. The estimation of health expectancies from cross-longitudinal studies. Math. Popul. Stud. 10 (4), 221–248. https://doi.org/10.1080/713644739.

Lopez, A., 1961. Problems in Stable Population Theory. Office of Population Research, Princeton University.

Luy, M., 2006. Mortality tempo-adjustment: an empirical application. Demogr. Res. 15 (21), 561–590. https://doi.org/10.4054/DemRes.2006.15.21.

Luy, M., Sauerberg, M., et al., 2018. Could the actual decrease in life expectancy be caused by "Tempo Effects"? In: REVES@30: Past, Present, and Future Trends in Population Health. https://globalagingandcommunity.com/reves-2018/program/.

Shih, Y. C., 2016. Living alone and subsequent living arrangement transitions among older Americans (Ph.D. thesis). University of Massachusetts Boston, Boston, MA.

Sullivan, D.F., 1971. A single index of mortality and morbidity. HSMHA Health Rep. 86 (4), 347–354.

Wolfbein, S.L., 1949. The length of working life. Popul. Stud. 3 (3), 286–294.

Chapter 11

Behavior of Stationary Population Identity in Two-Dimensions: Age and Proportion of Population Truncated in Follow-up

Arni S.R. Srinivasa Rao[*,1] **and James R. Carey**[†,‡]
[*]*Augusta University, Augusta, GA, United States*
[†]*Department of Entomology, University of California, Davis, CA, United States*
[‡]*Center for the Economics and Demography of Aging, University of California, Berkeley, CA, United States*
[1]*Corresponding author: e-mail: arrao@augusta.edu*

ABSTRACT
Stationary population identity arises while studying the average time left to die after forming a captive cohort of individuals in a stationary population. We extend the ideas of stationary population identity, which arises in stationary populations to two dimensions with the aim to answer questions related to aging, age structure of the captive cohort. These ideas found to be useful for identifying individuals by age group in a captive cohort.

Keywords: Captive cohort, Population age structure, Partition function

1 INTRODUCTION

Although one of the basic canons of population studies is that knowledge of age composition is fundamental to the understanding the dynamics of a population, information on the age structure of a natural population is usually extremely limited. Although there are some long-term animal studies in which most or even all newborn are marked at birth and thus are of known age throughout a study (Nussey et al., 2008), investigations with this information are the exception rather than the rule because of costs and logistics. Similar challenges exist for studies of natural arthropod populations where various

Handbook of Statistics, Vol. 40. https://doi.org/10.1016/bs.host.2018.12.001

methods are used to estimate the age of individuals including, for example, mechanical damage, biochemical assays, physiological markers, and gene expression (Cook and Sinkins, 2010). Because of the disconnect between the need for information on population age structure but the technical and budgetary challenges of obtaining the age of individuals for estimating age structure, Carey and his colleagues developed a new approach to obtaining information about the age structure of wild populations that involved neither mark-recapture or age-estimating technologies. Rather the method was based on the concept that the death patterns observed in captured or marked individuals of unknown age would yield information on the age structure of the population from which these individuals were sampled.

The starting point for this concept was to consider a closed stationary (zero growth) population in which the pre- and postcapture actuarial environments were identical and from which individuals were sampled at random and monitored through death. Using a simple table (see table 1 in Müller et al., 2004), Carey showed that, under the stationarity assumption, that the postcapture death distribution of the randomly sampled individuals equals the age structure of the population. In this same paper Müller and his colleagues expressed this relationship analytically. See Math Digest (2014) and PAA Affairs (2015) for a historical perspective on stationary population identity and related new developments. When population is stable, then the age structure of the population is sensitive to the momentum (Rao, 2014).

Stationary population identity (see, for example, Carey et al., 2018) and a generalized theorem on stationary population identity (Srinivasa Rao and Carey, 2015) are true when the population growth is stationary. This equality is obtained from the information on age of an individual and life left at the time of capture of individual while forming a captive cohort in various experiments in ecology and population biology (for example, see Carey et al., 2008; Müller et al., 2004). In one-dimensional stationary population identity, $f_1(a)=f_2(a)$ is true for any arbitrary age a of a population, where f_1 and f_2 are probability density functions of age composition and distribution of remaining lifespans in a stationary population (Carey et al., 2018). This result is fundamental in establishing the experimental observations by Carey (Carey et al., 2008; Müller et al., 2004). $f_1(a)=f_2(a)$ and other results describing the symmetries of life lived and remaining lifespans in Kim and Aron (1989) can be obtained as a direct consequences and applications of classical renewal theory frameworks (Feller, 1941, 1971; Lotka, 1939a,b). A generalized theorem of one-dimensional stationary population identity was proved for sufficiently large stationary population without standard life table and renewal theory framework (Srinivasa Rao and Carey, 2015). It was shown, in this generalized framework, that, there exists a graph from the family of exhaustive graphs constructed based on the coordinates of ages of individuals and their corresponding lives lived, will be equivalent to the graph obtained by the sequence of coordinates of the ages and corresponding lengths of life remaining (or life left) for each individual in a captive cohort (Srinivasa Rao and Carey, 2015). All these above results are helpful in explaining the pattern of remaining lifespans for certain captive cohorts observed by Carey. Less is known on determining an age structure of a captive cohort which is unknown at the time of formation, which will otherwise be helpful in understanding age wise or age-group wise remaining lifespans. Understanding age structure in a captive cohort formed for ecological and biodemographic experiments could be very helpful in epidemiology studies for understanding transmission dynamics of parasites in some vector borne diseases.

The purpose of this chapter is to introduce a variation on stationary population identity in which the postcapture period of monitoring is truncated to less than the maximal age of individuals sampled in the hypothetical population. We introduce and develop concepts of two-dimensional stationary population identity by considering age structure and proportion of population continuous in a range of entire age interval. We use these concepts for constructing partial age structure of the captured population which is unknown. We also consider truncation from right during the follow-up of captive cohort as an additional variable and use this variable in the partition function constructed in the process.

2 STRUCTURE OF THE TWO-DIMENSIONAL CAPTIVE COHORT

Let $x(a,i)$ be the ith individual who was captured at age a and $y(a,i)$ be the life left for ith individual who was captured at age a. Then we represent the total number of individuals captured at age a as $\int_0^\infty x(a,i)di$, who will have life left as $\int_0^\infty y(a,i)di$ units. Due to aging, $\int_0^\infty x(a,i)di$ will become $\int_0^\infty x(b,j)dj$ during the time period $b-a$ $(b>a)$. Suppose on the positive quadrant, we plot A, B, Q, and P where $A=(a,0)$, $B=(b,0)$, $Q=(b,\int_0^\infty x(b,j)dj)$, and $P=(a,\int_0^\infty x(a,i)di)$, respectively (see Fig. 1), then the area of the quadrilateral, $ABQ'P$, say, K for $BQ'>BQ$, satisfies the inequality (Eq. 1),

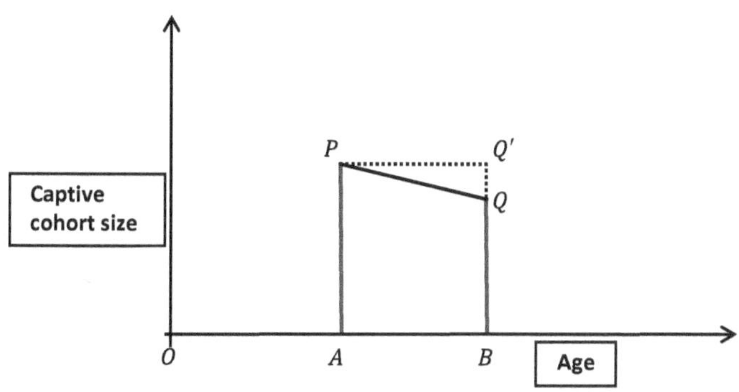

FIG. 1 Aging process in a captive cohort.

$$K > \frac{1}{2}(b-a)\left[\int_0^\infty x(a,i)di + \int_0^\infty x(b,j)dj\right] \quad (1)$$

and the death rate per $b-a$ time units is,

$$\frac{1}{(b-a)}\left[\int_0^\infty x(a,i)di - \int_0^\infty x(b,j)dj\right] \quad (2)$$

If death rate is *zero*, then K becomes,

$$K = (b-a)\int_0^\infty x(b,j)dj \quad (3)$$

In the event that a captive cohort is formed from a stationary population at time t_0, then $\int_0^\infty x(a,i,t_0)di$ is a part or a subcohort of the captive cohort whose size is $\int_0^\infty \int_0^\infty x(s,i,t_0)dids$ (here, s is a variable representing age). The subcohort $\int_0^\infty x(a,i)di$ will live for $a\int_0^\infty x(a,i,t_0)di$ time units if for each individual one-dimensional stationary population identity explained in, (Carey et al., 2018; Srinivasa Rao and Carey, 2015) is true. Then the relation between each age in a subcohort of a captive cohort, size of the subcohort and life lived by individuals in the subcohort is given in

$$a = \frac{\int_0^\infty y(a,i)di}{\int_0^\infty x(a,i)di} \quad (4)$$

The time left for a captive cohort describes the collective time that all the subcohorts of individuals of each age would live in the future. The time left for $\int_0^\infty \int_0^\infty x(s,i,t_0)dids$ is $\int_0^\infty \int_0^\infty y(s,i)dids$, and if a captive cohort consists of individuals of only one age, a, then the total time left for the captive cohort is also a.

We define a function of age structure in the captive cohort, H_1, and an associated set, A, such that,

$$H_1 : (0,\infty) \to J(k) \quad (5)$$

where $J(k) = \{1, 2, \ldots, k\}$ for sufficiently large k, and,

$$A = \{(a, H_1(a)) \forall a \in (0,\infty)\} \quad (6)$$

where $H_1(a) = \int_0^\infty x(a,i)di \in J(k)$. If H_1 is constant, then, the captive cohort is uniformly distributed over all ages or captive cohort consists of individuals of same age. If H_1 is not constant, then, there are three possibilities:

(i) $H_1(a) > H_1(a+h)$ for all $h>0$,
(ii) $H_1(a) < H_1(a+h)$ for all $h>0$,
(iii) H_1 is neither increasing nor decreasing for $0<a<\infty$.

We define a function of life left in a captive cohort, H_2, as follows:

$$H_2 : (t_1, t_2, \ldots, t_\infty) \rightarrow J(k) \tag{7}$$

for

$$t_m = mt_1 \in (a_{m-1}, a_m) \quad \text{for } m = 1, 2, \ldots \infty, t_\infty = a_\infty \tag{8}$$

and

$$\inf(a_{m-1}, \infty) = a_{m-1} \quad \text{for } m = 1, 2, \ldots \infty \tag{9}$$

The size of the captive cohort will start changing over the time t_1, t_2, ... The set C is computed as,

$$C = \{C(t_m) : 1 \leq m < \infty\} \tag{10}$$

describe the collection of individuals from the captive cohort who are available at various future time points. Here $C(t_m)$ is number of individuals from the original captive cohort (say at time t_0), who are surviving at time t_m, is given by,

$$C(t_m) = \int_0^\infty \int_0^\infty x(s, i) di ds - \sum_{s=0}^{m-1} \left[\int_0^\infty x(a_s, i) di \right] \tag{11}$$

We assume $C(t_\infty) = 0$.

3 TRUNCATION OF THE FOLLOW-UP DATA OF CAPTIVE COHORTS

The importance of the captive cohort method applied to the study of age structure in wild populations as described in papers by Carey and his colleagues (Carey et al., 2012a, 2018) is that the resulting population age structure estimates will include all age classes. This is because of the completeness of the database—i.e., the cohort of individuals captured in the wild are monitored until the last individuals die. Example applications in which databases on postcapture longevity are not only complete for single samples, but for multiple samples across entire field seasons, include the studies by Carey and his colleagues on medflies (Carey et al., 2008; Müller et al., 2004) and the study by Papadopoulos and his colleagues on mosquitoes (Papadopoulos et al., 2016). However, collecting complete postcapture data for studying wild populations has limitations. One limitation is that maintaining large numbers of flies, mosquitoes, or other insect species individually (i.e., in solitary confinement) from capture in the wild until the last individual dies can be both costly and logistically challenging (e.g., environmental chamber space). Therefore developing methods for estimating population structure using truncated data will have the potential to reduce costs as well as logistical demands. Researchers would then have the option of considering the trade-offs between gathering less complete vs complete postcapture data relative to their study needs. A second related limitation concerns the time required to gather complete postcapture longevity information. That is, the period required to monitor a captive cohort from time of capture through the death of the last individual. In theory as well as in practice this period will correspond to the maximum age for the species for individuals maintained in optimal laboratory conditions. For example, the period from the time of sampling to the date of death for the last individual in the medfly studies (Carey et al., 2012b) was over 200 days (≈7 months) and in the mosquito study (Papadopoulos et al., 2016) was over 120 days (≈4 months). A consequence of this potentially lengthy periods reduces, if not completely preempts, the usefulness of the information on population age structure for policy making in real time. For example, the use of information on population age structure for intervention decisions concerning the control of disease vector populations (Brady et al., 2016; Cook et al., 2008). Therefore methods for acquiring information on population age structure in real time, even if limited and partial, would be extremely useful for decision-making in the context of vector control.

4 CAPTIVE POPULATION AGE STRUCTURE, TRUNCATION, AND PARTITION FUNCTIONS

With the help of the properties of H_1 and H_2 described in this section, we brought an expression for the number of individuals from the original captive cohort who are surviving at some time in the future. An expression to compute the time left for the captive cohort for varying truncation times, t_m (say) for $m=1, 2, \ldots$ is derived. We have showed that we can construct partial age structure of the captive cohort up to certain age (say, a_s for $a_s < t_m$) by using the sequences of time lived and time left up to t_m.
Theorem 1 *$C(t_m)$, for $m=1, 2,$ is decreasing.*

Proof. We have,

$$\int_0^\infty \int_0^\infty x(s,i)dids - \int_0^\infty x(a_0,i)di > \int_0^\infty \int_0^\infty x(s,i)dids - \sum_{s=0}^{1} \int_0^\infty x(a_s,i)di$$

and in general,

$$\int_0^\infty \int_0^\infty x(s,i)dids - \sum_{s=0}^{m-1} \int_0^\infty x(a_s,i)di > \int_0^\infty \int_0^\infty x(s,i)dids - \sum_{s=0}^{m} \int_0^\infty x(a_s,i)di \quad \text{for each } m=1,2,\ldots,\infty \tag{12}$$

We will obtain $C(t_1)>C(t_2)>\ldots$ □

Remark 1 (i) Suppose, H_1 is decreasing, then, for $a_0, a_1, \ldots, a_\infty$ defined in Eq. (9) we have,

$$\int_0^\infty x(a_0,i)di > \int_0^\infty x(a_1,i)di > \cdots > \int_0^\infty x(a_\infty,i)di \Rightarrow \int_0^\infty\int_0^\infty x(s,i)dids - \int_0^\infty x(a_0,i)di < \int_0^\infty\int_0^\infty x(s,i)dids - \int_0^\infty x(a_1,i)di < \cdots$$

Remark 2 Suppose, $C(t_m)$ is decreasing, then we have,

$$\begin{aligned}\int_0^\infty\int_0^\infty x(s,i)dids - \int_0^\infty x(a_0,i)di &> \int_0^\infty\int_0^\infty x(s,i)dids - \sum_{s=0}^{1}\int_0^\infty x(a_s,i)di\\ &> \int_0^\infty\int_0^\infty x(s,i)dids - \sum_{s=0}^{2}\int_0^\infty x(a_s,i)di\\ &\vdots\\ &> \int_0^\infty\int_0^\infty x(s,i)dids - \lim_{m\to\infty}\sum_{s=0}^{m}\int_0^\infty x(a_s,i)di\end{aligned}$$

By using the relationship in Eq. (4), we can write, $C(t_m)$, in terms of life left as follows:

$$C(t_m) = \int_0^\infty\int_0^\infty x(s,i)dids - \sum_{s=0}^{m-1}\left[\frac{1}{a_s}\int_0^\infty y(a_s,i)di\right] \tag{13}$$

4.1 Time Left for a Captive Cohort and Right Truncation

We derive a formula for computing the time left for the captive cohort at time t_m, say, $S(t_m)$. Age distribution at cohort formation (say at t_0) is given in Table 1. The time left for the captive cohort at t_1, which was formed some time prior to t_1 (say, at t_0), is obtained as the difference between the time left at the time of formation of captive cohort and the time lived by captive individuals up to t_1 (Fig. 2).

TABLE 1 Age-Distribution in the Captive Cohort at t_0

Age	a_0	a_1	⋯	a_∞
Size	$\int_0^\infty x(a_0,i)di$	$\int_0^\infty x(a_1,i)di$	⋯	$\int_0^\infty x(a_\infty,i)di$

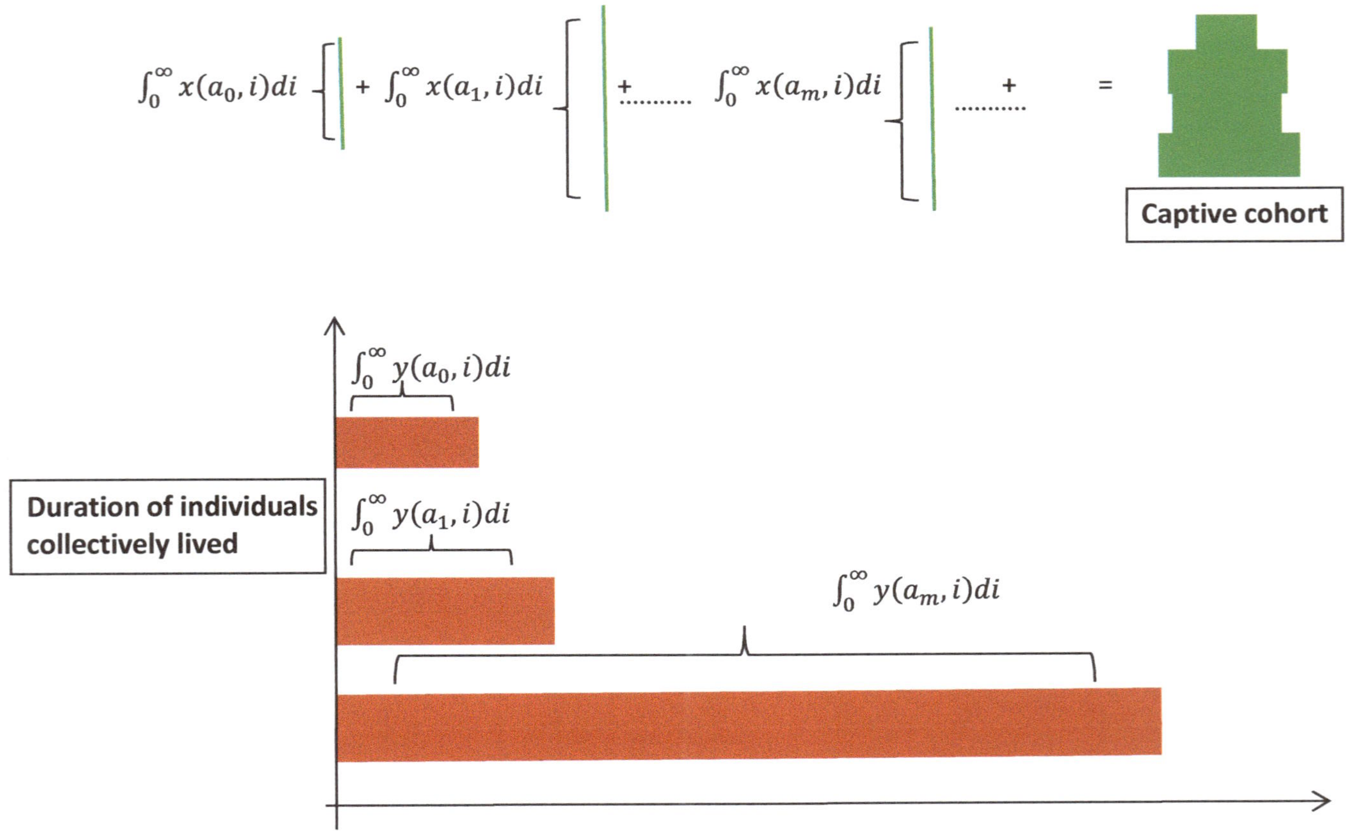

FIG. 2 Schematic structure of set *A* and collective life lived by *A*.

Let, $L(t_1)$ be the life lived by the captive cohort up to t_1 (including those lived up to $a_0 < t_1$), then,

$$\begin{aligned} L(t_1) &= a_0 \int_0^\infty x(a_0, i)di + t_1 \int_0^\infty x(a_1, i)di + t_1 \int_0^\infty x(a_2, i)di + \cdots + t_1 \int_0^\infty x(a_\infty, i)di \\ &= a_0 \int_0^\infty x(a_0, i)di + t_1 \left[\sum_{m=1}^{\infty} \int_0^\infty x(a_m, i)di \right] \end{aligned} \tag{14}$$

where $t_m \int_0^\infty x(a_m, i)di$ is life lived by individuals whose age is a_m for $m=1, 2, \ldots \infty$. We will obtain, $S(t_1)$, the time left by the captive cohort who are surviving at t_1, as,

$$S(t_1) = \int_0^\infty \int_0^\infty y(s, i)dids - L(t_1). \tag{15}$$

Since $a_0 < t_1 < a_1$, at t_1, the individuals with age a_1 would live for the duration $(a_1 - t_1)$ and similarly, at t_1, the individuals with age a_m would live for the duration $(a_m - t_1)$, for $m=2, 3, \ldots, \infty$. Hence,

$$\begin{aligned} S(t_1) &= (a_1 - t_1) \int_0^\infty x(a_1, i)di + (a_2 - t_1) \int_0^\infty x(a_2, i)di + \cdots + (a_\infty - t_1) \int x(a_\infty, i)di \\ &= \sum_{s=1}^{\infty} (a_s - t_1) \int_0^\infty x(a_s, i)di \end{aligned} \tag{16}$$

Suppose, we truncate information from right after t_1, then, the age structure that we are able to construct up to t_1 is $H_1(a_0)$. By Eqs. (14)–(16), we arrive at a relation between the total time left for a captive cohort at t_0 in terms of the time lived up to t_1 by the captive cohort and the time left for the surviving individuals at t_1 of the captive cohort as given in Eq. (17),

$$\int_0^\infty \int_0^\infty y(s, i)dids = a_0 \int_0^\infty x(a_0, i)di + t_1 \sum_{s=1}^{\infty} \int_0^\infty x(a_s, i)di + \sum_{s=1}^{\infty} (a_s - t_1) \int_0^\infty x(a_s, i)di \tag{17}$$

Theorem 2 *Suppose a captive cohort is formed at t_0 and follow-up information is right truncated for the period (t_m, ∞) for some $t_m > t_0$, then, the following relation hold:*

$$\begin{aligned} \int_0^\infty \int_0^\infty y(s, i)dids = &\sum_{s=0}^{m-1} a_s \int_0^\infty x(a_s, i)di + t_s \sum_{s=m}^{\infty} \int_0^\infty x(a_s, i)di \\ &+ \sum_{s=1}^{\infty} (a_s - t_m) \int_0^\infty x(a_s, i)di \end{aligned} \tag{18}$$

Proof. The number of individuals alive at t_m out of captive cohort are $\sum_{s=0}^{m-1}\int_0^\infty x(a_s,i)di$ and other individuals have died at various stages before t_m. Each subcohort of individuals whose age is a_s for $s<m$ have lived collectively for the period $\sum_{s=0}^{m-1}a_s\int_0^\infty x(a_s,i)di$, then, using the arguments in Eq. (14), $L(t_m)$, the life lived collectively up to t_m by the subcohort of individuals of age a_s is given by,

$$L(t_m)=\begin{cases}\sum_{s=0}^{m-1}a_s\int_0^\infty x(a_s,i)di, & \text{if } a_s<t_m\\ t_s\sum_{s=m}^{\infty}\int_0^\infty x(a_s,i)di, & \text{if } a_s>t_m\end{cases} \tag{19}$$

Individuals from the captive cohort whose age is $a_s>t_m$ will live for the period (a_s-t_m). By using the similar arguments of Eq. (16), we can compute, $S(t_m)$, the time left collectively for the subcohorts who are alive at t_m, as,

$$S(t_m)=\sum_{s=1}^{\infty}(a_s-t_m)\int_0^\infty x(a_s,i)di \quad \text{for } a_s>t_m \tag{20}$$

Hence, by using Eqs. (19) and (20), the time left for the captive cohort (which was formed at t_0), i.e., $\int_0^\infty\int_0^\infty y(s,i)dids$ is expressed as (Eq. 18). □

Corollary 1 *We can construct the age structure of the captive cohort up to age group a_s for $a_s<t_m$ by using the sequence of time lived and time left up to t_m in Theorem 2.*

θ_m is the proportion of individuals in the captive cohort who are followed until death up to t_m, is,

$$\theta_m=\frac{\sum_{s=0}^{m-1}\int_0^\infty x(a_s,i)di}{\int_0^\infty\int_0^\infty x(s,i)dids}$$

We have $\int_0^\infty x(a_0,i)di\le\theta_m\int_0^\infty\int_0^\infty x(s,i,t_0)dids\le\int_0^\infty\int_0^\infty x(s,i,t_0)dids$.

Suppose, H_1 is increasing, then, we have, $\theta_1<\theta_2<\theta_3<\cdots<\theta_m<\cdots$. Information available during (t_m,∞) is not observed, so there are options for the death distribution of remaining individuals who are surviving at t_m. Eventhough, all the terms of the sequence $(\theta_m)_{m\ge1}$ may not be possible to compute by t_m, we can see that this sequence is convergent. The probability that the remaining people will survive at least for the period t_1 $(=(m+1)t_1-mt_1)$ after t_m is one if Eq. (21) holds, else, there is a positive probability that the remaining cohort will finish within t_1 units after t_m. The information after t_m which we are unable to follow-up, we denote by $t_{m+b'}$ for $b=1, 2, \ldots$.

$$\int_0^\infty \int_0^\infty x(s,i)dids - \sum_{s=0}^{m-1} \int_0^\infty x(a_s,i)di > \int_0^\infty x(a_{m-1},i)di \tag{21}$$

For the posttruncation period, the remaining individuals whose age is not determined will survive for at least t_1 units after t_m, if Eq. (21) holds. The number of individuals who are likely to die by $t_{m+1'}$, say, $D(t_{m+1'})$, are at least α_1 and at most $\int_0^\infty x(a_{m-1},i)di - 1$ such that, the largest partition of the integer $\int_0^\infty \int_0^\infty x(s,i)dids - \sum_{s=0}^{m-1} \int_0^\infty x(a_s,i)di - \alpha_1$ is less than α_1. Here, we have considered only the partition of $\int_0^\infty \int_0^\infty x(s,i)dids - \sum_{s=0}^{m-1} \int_0^\infty x(a_s,i)di - \alpha_1$ expressing as sum of distinct parts and partitions such as (Eq. 22) are not considered.

$$\begin{aligned}
&\int_0^\infty \int_0^\infty x(s,i)dids - \sum_{s=0}^{m-1} \int_0^\infty x(a_s,i)di - \alpha_1 \\
&= 1+1+\cdots \left(\int_0^\infty \int_0^\infty x(s,i)dids - \sum_{s=0}^{m-1} \int_0^\infty x(a_s,i)di - 2\alpha_1\right) \text{ times} \\
&= \alpha_1 + 1+1+\cdots \left(\int_0^\infty \int_0^\infty x(s,i)dids - \sum_{s=0}^{m-1} \int_0^\infty x(a_s,i)di - 2\alpha_1\right) \text{ times}
\end{aligned} \tag{22}$$

For a general theory on integer partition, refer to Andrews and Eriksson (2004). When $D(t_{m+1'}) = \int_0^\infty x(a_{m-1},i)di - 1$, the slope of decline in H_1 is relatively lower than that of $d(t_{m+1'}) = \alpha_1$. Due to inverse relation between $d(t_{m+1'})$ and loss of information after t_m, the higher the value of $D(t_{m+1'})$ in the interval $(\alpha_1, \int_0^\infty x(a_{m-1},i)di - 1)$, the lesser is the loss of information due to truncation at t_m. If $D(t_{m+1'}) = \int_0^\infty x(a_{m-1},i)di - 1$, then, we will arrive at Eq. (12) and that ensures that the information loss after the time $t_{m+1'}$ is minimal.

$$\int_0^\infty \int_0^\infty x(s,i)dids - \left(\sum_{s=0}^{m-1} \int_0^\infty x(a_s,i)di + D(t_{m+1'})\right) < D(t_{m+1'}) \tag{23}$$

$q(t_{m+1'})$, the probability of dying of an individual during $(t_m, t_m + t_1)$ is,

$$q(t_{m+1'}) = \frac{D(t_{m+1'})}{\int_0^\infty \int_0^\infty x(s,i)dids - \sum_{s=0}^{m-1} \int_0^\infty x(a_s,i)di} \tag{24}$$

If $D(t_{m+1'}) = \alpha_1$, then, the inequality (Eq. 12) do not hold and we need to continue steps that we did after arriving at the inequality (Eq. 21). When Eq. (21) do not hold at the initial stage after truncation at t_m, then, the arguments provided to arrive at inequality (Eq. 12) hold.

Note 1 When age structure in the captive cohort does not follow any pattern then this situation can also be handled by using the technique of arranging the life lived in an ordered fashion explained for the one-dimensional stationary population identity in (Srinivasa Rao and Carey, 2015).

One of the motivations for developing this framework of two-dimensional Carey' s Equality is due to the absence of methods to estimate the age structure of wild insect populations. We wish attempt this by exploiting the information of lengths of life of captive individuals, population age distribution and time to death distribution.

5 DISCUSSION

Stationary population identity with two variables is structured in this chapter to suit analysis required to perform on a captive cohort when there exist very large number of individuals in each age-group of a population. Individuals in the captive cohort are constructed by life lived and size of the population in each age-group. Unlike, in one-dimensional stationary population identity (Srinivasa Rao and Carey, 2015), in this chapter, the collective lengths of remaining lifespans associated with the age at death during follow-up (through the construction of two functions, H_1 and H_2) and number surviving at the time of follow-up (as explained through Theorem 2 and Corollary 1) are simultaneously considered. Such considerations are new to the research in the stationary population identity which opened-up new doors in our theoretical understanding of the age structure of insect captive cohorts. The theory proposed is sensitive to the proportion of the population at each age and can be modified to adapt, even if there is no monotonic pattern for the two functions, H_1 and H_2 introduced. Truncation from right is taken as a variable to construct partial age structure of the population. We have constructed partial age structure in a deterministic way without any probabilistic assumptions. However, we have given a formula for computing probability of dying of an individual who is alive after truncation, by partitioning the number of individuals alive at the time of truncation. We have used distinct parts of the partition to suit monotonic nature of the age structure and one can consider all the parts of the partition function when there is no significant pattern of the age structure. We have not used any probabilistic assumptions on the remaining lifespans of the remaining individuals who are alive at the time of truncation in the captive cohort because such assumptions need to be justified and we have found no evidence in the published literature to start constructing partial age structure. The method proposed in this chapter can be also helpful in arranging age structure of a captive cohort on a continuous and flexible open intervals and geometric visualization of collective time lived and collective time remaining of a captive cohort.

6 CONCLUSIONS

We have essentially done following things in this chapter: (i) we have expressed captive cohort in a different way to previous works, (ii) each individual in every age is represented mathemtically in the formulations, (iii) sets of subcohots are analyzed using partition functions. Through this, we have achieved, (i) understanding internal structure of a captive cohort, (ii) construction of partial age structure of captive cohort.

REFERENCES

Andrews, G.E., Eriksson, K., 2004. Integer Partitions. Cambridge University Press. Special ICM2010 edition.

Brady, O.J., Godfray, H.C.J., Tatem, A.J., Gething, P.W., Cohen, J.M., McKenzie, F.E., Perkins, T.A., Reiner, R.C., Tusting, L.S., Sinka, M.E., Moyes, C.L., Eckhoff, P.A., Scott, T.W., Lindsay, S.W., Hay, S.I., Smith, D.L., 2016. Vectorial capacity and vector control: reconsidering sensitivity to parameters for malaria elimination. Trans. R. Soc. Trop. Med. Hyg. 110, 107–117.

Carey, J.R., Papadopoulos, N., Müller, H.-G., Katsoyannos, B., Kouloussis, N., Wang, J.-L., Wachter, K., Yu, W., Liedo, P., 2008. Age structure changes and extraordinary life span in wild medfly populations. Aging Cell 7, 426–437.

Carey, J.R., Müller, H.-G., Wang, J.-L., Papadopoulos, N.T., Diamantidis, A., Kouloussis, N.A., 2012a. Graphical and demographic synopsis of the captive cohort method for estimating population age structure in the wild. Exp. Gerontol. 47, 787–791.

Carey, J.R., Papadopoulos, N.T., Papanastasiou, S., Diamanditis, A., Nakas, C.T., 2012b. Estimating changes in mean population age using the death distributions of live-captured medflies. Ecol. Entomol. 37, 359–369.

Carey, J.R., Silverman, S., Rao, A.S.R.S., 2018. The life table population identity: discovery, formulations, proof, extensions, and applications. In: Srinivasa Rao, A.S.R., Rao, C.R. (Eds.), Handbook of Statistics. Integrated Population Biology and Modeling, Part A, 39. Elsevier, pp. 155–186.

Cook, P.E., Sinkins, S.P., 2010. Transcriptional profiling of *Anopheles gambiae* mosquitoes for adult age estimation. Insect Mol. Biol. 19 (6), 745–751.

Cook, P.E., McMeniman, C.J., O'Neil, S.L., 2008. Modifying insect population age structure to control vector-borne disease. Adv. Exp. Med. Biol. 627, 126–140.

Feller, W., 1941. On the integral equation of renewal theory. Ann. Math. Stat. 12 (3), 243–267.

Feller, W., 1971. An Introduction to the Probability Theory and its Applications, second ed. vol. II John Wiley & Sons.

Kim, Y.J., Aron, J.L., 1989. On the equality of average age and average expectation of remaining life in a stationary population. SIAM Rev. 31 (1), 110–113. https://doi.org/10.1137/1031005.

Lotka, A.J., 1939a. A contribution to the theory of self-renewing aggregates with special reference to industrial replacement. Ann. Math. Stat. 10, 1–25.

Lotka, A.J., 1939b. On an integral equation in population analysis. Ann. Math. Stat. 10, 144–161.

Math Digest, 2014. Age Tables Behind Bars, by Ben Polletta. http://www.ams.org/news/math-in-the-media/md-201410-toc#201410-populations. American Mathematical Society, (browsed on June, 15, 2015).

Müller, H.G., Wang, J.-L., Carey, J.R., Caswell-Chen, E.P., Chen, C., Papadopoulos, N., Yao, F., 2004. Demographic window to aging in the wild: constructing life tables and estimating survival functions from marked individuals of unknown age. Aging Cell 3, 125–131.

Nussey, D.H., Coulson, T., Festa-Bianchet, M., Gaillard, J.M., 2008. Measuring senescence in wild animal populations: towards a longitudinal approach. Funct. Ecol. 22 (3), 393–406.

PAA Affairs, 2015. Invited Essay: Stationary Populations and Implications of Carey's Equality by Arni S.R. Srinivasa Rao. (Spring 2015). http://www.populationassociation.org/wp-content/uploads/PAA-Spring15-Revised-2.pdf, Population Association of America, (browsed on June 1, 2015).

Papadopoulos, N., Carey, J.R., Ioannou, C., Ji, H., Müller, H.-G., Wang, J.-L., Luckhart, S., Lewis, E., 2016. Seasonality of post-capture longevity in a medically-important mosquito (*Culex pipiens*). Front. Ecol. Evol. 4, 63. https://doi.org/10.3389/fevo.20016.00063.

Rao, A.S.R.S., 2014. Population stability and momentum. Not. Am. Math. Soc. 61 (9), 1062–1065.

Arni S.R. Srinivasa Rao and James R. Carey (2015). Carey's equality and a theorem on stationary population, J. Math. Biol., 71, 583–594 *Placeholder Text.* https://doi.org/10.1007/s00285-014-0831-6

FURTHER READING

Rao, A.S.R.S., Carey, J.R., 2018. On the Three Properties of Stationary Populations and Knotting With Non-Stationary Populations. arXiv preprint arXiv:1811.03067.

Chapter 12

Demographic Situation of Manipur, India

Moirangthem Hemanta Meitei[1]
Department of Economics, Manipur University, Imphal, Manipur, India
[1]*Corresponding author: e-mail: mhmeitei@gmail.com*

ABSTRACT
Manipur abounds with different ethnic populations. Though small geographically, it resembles India in terms of the diversity of ethnic composition of population. Historical characteristics are important for a better understanding of demographic behavior of these different ethnic groups. From the available data majority population of valley the Meiteis exhibit a better performance in demographic indicators than the hill where the tribes predominantly settle. This dichotomy can show the need for differential approach to achieve a common demographic goal of reaching replacement-level fertility faster. This issue cannot be looked in isolation because demographic goals have to address other relevant health and development dimensions. In this regard the 2011 census, NFHS-4, and DLHS-4 can reveal the level and achievements of some sociodemographic goals targeted to achieve by 2010 listed in the National Population Policy 2000. These parameters show that districts have not fared well and the state as a whole also has failed to reach the targets. Not only unable to meet the sociodemographic target goals but it also does not show a kind of certainty when to reach the set goals in near future.

Keyword: Manipur, Ethnicity, Demographic goals, NPP 2000

1 OUTLINE

This chapter will discuss the origin of Manipur, the composition of Manipuri population from the angle of ethnic composition. Here how demographic transition takes place in the historical perspective and the impact of migration in the dynamics of population size and composition will be discussed. It also will dwell upon how the demographic parameters behave in the state over time.

Manipur is a small state in the far easternmost corner of India. In fact it belongs to the Northeastern region of India. It borders with Myanmar for

Handbook of Statistics, Vol. 40. https://doi.org/10.1016/bs.host.2018.11.002

about 350 km in the east and in the west lies Assam, Mizoram in the south, and Nagaland in the north. Fundamentally, the state has a small geographical area of 22,327 km^2. Topographically the state can be divided into two distinct regions as the hill and the plain (valley). The hill comprises of 90% of the total area, while the remaining 10% of the area belong to the plain. The entire plain area resembles bottom part of a bowl surrounded and flanked by walls of hill. Historically through folk tales and legendary stories the state is said to have nine layers of mountains surrounding the valley. But in terms of ethnic composition it is very rich having about 35 ethnic groups including scheduled tribes, Meiteis, and Meitei Pangals (Muslim). There are 33 recognized scheduled tribes as listed in 2011 census. The remaining major group belongs to the Meiteis and a smaller fraction comprises the Muslims (8%) in the state. In the past two decades Manipur was having 9 districts but very recently it increased to 16 districts: Bishnupur, Chandel, Churachandpur, Imphal East, Imphal West, Senapati, Tamenglong, Thoubal, Ukhrul, Kangpokpi, Tengnoupal, Pherzawl, Noney, Kamjong, Jiribam, and Kakching.

Manipur was an independent country ruled by the native kings over centuries. Later, in the eventful historical coincidence and process of expanding British empire it fell into the British rule in 1891. Historically the name Manipur was said given during the reign of a Hinduized king Garibaniwaza in 18th century, although the original names went by Kangleipak, Poireipak, and Meitrabak (Kamei, 2015). There are some mythical connections with regard to other names assigned to this part of world in the process of adapting Hindu culture. Most of these names are seemed associated with the introduction of Hinduism in this erstwhile country. It is thought that this country happened to be the land of Meiteis. However, the people of Manipur included the Meiteis and the people settling around the Meiteis also known with different names such as the hill people (*hao*). The Meiteis were named differently by the neighboring states: The Shans or Pong of Upper Burma called Cassay, the Burmese called Kathe, and Assamese called Mekhlee. These hill people in present day belong to the tribes. The Hinduized Meiteis tried at some point of history to link the Meiteis with the myths of Mahabharata to describe themselves as the descendants of the Aryans. But this is not scientifically proven. Kamei (2015) and Hodson (2011) are of the opinion that this could be under the influence of Hinduism. After the inclusion in the Indian dominion in 1949 there is a clear distinction of tribes belonging to scheduled and nonscheduled under the constitution of India. The topographical setting has helped to artificially differentiate the people of Manipur broadly into two as the Meities and people around the Meitei settling in the hill. Muslims are also a part of ethnic community in Manipur. They are known as Musalman or Meitei Pangals. Though the contacts with the Mayangs who were mostly the people of south-west side of country were made in the historical records during king Kyamba (invasion by Mayang in 1504 AD), the actual settlement of Muslims in present-day Manipur is cited during reign of king Khagemba (1597–1652).

Royal chronicle the "Cheitharol Kumbaba" made entry of invasion in the western part by a combined force of Kachari and Muslims in 1606. They were, however, defeated by the Meitei king and many of them were captured. They were made to settle in Manipur and allowed to marry Manipuri women but allowed to follow their own religion (Kamei, 2015). These Pangals have started settling in Manipur from early period of about 500 years back. Brown (1873/2001) took notice of these Bengali origin people with an estimated population of 4500 coming from the neighboring districts of Sylhet (part of modern Bangladesh) and Kachar (Southern Assam).

Manipur was an independent country before the British conquered in April 1891. The country however at that point of time did not have an exact boundary. But there was time and again mentioned that the boundary of the country drew deep inside the Burmese soil of Kachin area. Historically the people of Manipur comprise the Meiteis and other nonindigenous people living in the hills. The people in the hills comprise various dialects. In the historical period the Meiteis were migrants who initially migrated from the eastern side. It is believed that those migrated from the eastern side brought fire to Manipur. The strikingly interesting part which is still upheld is the practice of maintaining this fire till date burn in village called Andro, one of the native places of the aborigine Meitei ethnic group. The fire that is still kept alive is thought to have been brought by Poireiton. Later another powerful leader came up in the name of Pakhangba who started ruling this part of the world at Kangla defeating Poireiton. Later Pakhangba married sister of Poireiton and made an alliance. This incident was recorded right from 33 AD in the royal chronicle called "Cheitharol Kumbaba."

The Meitei is the larger ethnic group in Manipur compared to any other ethnic groups now. In the hill small tribes inhabit and they were different entities in early period. But after the unified administration of British these small tribes were given the nomenclature of Naga and Kukis. The uniqueness of the Meitei compared to the other smaller ethnic communities is faster development in evolution. The simplest example is the keeping of records in the form of royal chronicles and other valuable records in the form of books. Though this place may not have the paper as such, they invented some mechanisms to record the historical events. On top it must be remembered that these people used some kind of script which could make the record viable for further study among modern civilization. After Hinduization of the Meiteis there has been a breakthrough in the use of language and script of the Meiteilon. The connection with Hinduism started right from the days of Charairongngba (1697–1709) and he adopted Nimandi sect of Vaishnavism in 1703 (Kamei, 2015). This continued for a long time. By virtue of this connection people got educated in Sanskrit and Bengali script was adopted. It is also said that during the time of Garibniwaz some of the texts of ancient script were burnt at Kangla (palace). Luckily there are still copies of these old texts available (Kamei, 2015). Around 1980s there appeared a movement for revival to use and adopt the ancient script known as Meetei Mayek under the office order (published in Manipur Gazette No. 40 of 25 April 1980). Now this script is in full force among the younger generation in school education.

The people of the valley known as the Meiteis are thought to be a generic name in a later period. Before it there were various clans under different names. Till date there are basically seven clans in the Meitei society. These clans have a strong implication in Meitei life following rituals from birth, marriage, and death. These seven clans are Ningthouja (Mangang), Luwang, Angom, Khuman, Moirang, Kha-Nganba, and Sarang Leisangthem. The local name of clan is known as Salai (Sa = sagei—lineage; lai = god ancestor) (Kamei, 2015). Mythical interpretation of the origin of these clans is associated with identification of the Supreme Father (Shidaba Mapu—almighty God) by two of his sons Sanamahi and Pakhangba. While another parallel myth is origination of these seven clans with the seven days of the week. Later British anthropologist and ethnologists have contributed a lot to trace the people of Manipur. One of the theories (documented by Hodson, 2011, p. 10) as propounded by B.H. Hodgson in connection of Moy Tai (relating to present Meitei). But this could not pull further long. Later Hodson (2011) related Mei meaning man and Tei meaning a group of people. Here lies the connection of mythical world with origin of Meiteis because it is a belief that God Sanamahi created man from the image of almighty Mee meaning image and Tei (Teiba) meaning model. In combination it implies the image of God Meitei (Kamei, 2008). There is also a Hindu connection in such a way that the people of Manipur are the descendants of one of the Hindu epic heroes Arjuna of the Mahabharata. But this is refuted by Hodson (2011) as tainted because of Hinduism influence. On the basis of linguistic affinity Meiteis were grouped in the Kuki-Chin sub family (Grierson, 1904) and their ancestors "were among their kinsmen who migrated from China to upper water of the Irrawaddy and Chindwin rivers of upper Burma" (Kamei, 2008, p. 7). However, in the process of connection of people from the west Brahmins also settled in Manipur absorbing the local culture. From all these angles the origin of Meities may not be very definitive but it has the elements of the Shan, Tai, or Hindu ethos among the people of this country.

The other group of major ethnic groups in Manipur coresiding with the Meiteis are the hill people of the Nagas and the Kukis. Ethnographically the Nagas, Kukis, and Meiteis fall in Mongoloid group and belong to the Tibeto-Burmans of Sino-Tibetan group. The Kuki-Chin family consists of Kuki, Chin, Mizo, and the Meitei (Manipuri languages) linguistically (Grierson, 1904). Ethnically there is a divide between the Meiteis and the Kuki, Chin and Mizo (Zo, Zomi), and other cognate tribes and clans (Kamei, 2008).

There are various theories about the origin of the Nagas. The Nagas form an ethnic group of different tribes which settle around the states of Assam, Arunachal Pradesh, Manipur, and Nagaland. Noted historian Gangmumei

Kamei holds the view that the Nagas have been inhabiting in the hills of Manipur centuries before the Christian era (Kamei, 2008). The tribes belonging to Nagas in Manipur include: Anal, Maring, Moyon, Monsang, Lamkang, Chothe, Tangkul, Mao, Paomai, Maram, Thangal, Zeliangrong (Zeme, Liangmei, Rongmei, and Puimei), Chiru, Kharam, Tarao (Kamei, 2008). The settlement of Naga tribes in Manipur is quoted from Kamei (2008, p. 23):

> *These tribes occupy different area of Manipur hills; the Naga ethnic territory was well maintained till the migration of the Kuki-chin tribes during the historical period. The Zeliangrong tribes occupy the western hill of Manipur. The Rongmeis are a widely dispersed community. They had settled down in Imphal valley and the Cachar district of Assam. There are relics to show that the Zeliangrong habitat in the past, reached the edge of the Mizo hills. The Zeme and Liangmei are found in Nagaland; some Romgmeis had recently migrated to that state. The Zemes are also found in the North Cachar hills. The Mao, Maram and Thangal inhabit the north Manipur. The Tangkhuls are the most numerous tribe and occupy the hills of Ukhrul. The Tangkhul habitat spreads beyond the Indian border into upper Burma in Somra tract and Naga hills of Burma. Many Tangkhul have settled down in the foothills and in Imphal valley. The Marings occupy the Tengnoupal area of Chandel district, whereas the Moyon, Monshang, Lamkang, Anal, Tarao, Chothe are found in Chandel and Chakpikarong area of the same district.*

The Kuki-Chin Mizo was an outcome of the colonial imposition of a common identity and a search for common ethnic identity (Kamei, 2008). In other word it is a nomenclature assigned to this group of tribes for the convenience of administration. Further Kamei was of the view that the very word Kuki was not even known to the people to whom it was assigned. Historically it is mentioned that the first reference of the word Kuki was coined in 1777 when these tribesmen attacked the British during the time of Warren Hastings in Chittagong. The writings of the historical interpretation do not fit here. In the interest of the ethnic composition of Manipur the origin of this Kuki-Chin-Mizo group needs certain amount of deliberation.

The Kukis are known to the Meiteis as Khongjai or Khongsai. Even though the Kukis-Chin and Mizo (also Lushai) are falling in the same ethnic group, Kukis seemed to be distinct in Manipur away from the Lushais (Mizo) (Kamei, 2008). Kamei (2008, p. 31) further notes that "Reid's Lushai-Chin land (1893), and other writings of the British officers who served in Burma and north east India among these tribes, emphasized on the affinity and racial identity of the Chin, Lushai and Kukis." Depending heavily on the narration of Kamei some of the tribes of Kuki-Chin ethnic group such as Aimol have been recorded tribute paying tribes of Manipur in 12th century. Further he also noted the coexistence of tribes other than Nagas in Manipur for more than a thousand years speaking dialects of Kuki-Chin family which had great affinity to the migrants of 18th and 19th century (Kamei, 2008). Historically their civilizational evolution is lengthy and but one common thread among the Kuki tribes is their migratory behavior and their permanent settlement seemed to be quite a recent phenomenon. This can be traced from the historicity of their civilization. According to their tradition the Kuki tribes are originated from the hole or bowel of the earth. This group of tribes on the basis of historical linguistic are thought to move from south west China and Southeast Asia over centuries. The following are major tribes of Kuki-Chin group in Manipur. They include Thadou, Paite, Gangte, Simte, Vaiphei, Zou, Hmar (read as Mar), Mizo, Kom, Aimol, Baite, Mate, etc. (Kamei, 2008).

Here Kamei (2008) was quoting Lehman to establish the closeness of different ethnic groups in the state "The apparent close relationship among several Tibeto-Burman languages of Burma (including India) indicates clearly that in the not very distant past, the ancestors of such different people as the Nagas and the Chin belong relative to unitary speech group" (p. 38). The Chinese historical records reveal these groups were known as the Chiangs originally living in the north west of China. But because of constant war and hostility with the Chinese they migrated toward the south to the border of Tibet and Burma (Kamei, 2008). They moved toward Kabaw valley and further split into two groups. One moved to the south in Burma and the other into Manipur. This split was because of some forces. The Chins settled in Chindwin basin but later the Shan invaders pushed them out of Kabaw valley into hills. While trying to settle some groups were pushed to Manipur and Cachar around 1700 AD and some settled in the southern part of Chin hills. Their migration continued throughout the 19th and early 12th century. The earlier movement of Kuki-Chin-speaking tribe followed the waves of earlier settlers of Meiteis and the Naga from the upper Burma and Kabaw valley. So these early settlers of Tibeto-Burman people in Manipur sandwiched between the Naga and Kuki-Chins and they are often treated as bridge buffer tribe (Kamei, 2008).

In the later part of the settlement of the tribes of Naga group and Kuki groups in Manipur, the Meiteis name them as Hao/Hau, a generic name for all tribes of Manipur irrespective of their particular name or ethnicity.

> *... I can see no reason, and think there is far more ground to conclude them to be descendants of the surrounding hill tribes. The languages spoken by these tribes are in their pristine state: I conceive then, that in their spoken language, an indication of the descent of the Munniporees might be found. Tradition brings the Moirang tribe from the South, the direction of the Kookies, the Koomul form the East, the direction of the Murrings, and the Meithei and Looang form the North-west, the direction of the Koupooees. The languages of the Murrings, Kookies, and the Koupooees, are all very similar, and as the Koomul, etc., the offshoots of these tribes were, as before said, at different periods the dominant tribes in the valley, it might be expected that the present language of the people,*

united under the name of Meithei, would have a very apparent likeness to these languages, and such is the case. All these tribes have also traditions amongst themselves that the Munniporees are offshoots from them. These traditions then, and the composite nature of the language, appear to me to afford more reason for supposing the Munniporees to be descended from the surrounding hill tribes ...

Hodson (2011, p. 6)

As early as 1873 when Brown published *Statistical Account of Manipur*, the population of Manipur was 139,000 comprising of 65,000 in the hill and 74,000 in the valley. The people of Manipur as recorded by Brown consisted of various tribes that can be clubbed into two groups as Nagas and Kukis.

1.1 The Origin of Communities in Manipur

As a part of the study of the demographic events of Manipur one has to acquaint with historical accounts of the state. While it was still trying to consolidate its presence during the ages, during the 18th century some of the powerful kings of the kingdom made some kind of moves which led a great social change. In fact among the great kings of this kingdom was Garibaniwaz. He ruled during 1709–1748. During his time major invasions by the Burmese were recorded in 1718, 1723, 1725, and 1737. Along with defense the king also made several war expeditions to neighboring countries like Burma in the east and Takhel (modern Tripura) in the west. The important connection here is not about the war itself, rather how one can infer the size of population then. The Manipuri administration then used a system called "lallup" under which an abled bodied individual of a household must render military service. Though this system could not operate fully in the hill, it seems that it was quite effective in the valley areas. It is recorded that in the 1723 Burmese invasion the first wave of Manipuri defense was defeated and on hearing this the king sent a reinforcement of 4000 tribal soldiers, 5000 Meitei soldiers, and 300 cavalry (Kamei, 2015). Further Kamei (2015, p. 290) notes that in 1738 war expedition by Manipur to Burma the king raised 20,000 men and cavalry. This indicates that plausible size of population in Manipur could be around 100,000 with the idea that one man and cavalry was contributed from each household and there were 5 persons per household on average. Perhaps the size of population might be a little more also since those days early family formulation process must have prevailed though high level of mortality still persisted around the world.

After Garibaniwaz the Burmese attacked Manipur repeatedly resulting in mass exodus of people of Manipur taking shelter in the hills and far off places and leading to impoverishment of resources. Kamei (2015, pp. 313, 321) mentions that a large number of people of Manipur consisting of boatmen, silk workers, and silver smiths were taken as war captives to Burma. Due to incompetent administration with a strong leadership and internal conflicts within the royal family and a vengeance from the Burmese rulers Manipur experienced a crushing defeat and destruction from the hands of Burmese in the 19th century for a period of 7 years from 1819 to 1826. This incident is a historical importance to mark atrocities of the Burmese and known as "the seven years devastation." To quote Kamei (2015, pp. 338–339) "Manipur had never before faced such a national catastrophe brought about by the Burmese conquest. The oppression and cruelties with which the Burmese dealt with Manipur had no parallel in history of Manipur. It was a sort of genocide of the Meiteis that had been carried by the Burmese reign of terror. This is known in history of Manipur as the *Chahi Taret Khuntakpa*: Seven Years Devastation. The country was depopulated. One-third of the population accompanied Marjit Singh in his flight to Cachar, a sizeable section took refuge in the western hills of Manipur, and the rest lived in great fear of the Burmese and many thousands were taken as prisoner to Burma. The population of Imphal valley reduced to about 10,000 or about 2000 households only."

The above text and historical narrative clearly give the evidence that the state had been under so much pressure of accumulating and regenerating population after the great wars with the neighboring countries particularly Burma. The scar of Burmese invasions is still narrated by the older folks till date orally in the form of folksongs. The given size of population around Imphal was rather too small and thin. The true evidence of the scattered population of Manipur can still be traced from the language data of the census. The 2011 census reveals that there are 168,133 Manipuri-speaking population in Assam and among them more than half in Cachar district (105,254). The present growth of population in the state is taking place only after the great devastation. In other words the population is getting its level in the last nearly 200 years only.

1.2 Manipur at Crossroad

Historically prior to 1891 it was an independent kingdom, but after April 1891 the British India occupied it but fortunate enough to include in the early part of Indian census within the frame of Assam and adjoining areas. Therefore, official population data are available for Manipur right from 1901 census. The size of population was relatively very small of nearly 300,000 (3 lakhs). The growth of population in absolute number was also very small. Probably this must be due to high death rate in the population. However, just before the merging in the Indian dominion in 1949 or prior to 1951 census the net increase of population in the state was rather fluctuating in a close range of 60,000–67,000 leaving aside 1921. After coming into Indian domain the 1951 census recorded the population of the state to the tune of less than 6 lakhs. It implies that in the period of first half of 20th century population was able to double in a span of 50 years. But the doubling time got closer in later years. By 1981 census population reached the size of 14 lakhs which was even slightly more than the double the size of 1951 population. As India experienced population explosion there had been a rapid growth of population during the initial three decades since independence. Similarly Manipur also experienced a phenomenal growth rate until 1971 which can be observed in Table 1. But from 1981 onward population growth rate turned a different direction of slow but steady decline with an indication of below 3%. Net addition of population is a good indicator how population has been changing consistently over time since 1951 till 2011.

One interesting part in growth rate is the rural–urban growth differential. The concept of urban is more consistent from 1951 census. The growth rate of proportion urban population is higher than that of rural population in the state. However in 2001 census observed a slower growth of urban population than rural population. In the five hill districts there was no urban in 2001 census. The urban areas limited to the only four districts of the valley areas. Therefore, the proportion of urban population is low. But in 2011 it reverses and there appeared urban centers in all the districts. Therefore, the growth of urban is higher again in 2011. Summarily it implies that the concentration of population is higher in the urban areas. As the standard demographic operations of natural increase (difference between births and deaths), mobility of people to the urban centers can explain this phenomenon. Of course, the natural growth can happen in urban areas, but it is more so in the rural areas and the higher opportunities in urban areas can attract more persons to migrate to the urban centers. This amounts to say that urban growth is taking place in the state vibrantly.

In 1971 the number of districts of state was five and in 1981 it rose to six districts. Further the districts of Manipur were reorganized in 1991 census whereby the number of districts rose to eight. Again in 2001 census it further reached to nine districts and maintained in 2011 too. Fortunately, the 2011 census publication has included the population of these nine districts together since 1951. This table is reproduced here as Table 2. Among the nine districts five districts cover the hill part of the state, while the remaining four represent the valley area. The valley as mentioned earlier covers nearly 10% of the entire geographical area, whereas the remaining 90% of the land area are in the hill region. Naturally since agriculture is more successful in the valley area it was more convenient for settlement resulting in expansion of population and other related activities; therefore, the state capital right from the ancient time has been located in the central part of the valley (Kangla). The concentration of population is much higher in the valley. Some of the districts like Churachandpur and Chandel experienced an increase of six times of the 1951 population in 2011, whereas in all the valley districts the increase was well below fivefold close to fourfold (except in Imphal East). So in some way the increase in population in the valley areas is less than those of the hill districts.

TABLE 1 Population Growth in Manipur Since 1901

				Sex Ratio (M/F)			Net Addition and Growth Rate (%)			
Year	Total	Urban	Percent Urban Pop	*Total*	*Urban*	*Rural*	*Net Addition*	*Total*	*Urban*	*Rural*
1901	284,465	72,234	25.39	1037	1036	1038				
1911	346,222	74,650	21.56	1029	1051	1023	61,757	1.96	0.33	2.47
1921	384,016	80,003	20.83	1041	1091	1029	37,794	1.04	0.69	1.13
1931	445,606	85,804	19.26	1065	1137	1048	61,590	1.49	0.70	1.68
1941	512,069	99,716	19.47	1055	1102	1044	66,463	1.39	1.50	1.36
1951	577,635	2862	0.50	1036	633	1039	65,566	1.20	−35.51	3.32
1961	780,037	67,717	8.68	1015	985	1018	202,402	3.00	31.64	2.15
1971	1,072,753	141,492	13.19	980	980	980	292,716	3.19	7.37	2.68
1981	1,420,953	375,460	26.42	971	969	971	348,200	2.81	9.76	1.16
1991	1,837,149	505,645	27.52	958	975	951	416,196	2.57	2.98	2.42
2001	2,293,896	575,968	25.11	974	1009	963	456,747	2.22	1.30	2.55
2011	2,855,794	834,154	29.21	985	1026	969	561,898	2.19	3.70	1.63

Source: Statistical Handbook of Manipur 2017, pp. 6–7.

TABLE 2 Population of Manipur Over the Census Years

		1951	1961	1971	1981	1991	2001[a]	2011
Hill	Senapati	46,608	70,792	102,291	153,091	208,406	283,621	479,148
	Tamenglong	33,519	37,667	46,034	62,289	86,278	111,499	140,651
	Churachandpur	44,496	66,304	100,418	134,494	176,184	227,905	274,143
	Ukhrul	41,182	48,590	62,229	82,946	109,275	140,778	183,998
	Chandel	24,049	27,679	38,723	56,444	71,014	118,327	144,182
Valley	Bishnupur	57,340	79,005	108,785	141,760	180,773	208,368	237,399
	Thoubal	99,631	135,462	182,500	232,675	293,958	364,140	422,168
	Imphal East	89,821	135,594	189,713	255,365	330,460	394,876	456,113
	Imphal West	140,989	178,944	242,060	301,889	380,801	444,382	517,992
	Hill subtotal	189,854	251,032	349,695	489,264	651,157	882,130	1,222,122
	Valley subtotal	387,781	529,005	723,058	931,689	1,185,992	1,411,766	1,633,672
	Manipur	577,635	780,037	1,072,753	1,420,953	1,837,149	2,293,896	2,855,794

[a]*The population of Manipur state by sex includes the estimated population of Mao-Maram, Paomata, and Purul subdivisions of Senapati district for 2001 where census results were excluded.*

Source: Table A-2: Census of India, Manipur.

Over the years it can be seen that share of valley population in the state is declining since 1951. It maintained about two-thirds of the population in the valley districts till 1971. Since then there is a gradual decline in the share of population in the valley area. By the turn of the last millennium the share of population reduced to 60%, but in 2011 census it further shrinks down to 57%. No doubt the size of population is still larger in the valley area, but the increase in population is much higher in the hills. Of course the transition of population is taking place in the state but in a different manner. In the hill probably the birth rate is higher than the valley areas where the pressure of population is felt and people have oriented to the call of the small family norm of Indian family planning program. This will be discussed later.

Table 3 tries to give a comprehensive picture of the share of population by the districts. Historically valley areas have a higher concentration of population. Within the valley areas too Imphal, where capital is situated, as a whole abounds a large settlement. Before bifurcation of Imphal into Imphal East and Imphal West, Imphal had the highest concentration (share) of population. Kangla stands at the heart of Imphal itself and it was the main focus during the days of the kings of the country for most of the administrate functions. Otherwise, the Meitei principalities scattered around the valley area. For example, Moirang was the capital area of Moirang salai which situates around 45 km to the south of Imphal or Kangla.

TABLE 3 Share of Population by Different Districts

	1951	1961	1971	1981	1991	2001	2011
Senapati	8.1	9.1	9.5	10.8	11.3	12.4	16.8
Tamenglong	5.8	4.8	4.3	4.4	4.7	4.9	4.9
Churachandpur	7.7	8.5	9.4	9.5	9.6	9.9	9.6
Bishnupur	9.9	10.1	10.1	10.0	9.8	9.1	8.3
Thoubal	17.2	17.4	17.0	16.4	16.0	15.9	14.8
Imphal West	24.4	22.9	22.6	21.2	20.7	19.4	18.1
Imphal East	15.5	17.4	17.7	18.0	18.0	17.2	16.0
Ukhrul	7.1	6.2	5.8	5.8	5.9	6.1	6.4
Chandel	4.2	3.5	3.6	4.0	3.9	5.2	5.0
Manipur	100	100	100	100	100	100	100
Valley population	67.1	67.8	67.4	65.6	64.6	61.5	57.2

As per 2011 census the share of population is lowest in Tamenglong, one of the most inaccessible districts of the state followed by Chandel. Churachandpur, Tamenglong, and Ukhrul maintain somewhat a stable share of population in the state total. Meanwhile in the valley districts the share of population has been consistently declining particularly in Imphal West and Thoubal, while Bishnupur and Imphal East show declining share in last two censuses. The significance of highlighting the share of population in different districts relates to the seemingly polarization among the different ethnic groups of the state. Meanwhile since most of welfare schemes are streamlined on the basis of population, a sincere count of the population is expected and putting in suspect will refute the credibility of number.

Using the given population figures of the census years annual growth rate during the intercensal period is calculated using the formula $P_1 = P_0 \exp(rt)$ where P stands for population and subscripts refer to time period, r is the growth rate, and t is the time gap between two consecutive censuses (shown in Table 4). As evidenced by the share of population it is also observed that the growth rate has declined over time in almost all the districts, but the decline is more prominent in case of valley districts. However, it is not so well supported in the hill districts particularly and districts like Chandel, Ukhrul, Tamenglong, and Senapati have shown relatively higher rate of growth in the state. At the macrolevel aggregate growth is still high and in case of all the hill districts together maintaining around 3% which implies that if it goes at this rate the doubling time will be only 23 years. In the valley districts the growth rate has declined over time since 1951 census onward. During 1951–61 and 1961–71 the growth rate in the valley together was around 3%, but it reduced to 2.5% and 2.4% during 1971–81 and 1981–91 decades, respectively. By 2001–11 the valley population growth rate observed 1.5% implying about 50 years for doubling. Whether a higher growth rate of population in the state is good or bad is a separate issue, but it is important to understand whether a higher growth rate can warrant a better quality of life.

2 DEMOGRAPHIC COMPOSITION OF MANIPUR

Despite being a small state, Manipur has a wide diversity of population. Without having a good understanding of the social fabric of the composition of population it will be difficult to identify demographic situation of the state. As mentioned above there are 33 officially recognized scheduled tribes in the state in 2011. Recently four more tribes (Inpui, Liangmai, Rongmei, and Zeme) have been recognized by the government of India (Imphal Free Press, 10 May 2018). Besides the tribal population there are other groups which can be called as Meitei/Meetei which absorbs the Brahmans, a vast majority of local scheduled caste community (known as Loi), while the Muslims exist as a distinct community and nonlocals settlers who comprise mainly the Nepalis, Bengalis, Assamese, etc. However, in the hill districts the major

TABLE 4 Annual Growth Rate of Population

	1951–61	1961–71	1971–81	1981–91	1991–2001	2001–11
Senapati	4.2	3.7	4.0	3.1	3.1	5.2
Tamenglong	1.2	2.0	3.0	3.3	2.6	2.3
Churachandpur	4.0	4.2	2.9	2.7	2.6	1.8
Ukhrul	1.7	2.5	2.9	2.8	2.5	2.7
Chandel	1.4	3.4	3.8	2.3	5.1	2.0
Bishnupur	3.2	3.2	2.6	2.4	1.4	1.3
Thoubal	3.1	3.0	2.4	2.3	2.1	1.5
Imphal East	4.1	3.4	3.0	2.6	1.8	1.4
Imphal West	2.4	3.0	2.2	2.3	1.5	1.5
Hill subtotal	2.8	3.3	3.4	2.9	3.0	3.3
Valley subtotal	3.1	3.1	2.5	2.4	1.7	1.5
Manipur	3.0	3.2	2.8	2.6	2.2	2.2

population comprises the scheduled tribes. According to the 2011 census scheduled tribes constitute about 90% of the population in the hill districts. If the nature of growth of the scheduled tribe is known, it will suffice to learn the population dynamics of five districts of the hill. This may not be true in case of the valley area where most of the Meiteis have settled along with Muslims and other communities as well. However, settlement of other population in the hills is very much restricted because hills are earmarked more or less for the scheduled tribes. The list of the 33 recognized scheduled tribes in Manipur according 2011 census is reproduced here Aimol, Anal, Angami, Chiru, Chothe, Gangte, Hmar, Kabui, Kacha Naga, Koirao, Koireng, Kom, Lamgang, Mao, Maram, Maring, Any Mizo (Lushai) tribes, Monsang, Moyon, Paite, Purum, Ralte, Sema, Simte, Suhte, Tangkhul, Thadou, Vaiphei, Zou, Poumai Naga, Tarao, Kharam, and Any Kuki tribes.

On the other hand the composition of scheduled caste in the state is not brought to the forefront of the study. The scheduled caste population in the state is mainly the integral part of Meitei community. They are represented as Lois and Yaithibi in census. Historically the Lois were a group of indigenous people who were not in good terms with the king. Other communities belonging to scheduled caste in Manipur are Dhupi/Dhobi, Muchi/Ravidas, Namasudra, Patni, and Sutradhar. These communities of the scheduled caste do not form part of the Meitei or any tribe group of the state. In 2011 census almost 90% of the scheduled caste population belong to mainly Lois and Yaithibis.

Table 5 highlights the distribution of population of Manipur in 2011. The scheduled tribe population constitutes about 41% of the state population, while scheduled caste population makes only 3.4%. In some hill districts like Tamenglong and Ukhrul scheduled tribe population comprises 95%. In other hill districts tribe population settlement is quite large though slight less than observed in above two hill districts. Contrary to this the scheduled caste population mainly concentrates in the valley districts. In Bishnupur and Thoubal the scheduled caste population composition reaches to 9%, while in Imphal West and Imphal East it is slightly above 3%. However, in the hill districts their settlement is very scanty not even reaching 1% of the population of the hill districts.

From the above line of argument it is clear that the state has a distinct settlement system in the lines of the ethnic groups. For sure the hills are where mostly the tribes are settling and the valley abounds all communities but mostly the Meiteis and the Muslim population. This system of settlement is now cropping up some kinds of issues. One of the fundamental issues is the increasing density of population in the valley areas which demands more land space for settlement. Another issue associated with it is the loss of agricultural land. Cultivation on the hill slopes has not been successful for high yield at least as much as in the valley. Table 5 shows the differential density of population in the state by 2011. Since the hill districts are where the tribes inhabit,

TABLE 5 Composition of Scheduled Tribe and Scheduled Caste Population in Manipur 2011

	ST	SC	Total	Percentage Scheduled Tribe	Percentage Scheduled Caste	Density of Population per sq. km
Manipur	1,167,422	97,328	2,855,794	40.9	3.4	128
Senapati	419,210	1000	479,148	87.5	0.2	146
Tamenglong	134,626	22	140,651	95.7	0.0	32
Churachandpur	254,787	443	274,143	92.9	0.2	60
Bishnupur	3287	22,113	237,399	1.4	9.3	479
Thoubal	1808	40,593	422,168	0.4	9.6	821
Imphal West	24,161	16,530	517,992	4.7	3.2	998
Imphal East	27,657	15,839	456,113	6.1	3.5	643
Ukhrul	173,606	248	183,998	94.4	0.1	40
Chandel	128,280	540	144,182	89.0	0.4	44

Statistical Handbook, pp. 22–23.

the other communities are not allowed to settle in the hill. The landholding system in the hill is still under the customary laws. Therefore, private ownership of land in the hill areas is not practiced widely particularly among the Kuki tribes as land belongs of the Chief.

A dimension of composition of population in a greater detail among scheduled tribes is worth noting. Over time the list of scheduled tribes is ever-increasing though slowly. In some districts some tribes dominate others in terms of number. Only after the Naga-Kuki polarization the settlement is affected whereby people of similar ethnic groups try coreside in the adjacent places.

The distribution of scheduled tribes is presented in Table 6. Some of the major tribes in terms of population are listed below. An effort is made here to list the name of tribes to examine their growth. This table is specially prepared because it is hard to assemble the data on each tribe, although they are listed in the censuses. The growth of the listed scheduled tribes is examined in a uniform way. The 2001 census omitted some areas in Senapati district of Manipur. As a result the count of some tribes of the district was not reported. To maintain the growth a gap of 20 years is done here instead of usual intercensal gap of 10 years. Two growth rates from 1971–1991 and 1991–2011 are presented in Table 6. The growth rate in the 20-year gap between 1991 and 2011 censuses reveals that the growth rate of the scheduled tribe population was very high (3.1%). In the recent censuses between 2011 and 1991 the change in some tribes is negative, while some of the major tribes of the state such as Thadou, Tangkhul, Kabui, and Kacha Naga have had high growth rate. Of course, some of the smaller tribes also experienced much faster growth rate such as Maram, Koirao, Koireng, Kacha Naga, and Anal during 1991–2011. Further major scheduled tribes in each district are also highlighted in Table 7. Here the names of tribe having above 1000 population are displayed. There is no rule for selecting 1000, rather it is used only as to reflect larger population settled in the district. A greater constellation of tribes is observed in three districts, namely, Senapati, Chandel, and Churachandpur. Nonetheless in all the districts it is seen that tribes of different ethnic group settle side by side. This gives evidence that the narrower view of homelands for different major ethnic groups in the state will be in out of proposition and will be hard to materialize.

The distribution of scheduled tribe population is only a segment of the population in the state. A more complete picture, at a larger context, will be the examination of the population comprising other groups such as the scheduled tribe, Meiteis, Muslims, and other nonnative population. The language data spoken as mother tongue will be used to proceed further. Here the scheduled caste remains silent for reasons of being a component of Meitei population. As mentioned above scheduled caste population ("Loi" and "Yaithibi") is mostly concentrated in the valley areas but outskirt of capital, Imphal. Here a brief distribution is reproduced. Out of total scheduled caste population

TABLE 6 Scheduled Tribe Population in Manipur Recorded in Recent Census and Their Growth Rate

		Population					Growth Rate	
Sr. no.	Tribe	*1971*	*1981*	*1991*	*2001*	*2011*	*1971–91*	*1991–2011*
1	Aimol	836	1862	2108	2529	3190	4.62	2.07
2	Anal	6670	9349	10,642	21,242	23,509	2.34	3.96
3	Angami	70	566	308	132	95	7.41	−5.88
4	Chiru	2785	3744	6032	5622	8599	3.86	1.77
5	Chothe	1905	1687	2571	2762	3585	1.50	1.66
6	Gangte	6307	7891	12,793	9442	17,178	3.54	1.47
7	Hmar	23,312	29,216	35,767	42,933	48,375	2.14	1.51
8	Kabui	40,257	26,006	62,487	82,386	103,908	2.20	2.54
9	Kacha Naga	13,026	12,754	33,640	42,013	66,158	4.74	3.38
10	Koirao	1620	919	1716	2348	4475	0.29	4.79
11	Koireng	458	948	873	1410	1873	3.23	3.82
12	Kom	6550	9830	13,004	14,602	14,528	3.43	0.55
13	Lamgang	2622	3452	4031	5894	7770	2.15	3.28
14	Mao	33,379	50,715	76,972	4736	93,343	4.18	0.96
15	Maram	4539	6544	9592	1225	27,524	3.74	5.27
16	Maring	9825	11,910	15,698	23,238	26,424	2.34	2.60

17	Any Mizo Lushai	7483	6128	8240	15,164	8064	0.48	−0.11
18	Monsang	930	1139	1803	2130	2427	3.31	1.49
19	Moyon	1360	1642	2081	2970	2516	2.13	0.95
20	Paite	24,755	30,959	40,792	49,271	55,542	2.50	1.54
21	Purum		447	388	571	278	—	−1.67
22	Ralte	154	109	240	5	17	2.22	−13.24
23	Sema	3	24	111	13	40	18.05	−5.10
24	Simte	4177	5034	8833	11,065	6728	3.74	−1.36
25	Suhte	3	282	746	1905	804	27.58	0.37
26	Tangkhul	57,851	79,029	107,244	146,075	178,568	3.09	2.55
27	Thadou	59,955	56,467	121,994	182,594	215,913	3.55	2.85
28	Vaiphei	12,347	15,463	26,877	38,267	42,957	3.89	2.34
29	Zou	10,060	12,576	16,803	20,567	24,294	2.56	1.84
30	Poumai Naga					127,381		
31	Tarao					1066		
32	Kharam					1145		
33	Any Kuki tribes					28,342		
	Unclassified	1227	1287	7777	8030	20,806	9.23	4.92
	Total	334,466	387,979	632,163	741,141	1,167,422	3.18	3.07

TABLE 7 Distribution of Scheduled Tribes Above 1000 Population in District of Manipur as per 2011 Census

District	Tribes
Senapati	Any Kuki tribes, any Mizo (Lushai) tribes, Chiru, Kabui, Kacha Naga, Koirao, Koireng Kom, Mao, Maram, Maring, Poumai Naga, Tangkhul, Thadou, Vaiphei
Tamenglong	Chiru, Kabui, Kacha Naga, Gangte, Thadou
Churachandpur	Any Kuki tribes, any Mizo (Lushai) tribes, Chiru, Gangte, Hmar, Kabui, Kom, Paite, Simte, Thadou, Vaiphei, Zou
Bishnupur	Kabui
Imphal West	Kabui, Kacha Naga, Paite, Tangkhul, Thadou
Imphal East	Hmar, Kom, Kabui, Tangkhul, Thadou
Ukhrul	Tangkhul, Thadou
Chandel	Aimol, Anal, any Kuki tribes, Chothe, Gangte, Kom, Lamgang, Maring, Monsang, Moyon, Tangkhul, Thadou, Vaiphei, Zou

Source: Table A-11 (Appendix) District wise scheduled tribe population.

89.5% come from these two communities. Therefore, a separate analysis of scheduled caste population is omitted here (Table 8).

The census data of language are a useful source of information for a detailed study of ethnicity in terms of mother tongue spoken by the individuals. According to Manual on Vital Statistics (2009), "Mother tongue was defined as the language spoken in childhood by the person's mother to the person. If the mother died during the person's infancy, the language mainly spoken in the person's home in childhood will be the mother tongue." The Manual also mentions that the language data can shed important information on ethnicity of the population. Mention can be made that each tribe has its own dialect and most of them are reflected in the census also. The language data over time since 1981 are discussed here. This segregation of composition will enable to highlight how different ethnic groups in the state have grown over time. The understanding of the ethnic composition of the state may help to get a deeper look into myth of oversized migrants in the state.

Along with the mother tongue data, the data on religion are also used for arriving at some certainty of the Muslim population. The Muslim population is a group of population integrated to the Manipuri society apart from the Meiteis and the tribal ethnic groups. Historically they have come to Manipur some 400 years before and fully assimilated to the local way of life. One peculiarity of the Muslims in Manipur is their use of local Manipuri language.

TABLE 8 Scheduled Caste Composition in Manipur

All schedule castes	97,328
Dhupi, Dhobi	122
Lois	85,543
Muchi, Ravidas	190
Namasudra	3747
Patni	3664
Sutradhar	92
Yaithibi	1532
Generic castes, etc.	2438

SC-14 Scheduled caste population by religious community, 2011 census.

Therefore, when Manipuri-speaking population is accounted, Muslim population needs to be adjusted to get the number of the native Meiteis of the state on the basis of mother tongue data. A small exercise is done to get the number of different ethnic groups of population of the state. However, there is a problem in the case of 2001 census of Manipur where three subdivisions of Senapati district, namely, Mao-Maram, Paomata, and Purul did not have census count, although estimated total figure of the district was made available.

There are two important figures available in this regard. One of them is to use the proportion of scheduled tribes, and other is to use the religion data. In this regard it is worth mentioning how different faiths in Manipur came. A radical king (Pamheiba aka Garibaniwaz) in the mid-18th century made the people of the valley area (the Meiteis) converted into Hinduism, while almost one century later the tribes of the hill underwent conversion to Christianity (Kamei, 2008, pp. 52–53). Therefore, a distinct composition of population can be seen in such a manner that the tribes are synonymous with Christians, while Hindu can refer to the Meiteis among indigenous population. In recent times there is also a movement for revival of Meiteis to pre-Hinduized faith in different nomenclatures such as Apokpa, Sanamahism, etc. They all can be termed as the Meitei Marup (Kamei, 2008, pp. 53–54). Other third important dimension is the Muslims. Muslims or Meitei Pangals speak the local Manipuri language while they retain their faith. Therefore, from the Manipuri-speaking population Muslim population needs to be adjusted for obtaining the Meitei population. When attempt is made to adjust population of Senapati of 2001 (three subdivisions are excluded), the proportion of different religions is used from preceding and succeeding censuses

(1991 and 2011). The 1991 census data on religion do not provide classified data below district level, while in 2011 it comes down to subdivision level. While cross-checking the proportion of scheduled tribe in Mao-Maram subdivision it reaches to 88.67%, while Paomata and Purul get 98.28% and 98.1%, respectively. Meanwhile the corresponding proportion of Christian population of these three subdivisions are 92.71%, 98.05%, and 99.38%, respectively. From this finding it can now be safe to arrive at the idea that either scheduled tribe data or religion data can represent the composition of the ethnic population in Senapati district. Here only district-level information is used, since in 1991 census Purul sub-subdivision was not available. Applying linear interpolation, the estimated Christian and Hindu population for 2001 in Senapati district arrive at 244,481 and 34,885. Adjusting this population to the 2001 population it can be observed that the size of tribal population speaking both scheduled and nonscheduled as well as other languages turns out to be fairly large as much as Meitei population. Two important components of data put in Table 6 have come out. One is the religion data which provide the number of Muslim population in the state and the scheduled tribe population. As mentioned above Manipuri Muslims speak Manipuri as the Meiteis do. Therefore, if Meitei population is to be extracted from Manipuri-speaking population, it needs to adjust the Muslim population from the total Manipuri mother tongue speaking population. This is expressed in Table 9 where not only the Meitei population, scheduled tribe population but other language-based nonlocal populations are presented. Among the nonlocal population majority fall among Nepali-, Bengali-, and Hindi-speaking population. However, if critically examined the estimation of Meitei population may turn out to be somewhat an underestimate because among the nonlocals it is highly possible that some of them must be cross-classified to be Muslim. In other words those speaking other languages like Bengali or Hindi there could be Muslims (implying non-Meitei Pangals).

Table 9 also shows that the share of scheduled tribe population has been increasing over censuses since 1971 so also the share of Muslim population in the state. Population of each ethnic group has been increasing over time, but rate of increase is slowing down among the Meitei population. This has resulted in decrease in the share to the total population. In contrast to the experience of the Meitei population scheduled tribe and Muslim population in Manipur have been increasing faster. The growth of the nonlocal population is also increasing faster. Here some caution is needed for understanding the nonlocal population, some of the nonlocal population are settling in good number in the adjoining areas between the neighboring states, but such population are restricted in counting as the original people of Manipur. This is to be seen in the light of the classification of major ethnic groups of Manipur, namely, the Nagas, the Kukis, the Meiteis, and Meitei Pangals.

Table 9 further gives indirect information on the migrants to the state. There is a controversy in Manipur regarding the number of migrants. There

TABLE 9 Composition of Population by Community

	2011	2001	1991	1981	1971
Total population	2855,794	2293,896	1837,149	1418,949	1072,365
Scheduled tribe	1167,422	901,554	616,488	463,833	333,410
Estimated Meitei	1282,296	1075,159	976,599	774,816	607,433
Muslim	239,836	190,939	133,535	99,327	70,969
Nonlocal population by mother tongue					
Nepali	63,756	52,756	46,498	39,195	26,381
Bengali	30,611	27,100	19,385	19,393	15,025
Hindi	31,703	24,720	24,061	13,574	11,631
Other nonlocal	37,662	14,910	20,583	10,815	7904
Total nonlocal	163,732	119,486	110,527	82,977	60,941
Scheduled tribe	40.88	41.61	33.56	32.69	31.09
Muslim	8.40	8.81	7.27	7.00	6.62
Estimated Meitei	44.90	49.62	53.16	54.60	56.64
Nepali	2.23	2.43	2.53	2.76	2.46
Bengali	1.07	1.25	1.06	1.37	1.40
Hindi	1.11	1.14	1.31	0.96	1.08
Other nonlocal	1.32	0.69	1.12	0.76	0.74
All nonlocal	5.73	5.51	6.02	5.85	5.68

are reports of migrants outnumbering the local indigenous population very fast. Even though this is a speculation, it awakens the need for safeguarding the indigenous population of the state. There are movements for it for introduction of inner line permit in the state assembly. From the language data it can be seen that the plausible number of migrants over the years in Manipur constitutes 5.7% of the state population. Among them three ethnic groups are found to be promising. The Nepali-, Hindi-, and Bengali-speaking population as their mother tongue have come out to be in larger share. It is not sure these populations have taken their usual place of residence in Manipur as permanent migrants, but they are the persons counted on the soil of Manipur in the subsequent censuses.

2.1 Demographic Indicators

The basic demographic indicators are births, deaths, and migration. When migration does not occur, births and deaths fundamentally determine the size of a population. When there is a constant births and deaths over time, population tends to stable population where the age structure does not change. One of the important sources of demographic data is the census, but census does not provide demographic indicators in the form of rates such as birth rate and death rate. Fortunately in India the Sample Registration System (SRS) has been providing the birth and death rates (crude birth and death rates) for the state as done for India since 1971. The birth rate in early period was quite high in Manipur ranging around 30 per thousand in the first decade between 1970 and 1980. Later after 1980 the birth rate has declined consistently over time. The recent estimate from SRS put birth rate among the lowest in India. The 2015 SRS estimate of birth is 14.4 per thousand population which is fairly small. Similarly death rate in the state is very low. If the estimate is effective, the demographic performance of the state is not very poor. The trouble with the crude rate is inability to assign in which age births and deaths are occurring. Therefore, a more refined measure is desirable (Table 10).

2.2 Mortality Situation in Manipur

Mortality situation at the macrolevel can be assessed by crude death rate available over time for the state. However, crude death rate does not reveal the details of the mortality at the specific cases by age. Despite this drawback the available information of crude death rate has formed the basis for understanding the situation. This information is provided by SRS as presented in Table 11. The crude death rate in Manipur was not high ever since SRS has started generating this estimate. In 1971 the crude death rate was only 7 per 1000 population and urban estimates were lower than the rural estimates. In 1985 the CDR recorded the highest close to 8 per thousand population. Ever since 1985 till the latest estimate CDR has come down to a level which is

TABLE 10 Birth and Death Rates in Manipur Over Time

	Birth Rate			Death Rate		
Year	*Total*	*Rural*	*Urban*	*Total*	*Rural*	*Urban*
1971	33.3	34.0	26.4	6.9	7.1	5.5
1975	24.5	25.3	18.7	5.6	5.6	5.5
1980	30.9	31.6	24.1	6.3	6.4	4.6
1985	28.5	29.8	24.4	7.7	7.9	7.1
1990	21.1	22.2	17.4	5.9	5.8	6.3
1995	20.6	21.4	18.4	6.7	7.0	6.0
2000	18.3	19.1	16.2	5.6	5.4	6.0
2005	14.7	14.5	15.5	4.1	4.0	4.4
2010	14.9	14.8	15.3	4.2	4.3	4.0
2015	14.4	14.4	14.5	4.0	3.9	4.2

Source: SRS.

TABLE 11 Death Rate in Manipur

	SRS			NFHS			
Year	*Total*	*Rural*	*Urban*		*Total*	*Rural*	*Urban*
1971	6.9	7.1	5.5				
1975	5.6	5.6	5.5				
1980	6.3	6.4	4.6				
1985	7.7	7.9	7.1				
1990	5.9	5.8	6.3	NFHS-1	5.8	6.1	5.3
1995	6.7	7.0	6.0				
2000	5.6	5.4	6.0	NFHS-2	9.0	8.7	9.8
2005	4.1	4.0	4.4				
2010	4.2	4.3	4.0				
2015	4.0	3.9	4.2	NFHS-4	6.8	6.4	7.6

quite small. Unfortunately CDR is not used for comparison because it does not take care of age distribution. Therefore, there is no point for comparing the CDR of other states with the state CDR. In terms of rural–urban division the difference is very minimal and CDR of rural and CDR of urban do not deviate from each other. If the CDR values are taken at face value, they are quite low as found in other developed states. The estimates generated by SRS are reproduced against the findings provided by NFHS in different rounds. The NFHS estimates of CDR of first round were almost at par with the estimates of SRS, while NFHS-2 and NFHS-4 estimates of CDR saw a deviation from the estimates of SRS. Nevertheless, the CDR of Manipur either from SRS or NFHS is not high. From these estimates it can be seen that the mortality situation in Manipur has not been so high at the macrolevel.

To circumvent the inadequacy of crude death rate SRS-provided age-specific mortality rate is examined. From this information the mortality pattern of Manipur can be seen. It is observed that over time since 1991 till 2012 Manipur has been experiencing a declining trend of mortality for each age. The reduction in mortality is quite commendable. Even at the higher age of 70+ the mortality rate at the state macrolevel has gone down to 60.5 per 1000 in 2012 from a high of 81.7 per 1000 population in 1991. In other words the older population is also experiencing increasing longevity of life (life expectancy might be improved, but no evidence due to unavailability of life tables of Manipur). The trend among male and female also follows in similar fashion while there is a perceptible lower mortality for each age group till age 60 among females in 2012. But in the older age this advantage of lower mortality among women vanishes and becomes at par with that of menfolk (Table 12).

It is important to examine the level of infant mortality in the state. Infant mortality is often linked with the level of development of the society (Erdogan et al., 2013). In Manipur the infant mortality rate estimated by SRS is not high when compared with the other parts of the country. Estimates of infant mortality rates are also available in the NFHS for the state too. If there should be any comparison between these two estimates, the SRS estimates have been on the lower side. One observation that the SRS has made for the smaller states like Manipur is the inadequacy of sample sizes. Therefore, in SRS bulletins infant mortality rates classified by residence (rural–urban by sex) are not available (Table 13). If this statement is taken to the confidence, then the NFHS estimates are the next best option to rely on and presented in Table 14. The SRS infant mortality rate has come down to one-third from 1990 compared to the 2015 level, whereas in the NFHS it has come down to half since NFHS-1 to NFHS-4. Manipur also experienced a higher level of infant mortality among males than females in contrast to the SRS estimates where infant mortality rate is higher among the females than males.

TABLE 12 Three-Year Moving Average of Age-Specific Mortality Rate by Sex and Residence From 1990–92 to 2011–13

	Total			Male			Female		
Age Group	*1991*	*2001*	*2012*	*1991*	*2001*	*2012*	*1991*	*2001*	*2012*
0–4	6.99	4.92	2.20	7.6	5.1	2.1	6.4	4.7	2.4
5–9	1.60	1.01	0.27	1.8	1.5	0.3	1.4	0.6	0.2
10–14	0.88	0.77	0.49	1.1	1.3	0.3	0.6	0.2	0.7
15–19	1.42	1.02	0.58	1.9	1.2	0.7	0.9	0.8	0.5
20–24	1.76	1.46	0.78	1.5	1.9	1.4	2.0	1.1	0.2
25–29	1.60	2.21	1.07	2.4	3.8	1.6	0.9	0.8	0.6
30–34	2.73	3.01	1.23	3.7	3.9	1.8	1.8	2.0	0.7
35–39	3.34	3.52	2.21	3.3	5.1	3.2	3.4	1.9	1.2
40–44	3.90	4.28	2.90	4.6	5.2	4.3	3.1	3.4	1.4
45–49	5.68	5.43	3.23	7.4	7.4	4.9	3.9	3.4	1.6
50–54	8.67	6.87	5.35	7.9	7.6	7.6	9.4	6.0	2.9
55–59	13.71	9.59	5.21	14.6	9.1	7.0	12.8	10.1	3.5
60–64	21.61	13.09	7.56	22.5	12.2	7.6	20.8	13.9	7.5
65–69	34.09	25.58	13.55	37.2	28.9	13.4	31.0	22.0	13.8
70+	81.67	58.35	60.52	81.4	60.1	61.2	81.9	56.6	59.9
All ages	5.64	5.11	4.05	6.1	5.9	4.7	5.2	4.3	3.4

Source: SRS Compendium Sheet T8.

TABLE 13 Infant Mortality Rate, Manipur

Year	Total	Male	Female	Rural	Urban
1990	29.1	37.3	21.5	29.8	26.3
1995	35.7	36.9	34.2	41.3	17.4
2000	22.2	20.5	24.0	21.1	25.9
2005	9.0	8.0	10.0	10.0	8.0
2010	13.6	11.3	16.3	15.2	8.8
2015	9.0	8.0	10.0	10.0	8.0

Source: SRS Bulletins.

TABLE 14 NFHS Estimate of Infant and Child Mortality

	Infant Mortality Rate	Child Mortality Rate	Under-5 Mortality Rate	Infant Mortality Rate	
				Male	*Female*
NFHS-1 (1992–93)	42.4	20.2	61.7	—	—
NFHS-2 (1998–99)	37.0	19.9	56.1	40.0[a]	38.8[a]
NFHS-3 (2005–6)	29.7	12.6	41.9	42.0	29.5
NFHS-4 (2015–16)	21.7	4.3	25.9	24.1	19.3

[a]*Reference is for the 10-year period preceding the survey.*

The state-level estimates of infant mortality therefore provide a macrolevel position of the mortality condition. But for the planning purpose infant mortality rate at the district level is not available either in SRS or NFHS. In this regard indirect estimates generated by Brass technique from children ever born and children surviving are of great help (United Nations, 1983). On the basis of this technique NHSRC (n.d.) has provided district-level estimates of infant and under-5 mortality rates based on 2001 and 2011 census data. The table is reproduced here. There is substantial reduction of infant mortality rate in the state as a whole (9 points), while districts have also witnessed a similar pattern. During the intercensal period of 2011 and 2001 the hill districts have witnessed a larger decline in infant and under-5 mortality. This could perhaps be due to the improvement in health system due to the introduction of National Rural Health Mission (2005) in the country. This finding also provides a similar situation of having a higher infant mortality rate among males than the females in 2011 (Table 15).

So far the picture of mortality is about the presentation of the estimates of infant, child, and under-5 mortality rates and age-specific mortality rate from different sources. Manipur is a small state and SRS does not provide the life tables for such smaller states. Therefore, there is no composite indicator of mortality situation normally depicted by life expectancy from the life table. In other words the state does not have life table. However, one report by ICMR, PHFI, and IHME (2017) estimates the life expectancy of Manipur by 2016 for female and male to be 72.8 and 68, respectively, which has scored a marked increase from 1990 (corresponding figures of 65 and 62.9 for female and male, respectively).

Understanding what causes the death will provide an opportunity for its own sake and public health intervention through some program. Indeed the same report by ICMR, PHFI, and IHME (2017) also provides the causes of death for different age groups. For the young population below age 15 communicable diseases of diarrhea and lower respiratory infection and neonatal disorders are responsible for nearly three-fourths of deaths, while HIV still contributes 2.2% of total deaths. Among the economically productive and socially reproductive age 15–39 years, HIV accounts for 18% of deaths along with suicides and violence (15%). In case of the older population above age 40 major causes of death have become the cardiovascular diseases, cancer, and diarrhoeal and respiratory diseases. HIV still scores a high proportion of deaths among these groups too. A cause of concern regarding the cause of death is the menace of HIV in the state. As shown in the graph among the age 15–39 years the leading cause of death is HIV. According to NFHS-4 the prevalence of HIV in the three states of Manipur, Nagaland, and Mizoram was 1.49% (1.45% among women and 1.54% among male) among sample population of age 15–49. This estimate was by far the highest in the whole country (0.23% and 0.25% among female and male, respectively). The HIV prevalence by NFHS-4 (0.24%) for the country was quite comparable to the NACO estimate (0.26%) for year 2015 and that for Manipur stood 1.15%. Unfortunately, the situation of HIV prevalence has not improved since NFHS-3 10 years ago when the prevalence was still 1.13% among a population of 15–49 years. HIV prevalence in Manipur is still high within the country despite so many efforts by government and people. Earlier the route of HIV transmission was mainly from the injecting drug use, but recent report of Manipur State AIDS Control Society (2016–17) finds a gradual shift toward heterosexual route as the main transmission channel.

Among older population of age 40 years and above the degenerative diseases like cancer, cardiovascular diseases, and diabetes are prominent as the

TABLE 15 Indirect Estimates of Indicators of Infant and Child Mortality of Manipur by Sex, Based on 2011 Census

	Persons				Males				Females			
	Census (2011)		*Census (2001)*		*Census (2011)*		*Census (2001)*		*Census (2011)*		*Census (2001)*	
States and Districts	*IMR*	*U5*	*IMR*	*U5*	*IMR*	*U5*	*IMR*	*U5*	*IMR*	*U5*	*IMR*	*U5*
Manipur	43	57	52	70	45	60	47	62	41	54	57	78
1. Senapati	43	58	61	85	41	53	55	74	48	63	68	95
2. Tamenglong	42	55	78	112	43	56	77	111	41	53	78	114
3. Churachandpur	43	56	59	81	43	57	49	66	42	55	68	95
4. Bishnupur	40	53	52	71	42	56	48	64	38	50	57	78
5. Thoubal	43	57	46	61	46	61	42	55	40	52	51	68
6. Imphal West	42	56	39	51	48	65	38	49	36	47	41	53
7. Imphal East	42	55	49	65	46	61	44	59	37	49	53	72
8. Ukhrul	49	66	67	95	50	67	57	78	49	65	77	111
9. Chandel	51	69	61	84	55	75	54	74	47	63	67	95

Source: NHSRC, p. 39.

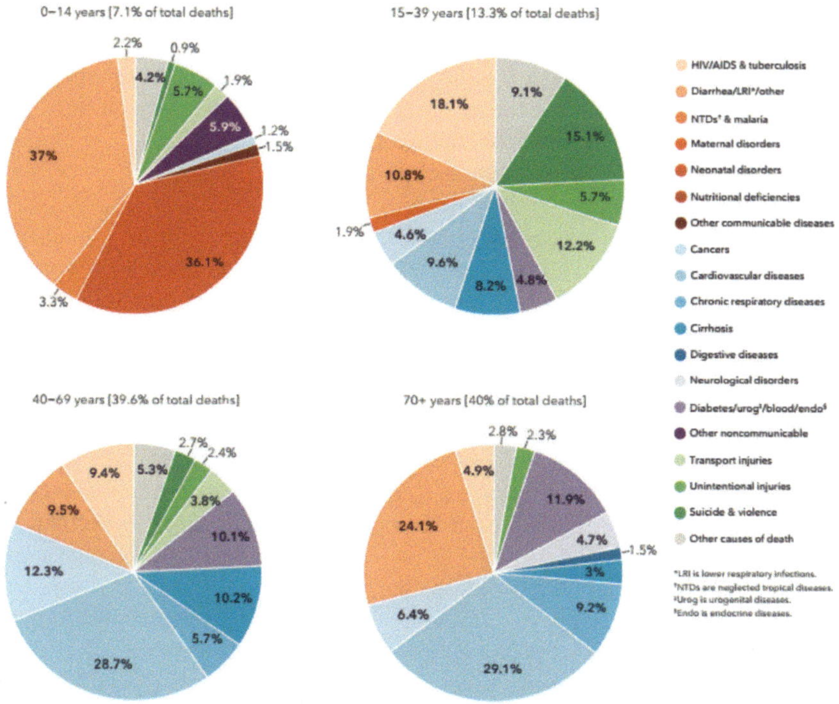

FIG. 1 Percent contribution of top 10 causes of death by age group, both sexes, 2016. *Source: Indian Council of Medical Research, Public Health Foundation of India and Institute for Health Metrics and Evaluation (2017, p. 143).*

cause of death. The share of cardiovascular disease is equally strong among 40–69 and 70+ population and so also diabetes in this age group. This report also suggests the return of diarrhoeal disease among the older population of 70+ years. In other words in Manipur communicable diseases still persist strongly. This may be explained by ignorance among people about the importance of healthy life, low standard of living due to poverty, and lack of infrastructure of health in the state (Fig. 1).

2.3 Fertility in Manipur

The macrolevel fertility situation in Manipur is already shown in Table 10. It is simply total number of births per 1000 population. Crude birth rate presented in Table 10 is estimated by SRS and fortunately NFHS also in various rounds published crude birth rate for the state. The CBR estimates by NFHS are 24.4, 26.1, 25.0, and 21.2 in the four rounds from NFHS-1 till NFHS-4, respectively (not shown in table). NFHS estimates were higher than the SRS estimates over time. Manipur has witnessed a consistent decline in fertility over time as per SRS. Around the time when it got statehood (1972) CBR was 33.3 per thousand but by 1990 after two decades it fell down to 21.1 and it further dipped down to 14.4 as published in SRS bulletins. On the other hand NFHS-1 in 1992–93 estimated CBR 24.4 which was very close to the SRS estimate but in the next two rounds of NFHS-2 and NFHS-3 witnessed a higher level of CBR compared to NFHS-1. Only NFHS-4 has seen a decline in CBR from NFHS-2.

SRS also generates a summary and standard measure of fertility, total fertility rate (TFR). Population Handbook of Population Reference Bureau (2004) defines "the total fertility rate (TFR) is the average number of children that would be born to a woman by the time she ended childbearing if she were to pass through all her childbearing years conforming to the age-specific fertility rates of a given year." TFR is a standardized measure and can compare with estimates from other places. If the SRS estimate is true, Manipur has got replacement fertility one decade earlier than what India set target by 2010 in its population policy 2000 (refer to Table 16). Table 16 also presents another measure of fertility in the form of gross reproduction rate which as defined by Population Handbook (2004): "The gross reproduction rate (GRR) is the average number of daughters that would be born to a woman (or group of women) during her lifetime if she passed through her childbearing years conforming to the age-specific fertility rates of a given year." Alternative estimate of TFR by NFHS-I made TFR at 2.8 and further went up to 3 in the second round further back to 2.8 and decreases to 2.6 in NFHS-4. The estimate by NFHS is much higher than the figures made by SRS.

TABLE 16 Total Fertility Rate and Gross Reproduction Rate

	TFR			GRR		
Indicators	*Total*	*Rural*	*Urban*	*Total*	*Rural*	*Urban*
1990–92	2.45	2.64	1.87	1.20	1.29	0.92
1994–96	2.59	2.79	2.10	1.21	1.29	1.00
1999–01	2.10	2.21	1.80	0.99	1.04	0.87
2004–06	1.61	1.64	1.56	0.73	0.75	0.68
2009–11	1.56	1.58	1.49	0.73	0.74	0.72
2011–13	1.47	1.47	1.46	0.70	0.69	0.71

Source: SRS.

The estimates of TFR by SRS and NFHS provide at the state level fertility measures, but it does not say anything about the below state-level or district-level estimates. This may be because of paucity of data. However, census provides some information on the fertility from two perspectives. One is the total number of children ever born to women and children surviving and the other is the number of children born in last 1 year. Though the census information is hardly used for generation of period measures, often the children even born can be useful to get the district-level estimates. Guilmoto and Rajan (2013) have made a study and brought out estimates of CBR and TFR at the district using the census data of 2001 and 2011. Here their estimates of CBR and TFR are reproduced in Table 17 to highlight the level and trend at the district level. Their estimates of fertility are still higher than that of SRS estimates but quite comparable with the estimates provided by NFHS at the state level. The interesting part is to get an idea of how the districts have fared which most of cases are missing. From their estimate in 2011 two districts Imphal West and Chandel got the lowest TFR, while others were well above replacement-level fertility. From this analysis it can be understood that Manipur may need a longer time to achieve the replacement fertility, even though the National Health Policy 2017 revised the target to achieve it by 2025.

While making the narratives of fertility in Manipur it is also thought that some parameters are needed to highlight. These parameters are essential for interpretation and understanding of fertility phenomenon in the state. Recall

TABLE 17 Estimates of Crude Birth Rate and Total Fertility Rate

	CBR, 2011	TFR, 2011	TFR, 2001
Manipur	20.0	2.5	2.6
Senapati	20.2	2.6	2.2
Tamenglong	20.4	2.7	2.8
Churachandpur	19.6	2.5	2.5
Bishnupur	18.9	2.4	2.5
Thoubal	24.2	3.1	3.3
Imphal West	17.2	2.0	2.2
Imphal East	20.4	2.5	2.6
Ukhrul	20.0	2.8	3.1
Chandel	17.8	2.1	2.8

the proximate determinants of Bongaarts and Potter (1983) in which they picked up four well-known parameters such as index of marriage, index of contraception, induced abortion, and duration of postpartum infecundability. Following evidences from the district-level household survey round-4 held during 2012–13 (DLHS-4) can shed some light of why fertility is high in the state. Table 18 is presented to convey some of the proximate determinants of fertility in Manipur from DLHS-4 survey results.

The mean age at marriage particularly of girls was 24 years, while for boys it was 26.7 years. In the context of other states of India age at marriage in Manipur is not low for both female and male. Within state except for Chandel (21.3 years) the average age at marriage among females was well above 22 and close to 23 years. This age at marriage is fundamentally above the legal age at marriage for girls in India (18 years). Despite relatively a higher average age at marriage a contrasting feature is the early marriage of women in the young age. This can be observed from the column of married women aged 20–24, among them a good proportion got married before age 18. In short, for the state nearly 23% of women in the 20–24 got married before they attained age 18. Only Imphal West and Thoubal district witnessed a lower proportion getting married before age 18 among this group of women. On the other hand higher level of fertility can also be linked with higher proportion of birth order. In the state 29% of births are of order 3 and above. This percentage is relatively higher in hill districts of Churachandpur, Ukhrul, Senapati, Tamenglong, and Chandel compared to the four remaining valley districts. High fertility level can also be corroborated with low use of family planning methods in the state.

The DLHS-4 found that only one-fifth of the currently married women were practicing family planning methods, while the modern scientific methods were much lower (11.3%). Naturally the unmet need for family planning service rose to more than half of the currently married women of 15–49 age. Indeed the unmet need for family planning services rose to two-thirds of the married women in some of districts, while the minimum was still close to half of the married women. Lastly the level of induced abortion in the state was quite small and low to only 3% of the total pregnancy outcome terminated as induced abortion and this was higher among the valley districts where better medical facilities and services are available. Abortion probably is one of the least preferred ways of practice to destroy a product of conception. The Manipuri society is still a traditional society where such an act is taken as immoral and a treachery within marriage. It is worth to mention that out of total fertility rate of 2.6 the wanted fertility rate was still 2.3 for the state (NFHS-4). This signifies the fact that most of the pregnancies are still wanted and scope for conducting an induced abortion remains very occasional among the women of the state. This attitude could be applicable in hills too where majority are Christians and leading a traditional way of life.

TABLE 18 Fertility-Related Indicators From DLHS-4

	Mean Age At Marriage		Women Aged 20–24 Married Before Age 18	Induced Abortion	Birth Order 3+	Current Use of Contraception	Current Use of Modern Contraception	Unmet Need for Family Planning Among Women of 15–49
	Boy	*Girl*						
Senapati	25.3	22.9	22.4	1.0	41.3	24.9	9.8	50.4
Tamenglong	26.4	22.7	21.2	0.0	36.1	7.9	4.5	67.8
Churachandpur	25.1	22.5	33.3	0.5	44.1	22.6	14.6	60.4
Bishnupur	26.2	24.1	23.0	7.7	18.7	30.2	18.5	48.1
Thoubal	26.6	23.1	19.8	5.2	21.4	18.6	10.2	49.3
Imphal West	27.2	26.0	16.3	3.5	12.2	22.8	12.5	54.2
Imphal East	27.8	25.4	25.6	4.1	21.5	22.4	12.5	50.6
Ukhrul	27.9	23.4	20.3	1.1	47.7	15.6	7.3	60.4
Chandel	25.9	21.3	25.3	1.3	32.1	8.2	5.5	66.3
DLHS-4	26.7	24.1	22.9	3.1	29.4	20.5	11.3	54.6
DLHS-3	27.3	23.6	24.7			43.1	19.5	25.6

2.4 Migration

Migration or the movement of a person from one place to another is when the mover makes the move of a permanent change in usual place of residence. Population Handbook defines "Migration is the geographic movement of people across a specified boundary for the purpose of establishing a new permanent or semipermanent residence." It is one of the important components of demographic process. Unlike fertility and mortality, usually migration contains a large element of conscious choice, whether to move or not (Gould, 2009, p. 155). Besides fertility and mortality which basically function as addition and depletion of a population, migration modifies the population causing addition or/and depletion. Migration involves a mobility which therefore requires two places as place of origin and place of destination. Migration strongly responds to the level of development of both the place of origin and place of destination. This concept was well founded from the early theorist like Ravenstein (1885) when he studied migration pattern in England in the late 19th century. Migration, therefore, is linked with development. Here mention may be made of the contribution by Lee (1966) regarding the development of theory of migration called "push and pull" which can accommodate the nuances of migration. Where development is more the pull factor operates stronger at the place of destination, while push factor works better where the place of origin is less developed (Gould, 2009).

Migration in Manipur has been a controversial issue. Right from the early days the people of the state particularly in the valley have been insisting to frame a policy which can protect the indigenous population. While the hill areas have been protected by the constitution since people in the hill areas belong to the scheduled tribe category, a similar law seems to be desirable for the population in the valley area too. In other words inmigration in the state has got no warm welcome unless the migrant has enough information of how long they will stay and particulars of his/her personal information and whereabouts. In this context migration has got a loud attention. However, from population context migration as an increase of population size may not be desirable, but the hind side of migration is benefits it accrues to the economy as it brings a cheaper labor supply with intrinsic skills coming with them—the efficiency of labor.

It is hard to find the reasons why people have to migrate to Manipur particularly interstate migration. Those who feel that it has a big potential for a future prospect (market) or opportunities may think of migrating to this part of country. This is a remote thinking because from market point of view the size of population is not so large (2.86 million population in 2011 census) and the effective demand is also not so attractive because the state has the third lowest per capita income in the country in 2016–17 year (PIB). Despite this consideration it is equally important to note that there is a vast area in the hill where some kind of suitable profitable activities such as organic farming are quite possible. And rhetoric or reality the Manipur is the corridor to the Southeast Asia to which India looks to and tries to open the gateway through the policy of Look East which got changed into Act East Policy of the country. Here lies the opportunities for those who dare to meet the challenges of migration.

Here it becomes quite significant to examine the size of migrants in the state. Since the valley area is quite convenient for agriculture, it has been a rice bowl of the state. Looking up in history the state was an exporter of rice and the uprising of women in 1939 (Nupi Lal) was to curb the unscrupulous rice export from the state by traders. Now the state is depending on the import of rice from outside the state. With the increase in population the cultivable areas of the valley is transforming into constructions for commercial and administrative purposes, reducing the land available for agriculture. While transforming the economy with greater degree of integration to globalization through the Act East Policy and connectivity through railway the people of Manipur feel that the larger population from other parts of the country will make Manipur a convenient destination. With this view it is strongly felt that there should be some policy which can protect and preserve the local traditions and cultural ethos of the indigenous population for their security of existence. There is always an apprehension that the valley people will not have even a plot of land to build a house. Meanwhile it seems that people have forgotten a hard fact that Manipuri people also move out of the state as other people can come to the state. The movement for protection of local indigenous population is perhaps well influenced by the concepts of the people who have seen bitter part of life in big cities of the country where a lot of people struggle for a secured roof to sleep peacefully and a handful of food grains for a square meal.

Table 19 observes the census figures of migrants on the basis of place of birth in Manipur. Although this methodology involves a great trouble of not

TABLE 19 Migration Experienced by Manipur

Type of Migrants	1951	1961	1971	1981	1991	2001
Internal	4204	10,295	25,068	27,728	17,381	18,529
International	2078	8058	12,480	5736	10,404	2148
Total incoming migrant	6282	18,353	37,548	33,464	27,785	20,677
Outmigrants	—	—	—	19,731	34,843	54,036

counting the return migrants, still it provides the evidence of the lifetime migrants. On this account it becomes quite clear that migration on the basis of place of birth does not seem to have an alarming impact over time since 1951 census. While examining the outmigrants from the Manipur to other parts of India, in 1981 Manipur accommodated more number of inmigrants to state. But contrary to experience of 1981, the 1991 and 2001 census proved the outmigrants from Manipur were much larger than the inmigrants. In other words the state is facing a negative net migration in the last few decades. Regrettably the 2011 migration tables of place of birth have not been published till this writing-up is complete.

From 1991 and 2001 census data a tabulation of inmigrants is done regarding where these migrants have made their destination in Manipur (Table 20). Majority of the migrants made their destination at Imphal West and Imphal East in 2001 as done in 1991 census at Imphal before bifurcation. This is no wonder because the migrants look for a convenient place where more opportunities are available, while Churachandpur has been experiencing more attraction of migrants more consistently than other districts. Chandel district also is found to attract the migrants next to Churachandpur. This could perhaps be linked with the strategic importance of Indo-Myanmar border town Moreh.

TABLE 20 Inmigration Destination in Manipur

	1991		2001	
District	*Number*	*Percentage*	*Number*	*Percentage*
Senapati	3990	14.4	1370	6.6
Tamenglong	225	0.8	243	1.2
Churachandpur	3867	13.9	2136	10.3
Ukhrul	208	0.8	1175	5.7
Chandel	3095	11.1	1769	8.6
Bishnupur	1196	4.3	754	3.7
Thoubal	1028	3.7	1170	5.7
Imphal West			5814	28.1
Imphal East			6246	30.2
Imphal[a]	14,176	51.0	12,060	58.3

[a]*In 2001 Imphal West and Imphal East were bifurcated from Imphal district of 1991.*

3 AGE COMPOSITION

So far the situation of births, deaths, and migration is placed for a deeper appreciation of the demographic process and historical perspective of the state. Now how the effect of these parameters can be manifested in a composite way has taken the prominence. How these indicators transform the demographic structure of the state can be assessed from the age and sex distribution of the population. The graphical representation of the age and sex structure of population is known as population pyramid. The population pyramid of the state is presented for last four censuses right from 1981 to 2011 in Fig. 2. Overall the shape of the pyramid is more or less the same with broad base and tapering tops. This shape is classic pyramid which implies that there is still high birth represented by broader base and relatively larger deaths among the older age population. This typical shape implies that there is a large concentration of population in young age. This is quite contrast to the shape of population pyramids of the developed countries where the shape is somewhat cylindrical implying a constant composition of population in different ages.

From these four pyramids of the age and sex structure of population, it depicts demographic transition of population in Manipur with a smooth decline of 0–14 population from 39% to 31%, respectively, since

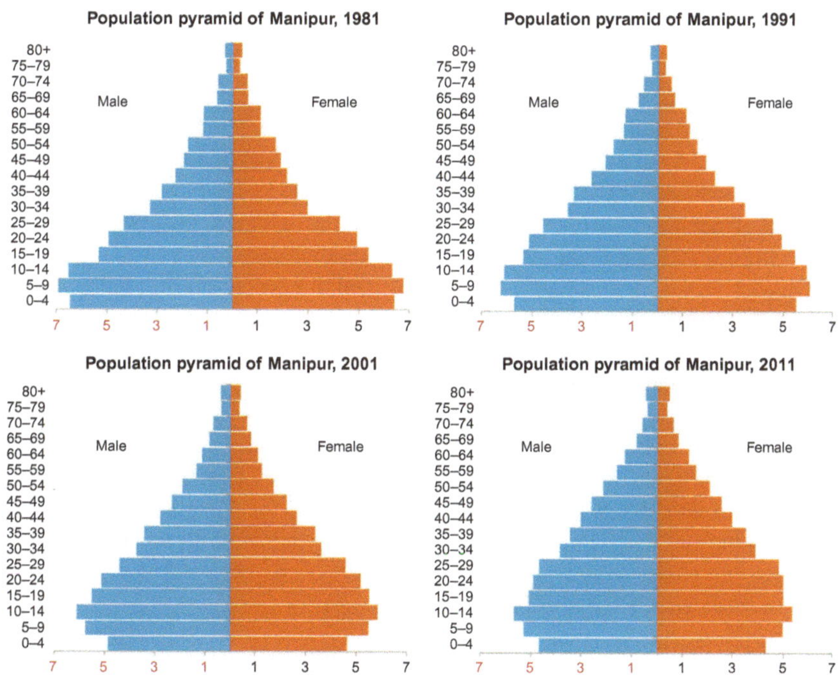

FIG. 2 Population pyramid of Manipur.

1981–2011 census. The decline of young age group is compensated by condensation of productive age group of 15–59 and also among the older population of 60 and above. A higher share of age 15–59 in the population structure from 55% to 63% in the last four censuses is an indication that if the society can spell out an appropriate policy to capitalize opportunities to benefit each individual at the right position and place, it only benefits the economy by reaping demographic dividend.

Fig. 2 is self-explanatory in such a way that there is a steady decline of young age group in the four censuses and this decline is compensated mainly by the working age group of 15–59 years. While these two age categories are compensating each other, one silently moving category is the population 60 and above years (60+). This age group of 60 years and above is generally known as aged population. The increase in the share of aged population is called population aging. Since the share of aged population has been steadily rising, it gives an indication that the state is slowly moving toward population aging. The median age of the state for these four censuses is 20, 21, 22, and 24, respectively. The median age clearly shows that Manipuri population is still very young. In other words this implies that there is a lot of population concentrating in the early period of life. This, therefore, conveys the message that fertility level is still high in the state and mortality decline is not fully anticipated by decline in fertility.

Another dimension which can be looked from age analysis is the dependency ratio. Dependency ratio is defined as the ratio between the dependent population—young population of less than 15 years—and older population of 60 years and above to economically productive age group (15–59 years). The old age dependency ratio is still low around 6%–7% in these four censuses. But there is a drastic decline in young age dependency in the same period from a high of 72% in 1981 to 58% in 2011. This change is mainly because of decline in birth rate while there is improvement in mortality situation whereby the younger population is experiencing survival to advanced adulthood. Other way round is the reading of potential support ratio which means the number of working age population per older population. This is another way of saying how many working age population is potentially able to support an old person in the state. This figure is quite consistent as a reverse of old age dependency ratio. In the state the support ratio is about 9 working age population per old age population. Being a traditional society the older persons can have a hope for a better support toward the end of life. This is all about the family support that matters a lot for the welfare of the aged population. The simple reason is that being underdeveloped the tax revenue that the working age is subscribing is not so effective and the tax base is also not large enough. The fundamental support for the old lies in the family. If mortality decline is at the pace of the national level, it is likely that the proportion of older population in the state will also be growing. Under such circumstances with the influence of globalization and structural change in the

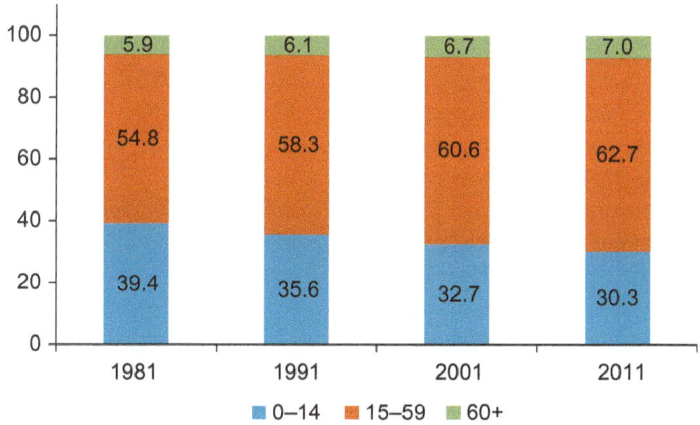

FIG. 3 Broad age composition of population in Manipur.

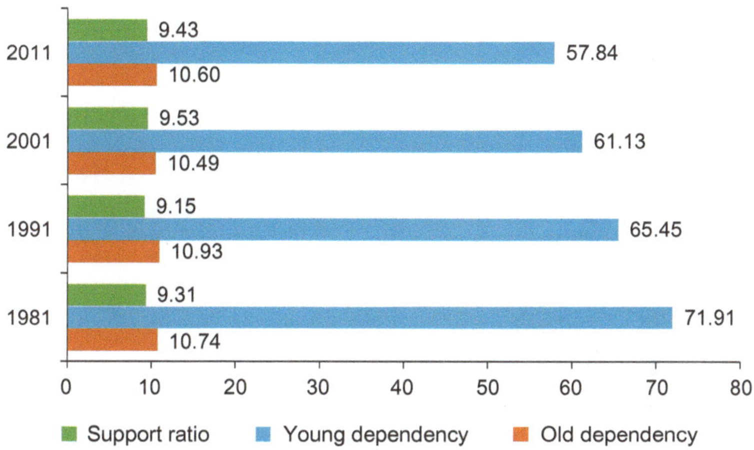

FIG. 4 Dependency ratio and potential support ratio in Manipur.

economic activities and social life the care and support of the older population may be in jeopardy unless the governance and state-specific proper policy is framed and implemented (Figs. 3 and 4).

3.1 Future Trajectory of Population

What will happen to Population of Manipur in due course of time? This indeed requires projection of population with realistic assumptions. In this text this part is not taken up. But what can be addressed is that how far the public policy will make a direction toward achieving certain goals. At this point of time there does not seem any specific state policy of population.

However, it is still mandatory for the state to fulfill national goals of population. The National Population Policy 2000 has made a strong objective that India should achieve a replacement-level fertility by 2010 and stabilization by 2045. The goal of achieving fertility by 2010 has not been met in the country and many northern and northeastern states. So the revised policy of achieving the replacement-level fertility is being postponed to 2025 as spelt out by the recently announced National Health Policy 2017. Other demographic parameter, infant mortality rate has to go down to 28 by 2019. In case of Manipur the first objective needs to be given more importance since the latter is already achieved by the state. The fertility goal is complicated and controversial since different estimates arrive at very diverse results. For instance the SRS estimate of TFR was 1.5 for the period of 2011–13, while NFHS-4 made it 2.6. If NFHS estimate is true, then at the pace of decline which has been observed during NFHS-3 and NFHS-4, achieving replacement-level fertility by 2025 seems to be a little hard unless there is a vigorous attempt from all the stakeholders. This also is difficult under the regime of reproductive and child health program (RCH) because RCH functions on the principle of choice and rights approach. In case of infant mortality if the SRS estimate is taken to confidence by 2016, Manipur already got IMR of 14 per 1000 live birth while NFHS-4's IMR estimate of 22 also showed much lower than the set goal. What needs to be done in the context of the state's demographic dynamics is to reduce fertility rate toward replacement level. This will not be easy either. If infant mortality rate has to go down further to the level achieved by the developed countries the role of the state is very big to expand the health facilities and services across the state equitably.

A quick critical review of the contextual situation of the villages of Manipur needs a revisit to evaluate how demographic indicators respond to the contextual realities. Setting target goals for demographic parameters will not help much unless other ancillary and facilitating parameters are taken into account. Table 21 gives the excerpts of selected indicators provided by district census handbook of 2011 census. For example, when a woman under labor pain has to walk for another 10 km for getting a transport facility, it does not make sense for a safe delivery. Here a few facts are placed for the purpose. The district handbook provides a list of village-level information. Selected indicators from 2011 census handbook table are extracted and analyzed for the purpose. Out of 2581 villages in the district handbook 66 villages did not have any household, and after removing them 2515 villages are retained for the analysis. The analysis reveals a large number of villages is far away from the district headquarter and the median distance of villages of hill districts like Tamenglong, Ukhrul, Senapati, Chandel, and Churachandpur is pretty far. The distance here is not only about physical distance, but the hill terrain that people have to go through for getting any effective health care or doing any official work or even for shopping. As usual the administrative inefficiencies that people have to tolerate for getting anything done is

TABLE 21 Situation of Villages in Manipur

District	Median Distance From District Headquarter (km)	Median Distance From Nearest Town (km)	% Villages Away From Black-Topped Road 5+ km	% Villages Away From All Weather Road 5+ km	% Villages Getting a Bus 5+ km	% Villages Distance From a Comm Bank 10+ km	% Villages Distance From Post Office 10+ km	% Villages Not Having a Primary School	% Villages Not Getting Tap Water	% Villages 10+ km From Primary Health Center	% Villages Having Anganwadi Center	% Village Having an ASHA	Number of Villages
Bishnupur	17	4	16.67	2.08	2.08	100	18.75	6.25	54.17	12.5	100	100	48
Chandel	35	20	64.65	77.91	56.51	100	58.84	60.23	100	68.37	56.74	71.63	430
Churachandpur	27	28	40.94	30.7	64.94	71.81	65.27	50.5	90.44	52.35	88.26	68.46	596
Imphal East	18	8	16.76	37.7	61.78	53.93	44.5	13.61	79.06	17.8	100	100	191
Imphal West	14	10	1.87	2.8	3.74	57.01	14.95	8.41	28.04	3.74	100	100	107
Senapati	57	27	51.56	9.56	42.00	83.71	47.38	33.18	99.85	31.39	95.07	96.71	669
Tamenglong	69.5	69	80.46	73.56	83.33	96.55	86.78	8.62	91.95	74.71	100	100	174
Thoubal	12	5	0	0	16.09	47.13	18.39	4.6	72.41	6.9	100	100	87
Ukhrul	69	57	39.91	0	43.19	99.06	99.06	18.78	94.37	83.1	88.73	91.08	213
Total	40	22	45.09	31.26	51.1	81.51	57.53	34.95	90.18	46.64	87.55	86.04	2515

unthinkable. Even for getting something for practical day-to-day needs the nearest town is very far for these districts. Imagine what kind of life people struggle in this age. Despite 70 years of development planning in districts like Tamenglong 80.5% of villages have to travel at least 5 km to get a black-topped road. There are still many villages lagging behind in other districts of Chandel (65%), Senapati (52%), Churachandpur (41%), and Ukhrul (40%), respectively in this count. The situation gets much better in the valley districts, but still among the villages of Bishnupur and Imphal East every one out of six villages face the similar situation. More than 80% of villages of Tamenglong and 43% of Ukhrul struggle to get a bus after initial effort of walking or otherwise at least 5 km. As of 2011 60% of villages in Chandel and half of villages in Churachandpur do not have even a primary school. What a disgrace to the national policy of education 1986 which spelt out free and compulsory primary education. With regard to health as per norm a qualified doctor is attached to a primary health center (PHC) in the rural areas. So PHC is taken as an appropriate proxy for an entry point contact with medical and health care. In the hill districts many villages do not have a PHC located nearby the village. In this context, 83% of villages of Ukhrul have to travel at least 10 kms to access health services from a nearest PHC. The corresponding situation in other districts is observed in Tamenglong (75%), Chandel (68%), Churachandpur (52%), Senapati (31%), and lowest in Imphal West (4%). It is also not sure whether the doctors and paramedics are available all throughout where the PHCs are nearby even. Only consolation and positive side is the installation of ICDS functionaries (Anganwadi centers) and ASHA workers in majority of the villages. Villages of Manipur also experience difficulty in accessing to modern credit and financial assistance. Even in this 21st century most of the villages are still financially deprived and excluded as evidence of localized banks are located at least 10 km away from the village.

Along with the demographic parameters a look at the process indicators will be of great help. These are the indicators which will facilitate achieving the demographic goals. These are the indicators that policy makers have to give adequate amount of attention to achieve the set demographic goals sensibly. Some of these indicators are presented in Table 22. The National Population Policy 2000 highlighted to achieve 14 socioeconomic goals and among them a few are listed (i) addressing the unmet needs of reproductive and child health services, (ii) reducing infant mortality rate to below 30 per 1000 live births, (iii) achieving universal immunization of children, (iv) achieving 80% institutional deliveries, and (v) promoting small family norms for achieving replacement-level fertility among others by 2010. A relook into these subgoals with the available data has shown that except for the infant mortality rate the process indicators of the reproductive and child health program have a long way to march. Since the state does not have a separate policy, the NPP 2000 also represents the state goals. Under such circumstances meeting the unmet for family planning or reproductive health component is far from

TABLE 22 Selected Health and Family Planning Indicators

	Family Planning			Women		Children		
District	*Use of Any Method*	*Use of Any Modern Method*	*Unmet Need*	*4+ Antenatal Care Visits*	*Delivery at Health Facility*	*Fully Vaccinated Children*	*Weight for Age % Below 2 SD*	*Anemia Among Children <11g/dL*
Chandel	15.4	8.1	29.7	42.9	38.2	50.6	13.2	26.6
Churachandpur	23.6	14.7	23.4	60.1	63.2	52.2	11.3	17.2
Senapati	17.5	11.5	30.6	39.5	38.2	58.7	14.7	19.8
Tamenglong	16.2	9.6	30.1	32.7	33.3	44.0	16.3	21.6
Ukhrul	12.7	6.6	30.7	27.0	33.7	42.4	11.1	16.7
Bishnupur	32.7	17.5	27.5	87.0	86.5	77.2	13.7	21.4
Imphal East	30.9	15.1	27.3	82.1	80.8	72.7	17.1	20.8
Imphal West	20.4	10.9	33.1	78.7	91.4	82.8	8.4	29.3
Thoubal	24.0	13.1	33.1	79.7	73.5	65.4	17.8	24.9
Manipur	23.6	12.7	30.1	69.0	69.1	65.8	13.8	22.8

Source: NFHS-4 Manipur Report.

reality. The unmet need for family planning reaches to a level of 30%, while current use of family planning method is still very poor for the state as a whole and very low for districts like Chandel (15.4%) or Ukhrul (12.7%) in the hill and even Imphal West had the lowest among the valley districts with 20.4% of currently married women reportedly using family planning method. The NPP 2000's subgoal of achieving 80% institutional deliveries by 2010 has not been able to meet the target till 2015–16 with only 69% deliveries had institutional delivery, while in the hill districts less than 40% of deliveries are taking place in health facilities in Chandel, Senapati, Tamenglong, and Ukhrul which is even less than half of what is targeted. With regard to vaccination of children of age 12–23 months the NPP 2000 made the goal that by 2010 universal immunization (vaccine preventable diseases) of children must have been attained, but in this case too only 65.8% of the children got fully vaccinated; while only Imphal West district crossed 80% of the children, otherwise in hill districts full immunization of even half of children has been very tough. In similar line about one in every five children was anemic and this was highest in Imphal West followed by Chandel district with 29.3% and 26.6%, respectively.

From the discussion made so far it becomes quite evident that Manipur needs to bring down fertility level to replacement level. But for this according to NFHS-4 the highest level of fertility (TFR) is among the scheduled tribe (3.19) who are mainly Christians (3.18) followed by Muslims with TFR of 2.6. Only infant mortality rate shows a fulfilling situation of mortality among the NPP 2000 goals but terribly fails in containing many communicable diseases particularly HIV. Therefore, more attention is directed toward reduction of fertility and impact of communicable diseases among the younger population. Hopefully by 2025 as anticipated by National Health Policy 2017 Manipur may be able to achieve replacement-level fertility. With the change of regime under NDA government the direction of reaching toward minimum government and maximum governance of the BJP-led government many opportunities may open to supplement the socioeconomic goals to faster achieve the demographic goals.

REFERENCES

Bongaarts, J., Potter, R.G., 1983. Fertility, Biology, and Behavior: An Analysis of the Proximate Determinants. Academic Press Inc, New York.

Brown, R., 1873/2001. Statistical Account of the Native State of Manipur and the Hill Territory Under Its Rule. Mittal Publications, New Delhi.

Erdogan, E., Ener, M., Arica, F., 2013. The strategic role of infant mortality in the process of economic growth: an application for high income OECD countries. Soc. Behav. Sci. 99, 19–25. Retrieved from, https://www.sciencedirect.com/science/article/pii/S1877042813039128.

Gould, W.T.S., 2009. Population and Development. Routledge, Abingdon.

Government of India, 2009. Manual on Vital Statistics, 2009. Ministry of Statistics and Programme Implementation, Central Statistical Organisation, New Delhi.

Grierson, G.A., 1904. Linguistic Survey of India, Vol. III, Pts. II & III. Office of the Superintendent, Government Printing, India, Calcutta.

Guilmoto, C.Z., Rajan, S.I., 2013. Fertility at the district level in India: lessons from the 2011 census. Economic & Political Weekly XLVIII (23), 59–70. (appendix, pp. 1–4).

Hapt, A., Kane, T.T., 2004. Population Handbook. Population Reference Bureau, Washington.

Hodson, T.C., 2011. The Meitheis. Low Price Publications, Delhi. (Original work published 1908).

Indian Council of Medical Research, Public Health Foundation of India and Institute for Health Metrics and Evaluation, 2017. India: Health of the Nation's States—The India State-Level Disease Burden Initiative. ICMR, PHFI and IHME, New Delhi. Retrieved from, https://www.healthdata.org/sites/default/files/files/policy_report/2017/India_Health_of_the_Nation%27s_States_Report_2017.pdf. (accessed 22/8/2018).

Kamei, G., 2008. Ethnicity and Social Change: An Anthology of Essays. Akansha Publishing House, New Delhi.

Kamei, G., 2015. History of Manipur: Pre-Colonial Period. Akansha Publishing House, New Delhi.

Lee, E.S., 1966. A theory of migration. Demography 3 (1), 47–57. http://links.jstor.org/sici?sici=0070-3370%281966%293%3A1%3C47%3AATOM%3E2.0.CO%3B2-B.

NHSRC. (n.d.). Indirect Estimates of District Wise IMR and Under 5 Mortality Using Census 2011 Data—Draft. NHSRC. Retrieved from, http://nhsrcindia.org/sites/default/files/Indirect%20Estimations%20of%20Districtwise%20%20IMR%20%20U5M%20with%20Census%202011%20Data_0.pdf

Ravenstein, E.G., 1885. The laws of migration. J. Stat. Soc. Lond. 48 (2), 167–235. http://www.jstor.org/stable/2979181. accessed 25/5/2012.

United Nations, 1983. Manual X: Indirect Techniques of Demographic Estimation. New York.

FURTHER READING

Annual Report NACO, 2016–17. Chapter 24. p. 340. Retrieved from, http://naco.gov.in/documents/annual-reports. (accessed 27/8/2018).

Brown, R., 2001. Statistical Accounts of Manipur. Mittal Publications, New Delhi. (Original work published 1874).

Census of India, 1971. Series 1 India Part ii-c (i): Social and Cultural Tables. 68–86, Registrar General & Census, Commissioner of India, New Delhi, pp. 8–20.

Census of India, 1981. SERIES-I India Part IV B(i): Population by Language/Mother-Tongue (Table C-7). Registrar General & Census, Commissioner of India, New Delhi.

Census of India, 1991a. Table C-7 Part B(i): Distribution of the 96 Non-scheduled Languages India, States and Union Territories—1991.

Census of India, 1991b. Table C-7 Part A(i): Distribution of the 18 Scheduled Languages India, States and Union Territories—1991.

Census of India, 2001. Table C-16 Population by Mother Tongue.

Census of India, 2011a. Table A-2 Decadal Variation in Population Since 1901.

Census of India, 2011b. Table A-11 State Primary Census Abstract for Individual Scheduled Tribes.

Census of India, 2011c. Table C -1 Population by Religious Community—2011.

Census of India, 2011d. Table District Census Hand Book: DCHB_Village_Release_1400.

Census of India, 2011e. Table C-16 Population by Mother Tongue.

Census of India, 2011f. Table SC-14 Scheduled Caste Population by Religious Community (for Each Caste Separately)—2011.

Directorate of Economics & Statistics, Government of Manipur, 2017. Statistical Handbook of Manipur 2017. Lamphelpat: Retrieved from, http://desmanipur.gov.in/files/NewsFiles/26Oct2017020620STATISTICAL%20HANDBOOK%20of%20Manipur%202017.pdf.
Government of Manipur (n.d.) Manipur Gazette of 33 and 40 on Approval of Meetei Mayek, April 22 & 25, 1980. Retrieved from, http://www.unicode.org/L2/L2008/08136-manipur.pdf accessed 26/9/2018.
International Institute for Population Sciences (IIPS), 2014. District Level Household and Facility Survey (DLHS-4), 2012-13: India. Manipur. Mumbai IIPS.
International Institute for Population Sciences (IIPS) & ICF, 2017a. National Family Health Survey (NFHS-4), 2015-16: India. IIPS, Mumbai.
International Institute for Population Sciences (IIPS) & ICF, 2017b. National Family Health Survey (NFHS-4), India, 2015-16: Manipur. IIPS, Mumbai.
Manipur State (n.d.) AIDS Control Society. SIMS Analysis Report, 2016-17. Retrieved from, https://manipursacs.nic.in/download/SIMS_Annual_report_2016-17.pdf.
Press Information Bureau, Government of India, Ministry of Statistics & Programme Implementation (n.d.) Per Capita Income of Manipur http://pib.nic.in/newsite/PrintRelease.aspx?relid=169546 (accessed 9/9/2018).
Registrar General of India, 2017. Sample Registration System Bulletin Volume 51 No.1. Retrieved from, http://www.censusindia.gov.in/vital_statistics/SRS_Bulletins/SRS%20Bulletin%20-Sep_2017-Rate-2016.pdf.
Registrar General of India, Sample Registration System: Compendium of India's Fertility and Mortality Indicators, 1971—2013. Retrieved from, http://www.censusindia.gov.in/vital_statistics/Compendium/Srs_data.html.
Tribes-of-Manipur (n.d.), Centre Recognizes 4 Tribes of Manipur, www.ifp.co.in/page/items/6059/6059-centre-recognizes-4-tribes-of-manipur (accessed 3/9/2018).

Section X

Collective Behaviors in Ecology

Chapter 13

Deriving Mesoscopic Models of Collective Behavior for Finite Populations

Jitesh Jhawar*[,1], Richard G. Morris[†] and Vishwesha Guttal*
[*]*Centre for Ecological Sciences, Indian Institute of Science, Bengaluru, India*
[†]*National Centre for Biological Sciences, TIFR, Bengaluru, India*
[1]*Corresponding author: e-mail: jiteshjhawar@iisc.ac.in*

ABSTRACT

Animal groups exhibit many emergent properties that are a consequence of local interactions. Linking individual-level behavior, which is often stochastic and local, to coarse-grained descriptions of animal groups has been a question of fundamental interest from both biological and mathematical perspectives. In this book chapter, we present two complementary approaches to derive *coarse-grained* descriptions of collective behavior at the so-called *mesoscopic* scales, which account for the stochasticity arising from the finite sizes of animal groups. We construct stochastic differential equations (SDEs) for a coarse-grained variable that describes the order/consensus within a group. The first method of construction is based on van Kampen's system-size expansion of transition rates. The second method employs Gillespie's chemical Langevin equations. We apply these two methods to two *microscopic* models from the literature, in which organisms stochastically interact and choose between two directions/choices of foraging. These "binary-choice" models differ only in the types of interactions between individuals, with one assuming simple pairwise interactions, and the other incorporating ternary effects. In both cases, the derived mesoscopic SDEs have multiplicative/state-dependent noise, i.e., the strength of the noise depends on the current state of the system. However, the different models demonstrate the contrasting effects of noise: *increasing* the order/consensus in the pairwise interaction model, while *reducing* the order/consensus in the higher-order interaction model. We verify the validity of such mesoscopic behavior by numerical simulations of the underlying microscopic models. Although both methods yield identical SDEs for *binary*-choice systems that are effectively one-dimensional, the relative tractability of the chemical Langevin approach is beneficial in generalizations to higher-dimensions. We hope that this book chapter provides a pedagogical review of two complementary methods to construct mesoscopic descriptions from microscopic rules, how the noise in mesoscopic models is often multiplicative/state-dependent, and finally, how such noise can have counter-intuitive effects on shaping collective behavior.

Handbook of Statistics, Vol. 40. https://doi.org/10.1016/bs.host.2018.10.002

Keywords: Demographic noise, Multiplicative noise, Nonlinear dynamics, Fokker–Planck equation, Langevin equation, Noise-induced transitions, Ecology, Collective motion, Population dynamics, Collective decision-making

NOMENCLATURE

- $\boldsymbol{N}$ state of the system, consisting of number of individuals (N_i) in different states/directions/food sources ($\boldsymbol{N} \equiv \{N_1, \ldots, N_n\}^{\mathsf{T}}$)
- $\boldsymbol{x}$ state of the system, consisting of proportions of individuals (x_i) in different states/directions/food sources ($\boldsymbol{x} \equiv \{x_1, \ldots, x_n\}^{\mathsf{T}}$)
- η uncorrelated Gaussian white noise process with mean zero and variance of one
- c rate of reaction pertaining to copying/feedback interaction among individuals
- h rate of reaction pertaining to higher-order interaction among individuals
- m order parameter representing the amount of consensus among individuals to choose a direction or food source
- N total number of individuals in the system or population, also referred as *system-size*
- n total number of states/directions/food sources in the system
- N_i number of individuals in i^{th} state/direction/food source
- P_s steady-state probability density function
- r total number of reactions/rules according to which individuals can change their state from X_i to X_j
- s rate of reaction of spontaneous switching between states/directions/food sources
- X_i an individual in a given state/direction/food source i
- x_i proportion of individuals in i^{th} state/direction/food source

1 INTRODUCTION

Collective behavior is widespread in the animal kingdom, where many properties of animal groups are emergent, i.e., they often arise from simple local interactions among individuals (Couzin et al., 2002; Parrish et al., 2002; Sumpter, 2006, 2010; Vicsek et al., 1995). These emergent properties include, for example, visually fantastic patterns of flocks of starlings swirling in synchrony (Attanasi et al., 2014; Cavagna et al., 2010), locusts on the march (Buhl et al., 2006; Simpson and Sword, 2008; Uvarov, 1977), conflict resolution and consensus decision-making in groups of animals (Beckers et al., 1990; Couzin et al., 2005; Deneubourg and Goss, 1989; Petit and Bon, 2010; Pratt et al., 2002), collective navigation (Berdahl et al., 2013; Grünbaum, 1998; Guttal and Couzin, 2010), or colonies of social insects performing complex tasks (Bonabeau et al., 1997; Dussutour et al., 2006; Franks et al., 1992; Gordon, 1996; Schultheiss and Cheng, 2012; Seeley, 1989).

Understanding these emergent properties has proved to be a topic of great interest in both biology and the physical sciences (Camazine et al., 2003; Flierl et al., 1999; Guttal, 2014; Sumpter, 2010). Broadly speaking, the focus of biologists has typically been to understand how organisms interact; processing local information from their neighbors. This information is often noisy and inaccurate, yet repeated interactions among individuals can result in the aforementioned collective properties. Physicists and mathematicians, by contrast, have tried to develop simple theories to describe how local interactions between such agents can combine to result in larger scale properties of groups and populations. These two approaches are almost complementary to each other, with theory able to make connections between the observed meso/macroscopic behaviors and the likely individual-level decisions of organisms; the latter being difficult to measure directly in the natural setting of a group.

Here, we will use the word microscopic to refer to behavior at the individual level ($N = 1$), and the word macroscopic to describe the behavior of infinite sized systems ($N \to \infty$). The mesoscopic scale is in between these two, and corresponds to intermediate (large but finite) group sizes. The aim of this book chapter is to introduce readers to a couple of complementary analytical methods that help us link individual behavior to emergent properties of the collective. In doing so, our focus will be on the mesoscopic scale: aiming to understand the important role of stochasticity resulting from finite group sizes.

2 BACKGROUND

For historical context, we begin by briefly presenting the classic *Vicsek model* of collective motion (Vicsek et al., 1995), which has had a profound impact on the field of collective behavior. It has inspired a large body of theoretical and empirical research in physics, mathematics, and biology. In this simple model, we consider a population of N self-propelled particles (SPP) moving with constant speed s in a continuous space of volume L^d, where d is the spatial dimension. (For the purpose of this chapter we restrict ourselves to one and two dimensions only.) Boundary conditions are assumed to be periodic, implying that a particle leaving the boundary at the right-extreme returns to the left-extreme edge of the space, etc. Topologically, a one-dimensional system with periodic boundaries is equivalent to a ring, whereas a two-dimensional system is equivalent to a torus.

The basic rule of the model is that each individual, i, moves in the average direction of its *neighbors*, plus some small error. More specifically, at any given time t, the directions of all individuals $\{\theta_i : i = 1,\ldots,N\}$ are updated synchronously according to

$$\theta_i(t+\delta t) = \overline{\theta(t)} + \Delta\theta(t), \tag{1}$$

where the overbar represents an average over individuals within a radius r from the location of individual i, the so-called *focal* individual. The first term on the right-hand side captures how the focal individual responds to interactions with its neighbors and corresponds to calculating $\arctan\left(\overline{\sin\theta}/\overline{\cos\theta}\right)$. The second term is a noise term representing the error a focal individual makes while copying the average neighborhood direction. $\Delta\theta(t)$ is therefore a random variable, typically chosen from a uniform distribution over the range $[-\eta/2, \eta/2]$, where $\eta \leq 2\pi$. Once the directions $\theta_i(t + \delta t)$ have been calculated, each individual moves in that direction with speed s for a small time interval δt, after which all angles are recomputed.

The Vicsek model is therefore a simple model of SPPs in continuous space, where each individual updates their direction of motion synchronously at discrete time steps based on Eq. (1). Nevertheless, the model exhibits a surprisingly rich variety of spatial patterns of collective movement. For example, it is evident from Fig. 1 that ordering is facilitated by both low levels of noise and

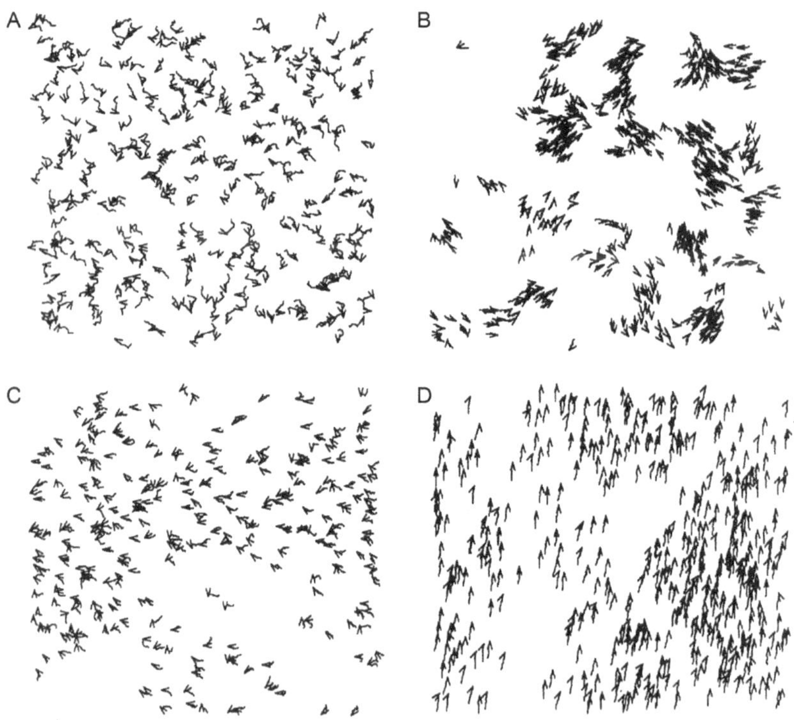

FIG. 1 Various patterns of collective motion produced by the classic Vicsek model of collective motion. There are 300 particles in each of these panels, with the *arrow* indicating the current direction of motion and the *short solid lines* preceding it indicates the trajectory of that particle over last 20 time steps. (A) High-density and large-noise ($L = 7$, $\eta = 2.0$), but initial stages of simulation. (B) Low-density and low-noise ($L = 25$, $\eta = 0.01$) lead to formation of groups that move in random directions. (C) High-density and large-noise (same as parameters in (A): $L = 7$, $\eta = 2.0$), but after some time and (D) high-density and low-noise ($L = 5$, $\eta = 0.1$) results in highly ordered motion. *Reproduced with permission from Vicsek, T., Czirók, A., Ben-Jacob, E., Cohen, I., Shochet, O., 1995. Novel type of phase transition in a system of self-driven particles. Phys. Rev. Lett. 75 (6), 1226.*

high densities of individuals. By calculating the average alignment of the individuals' motion, Vicsek defined a global "order parameter" in analogy with the polarization of a ferromagnet. Mathematically, this is given by the scalar quantity

$$m(t) = \frac{1}{N\,s}\left|\sum_{i=1}^{N} \mathbf{v}_\mathbf{i}(t)\right|, \tag{2}$$

where $\mathbf{v}_\mathbf{i}$ represents the velocity vector of individual i at time t. The normalization includes s, which is the speed of each individual. The quantity m is popularly known as the Vicsek order parameter in the physics literature; when close to 0 the group exhibits disordered motion, while when close to 1 the group is highly polarized, i.e., all individuals move in the same direction. Notably, Vicsek et al. (1995) showed via numerical simulations that the ensemble average of this quantity exhibits a phase transition in two dimensions as a function of increasing density, and decreasing noise (Fig. 2).

Since Vicsek's original paper, an array of similar SPP models of collective motion have been developed in continuous and discrete one-, two-, and three-dimensional space (Chaté et al., 2008; Chowdhury et al., 2002; Couzin et al., 2002; Czirók et al., 1999b; Grégoire and Chaté, 2004; Mogilner and Edelstein-Keshet, 1999; Nishinari et al., 2006; Strömbom, 2011). Several of these studies rely on numerical simulations to elucidate how simple microscopic interactions result in large scale collective motion. A number of analytical theories have been developed to describe the behavior of the order parameter m in Vicsek-like and other collective motion models (Czirók et al., 1999a; Ginelli, 2016; Jadbabaie et al., 2003; Toner and Tu, 1995). However, much of their focus is on understanding behavior at the macroscopic scale, i.e., in the limit of infinite population/group sizes, ignoring the role of stochastic effects arising from finite population/group sizes (Aldana et al., 2007; Baglietto and Albano, 2009; Kunwar et al., 2004; Marchetti et al., 2013; Ramaswamy, 2010; Romanczuk and Schimansky-Geier, 2012).

Here, our aim is to demonstrate a "first principles" approach to the derivation of *mesoscopic* models for the order parameter m, where noise is a function of both individual behaviors (which are stochastic) and finite size of groups. This is in contrast to the many ad hoc methods of including stochastic effects, such as simply appending a noise term to otherwise mean-field dynamics, for example. For analytical tractability, we will demonstrate the construction of such mesoscopic descriptions from microscopic rules that are simpler than the Vicsek model. Our aim is that this chapter be accessible to readers at the graduate, or even advanced undergraduate, level in numerate subjects like mathematics, physics, statistics, and quantitative biology. Although we endeavor to explain most concepts and techniques as they are introduced (occasionally referring to standard books for certain technical/detailed steps), basic familiarity of the reader with the following subjects will be helpful: dynamical-system-based analyses of ordinary differential

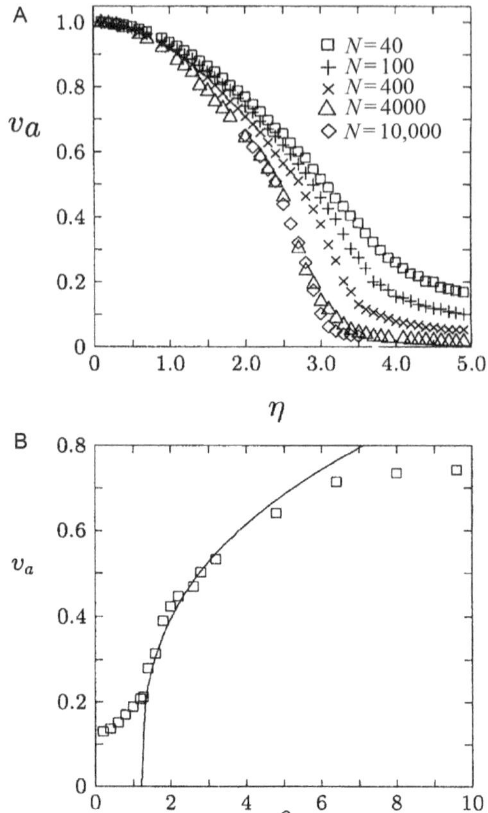

FIG. 2 Phase diagram of the Vicsek model. The order parameter (denoted as v_a in the figure; we adopt the notation m in this chapter for the equivalent quantity) which captures the degree of alignment or polarization of the entire population of particles. It shows features of phase transition from order to disorder as the strength of noise (η) increases in (A) and as the density of particles (ρ) increases in (B). See Vicsek et al. (1995) for parameter values and other details. *Reproduced with permission from Vicsek, T., Czirók, A., Ben-Jacob, E., Cohen, I., Shochet, O., 1995. Novel type of phase transition in a system of self-driven particles. Phys. Rev. Lett. 75 (6), 1226.*

equations (ODEs) (Strogatz, 2018), probability theory, stochastic differential equations (SDEs), Ito-calculus, master equations, and Fokker–Planck equations (Gardiner, 2009).

Before proceeding further, we state our goals in more specific terms. Given a microscopic model of interactions between individuals, we would like to derive an SDE for a coarse-variable that quantifies some aspect of the collective. If the coarse-variable is m, at mesoscopic scales where stochasticity is important, we expect the dynamical equation to be of the form

$$\frac{dm}{dt} = f(m) + g(m)\,\eta(t), \tag{3}$$

where $\eta(t)$ is a so-called delta-correlated Gaussian white noise with zero mean—i.e., $\langle \eta(t) \rangle = 0$ and $\langle \eta(t)\,\eta(t') \rangle = \delta(t-t')$ [see Gardiner (2009) for more details]. This is an example of an SDE and is sometimes also referred to as a Langevin equation. The function $f(m)$ is referred to as the *deterministic* or *drift* term, while $g(m)$ is often called the *stochastic* or *diffusion* term (Kolpas, 2008; Kolpas et al., 2007; Yates et al., 2009). If $g(m)$ is constant, the strength of noise is independent of the state of the system, and noise can be thought of as external, or *additive*, to the deterministic dynamics. On the other hand, if $g(m)$ is not constant, the strength of the noise depends on the current state $m(t)$. Such noise is called multiplicative, or state-dependent, and is known to produce unexpected results (Altschuler et al., 2008; Boettiger, 2018; Horsthemke and Lefever, 1984; Lawson et al., 2013). In Sections 3 and 4, we will not only describe how to derive equations of the form (3) from given probabilistic rules for individual-level behavior but also provide some heuristic insight into the different types of behaviors that can arise.

3 MESOSCOPIC DESCRIPTION OF A PAIRWISE BINARY-CHOICE MODEL

In this section, we illustrate two complementary methods for deriving mesoscopic dynamical descriptions of collective behavior from first principles. To do so, we choose a simple microscopic model of binary-choice, used to describe foraging animals that must choose between two food sources (Biancalani et al., 2014; Kirman, 1993), but also applied in several other diverse contexts, such as traders in financial markets (Alfarano et al., 2008; Kirman, 1993), and the recruitment of cell signaling molecules (Altschuler et al., 2008). Here, individuals can collect food from one of two sources, labeled by $i = 1, 2$, which can be thought of as being located at either end of a long line. The food source to which each individual is moving therefore determines its direction, either positive or negative, which is denoted by X_i. The proportion of the colony (total number of individuals, N) moving in a given direction is written using lowercase $x_i = N_i/N$, where N_i is simply the number of individuals with direction of motion X_i.

3.1 The Model

As with any model of collective decision-making, the choice that any individual makes is influenced by its *interactions* with the group (i.e., what others are doing) as well as external factors. In this section, we focus first on pairwise interactions, with higher-order interactions (specifically, ternary interactions) described in Section 4. For the case of our pairwise binary-choice model, we imagine a continuous-time stochastic protocol whereby individuals can interact in two different ways.

First, individuals can *copy* each other. For example, an interaction between a focal individual moving toward source 1 with another moving toward source 2 causes the focal individual to copy the other, and therefore switch to move toward the other source, 2. Likewise, an interaction between a focal individual of direction X_2 with another of direction X_1 causes the focal individual to switch to move toward source 1. These interactions can be represented in the same way as chemical reactions, writing:

$$X_1 + X_2 \xrightarrow{c} 2X_2, \tag{4a}$$

and

$$X_2 + X_1 \xrightarrow{c} 2X_1. \tag{4b}$$

Such copying interactions, occurring at a specific rate c, lead to an autocatalytic process wherein an individual's chance of switching directions depends on the proportion of individuals already moving in that direction.

Second, the model also incorporates a level of *random* choice, where individuals ignore the actions of the others, and switch from their current food source spontaneously at a rate s. The corresponding "chemical" reactions are

$$X_1 \xrightarrow{s} X_2, \tag{5a}$$

and

$$X_2 \xrightarrow{s} X_1. \tag{5b}$$

Since the proportions of individuals going toward either source sum to unity, i.e., $x_1 + x_2 = 1$, we can define a *coarse-variable* $m = x_1 - x_2 = 2x_1 - 1$, which takes values from -1 to 1 and encapsulates the propensity of the group to be moving to either of the food sources. The extreme values of m correspond to consensus decision-making, i.e., all individuals collectively forage from source 1 ($m = 1$) or source 2 ($m = -1$). The values of m close to zero imply a lack of consensus, or collective behavior, with equal proportions going toward either of the food sources. In-line with the previous section, we refer to m as the order parameter.

Indeed, let us compare the pairwise interaction binary-choice model (sometimes referred to as the "ant model" in the literature) with the Vicsek model. The pairwise interaction model is a highly simplified representation of collective behavior, analogs to both the classical Voter model (Cox and Griffeath, 1986) and the stepping stone model of population genetics (Kimura and Weiss, 1964). Unlike the Vicsek model, the rules are described via probabilistic reactions, rather than a difference equation. Further, here we ignore space and assume that any individual can interact with any other individual in the population. Additionally, the update protocol is asynchronous, i.e., individuals do not update their orientations simultaneously; this is in contrast to the synchronous update of orientations and positions of all individuals in the Vicsek model. Further, while the Vicsek model considers a continuous θ (the angle/direction of movement), this model assumes only two sources of food, and hence two directions. (For completeness, we present an extension to the pairwise interaction model in Appendix B, where we permit four-choices, in order to model movement in a two-dimensional space.) We also note that the order parameter of the Vicsek model ranges from 0 (disorder) to 1 (order), whereas here, it ranges from -1 to 1, with both extremes representing order (consensus) and the value $m = 0$ representing disorder (or lack of consensus).

Taken together, the aforementioned simplifications make the task of formally constructing mesoscopic equations relatively straightforward, whereas coarse-grained descriptions of the Vicsek model typically rely on symmetry arguments (Ramaswamy, 2010; Toner and Tu, 1995) and are harder to explicitly derive from the underlying microscopic model.

3.2 Constructing Mesoscopic SDEs

We describe two complementary approaches to constructing a mesoscopic dynamical equation for the coarse-grained variable m, of the form (3). The first method is based on van Kampen's system-size expansion method (Van Kampen, 1992) and is well known in the statistical physics community. Among other groups, the technique has been adopted extensively by McKane and colleagues for use in a variety of biological, ecological, and epidemiological contexts (Biancalani et al., 2014; McKane and Newman, 2004, 2005). The second approach is called the chemical Langevin equations (CLEs), developed by Gillespie (2000). The latter method is prevalent in chemistry (Sotiropoulos and Kaznessis, 2011) and to some extent in the biological literature concerning gene regulatory networks (El-Samad et al., 2006; Simpson et al., 2009), biochemical pathway analysis (Rüdiger, 2014), and epidemiology (Colizza et al., 2007; Wang et al., 2016; Yuan and Allen, 2011). However, aside from a couple of studies (Datta et al., 2010; Joshi and Guttal, 2018) it is, to our knowledge, rarely used in the ecology literature.

3.2.1 van Kampen's System-Size Expansion of Transition Rates

We begin by summarizing the step-by-step approach of this method: given a microscopic model, or protocol, the first step is to write how individual/microscopic rules translate into *transition rates* between different states of the system [denoted by the shorthand $\boldsymbol{x} = \{x_1, x_2\}^{\mathsf{T}}$]. The transition rates depend on the current state of the system and the system size (in this case, the number of individuals, N). In the second step, using these transition rates, we write a *master equation* for the temporal evolution of the probability density function (PDF) for the system's state, $P(\boldsymbol{x}, t)$. In the third step, we write the transition rates in terms of operators and Taylor-expand them for large but finite N, retaining only the two leading order terms. This provides a so-called *Fokker–Planck* equation for $P(\boldsymbol{x}, t)$, which approximates the aforementioned master equation. Then, we transform the state variables $\boldsymbol{x}$ to the coarse-variable of interest (in this case, m). Finally, we construct the SDE, or Langevin equation, that corresponds to the transformed Fokker–Planck equation.

Transition rates: For the model under consideration, since the individuals are indistinguishable, the state of the system is given by the variables x_1 and x_2: the proportion of individuals moving in the negative and positive directions, respectively. Both x_1 and x_2 take discrete values in the range $\{0, 1/N, \ldots, N - 1/N, 1\}$. Given an initial state (x_1, x_2), a single reaction of the type (4a) results in the state $(x_1 - 1/N, x_2 + 1/N)$, while (4b) results in $(x_1 + 1/N, x_2 - 1/N)$. Both transitions occur at a rate $c\, x_1 x_2$, i.e., the probability of picking two random individuals moving in different directions, multiplied by the rate at which they copy each other. Similarly, a single reaction of type (5a) also results in the state $(x_1 - 1/N, x_2 + 1/N)$, while (5b) results in $(x_1 + 1/N, x_2 - 1/N)$. Here the transitions occur at a rate $s\, x_1$ and $s\, x_2$, respectively, which is just the probability of picking a given individual, multiplied by the rate at which they switch direction. Combining the above, we write down the overall rate associated with a transition from state (x_1, x_2) to $(x_1 - 1/N, x_2 + 1/N)$, which is just

$$T_{21} = T\left(x_1 - \frac{1}{N}, x_2 + \frac{1}{N} \mid x_1, x_2\right) = c\, x_1\, x_2 + s\, x_1. \tag{6}$$

Similarly, for the transition (x_1, x_2) to $(x_1 + 1/N, x_2 - 1/N)$, we have

$$T_{12} = T\left(x_1 + \frac{1}{N}, x_2 - \frac{1}{N} \mid x_1, x_2\right) = c\, x_2\, x_1 + s\, x_2. \tag{7}$$

Here, the shorthand notation $T_{ij} = T(x_i + 1/N, x_j - 1/N \mid x_i, x_j)$ has been introduced for convenience in the coming manipulations. We also note that, due to the constraint $x_1 + x_2 = 1$, technically only one of x_1 or x_2 is required to fully determine the state of the system, a point to which we will return later.

Master equation: We describe the evolution, or rate of change, of $P(\boldsymbol{x}, t)$ by the influx of density associated with the state $\boldsymbol{x}$ from that of the other states $\boldsymbol{x}' \neq \boldsymbol{x}$, minus the outflow of density from state $\boldsymbol{x}$ to the other states $\boldsymbol{x}' \neq \boldsymbol{x}$. The resulting partial differential equation is called a master equation and is given by

$$\frac{\partial P(\boldsymbol{x},t)}{\partial t} = \sum_{\boldsymbol{x}' \neq \boldsymbol{x}} \left[T(\boldsymbol{x} \mid \boldsymbol{x}')\, P(\boldsymbol{x}',t) - T(\boldsymbol{x}' \mid \boldsymbol{x})\, P(\boldsymbol{x},t)\right] \tag{8}$$

$$\frac{\partial P(\boldsymbol{x},t)}{\partial t} = \sum_{x_1' \neq x_1} \sum_{x_2' \neq x_2} \left[T\left(x_1,x_2 \mid x_1',x_2'\right) P\left(x_1',x_2',t\right) - T\left(x_1',x_2' \mid x_1,x_2\right) P\left(x_1,x_2,t\right)\right] \tag{9}$$

$$\frac{\partial P(\boldsymbol{x},t)}{\partial t} = \sum_{x_1' \neq x_1} \sum_{x_2' \neq x_2} \left[T\left(x_1,x_2 \mid x_1',x_2'\right) P\left(x_1',x_2',t\right) - \left(T_{12} + T_{21}\right) P\left(x_1,x_2,t\right)\right]. \tag{10}$$

Step operators: The next step is to rewrite the master equation in terms of step operators, $\mathcal{E}_i^{\pm}$, which are defined as

$$\mathcal{E}_i^{\pm} f(x_i) = f\left(x_i \pm \frac{1}{N}\right). \tag{11}$$

The key idea is to use such step operators to compactly replace the sum that appears in Eq. (10). For example, the first term of Eq. (10) can be drastically simplified by using both the aforementioned step operators and the previously introduced shorthand for transition rates:

$$\begin{aligned}\sum_{x_1' \neq x_1} \sum_{x_2' \neq x_2} T(x_1,x_2 \mid x_1',x_2')\, P(x_1',x_2',t) &= T\left(x_1,x_2 \mid x_1 - \frac{1}{N}, x_2 + \frac{1}{N}\right) P\left(x_1 - \frac{1}{N}, x_2 + \frac{1}{N}, t\right) \\ &\quad + T\left(x_1,x_2 \mid x_1 + \frac{1}{N}, x_2 - \frac{1}{N}\right) P\left(x_1 + \frac{1}{N}, x_2 - \frac{1}{N}, t\right) \\ &= \mathcal{E}_1^{-} \mathcal{E}_2^{+}\, T_{12}(x_1,x_2)\, P(x_1,x_2,t) \\ &\quad + \mathcal{E}_1^{+} \mathcal{E}_2^{-}\, T_{21}(x_1,x_2)\, P(x_1,x_2,t).\end{aligned} \tag{12}$$

Note that the operators act on all functions, and all arguments therein, that appear to its right-hand side. The full master equation, now in the form of operators, is given by

$$\frac{\partial P(x_1,x_2,t)}{\partial t} = (\mathcal{E}_1^{-}\mathcal{E}_2^{+} - 1) T_{12} P(x_1,x_2,t) + (\mathcal{E}_1^{+}\mathcal{E}_2^{-} - 1) T_{21} P(x_1,x_2,t), \tag{13}$$

where the functional dependence of the rates T_{12} and T_{21} on the state (x_1, x_2) have been dropped for brevity.

Taylor expansion: We now recognize that for large N the action of the step operators may be approximated by a Taylor series. In doing so, we implicitly consider a continuous approximation of the variables x_i. Retaining only the first two terms in the expansion, we use the generic form

$$\mathcal{E}_i^{\pm} f(x_i) = f(x_i \pm \frac{1}{N}) = \left[1 \pm \frac{1}{N}\frac{\partial}{\partial x_i} + \frac{1}{2N^2}\frac{\partial^2}{\partial x_i^2}\right] f(x_i). \tag{14}$$

Applying this expansion scheme to Eq. (13) and simplifying, we get

$$\begin{aligned}\frac{\partial P(x_1,x_2,t)}{\partial t} &= \frac{1}{N}\left(\frac{\partial}{\partial x_2} - \frac{\partial}{\partial x_1}\right)[T_{12} P(x_1,x_2,t)] + \frac{1}{N}\left(\frac{\partial}{\partial x_1} - \frac{\partial}{\partial x_2}\right)[T_{21} P(x_1,x_2,t)] \\ &\quad + \frac{1}{2N^2}\left(\frac{\partial}{\partial x_1} - \frac{\partial}{\partial x_2}\right)^2 [(T_{12} + T_{21}) P(x_1,x_2,t)].\end{aligned} \tag{15}$$

Rearranging various terms, rescaling time according to $t' = t/N$ and then dropping the prime for ease of notation, we get the canonical form of a Fokker–Planck equation

$$\frac{\partial P(x_1,x_2,t)}{\partial t} = \left[-\sum_{i=1}^{2} \frac{\partial(\mathcal{A}_i P(x_1,x_2,t))}{\partial x_i} + \frac{1}{2N} \sum_{i,j=1}^{2} \frac{\partial^2 (\mathcal{B}_{ij} P(x_1,x_2,t))}{\partial x_i x_j} \right], \tag{16}$$

where $\mathcal{A}$ is vector and $\mathcal{B}$ is a matrix whose elements are given, respectively, by

$$\mathcal{A}_i = T_{ij} - T_{ji}, \quad \text{and} \quad \mathcal{B}_{ij} = (T_{ij} + T_{ji})(-1)^{i+j}. \tag{17}$$

Notice that our derivation of the Fokker–Planck equation (16) and the associated terms in Eqs. (17) make little reference to the pairwise interaction model. In other words, the above equations describing the time evolution of the PDF are based on very general features, and largely independent of the specific model under consideration. As a result, we may first apply this to the pairwise binary-choice model, before returning to reuse the same expression in Section 4 for the extension to ternary interactions. For the pairwise model, substituting Eqs. (6) and (7) into Eqs. (17), we get

$$\mathcal{A}_1 = s(x_2 - x_1) = -\mathcal{A}_2, \quad \text{and} \quad \mathcal{B}_{ij} = [2cx_1x_2 + s(x_1 + x_2)](-1)^{i+j}. \tag{18}$$

Mesoscopic SDE: Given a Fokker–Planck equation for a PDF, it is, in principle, always possible to construct a system of corresponding SDEs in order to describe the stochastic variables whose statistics are described by that PDF. For details, we refer the reader to the classic book on stochastic processes by Gardiner (2009). For the pairwise interaction model, the system of SDEs that are equivalent to a Fokker–Planck equation with the terms (18) is

$$\frac{dx_1}{dt} = \mathcal{A}_1 + \frac{1}{\sqrt{N}} \xi_1(t), \text{ and } \frac{dx_2}{dt} = \mathcal{A}_2 + \frac{1}{\sqrt{N}} \xi_2(t), \text{ where } \langle \xi_i(t) \xi_j(t') \rangle = \mathcal{B}_{ij} \delta(t - t'). \tag{19}$$

In the above SDEs, ξ_1 and ξ_2 represent noise sources that are correlated. Moreover, x_1 and x_2 are constrained by the fact that $x_1 + x_2 = 1$, and it is therefore possible to eliminate one of the two variables. To obtain a simpler SDE, which is still consistent with a Fokker–Planck description involving the coefficients (18), we therefore invoke the transformation $\xi_i = \sum_{j=1}^{2} \mathcal{G}_{ij} \eta_j$, where $\mathcal{B} = \mathcal{G}\mathcal{G}^T$ and the η_j ($j = 1, 2$) each correspond to independent delta-correlated Gaussian white noises. Because $\mathcal{B}$ and $\mathcal{G}$ are square 2×2 matrices, it is easy to see that $\mathcal{G}_{ij} = (-1)^{i+j+1} \sqrt{2cx_1x_2 + s(x_1 + x_2)}/\sqrt{2}$. Substituting this into Eqs. (19) we obtain

$$\frac{dx_1}{dt} = s(x_2 - x_1) + \sqrt{\frac{2cx_1x_2 + s(x_1 + x_2)}{2N}} [\eta_2(t) - \eta_1(t)] = -\frac{dx_2}{dt}. \tag{20}$$

Introducing $m = x_1 - x_2$ then yields

$$\frac{dm}{dt} = -2sm + \frac{1}{\sqrt{N}} \sqrt{c(1 - m^2) + 2s} [\eta_2(t) - \eta_1(t)]. \tag{21}$$

We use the fact that the difference between two variables, each drawn from independent but identical Gaussian distributions, is statistically equivalent to simply drawing from a single Gaussian distribution. Therefore, the above SDE simplifies to

$$\frac{dm}{dt} = -2sm + \sqrt{\frac{2}{N}}\sqrt{c(1-m^2)+2s}\,\eta(t). \tag{22}$$

Finally, it is helpful to move to a rescaled time $t' = 2st$. However, since the variance of the noise term depends on the timescale, care must be taken. Without wanting to digress too far on the subject of SDEs, the easiest course of action is to rescale the Fokker–Planck equation corresponding to (22) by multiplying it with $dt/dt' = 1/2s$. Relabeling t' as t and writing-down a new SDE, we get

$$\frac{dm}{dt} = -m + \sqrt{\frac{N_c}{N}}\sqrt{1-m^2+2s}\,\eta(t), \tag{23}$$

where $s/c = s$ and $N_c = 1/s$ (Biancalani et al., 2014). As before, $\eta(t)$ is delta-correlated Gaussian white noise, i.e., $\langle\eta(t)\rangle = 0$ and $\langle\eta(t)\,\eta(t')\rangle = \delta(t-t')$.

In the Section 3.3, we continue the analyses of Eq. (23) by solving for the steady-state solution of the corresponding Fokker–Planck equation and by generating sample stochastic trajectories numerically. Before doing so, however, we demonstrate an alternative, and often complementary approach to the derivation of mesoscopic SDEs, known as the chemical Langevin equation approach (Gillespie, 2000).

3.2.2 Gillespie's Chemical Langevin Approach

An alternative method for the construction of mesoscopic equations was developed by Gillespie (2000) in the context of many interacting chemical species, which provides a neat heuristic to write down SDEs directly from a set of chemical reactions. Here, as with van Kampen's approach, the method accounts for the stochasticity that arises from finite population sizes. The mathematical arguments for the approach are described in detail by Gillespie (2000), where both the similarities and differences to van Kampen's method are discussed at length. For completeness, we tersely recapitulate some of the arguments/steps in Appendix A. Given certain assumptions are satisfied, the procedure involves constructing the so-called *propensity functions* and *state change vectors/matrices*, and then simply assembling the SDE accordingly. We demonstrate this approach in the context of the pairwise interaction model described in Section 3.1.

Propensity functions: The propensity function, denoted by $a_j(\boldsymbol{x})$, represents the probability per unit time of reaction j. Here, as before, $\boldsymbol{x} = \{x_1, x_2\}^{\mathsf{T}}$, which represents the state of the system in terms of the concentration/proportion of individuals of different species. [Note that in Appendix A we follow Gillespie

(2000) and define the propensity function in terms of the absolute number of individuals (N_i) of each species present in the system, rather than concentrations, which are used here for comparison with van Kampen's approach.] The reaction propensities are computed based on the principle of mass action, i.e., that the probability of a reaction occurring at a given instant is equal to a specific rate multiplied by the product of the concentrations of all the constituent chemicals of that reaction at that time. Let us recall that the pairwise interaction model is described by the four microscopic rules or 'chemical reactions' set out in Eqs. (4a)–(5b). Consider the first reaction, which represents an individual originally moving toward source 1, who copies the behavior of another individual and changes to move toward source 2. The corresponding propensity function is just the copying rate c multiplied by the product of the proportion of individuals moving in direction 1 (x_1) and the proportion of individuals moving in direction 2 (x_2); thus, $a_1(\boldsymbol{x}) = c\,x_1x_2$. Applying this procedure to the remaining reactions [(4b), (5a), and (5b)] yields the following propensity functions:

$$
\begin{aligned}
a_2(\boldsymbol{x}) &= c\,x_1x_2,\\
a_3(\boldsymbol{x}) &= s\,x_1,\\
a_4(\boldsymbol{x}) &= s\,x_2.
\end{aligned}
$$

State change matrix: We then construct the state change, or stoichiometry, matrix $\boldsymbol{\nu}$, of which an element ν_{ji} represents the change in the number of individuals of species i when the reaction j occurs. The rows of $\boldsymbol{\nu}$ therefore correspond to the reactions and the columns represent the different species of the individuals. For the pairwise interaction model, for example, the first element in the second row represents a loss of one individual moving toward food source 1, and hence $\nu_{21} = -1$. Conversely, the second element in the second row represents a gain of an individual moving toward source 2, and hence $\nu_{22} = 1$. Accounting for all four reactions, $\boldsymbol{\nu}$ is a 4×2 matrix, given by

$$
\boldsymbol{\nu} = \begin{bmatrix} -1 & 1 \\ 1 & -1 \\ -1 & 1 \\ 1 & -1 \end{bmatrix}. \tag{24}
$$

Chemical Langevin equations: Using the propensity functions and the state change matrix, we can follow Gillespie's recipe and write down the chemical Langevin equation(s). While we refer the reader to the mathematical arguments in Gillespie (2000) and Appendix A, the basic idea is as follows. Each reaction j causes a change in the concentration of species i per unit time, which is given by the product of rate of that reaction [i.e., the propensity function $a_j(\boldsymbol{x})$] with the associated change in that species numbers (i.e., the stoichiometry matrix entry ν_{ji}). Thus, the expected rate of change in the concentration of species i due to reaction j will be $\nu_{ji}a_j(\boldsymbol{x})$. However, chemical reactions are stochastic events and the above term yields only an expected number of reactions per unit time. To take account of stochastic fluctuations, Gillespie uses the fact that, for large enough N, the change in the concentration of species i due to reaction j can be approximated by a normally distributed random variable for which the variance and the mean are both equal to the propensity of the reaction j. Applying this principle to all reactions j that result in a change in concentration of chemical i, we obtain the system of "chemical" Langevin equations

$$\frac{dx_i(t)}{dt} = \sum_{j=1}^{r} v_{ji} a_j(\boldsymbol{x}) + \frac{1}{\sqrt{N}} \sum_{j=1}^{r} v_{ji} [a_j(\boldsymbol{x}(t))]^{1/2} \eta_j(t), \quad \text{for } i = 1, \ldots, n. \tag{25}$$

where n is the total number of species/chemicals, r is the total number of reactions, N is the total number of individuals in the entire system, and η_j are the usual delta-correlated Gaussian noise sources with zero mean and unit variance.

Applying this scheme to the pairwise interaction model, we set $i = 1$, which represents the proportion of ants moving in the direction of food source 1. The first, a so-called *deterministic*, term of the Langevin equation for dx_1/dt, is found to be

$$\begin{aligned} \sum_{j=1}^{r} v_{j1} a_j(\boldsymbol{x}) &= v_{11} a_1 + v_{21} a_2 + v_{31} a_3 + v_{41} a_4, \\ &= -c x_1 x_2 + c x_1 x_2 - s x_1 + s x_2, \\ &= s(x_2 - x_1). \end{aligned}$$

Likewise, the second, *stochastic*, term can be computed as

$$\sum_{j=1}^{r} v_{j1} [a_j(\boldsymbol{x})]^{1/2} \eta_j(t) = v_{11} \sqrt{a_1} \eta_1(t) + v_{21} \sqrt{a_2} \eta_2(t) + v_{31} \sqrt{a_3} \eta_3(t) + v_{41} \sqrt{a_4} \eta_4(t), \tag{26}$$

$$\sum_{j=1}^{r} v_{j1} \left[a_j(\boldsymbol{x})\right]^{1/2} \eta_j(t) = -\sqrt{c\, x_1 x_2}\, \eta_1(t) + \sqrt{c\, x_1 x_2}\, \eta_2(t) - \sqrt{s\, x_1}\, \eta_3(t) + \sqrt{s\, x_2}\, \eta_4(t). \tag{27}$$

Putting these two expressions together, the CLE for the dynamics of the variable x_1 is given by

$$\frac{dx_1}{dt} = s\,(x_2 - x_1) + \frac{1}{\sqrt{N}} \left[-\sqrt{c\, x_1 x_2}\, \eta_1(t) + \sqrt{c\, x_1 x_2}\, \eta_2(t) - \sqrt{s x_1}\, \eta_3(t) + \sqrt{s\, x_2}\, \eta_4(t) \right]. \tag{28}$$

It is easy to see that the analogs expression for x_2 satisfies $dx_2/dt = -dx_1/dt$, which is consistent with the constraint $x_1 + x_2 = 1$. To obtain the equation for the order parameter m, we note that $m = x_1 - x_2$ and find

$$\frac{dm}{dt} = -2sm + \frac{2}{\sqrt{N}}\left[\sqrt{\frac{c(1-m^2)}{4}}(\eta_2(t)-\eta_1(t)) - \sqrt{s\frac{1+m}{2}}\eta_3(t) + \sqrt{s\frac{1-m}{2}}\eta_4(t)\right]. \tag{29}$$

By substituting c = 1, rescaling time $t' \rightarrow 2st$ and writing $t' = t$, we get

$$\frac{dm}{dt} = -m + \sqrt{\frac{N_c}{N}}\left[\sqrt{1-m^2}\eta_1'(t) - \sqrt{s(1+m)}\eta_2'(t) + \sqrt{s(1-m)}\eta_3'(t)\right]. \tag{30}$$

Since the noise sources are Gaussian and independent, the term in the square brackets is equivalent to a single noise source with a variance equal to the ℓ^2-norm of the variances of the individual noise sources, η_1', η_2', and η_3'. The result is that

$$\frac{dm}{dt} = -m + \sqrt{\frac{N_c}{N}}\sqrt{1-m^2+2s}\,\eta(t), \tag{31}$$

where $N_c = 1/s$. This is the same SDE that we derived using the system-size expansion approach (23), thus both approximations/methods yield the same mesoscopic description in this case. In Section 5, we comment on how the two methods are related, and how they do not necessarily yield the same equations for the extended version of a pairwise interaction model in two spatial dimensions—i.e., with more than two choices—as set out in Appendix B.

3.3 Characterizing Mesoscopic Dynamics

In this section, we employ three different methods to characterize the previously obtained mesoscopic SDE describing the dynamical behavior of the order parameter, m [Eqs. (23) or (31)]. Before doing so, we make some remarks on the qualitative features of the equation.

First, in the macroscopic ($N \rightarrow \infty$) limit, the system exhibits deterministic dynamics given by $dm/dt = -m$. This implies that order decays exponentially with time, reaching the deterministic stable state $m^* = 0$ asymptotically. In this state, an equal number of individuals choose food sources 1 and 2. Hence, the macroscopic system is disordered with no consensus or alignment among individuals.

Second, in the mesoscopic description (N large but finite) both the deterministic and stochastic terms of Eq. (23) are required to govern the dynamics of m. Here, the noise prefactor depends on the current state, m; such a feature is called multiplicative or state-dependent noise. In this case, the noise strength is maximum when the system is disordered (i.e., $m = 0$) and minimum when system is ordered (i.e., $m = \pm 1$). Furthermore, the strength of this noise becomes larger for decreasing system sizes.

As we show below, the interplay between a deterministic term that pulls the system toward disorder and a stochastic term that does the opposite can produce interesting and often counter-intuitive dynamics.

3.3.1 Steady-State Solution of the Fokker–Planck Equation

SDEs describe how to realize trajectories of stochastic variable(s). While often preferable for the purposes of gaining heuristic insight, or numerical implementation, they are equivalent to equations of the Fokker–Planck type, which are *partial* differential equations (PDEs) for the time-dependent probability distribution that governs the behavior of such stochastic variable(s). For example, for an SDE of the form $dm/dt = f(m) + g(m)\eta(t)$ (understood according to the Itô convention), the corresponding Fokker–Planck equation is given by Gardiner (2009)

$$\frac{\partial P(m,t)}{\partial t} = -\frac{\partial}{\partial m}[f(m)P(m,t)] + \frac{1}{2}\frac{\partial^2}{\partial m^2}\left[g^2(m)P(m,t)\right]. \tag{32}$$

In most cases, it is not possible to solve this equation analytically. However, it is often possible to solve this equation for the time-independent steady state. We set $\partial P(m, t)/\partial t = 0$ and denote the steady-state PDF by $P_s(m)$, giving

$$\frac{\partial}{\partial m}[f(m)P_s(m)] = \frac{1}{2}\frac{\partial^2}{\partial m^2}\left[g^2(m)P_s(m)\right]. \tag{33}$$

Integrating both sides with respect to m, we get

$$f(m)P_s(m) = \frac{1}{2}\frac{\partial}{\partial m}\left[g^2(m)P_s(m)\right] + c_0, \tag{34}$$

where c_0 is the integration constant. Under the assumption of no-flux boundary conditions, $c_0 = 0$. We integrate the above equation again and find that

$$P_s(m) = \frac{1}{P_0}\exp\left[2\int_{m0}^{m}\frac{f(m') - g(m')g'(m')}{g^2(m')}dm'\right], \tag{35}$$

where P_0 is the normalization constant. Applying this to the mesoscopic description of the pairwise interaction model (23) where $f(m) = -m$ and $g(m) = \sqrt{N_c/N}\sqrt{1+2s-m^2}$, we get

$$\frac{\partial P(m,t)}{\partial t} = \frac{\partial}{\partial m}[mP(m,t)] + \frac{N_c}{2N}\frac{\partial^2}{\partial m^2}\left[(1+2s-m^2)P(m,t)\right], \tag{36}$$

whose steady-state solution is given by Biancalani et al. (2014)

$$P_s(m) = \frac{1}{\mathcal{P}_0}\frac{1}{(1+2s-m^2)^{1-N/N_c}}, \tag{37}$$

where $N_c = 1/s$ and $\mathcal{P}_0$ is the normalization constant. In Fig. 3, we plot the steady-state distribution as a function of N.

3.3.2 Numerical Integration of the Mesoscopic SDE

Realizations of the stochastic trajectories of $m(t)$, governed by $P(m, t)$, can be obtained most easily by numerical integration of the SDE (23). There exist

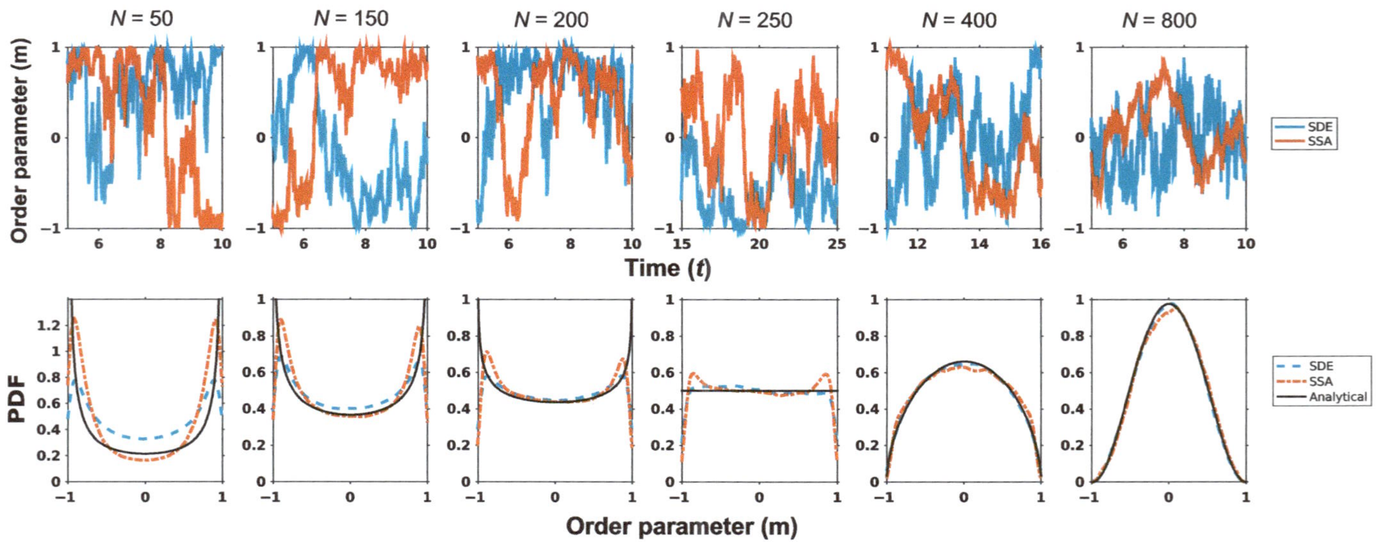

FIG. 3 Dynamics and steady-state properties of the order parameter or collective behavior (m) for the pairwise interaction model. Top row: *Blue line* indicates results from numerical integration of the analytically derived SDE Eq. (23), whereas the *orange line* shows Gillespie simulations of the model. Bottom row: Steady-state distribution of the order parameter m obtained by numerical integration of SDE (*blue* hyphenated lines), Gillespie simulations (*red* dotted lines), and the analytical solutions of the Fokker–Planck equation for the model (solid *black line*). $N_c = 250$.

more complicated schemes, but here we choose an initial value $m(t = 0) = m_0$ and employ a simple Euler–Maruyama approach (Press et al., 1996), which corresponds to the difference equation

$$\Delta m = -m\,\Delta t + \sqrt{\frac{N_c}{N}}\,\sqrt{1+2s-m^2}\,\sqrt{\Delta t}\,\eta(t), \tag{38}$$

where Δt is a small fixed time interval. In the cases where $P(m, t)$ cannot be obtained from the Fokker–Planck equation, we may repeat the above process a large number of times with different initial conditions and then plot the normalized histogram of m. The result converges, at long times, to the steady-state $P_s(m)$. In Fig. 3, we display the time series and the PDF of m for different values of system size N.

3.3.3 *Gillespie Simulations of the Microscopic Model*

In both the system-size expansion and the chemical Langevin approach, we assume finite but large systems for deriving Fokker–Planck and hence SDEs. To verify these approaches, we employ stochastic simulations (Gillespie, 1976, 1977) that are an exact representation of the master equation which describes the underlying microscopic process. While Fig. 3 presents results using the original scheme presented by Gillespie, there have since been many studies that improve both the speed and accuracy of the method. Interested readers are referred to studies such as Erban and Chapman (2009), which concerns improving the accuracy of the algorithm, and methods like τ-leaping (Gillespie, 2001) for accelerating the speed of simulations.

3.4 Results for the Pairwise Pairwise Interaction Model

We now present the results from all three approaches in order to understand both the dynamics and steady-state properties of the order parameter for the pairwise interaction model.

The steady-state solution of the Fokker–Planck equation predicts a group-size-dependent PDF of the order parameter m (bottom row of Fig. 3). Specifically, for large group sizes ($N > N_c$) the PDF $P_s(m)$ is unimodal with the mode occurring at $m = 0$. This implies that the most likely state is a disordered state consistent with the deterministic fixed point of the macroscopic/deterministic description. By contrast, for small group sizes $N < N_c$, the steady-state PDF $P_s(m)$ is *bi*modal, with modes at $m = \pm 1$. Therefore, the system spends most of its time in highly ordered states. These two states correspond to a consensus decision that results in clear collective movement of individuals in a single direction.

The PDF obtained from repeated numerical integration of the SDE (23) produces qualitatively similar results as the analytical solution of the Fokker–Planck equation indicating minimal transient dynamics before reaching the steady state. The advantage of numerical simulation of the SDE is that we can also visualize the time series of the order parameter m (blue line in top panel, in Fig. 3). We find that, for smaller group sizes, m frequently switches back and forth between two ordered states at $m = 1$ and $m = -1$. However, this is clearly not the case with larger groups, which spend large amounts of time in a disordered $m = 0$ state.

As expected, both the steady-state solution to the Fokker–Planck equation and repeated numerical integration of the SDE qualitatively match with the Gillespie simulations. Of course, as system size decreases, the results diverge slightly, which highlights the fact that the mesoscopic descriptions are only approximations of the underlying microscopic behavior.

4 TERNARY INTERACTION MODEL FOR BINARY CHOICE

We now present a simple extension of the pairwise interaction model that includes higher-order interactions. This model of collective behavior was originally developed by Dyson et al. (2015) to explain the quantitative behavior of a march of locusts in a one-dimensional arena (Buhl et al., 2006) and incorporates interactions between three individuals in addition to the previously discussed pairwise interactions; we refer to this model as the "ternary interaction model." Once again, we apply both the system-size expansion and chemical Langevin approaches to derive a mesoscopic description.

As in the pairwise model, individuals move toward food source 1 or 2 depending on both their interactions with others and random factors. We once again assume that individuals spontaneously switch at a rate s between food sources/directions

$$X_1 \xrightarrow{s} X_2, \tag{39a}$$

$$X_2 \xrightarrow{s} X_1. \tag{39b}$$

Likewise, we also retain the copying interactions, occurring at a rate c, of the pairwise interaction model, given by

$$X_1 + X_2 \xrightarrow{c} 2X_1, \tag{40a}$$

$$X_2 + X_1 \xrightarrow{c} 2X_2. \tag{40b}$$

The new component of this model is a higher-order interaction, where three individuals interact. The reaction occurs at a specific rate h, and results in all individuals moving in the direction of the majority. Depending on whether X_1 or X_2 was in the majority at the time of interactions, one of the two following reactions occur

$$2X_1 + X_2 \xrightarrow{h} 3X_1, \tag{41a}$$

$$2X_2 + X_1 \xrightarrow{h} 3X_2. \tag{41b}$$

We note that Schulze and Stauffer (2005) use a similar model to study evolution of language. In that model, a focal individual modifies its language spontaneously at a rate p, copies the language of a randomly chosen individual at a rate q, and finally, with a probability $(1-x)^2 r$, the individual chooses the language spoken by the fraction x of the population. This last reaction is identical to the ternary interaction set out in Eqs. (41).

4.1 Constructing a Mesoscopic SDE for the Ternary Interaction Model

As we did before with the pairwise interaction model, we now apply both the system-size expansion and the chemical Langevin equation methods in order to derive a mesoscopic equation for the order parameter $m = x_1 - x_2$.

4.1.1 van Kampen's System-Size Expansion of Transition Rates

Transition rates: The state variables of the system are again x_1 and x_2: the proportion of individuals moving in the direction of food sources 1 and 2, respectively. Similarly, since $x_1 + x_2 = 1$, one of either x_1 or x_2 can completely describe the state of the system. The transition rates T_{21} and T_{12} are the same as in the pairwise interaction model, except they must also incorporate reactions (41a) and (41b). The additional rates follow the "mass action" rationale set out previously, giving

$$T_{21} = T\left(x_1 - \frac{1}{N}, x_2 + \frac{1}{N} | x_1, x_2\right) = s x_1 + c x_1 x_2 + h x_1 x_2^2, \tag{42}$$

and

$$T_{12} = T\left(x_1 + \frac{1}{N}, x_2 - \frac{1}{N} | x_1, x_2\right) = s x_2 + c x_1 x_2 + h x_1^2 x_2. \tag{43}$$

Fokker–Planck equation: Next, we substitute these transition rates into the general form of the Fokker–Planck equation for binary-choice models (15), where we note that the steps from Eq. (10) to (15) are independent of model interaction details. Therefore, we get

$$\frac{\partial P(x_1,x_2,t)}{\partial t} = \left[-\sum_{i=1}^{2} \frac{\partial(\mathcal{A}_i P(x_1,x_2,t))}{\partial x_i} + \frac{1}{2N}\sum_{i,j=1}^{2} \frac{\partial^2(\mathcal{B}_{ij} P(x_1,x_2,t))}{\partial x_i x_j}\right], \tag{44}$$

where,

$$\mathcal{A}_1 = T_{12} - T_{21} = s(x_2 - x_1) + h x_1 x_2 (x_1 - x_2), \tag{45}$$

$$\mathcal{A}_2 = -\mathcal{A}_1, \tag{46}$$

$$\mathcal{B}_{ij} = (T_{ij} + T_{ji})(-1)^{i+j} = (-1)^{i+j}[s(x_1 + x_2) + 2c x_1 x_2 + h x_1 x_2 (x_1 + x_2)]. \tag{47}$$

Mesoscopic SDE: The coupled SDEs that correspond to realizations of the above Fokker–Planck equation are given by

$$\frac{dx_1}{dt} = \mathcal{A}_1 + \frac{1}{\sqrt{N}}\xi_1(t), \text{ and } \frac{dx_2}{dt} = \mathcal{A}_2 + \frac{1}{\sqrt{N}}\xi_2(t), \text{ where } \langle \xi_i(t)\xi_j(t')\rangle = \mathcal{B}_{ij}\delta(t-t'). \quad (48)$$

Using the same set of manipulations that we followed while analyzing the pairwise interaction model, we obtain the mesoscopic equation for the order parameter m for the ternary interaction model. To do so, we use the transformation: $\xi_i = \sum_{j=1}^{2} \mathcal{G}_{ij}\eta_j$, such that $\mathcal{G}$ satisfies $\mathcal{B} = \mathcal{G}\mathcal{G}^T$. We find that

$$\mathcal{G}_{ij} = (-1)^{i+j+1}\sqrt{s(x_1+x_2) + 2cx_1x_2 + hx_1x_2(x_1+x_2)}/\sqrt{2}.$$

After further transforming to $m = x_1 - x_2$, we obtain the mesoscopic dynamical equation (Dyson et al., 2015)

$$\frac{dm}{dt} = -2sm + \frac{h}{2}m(1-m^2) + \frac{2}{\sqrt{N}}\sqrt{s + \frac{2c+h}{4}(1-m^2)}\,\eta(t), \quad (49)$$

which, under the time rescaling $t' = 2st$, recovers the mesoscopic equation of the pairwise interaction model on setting $h = 0$.

4.1.2 Chemical Langevin Approach

As in Section 3.2.2, we define both propensity functions $a_j(\boldsymbol{x})$ and the state change, or stoichiometry, vector $\boldsymbol{\nu}$ for the ternary interaction model (Table 1). To derive the relevant ternary mesoscopic equation, we substitute both $a_j(\boldsymbol{x})$ and $\boldsymbol{\nu}$ from Table 1 into the chemical Langevin equations,

$$\frac{dx_i(t)}{dt} = \sum_{j=1}^{r} \nu_{ji}a_j(\boldsymbol{x}) + \frac{1}{\sqrt{N}}\sum_{j=1}^{r} \nu_{ji}[a_j(\boldsymbol{x})]^{1/2}\eta_j(t), \quad \text{for } i = 1, \ldots, n, \quad (50)$$

TABLE 1 Reactions in the Higher-Order Interaction Model, Their Propensities, and the State Change Vector

Serial No. (j)	Reaction	Propensity	ν_{j1}	ν_{j2}
1	$X_1 \xrightarrow{s} X_2$	sx_1	−1	1
2	$X_2 \xrightarrow{s} X_1$	sx_2	1	−1
3	$X_1 + X_2 \xrightarrow{c} 2X_1$	cx_1x_2	1	−1
4	$X_1 + X_2 \xrightarrow{c} 2X_2$	cx_1x_2	−1	1
5	$2X_1 + X_2 \xrightarrow{h} 3X_1$	$hx_1^2x_2$	1	−1
6	$2X_2 + X_1 \xrightarrow{h} 3X_2$	$hx_1x_2^2$	−1	1

where n is the number of species, r is the number of reactions, N is the total number of individuals, and η_j are Gaussian white noise sources with zero mean and correlation $\langle \eta_i(t)\eta_j(t') \rangle = \delta_{ij}\delta(t-t')$.

As an example, consider the variable x_1. The deterministic term of the CLE is given by

$$\sum_{j=1}^{r} \nu_{j1} a_j(\boldsymbol{x}) = \nu_{11}a_1 + \nu_{21}a_2 + \nu_{31}a_3 + \nu_{41}a_4 + \nu_{51}a_5 + \nu_{61}a_6$$

$$= -sx_1 + sx_2 + cx_1x_2 - cx_1x_2 + hx_1^2x_2 - hx_1x_2^2$$

$$= s(x_2 - x_1) + hx_1x_2(x_1 - x_2).$$

The stochastic term is given by

$$\sum_{j=1}^{r} \nu_{j1}[a_j(\boldsymbol{x})]^{1/2} = \nu_{11}\sqrt{a_1}\eta_1(t) + \nu_{21}\sqrt{a_2}\eta_2(t) + \nu_{31}\sqrt{a_3}\eta_3(t) + \nu_{41}\sqrt{a_4}\eta_4(t)$$

$$+ \nu_{51}\sqrt{a_5}\eta_5(t) + \nu_{61}\sqrt{a_6}\eta_6(t)$$

$$= -\sqrt{sx_1}\eta_1 + \sqrt{sx_2}\eta_2 + \sqrt{cx_1x_2}\eta_3 - \sqrt{cx_1x_2}\eta_4 + \sqrt{hx_1^2x_2}\eta_5 - \sqrt{hx_1x_2^2}\eta_6.$$

As before, it is easy to show that $dx_1/dt = -dx_2/dt$, because $\nu_{j1} = -\nu_{j2}$, for all j. As noise sources η_1 to η_6 are independent we may combine them into a single noise source to obtain

$$\sum_{j=1}^{r} \nu_{j1}[a_j(\boldsymbol{x})]^{1/2} = \sqrt{s(x_1+x_2) + 2cx_1x_2 + hx_1x_2(x_1+x_2)}\eta(t).$$

Hence, the CLE for the higher-order interaction model in terms of variable x_1 and x_2 is given by

$$\frac{dx_1}{dt} = s(x_2 - x_1) + hx_1x_2(x_1 - x_2) + \frac{1}{\sqrt{N}}\sqrt{s(x_1+x_2) + 2cx_1x_2 + hx_1x_2(x_1+x_2)}$$

$$\eta(t) = -\frac{dx_2}{dt}. \tag{51}$$

Substituting $m = x_1 - x_2$ and using the constraint $x_1 + x_2 = 1$, we get

$$\frac{dm}{dt} = -2sm + \frac{h}{2}m(1-m^2) + \frac{2}{\sqrt{N}}\sqrt{s + \frac{2c+h}{4}(1-m^2)}\eta(t). \tag{52}$$

This mesoscopic dynamical equation in m, for the one-dimensional ternary interaction model, derived using the chemical Langevin approach is exactly the same equation derived using the system-size expansion method. Therefore, we again see that both methods give exactly same results for one-dimensional/univariate models.

4.2 Characterizing Mesoscopic Dynamics

We now turn to characterizing the mesoscopic SDE (49) of the ternary interaction model both numerically and analytically. The methods are similar to the pairwise interaction model (Section 3.3).

4.2.1 Steady-State Solution of the Fokker–Planck Equation

The deterministic and the stochastic terms of the mesoscopic equation for the ternary interaction model are $f(m) = -2s\ m + h\ m(1 - m^2)/2$ and $g(m) = 2\sqrt{(s + (2c+h)(1-m^2)/4)/N}$, respectively. Therefore, the Fokker–Planck equation for the ternary interaction model is given by

$$\frac{\partial P(m,t)}{\partial t} = \frac{\partial}{\partial m}\left\{\left[2sm - \frac{h}{2}m(1-m^2)\right]P(m,t)\right\} + \frac{2}{N}\frac{\partial^2}{\partial m^2}\left\{\left[s + \frac{2c+h}{4}(1-m^2)\right]P(m,t)\right\}. \tag{53}$$

Using the general expression for steady-state PDF given by Eq. (35) (Gardiner, 2009), we get

$$P_s(m) = \frac{1}{P_0}[4s + (2c+h)(1-m^2)]^{\frac{4Ns(c+h)}{(2c+h)^2}-1} e^{\frac{hm^2N}{2(2c+h)}}, \tag{54}$$

where P_0 is the normalization constant (Dyson et al., 2015). In Fig. 4, we plot this expression for different values of N.

4.2.2 Numerical Integration of the Mesoscopic SDE

We applied the Euler–Maruyama scheme to the SDE (49) in exactly the same way as Section 3.3.2. The time series and the probability density function thus obtained are shown in Fig. 4 for different values of N.

4.2.3 Gillespie Simulations of the Microscopic Model

We implement the Gillespie algorithm in the same way as described in Section 3.3.3. We choose the sampling time as $1/s$, independent of system size. We later rescaled the time $t = t'/N$ to match the timescale of the mesoscopic equation (49). The results of these simulations are presented in Fig. 4.

4.3 Results for the Ternary Interaction Model

The steady-state solution of the Fokker–Planck equation is bimodal for all group sizes (black line in bottom row, Fig. 4). The two modes of the distribution are, however, less distinct for small system sizes, becoming more prominent with increasing system size. Increasing the rate of higher-order interactions (h) moves the location of the distribution modes to larger values

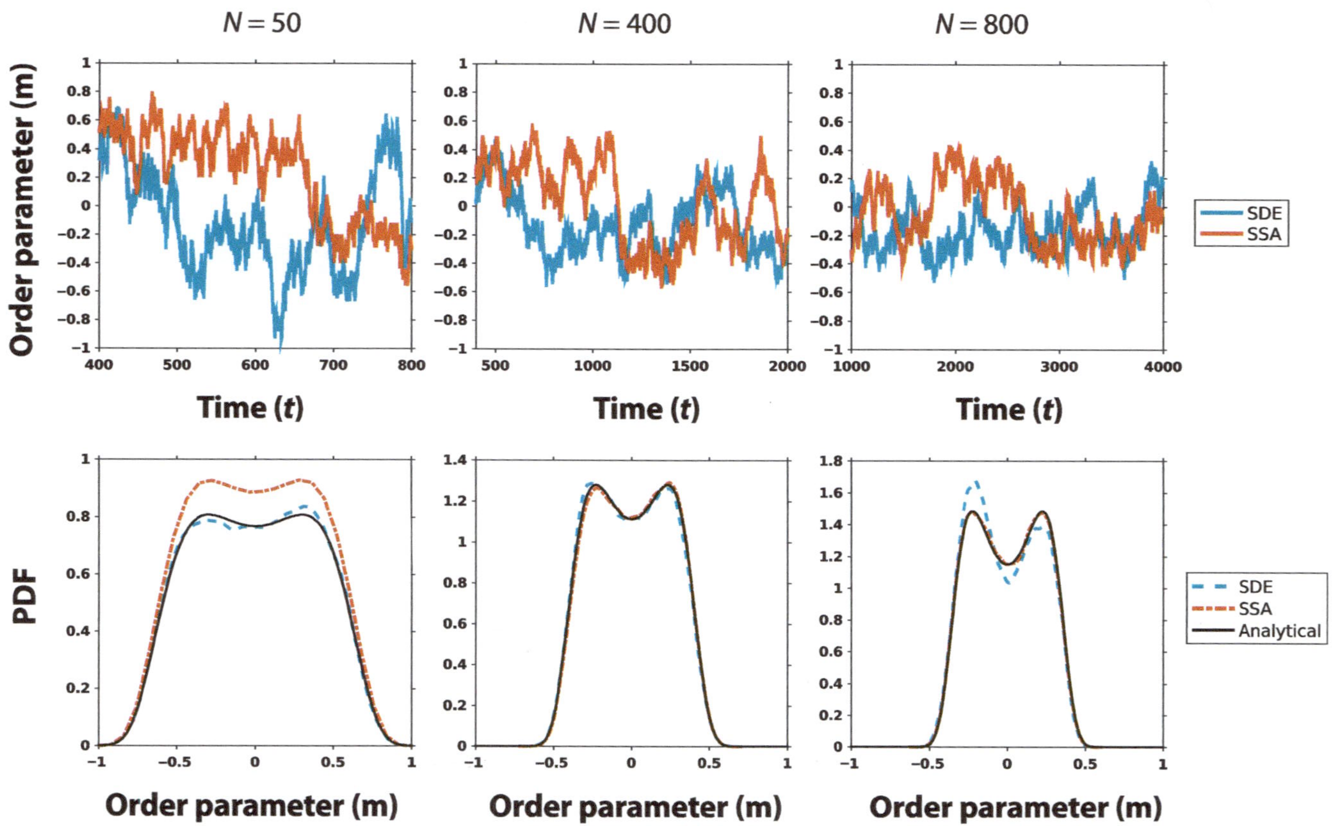

FIG. 4 Dynamics and steady-state distributions of order parameter for the higher-order interaction model Top row: Time series of order parameter (m) for three group size ($N = 50, 400, 800$) from the numerical integration of the stochastic differential equation (SDE) (49), shown in and from the stochastic simulation algorithm (SSA) method (). Bottom row: Corresponding steady-state probability density functions of m obtained from numerical solutions of SDE, SSA, and the analytical solution of the Fokker–Planck Eq. (53), shown in. Parameter values: $s = 0.05$, $c = 0.005$, $h = 0.21$.

of m. This implies that the system size only has quantitative but not qualitative influence on the steady-state distribution of the order parameter. The effect is therefore in sharp contrast to the pairwise interaction model, where increasing system size resulted in a transition from bimodal to unimodal distribution of m. Specifically, in the pairwise interaction model we observed two clear modes for small systems whereas, for the ternary interaction model, this corresponds to large system sizes.

To understand these results, consider the deterministic limit of the ternary interaction model's SDE. The $N \to \infty$ limit of Eq. (49) has three fixed points $(0, \pm\sqrt{(h-4s)/h})$. For $h > 4s$, three real roots exist, of which $m^* = \pm\sqrt{(h-4s)/h}$ are stable since $f'(m^*) < 0$. Visual inspection of two modes in the steady-state PDF reveals that they are indeed near the location of these two deterministic stable fixed points of the dynamical equation for all values of the system size N. This is unlike the pairwise interaction model where the equation has only one stable fixed point at $m = 0$, but nevertheless has a bimodal PDF whose modes are away from $m = 0$; an effect referred to as noise-induced bistability (Horsthemke and Lefever, 1984).

The effect of system size N in the mesoscopic description is captured in the stochastic term of (49), which is $\mathcal{O}(1/\sqrt{N})$. At small system sizes, the strength of stochasticity is relatively large and therefore, the order parameter is constantly pushed away from the stable equilibria $m^* = \pm\sqrt{(h-4s)/h}$. As a result, when $N = 50$ we observe a relatively large spread around the two stable states in the distribution of m (see bottom row in Fig. 4). On the other hand, for large system size the strength of stochasticity is less and thus system resides longer in the stable states. Consequently, the distribution shows two clear modes corresponding to deterministic stable equilibria in the system in low-noise/large system size case.

We can explore the temporal evolution of the order parameter, m, by numerically integrating Eq (49) (blue line in Fig. 4) or by Gillespie simulations (orange line in Fig. 4). For small group sizes m does not reside evidently in the stable states, moving constantly across different states (top row of Fig. 4). As the group size increases, the system resides longer in the two stable states ($m^* = \pm\sqrt{\frac{h-4s}{h}}$), exhibiting occasional transitions between them. This effect was further quantified by Dyson et al. (2015) by calculating the mean residence time in the two stable states, which increases with group size. The results from all three approaches (the analytical approach, the SDE simulations and the Gillespie simulations) to solving the mesoscopic dynamics of the ternary interaction model qualitatively match with each other, with slight discrepancy for small groups sizes (bottom row in Fig. 4). This is not surprising since the Langevin and the Fokker–Planck equations approximates the dynamics for large but finite systems, but Gillespie simulations are exact representation of the underlying process.

5 DISCUSSION

In this book chapter we presented two approaches from the literature, the system-size expansion method (Van Kampen, 1992) and the chemical Langevin equation method (Gillespie, 2000), to construct mesoscopic dynamical equations of collective behavior from microscopic rules. These methods allow us to capture how, for finite population sizes, stochasticity at the microscopic scale manifests at the mesoscopic scale. The resulting behavior is captured by SDEs, sometimes referred to as Langevin equations. We demonstrated applications of these approaches to two simple models of collective decision-making involving individuals that must make a binary choice; in both cases, both methods yield identical Langevin equations. In the following Section 5.1 we therefore compare various technical aspects of the two complementary methods. In Section 5.2, we discuss how the aforementioned techniques reveal important differences regarding the two models we considered; one considers only pairwise interactions between individuals, while the other incorporates higher-order interactions. Finally, in Section 5.3, we discuss some extensions of the model.

5.1 Comparison of System-Size Expansion With Chemical Langevin Approach

Both methods of constructing coarse-grained descriptions at mesoscopic scales are based on similar sets of assumptions. The system-size expansion method involves a formal procedure that begins with the master equation and then expands its terms in a small quantity: the inverse of the system size, $1/N$. By assuming N large and thereby keeping only two leading order terms, we obtain a Fokker–Planck equation, which describes the temporal evolution of the PDF of the order parameter, m. This is equivalent to assuming that the rate of change in the probability distribution for m is fully determined by its first and second jump moments, and hence that the noise is Gaussian in character. In contrast to this formal approach, the chemical Langevin method is based on heuristics that keep track of both the probability of each "chemical reaction" (or interaction), and how each reaction changes the state of the system. To do this, the timescale for reactions is so chosen that the changes in the number of individuals of each species can be approximated by a Gaussian/normal distribution. Thus, the two methods effectively make the same set of assumptions about the noise process that governs mesoscopic dynamics of the coarse-variable. [A more detailed discussion is provided in Gillespie (2002)]. Given this equivalence, it is not surprising that two methods do indeed yield identical Fokker–Planck equations and hence SDEs for the two univariate models under consideration.

More generally, the system-size expansion yields a Fokker–Planck equation, while the chemical Langevin approach results in one or more coupled SDEs. There is, of course, a formal equivalence between Fokker–Planck equations and SDEs. However, the decomposition of a Fokker–Planck equation into SDEs is not unique for multivariate systems. Specifically, this requires the decomposition of a known matrix—the matrix of second jump moments—into the product of an unknown matrix and its transpose. There exist many methods for such a decomposition, the Cholesky decomposition being a popular choice that ensures the unknown matrix is square but requires that the matrix of second jump moments is positive definite. In this case, the result is a set of coupled SDEs that involve the minimum number of possible noise sources (equal to the number of independent degrees-of-freedom required to fully describe the state of the system). By contrast, the chemical Langevin approach results in a system of coupled SDEs which involve as many noise sources as there are "chemical reactions." Typically, the fewer noise terms of the Cholesky decomposition involve cumbersome and hard-to-simplify prefactors, while the many terms of the chemical Langevin equations have relatively straightforward prefactors. We demonstrate this point in Appendix B where we consider an extension of the binary-choice to four-choices, representing collective movement in two dimensions rather than one. Using the system-size expansion to obtain a Fokker–Planck equation and then employing a Cholesky decomposition results in SDEs that are, for all intents and purposes, intractable, due to the complexity of the noise prefactors. However, by comparison, the chemical Langevin equations are significantly easier to construct and analyze.

We nevertheless stress that the two descriptions, as presented, *are* statistically the same, since they correspond to the same Fokker–Planck equation. The difference is purely esthetic, and arises in multivariate systems due to the nonuniqueness of mapping a Fokker–Planck equation to a set of SDEs. However, one area, so far not discussed, where the two approaches do differ is the so-called linear noise approximation (LNA) (Van Kampen, 1992). The LNA is attributed to van Kampen and typically used in conjunction with the expansion of transition rates in inverse system-size. It recognizes that, formally, expansions of this type should be accompanied by an ansatz regarding the N-dependence of the underlying variables. If the variance of the underlying variables is expected be proportional to $\sqrt{N}$, then this affects how the various terms of the expansion equate at lowest order, and results in mesoscopic descriptions that are *additive* in noise, rather than multiplicative. By contrast, if the variance is not expected to be Gaussian-like, then it does not affect the expansion at lowest order, and hence the LNA is not required.

5.2 Multiplicative Noise at Mesoscopic Scales

To better understand the mesoscopic descriptions that result from the aforementioned approaches, we considered two simple models of collective behavior from the literature, where individuals must make a binary choice between two alternatives. The models differ only in the way individuals interact. In the first case we assumed pairwise interactions (Biancalani et al., 2014;

Kirman, 1993; Lux, 1995), while in the second case three individuals were permitted to interact at any given time (Dyson et al., 2015). In the literature, Biancalani et al. (2014) and Dyson et al. (2015) used the system-size expansion approach to derive mesoscopic equations for such models. Here, we recapitulate their results and, in addition, demonstrate the chemical Langevin equation approach; the latter having been used extensively for analyzing chemical reactions and biochemical networks, but much less in the ecology literature (Datta et al., 2010).

From this exercise, we may learn a number of interesting points about noise at mesoscopic scales. First and foremost, for the simple binary-choice models considered, the SDEs for the coarse-variable (consensus/order) contain multiplicative noise, i.e., where the strength of stochasticity depends on the current state of the system. In other words, irrespective of the method of derivation, the resulting SDE is of the form $dm/dt = f(m) + g(m)\eta(t)$, with $g(m) \neq$ constant. Second, studying these two models (both pairwise and ternary interactions) together provided an opportunity to highlight how the nature of interactions can subtly influence collective behavior at mesoscopic scales. In both models, the noise term $g(m)$ decreases with group size by $\mathcal{O}(1/\sqrt{N})$, however, the effect of noise on collective behavior contrasts sharply between the two. Perhaps counter-intuitively, increasing the strength of noise in the SDE of the pairwise interaction model, actually increases order, despite the ostensibly randomizing effects of noise. This is exactly the opposite for the ternary interaction model, where increasing noise reduces the order, or level of consensus, among individuals.

To understand this quantitatively, consider the deterministic limit ($N \rightarrow \infty$), in which $dm/dt = f(m)$. Here, we expect the system to asymptotically reach one of the deterministic stable states, m^*, given by $f(m^*) = 0$ and $f'(m^*) < 0$. However, when the system is driven by multiplicative noise we have $dm/dt = f(m) + g(m)\eta(t)$, where the form of $g(m)$ determines the most probable states of the system (Horsthemke and Lefever, 1984). Crucially, these do *not* necessarily correspond to the deterministic stable states, as they would if the noise were simply additive [i.e., $g(m) =$ constant]. This is precisely the case for the pairwise interaction model; the deterministic stable state corresponds to $m^* = 0$, but the most likely states (i.e., the modes of the distribution) are close to $m^* \pm 1$. Thus, the multiplicative noise not only moves the system away from its deterministic stable state but also increases order. By contrast, in the higher-order (ternary) interaction model, the most likely states are indeed around or close to deterministic stable states.

5.3 Extensions and Concluding Remarks

In this chapter, we focused only on two simple models of collective behavior where interactions rules were either pairwise or at most ternary. Further, we focused on nonspatial models and assumed that the system is fully connected i.e. any individual can interact with any other individual. We deliberately chose these simple models because the main purpose of this chapter was to illustrate two complementary mathematical methods to deriving mesoscopic dynamics of the collective behavior; elucidating how stochasticity arising from individual behaviors is amplified at mesoscopic scales by small group sizes.

An obvious extension of this work includes considering more complicated higher-order interactions; many empirical studies quantify nature of interactions suggest that interactions (Bialek et al., 2012; Biro et al., 2006; Gautrais et al., 2012; Herbert-Read et al., 2011; Jiang et al., 2017; Katz et al., 2011; Mann et al., 2013) with some suggesting that interactions among flock-members can extend to several members, for example, up to seven in starling flocks (Ballerini et al., 2008). For better representation of realistic flock dynamics, a continuous state/direction, which is in contrast to discrete states we have considered so far, could be incorporated (Gavagnin and Yates, 2018; Othmer et al., 1988). However, analysis of continuous state models may require more sophisticated analytical tools than presented in this chapter. Further, the emergence of collective behavior at mesoscopic scales, which is absent at macroscopic scales, is also of relevance to understand the evolution of collective behavior. Much of the previous work concerning the evolution of social behavior in animals has involved either deterministic game theory, which assumes infinite population sizes (Krause et al., 2002; Sumpter, 2010; Torney et al., 2010), or simulations of large numbers of particles (Guttal and Couzin, 2010; Guttal et al., 2012; Reluga and Viscido, 2005; Spector et al., 2005; Wood and Ackland, 2007). Recent studies have indeed begun to highlight the role of stochasticity arising from small group sizes and finite population sizes (Joshi and Guttal, 2018; Joshi et al., 2017) where mesoscopic descriptions are important.

Further, our assumption of fully connected systems may be valid for small groups of animals, which is the focus of our paper, but for large groups, accounting for the local nature of interactions becomes important. Mesoscopic descriptions of such large but finite systems have been described for a small number of cases, and result in stochastic partial differential equations (Biancalani et al., 2010; Dean, 1996; McKane et al., 2014). On the other hand, the macroscopic or hydrodynamic descriptions of the collective motion models discussed in the first section have been well studied (Marchetti et al., 2013; Ramaswamy, 2010; Toner and Tu, 1995). It would therefore be interesting to understand how mesoscopic descriptions, which focus on capturing stochasticity arising from finite sized systems, dovetail with hydrodynamic descriptions at ostensibly larger length scales.

In summary, we presented two complementary approaches to derive mesoscopic descriptions of simple collective behavior models. It is worth noting, however, that the methods we presented are applicable to any ecological model that are based on asynchronous stochastic update rules and where spatial structure is not important. For example, these methods can also be useful to deriving mesoscopic description of population and community dynamics which captures demographic stochasticity (Majumder, 2018). While van Kampen's system-size expansion has indeed been applied to study simple ecological and evolutionary scenarios (Black and McKane, 2012; McKane and Newman, 2005; Traulsen et al., 2012), we have highlighted the difficulty of employing this method for studying multispecies interactions. In such cases, we suggest that the method of chemical Langevin approach could be powerful. Therefore, we hope that our book chapter provides a pedagogical review of mathematical methods for describing mesoscopic dynamics that are useful not only for collective behavior models but also for other biological dynamics where finite sizes of populations/groups is important.

5.4 Resources

The codes used for simulating the pairwise interaction and the ternary interaction models using Gillespie simulations and the corresponding mesoscopic stochastic differential equations can be found at GitHub (https://tinyurl.com/y8zvauqx).

ACKNOWLEDGMENTS

We thank Sabiha Majumder, Sumithra Sankaran, Jaideep Joshi, Jeffrey Phillippson, and Appilineni Kushal for various insightful discussions. We are thankful to Shubham Rana for useful comments on the chapter. J.J. acknowledges support by the Council for Scientific and Industrial Research, India for research scholarship. R.G.M. acknowledges support from Simons Foundation. V.G. acknowledges support from DBT-IISc partnership program, Science and Engineering Research Board (DST, Government of India) and DST-FIST for infrastructure support.

APPENDICES

APPENDIX A. THE CHEMICAL LANGEVIN EQUATION

In this appendix, we briefly recapitulate the approach developed by Gillespie (2000). We consider n chemicals/species $\{X_1, \ldots, X_n\}$ whose interactions are represented via r chemical reactions $\{R_1, \ldots, R_r\}$. We define the state of the system by $\boldsymbol{N}(t) \equiv \{N_1(t), \ldots, N_n(t)\}^{\mathrm{T}}$, where $N_i(t)$ denotes the number of X_i molecules in the system at time t. Next, we define a propensity function $a_j(\boldsymbol{N}(t))$, which represents the probability that for a given state $\boldsymbol{x}(t) = \boldsymbol{N}(t)/N$, the reaction R_j will occur within the next infinitesimal time interval $t + dt$. Here, $N = \sum_i N_i$ is the total number of individuals in the system. We also define the state change matrix $\boldsymbol{\nu}_j$ whose i^{th} component ν_{ji} is defined by the change in the number of X_i molecules due to the occurrence of reaction R_j. For example, consider the reaction $R_1 : X_1 + X_2 \rightarrow 2X_1$ and the reverse reaction $R_2 : 2X_1 \rightarrow X_1 + X_2$. The propensity function for the reaction R_1 would be $c_1N_1N_2/N^2$, and for the reaction R_2 it would be $c_2N_1(N_1 - 1)/2N^2$. The state change matrix would be $\boldsymbol{\nu}_1 = (+1, -1, 0, \ldots, 0)$, such that $\boldsymbol{\nu}_2 = -\boldsymbol{\nu}_1$.

Gillespie (2000) points out that when two fairly generic dynamical conditions are satisfied, the propensity functions can be used to write down SDEs that describe the temporal evolution of the state variables $\boldsymbol{x}$, and hence a Fokker–Planck equation for $P(\boldsymbol{x}, t)$. Here, following Gillespie (2000), we present the most crucial steps and approximations for this derivation.

To write the time evolution of the PDF of $\boldsymbol{N}$—i.e., $P(\boldsymbol{N}, t \mid \boldsymbol{N}_0, t_0)$—we consider a timescale sufficiently small that the probability of two or more reactions occurring is negligible compared to that of single reaction. We can then simply account for all the mutually exclusive ways of transitioning into or out-of state $\boldsymbol{N}$ via zero or one reaction. The result is that the probability of the system being in state $\boldsymbol{N}$ at time $t + dt$ is given by:

$$P(\boldsymbol{N},t+dt\,|,\boldsymbol{N}_0,t_0) = P(\boldsymbol{N},t\,|\,\boldsymbol{N}_0,t_0)\left[1-\sum_{j=1}^{r}a_j(\boldsymbol{N})\,dt\right] + \sum_{j=1}^{r}\left[P(\boldsymbol{N}-\boldsymbol{\nu}_j,t\,|\,\boldsymbol{N}_0,t_0)\,a_j(\boldsymbol{N}-\boldsymbol{\nu}_j)\,dt\right]. \quad \text{(A.1)}$$

The first term in the above equation represents the probability that the system is already in state $\boldsymbol{N}$ and no change occurs; this happens when none of the reactions, R_j, take place in the time interval $[t,t+dt]$, and corresponds to the probability $1-\sum_{j=1}^{r}a_j(\boldsymbol{N})dt$. The second term represents the probability that the system is in the state $\boldsymbol{N}$ due to a jump from one of the states $\boldsymbol{N} - \nu_j$ during the interval $[t,t+dt]$.

Rearranging the terms in Eq. (A.1) and taking the limit $dt \rightarrow 0$ gives rise to the *chemical master equation*:

$$\frac{\partial}{\partial t}P(\boldsymbol{N},t\,|\,\boldsymbol{N}_0,t_0) = \sum_{j=1}^{r}\left[P(\boldsymbol{N}-\boldsymbol{\nu}_j,t\,|\,\boldsymbol{N}_0,t_0)\,a_j(\boldsymbol{N}-\boldsymbol{\nu}_j) - P(\boldsymbol{N},t\,|\,\boldsymbol{N}_0,t_0)\,a_j(\boldsymbol{N})\right]. \quad \text{(A.2)}$$

In general, this integro-differential equation for the temporal evolution of $P(\boldsymbol{N}, t \mid \boldsymbol{N}_0, t_0)$ is difficult to solve, which is why we seek a method to instead write down approximate (chemical) Langevin equations.

Langevin equations: The state of the system after a time τ has elapsed will depend on the number of reactions that have taken place during that time. Let $K_j(\boldsymbol{N}, \tau)$ be the number of R_j reactions that occur in the time interval $[t, t + \tau]$. Since each R_j reaction contributes a change ν_{ji} to the species X_i, the number of X_i molecules at time $t + \tau$ will be

$$N_i(t+\tau) = N_i(t) + \sum_{j=1}^{r}K_j(\boldsymbol{N},\tau)\nu_{ji}, \quad \text{for } i=1,\ldots,n. \quad \text{(A.3)}$$

It is possible to approximate $K_j(\boldsymbol{N}, \tau)$ for any $\tau > 0$ when the following two conditions are satisfied: *Condition (i)*: The first condition requires τ to be sufficiently small that none of the propensity functions change considerably, i.e.,

$$a_j(\boldsymbol{N}(t')) \cong a_j(\boldsymbol{N}(t)), \quad \forall\, t' \in [t, t+\tau], \quad \forall\, j \in [1, r]. \tag{A.4}$$

This implies that we choose τ such that none of the species concentrations change appreciably due to occurrence of any reaction, and that all the reaction events occur independently of each other in the time interval $[t,\, t+\tau]$. Moreover, the quantities $K_j(\boldsymbol{N}, \tau)$, the number of times each reaction occurs in this time interval, must be statistically independent Poisson random variables, denoted by $\mathcal{P}_j(a_j(\boldsymbol{N}), \tau)$. As a result, Eq. (A.3) can be approximated by

$$N_i(t+\tau) = N_i(t) + \sum_{j=1}^{r} \mathcal{P}_j(a_j(\boldsymbol{N}), \tau) v_{ji}, \quad \text{for } i = 1, \ldots, n. \tag{A.5}$$

Condition (ii): In addition to the condition of τ being sufficiently small, we require τ to be large enough so that

$$\langle \mathcal{P}_j(a_j(\boldsymbol{N}), \tau) \rangle = a_j(\boldsymbol{N})\tau \gg 1. \tag{A.6}$$

As a result, each Poisson random variable, $\mathcal{P}_j(a_j(\boldsymbol{N}), \tau)$, can be approximated by a *normal* random variable with a mean and a standard deviation both equal to $a_j(\boldsymbol{N})\tau$. We denote such a normal random variable by $\mathcal{N}_j(a_j(\boldsymbol{N})\tau, a_j(\boldsymbol{N})\tau)$, where the first argument indicates the mean and the second argument the standard deviation. Therefore, Eq. (A.5) reduces to

$$N_i(t+\tau) = N_i(t) + \sum_{j=1}^{r} \mathcal{N}_j(a_j(\boldsymbol{N})\tau, a_j(\boldsymbol{N})\tau)\, v_{ji}, \quad \text{for } i = 1, \ldots, n. \tag{A.7}$$

Using the property that $\mathcal{N}(m, \sigma^2) = m + \sigma\mathcal{N}(0,1)$ and writing $\mathcal{N}(0,1) = \eta(t)$ for any t, we have

$$N_i(t+\tau) = N_i(t) + \sum_{j=1}^{r} v_{ji} a_j(\boldsymbol{N})\, \tau + \sum_{j=1}^{r} v_{ji} [a_j(\boldsymbol{N})\, \tau]^{1/2} \eta_j(t), \quad \text{for } i = 1, \ldots, n. \tag{A.8}$$

Let us denote τ by dt, and formally consider the infinitesimal limit $dt \to 0$. We also note that $\eta_j(t)$ and $\eta_{j'}(t')$ are statistically uncorrelated for $t \neq t'$ and $j \neq j'$, therefore,

$$\frac{dN_i(t)}{dt} = \sum_{j=1}^{r} v_{ji}\, a_j(\boldsymbol{N}) + \sum_{j=1}^{r} v_{ji}\, [a_j(\boldsymbol{N})]^{1/2} \eta_j(t), \quad \text{for } i = 1, \ldots, n. \tag{A.9}$$

Dividing the whole equation by N (total number of individuals in the system) and using the property that $a_j(\boldsymbol{N}) = N a_j(\boldsymbol{x})$, where $\boldsymbol{x} = \boldsymbol{N}/N$ we finally arrive at the Langevin equations used in the main body of the chapter:

$$\frac{dx_i}{dt} = \sum_{j=1}^{r} v_{ji} a_j(\boldsymbol{x}) + \frac{1}{\sqrt{N}} \sum_{j=1}^{r} v_{ji} [a_j(\boldsymbol{x})]^{1/2} \eta_j(t), \quad \text{for } i = 1, \ldots, n. \tag{A.10}$$

APPENDIX B. PAIRWISE INTERACTION MODEL IN TWO SPATIAL DIMENSIONS

In this appendix, we extend the pairwise interaction model to two spatial dimensions. As in the main body of the chapter, we show how to construct coarse-grained equations for the dynamics of an order parameter from microscopic rules via two methods: (*i*) the system-size expansion method and (*ii*) the chemical Langevin equation approach. We show that, in contrast to the one-dimensional case, two spatial dimensions are enough to differentiate between the two approaches; the system-size expansion has significant limitations, whereas the CLE approach yields coarse-grained equations with relative ease.

Our starting point is to assume that individuals forage from four sources, located along the positive x, positive y, negative x, and negative y axes. We label these four perpendicular directions 1–4, respectively. We chose this discretization in lieu of a continuous two-dimensional space for reasons of analytical tractability; in a continuous two-dimensional system there are infinitely many directions along which individuals can move, and thus the number of states to consider are also infinite, prohibiting a straightforward extension of the methods we have discussed for the one-dimensional pairwise interaction model. [For examples of so-called off-lattice techniques, see Dyson et al., 2012, 2015; Othmer et al., 1988.]

As before, an individual moving in direction j is denoted by X_j. It changes its direction based on two types of reaction: a copying interaction

$$X_i + X_j \xrightarrow{c} 2X_{j/i}, \quad \text{for } i \neq j \text{ and } i,j \in \{1,4\}, \tag{B.1}$$

where c denotes the specific copying rate; and a spontaneous change in direction

$$X_i \xrightarrow{s} X_j, \quad \text{for } i \neq j \text{ and } i,j \in \{1,4\}, \tag{B.2}$$

where s denotes the specific spontaneous direction switching rate. The proportion of ants in each of the four directions, denoted by x_i for $i \in \{1, 2, 3, 4\}$, satisfy the constraint $\sum_{i=1}^{4} x_i = 1$. We construct equations for the variable $\boldsymbol{x} = \{x_1, x_2, x_3, x_4\}^{\mathsf{T}}$, representing the proportion of ants moving/foraging in the four different directions 1, …, 4. We then calculate the group polarization, or order parameter $m = \sqrt{(x_1 - x_3)^2 + (x_2 - x_4)^2}$. Each of the state variables x_i can take values in the range $[0, 1]$. The order parameter too, takes values from 0 to 1, with 0 representing disordered motion and 1 representing highly aligned motion, or consensus among ants for the foraging source.

B.1 van Kampen's System-Size Expansion of Transition Rates

Transition rates: Using the shorthand notation $x_i^+ = x_i + 1/N$ and $x_j^- = x_j - 1/N$, the transition rates are the same as for the one-dimensional case

$$T_{ij}(\boldsymbol{x}) = T(x_i^+, x_j^- \mid x_i, x_j) = cx_ix_j + sx_j, \tag{B.3}$$

except the indices can now take the values 1, 2, 3, 4.

Master equation: The corresponding master equation is then just

$$\frac{\partial P(\boldsymbol{x},t)}{\partial t} = \sum_i \sum_{j \neq i} \left[P(x_i^+, x_j^-)\, T(x_i, x_j \mid x_i^+, x_j^-) - P(x_i, x_j)\, T(x_i^+, x_j^- \mid x_i, x_j) \right]. \tag{B.4}$$

As before, this can be simplified by using the step operators $\mathcal{E}_i^+$ and $\mathcal{E}_i^-$, such that

$$\mathcal{E}_i^{\pm}(x_i) = f(x_i \pm 1/N) = f(x_i^+), \tag{B.5}$$

whereby the master equation may be rewritten to give

$$\frac{\partial P(\boldsymbol{x},t)}{\partial t} = \sum_i \sum_{j \neq i} (\mathcal{E}_i^- \mathcal{E}_j^+ - 1) P(\boldsymbol{x},t)\, T_{ij}. \tag{B.6}$$

Fokker–Planck equation: Assuming a large but finite N, we approximate the action of the step operators via Taylor expansion. Ignoring terms at $O(1/N^3)$, we have

$$\mathcal{E}_i^+ f(x_i) = f(x_i + \frac{1}{N}) = \left(1 + \frac{1}{N}\frac{\partial}{\partial x_i} + \frac{1}{2N^2}\frac{\partial^2}{\partial x_i^2}\right) f(x_i), \tag{B.7a}$$

$$\mathcal{E}_i^- f(x_i) = f(x_i - \frac{1}{N}) = \left(1 - \frac{1}{N}\frac{\partial}{\partial x_i} + \frac{1}{2N^2}\frac{\partial^2}{\partial x_i^2}\right) f(x_i), \tag{B.7b}$$

and therefore

$$\mathcal{E}_i^- \mathcal{E}_j^+ = 1 + \frac{1}{N}\left(\frac{\partial}{\partial x_j} - \frac{\partial}{\partial x_i}\right) + \frac{1}{2N^2}\left(\frac{\partial}{\partial x_j} - \frac{\partial}{\partial x_i}\right)^2. \tag{B.8}$$

Substituting into (B.6), rearranging and making the substitution $t = N\tau$, we get the generic Fokker–Planck equation

$$\frac{\partial P(\boldsymbol{x})}{\partial \tau} = -\sum_{i=1}^{n-1} \frac{\partial}{\partial x_i} P(\boldsymbol{x}) \mathcal{A}_i(\boldsymbol{x}) + \frac{1}{2}\sum_{i,j=1}^{n-1} \frac{\partial^2}{\partial x_i \partial x_j} P(\boldsymbol{x}) \mathcal{B}_{ij}(\boldsymbol{x}), \tag{B.9}$$

where $n = 4$ is the total number of states/directions/food sources. Due to the constraint $\sum_{i=1}^4 x_i = 1$, we require only three variables to determine the state of the system, and the deterministic vector $\mathcal{A}(\boldsymbol{x})$ therefore has three elements

$$\mathcal{A}_i(\boldsymbol{x}) = \sum_{j \neq i}^{3} [T_{ij}(\boldsymbol{x}) - T_{ji}(\boldsymbol{x})], \quad \text{for } i = 1, \ldots, 3. \tag{B.10}$$

Similarly, the matrix $\mathcal{B}(\boldsymbol{x})$ is a 3×3, whose off-diagonal and diagonal elements, respectively, are given by

$$\mathcal{B}_{ij}(\boldsymbol{x}) = -\frac{1}{N}[T_{ij}(\boldsymbol{x}) + T_{ji}(\boldsymbol{x})], \quad \text{for } i \neq j, \tag{B.11}$$

$$\mathcal{B}_{ii}(\boldsymbol{x}) = \frac{1}{N}\sum_{j\neq i}[T_{ij}(\boldsymbol{x}) + T_{ji}(\boldsymbol{x})]. \tag{B.12}$$

For the pairwise interaction model in two dimensions represented by (B.1) and (B.2), $\mathcal{A}$ and $\mathcal{B}$ are given by

$$\mathcal{A}_i(\boldsymbol{x}) = s(1-4x_i),$$

$$\mathcal{B}_{ij}(\boldsymbol{x}) = -\frac{1}{N}[s(x_i + x_j) + 2cx_ix_j],$$

and

$$\mathcal{B}_{ii}(\boldsymbol{x}) = \frac{1}{N}[s(1+2x_i) + 2c\sum_{j\neq i}x_ix_j].$$

Mesoscopic SDEs: Given a multivariate Fokker–Planck equation of the form (B.9), we can write the corresponding SDEs as follows (Gardiner, 2009)

$$\frac{dx_i}{dt} = \mathcal{A}_i(\boldsymbol{x}) + \sum_{j=1}^{n-1}\mathcal{G}_{ij}(\boldsymbol{x})\eta_j(\tau), \tag{B.13}$$

where the η_j ($j = 1, 2, 3$) represent uncorrelated Gaussian white noise sources, and the matrix $\mathcal{G}$ is defined by the relation $\mathcal{G}\mathcal{G}^T = \mathcal{B}$. Since all x_i are real, the matrix $\mathcal{G}$ must necessarily be real; the condition for which is that $\mathcal{B}$ is positive definite—i.e., all the eigenvalues of $\mathcal{B}$ are real and positive. For the pairwise interaction model in two dimensions, we find that this condition does not hold. Numerically, we find that one of the eigenvalues of $\mathcal{B}$ is always zero, suggesting that $\mathcal{B}$ is positive *semi*-definite, and therefore $\mathcal{G}$ is not unique. We conclude, therefore, that deriving analytically tractable mesoscopic equations is not feasible for the two-dimensional pairwise interaction model using the van Kampen's method of system-size expansion. In the next section we describe the chemical Langevin equation method, which allows us to write down a coupled set of SDEs directly from the microscopic reactions.

B.2 Chemical Langevin Approach

Based on the scheme of reactions in Eqs. (B.1) and (B.2), we have 12 copying reactions and 12 spontaneous switching reactions. Each of these reactions result in a change in state, such as $x_i \to x_i + 1/N$ and $x_j \to x_j - 1/N$, for various combinations of $i \neq j$. For these 24 different reactions, we can write both propensity functions for the probability of each reaction and a state change vector v_{ji}, to capture the "stoichiometry" of the reactions. We show all the reactions with their propensity functions and corresponding state change vector entries in Table B.1. Using these details, we show how to write the CLE for the

TABLE B.1 Reactions in the Pairwise Interaction Model in Two Dimensions and Their Propensities and State Change Vector

Serial No. (j)	Reaction	Propensity	v_{j1}	v_{j2}	v_{j3}	v_{j4}
1	$X_1 + X_2 \xrightarrow{c} 2X_1$	cx_1x_2	1	−1	0	0
2	$X_2 + X_1 \xrightarrow{c} 2X_2$	cx_1x_2	−1	1	0	0
3	$X_1 + X_3 \xrightarrow{c} 2X_1$	cx_1x_3	1	0	−1	0
4	$X_3 + X_1 \xrightarrow{c} 2X_3$	cx_1x_3	−1	0	1	0
5	$X_1 + X_4 \xrightarrow{c} 2X_1$	cx_1x_4	1	0	0	−1
6	$X_4 + X_1 \xrightarrow{c} 2X_4$	cx_1x_4	−1	0	0	1
7	$X_2 + X_3 \xrightarrow{c} 2X_2$	cx_2x_3	0	1	−1	0
8	$X_3 + X_2 \xrightarrow{c} 2X_3$	cx_2x_3	0	−1	1	0
9	$X_2 + X_4 \xrightarrow{c} 2X_2$	cx_2x_4	0	1	0	−1
10	$X_4 + X_2 \xrightarrow{c} 2X_4$	cx_2x_4	0	−1	0	1
11	$X_3 + X_4 \xrightarrow{c} 2X_3$	cx_3x_4	0	0	1	−1
12	$X_4 + X_3 \xrightarrow{c} 2X_4$	cx_3x_4	0	0	−1	1
13	$X_1 \xrightarrow{s} X_2$	sx_1	−1	1	0	0
14	$X_2 \xrightarrow{s} X_1$	sx_2	1	−1	0	0
15	$X_1 \xrightarrow{s} X_3$	sx_1	−1	0	1	0
16	$X_3 \xrightarrow{s} X_1$	sx_3	1	0	−1	0
17	$X_1 \xrightarrow{s} X_4$	sx_1	−1	0	0	1
18	$X_4 \xrightarrow{s} X_1$	sx_4	1	0	0	−1
19	$X_2 \xrightarrow{s} X_3$	sx_2	0	−1	1	0
20	$X_3 \xrightarrow{s} X_2$	sx_3	0	1	−1	0
21	$X_2 \xrightarrow{s} X_4$	sx_2	0	−1	0	1
22	$X_4 \xrightarrow{s} X_2$	sx_4	0	1	0	−1
23	$X_3 \xrightarrow{s} X_4$	sx_3	0	0	−1	1
24	$X_4 \xrightarrow{s} X_3$	sx_4	0	0	1	−1

state x_1, such that the method can easily be applied to the remaining states. We recall that the generic form of a CLE is given by

$$\frac{dx_i}{dt} = \sum_{j=1}^{r} v_{ji} a_j(\boldsymbol{x}) + \frac{1}{\sqrt{N}} \sum_{j=1}^{r} v_{ji} a_j(\boldsymbol{x})^{1/2} \eta_j(t), \tag{B.14}$$

Substituting the values of v_{ji} and $a_j(\boldsymbol{x})$ for $i = 1$ from Table B.1 into the above equation we see that

$$\begin{aligned}\sum_{j=1}^{r} v_{j1} a_j(\boldsymbol{x}(t)) &= cx_1x_2 - cx_1x_2 + cx_1x_3 - cx_1x_3 + cx_1x_4 - cx_1x_4 + cx_2x_3 - cx_2x_3 \\ &\quad + cx_2x_4 - cx_2x_4 + cx_3x_4 - cx_3x_4 - sx_1 + sx_2 - sx_1 + sx_3 - sx_1 + sx_4, \\ &= s(x_2 - x_1) + s(x_3 - x_1) + s(x_4 - x_1), \\ &= s(1 - 4x_1).\end{aligned}$$

And similarly,

$$\begin{aligned}\sum_{j=1}^{r} v_{j1} [a_j(\boldsymbol{x}(t))]^{1/2} &= \sqrt{cx_1x_2}\eta_1 - \sqrt{cx_1x_2}\eta_2 + \sqrt{cx_1x_3}\eta_3 - \sqrt{cx_1x_3}\eta_4 \\ &\quad + \sqrt{cx_1x_4}\eta_5 - \sqrt{cx_1x_4}\eta_6 - \sqrt{sx_1}\eta_{13} + \sqrt{sx_2}\eta_{14} \\ &\quad - \sqrt{sx_1}\eta_{15} + \sqrt{sx_3}\eta_{16} - \sqrt{sx_1}\eta_{17} + \sqrt{sx_4}\eta_{18},\end{aligned}$$

which results in the following SDE for the variable x_1:

$$\begin{aligned}\frac{dx_1}{dt} = s(1 - 4x_1) + \frac{1}{\sqrt{N}}\Big(&\sqrt{cx_1x_2}\eta_1 - \sqrt{cx_1x_2}\eta_2 + \sqrt{cx_1x_3}\eta_3 - \sqrt{cx_1x_3}\eta_4 \\ &+ \sqrt{cx_1x_4}\eta_5 - \sqrt{cx_1x_4}\eta_6 - \sqrt{sx_1}\eta_{13} + \sqrt{sx_2}\eta_{14} - \sqrt{sx_1}\eta_{15} \\ &+ \sqrt{sx_3}\eta_{16} - \sqrt{sx_1}\eta_{17} + \sqrt{sx_4}\eta_{18}\Big),\end{aligned} \tag{B.15}$$

Performing the similar substitutions for the remaining variables results in

$$\begin{aligned}\frac{dx_2}{dt} = s(1 - 4x_2) + \frac{1}{\sqrt{N}}\Big(&\sqrt{cx_1x_2}\eta_2 - \sqrt{cx_1x_2}\eta_1 + \sqrt{cx_2x_3}\eta_7 - \sqrt{cx_2x_3}\eta_8 \\ &+ \sqrt{cx_2x_4}\eta_9 - \sqrt{cx_2x_4}\eta_{10} + \sqrt{sx_1}\eta_{13} - \sqrt{sx_2}\eta_{14} - \sqrt{sx_2}\eta_{19} \\ &+ \sqrt{sx_3}\eta_{20} - \sqrt{sx_2}\eta_{21} + \sqrt{sx_4}\eta_{22}\Big),\end{aligned} \tag{B.16a}$$

$$\begin{aligned}\frac{dx_3}{dt} = s(1 - 4x_3) + \frac{1}{\sqrt{N}}\Big(&\sqrt{cx_1x_3}\eta_4 - \sqrt{cx_1x_3}\eta_3 + \sqrt{cx_2x_3}\eta_8 - \sqrt{cx_2x_3}\eta_7 \\ &+ \sqrt{cx_3x_4}\eta_{11} - \sqrt{cx_3x_4}\eta_{12} + \sqrt{sx_1}\eta_{15} - \sqrt{sx_3}\eta_{16} \\ &+ \sqrt{sx_2}\eta_{19} - \sqrt{sx_3}\eta_{20} - \sqrt{sx_3}\eta_{23} + \sqrt{sx_4}\eta_{24}\Big),\end{aligned} \tag{B.16b}$$

$$\frac{dx_4}{dt} = s(1-4x_4) + \frac{1}{\sqrt{N}}(\sqrt{cx_1x_4}\eta_6 - \sqrt{cx_1x_4}\eta_5 + \sqrt{cx_2x_4}\eta_{10} - \sqrt{cx_2x_4}\eta_9$$
$$+ \sqrt{cx_3x_4}\eta_{12} - \sqrt{cx_3x_4}\eta_{11} + \sqrt{sx_1}\eta_{17} - \sqrt{sx_4}\eta_{18} \tag{B.16c}$$
$$+ \sqrt{sx_2}\eta_{21} - \sqrt{sx_4}\eta_{22} + \sqrt{sx_3}\eta_{23} - \sqrt{sx_4}\eta_{24}),$$

where, η_1, η_2, ..., η_{24} are delta-correlated Gaussian white noise sources. Unfortunately, further simplification of the stochastic terms is not possible in this case because the coupled nature of the equations. Nevertheless, the corresponding multivariate Fokker–Planck equation can be derived by following (Gillespie, 2002).

$$\frac{\partial P(\boldsymbol{x},t|\boldsymbol{x}_0,t_0)}{\partial t} = -\sum_{i=1}^{n}\frac{\partial}{\partial x_i}\left[\left(\sum_{j=1}^{r} v_{ji}a_j(\boldsymbol{x})\right)P(\boldsymbol{x},t|\boldsymbol{x}_0,t_0)\right]$$
$$+\frac{1}{2N}\sum_{i=1}^{n}\frac{\partial^2}{\partial x_i^2}\left[\left(\sum_{j=1}^{r} v_{ji}^2 a_j(\boldsymbol{x})\right)P(\boldsymbol{x},t|\boldsymbol{x}_0,t_0)\right] \tag{B.17}$$
$$+\frac{1}{N}\sum_{i'=1}^{n}\sum_{i<i'}\frac{\partial^2}{\partial x_i \partial x_{i'}}\left[\left(\sum_{j=1}^{r} v_{ji}v_{ji'}a_j(\boldsymbol{x})\right)P(\boldsymbol{x},t|\boldsymbol{x}_0,t_0)\right].$$

Substituting the values of v_{ji} and $a_j(\boldsymbol{x})$ from Table B.1 recovers a Fokker–Planck equation, of the form

$$\frac{\partial P(\boldsymbol{x})}{\partial \tau} = -\sum_{i=1}^{n}\frac{\partial}{\partial x_i}[P(\boldsymbol{x})\mathcal{A}_i(\boldsymbol{x})] + \frac{1}{2}\sum_{i,j=1}^{n}\frac{\partial^2}{\partial x_i \partial x_j}\left[P(\boldsymbol{x})\mathcal{B}_{ij}(\boldsymbol{x})\right], \tag{B.18}$$

where $\mathcal{A}(\boldsymbol{x})$ is given by

$$\mathcal{A}_i(\boldsymbol{x}) = s(1-4x_i),$$

while the diagonal and off-diagonal elements of $\mathcal{B}(\mathbf{x})$ are given by

$$\mathcal{B}_{ii}(\boldsymbol{x}) = \frac{1}{N}[s(1+2x_i) + 2c\sum_{j\neq i} x_i x_j],$$

and

$$\mathcal{B}_{ij}(\boldsymbol{x}) = -\frac{1}{N}[s(x_i+x_j) + 2cx_ix_j],$$

respectively. This is the same Fokker–Planck equation as derived using system-size expansion.

REFERENCES

Aldana, M., Dossetti, V., Huepe, C., Kenkre, V.M., Larralde, H., 2007. Phase transitions in systems of self-propelled agents and related network models. Phys. Rev. Lett. 98 (9), 095702.

Alfarano, S., Lux, T., Wagner, F., 2008. Time variation of higher moments in a financial market with heterogeneous agents: an analytical approach. J. Econ. Dyn. Control. 32 (1), 101–136.

Altschuler, S.J., Angenent, S.B., Wang, Y., Wu, L.F., 2008. On the spontaneous emergence of cell polarity. Nature 454 (7206), 886.

Attanasi, A., Cavagna, A., Del Castello, L., Giardina, I., Grigera, T.S., Jelić, A., Melillo, S., Parisi, L., Pohl, O., Shen, E., et al., 2014. Information transfer and behavioural inertia in starling flocks. Nat. Phys. 10 (9), 691.

Baglietto, G., Albano, E.V., 2009. Nature of the order-disorder transition in the Vicsek model for the collective motion of self-propelled particles. Phys. Rev. E 80 (5), 050103.

Ballerini, M., Cabibbo, N., Candelier, R., Cavagna, A., Cisbani, E., Giardina, I., Lecomte, V., Orlandi, A., Parisi, G., Procaccini, A., et al., 2008. Interaction ruling animal collective behavior depends on topological rather than metric distance: evidence from a field study. Proc. Natl. Acad. Sci. 105 (4), 1232–1237.

Beckers, R., Deneubourg, J.-L., Goss, S., Pasteels, J.M., 1990. Collective decision making through food recruitment. Insect. Soc. 37 (3), 258–267.

Berdahl, A., Torney, C.J., Ioannou, C.C., Faria, J.J., Couzin, I.D., 2013. Emergent sensing of complex environments by mobile animal groups. Science 339 (6119), 574–576.

Bialek, W., Cavagna, A., Giardina, I., Mora, T., Silvestri, E., Viale, M., Walczak, A.M., 2012. Statistical mechanics for natural flocks of birds. Proc. Natl. Acad. Sci. 109 (13), 4786–4791.

Biancalani, T., Fanelli, D., Di Patti, F., 2010. Stochastic Turing patterns in the Brusselator model. Phys. Rev. E 81 (4), 046215.

Biancalani, T., Dyson, L., McKane, A.J., 2014. Noise-induced bistable states and their mean switching time in foraging colonies. Phys. Rev. Lett. 112 (3), 038101.

Biro, D., Sumpter, D.J.T., Meade, J., Guilford, T., 2006. From compromise to leadership in pigeon homing. Curr. Biol. 16 (21), 2123–2128.

Black, A.J., McKane, A.J., 2012. Stochastic formulation of ecological models and their applications. Trends Ecol. Evol. 27 (6), 337–345.

Boettiger, C., 2018. From noise to knowledge: how randomness generates novel phenomena and reveals information. Ecol. Lett. 21 (8), 1255–1267.

Bonabeau, E., Theraulaz, G., Deneubourg, J.-L., Aron, S., Camazine, S., 1997. Self-organization in social insects. Trends Ecol. Evol. 12 (5), 188–193.

Buhl, J., Sumpter, D.J.T., Couzin, I.D., Hale, J.J., Despland, E., Miller, E.R., Simpson, S.J., 2006. From disorder to order in marching locusts. Science 312 (5778), 1402–1406.

Camazine, S., Deneubourg, J.-L., Franks, N.R., Sneyd, J., Bonabeau, E., Theraula, G., 2003. Self-Organization in Biological Systems. vol. 7. Princeton University Press.

Cavagna, A., Cimarelli, A., Giardina, I., Parisi, G., Santagati, R., Stefanini, F., Viale, M., 2010. Scale-free correlations in starling flocks. Proc. Natl. Acad. Sci. 107 (26), 11865–11870.

Chaté, H., Ginelli, F., Grégoire, G., Raynaud, F., 2008. Collective motion of self-propelled particles interacting without cohesion. Phys. Rev. E 77 (4), 046113.

Chowdhury, D., Guttal, V., Nishinari, K., Schadschneider, A., 2002. A cellular-automata model of flow in ant trails: non-monotonic variation of speed with density. J. Phys. A Math. Gen. 35 (41), L573.

Colizza, V., Barrat, A., Barthelemy, M., Valleron, A.-J., Vespignani, A., 2007. Modeling the worldwide spread of pandemic influenza: baseline case and containment interventions. PLoS Med. 4 (1), e13.

Couzin, I.D., Krause, J., James, R., Ruxton, G.D., Franks, N.R., 2002. Collective memory and spatial sorting in animal groups. J. Theor. Biol. 218 (1), 1–11.
Couzin, I.D., Krause, J., Franks, N.R., Levin, S.A., 2005. Effective leadership and decision-making in animal groups on the move. Nature 433 (7025), 513.
Cox, J.T., Griffeath, D., 1986. Diffusive clustering in the two dimensional voter model. Ann. Probab. 14 (2), 347–370.
Czirók, A., Barabási, A.-L., Vicsek, T., 1999. Collective motion of self-propelled particles: kinetic phase transition in one dimension. Phys. Rev. Lett. 82 (1), 209.
Czirók, A., Vicsek, M., Vicsek, T., 1999. Collective motion of organisms in three dimensions. Phys. A Stat. Mech. Appl. 264 (1–2), 299–304.
Datta, S., Delius, G.W., Law, R., 2010. A jump-growth model for predator-prey dynamics: derivation and application to marine ecosystems. Bull. Math. Biol. 72 (6), 1361–1382.
Dean, D.S., 1996. Langevin equation for the density of a system of interacting Langevin processes. J. Phys. A Math. Gen. 29 (24), L613.
Deneubourg, J.-L., Goss, S., 1989. Collective patterns and decision-making. Ethol. Ecol. Evol. 1 (4), 295–311.
Dussutour, A., Nicolis, S.C., Deneubourg, J.-L., Fourcassié, V., 2006. Collective decisions in ants when foraging under crowded conditions. Behav. Ecol. Sociobiol. 61 (1), 17–30.
Dyson, L., Maini, P.K., Baker, R.E., 2012. Macroscopic limits of individual-based models for motile cell populations with volume exclusion. Phys. Rev. E 86 (3), 031903.
Dyson, L., Yates, C.A., Buhl, J., McKane, A.J., 2015. Onset of collective motion in locusts is captured by a minimal model. Phys. Rev. E 92 (5), 052708.
El-Samad, H., Prajna, S., Papachristodoulou, A., Doyle, J., Khammash, M., 2006. Advanced methods and algorithms for biological networks analysis. Proc. IEEE 94 (4), 832–853.
Erban, R., Chapman, S.J., 2009. Stochastic modelling of reaction-diffusion processes: algorithms for bimolecular reactions. Phys. Biol. 6 (4), 046001.
Flierl, G., Grünbaum, D., Levins, S., Olson, D., 1999. From individuals to aggregations: the interplay between behavior and physics. J. Theor. Biol. 196 (4), 397–454.
Franks, N.R., Wilby, A., Silverman, B.W., Tofts, C., 1992. Self-organizing nest construction in ants: sophisticated building by blind bulldozing. Anim. Behav. 44, 357–375.
Gardiner, C., 2009. Stochastic Methods. vol. 4. Springer, Berlin.
Gautrais, J., Ginelli, F., Fournier, R., Blanco, S., Soria, M., Chaté, H., Theraulaz, G., 2012. Deciphering interactions in moving animal groups. PLoS Comput. Biol. 8 (9), e1002678.
Gavagnin, E., Yates, C.A., 2018. Stochastic and deterministic modeling of cell migration. In: Srinivasa Rao, A.S.R., Rao, C.R. (Eds.), Handbook of Statistics: Integrated Population Biology and Modeling, Part A, 39, Elsevier, pp. 37–91.
Gillespie, D.T., 1976. A general method for numerically simulating the stochastic time evolution of coupled chemical reactions. J. Comput. Phys. 22 (4), 403–434.
Gillespie, D.T., 1977. Exact stochastic simulation of coupled chemical reactions. J. Phys. Chem. 81 (25), 2340–2361.
Gillespie, D.T., 2000. The chemical Langevin equation. J. Chem. Phys. 113 (1), 297–306.
Gillespie, D.T., 2001. Approximate accelerated stochastic simulation of chemically reacting systems. J. Chem. Phys. 115 (4), 1716–1733.
Gillespie, D.T., 2002. The chemical Langevin and Fokker-Planck equations for the reversible isomerization reaction. J. Phys. Chem. A 106 (20), 5063–5071.
Ginelli, F., 2016. The physics of the Vicsek model. Eur. Phys. J. Spec. Top. 225 (11–12), 2099–2117.
Gordon, D.M., 1996. The organization of work in social insect colonies. Nature 380 (6570), 121.
Grégoire, G., Chaté, H., 2004. Onset of collective and cohesive motion. Phys. Rev. Lett. 92 (2), 025702.

Grünbaum, D., 1998. Schooling as a strategy for taxis in a noisy environment. Evol. Ecol. 12 (5), 503–522.

Guttal, V., 2014. Ecology: from individuals to collectives. Resonance 19 (4), 368–375.

Guttal, V., Couzin, I.D., 2010. Social interactions, information use, and the evolution of collective migration. Proc. Natl. Acad. Sci. 107 (37), 16172–16177.

Guttal, V., Romanczuk, P., Simpson, S.J., Sword, G.A., Couzin, I.D., 2012. Cannibalism can drive the evolution of behavioural phase polyphenism in locusts. Ecol. Lett. 15 (10), 1158–1166.

Herbert-Read, J.E., Perna, A., Mann, R.P., Schaerf, T.M., Sumpter, D.J.T., Ward, A.J.W., 2011. Inferring the rules of interaction of shoaling fish. Proc. Natl. Acad. Sci. 108 (46), 18726–18731.

Horsthemke, W., Lefever, R., 1984. Noise-induced transitions: theory and application in physics, chemistry, and biology. Springer Series in Synergetics, 15, Springer-Verlag, Berlin.

Jadbabaie, A., Lin, J., Morse, A.S., 2003. Coordination of groups of mobile autonomous agents using nearest neighbor rules. IEEE Trans. Autom. Control 48 (6), 988–1001.

Jiang, L., Giuggioli, L., Perna, A., Escobedo, R., Lecheval, V., Sire, C., Han, Z., Theraulaz, G., 2017. Identifying influential neighbors in animal flocking. PLoS Comput. Biol. 13 (11), 1–32.

Joshi, J., Guttal, V., 2018. Demographic noise and cost of greenbeard can facilitate greenbeard cooperation. Evolution. https://doi.org/10.1111/evo.13615.

Joshi, J., Couzin, I.D., Levin, S.A., Guttal, V., 2017. Mobility can promote the evolution of cooperation via emergent self-assortment dynamics. PLoS Comput. Biol. 13 (9), e1005732.

Katz, Y., Tunstrøm, K., Ioannou, C.C., Huepe, C., Couzin, I.D., 2011. Inferring the structure and dynamics of interactions in schooling fish. Proc. Natl. Acad. Sci. 108 (46), 18720–18725.

Kimura, M., Weiss, G.H., 1964. The stepping stone model of population structure and the decrease of genetic correlation with distance. Genetics 49 (4), 561–576.

Kirman, A., 1993. Ants, rationality, and recruitment. Q. J. Econ. 108 (1), 137–156.

Kolpas, A., 2008. Coarse-grained analysis of collective motion in animal groups (Dissertation). University of California, Santa Barbara.

Kolpas, A., Moehlis, J., Kevrekidis, I.G., 2007. Coarse-grained analysis of stochasticity-induced switching between collective motion states. Proc. Natl. Acad. Sci. 104 (14), 5931–5935.

Krause, J., Ruxton, G.D., Ruxton, G.D., 2002. Living in Groups. Oxford University Press.

Kunwar, A., John, A., Nishinari, K., Schadschneider, A., Chowdhury, D., 2004. Collective traffic-like movement of ants on a trail: dynamical phases and phase transitions. J. Phys. Soc. Jpn 73 (11), 2979–2985.

Lawson, M.J., Drawert, B., Khammash, M., Petzold, L., Yi, T.-M., 2013. Spatial stochastic dynamics enable robust cell polarization. PLoS Comput. Biol. 9 (7), e1003139.

Lux, T., 1995. Herd behaviour, bubbles and crashes. Econ. J. 105 (431), 881–896.

Majumder, S., 2018. Multiple stable states and abrupt transitions in spatial ecosystems (Dissertation). Indian Institute of Science, Bangalore.

Mann, R.P., Perna, A., Strömbom, D., Garnett, R., Herbert-Read, J.E., Sumpter, D.J.T., Ward, A.J. W., 2013. Multi-scale inference of interaction rules in animal groups using Bayesian model selection. PLoS Comput. Biol. 9 (3), e1002961.

Marchetti, M.C., Joanny, J.-F., Ramaswamy, S., Liverpool, T.B., Prost, J., Rao, M., Simha, R.A., 2013. Hydrodynamics of soft active matter. Rev. Mod. Phys. 85 (3), 1143.

McKane, A.J., Newman, T.J., 2004. Stochastic models in population biology and their deterministic analogs. Phys. Rev. E 70 (4), 041902.

McKane, A.J., Newman, T.J., 2005. Predator-prey cycles from resonant amplification of demographic stochasticity. Phys. Rev. Lett. 94 (21), 218102.

McKane, A.J., Biancalani, T., Rogers, T., 2014. Stochastic pattern formation and spontaneous polarisation: the linear noise approximation and beyond. Bull. Math. Biol. 76 (4), 895–921.

Mogilner, A., Edelstein-Keshet, L., 1999. A non-local model for a swarm. GAMS J. Math. Biosci. 38 (6), 534–570.

Nishinari, K., Sugawara, K., Kazama, T., Schadschneider, A., Chowdhury, D., 2006. Modelling of self-driven particles: foraging ants and pedestrians. Phys. A Stat. Mech. Appl. 372 (1), 132–141.

Othmer, H.G., Dunbar, S.R., Alt, W., 1988. Models of dispersal in biological systems. J. Math. Biol. 26 (3), 263–298.

Parrish, J.K., Viscido, S.V., Grunbaum, D., 2002. Self-organized fish schools: an examination of emergent properties. Biol. Bull. 202 (3), 296–305.

Petit, O., Bon, R., 2010. Decision-making processes: the case of collective movements. Behav. Process. 84 (3), 635–647.

Pratt, S.C., Mallon, E.B., Sumpter, D.J., Franks, N.R., 2002. Quorum sensing, recruitment, and collective decision-making during colony emigration by the ant Leptothorax albipennis. Behav. Ecol. Sociobiol. 52 (2), 117–127.

Press, W.H., Teukolsky, S.A., Vetterling, W.T., Flannery, B.P., 1996. Numerical Recipes in C. vol. 2. Cambridge University Press, Cambridge.

Ramaswamy, S., 2010. The mechanics and statistics of active matter. Ann. Rev. Condens. Matter Phys. 1, 323–345.

Reluga, T.C., Viscido, S., 2005. Simulated evolution of selfish herd behavior. J. Theor. Biol. 234 (2), 213–225.

Romanczuk, P., Schimansky-Geier, L., 2012. Mean-field theory of collective motion due to velocity alignment. Ecol. Complex. 10, 83–92.

Rüdiger, S., 2014. Stochastic models of intracellular calcium signals. Phys. Rep. 534 (2), 39–87.

Schultheiss, P., Cheng, K., 2012. Finding food: outbound searching behavior in the Australian desert ant Melophorus bagoti. Behav. Ecol. 24 (1), 128–135.

Schulze, C., Stauffer, D., 2005. Monte Carlo simulation of the rise and the fall of languages. Int. J. Mod. Phys. C 16 (05), 781–787.

Seeley, T.D., 1989. The honey bee colony as a superorganism. Am. Sci. 77 (6), 546–553.

Simpson, S.J., Sword, G.A., 2008. Locusts. Curr. Biol. 18 (9), R364–R366.

Simpson, M.L., Cox, C.D., Allen, M.S., McCollum, J.M., Dar, R.D., Karig, D.K., Cooke, J.F., 2009. Noise in biological circuits. Wiley Interdiscip. Rev. Nanomed. Nanobiotechnol. 1 (2), 214–225.

Sotiropoulos, V., Kaznessis, Y.N., 2011. Analytical derivation of moment equations in stochastic chemical kinetics. Chem. Eng. Sci. 66 (3), 268–277.

Spector, L., Klein, J., Perry, C., Feinstein, M., 2005. Emergence of collective behavior in evolving populations of flying agents. Genet. Program Evolvable Mach. 6 (1), 111–125.

Strogatz, S.H., 2018. Nonlinear Dynamics and Chaos: With Applications to Physics, Biology, Chemistry, and Engineering. CRC Press.

Strömbom, D., 2011. Collective motion from local attraction. J. Theor. Biol. 283 (1), 145–151.

Sumpter, D.J.T., 2006. The principles of collective animal behaviour. Philos. Trans. R. Soc. Lond. B Biolog. Sci. 361 (1465), 5–22.

Sumpter, D.J.T., 2010. Collective Animal Behavior. Princeton University Press.

Toner, J., Tu, Y., 1995. Long-range order in a two-dimensional dynamical XY model: how birds fly together. Phys. Rev. Lett. 75 (23), 4326.

Torney, C.J., Levin, S.A., Couzin, I.D., 2010. Specialization and evolutionary branching within migratory populations. Proc. Natl. Acad. Sci. 107 (47), 20394–20399.

Traulsen, A., Claussen, J.C., Hauert, C., 2012. Stochastic differential equations for evolutionary dynamics with demographic noise and mutations. Phys. Rev. E 85 (4), 041901.

Uvarov, B.P., 1977. Grasshopper and Locust: A Handbook of General Acridology. Vol. II: Behaviour, Ecology, Biogeography, Population Dynamics. Cambridge University Press, Cambridge.

van Kampen, N.G., 1992. Stochastic Processes in Physics and Chemistry. vol. 1. Elsevier.

Vicsek, T., Czirók, A., Ben-Jacob, E., Cohen, I., Shochet, O., 1995. Novel type of phase transition in a system of self-driven particles. Phys. Rev. Lett. 75 (6), 1226.

Wang, Z., Bauch, C.T., Bhattacharyya, S., d'Onofrio, A., Manfredi, P., Perc, M., Perra, N., Salathé, M., Zhao, D., 2016. Statistical physics of vaccination. Phys. Rep. 664, 1–113.

Wood, A.J., Ackland, G.J., 2007. Evolving the selfish herd: emergence of distinct aggregating strategies in an individual-based model. Proc. R. Soc. Lond. B Biol. Sci. 274 (1618), 1637–1642.

Yates, C.A., Erban, R., Escudero, C., Couzin, I.D., Buhl, J., Kevrekidis, I.G., Maini, P.K., Sumpter, D.J.T., 2009. Inherent noise can facilitate coherence in collective swarm motion. Proc. Natl. Acad. Sci. 106 (14), 5464–5469.

Yuan, Y., Allen, L.J.S., 2011. Stochastic models for virus and immune system dynamics. Math. Biosci. 234 (2), 84–94.

Chapter 14

Collective Behavior and Ecology

Glenn R. Flierl[1]
Department of Earth, Atmospheric, and Planetary Science, MIT, Cambridge, MA, United States
[1]*Corresponding author: e-mail: glenn@lake.mit.edu*

ABSTRACT
We explore individual-based and continuum models of the distribution of organisms which respond to their environment by altering their movement relative to the fluid they live in. Aggregations can occur either through sensing of gradients or by simply altering the random motions based on local cues. These may include information about positions and perhaps velocities of neighbors as well as environmental characteristics. Once aggregations form, the ecological interactions will also change: a patch of organisms can locally deplete their resource so that the average or the nonlinear terms representing uptake end up being smaller than they would be with a uniform distribution. We examine a number of cases of both this kind of interaction and predator–prey dynamics; the influence on the mean population is sensitive to the details of the functional forms of the growth and mortality processes. The impact can be quite significant.

Keywords: Collective behavior, Aggregation, Population dynamics, Theoretical ecology

1 INTRODUCTION

The spatial distribution of many organisms does not look like the Poisson distribution which would arise from randomly scattering them over an area. Instead, they clearly exhibit patchiness or aggregation. On land, one may be able to locate the individuals and compute statistics such a Lloyd's patchiness index which compare the number of neighbors in a circle around each organism to the expected number if they were distributed randomly. In the ocean, we could do this for sessile organisms, but for swimming animals or planktonic organisms carried by the currents, it is not possible currently to make a similar census. But patchiness is still evident from comparing plankton tows from nearby sites and times: the r.m.s. values can be large compared to the mean. If, for example, the organisms cluster into about 10% of the area, the r.m.s. amplitude is over three times the mean.

Handbook of Statistics, Vol. 40. https://doi.org/10.1016/bs.host.2018.11.010

When we look at interactions between organisms and their resources, whether those be nutrients or prey, and with their predators, we expect that they will be altered by patchiness. For example, a patch of primary producers will exhaust the local nutrient available much more quickly than if they were spread randomly. They then must rely on other processes or their own movements to restore the supply. Likewise the encounter rates between predators and prey will be altered if one or both are not evenly distributed. However, the common ecological models deal with only mean quantities and assume that the functions expressing quantities such as uptake of nutrients or grazing by herbivores can be expressed as functions of the means. But for global or regional models, we necessarily are trying to represent changes in time of spatially low-pass filtered values. But the interactions commonly are nonlinear (e.g., the predator–prey grazing PZ in the Lotka–Volterra equations), so that the filtered values of these will not match the result of calculating them with the already filtered fields. Of course, the models generally have poorly determined parameters so effects of small-scale patchiness can be partially compensated for by adjusting them. But there is no assurance that the degree of patchiness and its influence on the dynamics will not change because of alternations in the various biomasses or factors such as the amount of turbulence.

We shall explore the mechanisms by which aggregations form and examine some simple models of ecological interactions when these are active. We consider organisms in a fluid environment by capable of moving relative to the fluid. This adds some extra possibilities: patches can be pushed together or torn apart by the flow.

Here, we are trying to develop the ideas in a sequential, consistent, and pedagogical fashion; the reader should be aware that there are numerous articles and books (e.g., Okubo and Levin, 2001) that develop and were first to develop many of the ideas herein. Those wishing to follow-up on the material herein should treat this as an introduction and should dig into the literature. That said, Sections 1–4 are parallel to Flierl et al. (1999). They discussed the effects of turbulence in some detail; here we concentrate more on examples of ecological impacts of collective behavior.

2 INDIVIDUAL-BASED MODELS

Individual-based models (IBMs) represent each organism as an object with various characteristics including position and velocity. Its velocity may be altered by behavior: sinking, floating, swimming; it may also be subject to drag when moving relative to the water. They may also have other characteristics such as size which could change and alter their behavior.

Conceptually, we can think about the movement of organisms subject to random accelerations using Itô calculus. The position of the individual $\mathbf{X}$ and its velocity relative to the fluid (which is moving at $\mathbf{v}_{fluid}$) $\mathbf{U}$ satisfy

$$\begin{aligned} d\mathbf{X} &= [\mathbf{v}(\mathbf{X},t)+\mathbf{U}]dt \\ d\mathbf{U} &= \mathbf{A}(\mathbf{X},\mathbf{U},t)\,dt+\sqrt{2\sigma(\mathbf{X},\mathbf{U},t)}\,d\mathbf{W} \end{aligned} \tag{1}$$

with $\langle d\mathbf{W}\rangle$ representing the random process with $\langle d\mathbf{W}\rangle=0$ and $\langle dW_i(t_m)dW_j(t_n)\rangle=\delta_{t_nt_m}dt$. As we shall see, the $\sqrt{dt}$ factor brings the random accelerations in at the right order for well-behaved variance. To express more general dynamics, we can think of collecting **X** and **U** into a single six-dimensional vector **z** satisfying

$$dz_i=F_i(\mathbf{z},t)dt+S_{ij}(\mathbf{z},t)dW_j \tag{2}$$

We can easily conceive of expanding the vector **z** to include other characteristics such as size which can affect movement; these other individual characteristics may have both deterministic and random variations. For example, the growth rate for the size has a mean for as well as random fluctuations. While there is an extensive body of literature on such equations, we will only use the equations for the mean and covariance. The mean is simple

$$\frac{\partial}{\partial t}\langle z_i\rangle=\langle F_i(\mathbf{z},t)\rangle \tag{3}$$

but deceptively so: if F is a nonlinear function of **z**, this equation will not be closed. That is, $\langle F_i(\mathbf{z},t)\rangle$ is not expressible as some $\widetilde{F}_i(\langle\mathbf{z}\rangle,t)$. However, proceeding onwards, we find

$$\frac{\partial}{\partial t}\langle z_iz_j\rangle=\langle z_iF_j\rangle+\langle z_jF_i\rangle+\langle S_{ik}S_{jk}\rangle \tag{4}$$

2.1 Dispersion/Diffusion

For the first example, return to Eq. (1) and suppose that $\mathbf{u}=0$, σ is constant, and that the random velocity arises from an Ornstein–Uhlenbeck process: the deterministic acceleration damps the movement relative to the water, $\mathbf{A}=-\gamma\mathbf{U}$. Splitting the **z** vector into two three-vectors

$$\mathbf{z}=\begin{bmatrix}X_i\\U_i\end{bmatrix},\quad \mathbf{F}=\begin{bmatrix}U_i\\-\gamma U_i\end{bmatrix},\quad \mathbf{R}=\begin{bmatrix}0&0\\0&\sqrt{2\sigma}\end{bmatrix}$$

leads to

$$\frac{\partial}{\partial t}\langle\mathbf{X}\rangle=\langle\mathbf{U}\rangle,\quad \frac{\partial}{\partial t}\langle\mathbf{U}\rangle=-\gamma\langle\mathbf{U}\rangle$$

so that asymptotically the mean velocity vanishes and the position becomes constant. If we choose these to be zero initially, they stay zero. The variance, however, has nontrivial long-term values; according to Eq. (3)

$$\frac{\partial}{\partial t}\langle X_i X_j\rangle = \langle X_i U_j\rangle + \langle X_j U_i\rangle$$
$$\frac{\partial}{\partial t}\langle X_i U_j\rangle = -\gamma\langle X_i U_j\rangle + \langle U_i U_j\rangle$$
$$\frac{\partial}{\partial t}\langle U_i U_j\rangle = -2\gamma\langle U_i U_j\rangle + 2\sigma\delta_{ij}$$

The velocity variance limits to $\sigma\delta_{ij}/\gamma$, $\langle X_i U_j\rangle \to \sigma\delta_{ij}/\gamma^2$, and the position variance grows linearly in time $\langle X_i X_j\rangle \to 2t\,\sigma\delta_{ij}/\gamma^2$. With these rather severe restrictions on the parameters, the organisms will disperse with the variance in position growing linearly in time.

The ordinary equation for diffusion of a tracer

$$\frac{\partial}{\partial t}b = \kappa\nabla^2 b$$

has

$$\frac{\partial}{\partial t}\int d\mathbf{x}\; b(\mathbf{x}) = 0,\quad \frac{\partial}{\partial t}\int d\mathbf{x}\; x_i b(\mathbf{x}) = 0,\quad \frac{\partial}{\partial t}\int d\mathbf{x}\; x_i x_j b(\mathbf{x}) = 2\kappa\delta_{ij}\int d\mathbf{x}\; b(\mathbf{x})$$

so that $\int \dot{x}\; x^2 b(x)/\int d\mathbf{x}\; b(x)$ grows linearly with time as $2\kappa t$. By analogy, then $K \equiv \sigma/\gamma^2$ is identified at the effective diffusivity associated with the random motion. We shall see that the PDF for the position obeys the diffusion law under the restrictions above.

2.2 Taxis

A more interesting example is "taxis" where the organisms respond to gradients in a resource R by preferring upgradient swimming behavior proportional to ∇R. For collective aggregation, linear dependence on gradients of the cue can lead to velocities which are too large, so that we have to limit the velocities in some appropriate way; however, near the peak of the resource where $|\nabla R| = 0$, the linear assumption is adequate to illustrate the point. For this form, the F_i in the velocity part of $\mathbf{z}$ near the maximum is

$$F_i = -\gamma\left(U_i - \alpha\frac{\partial}{\partial x_i}R\right) \simeq -\gamma U_i + \gamma\alpha\frac{\partial^2 R}{\partial x_i \partial x_j}X_j$$

(with summation convention that repeated indices on the right side not appearing on the left are summed over). We can regard the second derivatives as constant so that

$$F_i = -\gamma U_i + \gamma\alpha R_{ij} X_j$$

If we rotate the coordinates to the principal axes of R_{ij}, we can replace this matrix by $\lambda_i\delta_{ij}$ with the λ_i's being the eigenvalues of the matrix; they must be negative. From

$$F_i = -\gamma U_i + \gamma\alpha\lambda_i X_i$$

we can find the equation for the mean position and velocity

$$\frac{\partial}{\partial t}\begin{bmatrix}\langle X_i\rangle\\ \langle U_i\rangle\end{bmatrix}=\begin{bmatrix}0 & 1\\ \gamma\alpha\lambda_i & -\gamma\end{bmatrix}\begin{bmatrix}\langle X_i\rangle\\ \langle U_i\rangle\end{bmatrix}$$

The solutions have the velocity and position decaying toward zero as $e^{\Lambda t}$ with Λ being an eigenvalue of the matrix

$$\Lambda^2+\gamma\Lambda+\alpha\gamma|\lambda_i|=0$$

Since the sum of the roots is negative and the product is positive, both Λ's are less than zero, and the solutions indeed decay, albeit at different rates along the different axes.

Now consider the variance around the final zero mean position and velocity. We have

$$\begin{aligned}
\frac{\partial}{\partial t}\langle X_iX_j\rangle &= \langle X_iU_j\rangle+\langle U_iX_j\rangle\\
\frac{\partial}{\partial t}\langle X_iU_j\rangle &= -\gamma\langle X_iU_j\rangle+\alpha\gamma\lambda_j\langle X_iX_j\rangle+\langle U_iU_j\rangle\\
\frac{\partial}{\partial t}\langle U_iU_j\rangle &= -2\gamma\langle U_iU_j\rangle+\alpha\gamma\lambda_i\langle X_iU_j\rangle+\alpha\gamma\lambda_j\langle X_jU_i\rangle+2\sigma\delta_{ij}
\end{aligned}$$

Unlike the free dispersal problem, these equations have a steady-state solution,

$$\langle X_iX_j\rangle=\frac{2}{\alpha\gamma|\lambda_i+\lambda_j|}\langle U_iU_j\rangle \quad\text{and}\quad \langle U_iU_j\rangle=\frac{2\sigma|\lambda_i+\lambda_j|}{\alpha(\lambda_i-\lambda_j)^2+2\gamma|\lambda_i+\lambda_j|}\,\delta_{ij}$$

which reduces to

$$\langle U_i^2\rangle=\frac{\sigma}{\gamma},\quad \langle X_i^2\rangle=\frac{\langle U_i^2\rangle}{\alpha\gamma|\lambda_i|}=\frac{\sigma}{\gamma^2}\frac{1}{\alpha|\lambda_i|}$$

The organisms will cluster around the peak in the resource, with the spread smaller in the directions where the peak is sharper (larger second derivative and therefore $|\lambda_i|$). We have talked about taxis as ability to swim up the gradient of the resource R; we will describe another option for sensing gradients using temporal memory.

Migration can be thought or as a kind of taxis, with the velocity relaxing toward some preferred value

$$F_i=-\gamma[\mathbf{U}-\mathbf{v}(\mathbf{x},t)]$$

The caveats mentioned above apply to this case

$$\frac{\partial^2}{\partial t^2}\langle X_i\rangle=-\gamma\frac{\partial}{\partial t}\langle X_i\rangle+\gamma\langle v_i(\mathbf{X},t)\rangle$$

and we can only use this to find the trajectory if the spread is negligible so that $\langle v_i(\mathbf{X},t)\rangle\simeq v_i(\langle X_i\rangle,t)$. We will discuss the effects of the spread later.

We should stress that taxis can be accomplished without directly sensing gradients. Making decisions based on a comparison of current conditions with some memory of the past can also be an efficient method of moving upgradient. We illustrate this point with a simple system of an animal which swims at a constant speed at an angle θ which has random variations with a variable amplitude. The "memory" M records the history of the resource concentration, but decays with a time scale T; i.e., it allows decisions to be made based on whether the current perceived resource value is better or worse than the recent past. If conditions are improving or not much different, the turns will be relatively small; otherwise they will be larger (Fig. 1).

$$
\begin{aligned}
dX &= s\cos\theta dt \\
dY &= s\sin\theta dt \\
d\theta &= F(M-R)dW \\
dM &= \frac{1}{T}(R-M)dt \\
F(M-R) &= \frac{\alpha}{2}[1-\tanh(R-M)]
\end{aligned}
\tag{5}
$$

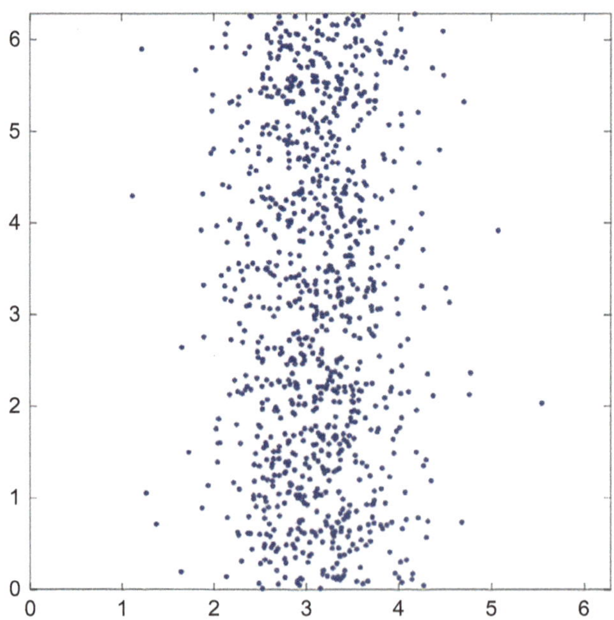

FIG. 1 Simulation with 1000 particles. $s=1/8$, $\alpha=5$, $1/T=5$, $R=5[1-\cos(x)]$.

2.3 Kinesis

But organisms can also find a high resource center even without being able to sense gradients either directly or via memory. This mechanism is known as "kinesis" and involves reducing random acceleration in regions of high R. This can also be thought of as spending more time feeding and less time moving through the water. We model this by letting

$$\sigma(\mathbf{x}) = \sigma_0 + \lambda_i \mathbf{x}_i^2$$

in the principal axes coordinate system. We will analyze this using the equation for the probability density function.

3 COLLECTIVE BEHAVIOR (USING IBMs)

Collective behavior results when the environmental cues determining $\mathbf{v}$ and σ arise from perception of neighboring organisms and their state. This could involve just the position of the near-neighbors, with organisms moving toward like organisms (or away if too crowded); this is "social taxis." Or they could respond by reducing random motion or accelerations (social kinesis). Finally, some organisms clearly sense the velocity of neighbors and try to match their movement.

We define a perceived neighbor density for the particular organism we are following (having position $\mathbf{X}$ and velocity $\mathbf{U}$) by a weighted sum over the positions of the other organisms at positions $\mathbf{X}_j$

$$R(\mathbf{X}) = \sum_j w(\mathbf{X}_j - \mathbf{X})$$

with the weighting kernel w having compact support corresponding to the sensing distance, e.g.,

$$w(Z) = \begin{cases} \frac{2}{\pi}(1 - |Z|^2) & |Z| < 1 \\ 0 & |Z| > 1 \end{cases}$$

where distances are measured in terms of the perception scale. The neighbor density is the "resource" to which the organisms respond.

Social kinesis and taxis correspond to $K(R)$ or $\mathbf{v} = \alpha \nabla R$, as before (Fig. 2).

Details of the clustering depend, of course, on the various parameters as we shall discuss below. But the experiments in Fig. 2 certainly suggest that social kinesis leads to less coherent aggregations which are more easily disrupted than social taxis.

3.1 Schooling

Animals in schools or flocks move to a good approximation in the same direction and speed. To model this with the IBM, we can take the preferred velocity $\mathbf{v}$ to be a function of the near-neighbor distances and velocities

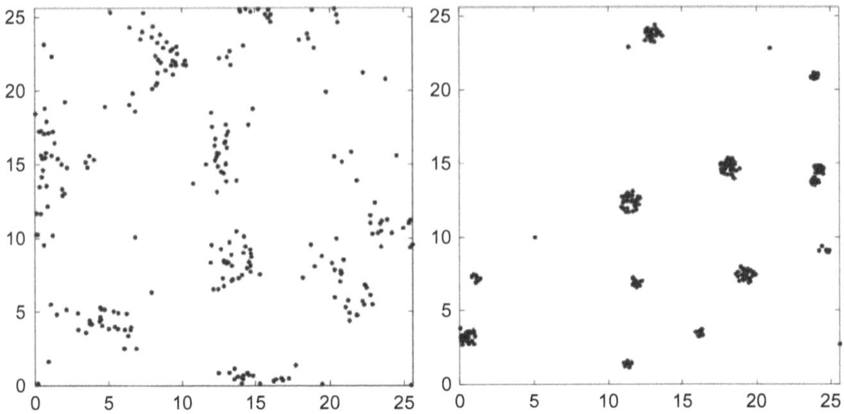

FIG. 2 Examples with 256 organisms and $\gamma = 0.1$ of social kinesis (*left*), $K(R) = 0.01/(1 + R)$ and taxis (*right*) $\alpha = 1$.

$$\mathbf{v}(\mathbf{X}) = s\frac{\mathbf{v}^*}{|\mathbf{v}^*|}, \quad \mathbf{v}^*(\mathbf{X}) = \sum_i w_0(\mathbf{X}_i - \mathbf{X}) + w_1(\mathbf{U}_i - \mathbf{U})$$

The weighting functions have the same spatial form but different prefactors representing the relative importance of the two terms. For weak attraction, w_0 small, the schools might better be described as "flocks": they are large and amorphous with individuals oriented coherently because of multiple encounters. For strong attraction ($w_0 \sim 4w_1 0$, we now see clumps of organisms moving together (Fig. 3). These are different from the social kinesis case, though, because the preferred speed is constant s, which makes it harder to form a tight group.

4 FOKKER–PLANCK EQUATION

For a single individual moving in response to a specified environment R (as in nonsocial taxis/kinesis), we can write an equation for the probability density $\mathcal{P}(\mathbf{z},t)$ which, when integrated over some volume in $\mathbf{z}$ space gives the probability that the organism's state $\mathbf{Z}$ is in the volume. That is, if we think of a problem with one space dimension, $\mathcal{P}([x,u],t)dxdu$ is the probability that it is spatially located in $x \pm dx/2$ and moving at velocity $u \pm du/2$.

The probability density evolves by fluxes into and out of the small volume around $\mathbf{z}$. This implies that $\frac{\partial}{\partial t}\mathcal{P}$ will be the divergence of a flux. The two contributions to the flux are "advection" by the F_i and "diffusion" by the random hops into or out of the volume. In our case the hops are in velocity rather than position—the $S_{ij}dW_j$. Note that in the Itô formalism, this is not the same as diffusivities used in fluid equations. The latter, conceptually, are represented by hops with probabilities determined on the faces of the volume so that the probability of crossing the line at $x = 0$ is the same whether coming from

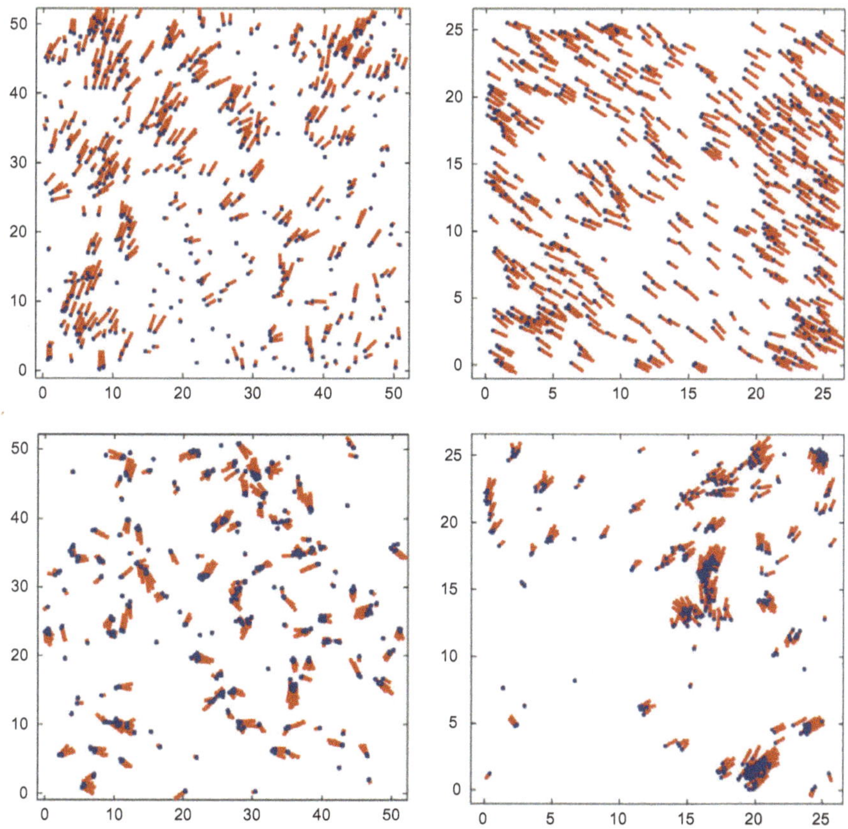

FIG. 3 Schooling in large (*left*) and small (*right*) domain corresponding to lower and higher densities. The upper figures have only orientation on the neighboring vectors while the lower also have a component toward others (social taxis). The preferred speed is $s = 1$, but the random accelerations and rapid changes in direction of $\mathbf{v}$ lead to variations (e.g., the upper plots).

the left or right. In contrast, Eq. (2) determines the probabilities based on the starting point of the particle. As a result, the Fokker–Planck (FP) equation

$$\frac{\partial}{\partial t}\mathcal{P}(\mathbf{z},t) = -\frac{\partial}{\partial z_i}\left[F_i\,\mathcal{P} - \frac{\partial}{\partial z_j}\sigma_{ij}\,\mathcal{P}\right], \quad \sigma_{ij} = \frac{1}{2}S_{ik}S_{jk} \tag{6}$$

has K within the $\frac{\partial}{\partial z_j}$ rather than outside.

4.1 Common Behavior

For organisms responding independently to a common signal, each will have a probability density governed by the same Eq. (7), and we can compute the biomass simply by multiplying $\mathcal{P}$ by the number and individual mass. Because the Fokker–Planck equation is linear, we can just regard $\mathcal{P}$ as the biomass. Implicitly, we are ignoring individual differences which alter their characteristics. In some cases, the differences may not be very significant; for example, making the swimming speed vary by 50% for searching problem (Eq. 5) makes the distribution wider and about 10% lower.

Given that the individuals are not influenced by each other but only by some common environment, we can use the Fokker–Planck equation; for simplicity, the case with relaxation of the velocities and spatially dependent random accelerations will be considered.

$$\frac{\partial}{\partial t}\mathcal{P} = -\frac{\partial}{\partial x_i}u_i\mathcal{P} + \frac{\partial}{\partial u_i}\left[\gamma(u_i - v_i(\mathbf{x}))\mathcal{P} + \sigma(\mathbf{x})\frac{\partial}{\partial u_i}\mathcal{P}\right] \tag{7}$$

We can write an explicit solution for the case of taxis when $K = \text{const.}$ and $\mathbf{v} = -\alpha\nabla R$:

$$\mathcal{P} = p_0 \exp\left(-\frac{\gamma^2}{\sigma}[R_0 - R(\mathbf{x})]\right)\exp\left(-\frac{\gamma|\mathbf{u}|^2}{2\sigma}\right)$$

with R_0 the maximum of R in the domain. The density of organisms is the zeroth moment of $\mathcal{P}$ with respect to the velocity

$$\rho(\mathbf{x},t) = P_0(\mathbf{x},t) = \int d\mathbf{u}\mathcal{P}(\mathbf{x},\mathbf{u},t)$$

and is large at the point where R is maximal and then decreases rapidly away

$$\rho = \rho_0 \exp\left(-\frac{\gamma^2}{\sigma}[R_0 - R(\mathbf{x})]\right)$$

For the case of kinesis in one-dimension, we can solve the FP equation numerically; the resulting steady-state distribution is no longer symmetric, and, as we shall see, $\rho \simeq \text{const.}/K$ (Fig. 4).

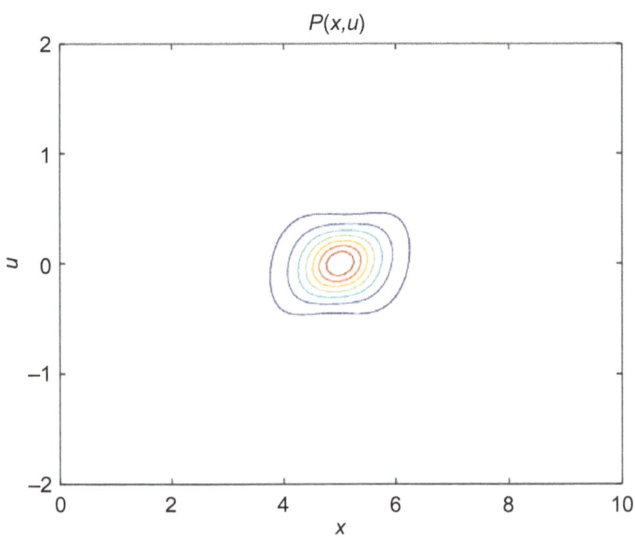

FIG. 4 $\mathcal{P}(x,u)$ with $K = (2.7 + 2.4\cos(\pi x/10))^2$.

In general, however, we cannot find analytical solutions for the particular problem at hand, and numerical solutions rapidly get unwieldy as we move to two or three dimensions since the PDE becomes a four- or six-dimensional advection diffusion problem with the velocities growing linearly with the **u** coordinates.

Instead, we derive an approximate equation by moment truncation. The first two moments of $\mathcal{P}$ are defined by

$$P_0(\mathbf{x},t) = \int d\mathbf{u}\,\mathcal{P} \equiv \rho, \quad P_{1,i}(\mathbf{x},t) = \int d\mathbf{u}\, u_i\mathcal{P}$$

From the moments of the FP equation, we find

$$\frac{\partial}{\partial t}\rho = -\frac{\partial}{\partial x_i}\int u_i\mathcal{P} = -\frac{\partial}{\partial x_i}P_{1,i}$$

$$\frac{\partial}{\partial t}P_{1,i} = -\frac{\partial}{\partial x_j}\int u_iu_j\mathcal{P} - \gamma P_{1,i} + \gamma v_i\rho$$

If the relaxation is fast because drag by the fluid is a dominant factor, then γ is large, and

$$P_{1,i} \simeq v_i\rho - \frac{1}{\gamma}\frac{\partial}{\partial x_j}\int u_iu_j\mathcal{P} \tag{8}$$

with $\mathbf{v}$ a combination of the fluid velocity $\mathbf{v}_{fluid}$ and the preferred swimming relative to the fluid. We can estimate the second-order moment when $\sigma \sim \gamma^2$. Then $\mathbf{u} \sim \sqrt{\sigma/\gamma}$ and the largest terms in the FP equation are just associated with redistributing the velocities

$$\frac{\partial}{\partial u_i}\left[\gamma u_i\mathcal{P} + \sigma\frac{\partial}{\partial u_i}\mathcal{P}\right] \simeq 0$$

so that

$$\mathcal{P} \simeq A(\mathbf{x},t)\exp\left(-\frac{\gamma}{2\sigma}\mathbf{u}^2\right) \quad \Rightarrow \quad \int u_iu_j\mathcal{P} = \delta_{ij}\frac{\sigma}{\gamma}\int\mathcal{P}$$

Thus the flux of ρ is approximately

$$F_i = P_{1,i} = v_i\rho - \frac{1}{\gamma}\frac{\partial}{\partial x_i}\frac{\sigma}{\gamma}\rho$$

or

$$\frac{\partial}{\partial t}\rho = -\nabla\cdot\mathbf{F}, \quad \mathbf{F} = \mathbf{v}\rho - \frac{1}{\gamma}\nabla(\gamma K\rho), \quad K = \sigma/\gamma^2 \tag{9}$$

Pulling the γK out yields

$$\mathbf{F} = \left[\mathbf{v} - \frac{1}{\gamma}\nabla(\gamma K)\right]\rho - K\nabla\rho$$

showing both the convergent velocities associated with taxis ($\mathbf{v}$) and kinesis ($-\frac{1}{\gamma}\nabla[\gamma K]$) as well at the down-gradient diffusive flux. We shall take γ to be constant and define an advective velocity $\mathbf{v}^* = \mathbf{v} - \nabla K$

$$\frac{\partial}{\partial t}\rho = -\nabla \cdot \mathbf{F}, \quad \mathbf{F} = \mathbf{v}^*\rho - K\nabla\rho \tag{10}$$

As an alternative approach, we can also simplify the equations by an expansion in Hermite functions (not normalized); in 1D, these look like

$$H_0 = \exp(-\gamma(u-v)^2/2\sigma), \quad H_1 = (u-v)H_0, \quad H_2 = [(u-v)^2 - \sigma/\gamma]H_0$$

They have the property that

$$\frac{\partial}{\partial u}\left[\gamma(u-v)H_n + \sigma\frac{\partial}{\partial u}H_n\right] = -n\gamma H_n$$

Substituting $\mathcal{P} = h_n(\mathbf{x}, t)H_n(u)$ in

$$\frac{\partial}{\partial t}\mathcal{P} + \frac{\partial}{\partial x}(u-v)\mathcal{P} + \frac{\partial}{\partial x}v\mathcal{P} = \frac{\partial}{\partial u}\left[\gamma(u-v)\mathcal{P} + \sigma\frac{\partial}{\partial u}\mathcal{P}\right]$$

gives

$$\frac{\partial}{\partial t}h_0 + \frac{\partial}{\partial x}h_1 + \frac{\partial}{\partial x}vh_0 = 0$$

$$\frac{\partial}{\partial t}h_1 + \frac{\partial}{\partial x}\left[h_2 + \frac{\sigma}{\gamma}h_0\right] + \frac{\partial}{\partial x}vh_1 = -\gamma h_1$$

Truncating to just h_0 and h_1 gives a pair of time- and space-dependent equations for the flux and the density (proportional to h_1 and h_0, respectively). The steady solution to the two equation system is

$$h_0 = \frac{\gamma C}{\gamma K - v^2}\exp\int^x \frac{\gamma v}{\gamma K - v^2}$$

while the large γ limit (equivalent to Eq. 8) just drops the v^2 terms.

$$h_0 = \frac{C}{K}\exp\int^x \frac{v}{K} \tag{11}$$

More formally, we assume $\sigma = K\gamma^2$ and $\gamma >> 1$ so that $h_1 \simeq -\frac{\partial}{\partial x}Kh_0$ from the h_1 equation.

4.2 Steady Solutions

Steady solutions in higher dimensional problems are not as simple; however, they must have a nondivergent flux; we begin with the simple case in which the flux has no rotational part and therefore vanishes

$$\mathbf{v}^*\rho - K\nabla\rho = 0$$

For taxis with $\mathbf{v}^* = \alpha\nabla R$, $K = \text{const.}$

$$\rho = p_0 \exp(\alpha R/K)$$

while for kinesis having $\mathbf{v} = 0$, $K = K(\mathbf{x})$

$$\rho = \frac{p_0}{K}$$

If both are active $\mathbf{v} = \alpha\nabla R$, $K = K_0/(1+\beta R)^2$ then

$$\rho = p_0(1+\beta R)^2 \exp\left(\alpha\left[R+\beta R^2+\beta^2 R^3/3\right]/K_0\right)$$

These solutions, which are just explicit, multidimensional versions of (11), illustrate the important differences between taxis and kinesis: the latter, with its algebraic enhancement of the concentration, leads to less intense and robust aggregations compared to the exponential form (Fig. 5).

But the flow need not be irrotational, especially if the fluid velocity is involved. We can write the general form of a nondivergent flux as

$$\mathbf{v}^*\rho - K\nabla\rho = -\nabla \times \rho\Psi$$

with Ψ some vector field. This gives an algebraic system

$$K\nabla \ln\rho + \Psi \times \nabla \ln\rho = \mathbf{v}^* + \nabla \times \Psi = \mathbf{F} \quad (12)$$

The explicit solution is

$$\nabla \ln\rho = \frac{1}{K(1+|\tilde{\Psi}|^2)}\left[\mathbf{F} - \tilde{\Psi} \times \mathbf{F} + (\tilde{\Psi}\cdot\mathbf{F})\tilde{\Psi}\right]$$

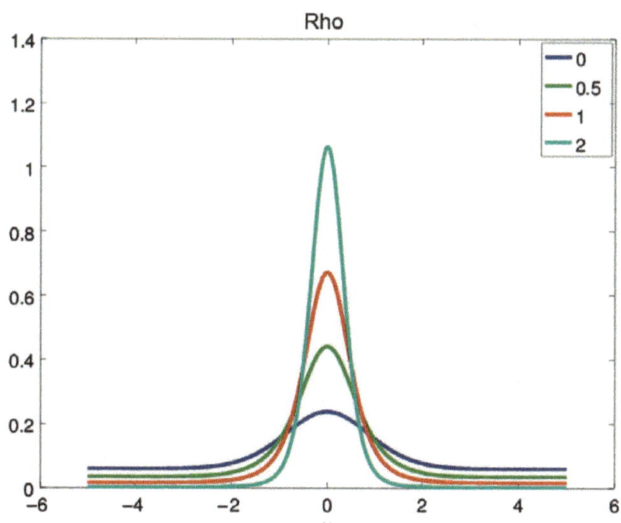

FIG. 5 Steady solutions with different values of $\alpha R_0/K_0$ normalized to have the same total number of organisms.

with $\widetilde{\Psi}=\Psi/K$. The curl of the right-hand side must be zero, placing constraints on Ψ which may or may not be possible to satisfy. If $\mathbf{v}^*$ is nondivergent, $\mathbf{F}=\nabla\times(\Psi-\psi)$ with ψ the vector streamfunction $\mathbf{v}^*=-\nabla\times\psi$, and we can choose $\Psi=\psi$ so that $\mathbf{F}=0$ and ρ is constant. As is obvious, convergence in $\mathbf{v}^*$ associated with behavior is necessary for persistent patchiness in a steady flow. The forms used above for taxis and kinesis are convergent, but this can also occur if the organisms are moving relative to the fluid; i.e., if $\mathbf{v}=\mathbf{v}_{fluid}+\mathbf{v}_{bio}$, the first term is nondivergent, but the second term need not be. Swimming toward the surface, for example, gives a vertical convergence. For depth-keeping behavior where w_{bio} cancels w_{fluid}, convergence of the horizontal fluid velocities can aggregate otherwise planktonic organisms: $\nabla\cdot\mathbf{v}=\frac{\partial}{\partial z}w_{bio}=-\frac{\partial}{\partial z}w_{fluid}\neq 0$. However, such depth-keeping behavior (with light often presumed to be the cue) is just another form of taxis, as evident in the term "phototaxis."

The fact that the result must be the gradient of a scalar places constraints on the kind of flow and K which can be accommodated. Even when $\psi=\Psi 0$ so that $\mathbf{v}^*=-\nabla\phi=\mathbf{F}$, the form

$$K\nabla\ln\rho=-\nabla\phi$$

implies $\nabla K\times\nabla\phi=0$ so that $K=K(\phi)$; the forms used in the case above with both taxis and kinesis satisfy this requirement.

Assuming an aggregation exists, we can examine the conditions near the peak to see when the velocity and diffusivity permit a local maximum in ρ.

- $\mathbf{v}^*=\mathbf{v}-\nabla K$ must be zero at the center; let this be $\mathbf{x}=0$.
- the diffusive term only spreads the distribution so we require $\nabla\cdot\mathbf{v}^*<0$ to maintain a peak in ρ.
- If we expand around $\mathbf{x}=0$

$$v_i^*\simeq x_j\frac{\partial v_i^*}{\partial x_j}=V_{ij}x_j$$

and we want particle trajectories

$$\frac{d}{dt}x_i=V_{ij}x_j$$

to move inwards either directly or in a spiral. If $\mathbf{v}^*$ is irrotational, we can diagonalize $\mathbf{V}$ corresponding to the principal axes of strain; if any of the strain rates are positive (meaning $\mathbf{V}$ has a positive eigenvalue), particles will escape along that axis, and the aggregation will not be maintained. If, however, $\mathbf{v}$ has a rotational component—vorticity $\boldsymbol{\zeta}=\nabla\times\mathbf{v}$ which is not zero—the swirling flow can move them around from the regions of outflow back into the regions of stronger inflow, so that they remain in the favorable region. This could occur if the relaxation is toward the fluid velocity plus a swimming velocity; the former often has vorticity. For particles to move into the origin and not away, all the eigenvalues of $\mathbf{V}$ must have negative real parts. This is a stronger requirement than negative divergence which only states that the inward velocity averaged around a contour enclosing $\mathbf{x} = 0$ is negative—that only requires the sum of the eigenvalues to be negative.

For the 2D movement example, we split **V** into symmetric and antisymmetric components,

$$\mathbf{V} = \begin{bmatrix} u_x & \dfrac{v_x + u_y}{2} - \dfrac{\zeta}{2} \\ \dfrac{v_x + u_y}{2} + \dfrac{\zeta}{2} & v_y \end{bmatrix}$$

with ζ the vertical vorticity $v_x - u_y$, and then rotate to the principal axes of strain

$$\mathbf{V} = \begin{bmatrix} s_1 & -\dfrac{\zeta}{2} \\ \dfrac{\zeta}{2} & s_2 \end{bmatrix}$$

with s_1 and s_2 the eigenvalues of the symmetric part of **V**

$$s_{1,2} = \frac{1}{2}\left[(u_x + v_y) \pm \sqrt{(u_y + v_x)^2 + (u_x - v_y)^2}\right] \tag{13}$$

The sum of the eigenvalues, $s_1 + s_2$ is the divergence of $\mathbf{v}^*$ and must be negative. To ensure negative real parts, the determinant must be positive

$$\zeta^2 > -4s_1s_2$$

This, then, indicates when the swirl is strong enough to overcome the weaker outflows.

Does diffusion alter these results? In the rotated coordinates the steady-state condition is (12); taking the density locally to be of the form

$$\rho = \rho_0 \exp\left(-\frac{1}{2}a_{11}x^2 - a_{12}xy - \frac{1}{2}a_{22}y^2\right)$$

and expanding around $\mathbf{x} = 0$ gives

$$s_1x - \frac{\zeta}{2}y + Ka_{11}x + Ka_{12}y = -\psi(a_{12}x + a_{22}y)$$

$$s_2y + \frac{\zeta}{2}y + Ka_{21}x + Ka_{22}y = \psi(a_{11}x + a_{12}y)$$

Set each coefficient to zero and solve the four equations for a_{11}, a_{12}, a_{22}, ψ and require $a_{11} > 0$, $a_{22} > 0$ and $a_{11}a_{22} > a_{12}^2$ so that this is indeed a local maximum in ρ. The result is $s_1 + s_2 < 0$ (which means negative divergence, i.e., convergence) and $\zeta^2 > -4s_1s_2$. Okubo(1970) finds this result by considering moments, although he uses the three combinations of $\frac{\partial}{\partial x_j}u_i$ in the parentheses in Eq. (13) rather than the principal components (mentioned, however, as an option).[a]

5 COLLECTIVE BEHAVIOR (USING FP)

If, however, we want to consider the joint PDF for all the organisms including social behavior, we have a problem: it is no longer just proportional to $\mathcal{P}$ for a single individual since the movements are not independent. Dealing with the joint PDF $\mathcal{P}(\mathbf{x}_1,\mathbf{v}_1,\mathbf{x}_2,\mathbf{v}_2,\mathbf{x}_3,\mathbf{v}_3,\ldots)$ for the positions and velocities raises problems reminiscent of the Boltzmann equation in gas kinetics. The strategy used there is

- Assume all the N individuals are the same and examine the single particle PDF

$$\mathcal{P}(\mathbf{x},\mathbf{v},t) = \int d\mathbf{x}_2 d\mathbf{v}_2 d\mathbf{x}_3 d\mathbf{v}_3 \ldots \mathcal{P}(\mathbf{x},\mathbf{v},\mathbf{x}_2,\mathbf{v}_2,\mathbf{x}_3,\mathbf{v}_3\ldots)$$

with the density then being $\rho(\mathbf{x},t) = N\int d\mathbf{v}\,\mathcal{P}(\mathbf{x},\mathbf{v},t)$.
- Assume that time during which an interaction takes place—the collision of molecules—is very short compared to the time scales of the evolution of $\mathcal{P}$ so that only the pairwise term $\mathcal{P}(\mathbf{x},\mathbf{v},\mathbf{x}_2\mathbf{v}_2,t)$ is needed.
- Write the pairwise collision term under the assumption that the colliding particles can be regarded as independent before the collision. This implies it can be expressed as the product of $\mathcal{P}(\mathbf{x},\mathbf{v},t)$ and $\mathcal{P}(\mathbf{x},\mathbf{v}_2,t)$. The change in $\mathcal{P}(\mathbf{x},\mathbf{v},t)$ is then expressed as an weighted integral of the scattering into and out of $\mathbf{v}$ $d\mathbf{v}$.

While the first of these assumptions seems acceptable, it is much less likely that the interaction time is very small compared to the independent motion period or that the interaction is only pairwise. Thus, Eq. (14) has to be regarded as heuristic when we are dealing with social behavior. Our simplification is simply regarding the probability of finding a neighbor at distance $|\mathbf{x}_1 - \mathbf{x}|$ is just $\mathcal{P}(\mathbf{x}_1,t)$, thereby disregarding the correlations arising from the interaction. However, the results seem similar enough to the IBM, and given that the $w(|\mathbf{x} - \mathbf{x}'|)$ is ad hoc in any case, we can learn a fair amount about the dynamics.

In short, we replace the joint PDF with the single-organism one with R now determined by integrals over the $\mathcal{P}$:

$$\frac{\partial}{\partial t}\rho = -\nabla\left[\mathbf{v}_{fluid}\,\rho + (\mathbf{v} - \nabla K)\rho - K\nabla\rho\right] \tag{14}$$

with the preferred velocity $\mathbf{v}$ and/or the diffusivity K being functions of R which is itself a functional of ρ.

[a]The "Okubo–Weiss" parameter is often applied when the divergence is zero in which case K causes the local maximum to continue to spread and higher order terms must be invoked.

Schooling is the most obvious manifestation of collective behavior—when an animal's movement responds to neighbors of the same species. However, "social" aggregations can also take simpler forms without the approximate alignment of velocity vectors characterizing schools. For the IBM, we can compute the perceived number of neighbors

$$n(X_i) = \sum w(X_j - X_i)$$

where the weighting function would decay with distance and most-likely be zero outside of some perception range. The preferred velocity for taxis is taken to be related to ∇n so that organisms move in the direction where they perceive more neighbors. Or n may determine the strength of K when the organisms are using kinesis.

Such models are relatively simple to simulate, although efficiency can become an issue with large numbers of particles, but they are not simple to analyze. Instead, we shall examine the taxis/kinesis problem for ρ with the neighbor function being

$$n(\mathbf{x},t) = \int d\mathbf{x}'\ w(\mathbf{x}-\mathbf{x}')\rho(\mathbf{x}',t)$$

with the normalization $\int d\mathbf{x}'\ w(\mathbf{x}-\mathbf{x}') = 1$. To include social taxis and kinesis, we use preferred velocities and diffusivity induced by random motion in the same kind of form as in the IBM:

$$\mathbf{v}^* = \nabla\Phi(n), \quad K = K(n)$$

where we use Φ in place of $-\ \phi$ so that the maximum concentration of neighbors is a point of attraction; movement is upgradient. Now let us consider when a uniform state will break into patches. If the density is uniform $\rho = \overline{\rho}$, then n is also; if the weighting function has integral 1, then $\overline{n} = \overline{\rho}$. In that case, Φ and K are constant and the flux vanishes. If we perturb this sate with some small $\rho'(\mathbf{x}, t)$, the dynamics gives

$$\begin{aligned}\frac{\partial}{\partial t}\rho' &= -\nabla\cdot[(\overline{\rho}+\rho')\nabla\Phi(\overline{n}+n') - K(\overline{n}+n')\nabla\rho'] \\ &= -\nabla\cdot[(\overline{\rho}+\rho')\Phi_n(\overline{\rho}+n')\nabla n' - K(\overline{\rho}+n')\nabla\rho']\end{aligned}$$

with $\Phi_n = \frac{\partial}{\partial n}\Phi$. The linearized equation

$$\frac{\partial}{\partial t}\rho' = -\nabla\cdot[\overline{\rho}\Phi_n(\overline{\rho})\nabla n' - K(\overline{\rho})\nabla\rho'] = -\overline{\rho}\Phi_n(\overline{\rho})\nabla^2 n' + K(\overline{\rho})\nabla^2\rho'$$

makes it evident that instability can only occur when the first term is positive and dominates. If we Fourier-transform $\rho' = \int d\mathbf{k}\,\hat{\rho}'(\mathbf{k})\exp(\imath\mathbf{k}\cdot\mathbf{x})$, the mode with wavenumber $\mathbf{k}$ will grow or decay at a rate given by

$$\frac{\partial}{\partial t}\hat{\rho}' = \mathbf{k}^2[\overline{\rho}\Phi_n(\overline{\rho})\hat{w}(\mathbf{k}) - K(\overline{\rho})]\hat{\rho}'(\mathbf{k}), \quad \hat{w}(\mathbf{k}) = \int d\mathbf{x}\, w(\mathbf{x})\exp(\imath\mathbf{k}\cdot\mathbf{x})$$

Since $\hat{w}$ will decay as $|\mathbf{k}|$ increases, we expect diffusion will dominate as small scales. The growth rate also decreases at large scales because of the $\mathbf{k}^2$ term. However, it will be positive in some range if the term in square brackets is positive as $\mathbf{k} \to 0$; i.e., if $\overline{\rho}\Phi_n(\overline{\rho}) > K(\overline{\rho})$. Then the growth rate of patches will have a maximum at some intermediate scale.

In the case $\Phi = \alpha n$, the instability condition is simply $\alpha\overline{\rho} > K$; however, this form has problems as the patch density increases. The convergent velocities continue to increase until finally the animals crowd into patches with sizes on the order of the sensing range. To illustrate this, we show the evolution of the density in the one-dimensional case when the initial state is supercritical in the middle part of the domain (Fig. 6). The weighting function is $w(x) = 0.75(1 - x^2)$ or zero for $|x| > 1$. Broader and flatter initial distributions can break up into narrow patches with high concentration. These interact very, very slowly with peaks moving toward each other under the influence of each other's far-fields and biomass leaking through the regions with ρ nearly zero from the lower peaks into the higher ones.

We can estimate the width of the peaks by assuming

$$\rho \simeq \rho_0 \exp(-x^2/2W^2)$$

with $W \ll 1$; equating the advective and diffusive fluxes gives

$$W = \left(\frac{\sqrt{2}K}{3\sqrt{\pi}\alpha\rho_0}\right)^{1/3}$$

which can be significantly smaller than the sensing distance which is scaled to be 1. In two dimensions, the power is 1/4 rather than 1/3. Fluid flow can speed up the mergers somewhat but becomes less effective as the separation between the unmerged patches increases and as their size decreases because ρ_0 grows.

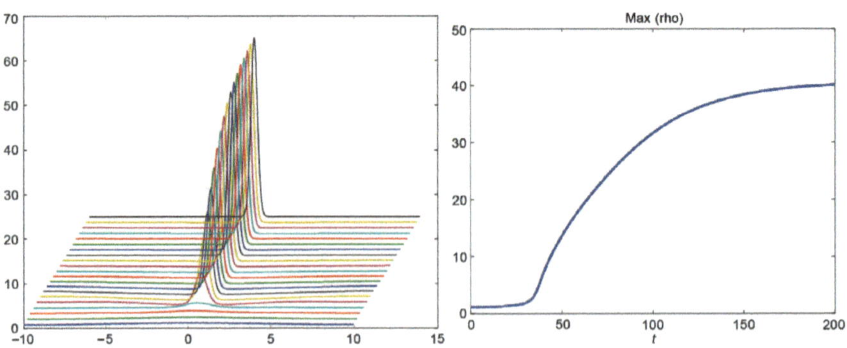

FIG. 6 Density vs x with time running into the page; the maximum density is shown on the right, indicating the leveling off.

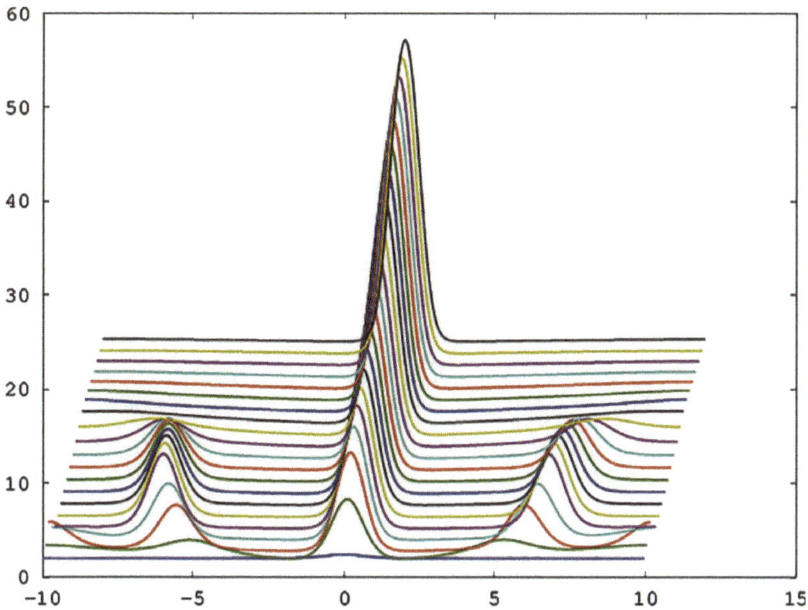

FIG. 7 Density-limited convergence with $\alpha\Phi = n/(1 + an)$, $a = 0.1$.

This rather unsatisfactory result stems from the fact that **v** crowds the animals together more and more strongly as the density increases. Flierl and Woods (2015) chose to resolve the issue by having the velocities saturate. We might expect Φ to increase initially as n increases (a larger number of neighbors to one side evokes the response to join the group), but then reasonably expect it to level off or even decrease as crowding occurs. In that case, $d\Phi/dn$ can decrease below the threshold $K/\overline{\rho}$ at large $\overline{\rho}$, and the uniform state will become stable again. We show this for $\alpha\Phi = n/(1 + an)$ in Fig. 7. This example, with $a = 0.1$, shows the initial growth of several peaks and then their merger. The instability criterion, $\overline{\rho}/(1+a\overline{\rho})^2 > K$ allows growth over a range of $\overline{\rho}$; for $a = 0.1$ and $K = 1$, the peak value has not reached the upper stable range ($\overline{\rho} > 78$) so the limitation is not very strong.

With $a = 0.2$, however, the unstable range is $1.91 < \overline{\rho} < 13.09$; as a result the peak stabilizes at much smaller values (comparable but less than the upper boundary of the unstable range) and is much wider (Fig. 8).

Kinesis has a rather different character. From the equation

$$\frac{\partial}{\partial t}\rho = \nabla^2(K(n)\rho)$$

we still have the steady solutions satisfying

$$K(n)\rho = K_0\rho_0 \tag{15}$$

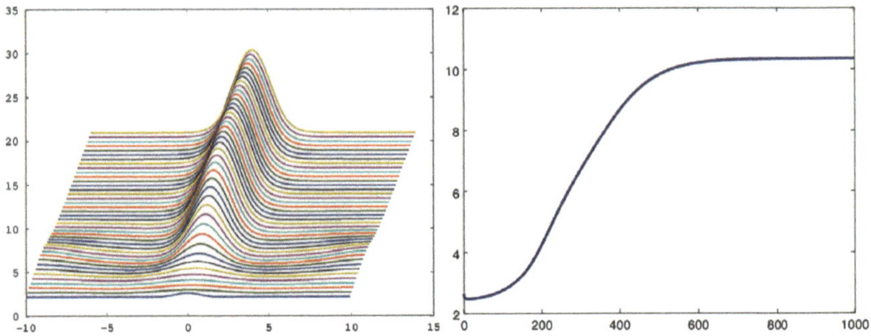

FIG. 8 Stronger limitation $a = 0.2$.

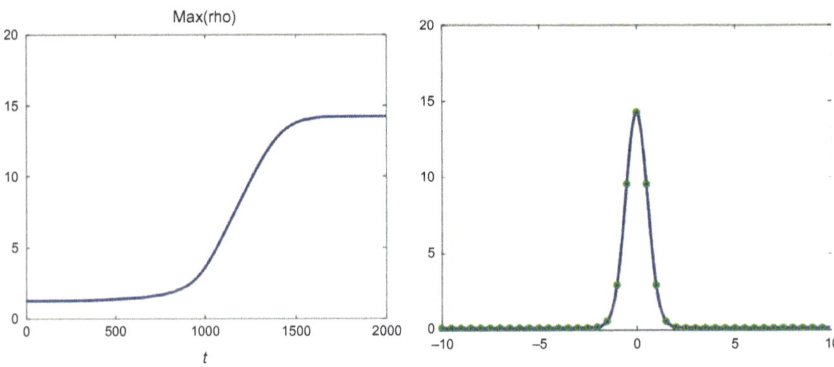

FIG. 9 Aggregation with $n_0 = 1$ and $\rho = 1 + 0.1\cos(\pi x/10)$ initially.

The uniform state will be unstable when

$$\frac{\partial}{\partial t}\rho' = \nabla^2[K(\overline{\rho})\rho' + \overline{\rho}K_n(\overline{\rho})n']$$

has exponentially growing eigenfunctions. In Fourier space, the growth rate is

$$\frac{\partial}{\partial t}\ln\hat{\rho}' = \mathbf{k}^2[-\overline{\rho}K_n(\overline{\rho})\hat{w}(\mathbf{k}) - K(\overline{\rho})]$$

with $\hat{w}$ the transform of the weighting functions. As expected, perturbations die out when $K_n = 0$, implying $\mathbf{v}^* = 0$. Instability requires the diffusivity to fall rapidly enough with increasing number of neighbors, such that

$$K_n(\overline{\rho}) < -K(\overline{\rho})/\overline{\rho}$$

For an exponential form $K(n) = K_0 \exp(-n/n_0)$, we just need $\overline{\rho} > n_0$ (Fig. 9).

6 COLLECTIVE BEHAVIOR AND ECOLOGY

Aggregation of the kind we have discussed above has, of course, implications on ecological interactions. To illustrate this, we consider standard advection–reaction–diffusion equations (see, for example, Franks, 2002)

$$\frac{\partial}{\partial t}b_i = -\nabla\cdot\left[\mathbf{v}_{fluid}\, b_i + (\mathbf{v}_i - \nabla K_i)b_i - K_i\nabla b_i\right] + \mathcal{B}_i(\mathbf{b}) \tag{16}$$

The $b_i(\mathbf{x}, t)$ are the biomasses in the different types (perhaps in terms of a "currency" like nitrogen), and the $\mathcal{B}_i$ represent the exchanges among the times by biological interactions. We want to understand how aggregation in the form of a convergent $\mathbf{v}_i - \nabla K_i$ affects the average values and other properties.

First of all, how dense to the organisms have to be before they will aggregate as in Fig. 1? We split the density into a uniform state $\overline{\mathbf{b}}$ in equilibrium

$$\mathcal{B}_i(\overline{\mathbf{b}}) = 0$$

and deviations from that state $\mathbf{b} = \overline{\mathbf{b}} + \mathbf{b}'$

$$\frac{\partial}{\partial t}b_i' = -\nabla\cdot\left[\mathbf{v}b_i' + (\mathbf{v}_i' - \nabla K_i')(\overline{b}_i + b_i') - K_i\nabla b_i'\right] + \mathcal{B}_i(\overline{\mathbf{b}} + \mathbf{b}')$$

For the stability problem, we drop quadratic and higher order terms in $\mathbf{b}'$

$$\frac{\partial}{\partial t}b_i' = -\nabla\cdot\left[\mathbf{v}b_i' + (\mathbf{v}_i' - \nabla K_i')\overline{b}_i - K_i(\overline{b})\nabla b_i'\right] + \mathcal{B}_{ij}b_j'$$

with the Jacobian matrix being

$$\mathcal{B}_{ij} = \frac{\partial^2\mathcal{B}(\mathbf{b})}{\partial b_i \partial b_j}$$

Its eigenvalues determine the stability to spatially uniform perturbations. We shall assume $\overline{\mathbf{b}}$ is what we will call "biostable": such perturbations (i.e., just involving the population dynamics) decay to the equilibrium $\overline{\mathbf{b}}$ so that all the eigenvalues of $\mathcal{B}_{ij}$ have negative real parts. For the stability calculation, we ignore the fluid velocity $\mathbf{v}$ and look at a Fourier mode $b_i'(\mathbf{x},t) = b_i'(t)\exp(\imath\mathbf{k}\cdot\mathbf{x})$. This results in with the flux convergence

$$D_i(\mathbf{b}) = \mathbf{k}^2\left[\overline{b}_i\Phi_{i,n}(\overline{\mathbf{b}})\hat{w}_i(\mathbf{k}) - K_i(\overline{\mathbf{b}})\right] \tag{17}$$

Instability will occur if one or more of the eigenvalues of $D_i\delta_{ij} + \mathcal{B}_{ij}$ has a positive real part.

6.1 One Variable

One of the simplest problems is logistic growth with aggregation

$$\mathcal{B}(P) = \mu P(1 - P/P_0) - dP$$

The quadratic term here represents competition for resources. The uniform state will be unstable to aggregation if

$$\mathbf{k}^2\left[\overline{P}\Phi_n(\overline{P})\hat{w}(\mathbf{k}) - K\right] + \mathcal{B}'(\overline{P}) > 0 \qquad \text{with} \qquad \overline{P} = P_0\left(1 - \frac{d}{\mu}\right)$$

We can readily see that for the simple doubly periodic domain, the mean value (spatial average) of P, denoted by $\langle P\rangle$, will be reduced. If we write $P = \langle P\rangle(t) + P'(\mathbf{x},t)$ then averaging the dynamical equation gives

$$\frac{\partial}{\partial t}\langle P\rangle = \mu\langle P\rangle(1-\langle P\rangle/P_0) - d\langle P\rangle - \mu\overline{P'^2}$$

In equilibrium

$$\langle P\rangle = P_0\left(1-\frac{d}{\mu}\right) - \frac{\overline{P'^2}}{\langle P\rangle}$$

The last term is negative definite, so that the solution to this quadratic clearly has $\langle P\rangle$ less that the first term, which would be the value in the absence of aggregation.

A numerical simulation is shown in Fig. 10. This includes a weak flow field as well as social taxis (Fig. 11). The fluid velocities are

$$\mathbf{v}_{fluid} = \hat{\mathbf{z}} \times \nabla\psi \quad \text{with} \quad \psi = \frac{0.1}{k}\cos(kx+\theta_x)\cos(ky+\theta_y)$$

The θ's undergo a random walk $d\theta = 0.15$ dW. We use simple upwind differencing to maintain positivity and a second-order Adams–Bashforth time-stepping. The sequence of images in Fig. 10 shows the stirring with no explicit diffusion; while it is slow, it can stir patches effectively.

More generally, the mean state would have

$$\langle \mathcal{B}(\langle P\rangle + P')\rangle = 0$$

If the fluctuations were small, which of course they are not, we could expand this around the uniform state P_u to find

$$\mathcal{B}'(P_u)(\langle P\rangle - P_u) + \frac{1}{2}\mathcal{B}''(P_u)\overline{P'^2} = 0$$

Since the P_u state is stable $\mathcal{B}'(P_u) < 0$ and

$$\langle P\rangle = P_u + \frac{\mathcal{B}''(P_u)}{2|\mathcal{B}'(P_u)|}\overline{P'^2}$$

For the logistic equation $\mathcal{B}'' < 0$ and the mean is reduced; however, other functional forms can have positive curvature and enhanced mean populations.

For example, $\mathcal{B} = b[\exp(-2b) - 0.05]$ (Fig. 12) gives a significant increase in biomass from the 1.825 value where $\mathcal{B}$ drops to zero, an increase associated with the aggregation since the curvature is positive and the penalty for large values of b is not strong. Behavior can also be effective at improving growth rates if the resource is intermittent and patchy with the organisms responding by upgradient movement (Fig. 13).

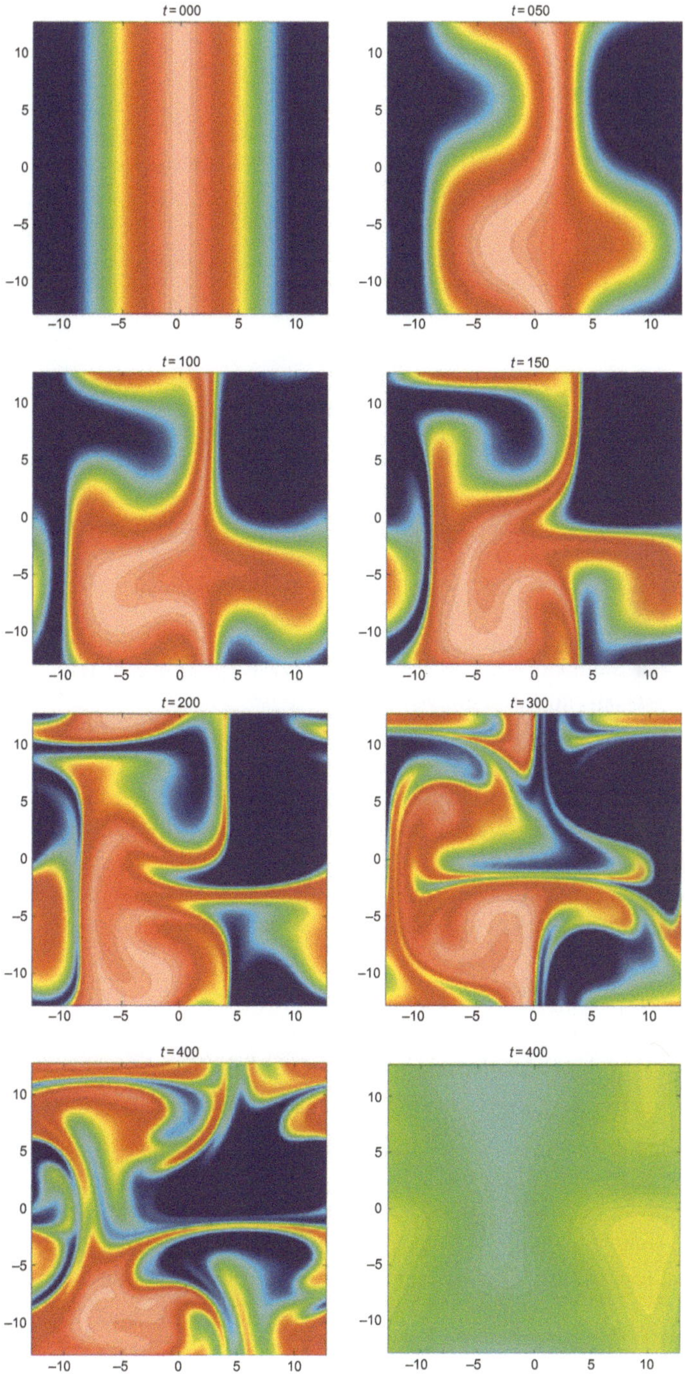

FIG. 10 Stirring by the velocity field with $K = 0$; the final figure shows the $K = 0.05$ case at the same time as the one to its left.

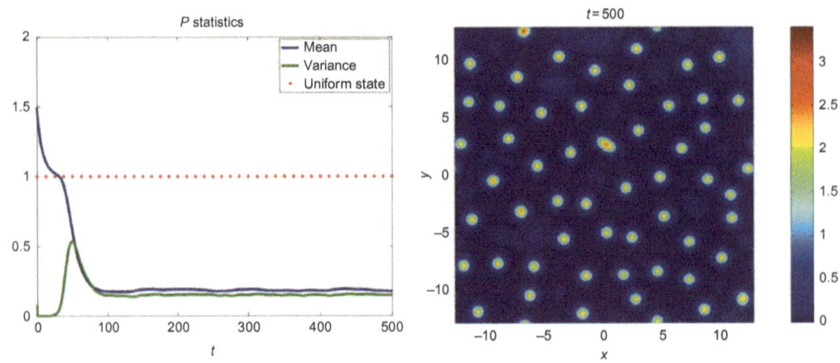

FIG. 11 Social taxis with $\alpha = 1$, $K = 0.05$. The mean (*blue*) and variance (*green*) are shown along with the equilibrium value for the uniform state (*red*, dotted).

6.2 Two-Variables

For the two-variable problem, instability requires either

$$D_1 + \mathcal{B}_{11} + D_2 + \mathcal{B}_{22} > 0 \quad \text{or} \quad det(\mathcal{B}_{ij}) + D_1\mathcal{B}_{22} + D_2\mathcal{B}_{11} + D_1D_2 < 0$$

If the uniform steady-state $\overline{\mathbf{b}}$ is biostable, then $\mathcal{B}_{11} + \mathcal{B}_{22} < 0$ and $det(\mathcal{B}_{ij}) > 0$; therefore we will need either $D_1 + D_2$ to be sufficiently positive or $D_1\mathcal{B}_{22} + D_2\mathcal{B}_{11}$ to be negative enough.

Turing instability deals with purely diffusive processes with $D_i = -\mathbf{k}^2K_i$ so that it is necessary for $K_1\mathcal{B}_{22} + K_2\mathcal{B}_{11}$ to be positive. This can only happen if one of the self-interaction terms is positive and one negative; suppose $\mathcal{B}_{11} > 0$ but $\mathcal{B}_{22}$ is negative and larger in magnitude. Then $K_1\mathcal{B}_{22} + K_1\mathcal{B}_{11} < 0$ and instability requires $K_2 > K_1$.

As a specific example of a two-variable system, we will consider the stabilized version of the Lotka–Volterra equations for a predator Z and prey P

$$\frac{\partial}{\partial t}P = -\nabla \cdot \left[\mathbf{v}P + (\mathbf{v}_p - \nabla K_p)P - K_p\nabla P\right] + \mu P(1 - P/P_0) - gPZ$$

$$\frac{\partial}{\partial t}Z = -\nabla \cdot [\mathbf{v}Z + (\mathbf{v}_z - \nabla K_z)Z - K_z\nabla Z] + agPZ - dZ$$

with

$$\mathcal{B}_{ij} = \begin{bmatrix} -\mu\langle P\rangle/P_0 & -g\langle P\rangle \\ ag\overline{Z} & 0 \end{bmatrix}, \quad \langle P\rangle = \frac{d}{ag}, \quad \overline{Z} = \frac{\mu}{g}\left(1 - \frac{\langle P\rangle}{P_0}\right) \tag{18}$$

For this model, the Turing instability cannot occur since $\mathcal{B}_{11} < 0$ and $\mathcal{B}_{22} = 0$. However, we can still find aggregation instabilities. The conditions become

$$D_1 + D_2 > \mu\langle P\rangle/P_0 \qquad \text{or} \qquad gd\overline{Z} + D_1D_2 < \mu\langle P\rangle D_2/P_0$$

with the D factors from Eq. (17). In the case of predator grouping ($D_2 > 0$), we need $D_2 > \mu\langle P\rangle/P_0$ or $D_2 > gd\overline{Z}P_0/\mu\langle P\rangle$. For prey aggregation, the single

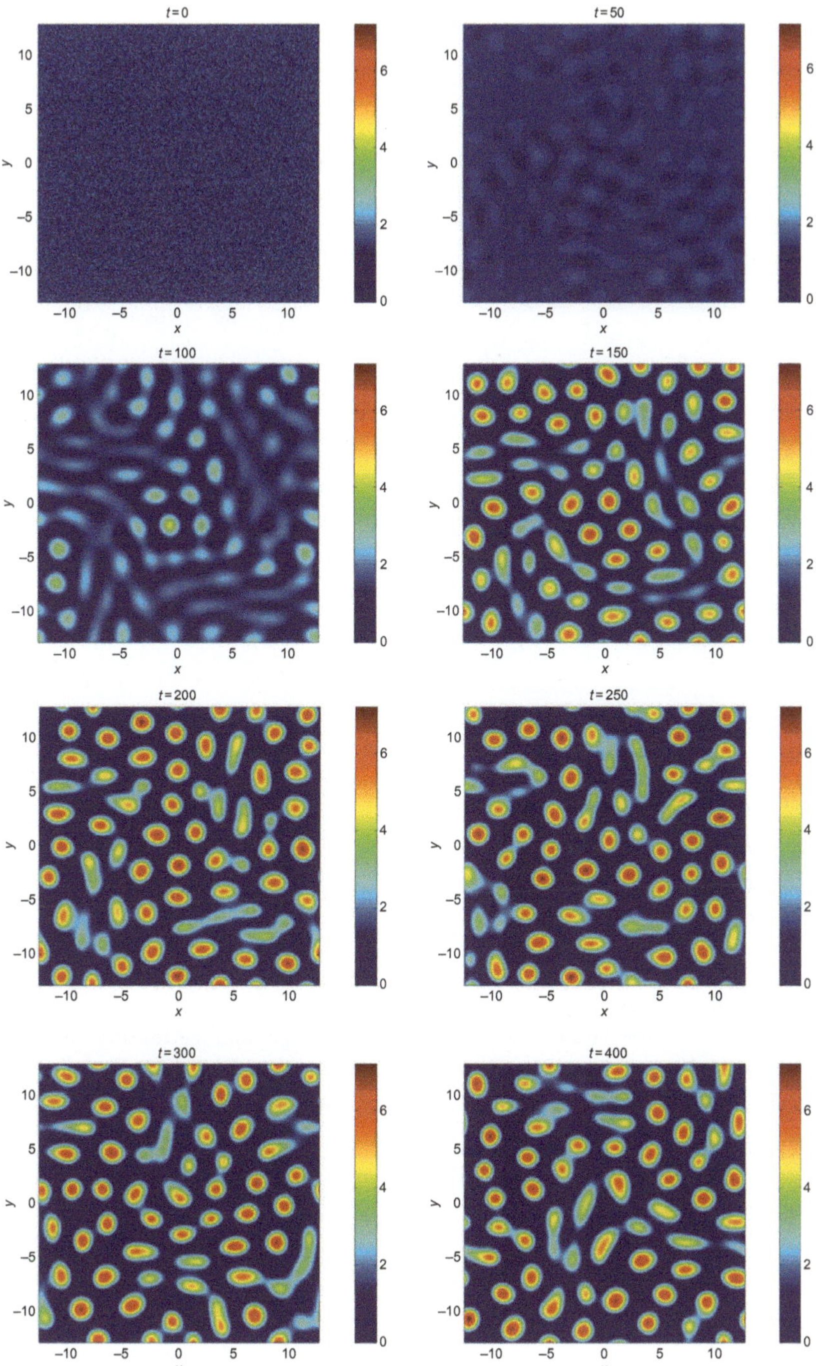

FIG. 12 A sequence from the positive curvature case. The stirring velocity can be seen elongating the patches.

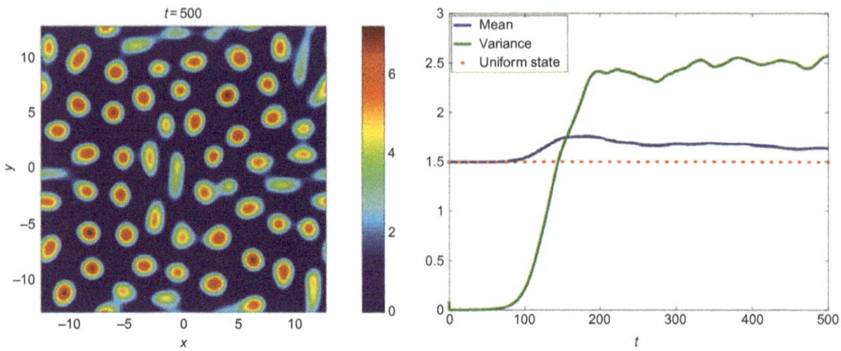

FIG. 13 The final state and statistics showing the elevation of the mean and the variance.

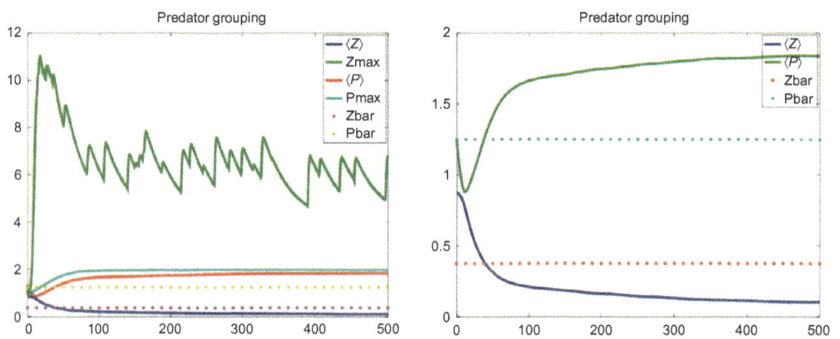

FIG. 14 Means, maximum values, and uniform state values. Parameters are $\mu = 0.1$, $g = 0.1$, $a = 0.4$, $d = 0.05$, $P_0 = 2$.

condition is $D_1 > \mu\langle P\rangle/P_0$. A numerical experiment solving this into the nonlinear regime shows that the mean of Z is lower than the uniform steady-state value $\overline{Z}$ and the mean of P is higher. The peak values show mergers of patches, followed by a decrease from over-grazing on the food supply (Figs. 14 and 15).

We can use the equations for the spatial means and variability to understand why the aggregations alter the mean values. We split P and Z into domain averages $\langle P\rangle$, $\langle Z\rangle$ and deviations P', Z'. The domain-averaged equations then give

$$\frac{\partial}{\partial t}\langle P\rangle = \mu\langle P\rangle(1-\langle P\rangle/P_0) - g\langle P\rangle\,\langle Z\rangle - \mu\langle P'^2\rangle/P_0 - g\langle P'Z'\rangle$$

$$\frac{\partial}{\partial t}\langle Z\rangle = ag\langle P\rangle\,\langle Z\rangle - d\langle Z\rangle + ag\langle P'Z'\rangle$$

If the system reaches steady state, we had have

$$\langle P\rangle = \frac{d}{ag} - \frac{ag\langle P'Z'\rangle}{\langle Z\rangle} \quad \text{and} \quad \langle Z\rangle = \frac{\mu a}{d}\langle P\rangle(1-\langle P\rangle/P_0) - \frac{\mu a}{dP_0}\langle P'^2\rangle$$

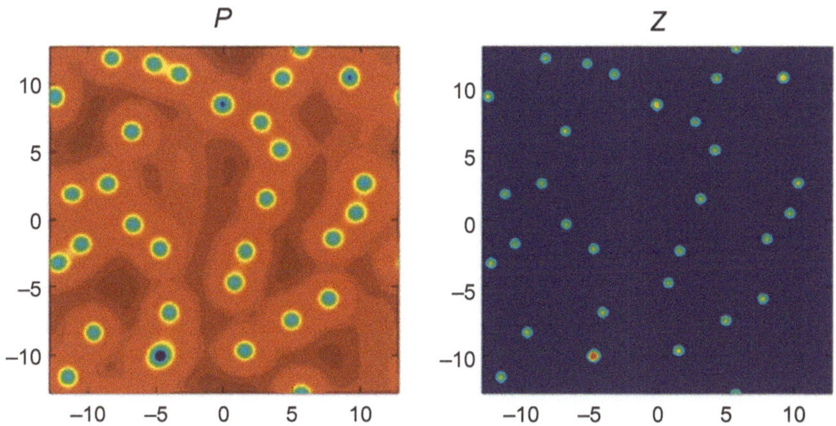

FIG. 15 Prey and predator distributions showing clear anticorrelation $\langle P'Z' \rangle < 0$.

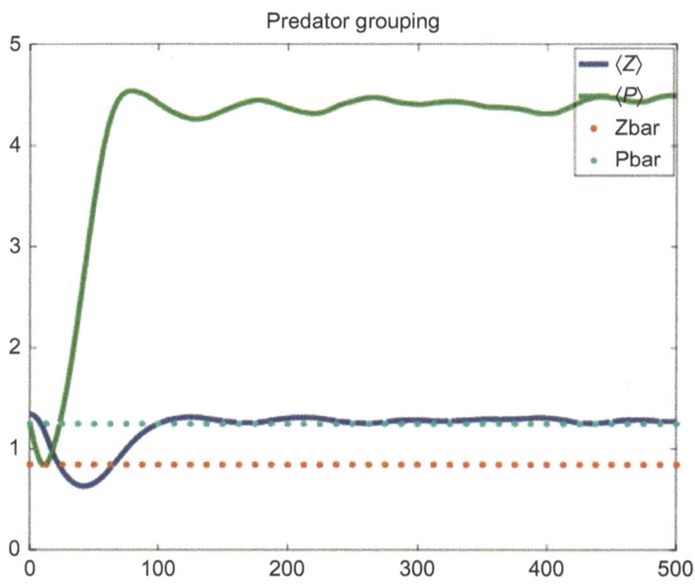

FIG.16 Mean Z and P with larger $P_0 = 8$.

Since $\langle P'Z' \rangle$ is negative; $\langle P \rangle > \overline{P}$. But the uptake term will decrease: the maximum of $P(1 - P/P_0)$ is at $P_0/2$ and $\overline{P}$ is higher than this. So the first term would give a smaller $\langle Z \rangle$, and the second term is negative definite. As a result, the average predator population is reduced by the grouping.

If we increase P_0 sufficiently, the uptake term will increase as $\langle P \rangle$ increases, and it turns out to be possible for the predator grouping to increase both $\langle P \rangle$ and $\langle Z \rangle$ as the logistic term is playing less of a role and the increase throughput for the prey allows higher predator levels (Fig. 16).

With prey aggregation ($D_1 > 0$)), the situation is rather different. If we keep the other parameters the same as in the previous case, the predator dies out. Essentially, the aggregation means that large areas do not have enough prey to support the predator. While growth of Z can occur in the prey patches, the growth rate is not sufficient to overcome diffusive losses. We can understand this by looking at a simple model of growth vs diffusion. We can solve

$$\frac{\partial}{\partial t}Z = [g_0\mathcal{H}(r_0 - r) - d]Z + \kappa\nabla^2 Z$$

(with $\mathcal{H}$ the Heaviside step-function) to find the conditions for exponentially growing solutions $Z = Z(r)\exp(\sigma t)$. In terms of the two nondimensional quantities

$$\Delta = \left(1 + \frac{\sigma}{d}\right)\frac{dr_0^2}{\kappa} \qquad \text{and} \qquad \Gamma = \frac{g_0 r_0^2}{\kappa}$$

involving the ratio of the diffusive time to the time scale for d and for growth, with additional contribution from the population growth rate. The solutions

$$Z = \begin{cases} Z_0 J_0(\sqrt{\Gamma - \Delta}\, r/r_0) & r < r_0 \\ Z_1 K_0(\sqrt{\Delta}\, r/r_0) & r > r_0 \end{cases} \tag{19}$$

where J_0 and K_0 are Bessel functions. Continuous Z and $\frac{\partial}{\partial r}Z$ implies

$$\frac{\sqrt{\Gamma - \Delta}\, J_1(\sqrt{\Gamma - \Delta})}{J_0(\sqrt{\Gamma - \Delta})} = \frac{\sqrt{\Delta}\, K_1(\sqrt{\Delta})}{K_0(\sqrt{\Delta})}$$

Growth will occur for $\Delta > \Delta_0 = dr_0^2/\kappa$; if we put this minimal Δ_0 in the equation above, we can find the critical value of Γ above which the population can grow and then be stabilized by the nonlinear terms. We show this as a plot of the ratio of the patch size to the distance diffused over a growth time scale $\sqrt{\Gamma}$ as a function of $d/g = \Delta_0/\Gamma$ (Fig. 17). Not surprisingly, when g_0 is not much larger than d, maintenance of the population requires a large area of growth.

If the predator is much more mobile than the prey, it can survive. As an example, let us consider Z exhibiting kinesis:

$$\frac{\partial}{\partial t}Z = \nabla^2 K_z Z + (agP - d)Z \tag{20}$$

and suppose K_z is large. Then the first approximation to a solution is simply

$$Z \simeq Z_0 = \frac{C}{K_z}$$

and we can iterate

$$\frac{\partial}{\partial t}Z_0 = \nabla^2 K_z Z = (agP - d)Z_0$$

Averaging this yields

$$\frac{\partial}{\partial t}\langle Z_0\rangle = ag\langle PZ_0\rangle - d\langle Z_0\rangle$$

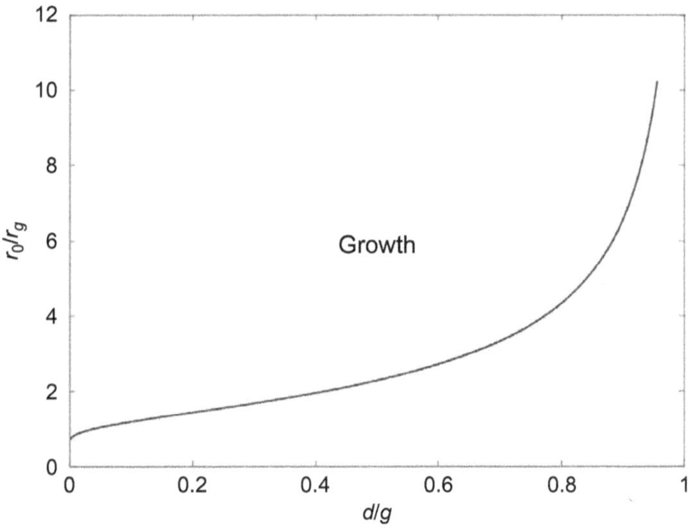

FIG. 17 Critical radius for instability r_0/r_g with $r_g = \sqrt{\kappa/g}$ as a function of d/g.

leading to an expression for the mean

$$\langle Z \rangle = \frac{ag}{d} \left\langle \frac{P}{K_z} \right\rangle C$$

But the average of the lowest order approximation

$$\langle Z \rangle = C \left\langle \frac{1}{K_z} \right\rangle$$

gives the final approximate expression

$$\left\langle \frac{P}{K_z} \right\rangle = \frac{d}{ag} \left\langle \frac{1}{K_z} \right\rangle = \overline{P} \left\langle \frac{1}{K_z} \right\rangle$$

Alternatively, we can just use Eqs. (18) and (20) to find the grazing

$$\langle PZ \rangle = \frac{d}{ag} \langle Z \rangle = \overline{P} \langle Z \rangle$$

This is, of course, consistent with the balance in the mean Z equation; however, we can now use it in the mean P equation

$$g\overline{P} \langle Z \rangle = \mu \langle P \rangle - \frac{\mu}{P_0} \langle P^2 \rangle$$

If we consider constant K_z so that $\langle P \rangle = \overline{P}$ and the prey patches as occupying a fraction α of the area, then $\langle P^2 \rangle \simeq \overline{P}^2/\alpha$ and

$$\langle Z \rangle \simeq \frac{\mu}{g} \left(1 - \frac{\overline{P}}{P_0 \alpha} \right)$$

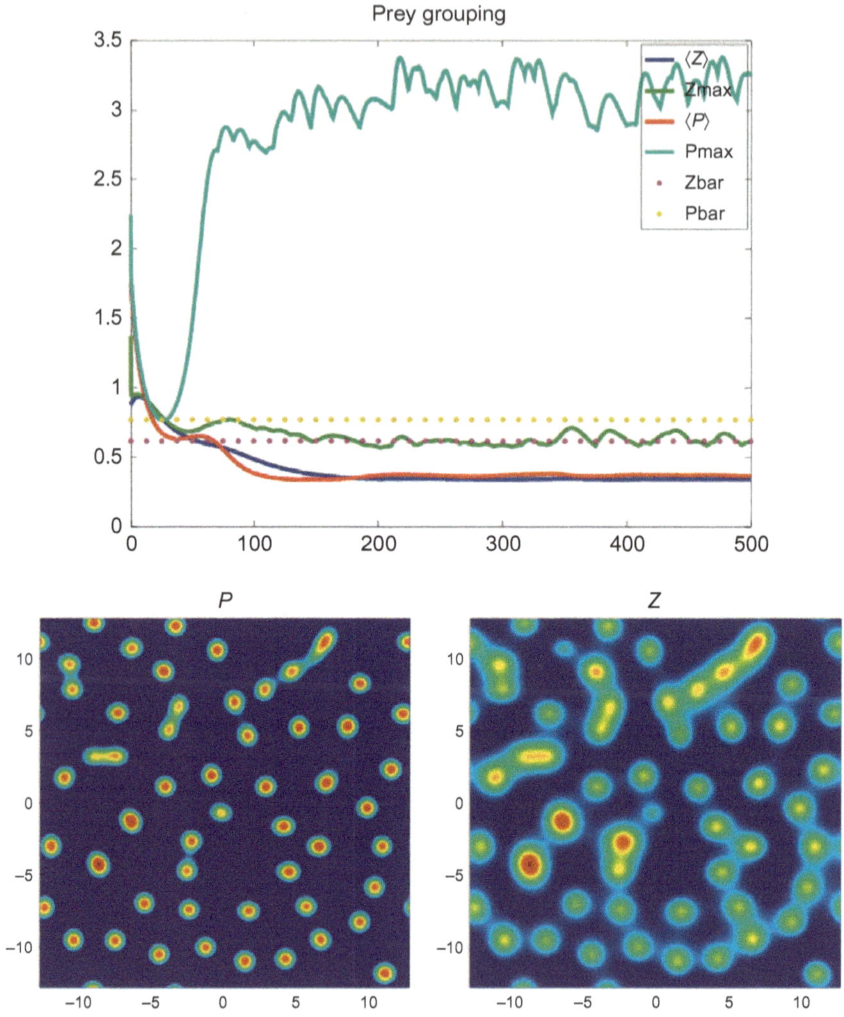

FIG. 18 Prey grouping with quadratic death rate.

and survival for the predator requires $\alpha > \overline{P}/P_0$. However, the high diffusivity K_z means that the predator does not remain in the patch very long, so survival remains unlikely.

However, altering the death term to quadratic as in Steele and Henderson (1981),

$$\frac{\partial}{\partial t}Z = \nabla \cdot [K_z \nabla Z] + agPZ - dZ^2$$

allows both to survive, albeit with reduced mean values (Fig. 18). There is now a significantly positive correlation $\langle P'Z' \rangle$ although the predator spreads more broadly, consistent with the solutions (19).

These results reinforce the conclusion that the net effects of collective behavior on the populations depend sensitively on the nature of the ecological dynamics.

7 CONCLUDING REMARKS

As the above should make clear, IBMs are fairly straightforward, although there are always concerns about the numerical approximations and interpolations needed when the velocities are only known on a grid. Also, care is needed if one is using random motions to represent inhomogeneous diffusivity; as Eq. (10) shows, simply adding random accelerations $\sqrt{2K}$ gives an extra term $-\nabla K$ in the advection; this must be corrected for by moving the particles with $\mathbf{v} + \nabla K$ rather than $\mathbf{v}$. But these kinds of issues should not obscure the fact that IBMs are often offer significant advantages. For example, one could consider a particle as representing a cohort and follow it through developmental stages (e.g., six naupliar stages, five copepodid stages, adults in the case of copepods, Davis, 1984). Since paths of other particles started nearby can diverge because of stochastic processes and turbulence, they can experience different environmental conditions and develop at different rates. As a result, a whole collection of particles can lead to a diverse and variable population structure in a region. In principle, this could be modeled with $\rho(\mathbf{x}, t, rmstage)$ with suitable randomness in the stage transitions, but the number of fields would make for a comparatively daunting computation.

However, IBMs have disadvantages compared to continuum models in other ways. Interactions between individuals or between predator and prey particles become a problem of finding which of the many particles are close enough to each other. This problem is not unique, so it may be possible to find approaches which are not brute force in order to make IBMs with large numbers of particles.

For independent individuals, the PDF function $\mathcal{P}(\mathbf{x},\mathbf{v},t)$ describes the behavior equally well, although the $\rho(\mathbf{x}, t)$ approximation Eq. (9) is not as precise. Here the difficulties become numerical: solving the advection equations in this compressible and highly nonlinear integral-differential equation or equation set. And, of course, there is the perennial issue of scale: if patch scales are the order of meters, resolving them will not be possible even in a regional ecosystem model with grid scales of a kilometer. But the interactions between different components is straightforward.

We have presented arguments and examples indicating that patchiness can significantly alter ecological interaction rates. As should be obvious, we still need further understanding of the impact in more complex models. The limited data available my help constrain the aggregation model to have appropriate amplitudes and spatial scale. The ways in which patchiness will vary with different environmental and ecological conditions will also be an important part of parameterizing this small-scale process in models. Many of the experiments here suggest that aggregation reduces the mean population; even though this is not the only measure of fitness, it does the raise the question of how such behavior became so common. Verdy and Flierl (2009) argue that evolution might favor schooling if the reproductive advantages offset the competition for food; that could happen when the food supply is replenished rapidly enough by turbulent mixing. Perhaps this argument or some of the other explanations such as higher success at searching out food or avoiding predation or a combination of multiple factors can lead to a better understanding of how evolution led to such behavior.

ACKNOWLEDGMENTS

My interest in this topic was sparked by Daniel Grünbaum and advanced through many discussion with him and with Simon Levin. I also thank Santiago Benevides for a careful reading and N.S.F. for support under OCE-1459702.

REFERENCES

Davis, C.S., 1984. Interaction of a copepod population with the mean circulation on georges bank. J. Mar. Res. 42, 573–590.

Flierl, G.R., Woods, N.W., 2015. Copepod aggregations: influences of physics and collective behavior. J. Stat. Phys. 158, 665–698.

Flierl, G., Grünbaum, D., Levin, S., Olson, D., 1999. From individuals to aggregations: the interplay between behavior and physics. J. Theoret. Bio. 196, 397–454.

Franks, P.J., 2002. NPZ models of plankton dynamics: their construction, coupling to physics, and application. J. Oceanogr. 58, 379–387.

Okubo, A., Levin, S., 2001. Diffusion and Ecological Problems: Modern Perspectives. Springer-Verlang, New York.

Steele, J.H., Henderson, E.W., 1981. A simple plankton model. Am. Nat. 117, 676–691.

Verdy, A., Flierl, G., 2009. Evolution and social behavior in krill. Deep Sea Res. II Top. Stud. Oceanogr. 55, 472–484.

FURTHER READING

Harris, S., 1971. An Introduction to the Theory of the Boltzmann Equation. Holt, Rinehart and Winston, New York.

Index

Note: Page numbers followed by "*f*" indicate figures, and "*t*" indicate tables.

N

Q

R

S

V

W

Z